From Riemann to Differential Geometry and Relativity

Lizhen Ji · Athanase Papadopoulos
Sumio Yamada
Editors

From Riemann to Differential Geometry and Relativity

Springer

Editors
Lizhen Ji
Department of Mathematics
University of Michigan
Ann Arbor, MI
USA

Sumio Yamada
Department of Mathematics
Gakushuin University
Tokyo
Japan

Athanase Papadopoulos
University of Strasbourg, CNRS
Institut de Recherche Mathématique
Avancée
Strasbourg Cedex
France

ISBN 978-3-319-60038-3 ISBN 978-3-319-60039-0 (eBook)
DOI 10.1007/978-3-319-60039-0

Library of Congress Control Number: 2017942965

Printed on acid-free paper

This Springer imprint is published by Springer Nature
The registered company is Springer International Publishing AG
The registered company address is: Gewerbestrasse 11, 6330 Cham, Switzerland

Preface

Bernhard Riemann is one of those few mathematicians whose work made a profound transformation of mathematics and physics. Not only his results are far-reaching, but his vision and approach to mathematics were directly felt and appreciated by all the later generations of mathematicians.

To say something original on Riemann's work is not easy, not because everything about him is known—far from it, but because it requires a profound reading and understanding of his mathematical writings, which are difficult, involving hidden geometric arguments, sometimes originating in physics and most of all relying on his broad intuitive vision. Besides a familiarity with the mathematical concepts involved, a reader of Riemann's works must be capable of following his very terse style. Anyone who has read his habilitation lecture, *Über die Hypothesen, welche der Geometrie zu Grunde liegen*, has felt its unusual tone. The mathematical ideas are expressed there in a broad and unusual language, and the results are generally stated without supporting proofs or calculations. Furthermore, these ideas are intertwined with philosophical and historical considerations, which may be incomprehensible to a reader who is not sensible to history and philosophy. André Weil mentions this memoir in a letter he wrote to his sister on March 26, 1940, and published in his *Collected Papers* (Springer Verlag, New York, Vol. 1, p. 244–255). He writes the following, talking about algebraic functions of one variable: "It is generally believed that there is nothing left to do on algebraic functions of one variable, because Riemann, who discovered almost everything we know about these functions (I am excepting the works of Poincaré and Klein on uniformization, and those of Hurwitz and Severi on correspondences) did not leave for us any statement of a big problem that concerns them. I am without doubt one of the most knowledgeable persons on this subject; certainly because I had the good fortune (in 1923) to learn it directly from Riemann's writings, whose memoir is of course one of the greatest things that a mathematician has ever written; there is not a single word there that is not of consequence."

Today, 150 years after Riemann's death, some of his highly original ideas are still poorly known to the mathematical community, in spite of the fact that a large number of books and articles were published on his work. The reason is that these books often concentrate on the results that are considered to lead to important developments, leaving in the dark some of Riemann's beautiful ideas that deserve to be contemplated and further exploited. Actualizing these ideas and including them in the context of current mathematics is a permanent necessity.

Several essays included in the present volume are the result of reading Riemann's writings, and the others are motivated by his ideas as they appear in the scientific literature.

The decision of editing this book was taken after two conferences held in Strasbourg, the first one on June 12–14, 2014, whose subject was "Riemann, topology and physics," and the second one on September 18–20, 2014, whose subject was "Riemann, Einstein and geometry," where Riemann's influence on relativity theory was emphasized. Consequently, this book contains several chapters on the latter theory.

Despite the variety of topics contained in this volume, there is one simple and common purpose, to highlight—hopefully in a new way—some of Riemann's original ideas and their subsequent development.

We would like to take this opportunity to thank Elena Griniari from Springer Verlag for her interest, support and efficient help in this edition, and Manfred Karbe for his invaluable advice.

Editing such a book required hard work. We consider it an expression of our gratitude for all that Riemann gave to human knowledge. His ghostly voice still inspires us all.

Ann Arbor, USA — Lizhen Ji
Strasbourg, France and Providence, USA — Athanase Papadopoulos
Tokyo, Japan — Sumio Yamada
April 2017

Bernard Riemann (Courtesy of the library of the University of Götttingen)

Riemann's wife and daughter (Courtesy of the library of the University of Götttingen)

Contents

Introduction

The present book is an addition to the living literature on Riemann. It contains a series of introductory essays in which the authors comment on some of Riemann's writings with the goal of making them more accessible, followed by surveys of some recent research topics rooted in Riemann's work or strongly motivated by his ideas. The overall goal is to give a comprehensive overview of Riemann's work, the origin of his ideas and their impact on mathematics, philosophy and physics. The various authors—each one with his own style—get into a great variety of subjects including Riemann surfaces, elliptic and Abelian integrals, the hypergeometric series, differential geometry, topology, integration theory, the zeta function, minimal surfaces, uniformization, trigonometric series, electromagnetism, heat propagation, Riemannian Brownian motion, and several other topics to which Riemann made essential contributions or that were greatly influenced by his work. One difference between this book and the existing books on Riemann is that it contains a significant part devoted to Riemann's impact on philosophy (there are three chapters on this subject out of a total of twenty chapters in the book), while another consequential part (again, three chapters) is concerned with the impact of Riemann's ideas on the theory of relativity. Let us add that even though part of the book deals with subjects that are treated in other books on Riemann, it is always useful to have, in the mathematical literature, surveys on the same subject written by different persons, each survey reflecting its author's interests and his ideas on what is important and what is only secondary material (although in Riemann's case secondary material is very rare).

Riemann's influential habilitation lecture, *Über die Hypothesen, welche der Geometrie zu Grunde liegen* (On the hypotheses that lie at the bases of geometry) is at the center of the discussion in several chapters of the book. The repercussion of this lecture in mathematics, physics and philosophy is immense. Occurrences of a single text that had such a profound influence on these three branches of human knowledge are very rare in history. Other examples may be found in the writings of very few thinkers: Aristotle, Newton, Leibniz, Descartes and Poincaré are some of them, and it is difficult to find more names.

The present introduction has several purposes. The first one is to provide the reader with a short summary of the topics that he will find in this book. Reading this summary will give him an idea of the great variety of themes on which Riemann worked and on the impact of his ideas. Another purpose, on which we stress in the last part of the introduction, is to transmit a few thoughts, beyond Riemann's work, on the intricate relation between mathematics, physics and philosophy.

The volume is organized into a preamble, four main parts, and a concluding chapter. Each of the four parts contains a series of essays, arranged in chapters.

Chapter 1, which constitutes the preamble, written by the author of the present introduction, is an overview of the prehistory of some of the main mathematical fields on which Riemann worked. In other words, the chapter concerns the predecessors of Riemann, more precisely, the mathematicians who started the fields in which he worked, and those who exerted a major influence on him. It appears clearly from this overview that for most of the questions which Riemann addressed, Euler stands in the background, as a dominating figure. This concerns the theory of functions (in particular algebraic functions and functions of a complex variable), of elliptic integrals, of Abelian integrals, of the hypergeometric series, of the zeta function and of Riemann's ideas on space, as well as his work on topology, differential geometry, trigonometric series, and integration, and his use of the techniques of the calculus of variations. Even though Euler was not the initiator of all these fields, he is, for most of them, the one who brought them to matureness. This applies in particular to the theories of algebraic and complex functions, to that of elliptic integrals and to the zeta function. Despite the fact that the history of topology can be traced back to the Greeks, and then to Leibniz and Descartes, Euler was the first to solve topological problems with the conviction that these problems are proper to this field, and that the classical method of analysis and algebra are insufficient for their solution. Another major figure to whom Riemann has an enormous debt for what concerns his intellectual and mathematical development, is his mentor Gauss, who worked on all the topics that Riemann tackled. We also know, from Riemann's writings and his correspondence, that he was a dedicated reader of Euler and Gauss's works. What we said about mathematics in Riemann's writings also holds for physics and philosophy, that is, it is possible to trace back several important ideas of Riemann in these fields to Euler and to Gauss. This chapter is also in some sense an essay on historical progression in mathematics and it is an occasion of revisiting the texts written by several pre-eminent mathematicians of which we are the heirs, and on whose shoulders we stand.

Part I, composed of Chapters 2 to 8, is an exploration of Riemann's works and their impact on mathematics and physics. Some of these chapters have a historical character, and others contain detailed reviews of some of Riemann's published works. Some relations with the works of Riemann's contemporaries are also highlighted.

Part II, containing Chapters 9 to 11, is more directed towards the philosophical aspects of Riemann's work. It focuses in particular on his ideas on space, making relations with conceptions of other thinkers on the same subject, and exploring the impact of these ideas on modern philosophy. The notion of *Mannigfaltigkeit* (usually translated as multiplicity, or manifoldness), which existed in the philosophical language, which Riemann introduced in mathematics, and which is an ancestor of the modern notion of manifold, is thoroughly discussed.

Part III, consisting of Chapters 12 to 16, is a series of five surveys of modern mathematical research topics that are based on ideas originating in Riemann's work. These topics belong to the fields of conformal geometry, algebraic geometry, the foundations of mathematics, integration, and probability theory.

Part IV, consisting of Chapters 17 to 19, is a collection of surveys on the theory of relativity and more especially on questions in relativity that are directly inspired or that rely on Riemann's work.

Chapter 20, the concluding chapter in this volume, is written by Lizhen Ji and is meant to give a quick overview of the life and works of Riemann. It contains in particular a brief summary of each of Riemann's published articles, together with a list of notions that Riemann introduced and that are named after him.

We now present in more detail the content of Parts I to IV, chapter by chapter.

Part I

Chapter 2 is written by Jeremy Gray and it starts with a review of some important aspects of Riemann's habilitation lecture, *On the hypotheses that lie at the bases of geometry*, from the points of view of mathematics, physics, and philosophy, highlighting the consequences of Riemann's conception of space. It is a matter of fact that the philosophical notion of space has also implications on Riemann's mathematical works. According to Gray, Riemann's conception of space is related to the question of whether objects of geometry are described by coordinates or not. Following the line of work started by Gauss on the differential geometry of surfaces, Riemann formulated in a novel way the question of "determination of position" in a manifoldness, and that of the difference between intrinsic and extrinsic properties.

Gray then turns to the more general question of the interaction between the mathematical ideas expressed in Riemann's lecture and physics and philosophy. Riemann claims in his memoir that he was influenced by Herbart, without being explicit on that. Based on passages from Riemann's notes on philosophy, Gray presents some ideas of Herbart's on space, time, and motion, and he discusses the way they were received by Riemann. For instance, Herbart addressed the question of whether or not our knowledge of space, time, and motion is generated by our experience. Gray explains where Riemann agrees or disagrees with Herbart's ideas. Like Newton before him, Riemann disputed the idea of an action at a distance. He imagined, like Euler did before him, that space is filled with a substance—ether—whose properties are responsible for the transmission of the forces of nature. This philosophico-physical idea was the driving force that led Riemann to the discovery of what became known as Riemannian geometry. From the physical viewpoint, this geometry is seen as the study of spaces with infinitesimal physical forces that are responsible for curvature.

Gray then considers another subject, namely, Riemann's theory of electrodynamics, formulated in his article *Ein Beitrag zur Elektrodynamik* (A contribution to electrodynamics). The paper was presented to the Royal Academy of Sciences at Göttingen in 1858, but subsequently withdrawn, and it was published posthumously in 1867. This article by Riemann is also analyzed in Chapter 3 of the present volume. It is motivated by a question concerning the velocity of electrical interaction. Riemann argues that this velocity, "within the limits of errors of observation, is equal to that of light."

After electrodynamics, Gray comments on Riemann's theory of heat diffusion, expressed in another essay known as the *Commentatio*, whose aim is to find conditions on the distribution of heat in an infinite, homogeneous, solid body under which a system of curves remains isothermal for an indefinite period of time. Riemann formulated this problem in terms of a positive definite quadratic form with constant coefficients at each point on the solid body, governing the heat flow. He then interprets the physical problem mathematically, as a problem concerning the reduction of a quadratic form. Thus, Riemann again places a physical problem at the heart of Riemannian geometry. In the same chapter, Gray addresses the question of the influence of the *Commentatio* on the works of later mathematicians.

Chapter 3, written by Hubert Goenner, deals with Riemann's work on electromagnetism. It is built around Riemann's paper *Ein Beitrag zur Elektrodynamik* (A contribution to electrodynamics) which is also considered in Chapter 2. Goenner analyzes this paper in some detail. He reminds the reader that the idea that electrical interaction is not instantaneous was voiced by Gauss already in 1845. Based on it, Riemann deduced an explanation of the electrodynamic actions of galvanic currents. Goenner highlights several important points in Riemann's paper, explaining how Riemann's theory anticipates that of Maxwell. He also mentions connections with later discoveries of Riemann that led him to change some important points in his theory. Riemann later on addressed these questions in more detail, in a course entitled "The mathematical theory of gravitation, electricity and magnetism," that he gave in summer 1861.

Goenner's commentary is a useful reading guide to Riemann's paper. It presents Riemann's work in a large perspective comprising the works of Gauss, Weber, and others. Incidentally, Goenner provides an answer to the question of why Riemann withdrew his paper, namely, a wrong factor that Riemann included in a function under an integral sign. According to Goenner, Riemann realized, after submitting this paper, that a different factor should be there and this led him to withdraw the paper. There are other conjectural reasons for Riemann's withdrawal of the paper, for instance, the fact that Riemann realized that he used a trivially forbidden interchange of integration—this explanation is the one given by Gray in Chapter 2 of the present volume.

Chapter 4, by Christian Houzel, concerns Riemann's solution of Jacobi's problem of inversion of Abelian integrals. These are integrals of the form $\int_{z_0}^{z} R(w,z)dz$ where R(w, z) is a rational function of the two variables w and z that are related by an algebraic equation f(w,z) = 0. The Jacobi inversion problem, which was formulated by Jacobi in 1832, generalizes the inversion problem for elliptic integrals to which Riemann also contributed in an essential way. The inversion of elliptic integrals leads to the so-called doubly periodic functions, that is, holomorphic functions defined on the torus. The inversion of Abelian integrals leads to what became known later on as automorphic functions, on more general surfaces. Elliptic integrals are in some sense generalizations of inverse trigonometric functions ($\int_0^x \frac{dt}{\sqrt{1-t^2}}$ represents the arcsine function, a special case of the class of elliptic integrals of the form $\int \frac{dx}{\sqrt{1-x^n}}$) and a major idea behind this study is that inverses of elliptic integrals may behave in some sense like trigonometric functions, having periods, addition formulae, etc. Weierstrass also worked on the Jacobi inversion problem. Riemann sketched a solution of this problem in his famous memoir *Theorie der Abel'schen Functionen* (Theory of Abelian functions), written in 1857, without giving a complete proof. He completed the proof in his memoir *Über das Verschwinden der ϑ-Functionen* (On the vanishing of ϑ functions), published 1866. Houzel makes a historical survey of this inversion problem and gives an outline of Riemann's proof. This proof uses all the concepts that Riemann introduced, including the representation of algebraic functions by Riemann surfaces that are ramified coverings of the Riemann sphere, his formulation of the problem in terms of periods of differential forms of the first kind on the associated Riemann surface, and his use of what became known later on as Riemann's theta functions. His 1857 memoir concludes with a proof of the fact that integrals of the algebraic differential forms on a Riemann surface may be expressed as quotients of products of translated theta functions. Riemann also contributed to the classification and the study of moduli of Abelian integrals. In the last section, Houzel indicates some later developments of Riemann's results by A. Weil (1948) and G. Kempe (1971–73).

Chapter 5 by Sumio Yamada concerns Riemann's work on minimal surfaces. It consists of an overview, from a modern viewpoint, of Riemann's two papers on the subject, *Über die Fläche vom kleinsten Inhalt bei gegebener Begrenzung* (On the surface of least area with a given boundary) and *Beispiele von Flächen kleinsten Inhalts bei gegebener Begrenzung* (Examples of surfaces of least area with a given boundary). Both papers were finalized after Riemann's death by K. Hattendorff to whom Riemann had left a set of notes on the subject. At the same time, Yamada makes a comparison between Riemann's work and that of Weierstrass on the same subject. He shows that Riemann's notes contain several results on minimal surfaces which are now classical, including the Weierstrass-Enneper representation, Schwarz's explicit construction of minimal surfaces, as well as the Schwarz–Christoffel transformation. He also mentions relations with the works of Euler and Lagrange and with Riemann's own work on the Riemann mapping theorem.

Chapter 6, by the author of this introduction, is a survey of the ideas from physics that are contained in Riemann's mathematical papers, and on the mathematical problems that he tackled that were motivated by physics. In fact, it is sometimes not easy to separate Riemann's mathematical ideas from physics, and it is clear that for certain topics, Riemann did not make any difference between mathematics and physics. Furthermore, Riemann's philosophical ideas are often in the background of his work on mathematics and physics. The main goal of Chapter 6 is to try to convey this general theory, by analyzing several writings of Riemann. These include his habilitation lecture, his habilitation text on trigonometric series, the *Commentatio*, a paper on differential geometry motivated by the problem of expressing the temperature at a point of a homogeneous solid body in terms of time and a system of coordinates on the body, his paper on the equilibrium of electricity, his paper on the propagation of planar air waves, his paper on the functions representable by Gauss's hypergeometric series $F(\alpha, \beta, \gamma, x)$, and a few others. The chapter also contains a discussion of some of Riemann's philosophical ideas, mentioning several of Riemann's predecessors in this domain, in particular the Greek philosophers. The intricate relation between physics and mathematics in Riemann's work that is surveyed in this chapter is a vast field.

Chapter 7, also written by the author of the present introduction, is an essay on the works of Cauchy and Puiseux, the two French predecessors of Riemann on the theory of functions of a complex variable.

Cauchy started working in this field in 1814, that is, twelve years before Riemann was born. He introduced several concepts which were useful to Riemann, including line integrals, the dependence of such an integral on the homotopy class of the path of integration, and the calculus of residues. Cauchy wrote a large number of papers on this subject. In 1851, he discovered, for a complex function of a complex variable, the notion of derivative independent of direction, and he showed that the real and imaginary parts of such a function must satisfy the two partial differential equations that became known as the Cauchy–Riemann equations. At the end of the same year, Riemann submitted his doctoral dissertation, which contains the same concept of derivative, with the same characterization.

Puiseux was much younger than Cauchy and he followed his lectures on the theory of functions of a complex variable. He wrote two remarkable papers on this subject, and in particular on the question of uniformizing (that is, making single-valued) a multi-valued function defined implicitly by an algebraic equation. Puiseux' first paper was published in 1850, that is, one year before Riemann defended his doctoral dissertation in which he introduced the concept of Riemann surface. Interpreted in the right setting, this work of Puiseux inaugurates a group-theoretic point of view on the theory of Riemann surfaces. The second paper by Puiseux was published in 1851. The notions that Puiseux discovered constitute a combinatorial version of Riemann surfaces. Hermite made the relation between uniformization and Galois theory, based on the work of Puiseux. This is also reported on in Chapter 7.

Chapter 7 and the next one are also an occasion for the reader to learn about the lives of several pre-eminent mathematicians who florished at the epoch of Riemann and who had ideas close to his. Gaining insight into the life of a great mathematician is interesting even if this life has nothing exceptional. It often makes us understand his motivations and makes his work more familiar.

Chapter 8, again by the author of the present introduction, is concerned with the reception of the concept of Riemann surface by the French school and how this concept is presented in the French treatises on analysis published in the few decades that followed Riemann's work on this subject.

It took several years for the mathematical community to understand the concept of Riemann surface that was conceived by Riemann as the base ground for general meromorphic functions and on which a multi-valued function becomes uniform, and to accept the validity of some major results that Riemann proved regarding these surfaces—his theorem saying that a meromorphic function is determined by its singularities, and other results in the same vein.

At the same time, Chapter 8 is a survey of the remarkable French school of analysis that started with Lagrange, then Cauchy, and attained a high degree of maturity in the second half of the nineteenth century. We also comment on the relations between this school and the German one. The reader will find in this chapter a survey of works related to the concept of Riemann surface and related matters (elliptic and Abelian integrals, the topology of surfaces, the uniformization of multi-valued functions, etc.) by various French authors including Briot, Bouquet, Appell, Goursat, Picard, Simart, Fatou, Jordan, Halphen, Tannery, Molk, Lacour, and Hermite.

Part II

The question of space, which was already addressed several times in Part I, is thoroughly studied in the three chapters that constitute Part II of the present volume.

Elaborate psychological, physical, philosophical and mathematical theories of space were developed by various thinkers since the birth of philosophy. We live in a space, even in the few months before we are officially born. Everyone has a feeling of space, and we are supposed to have the impression that this space is Euclidean. Making a philosophy of space implies going a step further than these primary feelings. The Greeks, since the early Pythagoreans, wondered about the properties of space that go beyond our immediate senses, addressing the questions of whether space is infinite, whether it is full of matter, whether void exists, etc. Riemann had his own ideas on space, and these are contained in his habilitation lecture, and also in unfinished notes that were published pusthumously. The mathematical notion of manifold was born from Riemann's reflections on space. This is the major theme addressed in Chapters 9 to 11.

Chapter 9, by Ken'ichi Ohshika, is essentially concerned with the notion of manifold, starting from the first introduction by Riemann, in his habilitation lecture, in a mathematical context (but still with a high philosophical flavor), of the word *Mannigfaltigkeit*, usually translated by *multiplicity*. The survey takes us until the modern notion of manifold, developed in the twentieth century, including the introduction of the specialized notions of Hausdorffness and differentiability. The contributions of Hilbert, Weyl, Kneser, Veblen–Whitehead, and Whitney are surveyed. Poincaré's two definitions of a manifold, formulated at the turn of the nineteenth century, are also presented. The first definition is close to that of a submanifold, and the other one, using the notion of analytic continuation, is closer to the modern definition of a manifold using charts. The philosophical background of Riemann is also discussed, including the influences of Kant and Herbart on his ideas. Ohshika explains how Riemann's point of view differs from that of Kant, who regarded Euclidean space and its geometry as given a priori, thus excluding in principle the concepts of non-Euclidean geometries, and who apparently never thought of a possibility of alternative views on space and time.

Chapter 10 by Franck Jedrzejewski is mainly devoted to the influence of Riemann on two pre-eminent twentieth-century French thinkers, Gilles Deleuze and Félix Guattari.

Deleuze was a philosopher with a large spectrum of themes, including literature, politics, psychoanalysis and art. Guattari was a philosopher and a psychoanalyst who followed during several years the famous seminar led by Lacan, who at the same time was Guattari's psychoanalyst. Deleuze and Guattari had a long and fruitful collaboration which culminated in their joint work *Capitalism and Schizophrenia*, a complex philosophical essay in two parts entitled *Anti-Oedipus*[1] (1972) and *A Thousand Plateaus*[2] (1980). In this work, the authors address various questions concerning political action, desire, psychology, economics, society, history, and culture. As the name of the first volume suggests, the work is critical of psychoanalysis as it was conceived by Freud. In fact, throughout his relation with Deleuze, Guattari distanced himself from Lacan, and Deleuze and Guattari expressed their disagreement with the fact of reducing the unconscious mind to the family circle of the individual (his relation with his parents). The second volume of *Capitalism and Schizophrenia* contains a discussion, evaluation, and critique of works of Freud, Jung, Reich and Francis Scott Fitzgerald. The publication of the two volumes generated a large debate in the intellectual milieu in France and sometimes beyond, and the ideas formulated by Deleuze and Guattari had a non-negligible political influence in the last quarter of the twentieth century. Their work belongs to the so-called post-structuralist and transcendental empiricism postmodernist currents.

At this point, the reader may rightly ask: *What does all this have to do with Riemann and with mathematics?* Another question will also soon be addressed: *Why were some twentieth-century French philosophers interested in Riemann and*

---

[1]G. Deleuze and F. Guattari, L'anti-Œdipe, Paris, Éd. de Minuit, 1972. English translation by R. Hurley, M. Seem and H. R. Lane: Anti-Œdipus, London and New York: Continuum, 2004.

[2]G. Deleuze and F. Guattari, Mille Plateaux, Paris, Éd. de Minuit, 1980. English translation by B. Massumi: A Thousand Plateaus, London and New York: Continuum, 2004.

how were they influenced by him? In Chapter 10, Jedrzejewski brings some answers to these questions. As a preliminary attitude, the reader has to realize that in the same way as there are mathematicians interested in philosophy, there are philosophers interested in mathematics, and this has been so since antiquity. Not only these philosophers were interested in mathematics, but they brought mathematical notions and ideas into the realm of philosophy, and they used them in their works, sometimes as essential elements in formulating systems of thought which they wanted to be coherent and built on a logical basis. We can quote here Jules Villemin (1920–2001), another pre-eminent French philosopher, from his major work *La Philosophie de l'algèbre*[3]:

> There exists an intimate—although less apparent and more uncertain—relationship between pure mathematics and theoretical philosophy. History of mathematics and of philosophy shows that a renewal of the methods of one of them, each time had an impact on the other one.[4]

There are many instances in the history of ideas of works on philosophy having an impact on the development of mathematics. We recall for example Plato's influence on the development of geometry and Aristotle's influence on axiomatics and the foundations of mathematics. The various views from which philosophers considered the notion of space had also a certain impact on mathematics, and this theme is considered in the various chapters that constitute the second part of the present volume. One may also mention the enormous influence of such views on the research conducted during several centuries on Euclid's parallel axiom that led eventually to the discovery of non-Euclidean geometry.

Deleuze's philosophical theories are rooted in the works of mathematicians like Riemann, Leibniz, Whitehead, Albert Lautman, and Gilles Châtelet. Already in 1968, in an essay entitled *Différence et répétition*, he expressed the fact that an idea, from the point of view of its organization, is the philosophical analogue of a continuous multiplicity in the sense of Riemann. Guattari developed a philosophical concept of multiplicity, based on Riemann's *Mannigfaltigkeit*, as an alternative to the notion of *substance*, which is one of the key concepts in metaphysics. Many other mathematical terms, like dimension, continuity, variability, order, and metric, acquired a philosophical significance in Deleuze's work.

In Chapter 10, Jedrzejewski makes a detailed comparison between some texts of Riemann and those of Deleuze and Guattari. In fact, many twentieth-century philosophers addressed questions related to or arising from mathematics, its logic and its language. Deleuze was particularly fascinated by topology. He was influenced of Leibniz, relying at the same time on his metaphysics, his differential

[3]Presses universitaires de France, 1962.

[4]Il existe un rapport intime quoique moins apparent et plus incertain entre les Mathématiques pures et la Philosophie théorique. L'histoire des mathématiques et de la philosophie montre qu'un renouvellement des méthodes de celles-là a, chaque fois, des répercussions sur celles-ci.

calculus and his ideas on topology. Jedrzejewski also mentions the work of another French philosopher, Henri Bergson.[5]

Beyond its relation with Riemann, this chapter by Jedrzejewski is an interesting example of how mathematics meets philosophy. The chapter is written in French, the reason being that Jedrzejewski wanted to use the original Deleuzian terms. An extended English summary of the chapter is provided by the author.

Chapter 11, by Arkady Plotnistky, also has a philosophical character. It concerns the "conceptual" nature of Riemann's thinking and its implications in mathematics, physics, and philosophy. The word "concept" is used here in a technical sense explained by Plotnitsky, who relies on another philosophical essay by Deleuze and Guattari, *What is philosophy?* (1994), in which these authors view thought ("la pensée"), with its creative nature, as a confrontation between the brain and chaos. Plotnitsky's discourse is at the level of "concepts of concept," promoted by Deleuze and Guattari in the realm of philosophical thinking, transferred (by Plotnitsky) to the physical and mathematical worlds as well, despite the fact that these authors claim that their concept of concept pertains uniquely to philosophical thinking. The discussion around this concept of concept and the confrontation between the ideas of Riemann, Hegel, Deleuze, and Guattari and others makes Plotnitsky's essay an original contribution to the realm of Riemannian philosophy. Understanding the difference between a philosophical and a mathematical concept is at the center of this essay, like in the previous essay by Jedrezejewski (Chapter 10). Riemann's habilitation lecture, *On the hypotheses that lie at the bases of geometry*, in which mathematics, physics, and philosophy are merged, is in the background and provides Plotnitsky with the main material for his argumentation. The question of whether Riemann's notion of space belongs to mathematics or to philosophy is central. A notion like the "plane of immanence" (*plan d'immanence*) as a plane of the movement of thought, in Riemann's approach, is characterized by its multi-component factor, and it is one of the main ways in which Plotnitsky approaches Riemann's work. His essay sheds a new light on Riemann's dissertation and in particular his rethinking of geometry in terms of manifoldness. Connections with works of several philosophers, artists and scientists are highlighted in this chapter. The themes discussed include Leibniz's monads, Grothendieck's topoi, and quantum physics.

Part III

Part III of this volume, consisting of Chapters 12 to 16, is mathematical. It covers recent developments in mathematics that are closely related to ideas of Riemann.

[5]The French philosopher and teacher Henri Bergson (1859–1941) was awarded in 1927 the Nobel prize for literature. He had a mathematical background and there is a famous controversy between him and Einstein concerning the philosophical notion of time, which might be interesting for the reader of this book.

Chapter 12, by Feng Luo, is a variation on the Riemann mapping theorem and, its generalization, the uniformization theorem. More precisely, it concerns the discrete version of these theorems.

The interest in a discrete version of the Riemann mapping theorem was given a strong impetus by W.P. Thurston who, in the 1980s, advertised this subject in several lectures and made the relation with circle packings. The idea behind this relation is that a conformal mapping (like the Riemann mapping) is characterized by the fact that it sends infinitesimal circles to infinitesimal circles. Circle packings involve smaller and smaller circles, therefore they should give information on conformal mappings. An idea that emerged was that studying circle packings might give a new point of view on the Riemann mapping theorem, or even a new proof of it. In this setting, a precise question concerning the convergence of circle packings to the Riemann mapping theorem was raised and was eventually solved by Rodin and Sullivan.

In Chapter 12, after a presentation of Thurston's ideas on a circle packing version of the Riemann mapping theorem, Luo reviews his own recent work on the discrete uniformization theorem for polyhedral surfaces. The proof is variational. The author highlights relations with approximation theory and with algorithmic and digitalization techniques.

The material discussed in Chapter 12 may be considered as an illustration of Riemann's ideas on the relation between the discrete and the continuous, one of the major themes in his habilitation lecture.

The next chapter concerns the Riemann–Roch theorem.

The history of the Riemann–Roch Theorem starts with the so-called Riemann existence theorem, which asserts the existence of meromorphic functions on Riemann surfaces. The classical Riemann–Roch theorem gives more precise information. It concerns the dimension of the space of meromorphic functions on a compact surface having poles of (at most) a certain order at some prescribed set of points. The theorem is a formula, expressing this dimension in terms of the genus of the surface and the total order at the ramified points, thus establishing a fundamental relation between topological and analytical notions.

There are several classical proofs of this theorem, some of them topological, others geometric and there are proofs involving abstract algebra, adapted to the case where the ground field (the field of scalars) is more general than that of the complex numbers. The result has many applications, and there are several versions and generalizations of the Riemann–Roch Theorem. Brill and Noether, back in 1874, already gave an algebro-geometric version of this theorem,[6] a version which is sometimes called the Riemann-Brill-Noether Theorem and which has vast modern developments. The Riemann-Roch Theorem was widely generalized by Hirzebruch in 1953, from Riemann surfaces to the setting of projective varieties over complex numbers. The modern version of this theorem is expressed in the setting of

[6]A. Brill and M. Noether, Über die algebraischen Functionen und ihre Anwendung in der Geometrie, Math. Ann. 7 (1874,) No. 2, 269– 316.

schemes. Grothendieck obtained a very general version of the Riemann–Roch theorem, formulated in the language of categories and functors, which holds for algebraic varieties defined over arbitrary ground fields. This was one of Grothendieck's major discoveries. In Chapter 2, §2.8 of his *Récoltes et semailles*,[7] he writes: "The year 1957 is the one where I was led to extract the theme 'Riemann–Roch' (Grothendieck's version) which overnight consecrated me 'great star'."[8] Grothendieck's version of the Riemann–Roch theorem was the starting point of K-theory. In the section called *La vision—ou douze thèmes pour une harmonie* (The vision—or twelve themes for a harmony) of *Récoltes et semailles* (Chapter 2, §2.8), Grothendieck considers the theme he calls the *Riemann–Roch-Grothendieck Yoga* as one of the twelve themes which he describes as his "great ideas" (*grandes idées*). There is also a discrete Grothendieck-Riemann–Roch theorem. The famous Atiyah–Singer index theorem, discovered in 1963, can be considered as another generalization of the Riemann–Roch theorem.

All this justifies the inclusion of a chapter on the Riemann–Roch theorem, whose original idea started with Riemann.

Thus, in Chapter 13, Norbert A'Campo, Vincent Alberge and Elena Frenkel present a modern version of the Riemann–Roch theorem. It concerns the space of sections of holomorphic line bundles over a Riemann surfaces. The proof uses Dolbeault cohomology, Serre duality for line bundles, and functional analysis (Fredholm operator theory). The chapter is intended to be a self-contained proof of this cohomological version of Riemann–Roch. All the required notions (holomorphic line bundle, degree, the Poincaré-Hopf index formula, the Picard group, sheaves, sheaf cohomology, Chern class, the Cauchy-Riemann operator, Dolbeault cohomology, Serre duality, the index of a Fredholm operator and divisor) are introduced and clearly explained, in a concise but sufficiently detailed manner so that the reader can understand the theorem and its proof.

The modern version of the Riemann–Roch theorem is an important monument of twentieth-century mathematics.

Chapter 14, by Victor Pambuccian, Horst Struve and Rolf Struve, concerns the foundations of mathematics. The reader might wonder about the existence of a relation between Riemann and the foundations of mathematics. This relation is hinted on by Riemann in his 1854 habilitation lecture. At the beginning of this lecture, Riemann mentions the axiomatic approach as one of the possible approaches to geometry (the other one, to which he will stick soon after, being the metrical approach). He does not further develop this idea, but he raises the issue of the necessity of having a *solid foundation* of geometry. Riemann's immediate successors knew about his interest in axiomatics. W. A. Clifford, one of the earliest commentators of Riemann's works, writes, in a text titled *The postulates of space*,

[7]A. Grothendieck, Récoltes et semailles, Réflexions et témoignage sur un passé de mathématicien, unpublished manuscript, 1985–1986, 929 p.

[8]L'année 1957 est celle où je suis amené à dégager le thème "Riemann-Roch" (version Grothendieck) – qui, du jour au lendemain, me consacre "grande vedette".

p. 565[9]: "It was Riemann, however, who first accomplished the task of analysing all the assumptions of geometry, and showing which of them were independent." Helmholtz, in his lecture *On the origin and significance of of geometrical axioms*[10] mentions several times Riemann's ideas on the axiomatic foundation of mathematics. It is also useful to recall that Hilbert, in an appendix of his *Foundations of Geometry*, mentions Riemann. He writes: "The investigations by Riemann and Helmholtz for the foundations of geometry led Lie to take up the problem of the *axiomatic*[11] treatment of geometry as introductory to the study of groups." Although Riemann did not develop the axiomatic point of view in any of his own writings, his heirs did, and in particular there were several attempts to axiomatize Riemannian geometry. This is the subject of Chapter 14 of the present volume.

As a matter of fact, the question of the foundations of geometry, like many other foundational questions, can be traced back to Aristotle, developed in his *Posterior analytics* and his other essays. Geometry, at that time, meant mostly Euclidean geometry, although spherical geometry was also known. In any case, the question of the foundations of Riemannian geometry naturally stems from Riemann's work. With this idea in mind, in Chapter 14, the authors present a set of approaches to the axiomatization of metric spaces, developed by several authors, some of them motivated by Riemann's work. These authors used new notions from various fields that were developed in the few decades that followed Riemann's work: transformation groups, Lie groups, the foundations of arithmetics, mathematical logic, and metric geometry. Although the abstract notion of group is absent from Riemann's writings, the ideas of homogeneity and symmetry are present at several places in his work. The discussion involving group theory that is done in the chapter by Pambuccian, H. Struve and R. Struve is welcome as an important element in the development of Riemann's ideas.

In Chapter 15, Toshikazu Sunada surveys some of the impact of the idea of a Riemann sum—the basic element of Riemann's integration theory—in various branches of mathematics. He reviews in particular how Riemann sums are used in some counting problems in elementary number theory and in the theory of quasicrystals. The chapter contains illuminating examples, and the author makes interesting connections between works of Riemann Fermat, Dirichlet, Gauss, Siegel, Delone, and others.

In Chapter 16, which is the last chapter of Part III, Jacques Franchi gives an exposition of the extension of the theory of Brownian motion to the setting of Riemannian manifolds and of recent work on relativistic Brownian motion.

We recall that the concept of Brownian motion was introduced initially as a description of the (random) motion of a particle subject to the action of a multitude of other particles in a fluid. Einstein published in 1905 a paper on this subject, in the

---

[9]Cf. The World of Mathematics, edited by J.R. Newman, Volume 1, Simon and Schuster, 1956, New York, 552–557.

[10]*Ibid.*p. 647–668.

[11]The emphasis is in the original text.

setting of his kinetic theory of gases. A rigorous mathematical theory of Brownian motion was developed later, in particular by N. Wiener, around the 1920s, on a probabilistic basis and in terms of stochastic processes. We note incidentally that Brownian motion is closely related to the theory of Riemann surfaces. In particular, the Riemann mapping theorem can be proved using Brownian motion. Such an approach was promoted by Sullivan and Thurston. One can also mention, in the same vein, a probabilistic proof of the Riemann mapping theorem by Patodi (1970) and two other proofs by Bismut (1984 and 1985) of the Atiyah–Singer theorem. These proofs are simpler than the original, using only the Gauss–Bonnet theorem. There is also a probabilistic proof of Picard's small theorem by B. Davis (1975).[12] These are only a few of the instances where probability is used to prove results in Riemannian geometry. The theory of Brownian motion on Riemannian manifolds was developed around 1970 by probabilists. This topic, which is the subject of Chapter 16, is another occasion for understanding the strong relation between Riemannian geometry and probability theory.

After his exposition of Brownian motion in Riemannian geometry, Franchi moves on to the extension of Brownian motion to a relativistic framework. This makes a new relation between Riemannian geometry and relativity theory, and it adds an element to explain Einstein's strong interest in this field. As Franchi explains, the relativistic extension of Brownian motion is a non-trivial theory, especially because of the relativistic constraint that the particle's velocity cannot exceed that of light.

Chapter 16 is also the occasion of following the history of the interesting theory of diffusion, where the first (negative) results were obtained by Dudley in 1965, who proved that a Lorentz-covariant Markov diffusion process cannot exist in the framework of special relativity, in particular because of the same problem of large velocities. At the same time, Dudley proposed a construction of a relativistic diffusion at the level of the tangent bundle of Minkowski space. He specified the asymptotic behavior of that diffusion and he showed that it is canonical, given the constraints of being covariant under the action of the Lorentz group. A similar approach on the unit tangent bundle of a generic Lorentzian manifold, that is, in the setting of general relativity, was made by Franchi and Le Jan in 2007. In this setting, relativistic diffusion becomes a random perturbation of the geodesic flow over a Lorentzian manifold. Some basic examples are then analyzed to some extent.

The exposition in Chapter 16 follows the gradual move from the Euclidean to the Riemannian and then to the relativistic worlds. This theory is another instance of the intricate interaction between geometry, analysis, probability, and physics relying heavily on Riemann's ideas. Thus, this chapter makes a natural transition between Part III and the next part of the book.

---

[12] I owe the last two examples to Jacques Franchi.

Part IV

Part IV of this volume concerns physics. It contains three chapters on the extension of Riemann's ideas to modern physics, mainly, to relativity theory. Riemann, in his habilitation lecture *Über die Hypothesen, welche der Geometrie zu Grunde liegen*, expressed the fact that physical space might not satisfy the axioms of Euclidean geometry. This, is, from the philosophical point of view, the starting point of Riemann's position as a predecessor of modern physics. At a more practical level, the mathematical development of Einstein's theory of general relativity is, fundamentally, in the tradition of Riemann's differential geometry; this is one of the themes of Part IV of the present volume.

Minkowski geometry is a semi-Riemannian geometry where the metric tensor is not positive-definite—a mathematical consequence of the physical fact that particles cannot move at a speed larger than that of light. Although the geometric setting of special relativity is Minkowski geometry, which is not Riemannian, the basic mathematical ideas that are used in the development of this geometry are similar to those introduced by Riemann. In other words, the fact that Minkowski geometry differs from Riemannian geometry does not affect the fact that it is in the lineage of Riemann's ideas on geometry and space. Riemann's discussion of the invariance properties of a metric, which he carries in his habilitation lecture, have their analogues as invariance properties of the Lorentz transformations of special relativity. In fact, many important features of a four-manifold equipped with a Riemannian metric together with its Riemann's curvature tensor have their analogue in Minkowski spacetime. In general relativity, the metric tensor that describes the local geometry of space is the mathematical representation of the gravitational potential. This is again in the tradition of Riemann, who conceived his infinitesimal metric tensor in close relation with physics, the curvature of the space being seen as a consequence of the physical infinitesimal forces.

Andreas Hermann and Emmanuel Humbert, in Chapter 17 of this volume, study a variant of the so-called Positive Mass Conjecture for closed Riemannian manifolds. The conjecture is a statement in general relativity which gives conditions under which the mass of an asymptotically flat spacetime is non-negative. The relevance of this theory to the subject of this book is that it is an important instance of a purely physical problem that can be formulated in terms of Riemannian geometry. The subject discussed has a long history. Using minimal surfaces and variational methods, Schoen and Yau proved in 1979 the positive mass conjecture for 3-dimensional Riemannian manifolds, in the setting of the Hamiltonian formulation of general relativity. Witten gave a later proof (1981) which holds in any dimension where the mainfold is spin. Roughly speaking, the positive mass theorem says that if the scalar curvature of a spacetime is everywhere positive, then its mass is positive. An inequality attributed to Penrose says that the mass of a spacetime can be estimated by the total area of the black holes contained in it, and that equality is attained only for a simple model of a black hole, the so-called Schwarzschild model.

Chapter 18, by Marc Mars, focuses on an important aspect of the rich interaction between mathematics and physics based on the interplay between differential geometry, in the tradition of Riemann, and gravity, in the setting of the theory of relativity. This aspect is the local characterization of pseudo-Riemannian manifolds, which is central in general relativity in order to identify spacetime geometries independently of the specific set of coordinates used to describe them. One of the many groundbreaking contributions of Riemann to geometry was the introduction of his tensor (the so-called Riemann tensor) which vanishes if and only if the metric is locally flat. It turns out that this fundamental local characterization result holds independently of the signature of the metric, and is the motivation of many other similar characterization theorems.

After reviewing the classical results on the subject, including Élie Cartan's characterization of Riemannian locally symmetric spaces in terms of the derivative of the Riemann tensor and Weyl's characterization of locally conformally flat spaces in terms of the vanishing of the conformal curvature (i.e., Weyl's) tensor, the chapter discusses a selection of various characterization results of physically relevant spacetimes. The emphasis is primarily on spacetimes describing stationary black holes, both in the static and in the rotating case. Thus, several characterization results are presented for the Schwarzschild spacetimes, as well as for the Kerr metric, and its charged and cosmological constant generalizations. Local characterization of other spacetimes, such as for instance *pp*-waves and related spacetimes, are also described.

The last chapter of Part IV, Chapter 19, by Jean-Philippe Nicolas, contains an exposition of Penrose's conformal technique and its application to asymptotic analysis in general relativity. The setting is again that of Lorentzian geometry considered as an extension of Riemannian geometry in which space and time are united by an indefinite metric of signature (1, 3). The author presents Penrose's approach to general relativity with the central role played by the light cone structure and he explains its relation with Riemannian geometry and with Einstein's theory. The focus is on Penrose's use of conformal compactifications in the study he made of the asymptotic properties of spacetimes and fields. Indeed, Penrose introduced in 1963 a basic geometrical construction which is termed in Chapter 19 a "compactified unphysical" spacetime. This is a manifold with boundary to which the conformal metric extends smoothly, i.e., there is a metric in the conformal class that extends as a smooth non-degenerate Lorentzian metric. Spacetimes that admit smooth conformal compactifications are characterized by a decay property of their Weyl curvature at infinity. When such a compactification exists, the boundary of the manifold is equipped with a nice geometric stratified structure.

After surveying Penrose's theory, Nicolas reviews some of its applications to questions of scattering and peeling. Scattering theory is a way of studying the evolution of solutions of a certain equation by a so-called scattering operator, an operator which associates to the asymptotic behavior of the solutions in the distant past their asymptotic behavior in the distant future. Peeling is a generic asymptotic behavior discovered by R. Sachs in the beginning of the 1960s. In the mid 1960s, Penrose proved that this behavior is equivalent to the boundedness of the rescaled field at infinity, using the conformal method and the 2-spinor formalism. The question of the genericity of the peeling behavior is discussed. A new approach to these questions together with results by L.J. Mason and Nicolas are presented. The two approaches to asymptotic analysis described in Chapter 19 make a fundamental use of the notion of conformal compactification.

Beyond the results presented, the true focus of the essay is on the nature of spacetime: whereas many modern approaches to general relativity break the symmetry between time and space by performing a 3 + 1 splitting of the geometry, Penrose's approach truly deals with the 4-dimensional manifold and relies on causal objects like lightcones instead of Cauchy hypersurfaces.

Reading the texts of the ancient mathematicians always sheds a new light on the problems that nurture us every day. Regarding Riemann, Weil writes the following in his *Apprenticeship of a Mathematician*[13]:

> [...] In the same year, I began to read Riemann. Some time earlier, and first of all in reading Greek poets, I had become convinced that what really counts in the history of humanity are the truly great minds, and that the only way to get to know these minds was through direct contact with their works. I have since learned to modify this judgement quite a bit, though I have never really let go of it completely. My sister, however, who had come to a similar viewpoint—either on her own or perhaps partly under my influence—held on to it until the very end of her too short life. During my year of instruction in philosophy, I had also been struck by a phrase of Poincaré's which expresses no less an extreme position: "The value of civilizations lies only in their sciences and arts." With such ideas in my mind I had no choice but to dive headlong into the works of the great mathematicians of the past, as soon as they were materially and intellectually within my grasp. Riemann was the first; I read his inaugural dissertation and his major work on Abelian functions. Starting out thus was a stroke of luck of which I have always been grateful. These are not hard to read as long as one realizes that every word is loaded with meaning: there is perhaps no other mathematician whose writing matches Riemann's for density. Jordan's second volume was good preparation for studying Riemann. Moreover, the library[14] had a good collection of Felix Klein's mimeographed lecture notes, a large part of which is simply a rather discursive, but intelligent, commentary fleshing out of the extreme concision of Riemann's work.

The theme of the present volume, beyond the reference to Riemann's work, belongs to the more general profound interrelation between mathematics, physics and philosophy. The relation is multiple. Physics may exert an influence on mathematics and vice versa. Physics has also an impact on philosophy, and philosophy on mathematics. The ancient Greeks, the founders of mathematics as a deductive science in the way we intend it today, were completely aware of these interrelations. One may mention here Archimedes, Ptolemy and many others great figures. Euler, Poincaré and Cauchy were also physicists and philosophers, and they also wrote on the interrelations and the impact of these fields, each of them with his own style and according to his own interests. The subject of Euler's first public lecture, delivered in Basel in 1724 (the year Immanuel Kant was born), on the occasion of his obtention of his philosophy diploma, was the comparison between the philosophical systems of Newton and Descartes. Euler's philosophy is at some places religiously oriented and some of his philosophical writings are permeated with theological considerations. They were influential to his approach to physics. We allude to this and we give some examples in Chapter 1 of the present volume. The philosophical writings of Cauchy, who like Euler, was a devote Christian, and

[13]A. Weil, The Apprenticeship of a Mathematician, Springer, Basel, 1991. Translated from the French: Souvenirs d'apprentissage, Basel, Birkhäuser, 1991, p. 40.

[14]The library is meant to be that of the École Normale Supérieure.

who was furthermore involved in several charities, are also infused with religion. In several passages, he mentions the limitation of mathematics. For instance, in the introduction to his *Cours d'analyse*,[15] he writes the following:

> [...] Thus, let us be persuaded that there are other truths than those of algebra, realities other than sensible objects. Let us cultivate ardently the mathematical sciences, without trying to extend them beyond their domain; and let us not imagine that one can tackle history with formulae, or use theorems of algebra or of differential calculus as an assent to morals.[16]

Regarding physics and its development, we quote the following passage, from lectures that Cauchy gave in Turin in 1833, *Sept leçons de physique générale*, (Seven lessons on general physics), p. 5[17]:

> [...] Among these sciences, there is one in which all the power of analysis is manifested, and in which calculus, created by man, takes care of teaching him, through a mysterious language, the links that exist between phenomena which apparently are very different, and between the particular and the general laws of creation. This science, which we can trace back to the discovery of the principle of universal gravitation, was successively enriched by the immortal works of people like Descartes, Huygens, Newton, and Euler. But it is particularly since twenty years that the rapid improvement of mathematical analysis allowed him to make huge progress. It is since that epoch that we were able to apply calculus to the theory of elasticity, to that of heat propagation in solids or in space, of the propagation of waves on the surface of a heavy fluid, of the transmission of sound through solid bodies; to the theory of dynamical elasticity, to that of vibration of plates or elastic lamina; and finally to the theory of light including the various reflection phenomena, simple refraction, double refraction, polarization, colors, etc. Finally, it is since that epoch that important works of people like Ampère, Fourier, Poisson and of some others of which I do not need to remind you the names, were published.[18]

---

[15] A.-L. Cauchy, Cours d'analyse de l'École Royale Polytechnique, 1[re]partie. Analyse algébrique. Imprimerie royale, Paris, 1821. Œuvres complètes, série 2, tome III.

[16] [...] Soyons donc persuadés qu'il existe des vérités autres que les vérités de l'algèbre, des réalités autres que les objets sensibles. Cultivons avec ardeur les sciences mathématiques, sans vouloir les étendre au-delà de leur domaine ; et n'allons pas nous imaginer qu'on puisse attaquer l'histoire avec des formules, ni donner pour sanction à la morale des théorèmes d'algèbre ou de calcul intégral.

[17] A.-L. Cauchy, Sept leçons de physique générale, Paris, bureau du journal Les Mondes et Gauthier-Villars, 1868.

[18] [...] Parmi ces sciences, il est une où se manifeste toute la puissance de l'analyse, et dans laquelle le calcul créé par l'homme se charge de lui apprendre, par un mystérieux langage, les liaisons qui existent entre des phénomènes en apparence très divers, entre les lois particulières et les lois générales de la création. Cette science, qu'on peut faire monter à la découverte du principe de la gravitation universelle, a été successivement enrichie des immortels travaux des Descartes, des Huyghens, des Newton, des Euler. Mais c'est particulièrement depuis vingt ans que le perfectionnement rapide de l'analyse mathématique lui a permis de faire d'immenses progrès. C'est depuis cette époque qu'on a pu appliquer le calcul à la théorie de l'élasticité, de la propagation de la chaleur dans des corps ou dans l'espace, de la propagation des ondes à la surface d'un fluide pesant, de la transmission du son à travers les corps solides ; à la théorie de l'élasticité dynamique, à celle des vibrations des plaques ou des lames élastiques ; enfin à la théorie de la lumière comprenant les phénomènes divers de la réflexion, de la réfraction simple, de la double réfraction, de la polarisation, de la coloration, etc... C'est enfin depuis cette époque qu'ont été publiés les importants travaux des Ampère, des Fourier, des Poisson et de quelques autres dont il est inutile de vous rappeler les noms.

To stay close to the epoch of Riemann, we quote another one of his close predecessors, Joseph Fourier, from his *Théorie analytique de la chaleur* (Analytic theory of heat), published in 1822, a text which was very important for Riemann who refers to it in his habilitation memoir on trigonometric functions. On the relation between mathematics and the study of nature, Fourier writes in the Introduction to his work[19]:

> The thorough study of nature is the most profound productive source of mathematical discoveries. Not only this study, offering to the researches a specific purpose, has the advantage of excluding fuzzy questions and dead-end calculations; it is also a secure way of forming the heart of analysis, and of discovering there the elements whose knowledge is the most important to us, and which this science must always preserve: these fundamental elements are those which reproduce themselves in every natural effect. One can see, for instance that the same expression, which geometers had considered as an abstract property, and which from this respect belong to general analysis, also represents the motion of light in the atmosphere, that it determines the laws of diffusion of heat in solid matter, and that it enters in the main questions of the theory of probability.[20]

Poincaré was a prototype of the scientist-philosopher, and it was probably under his influence that most of the pre-eminent French mathematicians of his epoch became deeply interested in physics and philosophy. We quote him from his 1908 ICM talk (Rome)[21]:

> We cannot forget what our goal should be. As I see it, it is twofold. Our science borders at the same time on philosophy and on physics, and it is for our two neighbors that we are working. On the other hand, we have always seen, and we shall also see the mathematicians walking in two opposite directions. On the one hand, mathematical science must reflect on itself, and this is useful, because reflecting on itself means reflecting on the human mind that created it, all the more since this is his creation for which he borrowed the less from outside. This is why certain mathematical speculations are useful, like the ones which aim at the study of postulates, of unusual geometries, of functions with strange behavior. The more these speculations deviate from the most common conceptions, and consequently of the nature of their applications, the better they will show what human mind is able to do, when it avoids more and more the tyranny of the external world, and consequently, the

[19] J. Fourier, Théorie analytique de la chaleur, Paris, Firmin Didot, 1822, Discours préliminaire, p. xiii.

[20] L'étude approfondie de la nature est la source la plus féconde des découvertes mathématiques. Non seulement cette étude, offrant aux recherches un but déterminé, a l'avantage d'exclure les questions vagues et les calculs sans issue ; elle est encore un moyen assuré de former l'analyse elle-même, et d'en découvrir les éléments qu'il nous importe le plus de connaître, et que cette science doit toujours conserver : ces éléments fondamentaux sont ceux qui se reproduisent dans tous les effets naturels. On voit, par exemple, qu'une même expression, dont les géomètres avaient considéré les propriétés abstraites, et qui sous ce rapport appartient à l'analyse générale, représente aussi le mouvement de la lumière dans l'atmosphère, qu'elle détermine les lois de la diffusion de la chaleur dans la matière solide, et qu'elle entre dans les questions principales de la théorie des probabilités.

[21] H. Poincaré, l'Avenir des mathématiques, Atti del IV congresso internazionale dei matematici, Volume 1, Accademia dei Lincei, Rome, 1909, p. 167–182.

> more they will let us know it itself. But on the other hand, it is on the side of nature that we must direct the greater part of our army.[22]

Closer to us, Grothendieck,[23] who is quoted several times in the present volume and who at several places declared that he was a heir of Riemann, has a huge amount of still unpublished philosophical writings. In his *Récoltes et semailles*, which we already mentioned in this introduction, expressing his ideas about a "unitary theory" in physics, and after a long digression involving Euclid, Newton, Riemann and Einstein, Grothendieck writes (§2.20, Note 71):

> To summarize, I foresee that the long-awaited renewal (if ever it comes…) will rather come from someone who has the soul of a mathematician, who is well informed about the great problems of physics, rather than from a physicist. But above all, we need a man having the "philosophical openness" that is required to grasp the crux of the problem. The latter is not at all of a technical nature, but it is a fundamental problem of "natural philosophy."[24]

The present volume is a modest tribute to all those who taught us creative science.

Athanase Papadopoulos

[22]Nous ne pouvons oublier quel doit être notre but ; selon moi ce but est double ; notre science confine à la fois à la philosophie et à la physique, et c'est pour nos deux voisines que nous travaillons ; aussi nous avons toujours vu et nous verrons encore les mathématiciens marcher dans deux directions opposées. D'une part, la science mathématique doit réfléchir sur elle-même et cela est utile, parce que réfléchir sur elle-même, c'est réfléchir sur l'esprit humain qui l'a créée, d'autant plus que c'est celle de ses créations pour laquelle il a fait le moins d'emprunts au dehors. C'est pourquoi certaines spéculations mathématiques sont utiles, comme celles qui visent l'étude des postulats, des géométries inaccoutumées, des fonctions à allures étranges. Plus ces spéculations s'écarteront des conceptions les plus communes, et par conséquent de la nature et des applications, mieux elles nous montreront ce que l'esprit humain peut faire, quand il se soustrait de plus en plus à la tyrannie du monde extérieur, mieux par conséquent elles nous le feront connaître en lui-même. Mais c'est du côté opposé, du côté de la nature, qu'il faut diriger le gros de notre armée.

[23]The fact that Grothendieck was not interested in physics is a myth. It suffices to read his non-mathematical writings to be convinced of the contrary.

[24]Pour résumer, je prévois que le renouvellement attendu (s'il doit encore venir …) viendra plutôt d'un mathématicien dans l'âme, bien informé des grands problèmes de la physique, que d'un physicien. Mais surtout, il y faudra un homme ayant "l'ouverture philosophique" pour saisir le nœud du problème. Celui-ci n'est nullement de nature technique, mais bien un problème fondamental de "philosophie de la nature."

Looking Backward: From Euler to Riemann

Athanase Papadopoulos

Il est des hommes auxquels on ne doit pas adresser d'éloges,
si l'on ne suppose pas qu'ils ont le goût assez peu délicat
pour goûter les louanges qui viennent d'en bas.
(Jules Tannery, [240] p. 102)

Abstract We survey the main ideas in the early history of the subjects on which Riemann worked and that led to some of his most important discoveries. The subjects discussed include the theory of functions of a complex variable, elliptic and Abelian integrals, the hypergeometric series, the zeta function, topology, differential geometry, integration, and the notion of space. We shall see that among Riemann's predecessors in all these fields, one name occupies a prominent place, this is Leonhard Euler.

Keywords Bernhard Riemann · Function of a complex variable · Space · Riemannian geometry · Trigonometric series · Zeta function · Differential geometry · Elliptic integral · Elliptic function · Abelian integral · Abelian function · Hypergeometric function · Topology · Riemann surface · Leonhard Euler · Space · Integration

AMS Mathematics Subject Classification: 01-02 · 01A55 · 01A67 · 26A42 · 30-03 · 33C05 · 00A30.

A. Papadopoulos (✉)
Institut de Recherche Mathématique Avancée, Université de Strasbourg and CNRS,
7 rue René Descartes, 67084 Strasbourg Cedex, France
e-mail: athanase.papadopoulos@math.unistra.fr

A. Papadopoulos
Brown University, Mathematics Department, 151 Thayer Street, Providence, RI 02912, USA

L. Ji et al. (eds.), *From Riemann to Differential Geometry and Relativity*,
DOI 10.1007/978-3-319-60039-0_1

1 Introduction

More than any other branch of knowledge, mathematics is a science in which every generation builds on the accomplishments of the preceding ones, and where reading the old masters has always been a ferment for new discoveries. Examining the roots of Riemann's ideas takes us into the history of complex analysis, topology, integration, differential geometry and other mathematical fields, not to speak of physics and philosophy, two domains in which Riemann was also the heir of a long tradition of scholarship.

Riemann himself was aware of the classical mathematical literature, and he often quoted his predecessors. For instance, in the last part of his Habilitation lecture, *Über die Hypothesen, welche der Geometrie zu Grunde liegen* [230] (1854), he writes[1]:

> The progress of recent centuries in the knowledge of mechanics depends almost entirely on the exactness of the construction which has become possible through the invention of the infinitesimal calculus, and through the simple principles discovered by Archimedes, Galileo and Newton, and used by modern physics.

The references are eloquent: Archimedes, who developed the first differential calculus, with his computations of length, area and volume, Galileo, who introduced the modern notions of motion, velocity and acceleration, and Newton, who was the first to give a mathematical expression to the forces of nature, describing in particular the motion of bodies in resisting media, and most of all, to whom is attributed a celebrated notion of space, the "Newtonian space." As a matter of fact, the subject of Riemann's habilitation lecture includes the three domains of Newton's *Principia*: mathematics, physics and philosophy. It is interesting to note also that Archimedes, Galileo and Newton are mentioned as the three founders of mechanics in the introduction (Discours préliminaire) of Fourier's *Théorie analytique de la chaleur* ([116], pp. i–ii), a work in which the latter lays down the rigorous foundations of the theory of trigonometric series. Fourier's quote and its English translation are given in Sect. 10 of the present chapter. In the historical part of his Habilitation dissertation, *Über die Darstellbarkeit einer Function durch eine trigonometrische Reihe* (On the representability of a function by a trigonometric series) [215], a memoir which precisely concerns trigonometric series, Riemann gives a detailed presentation of the history of the subject, reporting on results and conjectures by Euler, d'Alembert, Lagrange, Daniel Bernoulli, Dirichlet, Fourier and others. The care with which Riemann analyses the evolution of this field, and the wealth of historical details he gives, is another indication of the fact that he valued to a high degree the history of ideas and was aware of the first developments of the subjects he worked on. In the field of trigonometric series and in others, he was familiar with the important paths and sometimes the wrong tracks that his predecessors took for the solutions of the problems he tackled. Riemann's sense of history is also manifest in the announcement of his memoir *Beiträge zur Theorie der durch die Gauss'sche Reihe* $F(\alpha, \beta, \gamma, x)$ *darstellbaren Functionen* (Contribution to the theory of functions representable by Gauss's series

[1] In all this chapter, for Riemann's habilitation, we use Clifford's translation [231].

$F(\alpha, \beta, \gamma, x)$), published in the *Göttinger Nachrichten*, No. 1, 1857, in which he explains the origin of the problems considered, mentioning works of Wallis, Euler, Pfaff, Gauss and Kummer. There are many other examples.

Among Riemann's forerunners in all the fields that we discuss in this chapter, one man fills almost all the background; this is Leonhard Euler. Riemann was an heir of Euler for what concerns functions of a complex variable, elliptic integrals, the zeta function, the topology of surfaces, the differential geometry of surfaces, the calculus of variations, and several topics in physics.

Riemann refers to Euler at several places of his work, and Euler was himself a diligent reader of the classical literature: Euclid, Pappus, Diophantus, Theodosius, Descartes, Fermat, Newton, etc. All these authors are mentioned all along his writings, and many of Euler's works were motivated by questions that grew out of his reading of them.[2] Before going into more details, I would like to say a few words about the lives of Euler and Riemann, highlighting analogies, but also differences between them.

Both Euler and Riemann received their early education at home, from their fathers, who were protestant ministers, and who both were hoping that their sons will become like them, pastors. At the age of fourteen, Euler attended a Gymnasium in Basel, while his parents lived in Riehen, a village near the city of Basel.[3] At about the same age, Riemann was sent to a Gymnasium in Hanover, away from his parents. During their Gymnasium years, both Euler and Riemann lived with their grandmothers.[4] They both enrolled a theological curriculum (at the Universities of Basel and Göttingen respectively), before they obtain their fathers' approval to shift to mathematics.

There are also major differences between the lives of the two men. Euler's productive period lasted 57 years (from the age of 19, when he wrote his first paper, until his death at the age of 76). His written production comprises more than 800 memoirs and 50 books. He worked on all domains of mathematics (pure and applied) and physics (theoretical and practical) that existed at his epoch. He also published on geography, navigation, machine theory, ship building, telescopes, the making of optical instruments, philosophy, theology and music theory. Besides his research books, he wrote elementary schoolbooks, including a well-known book on the art of reckoning [64]. The publication of his collected works was decided in 1907, the year of his bicentenary, the first volumes appeared in 1911, and the edition is still in progress (two volumes appeared in 2015), filling up to now more than 80 large volumes. Unlike Euler's Riemann's life was short. He published his first work at the age of 25 and he died at the age of 39. Thus, his productive period lasted only 15 years. His collected works stand in a single slim volume. Yet, from the points of view of the originality and the impact of their ideas, it would be unfair to affirm that either of them stands before the other. They both had an intimate and permanent

[2]Cf. for instance Euler's *Problematis cuiusdam Pappi Alexandrini constructio* (On a problem posed by Pappus of Alexandria), [96], 1780.

[3]Today, Riehen belongs to the Canton of the city of Basel, and it hosts the famous Beyeler foundation.

[4]In 1842, at the death of his grandmother, Riemann quitted Hanover and attended the Gymnasium at the Johanneum Lüneburg.

relation to mathematics and to science in general. Klein writes in his *Development of mathematics in the 19th century* ([162], p. 231 of the English translation):

> After a quiet preparation Riemann came forward like a bright meteor, only to be extinguished soon afterwards.

On Euler, I would like to quote André Weil, from his book on the history of number theory, *Number Theory: An approach through history from Hammurapi to Legendre* [255]. He writes, in the concluding section:

> [...] Hardly less striking is the fact that Euler never abandoned a problem after it has once aroused his insatiable curiosity. Other mathematicians, Hilbert for instance, have had their lives neatly divided into periods, each one devoted to a separate topic. Not so Euler. All his life, even after the loss of his eyesight, he seems to have carried in his head the whole of the mathematics of his day, both pure and applied. Once he has taken up a question, not only did he come back to it again and again, little caring if at times he was merely repeating himself, but also he loved to cast his net wider and wider with never failing enthusiasm, always expecting to uncover more and more mysteries, more and more "herrliche proprietates" lurching just around the next corner. Nor did it matter to him whether he or another made the discovery. "*Penitus obstupui*", he writes ("I was quite flabberggasted": *Eu*.I-21.1 in E 506|1777, cf. his last letter to Lagrange, *Eu*.IV A-5.505|1775) on learning Lagrange's additions to his own work on elliptic integrals; after which he proceeds to improve upon Lagrange's achievement. Even when a problem seemed to have been solved to his own satisfaction (as happened with his first proof of Fermat's theorem $a^p \equiv a \bmod p$, or in 1749 with sums of two squares) he never rested in his search for better proofs, "more natural" (*Eu*. I-2.510 in E 262|1755; cf. Sect. VI), "easy" (*Eu*.1-3.504 in E 522|1772; cf. Sect.VI), "direct" (*Eu*.I-2.363 in E 242|1751; cf. Sect. VI); and repeatedly he found them.[5]

Let us say in conclusion that if we had to mention a single mathematician of the eighteenth century, Euler would probably be the right choice. For the nineteenth century, it would be Riemann. Gauss, who will also be mentioned many times in the present chapter, is the main figure astride the two centuries.

Euler's results are contained in his published and posthumous writings, but also in his large correspondence, available in several volumes of his *Opera Omnia*. We shall mention several times this correspondence in the present chapter. It may be useful to remind the reader that at the epoch we are considering here, there were very few mathematical journals (essentially the publications of the few existing Academies of Sciences). The transmission of open problems and results among mathematicians was done largely through correspondence. On this question, let us quote the mathematician Paul Heinrich Fuss, who published the first set of letters of Euler, and who was his great-grandson. He writes in the introduction to his *Correspondence* [118], p. xxv[6]:

> Since sciences ceased to be the exclusive property of a small number of initiates, correspondence between scholars was taken over by the periodical publications. The progress is undeniable. However, this freeness with which ideas and discoveries were communicated in the past, in private and very confidential letters, we do not find it any more in the ripe and

[5]In Weil's book, every piece of historical information is accompanied by a precise reference. Works that attain this level of scholarship are very rare.

[6]Unless otherwise stated, the translations from the French in this chapter are mine.

> printed pieces of work. At that time, the life of a scholar was, in some way, all reflected in that correspondence. We see there the great discoveries being prepared and gradually developed; no link and no transition is missing; the path which led to these discoveries is followed step by step, and we can draw there some information even in the errors committed by these great geniuses who were the authors. This is sufficient to explain the interest tied to this kind of correspondence.[7]

In the case of Euler, particularly interesting is his correspondence with Christian Goldbach, published recently in two volumes of the *Opera Omnia* [109]. It contains valuable information on Euler's motivations and progress in several of the domains that are surveyed in the sections that follow, in particular, topology, the theory of elliptic functions and the zeta function. A few lines of biography on this atypical person are in order.

Christian Goldbach (1690–1764) was one of the first German scholars whom Euler met at the Saint Petersburg Academy of Sciences when he arrived there in 1827. He was very knowledgeable in mathematics, although he was interested in this field only in an amateurish fashion, encouraging others' works rather than working himself on specific problems. He was also a linguist and thoroughly involved in politics. Goldbach studied law at the University of Königsberg. In Russia, he became closely related to the Imperial family. In 1732, he was appointed secretary of the Saint Petersburg Academy of Sciences and in 1737 he became the administrator of that institution. In 1740, he held an important position at the Russian ministry of foreign affairs and became the official cryptographer there. Goldbach had a tremendous influence on Euler, by being attentive to his progress, by the questions he asked him on number theory, and also by motivating him to read Diophantus and Fermat. Goldbach, who was seventeen years older than Euler, became later on one of his closest friends and the godfather of his oldest son, Johann Albrecht, the only one among Euler's thirteen children who became a mathematician. Paul Heinrich Fuss writes in the introduction to his *Correspondence* [118], p. xxii:

> It is more than probable that if this intimate relationship between Euler and this scholar, a relationship that lasted 36 years without interruption, hadn't been there, then the science of numbers would have never attained the degree of perfection which it owes to the immortal discoveries of Euler.[8]

Goldbach kept a regular correspondence with Euler, Nicolas and Daniel Bernoulli, Leibniz (in particular on music theory) and many other mathematicians.

---

[7]Depuis que les sciences ont cessé d'être la propriété exclusive d'un petit nombre d'initiés, ce commerce épistolaire des savants a été absorbé par la presse périodique. Le progrès est incontestable. Cependant, cet abandon avec lequel on se communiquait autrefois ses idées et ses découvertes, dans des lettres toutes confidentielles et privées, on ne le retrouve plus dans les pièces mûries et imprimées. Alors, la vie du savant se reflétait, pour ainsi dire, tout entière dans cette correspondance. On y voit les grandes découvertes se préparer et se développer graduellement ; pas un chaînon, pas une transition n'y manque ; on suit pas à pas la marche qui a conduit à ces découvertes, et l'on puise de l'instruction jusque dans les erreurs des grands génies qui en furent les auteurs. Cela explique suffisamment l'intérêt qui se rattache à ces sortes de correspondances.

[8]Il me semble plus que probable que si cette liaison intime entre Euler et ce savant, liaison qui dura 36 ans sans interruption, n'eût pas lieu, la science des nombres n'aurait guère atteint ce degré de perfection dont elle est redevable aux immortelles découvertes d'Euler.

After Goldbach and his influence on Euler, we turn to Gauss, who, among the large number of mathematicians with whom Riemann was in contact, was certainly the most influential on him.[9] We shall see in the various sections of the present chapter that this influence was crucial for what concerns the fields of complex analysis, elliptic integrals, topology, differential geometry—the same list as for Euler's influence on Riemann—and also for what concerns his ideas on space. There are other topics in mathematics and physics which were central in the work of Riemann and where he used ideas he learned from Gauss: the Dirichlet principle, magnetism, etc.; they are addressed in several other chapters of the present book.

The first contact between Gauss and Riemann took place probably in 1846, before Gauss became officially Riemann's mentor. In that year, in a letter to his father dated November 5 and translated in [233], Riemann informs the latter about the courses he plans to follow, and among them he mentions a course by Gauss on "the theory of least squares."[10] During his two year stay in Berlin (1847–1849), Riemann continued to study thoroughly Gauss's papers. In another letter to his father, dated May 30, 1849, he writes (translation in [233]):

> Dirichlet has arranged to me to have access to the library. Without his assistance, I fear there would have been obstacles. I am usually in the reading room by nine in the morning, to read two papers by Gauss that are not available anywhere else. I have looked fruitlessly for a long time in the catalog of the royal library for another work of Gauss, which won the Copenhagen prize, and finally just got it through Dr. Dale of the Observatory. I am still studying it.

During the same stay in Berlin, Riemann followed lectures by Dirichlet on topics related to Gauss's works. He writes to his father (letter without date, quoted in [233]):

> My own course of specialization is the one with Dirichlet; he lectures on an area of mathematics to which Gauss owes his entire reputation. I have applied myself very seriously to this subject, not without success, I hope.

Regarding his written production, Riemann endorsed Gauss' principle: *pauca sed matura* (few but ripe).

Riemann, as a child, liked history. In a letter to his father, dated May 3, 1840 (he was 14), he complains about the fact that at his Gymnasium there were fewer

---

[9]Some historians of mathematics claimed that when Riemann enrolled the University of Göttingen, as a doctoral student of Gauss, the latter was old and in poor health, and that furthermore, he disliked teaching. From this, they deduced that Gauss's influence on Riemann was limited. This is in contradiction with the scope and the variety of the mathematical ideas of Riemann for which he stated, in one way or another, but often explicitly, that he got them under the direct influence of Gauss or by reading his works. The influence of a mathematician is not measured by the time spent talking with him or reading his works. Gauss died the year after Riemann obtained his habilitation, but his imprint on him was permanent.

[10]The other courses are on the Cultural History of Greece and Rome, Theology, Recent Church history, General Physiology and Definite Integrals. Riemann had also the possibility to choose courses among Probability, Mineralogy and General Natural History. He adds: "The most useful to me will be mineralogy. Unfortunately it conflicts with Gauss's lecture, since it is scheduled at 10 o'clock, and so I'd be able to attend only if Gauss moved his lecture forward, otherwise it looks like it won't be possible. General Natural History would be very interesting, and I would certainly attend, if along with everything else I had enough money.".

lessons on history than on *Rechnen* (computing), cf. [233]. On August 5, 1841, he writes, again to his father, that he is the best student in history in his class. Besides history, Riemann was doing very well in Greek, Latin, and German composition (letters of February 1, 1845 and March 8, 1845). According to another letter to his father, dated April 30, 1845, it is only in 1845 that Riemann started being really attracted by mathematics. In the same letter, Riemann declares that he plans to enroll the University of Göttingen to study theology, but that in reality he must decide for himself what to do, since otherwise he "will bring nothing good to any subject."

Besides Euler and Gauss, we shall mention several other mathematicians. Needless to say, it would have been unreasonable to try to be exhaustive in this chapter; the subject would need a book, and even several books. We have tried to present a few markers on the history of the major questions that were studied by Riemann, insisting only on the mathematicians whose works and ideas had an overwhelming impact on him.

The content of the rest of this chapter is the following.

Section 2 is essentially an excursion into the realm of Euler's ideas on the notion of function, with a stress on algebraic functions and functions of a complex variable. Algebraic functions are multivalued, and Euler included these functions as an important element of the foundations of the field of analysis, which he laid down in his famous treatise *Introductio in analysin infinitorum* (Introduction to the analysis of the infinite) [61]. Riemann's work on what became known as Riemann surfaces was largely motivated by the desire to find a domain of definition for an algebraic multi-valued function on which it becomes single-valued.[11] The study of functions of a complex variable, which includes as a special case that of algebraic functions, is one of the far-reaching subjects of Riemann's investigations, and its development is one of the few most important achievements of the nineteenth century (probably the most important one).

Section 3 is concerned with elliptic integrals. These integrals constitute a class of complex functions with new interesting properties, and the work described in this section is a natural sequel to that which is reviewed in Sect. 2. We shall mention works done on this subject by Johann Bernoulli, Fagnano, Euler (who published thirty-three memoirs on elliptic integrals), Legendre, Abel and Jacobi.

Section 4 focusses on Abelian functions, a vast generalization of elliptic functions, which led to an important problem in which Riemann became interested, namely, the Jacobi inversion problem, and which he eventually solved using ϑ functions. In fact, Abelian integrals constitute one of the major topics that Riemann worked on. He started his investigation on this subject in his doctoral dissertation [214] (1851), worked on it in his 1854 memoir [217] whose title is quite rightly "The theory of Abelian functions," and he never stopped working and lecturing on it during the few years that were left to him. Some lecture notes and memoirs by Riemann on Abelian functions were published posthumously. In particular, his memoir *Über das Verschwinden der ϑ-Functionen* (On the vanishing on theta functions) [224], in which he gives a solution to Jacobi's problem of inversion for the general case of

[11] As a matter of fact, this is the origin of the use of the word "uniformization" by Riemann.

integrals of algebraic functions, is analyzed in Chap. 4 of the present volume, written by Houzel [141].

Section 5 is concerned with the so-called Gauss hypergeometric series. These series, in various forms, were studied by Euler in his *Institutiones calculi integralis* (Foundations of integral calculus), a treatise in three volumes [92] (1768–1770), and in several other papers by him, and by Gauss. The hypergeometric series is a family of functions of the form

$$1+\frac{\alpha\beta}{1.\gamma}x+\frac{\alpha(\alpha+1)\beta(\beta+1)}{1.2\gamma(\gamma+1)}x^2+\frac{\alpha(\alpha+1)(\alpha+2)\beta(\beta+1)(\beta+2)}{1.2.3\gamma(\gamma+1)(\gamma+2)}x^3+\dots$$

where α, β, γ are parameters and where the variable is x.

Gauss considered that almost any transcendental function is obtained from a hypergeometric series by assigning special values to the parameters. By providing such a broad class of functions, the introduction of the hypergeometric series in the field of analysis opened up new paths. Besides Euler and Gauss, the predecessors of Riemann in this field include Pfaff and Kummer.

In Sect. 6, we deal with the zeta function. The history of this function is sometimes traced back to the work of Pietro Mengoli (1625–1686) on the problem of finding the value of the infinite series of inverses of squares of integers. Indeed, it is reasonable to assume that questions about this series were accompanied by questions about the series of inverses of cubes and other powers. But it was Euler again who studied $\sum_1^\infty \frac{1}{n^s}$ as a function of s (for s real), establishing the functional equation that it satisfies, and the relation with prime numbers. This was the starting point of Riemann's investigations on what became later known as the Riemann zeta function.

In Sect. 7, we make a quick review of some works done by Riemann's predecessors on the notion of space. This is essentially a philosophical debate, but it has a direct impact on mathematics and in particular on Riemann's work on geometry, more especially on his habilitation dissertation. It is in his reflections on space that Riemann introduced in mathematics the notion of Mannigfaltigkeit, which he borrowed from the philosophical literature. This notion reflects Riemann's multi-faced view on space, and it is an ancestor of the modern notion of manifold. Our review of space is necessarily very sketchy, since this notion is one of the most fundamental notions of philosophy, and talking seriously about it would require a whole essay. In particular, there is a lot to say on the philosophy of space in the works of Newton, Euler and Riemann and the comparison between them, but it is not possible to do it in the scope of the present chapter. Our intent here is just to indicate some aspects of the notion of space as it appears in the works of these authors and those of some other philosophers, including Aristotle, Descartes and Kant, and, as much as possible in a short survey, to give some hints on the context in which they emerge.

It is also important to say that the effect of this discussion on space goes far beyond the limits of philosophy. Euler's theories of physics are strongly permeated with his philosophical ideas on space. Gauss's differential geometry was motivated by his investigations on physical space, more precisely, on geodesy and astronomy, and, more generally, by his aspiration to understand the world around him. At a more philosophical level, Gauss was an enthusiastic reader of Kant, and he criticized the latter's views on space, showing that they do not agree with the recent discoveries—his own and others'—of geometry. Riemann, in this field, was an heir of Gauss. In his work, the curvature of space (geometric space) is the expression of the physical forces that act on it. These are some of the ideas that we try to convey in Sect. 7 and in other sections of this chapter.

Section 8 is concerned with topology. Riemann is one of the main founders of this field in the modern sense of the word, but several important topological notions may be traced back to Greek antiquity and to the later works of Descartes, Leibniz and Euler. We shall review the ideas of Leibniz, and consider in some detail the works of Descartes and Euler on the so-called Euler characteristic of a convex polyhedral surface, which in fact is nothing else but an invariant of the topological sphere, a question whose generalization is contained in Riemann's doctoral dissertation [214] and his paper on Abelian functions [217], from where one can deduce the invariants of surfaces of arbitrary genus.

Section 9 is concerned with the differential geometry of surfaces. We review essentially the works of Euler, Gauss and Riemann, but there was also a strong French school of differential geometry, operating between the times of Euler and Riemann, involving, among others, Monge and several of his students, and, closer to Riemann, Bonnet.

Section 10 is a review of the history of trigonometric series and the long controversy on the notion of function that preceded this notion. In his Habilitation memoir, Riemann describes at length this important episode of eighteenth and nineteenth century mathematics which also led to his discovery of the theory of integration, which we discuss in the next section.

In Sect. 11, we review some of the history of the Riemann integral. From the beginning of integral calculus until the times of Legendre, passing through Euler, integration was considered as an antiderivative. Cauchy defined the integral by limits of sums that we call now Riemann sums, taking smaller and smaller subdivisions of the interval of integration and showing convergence to make out of that a definition of the definite integral, but he considered only integrals of continuous functions, where convergence is always satisfied. It was Riemann who developed the first general theory of integration, leading to the notion of integrable and non-integrable function.

The concluding section, Sect. 12 contains a few remarks on the importance of returning to the texts of the old masters.

Some of the historical points in our presentation are described in more detail than others; this reflects our personal taste and intimate opinion on what is important in history and worth presenting in more detail in such a quick survey. The reader will find at the end of this chapter (before the bibliography) a table presenting in parallel some works of Euler and of Riemann on related matters.

2 Functions

Vito Volterra, in his 1900 Paris ICM plenary lecture [250], declared that the nineteenth century was "the century of function theory."[12] In the language of that epoch, the expression "function theory" refers, in the first place, to functions of a complex variable. One of the mottos, which was the result of a thorough experience in the domain, was that a function of a real variable acquires its full strength when it is complexified, that is, when it is extended to become a function of a complex variable. This idea was shared by Cauchy, Riemann Weierstrass, and others to whom we refer now.

On functions of a complex variable, we first quote a letter from Lagrange to Antonio Lorgna, an engineer and the governor of the military school at Verona who made important contributions to mathematics, physics and chemistry. The letter is dated December 20, 1777. Lagrange writes (cf. Lagrange's Œuvres, [166] t. 14, p. 261):

> I consider it as one of the most important steps made by Analysis in the last period, that of not being bothered any more by imaginary quantities, and to be able to submit them to calculus, in the same way as the real ones.[13]

Gauss, who, among other titles he carried, was one of the main founders of the theory of functions of a complex variable, was also responsible for the introduction of complex numbers in several theories. In particular, he realized their power in number theory, and he used this in his *Disquisitiones arithmeticae* (Arithmetical researches) [121] (1801), a masterpiece he wrote at the age of 24. In his second paper on biquadratic residues [120] (Section 30), he writes that "number theory is revealed in its entire simplicity and natural beauty when the field of arithmetic is extended to the imaginary numbers." He explains that this means admitting integers of the form $a + bi$. "Such numbers," he says, "will be called complex integers."

In the same vein, Riemann, who had a marked philosophical viewpoint on things, writes, regarding complex functions, in Section 20 of his doctoral dissertation, *Grundlagen für eine allgemeine Theorie der Functionen einer veränderlichen complexen Grösse* [214] (Foundations of a general theory of functions of a variable complex magnitude) [214] (1851): "Attributing complex values to the variable quantities reveals a harmony and a regularity which otherwise would remain hidden."

---

[12]The title of Volterra's lecture is: *Betti, Brioschi, Casorati : Trois analystes italiens et trois manières d'envisager les questions d'analyse* (Betti, Brioschi, Casorati: Three Italian analysts and three manners of addressing the analysis questions). In that lecture, Volterra presents three different ways of doing analysis, through the works of Betti, Brioschi and Casorati, who are considered as the founders of modern Italian mathematics. The three mathematicians had very different personalities, and contrasting approaches to analysis, but in some sense they were complementing each other. In particular, Brioschi was capable of doing very long calculations, Betti was a geometer repugnant to calculations, and Casorati was an excellent teacher and an applied mathematician.

[13]Je regarde comme un des pas les plus importants que l'Analyse ait faits dans ces derniers temps, de n'être plus embarrassée des quantités imaginaires et de pouvoir les soumettre au calcul comme les quantités réelles.

Finally, let us quote someone closer to us, Jacques Hadamard, from his *Psychology of invention in the mathematical field* [137]. A sentence by him which is often repeated is that "the shortest and the best way between two truths of the real domain often passes by the imaginary one." We quote the whole passage ([137] pp. 122–123):

> It is Cardan, who is not only the inventor of a well-known joint which is an essential part of automobiles, but who has also fundamentally transformed mathematical science by the invention of imaginaries. Let us recall what an imaginary quantity is. The rules of algebra show that the square of any number, whether positive or negative, is a positive number: therefore, to speak of the square root of a negative number is mere absurdity. Now, Cardan deliberately commits that absurdity and begins to calculate on such "imaginary" quantities.
>
> One would describe this as pure madness; and yet the whole development of algebra and analysis would have been impossible without that fundament—which, of course, was, in the nineteenth century, established on solid and rigorous bases. It has been written that the shortest and the best way between two truths of the real domain often passes by the imaginary one.

In the rest of this section, we review some markers in the history of functions, in particular functions of a complex variable and algebraic functions, two topics which are at the heart of Riemann's work on Riemann surfaces, on Abelian functions, on the zeta function, and on other topics. Before that, we make a digression on the origin of the general notion of function.

It is usually considered that Euler's *Introductio in analysin infinitorum* [61] is the first treatise in which one can find the definition of a function, according to modern standards, and where functions are studied in a systematic way. We take this opportunity to say a few words on Euler's treatise, to which we refer several times in the rest of this chapter.

The *Introductio* is a treatise in two volumes, first published in 1748, which is concerned with a variety of subjects, including (in the first volume) algebraic curves, trigonometry, logarithms, exponentials and their definitions by limits, continued fractions, infinite products, infinite series and integrals. The second volume is essentially concerned with the differential geometry of curves and surfaces. The importance of the *Introductio* lies above all in the fact that it made analysis the branch of mathematics where one studies functions. But the *Introductio* is more than a treatise with a historical value. Two hundred and thirty years after the first edition appeared in print, André Weil considered that it was more useful for a student in mathematics to study that treatise rather than any other book on analysis. This is reported on by John Blanton who writes, in his English edition of the *Introductio* [62]:

> In October, 1979, Professor André Weil spoke at the University of Rochester on the life and work of Leonhard Euler. One of his remarks was to the effect that he was trying to convince the mathematical community that students of mathematics would profit much more from a study of Euler's *Introductio in analysin infinitorum*, rather than the available modern textbooks.

The importance of this work has also been highlighted by several other mathematicians. C. B. Boyer, in his 1950 ICM communication (Cambridge, Mass.) [28], compares the impact of the *Introductio* to that of Euclid's *Elements* in geometry and to al-Khwārizmī's *Jabr* in algebra. He writes:

> The most influential mathematics textbooks of ancient times (or, for that matter, of all times) is easily named. The *Elements* of Euclid, appearing in over a thousand editions, has set the pattern in elementary geometry ever since it was composed more than two and a quarter millenia ago. The medieval textbook which most strongly influenced mathematical development is not so easily selected; but a good case can be made out of *Al-jabr wal muqābala* of al-Khwārizmī, just about half as old as the *Elements*. From this Arabic work, algebra took its name and, to a great extent, its origin. Is it possible to indicate a modern textbook of comparable influence and prestige? Some would mention the *Géométrie* of Descartes, or the *Principia* of Newton or the *Disquisitiones* of Gauss; but in pedagogical significance these classics fell short of a work less known. [...] over these well known textbooks there towers another, a work which appeared in the very middle of the great textbook age and to which virtually all later writers admitted indebtedness. This was the *Introductio in analysin infinitorum* of Euler, published in two volumes in 1748. Here in effect Euler accomplished for analysis what Euclid and al-Khwārizmī had done for synthetic geometry and elementary algebra, respectively.

Even though the *Introductio* is generally given as the main reference for the introduction of functions in analysis, regarding the *usage* of functions, one can go far back into history. Tables of functions exist since the Babylonians (some of their astronomical tables survive). Furthermore, in ancient Greece, mathematicians manipulated functions, not only in the form of tables. In particular, the chord function (an ancestor of our sine function)[14] is used extensively in some Greek treatises. For instance, Proposition 67 of Menelaus' *Spherics* (1st-2nd century A.D.) says the following [212]:

> Let ABC and DEG be two spherical triangles whose angles A and D are equal, and where C and G are either equal or their sum is equal to two right angles. Then,
>
> $$\frac{\text{crd } 2AB}{\text{crd } 2BC} = \frac{\text{crd } 2DE}{\text{crd } 2EG}.$$

Youschkevitch, in his interesting survey [260], argues that the general idea of a dependence of a quantity upon another one is absent from Greek geometry. The author of the present chapter declares that if in the above proposition of Menelaus one does not see the notion of function, and hence the general idea of a dependence of a quantity upon another one, then this author fails to know what mathematicians mean by the word function.

Leibniz and Johann I Bernoulli, who were closer to Euler, manipulated functions, even though the functions they considered were always associated with geometrical objects, generally, curves in the plane. For instance, in a memoir published in 1718 on the isoperimetry problem in the plane, [24] Bernoulli writes:

> Here, we call *function* of a variable magnitude, a quantity formed in whatever manner with that variable magnitude and constants.[15]

The functions that Bernoulli considers in this memoir are associated to *arbitrary* curves in the plane having the same perimeter, among which Bernoulli looks for the

[14]The relation between chord and sine is : $\sin x = \frac{1}{2}\text{crd } 2x$.

[15]On appelle ici *Fonction* d'une grandeur variable, une quantité composée de quelque manière que ce soit avec cette grandeur variable et des constantes. [The emphasis is Bernoulli's].

one which bounds the greatest area. This is an example of the general idea that before Euler, analysis was tightly linked to geometry, and the study of functions consisted essentially in the study of curves associated to some geometric properties. With the *Introductio*, things became different. Analysis started to release itself from geometry, and functions were studied for themselves. Let us now make a quick review of the part of this treatise which concerns us here.

The first chapter is called *On functions in general*. In this chapter, Euler states his general definition of a function, after a description of what is a variable quantity:

> A function of a variable quantity is an analytic expression composed in any way whatsoever of the variable quantity and numbers or constant quantities.

The word "analytic" means in this context that the function is obtained by some process that uses the four operations (addition, subtraction, multiplication and division), together with root extraction, exponentials, logarithms, trigonometric functions, derivatives and integrals. Analyticity in terms of being defined by a convergent power series is not intended by this definition. The meaning of the word "analytic function" rather is "a function used in (the field of) analysis." Concerning the notion of variable, Euler writes (Sect. 3)[16]:

> [...] Even zero and complex numbers are not excluded from the signification of a variable quantity.

Thus, functions of a complex variable are included in Euler's *Introductio*. We note however that in this treatise, Euler, in his examples, always deals with functions that are given by formulae: polynomials, exponentials, logarithms, trigonometric functions, etc. but also infinite products and infinite sums.

After the definition of a function, we find in the *Introductio* the definition of an algebraic function. In Sect. 7, Euler writes:

> Functions are divided into algebraic and transcendental. The former are those made up from only algebraic operations, the latter are those which involve transcendental operations.

And in Sect. 8:

> Algebraic functions are subdivided into non-irrational and irrational functions: the former are such that the variable quantity is in no way involved with irrationality; the latter are those in which the variable quantity is affected by radical signs.

Concerning irrational functions (Sect. 9), he writes:

> It is convenient to distinguish these into explicit and implicit irrational functions.
>
> The explicit functions are those expressed with radical signs, as in the given examples. The implicit are those irrational functions which arise from the solution of equations. Thus Z is an implicit irrational function of z if it is defined by an equation such as $Z^7 = az$ or $Z^2 = bz^5$. Indeed, an explicit value of Z may not be expressed even with radical signs, since common algebra has not yet developed to such a degree of perfection.

---

[16]We are using the translation from Latin in [61].

And in Sect. 10:

Finally, we must make a distinction between single-valued and multi-valued functions.

A single-valued function is one for which, no matter what value is assigned to the variable z, a single value of the function is determined. On the other hand, a multi-valued function is one such that, for some value substituted for the variable z, the function determines several values. Hence, all non-irrational functions, whether polynomial or rational, are single-valued functions, since expressions of this kind, whatever value be given to the variable z, produce a single value. However, irrational functions are all multi-valued, because the radical signs are ambiguous and give paired values. There are also among the transcendental functions, both single-valued and multi-valued functions; indeed, there are infinite-valued functions. Among these are the arcsine of z, since there are infinitely many circular arcs with the same sine.

Euler then gives examples of two-valued, three-valued and four-valued functions, and in Sect. 14 he writes:

Thus Z is a multi-valued function of z which for each value of z, exhibits n values of Z where n is a positive integer. If Z is defined by this equation

$$Z^n - PZ^n - 1 + QZ^{n-2} - RZ^{n-3} + SZ^{n-4} - \cdots = 0$$

[...] Further it should be kept in mind that the letters P, Q, R, S, etc. should denote single-valued functions of z. If any of them is already a multi-valued function, then the function Z will have many more values, corresponding to each value of z, than the exponent would indicate. It is always true that if some of the values are complex, then there will be an even number of them. From this we know that if n is an odd number, there will be at least one real value of Z.

He then makes the following remarks:

If Z is a multi-valued function of z such that it always exhibits a single real value, then Z imitates a single-valued function of z, and frequently can take the place of a single-valued function.

Functions of this kind are $P^{\frac{1}{3}}$, $P^{\frac{1}{5}}$, $P^{\frac{1}{7}}$, etc. which indeed give only one real value, the others all being complex, provided P is a single-valued function of z. For this reason, an expression of the form $P^{\frac{m}{n}}$, whenever n is odd, can be counted as a single-valued function, whether m is odd or even. However, if n is even then $P^{\frac{m}{n}}$ will have either no real value or two; for this reason, expressions of the form $P^{\frac{m}{n}}$, with n even, can be considered to be two-valued functions, provided the fraction $\frac{m}{n}$ cannot be reduced to lower terms.

From this discussion we single out the fact that algebraic functions are considered as functions, even though they are multi-valued. They are solutions of algebraic equations. Since we are talking about history, it is good to recall that the study of such equations is an old subject that can be traced back to the work done on algebraic curves by the Greeks. In fact, Diophantus (3rd century B.C.) thoroughly studied integral solutions of what is now called "Diophantine equations." They are examples of algebraic equations.[17] Algebraic equations are also present in the background of the

[17]For what concerns Diophantus' *Arithmetica*, we refer the interested reader to the recent and definitive editions [52–54, 210] by R. Rashed.

geometric work of Apollonius (3d–2d century B.C.) on conics. In that work, intersections of conics were used to find geometrical solutions of algebraic equations.[18] It is true however that in these works, there is no *definition* of an algebraic function as we intend it today, and in fact at that time there was no definition of function at all.

The multi-valuedness of algebraic functions gave rise to tremendous developments by Cauchy and Puiseux, and it was also a major theme in Riemann's work, in particular in his doctoral dissertation [214] (1851) and his memoir on Abelian functions [217] (1857). In fact, the main reason for which Riemann introduced the surfaces that we call today Riemann surfaces was to find ground spaces on which multi-valued functions are defined and become single-valued. We discuss the works of Cauchy and Puiseux in relation with that of Riemann in Chap. 7 of the present volume, [191].

We note for later use that a definition of "continuity" is given in Volume 2 of the *Introductio*, where Euler says that a curve is continuous if it represents "one determinate function," and discontinuous if it is decomposed into "portions that represent different continuous functions." We shall see that such a notion was criticized by Cauchy (regardless of the fact that it is called "continuity").[19]

We note finally that it is usually considered that the expression *analysin infinitorum* in the title of Euler's treatise does not refer to the field of infinitesimal analysis in the sense of Newton or Leibniz, but, rather, to the use of infinity (infinite series, infinite products, continued fractions expansions, integral representations, etc.) in analysis. Euler was also the first to highlight the zeta function, the gamma function and elliptic integrals as functions. However, it is good to recall that infinite sums were known long before Euler. For instance, Zeno of Elea (5th c. B.C.) had already addressed the question of convergence of infinite series, and to him are attributed several well-known paradoxes in which the role and the significance of infinite series and their convergence are emphasized (the paradox of Achilles and the tortoise, the arrow paradox, the paradox of the grain of millet, etc.). However, infinite series are not considered as functions in these works. Zeno's paradoxes are commented in detail in Aristotle's *Physics* [20], but also by mathematicians and philosophers from the modern period, including Bertrand Russell, Hermann Weyl, Paul Tannery and several others; cf. [236] pp. 346–354, and [241, 258].

We also recall that convergent series were used by Archimedes in his computations of areas and volumes.

Before leaving this book, let us mention that Euler establishes there a hierarchy among transcendental functions by introducing a notion close to what we call today the transcendence degree of a function.

In his later works, Euler dealt with much more general functions. For instance, in his 1755 memoir [103], entitled *Remarques sur les mémoires précédents de M. Bernoulli* (Remarks on the preceding memoirs by Mr. Bernoulli), any mechanical

---

[18]For a recent and definitive edition of Apollonius' *Conics*, we refer the reader to the volumes [11–15], again edited by R. Rashed.

[19]There are other imperfections in the *Introductio*, even though this book is one of the most interesting treatises ever written on elementary analysis.

curve (that is any curve drawn by hand) is associated with a function.[20] In his *Institutiones calculi differentialis cum eius usu in analysi finitorum ac doctrina serierum* (Foundations of differential calculus, with applications to finite analysis and series) [83], also published in 1755, Euler gave again a very general definition of a function (p. vi):

> Those quantities that depend on others in this way, namely, those that undergo a change when others change, are called functions of these quantities. This definition applies rather widely and includes all ways in which one quantity could be determined by another.

Likewise, in his memoir [104], *De repraesentatione superficiei sphaericae super plano* (On the representation of Spherical Surfaces onto the Plane) (1777), Euler dealt with "arbitrary mappings" between the sphere and the plane. He writes[21]:

> I take the word "mapping" in the widest possible sense; any point of the spherical surface is represented on the plane by any desired rule, so that every point of the sphere corresponds to a specified point in the plane, and inversely.

We shall consider again the question of functions, from the epoch of Euler and until the work of Riemann, in Sect. 10 concerned with trigonometric functions.

Riemann, in his doctoral dissertation, [214] (1851), also considers arbitrary functions. In fact, the dissertation starts as follows: "If we designate by z a variable magnitude, which may take successively all possible real values, then, when to each of these values corresponds a unique value of the indeterminate magnitude w, we say that w is a function of z [...]" One may also refer to the beginning of Sect. XIX of the same dissertation, where Riemann states that the principles he is presenting are the bases of a general theory of functions which is independent of any explicit expression.

The details of the seventeenth-century debate concerning functions are rather confusing if one does not include them in their historical context. For instance, the notion of "continuity" which we alluded to and which is referred to in the debate is different from what we intend today by this word. In fact, the word "continuity," even restricted to the works of Euler, varied in the course of time.

Cauchy, the major figure standing between Euler and Riemann for what concerns the notion of function, in his *Mémoire sur les fonctions continues* (Memoir on continuous functions) [37], starts as follows:

---

[20] One may recall here that the mathematicians of Greek antiquity (Archytas of Tarentum, Hippias, Archimedes, etc.) who examined curves formulated a mechanical definition. The curves with which they dealt were not necessarily defined by equations, they were "traced by a moving point," sometimes (in theory) using a specific mechanical device. Of some interest here would be the connections between this subject and the theory of mechanical linkages, which was extensively developed in the nineteenth century and became fashionable again in the twentieth century. A conjecture by Thurston says (roughly speaking) that any "topological curve" is drawable by a mechanical linkage. This is a vast generalization of a result of Kempe stating that any bounded piece of an algebraic curve is drawable by some linkage, cf. [160]. We refer the reader to Sossinsky's survey of this subject and its recent developments [238], in particular the solution of Thurston's conjecture.

[21] We are using George Heines' translation.

> In the writings of Euler and Lagrange, a function is termed *continuous* or *discontinuous* according to whether the various values of this function corresponding to various values of the variable follow or not the same law, or are given or not by only one equation. It is in these terms that the continuity of functions was defined by these famous geometers, when they used to say that "the arbitrary functions, introduced by the integration of partial differential equations, may be continuous or discontinuous functions." However, the definition which we just recalled is far from offering mathematical accuracy [...] A simple change in notation will often suffice to transform a continuous fonction into a discontinuous one, and conversely.[22]

In fact, one might consider that Euler's definition of continuity is just one definition that is different from the new definition which Cauchy had in mind (and which is the definition we use today). This would have been fine, and it would not be the only instance in mathematics where the same word is used for notions that are different, especially at different epochs. But Cauchy showed by an example that in this particular case Euler's definition is inconsistent, because the property it expresses depends on the parametrization that is used. Cauchy continues:

> But the non-determinacy will cease if we substitute to Euler's definition the one I gave in Chapter II of the *Analyse algébrique*. According to the new definition, a function of the variable x will be *continuous* between two limits a and b of this variable if between two limits the function has always a value which is unique and finite, in such a way that an infinitely small increment of this variable always produces an infinitely small increment of the function itself.[23]

We quoted these texts in order to give an idea of the progress of the notion of continuity. We now come to the study of functions of a complex variable.

In his memoir on Abelian functions, Riemann refers explicitly to Gauss for the fact that we represent a complex magnitude $z = x + iy$ by a point in the plane with coordinates x and y.

It is not easy to know when the theory of functions of a complex variable started, and, in fact, the answer depends on whether one studies holomorphic functions, and what properties of holomorphic functions are meant (before the epoch of Riemann, they were not known to be equivalent): angle-preservation, power series expansion, the Cauchy-Riemann equation, etc.

Euler used complex variables and the notion of conformality (angle-preservation) in his memoirs on geographical maps. He wrote three memoirs on this subject,

---

[22]Dans les ouvrages d'Euler et de Lagrange, une fonction est appelée *continue* ou *discontinue*, suivant que les diverses valeurs de cette fonction, correspondantes à diverses valeurs de la variable, sont ou ne sont pas assujetties à une même loi, sont ou ne sont pas fournies par une seule équation. C'est en ces termes que la continuité des fonctions se trouvait définie par ces illustres géomètres, lorsqu'ils disaient que "les fonctions arbitraires, introduites par l'intégration des équations aux dérivées partielles, peuvent être des fonctions continues ou discontinues." Toutefois, la définition que nous venons de rappeler est loin d'offrir une précision mathématique [...] un simple changement de notation suffira souvent pour transformer une fonction continue en fonction discontinue, et réciproquement.

[23]Mais l'indétermination cessera si à la définition d'Euler on substitue celle que j'ai donnée dans le chapitre II de l'*Analyse algébrique*. Suivant la nouvelle définition, une fonction de la variable réelle x sera continue entre deux limites a et b de cette variable, si, entre ces limites, la fonction acquiert constamment une valeur unique et finie, de telle sorte qu'un accroissement infiniment petit de la variable produise toujours un accroissement infiniment petit de la fonction elle-même.

De repraesentatione superficiei sphaericae super plano (On the representation of spherical surfaces on a plane) [104], *De proiectione geographica superficiei sphaericae* (On the geographical projections of spherical surfaces) [105], and *De proiectione geographica Deslisliana in mappa generali imperii russici usitata* (On Delisle's geographic projection used in the general map of the Russian empire) [106]. The three memoirs were published in 1777. In the development of the theory, he used complex numbers to represent angle-preserving maps. Lagrange also studied angle-preserving maps, in his memoir *Sur la construction des cartes géographiques* (On the construction of geographical maps) [165], published in 1779.

In fact, the notion of angle-preserving map can be traced back to Greek antiquity, see the survey [194]. We already recalled that Euler, in his didactical treatise *Introductio in analysin infinitorum*, refers explicitly to functions in which the variable is a complex number. De Moivre, already in 1730, considered polynomials defined on the complex plane, and it is conceivable that other mathematicians before him did the same [182]. Remmert, who, besides being a specialist of complex analysis, is a highly respected historian in this field, writes in his *Theory of complex variables* [213] that the theory was born at the moment when Gauss sent a letter to Bessel, dated December 18, 1811, in which he writes[24]:

> At the beginning I would ask anyone who wants to introduce a new function in analysis to clarify whether he intends to confine it to real magnitudes (real values of the argument) and regard the imaginary values as just vestigial—or whether he subscribes to my fundamental proposition that in the realm of magnitudes the imaginary ones $a + b\sqrt{-1} = a + bi$ have to be regarded as enjoying equal rights with the real ones. We are not talking about practical utility here; rather analysis is, to my mind, a self-sufficient science. It would lose immeasurably in beauty and symmetry from the rejection of any fictive magnitudes. At each stage truths, which otherwise are quite generally valid, would have to be encumbered with all sorts of qualifications.

In fact, the letter also shows that at that time Gauss was already aware of the concept of complex integration, including Cauchy's integral theorem; cf. [126] Vol. 8, p. 90–92.

Cauchy, in his *Cours d'analyse* [34] (1821), starts by defining functions of real variables (p. 19), and then passes to complex variables. There are two distinct definitions in the real case, for functions of one or several variables:

> When variable quantities are so tied to each other that, given the value of one of them, we can deduce the values of all the others, we usually conceive these various quantities expressed in terms of one of them, which then bears the name *independent variable*; and the other quantities expressed in terms of the independent variable are what we call functions of that variable.
>
> When variable quantities are so tied to each other that, given the values of some of them, we can deduce the values of all the others, we usually conceive these various quantities expressed in terms of several of them, which then bear the name *independent variables*; and the remaining quantities expressed in terms of the independent variables, are what we call functions of these same variables.[25]

[24]The translation is Remmert's; cf. [213] p. 1.

[25]Lorsque des quantités variables sont tellement liées entre elles que, la valeur de l'une d'elles étant donnée, on puisse en conclure les valeurs de toutes les autres, on conçoit d'ordinaire ces

Talking about Cauchy's work on functions of a complex variable, one should also mention the Cauchy–Riemann equation as a characterization of complex analycity, which Cauchy and Riemann introduced in the same year, 1851, Cauchy in his papers [38, 39] and Riemann in his doctoral dissertation [214]. It is important to note also that the Cauchy–Riemann equations,

$$\frac{\partial u}{\partial x} = \frac{\partial v}{\partial y} \text{ and } \frac{\partial u}{\partial y} = -\frac{\partial v}{\partial x},$$

without the complex character, were used by d'Alembert in 1752, in his works on fluid dynamics, *Essai d'une nouvelle théorie de la résistance des fluides* (Essay on a new theory of fluid resistance) [261] p. 27. D'Alembert showed later that functions u and v satisfying this pair of equations also satisfy Laplace's equation: $\Delta u = 0$ and $\Delta v = 0$.

The work of Cauchy is also reviewed in the chapter [191] in the present volume, written by the present author.

Riemann's doctoral dissertation [214] is in some sense an essay on functions of a complex variable. Right at the beginning of the dissertation, Riemann states explicitly what he means by a function. He starts with functions of a real variable:

> If we designate by z a variable magnitude, which may take successively all possible real values, then, if to each of these values corresponds a unique value of the indeterminate magnitude w, we say that w is a function of z.

He then talks about continuity of functions, in the modern sense of the word (as opposed to the sense that Euler gave to this word).[26] Then he writes:

> This definition does not stipulate any law between the isolated values of the function, this is evident, because after this function has been dealt with for a given interval, the way it is extended outside this interval remains quite arbitrary.

Riemann then recalls that the possibility of using some "mathematical law" that assigns to w a value for a given value of z was proper to the functions which Euler termed *functiones continuæ*. He writes that "modern research has shown that there exist analytic expressions by which any continuous function on a given interval can be represented." He then declares that the case of functions of a complex variable is treated differently. In fact, Riemann considers only functions of a complex variable

---

(Footnote 25 continued)
diverses quantités exprimées au moyen de l'une d'entre elles, qui prend alors le nom de *variable indépendante* ; et les autres quantités exprimées au moyen de la variable indépendante sont ce qu'on appelle des *fonctions* de cette variable.

Lorsque les quantités variables sont tellement liées entre elles que, les valeurs de quelques unes étant données, on puisse en conclure celles de toutes les autres, on conçoit ces diverses quantités exprimées au moyen de plusieurs d'entre elles, qui prennent alors le nom de *variables indépendantes*; et les quantités restantes, exprimées au moyen des variables indépendantes, sont ce qu'on appelle des *fonctions* de ces mêmes variables.

[26] In the *Introductio* Euler used the expression *continuous function* for a function that is "given by a formula." This is thoroughly discussed in Sect. 10 of the present chapter.

whose derivative does not depend on the direction, that is, holomorphic functions. He makes this property part of his definition of a function of a complex variable. Thus, when he talks about a function in the complex setting, he considers only conformal maps.

Regarding Riemann's dissertation, let us note that in a letter to his brother, dated November 26, 1851 [233], after he submitted his doctoral dissertation manuscript, he writes that Gauss took it home to examine it for a few days, and that before reading it, Gauss told him:

> [Riemann speaking] for years he had been preparing an essay, on which today he is still occupied, whose subject is the same or at least in part the same as that covered by me. Already in his doctoral dissertation now 52 years ago he actually expressed the intention to write on this subject.

This is an instance where Gauss was aware of a theory, or part of it, long before its author; we shall mention several other such instances in what follows.

3 Elliptic Integrals

In the huge class of integrals of functions, the integrals of algebraic functions constitute the simplest and the most natural class to work with. The class of elliptic integrals (and their Abelian generalizations) which deal with such functions soon turned out to be enough tractable and at the same time very rich from the point of view of the problems that they posed. These integrals led to a huge amount of work by several prominent mathematicians, as we shall see in this section.

Riemann had several reasons to work on Abelian integrals. Motivated by lectures by Dirichlet, Jacobi and others, he worked on the open problems that these functions presented, in particular the Jacobi inversion problem.

When Riemann started his work on integrals as functions of a complex variable, this subject was already well developed. An important challenging problem that he tackled was the so-called Jacobi inversion problem which we mention below. Most of all, these functions constituted for Riemann an interesting class of non-necessarily algebraic functions of a complex variable. The double periodicity of these integrals, the multi-valuedness of their inverses, the operations that one can perform on them, constituted a treasure of examples of new functions of a complex variable, and a context in which his theory of Riemann surfaces may naturally be used.

We start by summarizing some of the main ideas and problems that concern elliptic functions that were addressed since the time of Euler.

(1) The study of definite integrals representing arcs of conics and of lemniscates, and the comparison of their properties with those of integrals representing arcs of circles, which are computable in terms of the trigonometric functions or their inverses. We recall, by way of comparison, that whereas the integral $\int_0^x \frac{dt}{\sqrt{1-t^2}}$ represents arc length along a circle centered at the origin, the

integral $\int_0^x \frac{dt}{\sqrt{1-t^4}}$ represents arc length along the lemniscate of polar equation $r^2 = \cos 2\theta$.

(2) The search for sums and product formulae for such integrals, in the same way as there are formulae for sums and products of trigonometric functions.
(3) The study of periods, again, in analogy with those of trigonometric functions.

In fact, some of the first questions concerning elliptic integrals can be traced back to Johann I Bernoulli who tried to use the newly discovered integral calculus to obtain formulae for lengths of arcs of conic sections and some other curves. Bernoulli found the first addition formulae for such integrals. Finding general addition theorems for elliptic integrals remained one of the major problems for the following hundred years, involving the works of several major figures including Euler, Legendre, Abel, Jacobi and Riemann. Bernoulli also discovered that the lengths of some curves, expressed using integrals, may be expressed using infinite series [23].

Johann Bernoulli was Euler's teacher, and it is not surprising that the latter became interested in these problems early in his career. In his first paper on the subject, *Specimen de constructione aequationum differentialium sine indeterminatarum separatione* (Example of the construction of differential equations without separation of variables) [70] written in 1733, Euler gives a formula for arc lengths of ellipses. He obtains them by first writing a differential equation satisfied by these arcs. Generally speaking, Euler systematically searched for differential equations that describe the various situations that he was studying.

Between the work of Bernoulli and that of Euler, we must mention that of Fagnano, who, around the year 1716, in a study he was carrying on the lemniscate, discovered some results which Euler considered several years later as outstanding. These results included an addition formulae for a class of elliptic integrals [111], and the fact that on an ellipse or a hyperbola, one may find infinitely many pairs of arcs whose difference is expressible by algebraic means. The word used by Euler and others for such arcs (or differences of arcs) is that they are "rectifiable." Fagnano managed to reduce the rectifiability of the lemniscate to that of the ellipse and hyperbola. A few words on Fagnano one in order.

Giulio Carlo de' Toschi di Fagnano (1682–1766) was a noble Italian interested in science, who worked during several decades in isolation, away from any scientific environment. Weil's authoritative book on the history of number theory [255] starts with the following:

> Accroding to Jacobi, the theory of elliptic functions was born between the twenty-third of December 1751 and the twenty-seventh of January 1752. On the former date, the Berlin Academy of Sciences handed over to Euler the two volumes of Marchese Fagnano's *Produzioni Mathematiche*, published in Pesaro in 1750 and just received from the author; Euler was requested to examine the book and draft a suitable letter of thanks. On the latter date, Euler, referring explicitly to Fagnano's work on the lemniscate, read to the Academy the first of a series of papers, eventually proving in full generality the addition and multiplication theorems for elliptic integrals.

On p. 245 of the same treatise, Weil writes:

> On 23 December 1751 the two volumes of Fagnano's *produzioni Mathematiche*, just published, reached the Berlin Academy and were handed over to Euler; the second volume contained reprints of pieces on elliptic integrals which appeared between 1714 and 1720 in an obscure Italian journal and had remained totally unknown. On reading these few pages Euler caught fire instantly; on 27 January 1752 he was presenting to the Academy a memoir [88] with an exposition of Fagnano's main results, to which he was already adding some of his own.
>
> The most striking of Fagnano's results concerned transformations of the "lemniscate differential"
> $$w(z) = \frac{dz}{\sqrt{1-z^4}};$$
> how he had reached them was more than even Euler could guess. "Surely his discoveries would shed much light on the theory of transcendental functions," Euler wrote in 1753, "if only his procedure supplied a sure method for pursuing these investigations further; but it rests upon substitutions of a tentative character, almost haphazardly applied ..."[27]

In a letter dated October 17, 1730 ([109], p. 624), well before being aware of Fagnano's work, Euler informed Goldbach that "even admitting logarithms,"[28] he could by no means compute the integral $\int \frac{a^2dx}{\sqrt{a^4-x^4}}$, that "expresses the curve element of the rectangular elastic curve, or rectify this ellipse." Fagnano, instead of giving explicit values, established equalities between such integrals which paved the way to a new series of results by Euler and others. In a letter to Goldbach, dated May 30, 1752, that is, about six months after reading Fagnano's work, Euler writes (see [109] p. 1064): "Recently some curious integrations occurred to me." He first notes that three differential equations

$$\frac{dx}{\sqrt{1-x^2}} = \frac{dy}{\sqrt{1-y^2}},$$

$$\frac{dx}{\sqrt{1-x^4}} = \frac{dy}{\sqrt{1-y^4}},$$

and

$$\frac{dx}{\sqrt{1-x^3}} = \frac{dy}{\sqrt{1-y^3}}$$

can be integrated explicitly, and lead respectively to

$$y^2 + x^2 = c^2 + 2xy\sqrt{1-c^2},$$

---

[27] The reference is to Euler's memoir *Specimen novae methodi curvarum quadraturas et rectificationes aliasque quantitates transcendentes inter se comparandi* (An example of a new method for the quaduraturе and rectificaiton of curves and of comparing other quantities which are transcendentally related to each other) [89].

[28] The reference to logarithms comes from the fact that $\frac{dt}{t}$ and some more general rational functions can be integrated using logarithms.

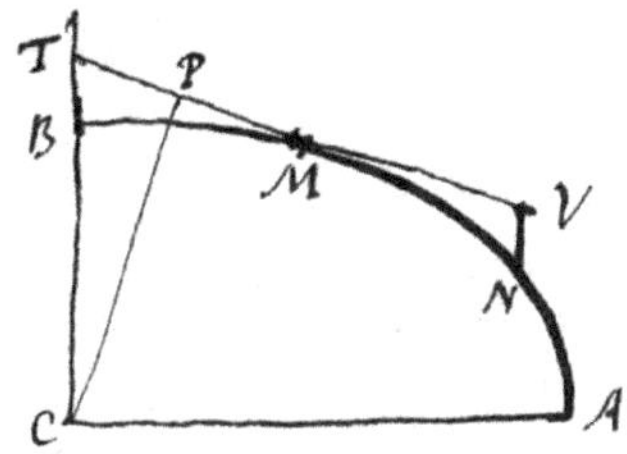

Fig. 1 A picture from Euler's calculation of the length of a segment of an ellipse, from Euler's letter to Goldbach dated May 30, 1752.

$$y^2 + x^2 = c^2 + 2xy\sqrt{1 - c^4} - c^2y^2$$

and

$$y^2 + x^2 + c^2x^2y^2 = 4c - 4c^2(x + y) + 2xy - 2cxy(x + y).$$

He adds that from these and other formulae of the same kind, he deduced the following theorem (see Fig. 1):

> If, in the quadrant ACB of an ellipse, the tangent VTM at an arbitrary point M is drawn which meets one of the axes, CB, at T, if TV is taken equal to CA and from V, VN is drawn parallel to CB, and if finally CP is the perpendicular on the tangent through the center C, then I say the difference of the arcs BM and AN will be rectifiable, namely, $BM - AN = MP$.

In the following letter to Goldbach, dated June 3rd, 1752, Euler gave a proof of this theorem and clarified a formula that Fagnano had given in his 1716 paper [110].

About five weeks after Euler received the work of Fagnano, he presented to the Berlin Academy a memoir entitled *Observationes de comparatione arcuum curvarum irrectificibilium* (Observations on the comparison of arcs of irrectifiable curves) [88] in which he expands on what he had announced in his correspondence with Goldbach, generalizing Fagnano's duplication result on the lemniscate to a general multiplication result and giving examples of arcs of an ellipse, hyperbola and lemniscate whose differences are rectifiable. This was the beginning of a systematic study by Euler of elliptic integrals. The year after, he presented to the Saint Petersburg Academy of Sciences a memoir entitled *De integratione aequationis differentialis* $\frac{mdx}{\sqrt{1-x^4}} = \frac{ndy}{\sqrt{1-y^4}}$ (On the integration of the differential equation $\frac{mdx}{\sqrt{1-x^4}} = \frac{ndy}{\sqrt{1-y^4}}$) [87] which starts with the sentence[29]:

> When, prompted by the illustrious Count Fagnano, I first considered this equation, I found indeed an algebraic relation between the variables x and y which satisfied the equation.

Several years later, in his famous treatise *Institutiones calculi integralis* [92], Euler included a section on the addition and multiplication of integrals of the form

[29]The translation is by S. G. Langton.

$$\int \frac{PdZ}{\sqrt{A + 2BZ + CZ^2 + 2DZ^3 + EZ^4}}.$$

Fagnano's works, in three volumes, were edited in 1911–1912 by Gambioli, Loria and Volterra [112].

Among the large number of memoirs that Euler wrote on elliptic integrals,[30] we mention the short memoir [82], *Problema, ad cuius solutionem geometrae invitantur; theorema, ad cuius demonstrationem geometrae invitantur* (A Problem, to which a geometric solution is solicited; a theorem, to which a geometric proof is solicited), published in 1754, containing his result on the rectification of the difference of two arcs of an ellipse. We also mention the memoir [90], *Demonstratio theorematis et solutio problematis in actis erud. Lipsiensibus propositorum* (Proof of a theorem and solution of a theorem proposed in the Acta Eruditorum of Leipzig) [90], in which he studies the division by 2 of an arc of ellipse. The memoir [93], entitled *Integratio aequationis* $\frac{dx}{\sqrt{\alpha+\beta x+\gamma x^2+\delta x^3+\epsilon x^4}} = \frac{dy}{\sqrt{\alpha+\beta y+\gamma y^2+\delta y^3+\epsilon y^4}}$ (The integration of the equation $\frac{dx}{\sqrt{\alpha+\beta x+\gamma x^2+\delta x^3+\epsilon x^4}} = \frac{dy}{\sqrt{\alpha+\beta y+\gamma y^2+\delta y^3+\epsilon y^4}}$), written in 1765 and published in 1768, is mentioned by Jacobi in a letter to Legendre which we quote below.

Besides Euler, one may mention d'Alembert. In a letter dated December 29, 1746, Euler writes to his Parisian colleague (see [107] p. 251):

> I read with as much profit as satisfaction your last piece with which you honored our Academy. [...] But what pleased me most in your piece is the reduction of several integral formulae to the rectification of the ellipse and the hyperbola; a matter to which I had also already given my thoughts, but I was not able to get entirely to the formula
>
> $$\frac{dx}{\sqrt{\alpha + \beta x + \gamma x^2 + \delta x^3 + \epsilon x^4}}$$
>
> and I regard your formula as a masterpiece of your expertise.[31]

Lagrange, whose name is associated with that of Euler in several contexts, studied elliptic integrals in his famous *Théorie des fonctions analytiques* (Theory of analytic

[30] The Euler archive lists thirty-three memoirs by him under the heading "Elliptic integrals," published between 1738 and 1882. It is sometimes hard to know exactly the year where Euler wrote his memoirs. For several of them, the date of publication was much later that the date of writing, and there are several reasons for that, including the huge backlog of the publication department of the Academies of Sciences of Saint Petersburg and Berlin, the main reason being that Euler used to send them too many papers.

[31] J'ai lu avec autant de fruit que de satisfaction votre dernière pièce dont vous avez honoré notre académie. [...] Mais ce qui m'a plu surtout dans votre pièce c'est la réduction de plusieurs formules intégrales à la rectification de l'ellipse et de l'hyperbole ; matière à laquelle j'avais aussi déjà pensé, mais je n'ai pu venir à bout de la formule

$$\frac{dx}{\sqrt{\alpha + \beta x + \gamma x^2 + \delta x^3 + \epsilon x^4}}$$

et je regarde votre formule comme un chef-d'œuvre de votre expertise.

functions) [164] (first edition 1797). In particular, he discovered a relation between Euler's addition formula and a problem in spherical trigonometry.

After Euler, d'Alembert and Lagrange, we must talk about Legendre, who investigated these integrals for almost forty years. He wrote two famous treatises on the subject, his *Exercices de calcul intégral sur divers ordres de transcendantes et sur les quadratures* (Exercises of integral calculus on various orders of transcendence and on the quadratures) [171] (1811–1816) and his *Traité des fonctions elliptiques et des intégrales eulériennes* (Treatise of elliptic functions and Eulerian integrals) [172] (1825–1828), both in three volumes. In the introduction to the latter (p. 1ff.), Legendre makes a brief history of the subject, from its birth until the moment he started working on it. According to his account, elliptic functions were first studied by MacLaurin and d'Alembert who found several formulae for integrals that represent arcs of ellipses or arcs of hyperbolas.[32] Legendre declares that their results were too disparate to form a theory. He then mentions Fagnano, recalling that his work was the starting point of the profound analogy between elliptic integrals and trigonometric functions. After describing Fagnano's work, Legendre talks about some of the main contributions of Euler, Lagrange and Landen on the subject. His treatise starts with a detailed study of integrals of the form $\int \frac{Pdx}{R}$ investigated by Euler, where P is an arbitrary rational function of x and $R = \sqrt{\alpha + \beta x + \gamma x^2 + \delta x^3 + \epsilon x^4}$. The expression *Eulerian integral* contained in the title of Legendre's treatise was coined by him. He writes:

> Although Euler's name is attached to almost all the important theories of integral calculus, I nevertheless thought that I was allowed to give more especially the name *Eulerian integral* to two sorts of transcendants whose properties constituted the subject of several beautiful memoirs of Euler, and form the most complete theory on definite integrals which exists up to now [...] [33]

After Legendre, and among the immediate predecessors of Riemann on elliptic functions, we find Abel, Jacobi, and Gauss. The last two were his teachers in Berlin and Göttingen respectively. With this work, the emphasis in the study of elliptic integrals shifted to that of their inverses. Considering inverses is naturally motivated by the analogy with trigonometric functions, as one may see by recalling that the integral $\int_0^x \frac{dt}{\sqrt{1-t^2}}$ represents the arcsine function, and therefore, its inverse is the more tractable sine function. The periodic behavior of inverses of elliptic integrals like $\int_0^x \frac{dt}{\sqrt{1-t^4}}$ and others, which became later one of the main questions in that theory, is in some sense a generalization of that of trigonometric functions.

---

[32]See e.g. [5, 179].

[33]Quoique le nom d'Euler soit attaché à presque toutes les théories importantes du calcul intégral, cependant j'ai cru qu'il me serait permis de donner plus spécialement le nom d'*Intégrales Eulériennes* à deux sortes de transcendantes dont les propriétés ont fait le sujet de plusieurs beaux mémoires d'Euler, et forment la théorie la plus complète que l'on connaisse jusqu'à présent sur les intagrales définies [...].

Abel and Jacobi developed simultaneously the theory of elliptic integrals, and separating their results has always been a difficult task. It is also well established that Gauss discovered several results of Abel and Jacobi before them, but never published them. This is attested in his notebook and in his correspondence, published in his Collected Works. Gauss started his notebook in 1796, at the age of 19, and he wrote his last note there in 1814. The notes consist of 146 statements, most of them very concise, and they fill up a total of 20 pages in his *Collected Works* (vol. 10). This edition of the notebook published in Gauss's Collected Works is accompanied by detailed comments by Bachmann, Brendel, Dedekind, Klein, Lœwy, Schlesinger and Stäckel. There is a French translation of the notebook [125]. Among the notes contained in this diary, several concern elliptic functions. For instance, in Notes 32 and 33, Gauss studies the inverse of the lemniscate integral $\int \frac{dx}{\sqrt{1-x^4}}$, as a particular case of the elliptic integral $\int \frac{dx}{\sqrt{1-x^n}}$. In Note 53, he mentions that he is studying the general integral $\int \frac{dx}{\sqrt[n]{1-x^n}}$, which was already considered by Euler in his *Institutiones calculi integralis*. In Note 54, he states that he has an easy method for determining the integral $\int \frac{x^n dx}{1+x^m}$, again an integral that was considered by Euler. There are several other notes on elliptic integrals in Gauss's notebook.

Jacobi read Euler's works while he was in high school. He obtained his PhD at the age of 21, and at the age of 22, he started a correspondence with Legendre, who was 74, informing him about his results on elliptic integrals. This correspondence became famous. It is reproduced in Crelle's Journal[34] and in Jacobi's *Collected Works*.[35] The beginning of this correspondence is touching. Jacobi sends his first letter to Legendre on August 5, 1827, expressing his great respect for the work of his older French colleague. He writes ([146] vol. 1, p. 390):

> A young geometer dares to present you a few discoveries in the theory of elliptic functions, to which he was led by a diligent study of your beautiful writings. It is to you, Sir, that this brilliant part of analysis owes the highest degree of perfection to which it has been elevated, and it is only in following the footsteps of such a great master that the geometers will be able to push it beyond limits which have been so far prescribed. Thus, it is to thee that I must offer the following, as a fair tribute of admiration and gratefulness.[36]

In his response, dated November 30, 1827, Legendre, referring to one of the theorems that Jacobi communicated to him, writes ([146] vol. 1, p. 396):

> I checked this theorem by my own methods and I found it perfectly correct. Even though I regret that this discovery escaped me, the joy I experienced was most vivid when I saw the

[34]Crelle's Journal, 80 (1875), p. 205–279.

[35]*Collected Works*, t. I, pp. 385–46.

[36]Un jeune géomètre ose vous présenter quelques découvertes faites dans la théorie des fonctions elliptiques, auxquelles il a été conduit par la lecture assidue de vos beaux écrits. C'est à vous, Monsieur, que cette partie brillante de l'analyse doit le haut degré de perfectionnement auquel elle a été portée, et ce n'est qu'en marchant sur les vestiges d'un si grand maître, que les géomètres pourront parvenir à la pousser au-delà des bornes qui lui ont été prescrites jusqu'ici. C'est donc à vous que je dois offrir ce qui suit comme un juste tribut d'admiration et de reconnaissance.

> significant improvement that was added to the beautiful theory of which I am the creator and which I developed almost alone during more than forty years.[37]

In another letter, sent on January 12, 1828, Jacobi informs Legendre about Abel's discoveries, in particular on the division of the lemniscate ([146], vol. 1, p. 401):

> Since my last letter, researches of the highest importance were published on elliptic functions by a young geometer, who may be personally known to you.[38]

Legendre sent his response on February 9, informing his correspondent that he knew about Abel's work, but that he was happy to see it summarized in a language which was closer to his own.[39]

Regarding Gauss's work on elliptic functions, we mention an excerpt of the first letter from Jacobi to Legendre [146] pp. 393–394:

> These researches were born only very recently. However, they are not the only ones that are conducted in Germany on the same object. Mr. Gauss, when he learned about them, informed me that he had developed, already in 1808, the cases of 3 sections, 5 sections and 7 sections, and that he found at the same time the corresponding new scales of modules. It seems to me that this information is very interesting.[40]

Legendre was outraged by Gauss's reaction. In his response to Jacobi, dated November 30, 1827, he writes ([146] p. 398):

> How is it possible that Mr. Gauss dared telling you that most of your theorems were known to him and that he discovered them back in 1808? This excess of impudence is unbelievable from a man who has enough personal merit so as he does not need to appropriate the discoveries of others... But this is the same man who, in 1801, wanted to attribute to himself the law of reciprocity published in 1785 and who wanted, in 1809, to take hold of the method of least squares that was published in 1805.[41]

---

[37] J'ai vérifié ce théorème par les méthodes qui me sont propres et je l'ai trouvé parfaitement exact. En regrettant que cette découverte m'ait échappée je n'en ai pas moins éprouvé une joie très vive de voir un perfectionnement si notable ajouté à la belle théorie, dont je suis le créateur, et que j'ai cultivé presque seul depuis plus de quarante ans.

[38] Depuis ma dernière lettre, des recherches de la plus grande importance ont été publiées sur les fonctions elliptiques de la part d'un jeune géomètre, qui peut-être vous sera connu personnellement.

[39] [J'avais déjà connaissance du beau travail de M. Abel inséré dans le *Journal de Crelle*. Mais vous m'avez fait beaucoup de plaisir de m'en donner une analyse dans votre langage qui est plus rapproché du mien.] ([146], t. 1, p. 407).

[40] Il n'y a que très peu de temps que ces recherches ont pris naissance. Cependant elles ne sont pas les seules entreprises en Allemagne sur le même objet. M. Gauss, ayant appris de celles-ci, m'a fait dire qu'il avait développé déjà en 1808 les cas de 3 sections, 5 sections et de 7 sections, et trouvé en même temps les nouvelles échelles de modules qui s'y rapportent. Cette nouvelle, à ce qui me paraît, est bien intéressante.

[41] Comment se fait-il que M. Gauss ait osé vous dire que la plupart de vos théorèmes lui étaient connus et qu'il en avait fait la découverte dès 1808 ? Cet excès d'impudence n'est pas croyable de la part d'un homme qui a assez de mérite personnel pour n'avoir pas besoin de s'approprier les découvertes des autres... Mais c'est le même homme qui en 1801 voulut s'attribuer la découverte de la loi de réciprocité publiée en 1785 et qui voulut s'emparer en 1809 de la méthode des moindres carrés publiée en 1805.

It was only at the publication of Gauss's *Collected Works*,[42] containing in particular his famous notebook, that it became clear that Gauss's assertion concerning the fact that he had discovered before Abel most of the properties of elliptic functions, including their double periodicity, was correct. One of the first results of Abel concerns integrals of arcs of lemniscate, a curve which he showed to be divisible by ruler and compass into n equal parts, for the same values of n for which the circle is divisible into n equal parts. The same result was stated without proof in Gauss's *Disquisitiones arithmeticae* [121].

Abel's first major results on elliptic functions are contained in his 1827 paper *Recherches sur les fonctions elliptiques* (Researches on elliptic functions) [1]. He explains there the double periodicity of these functions, as well as their multiplication and division properties. The analogy with circular functions is again highlighted. At the beginning of his paper, Abel talks about his famous predecessors, Euler, Lagrange and Legendre. He writes (p. 101):

> The first idea of these [elliptic] functions were given by the immortal Euler, who showed that the separable equation
>
> $$\frac{\partial x}{\sqrt{\alpha+\beta x+\gamma x^2+\delta x^3+\epsilon x^4}}+\frac{\partial y}{\sqrt{\alpha+\beta y+\gamma y^2+\delta y^3+\epsilon y^4}}=0$$
>
> is algebraically integrable. After Euler, Lagrange added something, when he gave his elegant theory of the transformation of the integral
>
> $$\int\frac{Rdx}{\sqrt{(1-p^2x^2)(1-q^2x^2)}},$$
>
> where R is a rational function of x. But the first, if I am not mistaken, who went thoroughly into the nature of these functions, is Mr. Legendre, who, first in a memoir on elliptic functions, and then in his excellent *Exercices de mathématiques*, developed numerous elegant properties of these functions, and showed their usefulness.[43]

Riemann was already interested in elliptic functions while he was a student in Berlin. Klein, in his *Development of mathematics in the 19th century* [162] (Chap-

---

[42] Gauss's collected works, *Carl Friedrich Gauss' Werke*, in twelve volumes, were published between 1863 and 1929.

[43] La première idée de ces fonctions a été donnée par l'immortel Euler, en démontrant que l'équation séparée

$$\frac{\partial x}{\sqrt{\alpha+\beta x+\gamma x^2+\delta x^3+\epsilon x^4}}+\frac{\partial y}{\sqrt{\alpha+\beta y+\gamma y^2+\delta y^3+\epsilon y^4}}=0$$

est intégrable algébriquement. Après Euler, Lagrange y a ajouté quelque chose, en donnant son élégante théorie de la transformation de l'intégrale

$$\int\frac{Rdx}{\sqrt{(1-p^2x^2)(1-q^2x^2)}},$$

où R est une fonction rationnelle de x. Mais le premier, et si je ne me trompe, le seul, qui ait approfondi la nature de ces fonctions, est M. Legendre, qui d'abord dans un mémoire sur les fonctions elliptiques, et ensuite dans ses excellents exercices de mathématiques, a développé nombre de propriétés élégantes de ces fonctions, et a montré leur application.

ter VI) writes that the latter, since the end of the 1840s, was interested in elliptic functions because this subject was fashionable in Germany. From a letter to his father, dated May 30, 1849, we know that Riemann was following in Berlin Jacobi's and Eisenstein's lectures on elliptic functions. He writes (cf. [233]): "Jacobi has just begun a series of lectures in which he leads off once again with the entire system of the theory of elliptical functions in the most advanced, but elementary way." In another letter (without date), also written in Berlin, Riemann writes: "I enrolled with five other students into a private class (Privatissumum) with Eisenstein, who was promoted in the course of this semester to a Privatdozent with a paper on the theory of elliptic functions."

We already mentioned Euler's impact on Jacobi. Eisenstein is another prominent mathematician on which Euler exerted a crucial influence. In his biography of Eisenstein [237], M. Schmitz writes that during the period 1837–1842, while he was a Gynmasium pupil, Eisenstein attended lectures by Dirichlet at the University of Berlin, and that he studied on his own Gauss's *Disquisitiones Arithmeticae* as well as papers and books by Euler and Lagrange. We quote Eisenstein, from his autobiography translated in [237]:

> After I had acquired the fundamentals by private study (I never had a private tutor) I proceeded to advanced mathematics and studied, besides other books containing advanced material, the brilliant work of Euler and Lagrange about differential and integral calculus. I was able to commit this material securely to my memory and to master it entirely, because I made it a rule to compose every theory in writing as soon as I understood it.

In his ICM communication [254], Weil declares (p. 233) that "Eisenstein fell in love with mathematics at an early age by reading Euler and Lagrange."

We shall conclude this section with two other quotes of Weil. Before that, let us recall that elliptic integrals are studied in number theory in relation with the theory of elliptic curves. Weil writes in an essay on the history of number theory, [253], p. 15, that Fermat, in his work on number theory, had already dealt with elliptic curves (without the name), in particular in his proof of the non-existence of integer solutions for the equation $x^4 - y^4 = z^2$. We quote him from his book on the history of number theory that we already mentioned ([255] p. 242):

> What we call now "elliptic curves" (i.e. algebraic curves of genus 1) were considered by Euler under two quite different aspects without ever showing an awareness of the connection between them, or rather of their substantial identity. On the one hand, he must surely have been familiar, from the very beginning of his career, with the traditional methods for handling Diophantine equations of genus 1. [...] On the other hand he had inherited from his predecessors, and notably from Johann Bernoulli, a keen interest in what we know as "elliptic integrals" because the rectification of the ellipse depends upon integrals of that type; they were perceived to come next to the integrals of rational functions in order of difficulty.

Eisenstein and Dirichlet were mostly interested in elliptic functions because of their use in number theory, contrary to Riemann, who, even though he was introduced to elliptic functions through Eisenstein's lectures, was not excited by that field. Weil writes in his essays [253], p. 21:

> [...] The case of Riemann is more curious. Of all the great mathematicians of the last century, he is outstanding for many things, but also, strangely enough, for his complete lack of interest for number theory and algebra. This is really striking, when one reflects how close he was, as a student, to Dirichlet and Eisenstein, and, at a later period, also to Gauss and to Dedekind who became his most intimate friend. During Riemann's student days in Berlin, Eisenstein tried (not without some success, he fancied) to attract him to number theory. In 1855, Dedekind was lecturing in Göttingen on Galois theory, and one might think that Riemann, interested as he was in algebraic functions, might have paid some attention. But there is not the slightest indication that he ever gave any serious thoughts to such matters.

We shall mention the work of Dirichlet on number theory (in particular on the prime number theorem) in Sect. 6 below. In Chap. 8 [192] of the present book, we report on several treatises on elliptic functions that were published in France during the few decades that followed Riemann's early work on the subject. In the next section, we review the more general Abelian functions.

4 Abelian Functions

A few years before Riemann started his work on elliptic functions and elliptic integrals, the general interest moved towards the more general Abelian integrals, and their inversion. The term *Abelian function*, first introduced by Jacobi in honor of Abel, is generally given to the functions obtained by inverting an arbitrary algebraic integral or a combination of such integrals. An algebraic integral is an integral of the form $\int R(x, y)dx$ where R is a rational function of the two variables x and y and where x and y satisfy furthermore a polynomial equation $f(x, y) = 0$. In his 1826 memoir submitted to the Paris Academy, Abel extended Euler's addition formula for elliptic integrals to Abelian integrals. He proved that the sum of an arbitrary number of such integrals can be written as the sum of p linearly independent integrals, to which is added an algebraic-logarithmic expression. Here p is the so-called *genus* of the algebraic curve defined by the equation $f(x, y) = 0$. After he learned about Abel's work, Jacobi formulated a generalized inversion problem for a system of p hyperelliptic integrals. His ideas were pursued by several mathematicians, and in particular by Riemann, who gave a solution to the inversion problem in terms of ϑ functions.

Abel also discovered that the inverse functions of elliptic integrals are doubly periodic functions defined on the complex plane. This property was at the basis of the later introduction of group theory in the theory of elliptic curves.

In the passage from elliptic functions to Abelian functions, one must also mention Galois. The day before his death, Galois sent a letter to his friend Auguste Chevalier in which he described his thoughts, saying that one could write a memoir based on his ideas on integrals. The letter is analyzed by Picard in his article [196].[44] Picard writes:

[44] This article constituted the preface to the Collected Works of Galois which were published shortly after.

> All what we know about these researches is contained in what he says in this letter. Several points remain obscure in some statements of Galois; however, we can have a precise idea of some of the results he reached in the theory of integrals of algebraic functions. We thus acquire the certainty that he possessed the most essential results on Abelian integrals that Riemann was led to obtain twenty-five years later. We see without surprise Galois talking about the periods of an Abelian integral relative to an arbitrary algebraic function [...] The statements are precise; the famous author makes the classification of Abelian integrals into three kinds, and he declares that if n denotes the number of linearly independent integrals of the first kind, the number of periods is $2n$. The theorem relative to the parameter inversion in the integrals of the third type is clearly marked, as well as the relations with the periods of Abelian integrals. Galois also talks about a generalization of Legendre's classical equation where the periods of elliptic integrals appear, a generalization which probably led him to the important relation that was discovered later on by Weierstrass and Mr. Fuchs.[45]

In his paper on Abelian functions [217], Riemann establishes existence results for Abelian functions and more generally their determination in terms of the points of discontinuity and the information on the ramification at these points. It is in that paper that Riemann introduces the notion of birational equivalence and number of moduli, both of which played an essential role in mathematics. In the same paper, he presents Abel's addition theorem for elliptic integrals, and he solves Jacobi's inversion problem in terms of p variable magnitudes, for a $(2p+2)$-connected surface. It is also in this paper that Riemann gives his well known classification of Abelian integrals into three types, a classification which depends on the existence and the nature of the singularities (poles or logarithmic). Riemann mentions in his paper several works on the inversion problem, in particular the successful attempt by Weierstrass in the case of hyperelliptic integrals.

On the work of Riemann on Abelian integrals, the reader is also referred to Chap. 4, by Houzel, in the present volume [141]. For a comprehensive survey on the work of Abel, the interested reader is referred to the article [142] by Houzel.

[45]Nous ne connaissons de ces recherches que ce qu'il en dit dans cette lettre ; plusieurs points restent obscurs dans quelques énoncés de Galois, mais on peut cependant se faire une idée précise de quelques-uns des résultats auxquels il était arrivé dans la théorie des intégrales de fonctions algébriques. On acquiert ainsi la conviction qu'il était en possession des résultats les plus essentiels sur les intégrales abéliennes que Riemann devait obtenir vingt-cinq ans plus tard. Nous voyons sans étonnement Galois parler des périodes d'une intégrale abélienne relative à une fonction algébrique quelconque [...] Les énoncés sont précis ; l'illustre auteur fait la classification en trois espèces des intégrales abéliennes, et affirme que, si n désigne le nombre des intégrales de première espèce linéairement indépendantes, les périodes seront en nombre $2n$. Le théorème relatif à l'inversion du paramètre dans les intégrales de troisième espèce est nettement indiqué, ainsi que les relations entre les périodes des intégrales abéliennes ; Galois parle aussi d'une généralisation de l'équation classique de Legendre, où figurent les périodes des intégrales elliptiques, généralisation qui l'avait probablement conduit à l'importante relation découverte depuis par Weierstrass et par M. Fuchs.

5 Hypergeometric Series

The theory of the hypergeometric series is another topic which Riemann tackled and whose roots involve in an essential way the works of Euler and Gauss. Riemann's main paper on the subject is *Beiträge zur Theorie der durch die Gauss'sche Reihe $F(\alpha, \beta, \gamma, x)$ darstellbaren Functionen* (Contribution to the theory of functions representable by Gauss's series $F(\alpha, \beta, \gamma, x)$) [222], published in 1857. The work in this paper was used by Riemann later in his development of the theory of analytic differential equations. There are also fragments on the same subject published in Riemann's *Collected works*.

The hypergeometric series is a function of the form

$$F(\alpha, \beta, \gamma, x) = 1 + \frac{\alpha\beta}{1.\gamma}x + \frac{\alpha(\alpha+1)\beta(\beta+1)}{1.2\gamma(\gamma+1)}x^2 + \frac{\alpha(\alpha+1)(\alpha+2)\beta(\beta+1)(\beta+2)}{1.2.3\gamma(\gamma+1)(\gamma+2)}x^3 + \dots$$

where x is the variable.

The term "hypergeometric series" appears in Euler's *Institutiones calculi integralis* [92] (1769), Chapter XI. The series is a solution of the so-called Euler hypergeometric differential equation which appears in Chapters VIII and XI of the same treatise. As a matter of fact, this name was given to several different but closely related objects. Euler, in one of his earliest memoir *De progressionibus transcendentibus seu quarum termini generales algebraice dari nequeunt* (On transcendental progressions, that is, those whose general terms cannot be given algebraically) [65], published in 1738, starts by mentioning Wallis's "hypergeometric series" $1! + 2! + 3! + 4! + \cdots$ (without the factorial notation). The terminology here refers to the fact that in analogy with the case of geometric progressions, where each term is obtained from the preceding one by multiplying it by a constant, one defined a hypergeometric progression as a progression in which each term is obtained from the preceding one by multiplying it by a factor which increases by a unit at each step. Wallis's papers on this subject include [251] (1, Scholium to Proposition 190) and [252] (p. 315).

Gauss mentions a hypergeometric series in his doctoral dissertation *Demonstratio nova theorematis omnem functionem algebraicam rationalem integram unius variabilis in factores reales primi vel secundi gradus resolvi posse* [119] (New proof of the theorem that every rational integral algebraic function of one variable can be resolved into real factors of the first or second degree) (1799).

We refer the reader to the paper [59] for a comprehensive history of the hypergeometric series.

At the beginning of his announcement of his memoir [222], Riemann states: "This memoir treats a class of functions which are useful to solve various problems in mathematical physics." As a matter of fact, these functions are still commonly used today in mathematical physics. Riemann notes that the name *hypergeometric series* was first proposed by Pfaff, for a more general series, whereas Euler, after Wallis, used such a name for a series which is slightly different. Pfaff was Gauss's friend, and had been his teacher. He studied this function in his book *Disquisitiones analyticae maxime ad calculum integralem et doctrinam serierum pertinentes* (Analytic investigations most relevant for integral calculus and the doctrine of series) [195] (1797). Gauss has a series of unpublished results on the hypergeometric series, which he communicated to the astronomer Bessel, who was also his friend, in a letter dated September 3, 1805. The results were used by Gauss in his later works. In his writings on the subject, Gauss used continued fractions in his study of the quotient of two hypergeometric series. He developed these ideas in his paper *Disquisitiones generales circa serium* $1+\frac{\alpha\beta}{1.\gamma}x+\frac{\alpha(\alpha+1)\beta(\beta+1)}{1.2\gamma(\gamma+1)}xx+\frac{\alpha(\alpha+1)(\alpha+2)\beta(\beta+1)(\beta+2)}{1.2.3\gamma(\gamma+1)(\gamma+2)}x^3+\ldots$ *etc.* (General investigations on the series $1+\frac{\alpha\beta}{1.\gamma}x+\frac{\alpha(\alpha+1)\beta(\beta+1)}{1.2\gamma(\gamma+1)}xx+\frac{\alpha(\alpha+1)(\alpha+2)\beta(\beta+1)(\beta+2)}{1.2.3\gamma(\gamma+1)(\gamma+2)}x^3+\ldots$ etc.) [122]. The same year, he wrote another paper on the same subject which he never published but which is contained in his *Collected Works* edition [123]. Riemann, in his paper [222], proved that these fractions converge in the complex plane cut along the subset $[2,+\infty]$ of the x-axis. In the same memoir, he introduced in the study of the hypergeometic functions a new method, which applies to all functions that satisfy linear differential equations with algebraic coefficients. He recalls in the announcement of that memoir, published in the *Göttinger Nachrichten*, No. 1, 1857, that Euler and Gauss made a thorough study of these functions from the theoretical point of view.

In the introduction to his paper [122] Gauss declares that practically any transcendental function that appears in analysis may be obtained as a special case of the hypergeometric series. In fact, it is known that functions like $\log(1+z)$, $\arcsin z$ and several orthogonal polynomials, including Legendre polynomials and Chebyshev polynomials, can be expressed using hypergeometric functions. The so-called confluent hypergeometric function (or Kummer's function) is a limit of the hypergeometric function.

The introduction of the hypergeometric series brought a whole new class of new functions to the field of analysis which, at least in the times of Euler, consisted in the study of functions.

6 The Zeta Function

This section is concerned with Riemann's article *Über die Anzahl der Primzahlen unter einer gegebenen Grösse* (On the number of primes less than a given magnitude) [219]. This memoir, which is only 8 pages long, changed the course of mathematics. Riemann wrote it at the occasion of his election to the Berlin Academy of Sciences, on August 11, 1859. Every newly elected member at that academy was asked to report on his most recent research, and Riemann chose this topic. A short history of the subject will show that the list of predecessors of Riemann in this field includes names which are familiar to us now: Euler, as always, then Legendre, Dirichlet and Gauss.

Riemann starts his memoir by recalling that Gauss and Dirichlet had been interested in this subject several years before him. He displays the following formula, which he recalls was noted by Euler, and which was his own departure point:

$$\prod \frac{1}{1-\frac{1}{p^s}} = \sum \frac{1}{n^s}.$$

Here, p takes all the prime values and n all the integer values. Riemann considers the function represented by these two expressions as a function of a complex variable s as long as the two series converge, and he denotes this function by $\zeta(s)$.[46] He then gives an integral formula for this function, and he notes that this integral is "uniform" (uni-valued), that it is defined and finite for any value of s except for $s = 1$ and that it vanishes when s is a negative odd integer.

The distribution of primes, which is the subject of Riemann's paper, may be traced back to Greek antiquity. The reader may recall that there are several results on prime numbers in Euclid's *Elements*. In particular, Proposition 20 of Book IX says that there are infinitely many primes. It is also known that the Greeks had a method to list effectively the sequence of primes (Eratosthenes sieve). Without any doubt, the general question of the distribution of primes kept busy the mathematicians of that epoch. It is also good to recall, right at the beginning, that Euler, in his paper *Variae observationes circa series infinitas* (Various observations about infinite series) [76], showed that the series of inverses of primes,

$$\frac{1}{2} + \frac{1}{3} + \frac{1}{5} + \frac{1}{7} + \frac{1}{11} + \ldots,$$

diverges, which in some sense is a wide generalization of the fact that the number of primes is infinite.

Euler was fascinated by the question of the distribution of primes. We quote him from a paper entitled *Découverte d'une loi tout extraordinaire des nombres par rapport à la somme de leurs diviseurs* (Discovery of a very extraordinary law of numbers in relation to the sum of their divisors) [80], written in 1747 and published in 1751:

> Mathematicians tried in vain, until now, to discover some or other order in the sequence of prime numbers, and we have reasons to think that this is a mystery which human mind will never be able to penetrate. To be convinced, it suffices to take a look at the tables of prime numbers, that a few persons have taken the trouble to continue beyond one hundred thousand: one will primarily notice that there is no order and no rule there.[47]

Let us return now to the zeta function.

---

[46]Even though the notation $\zeta(s)$ and the name zeta function first appear in Riemann's paper, we shall commit the usual anachronism of using the notation $\zeta(s)$ for the series $\sum_{n=1}^{\infty} \frac{1}{n^s}$ even when we talk about the work done on this series before Riemann.

[47]Les mathématiciens ont tâché jusqu'ici en vain à découvrir un ordre quelconque dans la progression des nombres premiers, et on a lieu de croire, que c'est un mystère auquel l'esprit humain ne saurait jamais pénétrer. Pour s'en convaincre, on n'a qu'à jeter les yeux sur les tables des nombres premiers, que quelques personnes se sont donné la peine de continuer au-delà de cent mille : et on s'apercevra d'abord qu'il ne règne aucun ordre ni règle.

The history of the zeta function in Euler's works naturally starts with the question of the value of the sum of the series of reciprocals of squares, $\zeta(2) = \sum_1^\infty \frac{1}{n^2}$. Before Euler, this series was known to be convergent, and the determination of its value was an open question whose formulation can be traced back at least to Pietro Mengoli in his treatise *Novae quadrature arithmeticae, seu de additione fractionum*[48] (New arithmetic quadratures, or the addition of fractions) [180] (1650). Several mathematicians worked on the problem, including Wallis, Leibniz, Stirling, de Moivre, Goldbach and several Bernoullis. In fact, the question of computing infinite sums was already a fashionable subject at that epoch. Mengoli, Huygens and Leibniz independently computed the sum of reciprocals of the triangular numbers, that is, numbers of the form $\frac{(n)(n+1)}{2}$. Leibniz's computation of the series of inverses of triangular numbers uses the classical "telescopic method" known to students, so its level of difficulty has nothing to do with Euler's computation of $\zeta(2)$. The problem of finding the value of $\zeta(2)$ became widely known among mathematicians after it was asked explicitly by Jakob Bernoulli in his series of papers *Positiones de seriebus infinitis* (Positions of an infinite series) (1689).[49] In the same work, Bernoulli considered the series for an arbitrary rational number s.

Euler published several papers on various aspects of the zeta function. In particular, he was the first to discover a formula establishing a relation between this series and prime numbers. It is interesting to recall that Euler has been investigating the convergence of infinite series and infinite products since his early days as a mathematician.[50] His first letter addressed to Goldbach, dated October 13, 1729, concerns the Γ function, a function that interpolates the factorials. Goldbach had asked the opinion of several mathematicians on that problem. Euler writes [109][51]:

> When lately I came across a few ideas that apparently could contribute to the interpolation of series having a variable law—as you are wont to call it—I took a closer look and discovered many things regarding that subject. As Mr. Bernoulli hinted that these results might please you, Sir, I decided to write to you and submit them to your judgment. For the series 1, 2, 6, 24, 120, ..., which you have treated extensively, as I see, I have found the general term [...]

The letter ends with:

> You, Sir, who have already enriched the theory of series by so many important discoveries, will therefore judge for yourself what else may be expected from this novel way to deal with series. It would certainly acquire its greatest utility and perfection if you could bring yourself to investigate how the differential calculus can be most conveniently applied to these questions. For up to now my method has the drawback that I cannot find what I want, but rather have to be content with wanting what I find.

---

[48]Mengoli's treatise is entirely devoted to the theory of infinite series, despite the word *quadrature* (that is, computation of areas) in the title.

[49]See the comments of this work of Bernoulli in Weil's article [256] p. 4.

[50]One should note that power series representations of functions already appear in the works of Newton, in the 1660s.

[51]In this volume of the *Opera Ominia*, the letters are translated into English.

In his paper *De summatione innumerabilium progressionum* (The summation of an innumerable progression) [66], Euler starts by giving a 7-digit approximate value of $\zeta(2)$, namely, 1.644934. Needless to say, such a computation needed from his part a large amount of computing, because the series converges very slowly. Before that, Wallis had given, in his *Arithmetica infinitorum* (Arithmetic of the infinite), 1655, a 3-digit approximation of that series. Goldbach and Daniel Bernoulli also gave 3-digit approximations, in 1728.[52] The reader may find interesting information on that subject in the correspondence between Euler, Bernoulli and Goldbach.

In 1735, Euler, who was 28 years old, obtained the summation formula for $\zeta(2)$ and, more generally, for the infinite series $\zeta(2\nu) = \sum_1^\infty \frac{1}{n^{2\nu}}$ for any positive integer ν. He found the values $\zeta(2) = \pi^2/6$ and $\zeta(2\nu) = r_\nu \pi^{2\nu}$, where r_ν are rational numbers which are closely related to the Bernoulli numbers. In the introduction to his memoir *De summis serierum reciprocarum* (On the sums of series of reciprocals) [72] (1735), he writes[53]:

> So much work has been done on the series $\zeta(n)$ that it seems hardly likely that anything new about them may still turn up ... I too, in spite of repeated efforts, could achieve nothing more than approximate values for their sums ... Now, however, quite unexpectedly, I have found an elegant formula for $\zeta(2)$, depending upon the quadrature of the circle.[54]

Euler's discovery made him famous, perhaps for the first time, among mathematicians in all Europe. When the news of Euler's discovery reached the city of Basel, the first reaction of his teacher, Johann Bernoulli, was to exclaim that the most burning desire of his deceased older brother Jakob was now fulfilled. Seen all the work he has done on the subject, there is no doubt that throughout his life, Euler tried (without success) to find a formula for $\zeta(s)$ for s an odd integer.

It was not unusual for Euler to publish several proofs of the same result, and his result on the convergence on $\zeta(2)$ is one instance of this fact. In particular, there are proofs of this fact in his memoirs [72] (presented to the Saint Petersburg Academy on December 5, 1735 and published in 1740) and [76] (presented to the Saint Petersburg Academy on April 25, 1727 and published in 1744), and an account is given in his *Introductio* [61] (first edition 1748).

In a letter to Goldbach dated August 28, 1742 (Letter 54 in [109]), Euler expresses $\zeta(2)$ in terms of dilogarithms. We recall that the dilogarithm function[55] is defined as

$$\mathrm{Li}(x) = \sum_{k=1}^{\infty} \frac{x^k}{k^2}.$$

---

[52] In a letter to Goldbach, sent in 1728, Daniel Bernoulli writes that the value of the series $\zeta(2)$ "is very nearly 8/5," and Goldbach answers that $\zeta(2) - 1$ lies between 16233/25200 and 30197/46800; cf. Weil [255] p. 257 for more details on this history.

[53] The translation from the Latin is by André Weil, [255] p. 261.

[54] Weil adds: [i.e., upon π].

[55] This name was still not given to that function in the work of Euler mentioned.

We have $\mathrm{Li}(1) = \zeta(2)$. In his paper [66], presented to the Saint Petersburg Academy in 1731 and published in 1738, Euler had already used the dilogarithm function to find numerical approximations for $\zeta(2)$.

In his memoir *Remarques sur un beau rapport entre les séries des puissances tant directes que réciproques* (Remarks on a beautiful relation between direct as well as reciprocal power series), [94], written in 1749 and published in 1768, Euler found the functional equation satisfied by the zeta function. The relation is not explicitly written by Euler but it follows from a relation he writes, as pointed out by Weil in [253] p. 10, who deduces it immediately from the following formula which Euler writes:

$$\frac{1 - 2^{n-1} + 3^{n-1} - 4^{n-1} + 5^{n-1} - 6^{n-1} + \text{ etc.}}{1 - 2^{-n} + 3^{-n} - 4^{-n} + 5^{-n} - 6^{-n} + \text{ etc.}}$$

$$= \frac{-1.2.3...(n-1)(2^n - 1)}{(2^{n-1} - 1)\pi^n} \cos\frac{n\pi}{2}.$$

Weil comments on this formula:

> In the left hand side, we have formally the quotient $\zeta(1-n)\zeta(n)$, except that Euler had written alternating signs to make the series more tractable; the effect of this is merely to multiply $\zeta(n)$ by $1 - 2^{1-n}$, and $\zeta(1-n)$ by $1 - 2^n$. In the right hand side we have the gamma function, which Euler had invented. Euler proves the formula for every positive integer n (using the so-called Abel summation to give a meaning to the divergent series in the numerator of the left hand side), and conjectures its validity for all n.

It was Riemann who showed later on that this equation is valid for any real number $\neq 0, 1$.

In his paper *Variae observationes circa series infinitas* which we already mentioned, [76], Euler found, for $s > 1$, the formula

$$\zeta(s) = \frac{1}{\Gamma(s)} \int_0^\infty \frac{x^{s-1}dx}{e^x - 1}.$$

Here Γ is the Euler gamma function, which is an extension of the factorial:

$$\Gamma(s) = \int_0^\infty e^{-u}u^{s-1}du.$$

In the same paper, he obtained the following formula, valid for real $s > 1$:

$$\zeta(s) = \prod_p \frac{1}{1 - \frac{1}{p^s}}$$

where the product is over all prime numbers p. (Weil explains Euler's derivation of this formula in [255] pp. 265–266.) This equality was the starting point of Riemann's investigations in his paper [219], and it became at the basis of the field called "analytic number theory." Incidentally, it gives a new proof of the fact that there are infinitely many prime numbers (taking $s = 1$ in the formula). We note by the way that Euler gave another proof of the existence of infinitely many prime numbers, using the divergence of the harmonic series $\sum \frac{1}{n}$.

After Euler, the next substantial work on the zeta function, $\zeta(s)$, was done more than a century later, by Riemann. Indeed, in the history of number theory that he wrote, Weil considers (see [255] p. 278) that after Euler, the subject was dead, and that Riemann resurrected it. He conjectures that in 1859, Riemann started working on this subject after he seized a remark by Eisenstein, see [256] for the details. Let us summarize some of the major ideas that Riemann brought in his short paper:

(1) Using analytic continuation, Riemann showed that the zeta function can be extended to a holomorphic function defined on the complex plane, except at the point 1 where the function has a simple pole with residue 1.
(2) He discovered the relation between the zeros of the zeta function and the asymptotic distribution of prime numbers. In fact, Riemann gave the principal term in the asymptotic law of the so-called counting function $\pi(x)$ which measures the number of prime numbers $\leq x$. More precisely, Riemann gave the formula

$$\pi(x) \sim \frac{x}{\log x}, \quad x \to \infty$$

with a sketch of a proof. The result became known as the "prime number theorem." Complete proofs of this theorem were given later by Hadamard and de la Vallée Poussin in 1896.
(3) Starting from the functional equation discovered by Euler—and of which Riemann provided two new proofs adapted to the newly extended function—Riemann showed that the set of zeros of the zeta function contains the even negative integers, and conjectured that all the other zeros are situated on the line $\mathrm{Im}(s) = \frac{1}{2}$. This is the famous Riemann hypothesis.
(4) Riemann obtained a new functional equation satisfied by the zeta function:

$$\zeta(s) = 2^s \pi^{s-1} \sin(\frac{\pi s}{2}) \Gamma(1-s) \zeta(1-s)$$

for $s \neq 0, 1$.

Finding the asymptotic behavior of the prime counting function $\pi(x)$ was, at the epoch of Riemann, one of the major problems in number theory. Legendre, Gauss and Dirichlet had already investigated this problem, and more precisely, they worked on a conjecture saying that $\pi(x)$ is asymptotic to a function of the size of $\frac{x}{\ln x}$. Riemann's main contribution was the introduction of complex analysis in this study, and his intuition that the distribution of primes is related to the zeros of the zeta function extended to the complex plane. The works by de la Vallée Poussin and Hadamard rely heavily on Riemann's ideas, and the outlines of their proofs are based on his sketch. We talk about Hadamard's work on the zeta function and the prime number theorem in Chap. 8 of the present volume, [192]. Let us add here a few historical notes on the counting function; it will give us the occasion to mention again the work of Legendre.

In 1798, Legendre published his *Essai sur la théorie des nombres* (Essay on number theory) [170], a long essay (about 472 pages without the tables) in which, based on numerical evidence, he proposed a conjecture on the form of the counting function $\pi(x)$. He writes (p. 19):

> Moreover, it is likely that the rigorous formula which gives the value of b when a is very large is of the form $b = \dfrac{a}{A \log a + B}$, A and B being constant coefficients, and $\log a$ denoting a hyperbolic logarithm. The exact determination of these coefficients would be a curious problem, worth of training the expertise of the analysts.[56]

Legendre also gave an approximate value of the constant $A(x)$. Let us note incidentally that Legendre, in the preface to his essay, makes a short history of the development of number theory, starting with the Greeks (Euclid and Diophantus), and passing by Viète, Bachet, Fermat, Euler and Lagrange.

In the second edition of his essay (1808), Legendre formulated another conjecture, saying that there are infinitely many primes in any arithmetic progression, that is, primes of the form $l + kn$ for any natural integer n. This conjecture is at the foundations of the theory of Dirichlet series, and it was at the basis of several approaches on the prime number theorem. The conjecture was proved by Dirichlet in 1837 [56], in a paper which brought new tools on how to approach the prime number theorem. In particular, Dirichlet introduced in this paper his famous L-function.

Besides Dirichlet and Legendre, one has to mention Gauss, who, at the age of 15 or 16, started an extensive investigation on the distribution of prime numbers. Based mostly on empirical data (tables of prime numbers that he compiled), he observed that the density of prime numbers around a fixed number x is inversely proportional to $\log x$, and he deduced that the counting function $\pi(x)$ should be well approximated by the integral $\int_2^x \frac{dt}{\log t}$. Gauss never published this work, but he described it in an 1849 letter to his friend and former student, the astronomer J. F. Encke. Gauss, in that letter, makes a comparison between his results and those of Legendre. The letter is included in Gauss's correspondence, edited in his Complete Works, and it is also translated and commented in the article [128] by L. J. Goldstein.

Finally, one has to mention the work of Chebyshev in his two papers [42, 43], done slightly before Riemann (the papers are published in 1851 and 1852), in which he gave precise approximate values for the prime number counting function, making use of the zeta function in the study of the counting function, as Riemann did in his 1859 paper. Chebyshev's paper [43] contains the proof of the so-called Bertrand

---

[56] Au reste, il est vraisemblable que la formule rigoureuse qui donne la valeur de b lorsque a est très grand, est de la forme $b = \dfrac{a}{A \log a + B}$, A et B étant des coefficients constants, et $\log a$ désignant un logarithme hyperbolique. La détermination exacte de ces coefficients serait un problème curieux et digne d'exercer la sagacité des Analystes.

postulate stating that for any integer $n \geq 3$, there exists a prime number p satisfying $n < p < 2N$.[57]

The question of the zeros of the zeta function was proposed by Hilbert in one of the problems he offered at the Paris 1900 ICM.

Riemann's memoir [219] had a major influence on several later mathematicians, including Weil, Siegel, and Selberg.

We conclude this section by quoting Weil, from an obituary article by A. Knapp [163]:

> A substantial portion of Weil's research was motivated by an effort to prove the Riemann hypothesis concerning the zeroes of the Riemann zeta function. He was continually looking for new ideas from other fields that he could bring to bear on a proof. He commented on this matter in a 1979 interview[58]: Asked what theorem he most wished he had proved, he responded, "In the past it sometimes occurred to me that if I could prove the Riemann hypothesis, which was formulated in 1859, I would keep it secret in order to be able to reveal it only on the occasion of its centenary in 1959. Since in 1959, I have felt that I am quite far from it, I have gradually given up, not without regret."[59]

One of the famous Weil conjectures is known as the "Riemann hypothesis over finite fields."

7 On Space

Riemann's habilitation lecture contains a discussion on the nature of physical space and its relation with geometry. The concepts on which Riemann dwells there make it clear that the theme of space belongs to his profound thought. One of the main ideas on which he stresses is the possibility that physical space is different from the space of Euclidean geometry, a point of view that makes Riemann in some sense a predecessor of modern physics.

---

[57]The work of Chebyshev deserves to be much more developed than in these few lines. Like his famous Swiss-Russian predecessor, Leonhard Euler Chebyshev published on most of the fields of pure and applied mathematics. In 1852, he made a stay in France, whose aim was essentially to visit factories and industrial plants, but during his stay he also met several French mathematicians and discussed with them. The list includes Bienaymé, Cauchy, Liouville, Hermite, Lebesgue, Poulignac, Serret and others. A detailed report on this stay, written by Chebyshev himself, is contained in his *Collected works* [45]. Chebyshev used to published in French journals and his relations with French mathematicians remained constant over the years. In 1860, he was elected corresponding member of the Paris Academy of Sciences, and in 1874 foreign member. We learn from his report that at the end of his 1852 stay in France, on his way back to Russia, Chebyshev stopped in Berlin and had several discussions with Dirichlet. It is conceivable that during that meeting the two mathematicians talked about the problems related to the prime number counting function. We refer the reader to the article [193] where some of Chebyshev's works are compared with works of Euler.

[58]Pour la Science, November 1979.

[59]Autrefois, il m'est quelquefois venu à l'esprit que, si je pouvais démontrer l'hypothèse de Riemann, laquelle avait été formulée en 1859, je la garderais secrète pour ne la révéler qu'à l'occasion de son centenaire en 1959. Comme en 1959, je m'en sentais encore bien loin, j'y ai peu à peu renoncé, non sans regret.

In speculating on space, Riemann follows a long tradition which includes the Greeks, Newton, Descartes, Kant and many others, a tradition which survived until the modern period; one may mention, among the mathematicians of the post-Riemannian period, Hermann Weyl, René Thom, Alexandre Grothendieck, and there are many others. It is therefore natural to have, in this chapter, a section on space, in which, not only we review Riemann's ideas—this is done in several chapters of the present volume–but where we mention some of the ideas on this subject that were expressed by his predecessors. Our exposition will necessarily be succinct. Writing a serious essay on the notion of space needs a whole volume.

Space is one of the first very few basic philosophico-epistemological notions. It appears at several places in the works of Aristotle: there are sections on space in the *Categories*, [18], the *Physics* [20], the *Metaphysics* [19], the treatise *On the heavens* [21], etc. Furthermore, like for many other subjects, we learn from Aristotle's works the opinions of his predecessors on space: the Meletians, the Pythagoreans, Plato, etc.

In the *Categories* (5a, 8-14), Aristotle explains that space, like time, belongs to the category of *continuous quantity*.[60] In Book IV of his *Physics*, he writes about the difference between "space" and "place." This is a fundamental distinction, with an impact in physics, and it had a huge influence on later thinkers.[61] The question has also implications in the history of topology. The Greek origin for the word place is *topos* (τόπος), and is translated into Latin by *situs*. The expression *analysis situs*, which was used by Leibniz and the Western founders of topology, finds its origin there.

Among the Western thinkers whose work on the theme of space emerges amid the classical philosophical monuments, we mention Galileo, Newton, Descartes, Leibniz, Huygens and Kant. Most of them are quoted by Riemann.

We start by quoting a text from Greek antiquity. This is a fragment by Archytas of Tarentum which is often referred to in the literature on Pythagorean philosophy, to show the kind of questions on space and on place that the ancient Greeks addressed, e.g., whether space is bounded or not, and the paradoxes to which this question leads (see [143] p. 541):

---

[60]In the *Categories*, (4b 20-5b 11) Aristotle distinguishes seven different types of quantities, which he classifies as continuous and discrete. Discrete quantity comprises number and speech. Continuous quantity comprises the line, the surface, the body, time, and space. Needless to say, although this classification may appear limited from a modern point of view, it has the great merit of existing, may be for the first time. Aristotle asked the pertinent questions.

[61]This theme of space and its relation to place was particularly expanded by Aristotle's commentators. We mention in particular the medieval Andalusian polymath Averroes (1126–1198). The third chapter of Rashed's book *Les mathématiques infinitésimales du IXème au XIème siècle* [211] contains a critical edition together with a translation and commentaries of the treatise *On space* by the Arabic scientist Ibn al-Haytham (known in the West under the name al-Hazen) in which this author criticizes Aristotle's theory of space developed in his *Physics*, and where he defines subsets of space by metric properties. There is also a rich discussion on the notion of space in Greek philosophy in the multi-volume encyclopedic work of P. Duhem [57], see in particular vol. I, p. 197ff.

> "But Archytas," as Eudemus says, "used to propound the argument in this way: 'If I arrived at the outermost edge of the heaven [that is to say at the fixed heaven], could I extend my hand or staff into what is outside or not?' It would be paradoxical not to be able to extend it. But if I extend it, what is outside will be either body or place. It doesn't matter which, as we will learn. So then he will always go forward in the same fashion to the limit that is supposed in each case and will ask the same question, and if there will always be something else to which his staff [extends], it is clear that it is also unlimited. And if it is a body, what was proposed has been demonstrated. If it is place, place is that in which body is or could be, but what is potential must be regarded as really existing in the case of eternal things, and thus there would be unlimited body and space." (Eudemus, Fr. 65 Wehrli, Simplicius, In Ar. Phys. iii 4; 541)

The most basic question that was addressed by many of the philosophers of the modern period that we mentioned is probably the following: Does space have an objective existence or is it only a construction of human mind? Before trying to answer this question, or to have an opinion on it, it is helpful to make it precise what notion of space it refers to: three-dimensional physical space? the three-dimensional space of Euclidean geometry? an abstract notion of space? Other related questions are: Is Euclid's three-dimensional geometry a pure logical construction or is it a mathematical formulation of the properties of external nature? Is the space of (theoretical) physics the same as the mathematicians' space? Does void exist, and what function does it have? These are some of the questions which obviously obsessed Riemann, and before him, many others.

In Descartes' doctrine, space depends on matter, therefore void cannot exist. Leibniz and Euler after him shared the same opinion. Newton had a notion of "absolute space" and "relative space." Furthermore, following the ancient Greeks, Descartes made a difference between space and place. We quote some passages fom his *Principes de la philosophie* (Principles of philosophy) [50] (1644).

> *Principle XIV. How place and space differ:* However, place and space are different in names, because place indicates more expressly situation than magnitude or figure, and that on the contrary, we think about that one when we talk about space; for we say that a thing entered at the place of another, even though it does not have exactly neither the same magnitude nor figure, and for that we do not mean that it occupies the same space that this other thing occupies; and when the situation is changed, we say that the place has also changed, even though it has the same magnitude and figure than before: in this sort, if we say that a thing is in some place, we only mean that it is situated in such a way with respect to other things; but if we add that it occupies a certain space, or place, then we mean that it has such magnitude and figure that it can occupy it exactly.[62]

[62] *Principe XIV. Quelle difference il y a entre le lieu et l'espace:* Toutefois le lieu et l'espace sont différents en leurs noms, parce que le lieu nous marque plus expressément la situation que la grandeur ou la figure, et qu'au contraire nous pensons plutôt à celles-ci lorsqu'on nous parle de l'espace ; car nous disons qu'une chose est entrée en la place d'une autre, bien qu'elle n'en ait exactement ni la grandeur ni la figure, et n'entendons point qu'elle occupe pour cela le même espace qu'occupait cette autre chose ; et lorsque la situation est changée, nous disons que le lieu est aussi changé, quoiqu'il soit de même grandeur et de même figure qu'auparavant : de sorte que si nous disons qu'une chose est en un tel lieu, nous entendons seulement qu'elle est située de telle façon à l'égard de quelques autres choses ; mais si nous ajoutons qu'elle occupe un tel espace, ou un tel lieu, nous entendons outre cela qu'elle est de telle grandeur et de telle figure qu'elle peut le remplir tout justement.

> *Principle XV: How the surface surrounding a body can be taken as its exterior place:* Thus, we never make a distinction between space and extent, for what regards length, width and depth; but we sometimes consider place as if it were within the thing which is placed, and sometimes also as if it were outside it. By no means the interior differs from space; but sometimes we take the exterior to be either the surface surrounding immediately the thing which is placed (and one has to notice that by surface we must not intend any part of the body surrounding it but only the extremity which is between the body which surrounds and the one which is surrounded which is only a mode or a way), or to be the surface in general, which is not part of a body rather than another one, and which always seems to be the same, provided it has the same magnitude and the same figure; because even if we see that the body that surrounds another body passes somewhere else with its surface, we are not used to say that what was surrounded by it has changed its place for this reason, it stays at the same situation regarding the other bodies that we consider as still. Thus, we say that a boat which is carried away by the stream of a river, and which is at the same time pushed away by the wind by a force which is so equal that it does not change its situation regarding the shores, stays at the same place, even though we see that all the surface that surrounds it changes permanently.[63]

Euler had also a strong philosophical background and, needless to say, a tendency for abstraction. We recall that the subject of his first public lecture, delivered at the University of Basel at the occasion of his graduation, was the comparison between the philosophical systems of Newton and Descartes. The notions of space, of motion and of force are discussed in several of his papers on physics. His most important work related to these matters is his *Mechanica*, in two volumes of 500 pages each, [63] with its systematic use of analysis (differential equations) in the field of mechanics, as opposed to Newton's geometric point of view developed in his *Principia*. In his memoir *Recherches sur l'origine des forces* (Research on the origin of forces) [78] (1750), Euler uses an argument involving a notion of "impenetrability of bodies" from which he deduces the law of shock of bodies. We also mention his *Anleitung zur Naturlehre, worin die Gründe zu Erklärung aller in der Natur sich ereignenden Begebenheiten und Veränderungen festgesetzet wedren* (Introduction to natural science establishing the fundamentals for the explanation of the events and changes that occur in nature), [98], a long memoir written in 1745, but never completed and published in 1862. Hermann Weyl says ([258] p. 42) about this memoir that Euler "in magnificent clarity

---

[63] *Principe XV. Comment la superficie qui environne un corps peut être prise pour son lieu exterieur:* Ainsi nous ne distinguons jamais l'espace d'avec l'étendue en longueur, largeur et profondeur ; mais nous considérons quelquefois le lieu comme s'il était en la chose qui est placée, et quelquefois aussi comme s'il en était dehors. L'intérieur ne diffère en aucune façon de l'espace ; mais nous prenons quelquefois l'extérieur ou pour la superficie qui environne immédiatement la chose qui est placée (et il est à remarquer que par la superficie on ne doit entendre aucune partie du corps qui environne, mais seulement l'extrémité qui est entre le corps qui environne et celui qui est environné, qui n'est rien qu'un mode ou une façon), ou bien pour la superficie en général, qui n'est point partie d'un corps plutôt que d'un autre, et qui semble toujours la même, tant qu'elle est de même grandeur et de même figure ; car encore que nous voyions que le corps qui environne un autre corps passe ailleurs avec sa superficie, nous n'avons pas coutume de dire que celui qui en était environné ait pour cela changé de place lorsqu'il demeure en la même situation à l'égard des autres corps que nous considérons comme immobiles. Ainsi nous disons qu'un bateau qui est emporté par le cours d'une rivière, et qui en même temps est repoussé par le vent d'une force si égale qu'il ne change point de situation à l'égard des rivages, demeure en même lieu, bien que nous voyions que toute la superficie qui l'environne change incessamment.

summarizes the foundations of the philosophy of nature of his time." In this memoir, Euler discusses notions like the *extent* of material bodies, the infinite divisibility of these bodies, motion, space, place magnitude, aether and gravity. His memoir *Essai d'une démonstration métaphysique du principe général de l'équilibre* (Essay on a metaphysical demonstration of the general principle of equilibrium) [81] concerns again, force, equilibrium, motion and gravity. In his memoir *Réflexions sur l'espace et le temps* (Reflections on space and time) [79], he makes a comparison between the mathematicians' and the philosophers' (which he calls the "metaphysicians") points of view. He describes *position* as the relation of a body with other bodies around it. He declares that the metaphysicians are wrong in claiming that the notions of space and place are abstract constructions of the mind, and he argues to show the reality of space and time. He claims that both absolute space and time, as mathematicians represent them, are real and exist beyond human imagination. He discusses inertia and the relativity of motion, the ideas of place and position, supported by notions from mechanics.

Euler's philosophical ideas, and their impact on Riemann, have not yet been seriously discussed in the literature.

Immanuel Kant is among the commanding figures that preceded Riemann on the subject of philosophy of space. As a matter of fact, space was already a major theme in Kant's *Inaugural dissertation* (1770). Kant expresses there his doctrine of the a priori nature of space and of geometric objects, that is, the belief that they are not derived from an outside experience. The following excerpt contains an expression of this point of view, which, as we shall recall, Gauss criticized later ([159] Sect. 15, A–D):

> The concept of space is not abstracted from external sensations. For I am unable to conceive of anything posited without me unless by representing it as in a place different from that in which I am, and of things as mutually outside of each other unless by locating them in different places in space. Therefore the possibility of external perceptions, as such, presupposes and does not create the concept of space, so that, although what is in space affects the senses, space cannot itself be derived from the senses.
>
> The concept of space is a singular representation comprehending all things in itself, not an abstract and common notion containing them under itself. What are called several spaces are only parts of the same immense space mutually related by certain positions, nor can you conceive of a cubic foot except as being bounded in all directions by surrounding space.
>
> The concept of space, therefore, is a pure intuition, being a singular concept, not made up by sensations, but itself the fundamental form of all external sensation. This pure intuition is in fact easily perceived in geometrical axioms, and any mental construction of postulates or even problems. That in space there are no more than three dimensions, that between two points there is but one straight line, that in a plane surface from a given point with a given right line a circle is describable, are not conclusions from some universal notion of space, but only discernible in space as in the concrete. Which things in a given space lie toward one side and which are turned toward the other can by no acuteness of reasoning be described discursively or reduced to intellectual marks. There being in perfectly similar and equal but incongruous solids, such as the right and the left hand, conceived of solely as to extent, or spherical triangles in opposite hemispheres, a difference rendering impossible the coincidence of their limits of extension, although for all that can be stated in marks intelligible to the mind by speech they are interchangeable, it is patent that only by pure intuition can the difference, namely, incongruity, be noticed. Geometry, therefore, uses principles not

> only undoubted and discursive but falling under the mental view, and the obviousness of its demonstrations—which means the clearness of certain cognition in as far as assimilated to sensual knowledge—is not only greatest, but the only one which is given in the pure sciences, and the exemplar and medium of all obviousness in the others. For, since geometry considers the relations of space, the concept of which contains the very form of all sensual intuition, nothing that is perceived by the external sense can be clear and perspicuous unless by means of that intuition which it is the business of geometry to contemplate. Besides, this science does not demonstrate its universal propositions by thinking the object through the universal concept, as is done in intellectual disquisition, but by submitting it to the eyes in a single intuition, as is done in matters of sense.
>
> Space is not something objective and real, neither substance, nor accident, nor relation; but subjective and ideal, arising by fixed law from the nature of the mind like an outline for the mutual co-ordination of all external sensations whatsoever. Those who defend the reality of space either conceive of it as an absolute and immense receptacle of possible things, an opinion which, besides the English, pleases most geometricians, or they contend for its being the relation of existing things itself, which clearly vanishes in the removal of things and is thinkable only in actual things, as besides Leibniz, is maintained by most of our countrymen. The first inane fiction of the reason, imagining true infinite relation without any mutually related things, pertains to the world of fable. But the adherents of the second opinion fall into a much worse error. Whilst the former only cast an obstacle in the way of some rational or monumental concepts, otherwise most recondite, such as questions concerning the spiritual world, omnipresence, etc., the latter place themselves in fiat opposition to the very phenomena, and to the most faithful interpreter of all phenomena, to geometry. For, not to enlarge upon the obvious circle in which they become involved in defining space, they cast forth geometry, thrown down from the pinnacle of certitude, into the number of those sciences whose principles are empirical. If we have obtained all the properties of space by experience from external relations only, geometrical axioms have only comparative universality, such as is acquired by induction. They have universality evident as far as observed, but neither necessity, except as far as the laws of nature may be established, nor precision, except what is arbitrarily made. There is hope, as in empirical sciences, that a space may some time be discovered endowed with other primary properties, perchance even a rectilinear figure of two lines.

The reader will notice that Kant talks about "geometrical axioms," and mentions axioms of Euclidean geometry such as the fact that "between two points there is but one straight line." Kant was by no means a mathematician, but he had a sufficient knowledge, as a philosopher, of several basic principles of mathematics.

It appears from Gauss's correspondence, published in Volume VII of his *Collected Works* (p. 200ff.) that he was meditating on the nature of space since a very young age, probably from the age of 16. It is from these meditations that he became interested in the parallel postulate and in non-Euclidean geometry, spherical and (the hypothetical) hyperbolic. Unlike most of the geometers that preceded him, Gauss was convinced, at a very early stage of his life, that the parallel postulate was not a consequence of the others, and he spent a lot of time and energy pondering on the principles of hyperbolic geometry, a geometry resulting from the negation of this postulated.

Gauss was also thoroughly interested in philosophy, and, in particular, he read Kant. He became very critical of the latter's conception of space, exemplified in the text we just quoted as being "not something objective and real, neither substance, nor accident, nor relation, but subjective and ideal, arising by fixed law from the nature of the mind." On Kant, Gauss had the advantage of being a mathematician. In a letter to his friend Bessel, dated April 9, 1830, Gauss writes (translation from [30] p. 13):

> We must confess in all humility that a number is *solely* a product of our mind. Space, on the other hand, possesses also a reality outside of our minds, the laws of which we cannot fully prescribe a priori.

In another letter, sent to Wolfgang Bolyai on March 6, 1832 and published in his *Collected Works*, Gauss writes, concerning the two hypotheses on the angle sum in a triangle, that it is precisely in the difficulty of this decision that "lies the clearest proof that Kant was wrong in asserting that space is just a form of our perception."

Gauss was also very critical of Kant's argument based on symmetries in the text we quoted above ("There being in perfectly similar and equal but incongruous solids, such as the right and the left hand, conceived of solely as to extent,...it is patent that only by pure intuition can the difference, namely, incongruity, be noticed"). We further discuss this in Chap. 6 of the present volume [190].

It is not surprising that Riemann declares, in his habilitation lecture, that, concerning his ideas on space, he is influenced by Gauss.

Riemann's ideas on space were discussed by Clifford, the first mathematician who translated into English Riemann's habilitation text, cf. [46].

8 Topology

Poincaré, who is certainly the major founder of the modern field of topology,[64] declares in his "Analysis of his own works" (*Analyse des travaux scientifiques de Henri Poincaré faite par lui-même*), [202] p. 100, that he has two predecessors in the field, namely, Riemann and Betti. The latter, in his correspondence with his friend and colleague Placido Tardy reports on several conversations he had with Riemann on topology. Two letters from Betti to Tardy on this subject are reproduced and translated in the book [206] by Pont, in the article [257] by Weil, and prior to them, by Loria in his obituary on Tardy [178].

The first of these two letters by Betti, dated October 6, 1863, starts with the following (Weil's translation): "I have newly talked with Riemann about the connectivity of spaces, and have formed an accurate idea of the matter," and he goes on explaining to his friend the notion of connectivity and that of the order of connectivity. Betti then writes:

> What gave Riemann the idea of the cuts was that Gauss defined them to him, talking about other matters, in a private conversation. In his writings one finds that analysis situs, that is, this consideration of quantities independently from their measure, is "wichtig"; in the last

[64] We may quote P. S. Alexandrov, who declared in a talk he gave at a celebration of the centenary of Poincaré's birth [9]: "To the question of what is Poincaré's relationship to topology, one can reply in a single sentence: he created it." On Poincaré and Riemann, Alexandrov, in the same talk, says the following: "The close connection of the theory of functions of a complex variable, which Riemann has observed in embryonic form, was first understood in all its depth by Poincaré."

> years of his life he has been much concerned with a problem in analysis situs, namely: given a winding thread and knowing, at every one of its self-intersections, which part is above and which below, to find whether it can be unwound without making knots; this problem he did not succeed in solving except in special cases ...

The second letter, dated October 16, 1863, starts with: "Riemann proves quite easily that every space can be reduced to an SC space by means of 1-cuts and SC 2-cuts." In the same letter, Betti elaborates on this subject, giving many examples in n dimensions. He concludes the letter by noting that the number of line sections is equal to the number of periodicity moduli of an $(n-1)$-integral, the number of simply connected surface sections to the number of periodicity moduli of an $(n-2)$-integral, and so on.

This should make clear the parentage, for what concerns topology, from Riemann to Poincaré, potentially including Betti. In this section, we go further back in the history of topological ideas, and we review some of the important works done before Riemann in this field.

René Thom considers that topology was born in ancient Greece. He expanded on this idea in several articles, cf. [246, 247]. This is a perfectly reasonable theory. In fact, the question depends on what sense we give to the word "topology." If the matter concerns the notions of limit and convergence, then the roots of this field are indeed in Greek antiquity, and more especially, in the writings of Zeno, which do not survive, but which were quoted by his critics and commentators, including Plato, Aristotle and Simplicius. Likewise, if the question concerns the notion of space, and the related notion of place, then the roots also are in Greek science. We already alluded to this fact in the previous section. The Greeks made a distinction between space and place and the notion of place (situs) is at the basis of topology. The three words place, situs and $\tau \acute{o} \pi o \varsigma$ are synonyms. To the best of our knowledge, a systematic investigation of the origin of topology in Greek antiquity has never been conducted. A whole book may be written on that subject. Failing to do this now, we shall start our exposition of the roots of topology with Leibniz, as it is usually done. Indeed, it is commonly accepted that the first explicit mention of topology as a mathematical field was made by him.

Even though no purely topological result can be attributed to Leibniz, he had the privilege to express for the first time, back in the seventeenth century, the need for a new branch of mathematics, which would be "a geometry that is more general than the rigid Euclidean geometry and the analytic geometry of Descartes." Leibniz describes his geometry as purely qualitative and concerned with the study of figures independently of their metrical properties. In a letter to Christiaan Huygens, sent on September 8, 1679 (cf. [176] pp. 578–569 and [144] vol. VIII n° 2192), he writes:

> After all the progress I have made in these matters, I am still not happy with Algebra, because it provides neither the shortest ways nor the most beautiful constructions of Geometry. This is why when it comes to that, I think that we need another analysis which is properly geometric or linear, which expresses to us directly *situm*, in the same way as algebra expresses *magnitudinem*. And I think that I have the tools for that, and that we might represent figures

> and even engines and motion in character, in the same way as algebra represents numbers in magnitude.[65]

In the same letter ([176] p. 570), Leibniz adds:

> I found the elements of a new characteristic, completely different from Algebra and which will have great advantages for the exact and natural mental representation, although without figures, of everything that depends on the imagination. Algebra is nothing but the characteristic of undetermined numbers or magnitudes. But it does not directly express the place, angles and motions, from which it follows that it is often difficult to reduce, in a computation, what is in a figure, and that it is even more difficult to find geometrical proofs and constructions which are enough practical even when the Algebraic calculus is all done.[66]

Together with his letter to Huygens, Leibniz included the manuscript of an essay he wrote on the new subject. He writes, in the same letter ([176] p. 571):

> But since I don't see that anybody else has ever had the same thought, which makes me fear that it might be lost if I do not get enough time to complete it, I will add here an essay which seems to me important, and which will suffice at least to rendre my aim more credible and easier to conceive, so that if something prevents its realization now, it will serve as a monument for posterity and give the possibility to somebody else to finish it.[67]

He then explains in more detail his vision of this new domain of mathematics, and where it stands with respect to algebra and geometry, giving several examples of a formalism to denote loci, showing how this formalism expresses statements such that the intersection of two spherical surfaces is a circle, and the intersection of two planes is a line.

Leibniz' letter ends with the words ([174] p. 25):

---

[65]Après tous les progrès que j'ai faits en ces matières, je ne suis pas encore content de l'Algèbre, en ce qu'elle ne donne ni les plus courtes voies, ni les plus belles constructions de Géométrie. C'est pourquoi lorsqu'il s'agit de cela, je crois qu'il nous faut encore une autre analyse proprement géométrique ou linéaire, qui nous exprime directement *situm*, comme l'algèbre exprime *magnitudinem*. Et je crois d'en avoir le moyen, et qu'on pourrait représenter des figures et même des machines et mouvements en caractères, comme l'algèbre représente les nombres en grandeurs. [We have modernized the French.].

[66]J'ai trouvé quelques éléments d'une nouvelle caractéristique, tout à fait différente de l'Algèbre, et qui aura de grands avantages pour représenter à l'esprit exactement et au naturel, quoique sans figures, tout ce qui dépend de l'imagination. L'Algèbre n'est autre chose que la caractéristique des nombres indéterminés ou des grandeurs. Mais elle n'exprime pas directement la situation, les angles et les mouvements, d'où vient qu'il est souvent difficile de réduire dans un calcul ce qui est dans la figure, et qu'il est encore plus difficile de trouver des démonstrations et des constructions géométriques assez commodes lors même que le calcul d'Algèbre est tout fait.

[67]Mais comme je ne remarque pas que quelqu'autre ait jamais eu la même pensée, ce qui me fait craindre qu'elle ne se perde, si je n'y ai pas le temps de l'achever, j'ajouterai ici un essai qui me paraît considérable, et qui suffira au moins à rendre mon dessein plus croyable et plus aisé à concevoir, afin que si quelque hasard en empêche la perfection à présent, ceci serve de monument à la postérité, et donne lieu à quelque autre d'en venir à bout.

> I have only one remark to add, namely, that I see that it is possible to extend the characteristic to things which are not subject to imagination. But this is too important and it would lead us too far for me to be able to explain myself on that in a few words.[68]

When Leibniz started his correspondence with Huygens, the latter was already a well established scientist whose achievements were behind him, and it was not easy to convince him of the usefulness of a new theory. Huygens thought that the theory was too abstract and he remained skeptical about it. He was above all a geometer working on concrete geometrical problems.

One may recall that when Leibniz sent him the above letter, Huygens was considered as a world authority in geometry and physics. He was settled in Paris since 15 years, and he was a leading member of the *Académie Royale des Sciences*. Leibniz had studied mathematics with Huygens, who was seventeen years older than him, and he considered him as his mentor. Huygens responded to Leibniz in a letter dated November 22, 1679 ([176] p. 577):

> I have examined carefully what you are asking me regarding your new characteristic, and to be frank with you, I cannot not conceive the fact that you have so much expectations from what you spread on me. Because your example of places concerns only realities which were already perfectly known, and the proposition saying that the intersection of a plane and a spherical surface makes the circumference of a circle does not follow clearly. Finally, I cannot see in what way you can apply your characteristic to which you seem you want to reduce all these different matters, like the quadratures, the invention of curves by the properties of tangents, the irrational roots of equations, Diophantus' problems, the shortest and the most beautiful constructions of the geometric problems. And what still appears to me stranger than anything else, the invention and the explanation of machines. I say it to you unsuspiciously, in my opinion this is only wishful thinking, and I need other proofs in order to believe that there could be some reality in what you present. I would nevertheless restrain myself from saying that you are mistaken, knowing the subtlety and the deepness of your mind. I only beg you that the magnificence of the things you are searching won't let you postpone from giving us those which you already found, like this Arithmetic Quadrature you discovered, concerning the roots of the equations beyond the cubical, if you are still satisfied with it.[69]

---

[68] Je n'ai qu'une remarque à ajouter, c'est que je vois qu'il est possible d'étendre la caractéristique jusqu'aux choses, qui ne sont pas sujettes à l'imagination ; mais cela est trop important et va trop loin pour que je me puisse expliquer là-dessus en peu de paroles.

[69] J'ai examiné attentivement ce que vous me demandez touchant votre nouvelle caractéristique, mais pour vous l'avouer franchement, je ne conçois pas parce que vous m'en étalez, que vous y puissiez fonder de si grandes espérances. Car votre exemple des Lieux ne regarde que des vérités qui nous étaient déjà fort connues, et la proposition de ce que l'intersection d'un plan et d'une surface sphérique fait la circonférence d'un cercle, s'y conclut assez obscurément. Enfin, je ne vois point de quel biais vous pourriez appliquer votre caractéristique à toutes ces choses différentes qu'il semble que vous y vouliez réduire, comme les quadratures, l'invention des courbes par la propriétés des tangentes, les racines irrationnelles des Équations, les problèmes de Diophante, les plus courtes et plus belles constructions des problèmes géométriques. Et ce qui me paraît encore le plus étrange, l'invention et l'explication des machines. Je vous le dis ingénument, ce ne sont là à mon avis que de beaux souhaits, et il me faudrait d'autres preuves pour croire qu'il y eût de la réalité dans ce que vous avancez. Je n'ai pourtant garde de dire que vous vous abusiez, connaissant d'ailleurs la subtilité et profondeur de votre esprit. Je vous prie seulement que la grandeur des choses que vous cherchez ne vous fasse point différer de nous donner celles que vous avez déjà trouvées, comme est

In another letter dated January 11, 1680 ([176] p. 584) Huygens writes:

> For what concerns the effects of your characteristic, I see that you insist on being persuaded of them, but as you say yourself, the examples will be more important than reasonings. This is why I am asking you much simpler examples, but capable of overcoming my incredulity, because that of the places, I confess, does not seem to me of that sort.[70]

The essay that Leibniz sent did not obtain Huygens' backing and it remained hidden among other manuscripts in Huygens' estate. It was published for the first time in 1833, and drew the attention of several nineteenth-century mathematicians, including Grassmann (1809–1877), the founder of the theory of vector spaces, who realized its importance for the new field of topology. There are two recent editions of this text, both included in doctoral dissertations, by J. Acheverría [175] (1995), in France, and by de Risi, [234] (2007), in Germany. The two dissertations contain other texts by Leibniz on the same subject.

Leibniz used several names for the new field, including *analysis situs, geometria situs, characteristica situs, characteristica geometrica, analysis geometrica, speciosa situs*, etc.

The first mathematician who worked consciously on topological questions is Euler. These questions include the definition and the invariance of the Euler characteristic of a convex polyheron, the problem known as that of the Königsberg seven bridges, another question related to the Knight's tour on the chessboard, and a musical question concerning a graph known as the *speculum musicum*. This graph was introduced in Euler's *Tentamen novae theoriae musicae ex certissimis harmoniae principiis dilucide expositae* (A attempt at a new theory of music, exposed in all clearness according to the most well-founded principles of harmony) [71]. Its vertices are the twelve notes of the chromatic scale, and the edges connect two elements which differ by a fifth or a major third with the property that one may traverse all the edges of the graph passing exactly once by each note. The article [189] is a detailed survey of the work of Euler on these questions. In the present section, we start by reviewing the work of Euler on the question of the seven bridges of Königsberg. This work shows that Euler considered himself as the direct heir of Leibniz for what concerns the field of topology. We shall then describe in detail the works of Euler and Descartes on the Euler characteristic, a question which is directly related to the topological classification of surfaces, which was one of Riemann's major achievements in topology. We recall that Euler formulated this result for a surface which is the boundary of a convex polyhedron having F faces, A edges and S vertices; the formula is then:

$$F - A + S = 2.$$

(Footnote 69 continued)
cette Quadrature Arithmétique et que vous avez découvert pour les racines des équations au-delà du cube, si vous en êtes content vous-même.

[70]Pour ce qui est des effets de votre caractéristique, je vois que vous persistez à en être persuadé, mais, comme vous dites vous-même, les exemples toucheront plus que les raisonnements. C'est pourquoi je vous en demande des plus simples, mais propres à convaincre mon incrédulité, car celui des lieux, je l'avoue, ne me paraît pas de cette sorte.

We start with the problem of the Königsberg bridges.

In the eighteenth century, the city of Königsberg[71] consisted of four quarters separated by branches of the river Pregel and related by seven bridges. The famous "problem of the seven bridges of Königsberg" asks for a path in that city that starts at a given point and returns to the same point after crossing once and only once each of the seven bridges. At the time of Euler, this was a popular question among the inhabitants of Königsberg.

Euler showed that such a path does not exist. He presented his solution to the Saint Petersburg Academy of Sciences on August 26, 1735, and in the same year he wrote a memoir on the solution of a more general problem entitled *Solutio problematis ad geometriam situs pertinentis* (Solution of a problem relative to the geometry of position) [74]. Euler learned about the problem from a letter, dated March 7, 1736, sent to him by Carl Leonhard Gottlieb Ehler, one of his friends who was the mayor of Danzig[72] and who had worked as an astronomer in Berlin. Euler solved the problem just after he received the letter. In a letter dated March 13, 1736, written to Giovanni Marioni, an Italian astronomer working at the court of Vienna, Euler declares that he became interested in this question because he realized that the problem could not be solved using geometry, algebra or combinatorics, and that therefore he wondered whether "it belonged to the 'geometry of position,' (*geometria situs*) which Leibniz has so much sought for." In the same letter, Euler announced to Marioni that after some thought, he found a proof which applies not only to that case, but to all similar problems.

In the introduction of his paper, Euler writes (translation from [29]):

> In addition to that branch of geometry which is concerned with magnitudes, and which has always received the greatest attention, there is another branch, previously almost unknown, which Leibniz first mentioned, calling it the geometry of position. This branch is concerned only with the determination of position and its properties; it does not involve measurements, nor calculations made with them. It has not yet been satisfactorily determined what kind of problems are relevant to this geometry of position, or what methods should be used in solving them. Hence, when a problem was recently mentioned, which seemed geometrical but was so constructed that it did not require the measurement of distances, nor did calculation help at all, I had no doubt that it was concerned with the geometry of position, especially as its solution involved only position, and no calculation was of any use. I have therefore decided to give here the method which I have found for solving this kind of problem, as an example of the geometry of position.

Euler's work on this problem is commented in several articles and books.

We now come to Euler's polyhedron formula, and we start with Descartes.

Long before Euler came out with his formula $F - A + S = 2$ relating the faces (F), edges (A) and vertices (S) of a convex polyhedron, Descartes obtained an equivalent result, with a geometric proof, involving the solid angles and the dihedral angles between the faces. Described in modern terms, Descartes' proof consists in computing in two different manners the total curvature of the boundary of the polyhedron.

---

[71] Today, the city of Königsberg, called Kaliningrad, is part of a Russian exclave between Poland and Lithuania on the Baltic Sea.

[72] Today, Danzig is the city of Gdansk, in Poland.

Descartes wrote that proof at the age of 25, but did not publish it. The story of Descartes' manuscript is interesting and we recall it now.

Descartes' manuscript was discovered in Hanover, among Leibniz's estate. The latter had copied Descartes' proof during a stay in Paris, in 1675 or 1676, presumably with the intention of publishing it. The original manuscript of Descartes, which carries the title *Progymnasmata de solidorum elementis* (Preparatory exercises to the elements of solids) [47] is mentioned in an inventory of papers which Descartes left in some chests in Stockholm, the city where he died. The copy, made by Leibniz, carries the same title, with the additional mention *excerpta ex manuscripto Cartesii* (Excerpt from a manuscript of Descartes). After the manuscript was discovered, a French translation was published by Foucher de Careil in 1859, in a volume of unpublished works of Descartes. This publication contained errors, because Foucher, who did it, was not a mathematician. The edition is nevertheless interesting, and in the introduction to the volume [114], Foucher recalls the story of the discovery. The story is also told by Adam in the commentaries of the volume of the Adam-Tannery edition of Descartes' works containing this theorem ([51] tome X, pp. 257–263).

In 1890, Jonquières presented to the Paris Académie des Sciences two Comptes Rendus notes entitled *Sur un point fondamental de la théorie des polyèdres* (On a fundamental property of the theory of polyhedra) [148] and *Note sur le théorème d'Euler dans la théorie des polyèdres* (Note on the theorem of Euler on the theory of polyhedra) [149], without being aware of the work of Descartes on this subject. After Jordan pointed out the existence of the work of Descartes in Foucher's edition, Jonquières published other Comptes Rendus notes on the work of Descartes, cf. [150–152]. Poincaré, in his celebrated first memoir on *Analysis situs* [200] attributes to Jonquières the generalization of Euler's theorem to non-necessarily convex polyhedral surfaces.

There is a relatively recent (1987) critical edition of Descartes' *Progymnasmata* with a French translation, with notes and commentaries, by P. Costabel [47].

Euler reported on his work on polyhedra in his correspondence with Goldbach. In a letter dated November 14, 1750, Euler informs his friend of the following two results which he refers to as Theorems 6 and 11 respectively:

> 6. In any solid enclosed by plane surfaces the sum of the number of faces and the number of solid angles is greater by 2 than the number of edges.
>
> 11. The sum if all planar angles equals four times as many right angles as the number of solid angles, decreased by 8.

The term "solid enclosed by plane surfaces" refers to a convex polyhedron. The first result is the Euler characteristic formula, and the second one is a form of the Gauss-Bonnet theorem for the sphere. Euler writes:

> I am surprised that these general properties in stereometry have not been noticed by anybody, as far as I know, but still more that the most important of them, viz., Th. 6 and Th. 11, are so hard to prove; indeed I still cannot prove them in a way that satisfies me.

In the same letter, Euler gives several examples where the two theorems are satisfied.

In his memoir [85], entitled *Demonstratio nonnullarum insignium proprietatum, quibus solida hedris planis inclusa sunt praedita* (Proof of some of the properties of solid bodies enclosed by planes) and written one year after [84], Euler gave proofs of the two results. In the introduction of [85], he declares that his polyhedron formula is part of a more general research he is conducting on polyhedra. In fact, in the letter to Goldbach we mentioned, Euler announces a result on volumes of simplices in terms of their side lengths (a three-dimensional analogue of Heron's formula for the area of triangles), which he proves later in his paper *Demonstratio nonnullarum insignium proprietatum, quibus solida hedris planis inclusa sunt praedita* [85]. Euler writes[73]:

> Although I had uncovered many properties which are common to all bodies enclosed by plane faces and which seemed to be completely analogous to those which are commonly included among the first principles of rectilinear plane figures, still, not without a great deal of surprise did I realize that the most important of those principles were so recondite that all the time and effort spent looking for a proof of them had been fruitless. Nor, when I consulted my friends, who are otherwise extremely versed in these matters and with whom I had shared those properties, were they able to shed any light from which I could derive these missing proofs. After the consideration of many types of solids I came to the point where I understood that the properties which I had perceived in them clearly extended to all solids, even if it was not possible for me to show this in a rigorous proof. Thus, I thought that those properties should be included in that class of truths which we can, at any rate, acknowledge, but which it is not possible to prove.

One advantage of Euler's proof, compared to the one of Descartes, is that it shows in a clear way the combinatorial aspect of the problem, highlighting the notion of edges and faces of the polyhedron.

The proof that Euler gives in [85] is based on an induction on the number of solid angles, reducing them by one at each step. He writes:

> These proofs are in no way inferior to those proofs used in Geometry except that here due to the nature of solids one must use more imagination, in as much as solids are being depicted on a flat surface.

At the same time, Euler was laying down the foundations of combinatorial topology. He writes (Scholion to Proposition 4):

> I admit that I have thus brought to light only the first principles of Solid Geometry, on which this science should be built as it develops further. No doubt it contains many outstanding qualities of solids of which we are so far completely ignorant. [...]

Legendre, in his *Éléments de géométrie* (Note XII) published in 1794 [173], gave a complete proof of Euler's theorem based on geometry. This proof is considered as one of the simplest, and it is repeated in more modern works, e.g. in Hopf [140].

A large number of mathematicians commented on Euler's polyhedron formula, expanding some arguments in Euler's proofs, giving new proofs, and sometimes comparing Euler's work with that of Descartes. To show the diversity of these works, we mention the papers by Andreiev [10], Bertrand [26], Bougaïev [22], Brianchon [27], Catalan [31], Cauchy [32, 33], Feil [113], Gergonne, [130], Grunert, [135], Jonquières [148–152, 154–156], Jordan [157, 158], Lebesgue [168], Lhuillier [177],

[73]Translation by C. Frances and D. Richeson.

Poincaré [199, 200], Poinsot [203, 204], Prouhet [208, 209], Steiner [239], Valat [249] and Thiel [244]. We shall quote some of these works below. We also mention that in 1858, the Paris *Académie des Sciences* proposed as a subject for the 1861 *Grand prix*: "To improve, in some important point, the geometric theory of polyherda." Möbius participated and presented a memoir (but did not get the prize).

In 1811, Cauchy brought out a purely combinatorial proof of that theorem. In this proof, one starts by deleting a face of the polyhedron and reduces the problem to another one concerning a planar polygon.[74] In his article *Recherches sur les polyèdres* (Researches on polyhedra) [32], published in 1813, Cauchy writes:

> Euler has determined, in the Petersburg *Mémoires*, year 1758, the relation that exists between the various elements that compose the surface of a polyhedron; and Mr. Legendre, in his *Éléments de Géométrie*, proved in a much simpler manner Euler's theorem, by considerations of spherical polygons. Having been led by some researches to a new proof of that theorem, I reached a theorem which is much more general than the one of Euler, whose statement is the following:
> Theorem. If we decompose[75] a polyhedron in as many others as we wish, by taking at will new vertices in the interior, and if we represent by P the number of new polyhedra thus formed, by S the total number of vertices, including those of the initial polyhedron, by F the total number of faces, and by A the total number of edges, then we will have
>
> $$S + F = A + P - 1,$$
>
> that is, the sum of the number of vertices and that of faces will overpass by one the sum of the number of edges that of polyhedra.[76]

Poinsot [203], in 1858, published a proof of Euler's formula using some of Cauchy's arguments. He writes: "This relation, which Euler was the first to prove, does not hold only for convex polyhedra, as one might think, but for polyhedra of any kind." In fact, this statement needs some explanation. We are used today to the fact that Euler's formula is valid for polyhedra which are homeomorphic to a sphere. This notion did not exist at that time, neither the word, nor the idea. One had to

74 A similar proof is given by Hilbert and Cohn-Vossen [138] p. 290.

75 Cauchy "decomposes" the polyhedron by taking new vertices in the interior of the three-dimensional polyhedron (and not on the boundary surface).

76 Euler a déterminé, dans les *Mémoires* de Pétersbourg, année 1758, la relation qui existe entre les différents éléments qui composent la surface d'un polyèdre ; et M. Legendre, dans ses *Éléments de Géométrie*, a démontré d'une manière beaucoup plus simple le théorème d'Euler, par la considération des polygones sphériques. Ayant été conduit par quelques recherches à une nouvelle démonstration de ce théorème, je suis parvenu à un théorème plus général que celui d'Euler et dont voici l'énoncé :
Théorème. Si l'on décompose un polyèdre en tant d'autres que l'on voudra, en prenant à volonté dans l'intérieur de nouveaux sommets ; que l'on représente par P le nombre de nouveaux polyèdres ainsi formés, par S le nombre total de sommets, y compris ceux du premier polyèdre, par F le nombre total de faces, et par A le nombre total des arêtes, on aura

$$S + F = A + P - 1,$$

c'est-à-dire que la somme faite du nombre des sommets et de celui des faces surpassera d'une unité la somme faite du nombre des arêtes et de celui des polyèdres.

wait for that to the work of Jordan, who set up the precise hypotheses under which Euler's formula is valid. In his article [155], he writes that Euler's theorem is valid for polyhedra which he calls "simple," or "Eulerian," that is, polyhedra ([155] p. 35) "such that any contour drawn on the surface which does not traverse itself divides this surface into two separate regions; a category that contains as a particular case convex polyhedra."[77] A few pages later (p. 38), Jordan makes the following commentary: "It would have been easy to show that if we can draw on a polyhedron λ different contours which do not intersect each other and which do not divide the surface into separate parts, we would have $S + F = A + 2 - 2\lambda$."[78] In fact, Jordan had extracted the notion we call today "topological surface of finite type," to which the general theory applies, cf. [154] p. 86:

> A surface is said to be of type (m, n) if it is bounded by m closed contours and if furthermore we can draw on it n closed contours that do not intersect themselves nor mutually, without dividing it into two distinct regions.[79]

Then Jordan makes the relation with the polyhedra to which Euler's formula applies: "The polyherda of kind (0, 0) are nothing but those which I called *Eulerian*."[80]

It is interesting to read Lebesgues' comments on some proof of Euler's theorem, because it gives us some hints of how the subject of topology was viewed in those days. Lebesgues' comments are written in 1924 ([168] p. 319):

> I don't agree at all with those who pretend to attribute Euler's theorem to Descartes. Descartes did not state the theorem; he did not see it. Euler perceived it and he fully understood its character. For Euler, the description of the form of a polyhedron must precede the use of the measures of its elements, and this is why he set his theorem as a fundamental theorem. For him, like for us, this is a theorem of enumerative *Analysis situs*; therefore he tried to find it by considerations independent of any metrical theory, that in effect belong to what we call the field of *Analysis situs*. And this is why he left to Legendre the honor of finding a rigorous proof. None of us who had read a little bit of Euler and who were amazed by his prodigious technical masterliness will doubt, even for one second, that if Euler had thought of putting aside his theorem and deducing it from one of its metric corollaries, he would have easily succeeded. (It should be noted that Euler does not at all restrict his researches to convex polyhedra.) It seems to me, on the contrary, that the fact that Descartes passed so closely to the theorem without seeing it, emphasizes Euler's credit. (At least, this is what we believe, because Descartes employed in his notebook some algebraic characters which he used before knowing Viète's characters.) But Leibniz, who found Descartes' notebook enough interesting to copy it, who realized that Descartes' geometry does not apply to questions involving order and position relations, who dreamed of constructing the algebra of these relations and who in advance gave it the name *Analysis situs*, did not notice, in Descartes' notebook Euler's

[77][...] tels que tout contour fermé tracé sur leur surface et ne se traversant pas lui-même divise cette surface en deux régions séparées ; catégorie qui renferme comme cas particulier les polyèdres convexes.

[78]Il serait aisé de démontrer que si l'on peut tracer sur un polyèdre λ contours différents, ne se coupant pas mutuellement et ne divisant pas la surface en parties séparées, on aura $S + F = A + 2 - 2\lambda$.

[79]Une surface sera dite d'espèce (m, n) si elle est limitée par m contours fermés et si l'on peut d'autre part y tracer n contours fermés ne se coupant pas eux-mêmes ni mutuellement, sans la partager en deux régions distinctes.

[80]Les polyèdres de l'espèce (0, 0) ne sont autres que ceux que j'ai appelés *eulériens*.

> theorem which is so fundamental in *Analysis situs*. This theorem really belongs to Euler. As for the proof, one could, may be with a little bit of unfairness, call it the proof of Legendre and Descartes. This proof is metrical, and it is fair to blame it for the fact that it uses notions that are foreign to *Analysis situs*. But one should not exaggerate the value of this grievance.[81]

We now give a quick review of some work of Gauss on topology, another field in which his impact on Riemann was huge.

Gauss was interested in applications of *Geometria situs* (a term he used in his writings), in particular in astronomy, geodesy and electromagnetism. In astronomy, he addressed the question of whether orbits of celestial bodies may be linked (cf. his short treatise entitled *Über die Grenzen der geocentrischen Orter der Planeten*). From his work on geodesy, we mention his letter to Schumacher, 21 Nov. 1825, (from Gauss's *Werke* vol. VIII, p. 400):

> Some time ago I started to take up again a part of my general investigations on curved surfaces, which shall become the foundation of my projected work on higher geodesy. [...] Unfortunately, I find that I will have to go very far afield [...]. One has to follow the tree down to all its root threads, and some of this costs me week-long intense thought. Much of it even belongs to geometria situs, an almost unexploited field.

From Gauss's *Nachlaß*, we know that he worked on a combinatorial theory of knot projections, during the year 1825, and again in 1844. (Gauss's *Werke*, Vol. VIII, pp. 271–286). We already mentioned at the beginning of this section, that we learn from a letter sent by Betti to Tardy that the idea of analyzing a surface by performing successive cuts was given to Riemann by Gauss, in a private conversation. Besides Riemann, Gauss had two students who worked on topology and who were certainly influenced by him: Listing and Möbius.

---

[81] Je ne suis pas du tout d'accord avec ceux qui prétendent attribuer à Descartes le théorème d'Euler. Descartes n'a pas énoncé le théorème ; il ne l'a pas vu. Euler l'a aperçu et en a bien compris le caractère. Pour Euler, la description de la forme d'un polyèdre doit précéder l'utilisation des mesures de ses éléments et c'est pourquoi il a posé son théorème comme théorème fondamental. C'est, pour lui comme pour nous, un théorème d'*Analysis situs* énumérative ; aussi a-t-il cherché à le démontrer par des considérations indépendantes de toute propriété métrique, appartenant bien à ce que nous appelons le domaine de l'*Analysis situs*. Et c'est pourquoi il a laissé à Legendre l'honneur d'en trouver la preuve rigoureuse ; aucun de ceux qui ont quelque peu lu Euler, et qui ont été stupéfaits de sa prodigieuse virtuosité technique, ne doutera un seul instant que si Euler avait pensé à faire passer son théorème au second plan et à le déduire d'un de ses corollaires métriques, il n'y eût facilement réussi. (Il convient d'ajouter qu'Euler ne restreint nullement ses recherches aux polyèdres convexes.) Que Descartes soit passé si près du théorème sans le voir me paraît au contraire souligner le mérite d'Euler. Encore peut-on dire que Descartes était jeune quand il s'occupait de ces questions. (C'est du moins ce que l'on croit, parce que Descartes a employé dans son cahier certains caractères cossiques qu'il utilisait avant de connaître les notations de Viète.) Mais Leibniz qui a trouvé le cahier de Descartes assez intéressant pour le copier, qui a reconnu que la géométrie de Descartes ne s'appliquait pas aux questions où interviennent des relations d'ordre et de position, qui a rêvé de construire l'algèbre de ces relations et l'a nommée à l'avance *Analysis situs*, n'a pas aperçu, dans le cahier de Descartes, le théorème d'Euler si fondamental en *Analysis situs*. Le théorème appartient bien à Euler ; quant à la démonstration, on pourrait, un peu injustement peut-être, la dénommer démonstration de Legendre et Descartes. Cette démonstration est métrique ; il est juste de lui reprocher de faire appel à des notions étrangères à l'*Analysis situs*. Mais il ne faudrait pas s'exagérer la valeur de ce grief.

Riemann introduced the fundamental topological notions for surfaces: connectedness, degree of connectivity, the classification of closed surfaces by their genus. He developed this theory for the purpose of using it in his work on the theory of functions of a complex variable. In his memoir on Abelian functions, he talks about *analysis situs*, referring to Leibniz:

> In the study of functions obtained by the integration of exact differentials, a few theorems of analysis situs are almost essential. Under that name, which was used by Leibniz, although may be in a slightly different sense, it is permitted to designate the theory of continuous magnitudes which studies these magnitudes, not as independent of their position and measurable with respect to each other, but by disregarding all idea of measure and studying them only for what regards their relation of position and inclusion. I intend to treat this subject later, in a way that is completely independent of any measure.

In his habilitation dissertation, Riemann mentions the possibility of working in the new field of topology, talking about the notion of "place." We quote this cryptic passage:

> Measure consists in the superposition of the magnitudes to be compared; it therefore requires a means of using one magnitude as the standard for another. In the absence of this, two magnitudes can only be compared when one is a part of the other; in which case we can only determine the more or less and not the how much. The researches which can in this case be instituted about them form a general division of the science of magnitude in which magnitudes are regarded not as existing independently of position and not as expressible in terms of a unit, but as regions in a manifoldness. Such researches have become a necessity for many parts of mathematics, e.g., for the treatment of many-valued analytical functions; and the want of them is no doubt a chief reason for which the celebrated theorem of Abel and the achievements of Lagrange, Pfaff, Jacobi for the general theory of differential equations, have so long remained unfruitful. Out of this general part of the science of extended magnitude in which nothing is assumed but what is contained in the notion of it, it will suffice for the present purpose to bring into prominence two points; the first of which relates to the construction of the notion of a multiply extended manifoldness, the second relates to the reduction of determinations of place in a given manifoldness to determinations of quantity, and will make clear the true character of an n-fold extent.

He also describes the passage from one dimension to another:

> If in the case of a notion whose specialisations form a continuous manifoldness, one passes from a certain specialisation in a definite way to another, the specialisations passed over form a simply extended manifoldness, whose true character is that in it a continuous progress from a point is possible only on two sides, forward or backwards. If one now supposes that this manifoldness in its turn passes over into another entirely different, and again in a definite way, namely so that each point passes over into a definite point of the other, then all the specialisations so obtained form a doubly extended manifoldness. In a similar manner one obtains a triply extended manifoldness, if one imagines a doubly extended one passing over in a definite way to another entirely different; and it is easy to see how this construction may be continued. If one regards the variable object instead of the determinable notion of it, this construction may be described as a composition of a variability of $n + 1$ dimensions out of a variability of n dimensions and a variability of one dimension.

Riemann's ideas on topology are explained in some sections of Klein's booklet [161]. For instance, Sect. 8 carries the title *Classification of closed surfaces according to the value of the integer p*. Let us comment on a passage from Klein's booklet concerning the classification of surfaces, as an example of his point of view on topology. We know that topology, which was an emerging subject, plays an important role in Riemann surface theory. We already mentioned that Riemann introduced several major notions on surface topology. Klein tried to make a more systematic exposition of these ideas. His book [161] contains a chapter in which the classification of closed surfaces according to genera is presented. On p. 32, he writes:

> That it is impossible to represent surfaces having different p's[82] upon one another, the correspondence being uniform, seems evident. It is not meant, however, that this kind of geometrical certainty needs no further investigation; cf. the explanations of G. Cantor (*Crelle*, t. LXXXIV. pp. 242 *et seq.*). But these investigations are meanwhile excluded from consideration in the text, since the principle there insisted upon is to base all reasonings ultimately on intuitive relations.

Klein then states the converse: between any two surfaces of the same genus it is possible to find a "uniform correspondence." He declares that this statement is more difficult to prove, and he refers to the 1866 article by Jordan [153].[83] This paper is an important milestone in the history of topology, because it contains the first attempt to classify surfaces up to homeomorphism, although there was no precise definition of homeomorphism yet.[84] Jordan's aim, in his paper, is to prove the following theorem, which he states in the introduction:

> In order for two surfaces or pieces of flexible and inextensible surfaces to be be applied onto each other without tear and duplication, the following is necessary and sufficient:
>
> (1) That the number of separated contours that respectively bound these portions of surfaces be the same. (If the surfaces considered are closed, this number is zero.)
>
> (2) That the maximal number of closed contours which do not self-intersect or intersect each other that we can draw on each of the two surfaces without cutting it into two separate regions be the same on both sides.

Jordan gives the following "definition" of two surfaces being "applicable onto each other." For a modern reader, this definition may seem fuzzy, but one has to remember that this paper was written in the heroic epoch of the foundations of modern topology, that the notion of homeomorphism seems extremely natural for us today, but that it was not so in those times. Jordan writes:

---

[82] We recall that p denotes the genus.

[83] Camille Jordan (1838–1922), who is mostly known for his results on the topology of surfaces and on group theory, also worked on function theory in the sense of Riemann. The title of the second part of his doctoral thesis is: "On periods of inverse functions of integrals of algebraic differentials." The subject was proposed to him by Puiseux, whom we mention in this paper concerning uniformization. Jordan is among the first who tried to study the ideas of Galois, and he is also among the first who introduced group theory in the study of differential equations.

[84] The word "homeomorphism" was introduced by Poincaré in his article [200] but with a meaning that is different from the one it has today. There is a definition of homeomorphism in the 1909 article by Hadamard [136], as being a one-to-one continuous map. This is not correct, unless Hadamard meant, by "one-to-one continuous", "one-to-one bi-continuous." We refer the reader to the paper [183] on the rise and the development of the notion of homeomorphism. This paper contains several quotes, some of which are very intriguing.

> We shall rely on the following principle, which we can consider as evident, and take it if necessary as a definition: *Two surfaces S, S' are applicable onto each other if we can decompose them into infinitely small elements such that to any contiguous elements of S correspond contiguous elements of S'.*

Besides Klein's booklet, several books and treatises explaining Riemann's ideas appeared in the decades that followed Riemann's work. We mention as examples Neumann's *Vorlesungen über Riemann's Theorie der Abel'schen Integrale* (Lectures on Riemann's theory of Abelian integrals) [186], Picard's *Traité d'Analyse* [197], Appell and Goursat's *Théorie des fonctions algébriques et de leurs intégrales* (Theory of Algebraic functions and their integrals) [16], and there are others. The last two treatises, together with several other French books on the theory of functions of a complex variable, are reviewed in Chap. 8 of the present volume [192].

Among the other important topological notions that were introduced before Riemann and that were used by him, we must mention the notion of *homotopy of paths* and its use in complex analysis (in particular, in the theory of line integrals), especially by Cauchy and Puiseux. This is discussed in detail in Chap. 7 of the present volume [191]. Cauchy published his first work on the subject in 1825 [36]. This is also a topic on which Gauss was a forerunner, but he did not publish anything about it. This is attested in his letter to Bessel, December 18, 1811, published in Volume VIII of his *Collected works* (pp. 90–92), a letter in which Gauss makes the important remark that if one defines integrals along paths in the complex plane, then the value obtained may depend on the path.

Regarding the history of Riemann's ideas on topology, we could have also commented on his predecessors regarding the notion of the discreteness and continuity of space, but this would have taken us too far. We make a few remarks on this matter in Chap. 6 of the present volume [190] .

We end this section by quoting Alexander Grothendieck, from his *Récoltes et semailles* (Harvesting and Sowing),[85] commenting on Riemann's reflections on this theme ([133] Chap. 2 Sect. 2.20):

> It must be already fifteen or twenty years ago since, skimming the modest volume that constitutes Riemann's complete works, I was struck by a remark which he made "incidentally." He observes there that it might be that the ultimate structure of space is "discrete," and that the "continuous" representations which we make of it may be an oversimplification (which may turn out to be excessive on the long run...) of a more complex reality; that for human thought, "the continuous" was easier to grasp than "the discontinuous," and that it serves us, subsequently, as an "approximation" for apprehending the discontinuous. This is an amazingly penetrating remark expressed by a mathematician, at a time where the Euclidean model of physical space was not questioned. In a strictly logical sense, it was rather the discontinuous which, traditionally, served as a technical mode of approaching the continuous.
>
> Moreover, the mathematical developments of the last decades showed an even more intimate symbiosis between continuous and discontinuous structures, which was not yet visualized

[85] The complete title is: *Récoltes et semailles : Réflexions et témoignage sur un passé de mathématicien* (Harvesting and Sowing: Reflections and testimony on the past of a mathematician). This is a long manuscript by Grothendieck in which he meditates on his past as a mathematician and where he presents without any compliance his vision of the mathematical milieu in which he evolved, especially the decline in morals, for what concerns intellectual honesty.

in the first half of this century. The fact remains that finding a model which is "satisfactory" (or, if need be, a collection of such models, linked in the most possible satisfying way...), whether the latter is "continuous," "discrete," or of a "mixed" nature—such a task will surely involve a great conceptual imagination, and a consummate intuition, in order to apprehend and disclose mathematical structures of a new type. This kind of imagination and "intuition" seems to me a rare object, not only among physicists (where Einstein and Schrödinger seem to be among the rare exceptions), but even among mathematicians (and I am talking in full knowledge of the facts).

To summarize, I foresee that the long-awaited renewal (if ever it comes...) will rather come from someone who has the soul of a mathematician, who is well informed about the great problems of physics, rather than from a physicist. But above all, we need a man having the "philosophical openness" required to grasp the crux of the problem. The latter is not at all of a technical nature, but it is a fundamental problem of "natural philosophy."[86]

9 Differential Geometry

In this section, we shall review some milestones in the history of differential geometry, concerning especially the study of geodesics and of curvature, from its beginning until the work of Riemann.

Differential geometry starts with the study of differentiable curves. The notion of curvature of planar curves already appears in works of Newton and of Johann I and

---

[86]Il doit y avoir déjà quinze ou vingt ans, en feuilletant le modeste volume constituant l'œuvre complète de Riemann, j'avais été frappé par une remarque de lui "en passant." Il y fait observer qu'il se pourrait bien que la structure ultime de l'espace soit "discrète", et que les représentations "continues" que nous en faisons constituent peut-être une simplification (excessive, peut-être, à la longue...) d'une réalité plus complexe; que pour l'esprit humain, "le continu" était plus aisé à saisir que "le discontinu", et qu'il nous sert, par suite, comme une "approximation" pour appréhender le discontinu. C'est là une remarque d'une pénétration surprenante dans la bouche d'un mathématicien, à un moment où le modèle euclidien de l'espace physique n'avait jamais été mis en cause; au sens strictement logique, c'est plutôt le discontinu qui, traditionnellement, a servi comme mode d'approche technique vers le continu.

Les développements en mathématique des dernières décennies ont d'ailleurs montré une symbiose bien plus intime entre structures continues et discontinues, qu'on ne l'imaginait encore dans la première moitié de ce siècle. Toujours est-il que de trouver un modèle "satisfaisant" (ou, au besoin, un ensemble de tels modèles, se "raccordant" de façon aussi satisfaisante que possible...), que celui-ci soit "continu," "discret" ou de nature "mixte"—un tel travail mettra en jeu sûrement une grande imagination conceptuelle, et un flair consommé pour appréhender et mettre à jour des structures mathématiques de type nouveau. Ce genre d'imagination ou de "flair" me semble chose rare, non seulement parmi les physiciens (où Einstein et Schrödinger semblent avoir été parmi les rares exceptions), mais même parmi les mathématiciens (et là je parle en pleine connaissance de cause).

Pour résumer, je prévois que le renouvellement attendu (s'il doit encore venir ...) viendra plutôt d'un mathématicien dans l'âme, bien informé des grands problèmes de la physique, que d'un physicien. Mais surtout, il y faudra un homme ayant "l'ouverture philosophique" pour saisir le nœud du problème. Celui-ci n'est nullement de nature technique, mais bien un problème fondamental de "philosophie de la nature."

Jakob Bernoulli. We mentioned, in Sect. 2, Johann Bernoulli's paper [24] published in 1718 on the isoperimetry problem in the plane.

Differential geometry is also concerned with geodesics. In 1744, Euler published a book [73] in which he sets the bases of the calculus of variations. The title is *Methodus inveniendi lineas curvas maximi minimive proprietate gaudentes, sive solutio problematis isoperimetrici lattissimo sensu accepti* (Method for finding curved lines enjoying properties of maximum or minimum, or solution of isoperimetric problems in the broadest accepted sense). In that book, several applications of the new methods are presented, among them isoperimetry problems, the problem of finding the shape of the brachystochrone, the study of the catenoid, and that of finding geodesics between two points on a surface. With this work of Euler, the methods of differential calculus, more precisely those of finding minima and maxima, were suddenly generalized to the realm of a variable moving in an infinite dimensional space (even though the expression "infinite dimensional" was not there yet), namely, in the question of looking for curves of minimal length or satisfying other geometric properties, among all curves joining two points.

Riemann's differential geometry is essentially about curvature, actually, the curvature of space, and we must talk now about the history of curvature, which starts with curvature of curves and surfaces. The history starts again with Euler.

Volume II of Euler's *Introductio in analysin infinitorum* [61], published in 1748, is concerned with the differential geometry of space curves and surfaces. Curves are given there by parametric equations of the form $x = x(t)$, $y = y(t)$, $z = z(t)$, and surfaces by parametric equations of the form $x = x(u, v)$, $y = y(u, v)$, $z = z(u, v)$. It is possible that this is the first time where such a parametric representation of surfaces appears in print. About twenty years after the first edition of this treatise was published, Euler wrote a memoir entitled *Recherches sur la courbure des surfaces* (Researches on the curvature of surfaces) [91] (1767), another work which transformed the subject. The aim of this memoir was to introduce and study the curvature at a point on a surface. Euler's idea, which is very natural, was to introduce a notion of curvature at a point of a differentiable surface based on the curvature of curves that pass through that point. His intuition was that to understand curvature at a point of a surface, it suffices to study the curvature of curves that are intersections of that surface with Euclidean planes. Moreover, he showed that it is sufficient to consider the planes that are perpendicular to the surface, that is, the planes containing the normal vector to the surface at that point. Each such curve has an osculating circle, and the collection of radii of these circles contains all the information about the curvature of the surface at that point. Furthermore, Euler proved that at any given point on the surface, the maximal curvature and the minimal curvature associated to the normal planes determine all the other normal curvatures. To be more precise, given a point on the surface and a tangent vector v at that point, let us call *normal curvature* though v the curvature of a curve obtained by intersecting the surface with a plane containing the vector v and the normal vector at that point. The *maximal* and *minimal normal curvature* at the given point are the maximum and minimum of the normal curvatures taken over all the normal planes at that point. Likewise, the *normal curvature radius* at the given point in the direction of the vector v is the curvature radius of the associated curve. We have a similar notion of *maximal* and *minimal normal curvature radii* at the given point. Euler showed that the directions of the planes that realize these extremal curvatures (except in very special cases) are orthogonal to each other, and he proved that at a given point, if ρ_1 and ρ_2 are the maximal and minimal normal curvature radii respectively, then the normal curvature radius ρ of the normal section through an arbitrary tangent vector v is given by the equation

$$\rho = \frac{2\rho_1\rho_2}{(\rho_1+\rho_2)-(\rho_1-\rho_2)\cos(2\varphi)},$$

where φ is the angle between v and the tangent vector to the normal plane with maximal curvature radius.

It is usual to write Euler's equation in the following form:

$$\frac{1}{\rho} = \frac{\cos^2\varphi}{\rho_1} + \frac{\sin^2\varphi}{\rho_2}.$$

We note that ρ_1 and ρ_2 may also take negative values and that Euler's equation has also a meaning when ρ_1 or ρ_2 is infinite; in the latter case, the curvature $\frac{1}{\rho}$ is zero for all φ. There is a classical local classification of differentiable surfaces at a point in terms of the signs of ρ_1 and ρ_2.

Euler writes ([91], Réflexion VI, p. 143):

> Thus, the judgement of the curvature of surfaces, however complicated it seems at the beginning, is reduced for each point to the knowledge of two osculating radii, one of which is the largest and the other the smallest at that element. These two objects determine entirely the nature of the curvature, displaying for us the curvature of all the possible sections that are perpendicular to the proposed element.[87]

There are other memoirs by Euler on the differential geometry of surfaces. We mention his *Solutio trium problematum difficiliorum ad methodum tangentium inversam pertinentium* (Solution of three rather difficult problems pertaining to the inverse method of tangents) [75] published in 1826, that is, several years after Euler's death. In this memoir, Euler addresses "inverse problems" in differential geometry, e.g., to reconstruct curves from information on their tangents. We also mention Euler's *De solidis quorum superficiem in planum explicare licet* (On solids whose entire surface can be unfolded onto a plane) [95] in which for the first time the notion of a surface developable on the plane is introduced. This notion was thoroughly investigated in the later works of Monge and his students that we mention below, and much later by Eugenio Beltrami. This paper [95] on developable surfaces also addresses a so-called

[87] Ainsi le jugement sur la courbure des surface, quelque compliqué qu'il ait paru au commencement, se réduit pour chaque élément à la conaissance de deux rayons osculateurs, dont l'un est le plus grand et l'autre le plus petit dans cet élément ; ces deux choses déterminent entièrement la nature de la courbure en nous découvrant la courbure de toutes les sections possibles qui sont perpendiculaires sur l'élément proposé.

"inverse problem," namely, the question of giving a characterization of the surfaces that are developable.

Gauss's development of differential geometry attained a high degree of perfection in the 1820s, motivated by his works on geography, astronomy and geodesy. He was probably the first to formulate the question of finding the properties of surfaces which are independent of their embedding in 3-space. After Euler in [91] highlighted the role of the maximal and minimal curvature at a point of a surface, it was Gauss's idea to take the product of these quantities as a measure of the curvature at that point, and to show that the result is an isometry invariant of the surface. This is (expressed in modern terms) the content of Gauss's *Theorema egregium*. The result is contained in Gauss's *Disquisitiones generales circa superficies curvas* (General investigations on curved surfaces) [124] (1828). In the abstract he wrote for his *Disquisitiones*, Gauss says (translation from [124] p. 48):

> [...] These theorems lead to the consideration of the theory of curved surfaces from a new point of view, where a wide and still wholly uncultivated field is open to investigation. If we consider surfaces not as boundaries of bodies, but as bodies of which one dimension vanishes, and if at the same time we conceive them as flexible but not extensible, we see that two essentially different relations must be distinguished, namely, on the one hand, those that presuppose a definite form of the surface in space; on the other hand, those that are independent of the various forms that the surface may assume. This discussion is concerned with the latter. In accordance with what has been said, the measure of curvature belongs to this case. But it is easily seen that the consideration of figures constructed upon the surface, their angles, their areas and their integral curvatures, the joining of the points by means of shortest lines, and the like, also belong to this case. All such investigations must start from this, that the very nature of the curved surface is given by means of the expression of any linear element in the form $\sqrt{Edp^2 + 2Fdpdq + Gdq^2}$.

Gauss, in the same memoir, used the so-called Gauss map from a surface to the unit sphere, defined by sending the normal unit normal vector at a point to the corresponding point on the sphere and showing that one can recover the curvature of the surface, which he had defined as the product of the minimal and maximal curvatures. The curvature, using the Gauss map, is obtained by taking the ratio of the area of the image of the Gauss map by the area of the surface (the definition of the curvature at a point needs a passage to the infinitesimal level).

Riemann's most important articles on differential geometry are his habilitation lecture *Über die Hypothesen welche der Geometrie zu Grunde liegen* (1854) which is mentioned several times in the present chapter and in the other chapters of this book, and the *Commentatio mathematica, qua respondere tentatur quaestioni ab Ill*^{ma} *Academia Parisiensi propositae* (A mathematical note attempting to answer a question posed by the distinguished Paris Academy), a memoir which he wrote in 1861, at the occasion of his participation to a competition prize set by the Paris Academy of Sciences, and which we consider in some detail in Chap. 6 of the present volume [190]. These two memoirs are unusual for opposite reasons: the first one lacks of formulae, and the second one is full of them. The second memoir contains the explicit form of the object which we call today the Riemann tensor.

In his habilitation lecture, Riemann makes a clear reference to Gauss's *Disquisitiones* as one of his major sources of inspiration, a work which he includes however in a broad philosophical discussion on magnitude, measure, quantity and the possibility of geometric representation. It is always good to read Riemann:

> Having constructed the notion of a manifoldness of n dimensions, and found that its true character consists in the property that the determination of position in it may be reduced to n determinations of magnitude, we come to the second of the problems proposed above, viz. the study of the measure-relations of which such a manifoldness is capable, and of the conditions which suffice to determine them. These measure-relations can only be studied in abstract notions of quantity, and their dependence on one another can only be represented by formulæ. On certain assumptions, however, they are decomposable into relations which, taken separately, are capable of geometric representation; and thus it becomes possible to express geometrically the calculated results. In this way, to come to solid ground, we cannot, it is true, avoid abstract considerations in our formulæ, but at least the results of calculation may subsequently be presented in a geometric form. The foundations of these two parts of the question are established in the celebrated memoir of Gauss, *Disqusitiones generales circa superficies curvas*.

For the case of surfaces, he writes:

> In the idea of surfaces, together with the intrinsic measure-relations in which only the length of lines on the surfaces is considered, there is always mixed up the position of points lying out of the surface. We may, however, abstract from external relations if we consider such deformations as leave unaltered the length of lines—i.e., if we regard the surface as bent in any way without stretching, and treat all surfaces so related to each other as equivalent. Thus, for example, any cylindrical or conical surface counts as equivalent to a plane, since it may be made out of one by mere bending, in which the intrinsic measure-relations remain, and all theorems about a plane—therefore the whole of planimetry—retain their validity. On the other hand they count as essentially different from the sphere, which cannot be changed into a plane without stretching. According to our previous investigation the intrinsic measure-relations of a twofold extent in which the line-element may be expressed as the square root of a quadric differential, which is the case with surfaces, are characterized by the total curvature. Now this quantity in the case of surfaces is capable of a visible interpretation, viz., it is the product of the two curvatures of the surface, or multiplied by the area of a small geodesic triangle, it is equal to the spherical excess of the same. The first definition assumes the proposition that the product of the two radii of curvature is unaltered by mere bending; the second, that in the same place the area of a small triangle is proportional to its spherical excess. To give an intelligible meaning to the curvature of an n-fold extent at a given point and in a given surface-direction through it, we must start from the fact that a geodesic proceeding from a point is entirely determined when its initial direction is given. According to this we obtain a determinate surface if we prolong all the geodesics proceeding from the given point and lying initially in the given surface-direction; this surface has at the given point a definite curvature, which is also the curvature of the n-fold continuum at the given point in the given surface-direction.

In these passages, Riemann summarizes his ideas on the general metric on (what became known later on as) an n-dimensional differentiable manifold, defined by a quadratic form on each tangent space, a broad generalization of Gauss's investigations on surfaces in which the quadratic form determines the metric, permits to calculate distances, angles and the curvature at any point. The curvature is the product of two quantities and is invariant by bending. The quadratic form represents the square of the line element. With these tools, one can study geodesic triangles on surfaces, prove that a geodesic is determined by its initial vector, generalize these matters to immersed surfaces, etc.

One may also include in Riemann's list of works on differential geometry his two papers on minimal surfaces [220, 221]. They are reviewed in Chap. 5 of the present volume, [259]. We also note that in his doctoral dissertation defended in 1880 in Paris and written under the supervision of Bonnet (cf. [188]), Niewenglowski explains that Riemann, in his work on minimal surfaces, was inspired by Bonnet; cf. also the comments in Chap. 8 of the present volume [192]. Again, minimal surfaces first appear in the work of Euler (cf. Chapter V, Sect. 47 of Euler's first treatise on the calculus of variations, [73]).

In a longer survey on Riemann's predecessors in the field of differential geometry, one would have analyzed the works of several French mathematicians who stand between Euler and Gauss, in particular Gaspard Monge (1764–1818), Jean-Baptiste Meusnier (1754–1793), Siméon-Denis Poisson (1781–1840), Charles Dupin (1784–1873), Olinde Rodrigues (1795–1851) and there are others. We only mention some of these works.

Monge, who was the founder of a famous school on projective and differential geometry, continued Euler's work on developable surfaces, cf. [184, 185]. He worked in particular with two orthogonal line fields that are defined by Euler's minimal and maximal directions of curvature radii, and he coined the expression *umbilical point* for points where the two curvature radii have the same value. (On the sphere, every point is umbilical.) Monge expressed several times in his writings his debt to Euler. In [185], he writes:

> Having resumed this matter, at the occasion of a memoir that Mr. Euler gave in the 1771 volume of the Petersburg Academy on developable surfaces, in which this famous geometer gives formulae to determine whether a given surface may or may not be applied onto a plane, I reached results on the same subject which seem to me much simpler, and whose usage is much easier.[88]

Poisson was a student of Lagrange and Laplace. He wrote a memoir entitled *Mémoire sur la courbure des surfaces* (Memoir on the curvature of surfaces) [205] (1832) in which he studied singular points of the curvature. We mention by the way that there are several points in the work of Poisson that are related to Riemann's works, in particular, concerning definite integrals and Fourier series.

Meusnier was a student of Monge. He gave a formula for the curvature of a curve obtained by intersecting a surface by a non-normal section, in terms of that of the normal sections at the given point that were considered by Euler. His paper on the subject carries almost the same title as Euler's paper [91], *Mémoire sur la courbure des surfaces* [181] (1785). In this paper, Meusnier writes (p. 478):

> Mr. Euler treated the same matter in a very beautiful memoir, printed in 1760 among those of the Berlin Academy. This famous geometer addresses the question in a manner which is different from the one which we just presented. He makes the curvature of an element of

---

[88] Ayant repris cette matière, à l'occasion d'un mémoire que M. Euler a donné dans le volume 1771 de l'Académie de Pétersbourg, sur les surfaces développables, et dans lequel cet illustre géomètre donne des formules pour reconnaître si une surface courbe proposée jouit ou non de la propriété de pouvoir être appliquée sur un plan, je suis parvenu à des résultats qui me semblent beaucoup plus simples, et d'un usage bien plus facile pour le même sujet.

the surface dependent on that of the various sections that one can perform on it by cutting it with planes.[89]

Meusnier's work is surveyed by Darboux in the paper [49].

Dupin is mostly known for the so-called Dupin indicatrix, a geometric device which characterizes the shape of a surface at a given point and which turned out to be related to the Gaussian curvature at that point. His famous work on geometry bears the title *Développements de géométrie, avec des applications à la stabilité des vaisseaux, aux déblais et remblais, au défilement, à l'optique, etc. pour faire suite à la géométrie descriptive et à la géométrie analytique de M. Monge : Théorie.* (Developments in geometry, with applications to the stability of vessels, cuts and fills, scrolling, optics, etc. as a sequel to the descriptive and analytic geometry of Mr. Monge: Theory) [58], 1813.

Rodrigues introduced, before Gauss, the Gauss map, and showed that at a given point the ratio of the area of the image to the area on the surface is equal (at the infinitesimal level) to the product of the two principal curvatures (those defined by Euler), cf. [235]. This result pre-dates that of Gauss, but the fact that curvature is an isometry invariant is however absent from Rodrigues' work.

Finally, it is fair to mention that Riemann's work in high dimensions was prepared by works of other mathematicians done in higher dimensions, and we mention in particular Jacobi [145] on the reduction of pairs of quadratic forms, Grassmann on higher-dimensional linear algebra [131], and Cayley [41] on higher-dimensional analytic geometry.

10 Trigonometric Series

Riemann's habilitation dissertation (*Habilitationsschrift*) is concerned with trigonometric series. Its title is *Über die Darstellbarkeit einer Function durch eine trigonometrische Reihe* (On the representability of a function by a trigonometric series) [215]. The dissertation is divided into two parts. The first part is a survey of the history of the problem of representing by a trigonometric series a function which is arbitrary, but which—according to Riemann's words at the beginning of his memoir—is "given graphically." It is important to note the last fact, since, for instance, a function which is discontinuous at a dense set of points cannot be given graphically. Dealing with the most general functions is part of the subject of the second part of Riemann's memoir.

We shall report on the historical part of Riemann's memoir. It involves several mathematicians, in particular Euler, although not directly his work on trigonometric series. We therefore note right away that Euler uses trigonometric series in his two

[89]M. Euler a traité la même matière dans un fort beau mémoire, imprimé en 1760 parmi ceux de l'Académie de Berlin. Cet illustre géomètre envisage la question d'une manière différente de celle que nous venons d'exposer ; il fait dépendre la courbure d'un élément de surface de celle des différentes sections qu'on peut y faire en le coupant par des plans.

memoirs on astronomy *Recherches sur la question des inégalités du mouvement de Saturne et de Jupiter* (Researches on the question of the inequalities in the motion of Saturn and Jupiter) [77], and *De motu corporum coelestium a viribus quibuscunque perturbato* (On the movement of celestial bodies perturbed by any number of forces) [86], both presented for competitions proposed by the Paris Académie des Sciences, in 1748 and 1756 respectively. The two memoirs are analyzed in the paper [129] in which the authors show that Euler was much more concerned with convergence of series than what is claimed in several books and articles on his work.

The starting point of Riemann's historical survey is the controversy between Euler and d'Alembert which originated in the publication in 1747 of a memoir by the latter, *Recherches sur la courbe que forme une corde tendue mise en vibrations* (Researches on the curve formed by a taut string subject to vibrations) [8]. In this memoir, d'Alembert gave the partial differential equation that represents the motion of a point on a vibrating string subject to small vibrations:

$$\frac{\partial^2 y}{\partial t^2} = \alpha^2 \frac{\partial^2 y}{\partial x^2}.$$

In this equation, t represents time, α is a constant and $y = y(t)$ is the oscillation of the string at a point whose coordinate is x along the string. The main problem, after this discovery, was to characterize the functions that are solutions of that equation.

One first obvious (but wrong) guess for a necessary condition on the solution is that it should be order-two differentiable. However, it was soon realized that this condition is too restrictive. Understanding the exact nature of the solutions of the vibrating string equation led to a fierce controversy which involved some of the most brilliant mathematicians of the eighteenth century. Among them are Euler, d'Alembert, Lagrange and Daniel Bernoulli. There is a large amount of primary literature concerned with this debate, including several memoirs by each of these mathematicians and the correspondence between them. Let us recall a few points of the history of that controversy.

In the memoir [8], d'Alembert wrote that the general solution to the wave equation is a function of the form

$$y(x,t) = \frac{1}{2}\left(\phi(x+\alpha t) + \phi(x-\alpha t)\right),$$

where ϕ is an "arbitrary" periodic function whose period is the double of the length of the string. The problem was to give a meaning to the adjective "arbitrary."

At the beginning of his memoir [8], d'Alembert declares that he will show that the problem admits infinitely many other solutions than the usual one represented by the sine curve (which he calls *compagne de la cycloïde allongée*). But from his point of view (like from Euler's one) the only acceptable functions were those given by a formula (functions which, as we recall, were termed "analytic" by Euler). The reason is that it was considered that the powerful methods of analysis can be applied only to such functions.

Euler published an article in the next volume of the Memoirs (1748) [68] in which he gave an exposition of d'Alembert's results but where he expressed a different point of view on the nature of the solution of the wave equation.[90] He claimed that a solution is not necessarily given by a formula, but that it might be "discontinuous" in the sense that it could be a concatenation of functions defined on smaller intervals on which the restriction of the function is defined by formulae. We already mentioned this notion of "discontinuity" in Sect. 2 of the present chapter. His assertion was supported by physical evidence, more precisely, by the fact that the initial form of a string, in a musical instrument that is pinched in the usuel manner, is a concatenation of two segments with a corner at their intersection. More than that, Euler pointed out that one may give an arbitrary initial form to the string, and therefore the solution may be arbitrary. Euler's paper introduced some doubts concerning the assertion made by d'Alembert that the solution must be twice differentiable and given by a formula. D'Alembert, who disagreed with Euler's claim, published the following year a memoir in which he confirmed his initial ideas. The rest of the controversy on the notion of function is very interesting and there are several articles on this subject. We recommend in particular the introduction, by Youschkevitch and Taton, of Volume V of Series IV A of Euler's *Opera omnia* containing Euler's correspondence with Clairaut, d'Alembert and Lagrange [261].

The difficulty of defining a general notion of function is never too much emphasized. We mention in this respect that in 1787, that is, four years after Euler's death, the Academy of Sciences of Saint-Petersburg proposed, as a competition question, to write an essay on the nature of an arbitrary function.[91] The prize went to the Alsatian mathematician Louis-François-Antoine Arbogast (1759–1803), who, in his *Mémoire sur la nature des fonctions arbitraires qui entrent dans les intégrales des équations aux différentielles partielles* (Memoir on the nature of arbitrary functions that appear in the integrals of partial differential equations), [17] (1791), adopted Euler's point of view: he accepted discontinuous functions in the sense Euler defined them, as solutions of partial differential equations. It is interesting to note that in the description of that problem, the Academy starts with the physical problem of vibrating strings:

> The problem of the vibrating strings is without doubt one of the most famous problems of applied mathematics. The most celebrated geometers of our time, who solved it, have argued on the legitimacy of their solution, without ever being able to convince each other. It is not that it is difficult to reduce the problem itself to pure analysis, but as it has given the first occasion to treat three-variable differential equations which give, by integrating them, arbitrary and varying functions, the important question which divided the points of view of these great men is whether these functions are entirely arbitrary, whether they represent all the arbitrary curves and surfaces, formed by a voluntary motion of the hand, or whether they include only those that are comprised under an algebraic or transcendental equation. Besides the fact that on that decision depends the way of terminating the dispute on vibrating strings, the same question on the nature of arbitrary functions re-emerges each time an arbitrary

[90]Euler published a Latin and a French version of his memoir, which appeared in the years 1749 and 1748 respectively. (The title of the French version, *Sur la vibration des cordes, traduit du latin*, although it was published first, shows that it was written after the Latin one.).

[91]*Histoire de l'Académie Impériale des Sciences, année 1787*, p. 4.

> problem leads to differential equations with three or more variables: this happens even very often, not only when we treat subjects of sublime mechanics, but most of all in the theory of fluid motion: in such a way that one cannot rigorously sustain that such a problem has been solved before setting precisely the nature of of arbitrary functions. The Academy invites then all the geometers to decide:
>
> *Whether arbitrary functions, to which we are led by integrating equations with one or several variables, represent arbitrary curves or surfaces, either algebraic or transcendental, or mechanical, discontinuous or produced by a voluntary motion of the hand; or whether these functions only contain continuous curves represented by an algebraic or transcendental equation.*[92]

We now come to the problem of trigonometric series.

Brook Taylor, in his 1713 memoir entitled *De motu nervi tensi* (On the motion of a tense string) [242] (cf. also his *Methodus incrementorum directa et inversa* (Direct and Indirect Methods of Incrementation), [243] (first edition 1715), showed that the vibration problem admits as a solution the sine and cosine functions. For several reasons which we shall mention below, it was tempting to conjecture then that the general solution of the problem is obtained by taking an infinite sum of trigonometric functions. This was done by Daniel Bernoulli (1700–1782).

In 1753, Bernoulli wrote a memoir on the vibration of strings. Bernoulli had already thought about this question for several years. In his approach to it, like in the other physical problems he considered, Bernoulli was an adept of Leibniz' calculus, rather than Euler's geometric methods (which were adopted by d'Alembert). As a physicist, the mathematical notion of function was not a central theme in his research, and from his point of view, the function representing the solution of the question was simply identified with the shape of the vibrating string. While Taylor had considered each trigonometric solution individually, that is, he noticed that functions of the form

[92]Le problème des cordes vibrantes est sans contredit un des plus fameux problèmes de la mathématique appliquée. Les plus célèbres géomètres de notre temps, qui l'ont résolu, se sont disputés sur la légitimité de leurs solutions, sans avoir jamais pu se convaincre l'un l'autre. Ce n'est pas que le problème en lui-même ne soit pas facilement réduit à l'analyse pure ; mais comme il a été le premier qui ait donné occasion de traiter des équations différentielles à trois variables, par l'intégration desquelles on parvient à des fonctions arbitraires et variables, la question importante qui partagea les avis de ces grands hommes fut : si ces fonctions sont entièrement arbitraires ? si elles représentent toutes les courbes et surfaces quelconques, formées par un mouvement volontaire de la main ? ou si elles ne renferment que celles qui sont comprises sous une équation soit algébrique soit transcendante ? Outre que c'est de cette décision que dépend le moyen de terminer cette dispute sur les cordes vibrantes, la même question sur la nature des fonctions arbitraires renaît toutes les fois qu'un problème quelconque conduit à des équations différentielles à trois et plusieurs variables : ce qui arrive même bien souvent, non seulement lorsqu'on traite des sujets de la mécanique sublime, mais surtout des mouvements des fluides : de sorte qu'on ne saurait soutenir rigoureusement qu'un pareil problème ait été résolu, avant qu'on ait fixé exactement la nature des fonctions arbitraires. L'Académie invite donc tous les géomètres de décider:

Si les fonctions arbitraires, auxquelles on parvient par l'intermédiaire des équations à trois ou plusieurs variables, représentent des courbes ou surfaces quelconques, soit algébriques soit transcendantes, soit mécaniques, discontinues, ou produites par un mouvement volontaire de la main ; ou si ces fonctions renferment seulement des courbes continues représentées par une équation algébrique ou transcendante?

$$y(x,t) = \sin \frac{n\pi x}{l} \cos \frac{n\pi \alpha t}{l}$$

are solutions of the wave equation, Bernoulli stated that the general solution was an infinite sum of such functions. Thus, he added to the debate the question of the convergence of trigonometric series. In the meanwhile, d'Alembert published a first supplement to his memoir *Sur les vibrations des cordes sonores* (On the vibration of sonorous strings) [4] in which, referring to Daniel Bernoulli's work, he writes:

> The question is not to *conjecture*, but to *prove*, and it would be dangerous (although, to tell the truth, this misfortune is unlikely to happen) that this kind of proof which is so odd enters into geometry. The only thing which seems surprising is that such reasonings are used in way of a proof by a famous mathematician.[93]

After the publication of Bernoulli's memoir, Euler wrote a new memoir in which he generalizes Bernoulli's result, *Remarques sur les mémoires précédents de M. Bernoulli* (Remarks on the preceding memoirs by Mr. Bernoulli) [103] (1755). He also confirms his own intuition that a solution of the vibrating string equation may be an arbitrary function. Today, we know that, in some sense, the solution he proposed is identical to that of Bernoulli, but the relation between trigonometric series and arbitrary functions was not yet discovered.

In his memoir, Euler starts by declaring that, without any doubt, Bernoulli developed the theory of formation of sound infinitely better than any other scientist before him, that his predecessors stopped at the mechanical determination of the motion of a tight string without any thorough investigation of the nature of sound, and that it was still not understood how a single string can emit several sounds at the same time. He then expresses his doubts about the fact that Bernoulli's infinite series of sines could be the general solution of the problem. He writes that it is impossible for the curve made by a vibrating string to be constituted by an infinite number of trochoïds (which is the name he used for the sine curves). He declares that there are infinitely many curves that are not included in that solution.

The solution of this problem was given by Fourier and completed by Dirichlet and Riemann, in the following century, as we shall discuss below. In the same paper, Euler insists on the fact that the general solution of the equation of the vibrating string cannot be given by a formula. He mentions his own conflict with d'Alembert, saying that he wishes very much that the latter explains why he is mistaken. Based on partial differential calculus, Euler gives a new explanation of the fact that by varying the initial shape of a string, any function becomes admissible as a solution of the problem.

Between November 1, 1759 and the end of the same year, Euler presented three memoirs, [99–101], on the propagation of sound. In these memoirs, he studies respectively the propagation in one, two and three dimensions. The differential equations

[93]Il ne s'agit pas de *conjecturer*, mais de *démontrer*, et il serait dangereux (quoi qu'à la vérité ce malheur soit peu à craindre) qu'un genre de démonstration si singulier s'introduisît en Géométrie. Ce qui pourra seulement paraître surprenant, c'est que de pareils raisonnements soient employés comme démonstratifs par un mathématicien célèbre.

that describe this propagation are the same as those which describe the vibration of strings. Euler mentions the limitations of the works of "Taylor, Bernoulli and some others." Despite the fact that the debate on the vibrating string had already lasted many years, the relation between the scientists working on that subject was still very tense.

In a later memoir, titled *Mémoire sur les vibrations des cordes d'une épaisseur inégale* (Memoir of the vibration of strings of uneven width) [25] (1767), Bernoulli gave an additional reason for the use of an infinite sum (p. 283):

> When a string makes several vibrations at the same time, of whatever number, and in whatever order, the absolute curvature will always be expressed by the general equation
>
> $$y = \alpha \sin \frac{x}{l}\pi + \beta \sin \frac{2x}{l}\pi + \gamma \sin \frac{3x}{l}\pi + \text{etc.}$$
>
> and since the number of arbitrary coefficients is infinite, one can make the curve pass by whatever number of points of positions that we wish, which indicates that all the curves belong to this case, provided we do not oppose the hypotheses. And it would be opposing them if we do not treat the quantities y, dy and ddy as infinitely smaller, at every point of the curve, than the quantities x, dx and $\frac{dx^2}{l}$. [94]

The interested reader may skim the volume of Euler's collected works containing the correspondence between Euler and Lagrange, [108], and the volume containing the correspondence between Lagrange and d'Alembert, [167], not only in order to understand more deeply this multi-faced controversy, but also in order to feel the cultural and scientific atmosphere in Europe during that period. Let us quote, as examples, two excerpts related to the discussion around the solution of the wave equation. In a letter to Lagrange, dated Octobre 1759, Euler writes:

> I was pleased to learn that you agree with my solution relative to the vibrating strings, which d'Alembert tried hard to refute using various sophisms, for the only reason that he did not propose it himself. He announced that he will publish an overwhelming proof of it; I don't know whether he did it. He thinks he will be able to impress people by his half-scholar eloquence. I doubt that he can seriously play such a role, unless he is profoundly blinded by self-esteem. He wanted to insert in our *Memoirs*, not a proof, but a simple declaration according to which my solution was very deficient. On my side, I proposed a new proof which has all the required rigor.[95]

---

[94]Lorsque la corde fait à la fois plusieurs espèces de vibration, quel qu'en soit le nombre, et de quelque ordre qu'elles soient, la courbure absolue sera toujours exprimée par cette équation générale

$$y = \alpha \sin \frac{x}{l}\pi + \beta \sin \frac{2x}{l}\pi + \gamma \sin \frac{3x}{l}\pi + \text{etc.}$$

et comme le nombre des coefficients arbitraires est infini, on peut faire passer la courbe par tant de points donnés de position qu'on voudra, ce qui marque que toutes les courbes se trouvent dans ce cas, pourvu qu'on ne fasse pas violence aux hypothèses. Et ce serait leur faire violence, si on ne faisait pas les quantités y, dy et ddy comme infiniment plus petites dans tous les points de la courbe, que les quantités x, dx et $\frac{dx^2}{l}$.

[95]J'ai appris avec plaisir que vous approuviez ma solution relative aux cordes vibrantes, que d'Alembert s'est efforcé de réfuter par divers sophismes, et ceci pour l'unique raison qu'il ne l'a pas proposée lui-même. Il a annoncé qu'il en publierait une accablante réfutation ; j'ignore s'il

In a letter to d'Alembert, dated March 20, 1765, Lagrange writes:

> Concerning [our discussion] on vibrating strings, it is now reduced to a point which, it seems to me, escapes any analysis. Moreover, I found, by a completely direct way, that if we admit in the intial figure the conditions that you ask, the solution reduces to the one of Mr. Bernoulli, namely, $y = \alpha \sin \frac{\pi x}{a} + \beta \sin \frac{2\pi x}{a} + ...$, and it is difficult for me to believe that this is the only one that can be found in nature. Besides, the phenomena of sound propagation can be explained only if we admit discontinuous functions, as I proved in my second dissertation.[96]

Let us also quote Nicolaus Fuss, the famous biographer of Euler,[97] from his *Éloge* [117]:

> The controversy between Messrs. Euler, d'Alembert & Bernoulli regarding the motion of the vibrating strings can be of interest only to professional geometers. Mr. D. Bernoulli, who was the first to develop the physical part which concerns the production of sound generated by this motion, thought that Taylor's solution was sufficient to explain it. Messrs. Euler and d'Alembert, who had exhausted, in this difficult matter, everything exquisite and profound that an analytic mind may possess, showed that the solution of Mr. Bernoulli, extracted from Taylor's Trochoids, is not general, and that it is even deficient. This controversy, which lasted a long time, with all the consideration that such famous men owe to each other, gave rise to a quantity of excellent memoirs; it really ended only at the death of Bernoulli.[98]

D'Alembert eventually accepted functions that are discontinuous (in the sense of Euler) as solutions of partial differential equations; cf. his 1780 memoir entitled *Sur les fonctions discontinues* (On discontinuous functions), published in [3] (t. VIII, Sect. VI) in which he formulates a *Règle sur les fonctions discontinues qui peuvent entrer dans l'intégration des équations aux dérivées partielles* (Rule on discontinuous

---

(Footnote 95 continued)
l'a fait. Il croit qu'il pourra jeter de la poudre aux yeux avec son éloquence de demi-savant. Je doute qu'il joue ce rôle sérieusement, à moins qu'il ne soit profondément aveuglé par l'amour-propre. Il a voulu insérer dans nos Mémoires non une démonstration, mais une simple déclaration suivant laquelle ma solution était très défectueuse ; pour ma part, j'ai proposé une nouvelle démonstration possédant toute la rigueur voulue.

[96]à l'égard de [notre discussion] sur les cordes vibrantes, elle est maintenant réduite à un point qui échappe, ce me semble, à l'Analyse. Au reste, j'ai trouvé par une voie tout à fait directe qu'en admettant dans la figure initiale les conditions que vous y exigez, la solution se réduit à celle de M. Bernoulli, savoir : $y = \alpha \sin \frac{\pi x}{a} + \beta \sin \frac{2\pi x}{a} + ...$, et j'ai peine à croire que celle-ci soit la seule qui puisse avoir lieu dans la nature. D'ailleurs, les phénomènes de la propagation du son ne peuvent s'expliquer qu'en admettant les fonctions discontinues, comme je l'ai prouvé dans ma seconde dissertation.

[97]Nicolaus Fuss (1755–1826) was first hired as Euler's secretary, then he became successively his favorite student, his closest friend, his collaborator and colleague at the Saint Petersburg Academy of Sciences, and eventually his grandson-in-law (the husband of Euler's grand-daughter Albertine).

[98]La controverse entre MM. Euler, d'Alembert & Bernoulli au sujet du mouvement des cordes vibrantes ne peut intéresser proprement que les Géomètres de profession. M. D. Bernoulli, qui fut le premier à en développer la partie physique qui regarde la formation du son engendré par ce mouvement, crut la solution de Taylor suffisante de l'expliquer. MM. Euler et d'Alembert, qui avaient épuisé, dans cette matière difficile, tout ce que l'esprit analytique a de sublime & de profond, firent voir que la solution de M. Bernoulli, tirée des Trochoïdes Tayloriennes, n'est pas générale, qu'elle est même insuffisante. Cette controverse qui a été continuée longtemps, avec tous les égards que des hommes aussi illustres se doivent mutuellement, a donné naissance à quantité d'excellents mémoires ; elle n'a fini proprement qu'à la mort de M. Bernoulli.

functions that may enter into the integration of partial differential equations). We refer the interested reader to the papers [147, 245, 260, 261] for more on the history of the subject.

The confirmation of Bernoulli's conjecture followed from Fourier's manuscript *Théorie de la propagation de la chaleur dans les solides* (Theory of heat propagation in solids),[99] [115] read to the Academy in 1807, that is, twenty-five years after Bernoulli's death. The manuscript carries the subtitle: "Mémoire sur la propagation de la chaleur avec notes séparées sur cette propagation—sur la convergence des séries $\sin x - \frac{1}{2}\sin 2x + \frac{1}{3}\sin 3x$ &c." (Memoir on the propagation of heat, with separate notes on that propagation—on the convergence of the series $\sin x - \frac{1}{2}\sin 2x + \frac{1}{3}\sin 3x$ etc.). Let us quote an excerpt ([132] p. 183):

> It follows from my researches on this object that the arbitrary functions, even discontinuous, can always be represented by the sine or cosine expansions of multiple arcs, and that the integrals which contain these developments are precisely as general as those where arbitrary functions of multiple arcs enter. A conclusion that the celebrated Euler has always rejected.[100]

In 1811, the Paris *Académie des sciences* proposed a competition whose title was: *Donner la théorie mathématique des lois de la propagation de la chaleur et comparer les résultats de cette théorie à des expériences exactes* (To give the mathematical theory of the laws of propagation of heat and to compare the results of this theory with exact experiences). Fourier submitted for the prize a very extensive work which included his 1807 manuscript. The jury of the competition consisted of Lagrange, Laplace, Maus, Haüy and Legendre. Darboux, in his review [48], quotes part of the report on the work of Fourier:

> This piece contains genuine differential equations of heat transmission, either in the interior of bodies, or at their surface. And what is new in the subject, added to its importance, has led the Class to crown this treatise, while noting however that the manner with which the author arrives at his equation is not exempt of difficulties, and that his analysis, to integrate them, still leaves something to be desired, either relative to the generality, or even from the point of view of rigor.[101]

The sum of Fourier's work on the propagation of heat was collected in his masterpiece, *Théorie analytique de la chaleur* (Analytic theory of heat) [116], published in 1822. The following result is stated at the end of Chapter III of this memoir, as a summary of what has been done (art. 235, p. 258):

---

[99] This memoir, read to the Academy 1807, was never published, until it was edited with comments by Grattan-Guinness, see [132].

[100] Il résulte de mes recherches sur cet objet que les fonctions arbitraires même discontinues peuvent toujours être représentées par les développements en sinus ou cosinus d'arcs multiples, et que les intégrales qui contiennent ces développements sont précisément aussi générales que celles où entrent les fonctions arbitraires d'arcs multiples. Conclusion que le célèbre Euler a toujours repoussée.

[101] Cette pièce renferme de véritables équations différentielles de la transmission de la chaleur, soit à l'intérieur des corps, soit à leur surface ; et la nouveauté du sujet, jointe à son importance, a déterminé la Classe à couronner cet Ouvrage, en observant cependant que la manière dont l'Auteur parvient à ses équations n'est pas exempte de difficultés, et que son analyse, pour les intégrer, laisse encore quelque chose à désirer, soit relativement à la généralité, soit même du côté de la rigueur.

> It follows from all that was proved in this section, concerning the series expansion of trigonometric functions, that if we propose a function fx, whose value is represented on a given interval, from $x = 0$ to $x = X$, by the ordinate of a curves line drawn arbitrarily, one can always expand this function as a series which will contain only the sines, or the cosines, or the sines and the cosines of multiple arcs, or only the cosines of odd multiples.[102]

Trigonometric functions were essential for the solution of the heat equation, as they were for the wave equation at the time of Bernoulli. What is important for our topic here is that trigonometric series became an essential tool in the field of complex analysis, independently of the heat flow. Fourier writes in the same section (art. 235 p. 258):

> We cannot entirely solve the fundamental questions of the theory of heat without reducing to this form the functions that represent the initial state of temperatures.
>
> These trigonometric series, ordered according to the cosines or sines of the multiples of the arc, pertain to elementary analysis, like the series whose terms contain successive powers of the variable. The coefficients of the trigonometric series are definite areas, and those of power series are fractions given by differentiation, in which one also attributes to the variable a definite value.[103]

Fourier then summarizes several properties of these series, including the integral formulae for the coefficients, and he also states the following:

> The series, ordered according to the cosines or the sines of the multiple arcs, are always convergent, that is, when we give to the variable an arbitrary non-imaginary value, the sum of the terms converges more and more to a unique and fixed limit, which is the value of the expanded function.[104]

It is interesting to recall that Fourier, in his *Théorie analytique de la chaleur* quotes Archimedes, Galileo and Newton, the three scientists mentioned by Riemann in the last part of his Habilitation lecture, *Über die Hypothesen, welche der Geometrie zu Grunde liegen*. Fourier writes ([116], pp. i–ii):

> The knowledge that the most ancient people could have acquired in rational mechanics did not reach us, and the history of that science, if we except the first theorems on harmony, does

---

[102]Il résulte de tout ce qui a été démontré dans cette section, concernant le développement des fonctions en séries trigonométriques, que si l'on propose une fonction fx, dont la valeur est représentée dans un intervalle déterminé, depuis $x = 0$ jusqu'à $x = X$, par l'ordonnée d'une ligne courbe tracée arbitrairement on pourra toujours développer cette fonction en une série qui ne contiendra que les sinus, ou les cosinus, ou les sinus et les cosinus des arcs multiples, ou les seuls cosinus des multiples impairs.

[103]On ne peut résoudre entièrement les questions fondamentales de la théorie de la chaleur, sans réduire à cette forme les fonctions qui représentent l'état initial des températures.

Ces séries trigonométriques, ordonnées selon les cosinus ou les sinus des multiples de l'arc, appartiennent à l'analyse élémentaire, comme les séries dont les termes contiennent les puissances successives de la variable. Les coefficients des séries trigonométriques sont des aires définies, et ceux des séries de puissances sont des fonctions données par la différenciation, et dans lesquelles on attribue aussi à la variable une valeur définie.

[104]Les séries ordonnées selon les cosinus ou les sinus des arcs multiples sont toujours convergentes, c'est-à-dire qu'en donnant à la variable une valeur quelconque non imaginaire, la somme des termes converge de plus en plus vers une seule limite fixe, qui est la valeur de la fonction développée.

> not go back further than the the discoveries of Archimedes. This great geometer explained the mathematical principles of the equilibrium of solids and of fluids. About eighteen centuries passed before Galileo, the first inventor of the dynamical theories, discovered the laws of motions of massive bodies. Newton encompassed in that new science all the system of the universe.[105]

Riemann claims in his memoir on trigonometric series that the question of finding conditions under which a function can be represented by a trigonometric series was completely settled in the work of Dirichlet, "for all cases that one encounters in nature. [...] The questions to which Dirichlet's researches do not apply do not occur in nature." We quote Dirichlet, from his 1829 memoir *Sur la convergence des séries trigonométriques qui servent à représenter une fonction arbitraire entre des limites données* (On the convergence of trigonometric series used to represent an arbitrary function between two given bounds) [55] (1829), in which he gives a solution to the convergence problem, and where he also mentions the work of Cauchy on the same problem. His paper starts as follows:

> The series of sines and cosines, by means of which one can represent an arbitrary function on a given interval, enjoy among other remarkable properties the one of being convergent. This property has not escaped the attention of the famous geometer who opened a new field for the applications of analysis, introducing there the way of expressing the arbitrary functions which are our subject here. It is contained in the memoir containing his first researches on heat. But, to my knowledge, nobody up to now gave a general proof. I know only on this subject a work due to Mr. Cauchy and which is part of his *Mémoires de l'Académie des sciences de Paris pour l'année 1823*. The author of this work confesses himself that his proof fails for certain functions for which convergence is nevertheless indisputable. A careful examination of that memoir led me to the belief that the proof which is presented there is insufficient even for the cases to which the authors thinks it applies.[106]

Riemann, after quoting the work of Dirichlet, gives two reasons for investigating the cases to which Dirichlet's methods do not apply directly. The first reason is that sorting out these questions will bring more clarity and precision to the principles of infinitesimal calculus. The second reason is that Fourier series will be useful not only

---

[105] Les connaissances que les plus anciens peuples avaient pu acquérir dans la mécanique rationnelle ne nous sont pas parvenues, et l'histoire de cette science, si l'on excepte les premiers théorèmes sur l'harmonie, ne remonte point au-delà des découvertes d'Archimède. Ce grand géomètre expliqua les principes mathématiques de l'équilibre des solides et des fluides. Il s'écoula environ dix-huit siècles avant quc Galilée, premier inventeur des théories dynamiques, découvrit les lois du mouvement des corps graves. Newton embrassa dans cette science nouvelle tout le système de l'univers.

[106] Les séries de sinus et de cosinus, au moyen desquelles on peut représenter une fonction arbitraire dans un intervalle donné, jouissent entre autres propriétés remarquables aussi de celles d'être convergentes. Cette propriété n'avait pas échappé au géomètre illustre qui a ouvert une nouvelle carrière aux applications de l'analyse, en y introduisant la manière d'exprimer les fonctions arbitraires dont il est question ; elle se trouve énoncée dans le Mémoire qui contient ses premières recherches sur la chaleur. Mais personne, que je sache, n'en a donné jusqu'à présent une démonstration générale. Je ne connais sur cet objet qu'un travail dû à M. Cauchy et qui fait partie des Mémoires de l'Académie des sciences de Paris pour l'année 1823. L'auteur de ce travail avoue lui-même que sa démonstration tombe en défaut pour certaines fonctions pour lesquelles la convergence est pourtant incontestable. Un examen attentif du mémoire cité m'a porté à croire que la démonstration qui y est exposée n'est pas même suffisante pour les cas auxquels l'auteur la croit applicable.

in physics, but also in number theory. Riemann says that in this field, it is precisely the functions which Dirichlet did not consider that seem to be the most important.

The so-called Dirichlet conditions for a function defined on the interval $[0, 2\pi]$ to have a Fourier trigonometric expansion is now classical. Picard, in [198] (p. 8) writes that Dirichlet's memoir on Fourier series remained a model of rigor. We conclude this section by quoting Riemann. Talking about Dirichlet's work on trigonometric series, he writes:

> This work of Dirichlet gave a firm foundation to a great number of important analytic researches. Highlighting a point on which Euler was mistaken, he succeeded in clearing out a question that had occupied so many eminent geometers for more than seventy years.

Riemann, in the same memoir, developed his integration theory in order to build a general theory for Fourier series, in particular for functions which have an infinite number of discontinuity points. This is the subject of the next section.

11 Integration

The second part of Riemann's memoir on trigonometric functions [215] carries the title "On the notion of definite integral and on the scope of its applicability." The relation between integration and trigonometric series is based on Fourier's formulae which give the coefficients of a trigonometric series in the form of integrals. Riemann starts the second part of the memoir by formulating a question: "First of all, what do we mean by

$$\int_a^b f(x)dx?"$$

The rest of the memoir is the answer to this question.

We explained at length that one of the fundamental questions in eighteenth century mathematics was "*What is a function*," and how this question led to a celebrated controversy. Riemann had to deal with the same question in his memoir on trigonometric series, more than a century after the controversy started, and he gave it the definitive answer. The problem in Riemann's memoir was addressed in a new context, namely, his integration theory, which was developed in a few pages at the end of that memoir. More particularly, the question became in that context: "*What are the functions that can be integrated*?" and in particular, whether the known functions were sufficient for the theory that became known as the Riemann integral or whether a new class of functions was needed.

We recall that since Newton and Leibniz, and passing by Euler, integration was defined as an anti-derivarive. Cauchy started an approach to integrals as limits of sums associated to partitions of the interval of definition, that is, sums of the form

$$\sum_{k=1}^{\infty} f(x_{k-1})(x_k - x_{k-1}),$$

cf. e.g. Cauchy's *Résumé des leçons données à l'École Royale Polytechnique sur le calcul infinitésimal* (Summary of lectures on infinitesimal calculus given at the *École Royale Polytechnique*) (1823) [35]. In Cauchy's setting, the limit always exists because he considered only continuous functions. It was soon realized that the definition may apply to more general functions. Dirichlet, in his work on trigonometric functions, used Cauchy's theory applied to discontinuous functions. Riemann states in his memoir that Cauchy's integration theory involves some random definitions which cannot make it a universal theory.

Riemann answered the question of how far discontinuity is allowed. He was led to the most general functions, which he termed "integrable." In Sects. 7, 8 and 9 of his memoir, he applies his new integration theory to the problem of representing arbitrary functions by trigonometric series. The main results are stated in three theorems in Sect. 8, and the propositions concerning the representation of functions by trigonometric series are contained in Sect. 9. Sections 10 and 11 contain results on the behavior of the coefficients of a trigonometric series. The last sections (Sects. 12 and 13) concern particular cases, more precisely, cases where the Fourier series is not convergent.

It is curious that Riemann mentions Cauchy several times in this memoir on trigonometric series, but he never refers to him in his dissertation on the theory of functions of a complex variable.

There is a section on the history of integration in Lebesgue's book [169]. In particular, Lebesgue summarizes Cauchy's theory, as well as an unpublished work of Dirichlet on the subject, which reached us through a description of Lipschitz ([169] p. 9). Dirichlet's work applies to functions with an infinite number of discontinuity points, but forming a non-dense subset. Riemann, using series, constructed functions to which the preceding techniques do not apply and which may still be integrated. These functions of Riemann do not have a graphical representation. We are far from Euler's "arbitrary drawable function" which, indeed, he thought exceeded the power of the calculus (by not being differentiable).

Chapter 2 of Lebesgue's treatise is a survey of the Riemann integral. This theory allows one to prove theorems such as the fact that a uniformly convergent integrable sequence of functions is an integrable function (p. 30), and that a uniformly convergent series of integrable functions may be integrated term by term. Lebesgue also mentions the work of Darboux, involving the notions of upper and lower limits. He then presents his own geometric theory (as opposed to the analytic theory of Riemann), based on set theory and measure theory. There are more comments on Lebesgues' integration theory in Chap. 8 of the present volume [192].

To conclude this section, let us mention that Riemann's ideas about the general notion of function in relation with integration theory underwent several developments in the twentieth century (one may think about the difficulties in the introduction of general measurable functions).

Riemann's memoir on trigonometric series was published 13 years after it was written. It was translated into French by Darboux and Houël.

It is interesting to note that trigonometric series are used in the proof of the so-called Poincaré lemma, a lemma which plays an essential role in the proof of the modern version of the Riemann–Roch theorem which is presented in Chap. 13 of the present volume.

12 Conclusion

In the preceding sections, we reviewed part of the historical origins of Riemann's mathematical works. One should write another article about the roots of his ideas in physics and philosophy. The intermingling between the old and new ideas of physics and philosophy is yet another subject. In this respect, and since the present book is also about relativity, we quote Kurt Gödel from his article *A remark about the relationship between relativity theory and idealistic philosophy* [127]. Talking about the insight that this theory brings into the nature of time, he writes:

> In short, it seems that one obtains an unequivocal proof for the view of those philosophers who, like Parmenides, Kant and the modern idealists, deny the objectivity of change and consider change as an illusion or an appearance due to our special mode of perception.

It would be stating the obvious to say that mathematicians should read the works of mathematicians from the past, not only the recent past, but most of all the founders of the theories they are working on. Yet, very few do it. I would like to conclude the present chapter by quoting some pre-eminent mathematicians who expressed themselves on this question. I start with Chebyshev.

We learn from his biographer in [207] that Chebyshev's thoroughly studied the works of Euler, Lagrange, Gauss, Abel, and other pre-eminent mathematicians. The biographer also writes that, in general, Chebyshev was not interested in reading the mathematical works of his contemporaries, considering that spending time on that would prevent him of having original ideas.

On the importance of reading the old masters, we quote again André Weil, from his 1978 ICM talk ([254] p. 235):

> From my own experience I can testify about the value of suggestions found in Gauss and in Eisenstein. Kummer's congruences for Bernoulli numbers, after being regarded as little more than a curiosity for many years, have found a new life in the theory of p-adic and L-functions, and Fermat's ideas on the use of the infinite descent in the study of Diophantine equations of genus one have proved their worth in contemporary work on the same subject.

Among the more recent mathematicians, I would like to quote again Grothendieck. During his apprenticeship, like most of us, Grothendieck was not encouraged to read ancient authors. He writes, in *Récoltes et semailles* (Chap. 2, Sect. 2.10):

> In the teaching I received from my elders, historical references were extremely rare, and I was nurtured, not by reading authors which were slightly ancient, nor even contemporary, but only through communication, face to face or through correspondence with others mathematicians, and starting with those who were older than me.[107]

[107]Dans l'enseignement que j'ai reçu de mes aînés, les références historiques étaient rarissimes, et j'ai été nourri, non par la lecture d'auteurs tant soit peu anciens ni même contemporains, mais

In the same work, we read (Chap. 2, Sect. 2.5):

> I personally feel that I belong to a lineage of mathematicians whose spontaneous mission and joy is to constantly construct new houses. [...] I am not strong in history, and if I were asked to give names of mathematicians in that lineage, I can think spontaneously of Galois and Riemann (in the past century) and Hilbert (at the beginning of the present century).[108]

Grothendieck's attitude towards mathematics and mathematicians changed drastically at the time he decided to quit the mathematical milieu, in 1970, twenty years after he obtained his first job, putting an end to an extraordinarily productive working life and to his relation with his contemporary mathematicians. One thing which is not usually mentioned about him is that his writings, during the period that followed, contain many references to mathematicians of the past, to whom Grothendieck expresses his debt, and among them stands Riemann. In his *Récoles et semailles* [133], Grothendieck writes (Chap. 2, Sect. 2.5):

> Most mathematicians are led to confine themselves in a conceptual framework, in a "*Universe*," which is fixed once and for all—essentially, the one they found "ready-made" at the time they were students. They are like the heirs of a big and completely furnished beautiful house, with its living rooms, kitchens and workshops, its kitchen set and large equipment, with which there is, well, something to cook and to tinker. How this house was progressively constructed, over the generations, and why and how such and such tool (and not another) was conceived and shaped, why the rooms are fit out in such a manner here, and in another manner there—these are as many questions as these heirs will never think to ask. That is the "Universe," the "given" in which we must live, that's it! Something which will seem great (and, most often, we are far from having discovered all the rooms), **familiar**[109] at the same time, and, most of all, **unchanging**.[110]

(Footnote 107 continued)
surtout par la communication, de vive voix ou par lettres interposées, avec d'autres mathématiciens, à commencer par mes aînés.

[108] Je me sens faire partie, quant à moi, de la lignée des mathématiciens dont la vocation spontanée et la joie est de construire sans cesse des maisons nouvelles. [...] Moi qui ne suis pas fort en histoire, si je devais donner des noms de mathématiciens dans cette lignée-là, il me vient spontanément ceux de Galois et de Riemann (au siècle dernier) et celui de Hilbert (au début du siècle présent).

[109] The emphasis is Grothendieck's.

[110] La plupart des mathématiciens sont portés à se cantonner dans un cadre conceptuel, dans un "*Univers*" fixé une fois pour toutes—celui, essentiellement, qu'ils ont trouvé "tout fait" au moment où ils ont fait leurs études. Ils sont comme les héritiers d'une grande et belle maison toute installée, avec ses salles de séjour et ses cuisines et ses ateliers, et sa batterie de cuisine et un outillage à tout venant, avec lequel il y a, ma foi, de quoi cuisiner et bricoler. Comment cette maison s'est construite progressivement, au cours des générations, et pourquoi et comment ont été conçus et façonnés tels outils (et pas d'autres...), pourquoi les pièces sont aménagées de telle façon ici, et de telle autre là—voilà autant de questions que ces héritiers ne songeraient pas à se demander jamais. C'est ça "l'Univers", le "donné" dans lequel il faut vivre, un point c'est tout ! Quelque chose qui paraît grand (et on est loin, le plus souvent, d'avoir fait le tour de toutes ses pièces), mais **familier** en même temps, et surtout: **immuable**.

We conclude with Grothendieck's reference to Riemann. In his *Sketch of a program*, [134], he writes (p. 240 of the English translation)[111]:

> Whereas in my research before 1970, my attention was systematically directed towards objects of maximal generality, in order to uncover a general language adequate for the world of algebraic geometry [...] here I was brought back, via objects so simple that a child learns them while playing, to the beginnings and origins of algebraic geometry, familiar to Riemann and his followers!

Topic	**Euler**	**Riemann**
Functions of a complex variable	• Introductio in analysin infinitorum (1748) • De repraesentatione superficiei sphaericae super plano (1777)	• Grundlagen für eine allgemeine Theorie der Functionen einer veränderlichen complexen Grösse (1851) • Theorie der Abel'schen Functionen (1857)
Elliptic and Abelian integrals	• Specimen de constructione aequationum differentialium sine indeterminatarum separatione (1738) • Observationes de comparatione arcuum curvarum irrectificibilium (1761) • De integratione aequationis differentialis $\frac{mdx}{\sqrt{1-x^4}} = \frac{ndy}{\sqrt{1-y^4}}$ (1761)	• Grundlagen für eine allgemeine Theorie der Functionen einer veränderlichen complexen Grösse (1851) • Theorie der Abel'schen Functionen (1857) • Über das Verschwinden der ϑ-Functionen (1857)
Hypergeometric series	• De summatione innumerabilium progressionum (1738) • Institutionum calculi integralis volumen secundum (1769) • Specimen transformationis singularis serierum (1778)	• Beiträge zur Theorie der durch die Gauss'sche Reihe $F(\alpha, \beta, \gamma, x)$ darstellbaren Functionen (1857)
The zeta function	• Variae observationes circa series infinitas (1744) • Remarques sur un beau rapport entre les series des puissances tant directes que réciproques (1749)	• Über die Anzahl der Primzahlen unter einer gegebenen Grösse (1859)
Integration	• Institutionum calculi integralis (1768–1770)	• Über die Darstellbarkeit einer Function durch eine trigonometrische Reihe (1854)

[111] The English translation is by Lochak and Schneps.

Topic	Euler	Riemann
Space and philosophy of nature	• Anleitung zur Naturlehre (1745) • Reflexions sur l'espace et le temps (1748)	• Grundlagen für eine allgemeine Theorie der Functionen einer veränderlichen complexen Grösse (1851)
Topology	• Solutio problematis ad geometriam situs pertinentis (1741) • Elementa doctrinae solidorum (1758) • Demonstratio nonnullarum insignium proprietatum, quibus solida hedris planis inclusasunt praedita (1758)	• Grundlagen für eine allgemeine Theorie der Functionen einer veränderlichen complexen Grösse (1851) • Theorie der Abel'schen Functionen (1857) • Über die Hypothesen, welche der Geometrie zu Grunde liegen (1854)
Differential geometry	• Introductio in analysin infinitorum (1748) • Recherches sur la courbure des surfaces (1767)	• Über die Hypothesen, welche der Geometrie zu Grunde liegen (1854) • Commentatio mathematica, qua respondere tentatur quaestioni ab Illma Academia Parisiensi propositae (1861) • Ein beitrag zu den Untersuchungen über die flüssigen Bewegung eines gleichartigen Ellipsoides (1861) • Über die Fläche vom kleinsten Inhalt bei gegebener Begrenzung (1867)
Trigonometric series	• Recherches sur la question des inegalités du mouvement de Saturne et de Jupiter (1748)	• Über die Darstellbarkeit einer Function durch eine trigonometrische Reihe (1854)
Acoustics	• Dissertatio physica de sono (1727) • Sur la vibration des cordes (1748)	• Über die Fortpflanzung ebener Luftwellen von endlicher Schwingungsweite (1860)

Acknowledgements I would like to thank Vincent Alberge, Jeremy Gray and Marie Pascale Hautefeuille who read a preliminary version of this chapter and suggested corrections.

References

1. N.H. Abel, Recherches sur les fonctions elliptiques. J. für die Reine und Angew. Math. **2**, 101–181 (1827)
2. L.V. Ahlfors, Quasiconformal mappings and their applications. Lect. Modern Math. **2**, 151–164 (1964)
3. d'Alembert, Jean le Rond, Opuscules mathématiques, Paris, Claude-Antoine Jombert, 1780
4. d'Alembert, Jean le Rond, Recherches sur les vibrations des cordes sonores & Supplément, Opuscules mathématiques, Paris, 1761, tome 1, pp. 1–73

5. d'Alembert, Jean le Rond, Recherches sur le calcul intégral, *Mémoires de l'Académie des sciences de Berlin*, 2, 1746, pp. 182–192
6. d'Alembert, Jean le Rond, Suite des recherches sur la courbe que forme une corde tendue mise en vibration (1747), *Mémoires de l'Académie des sciences de Berlin* 3, 1749, pp. 220–249
7. d'Alembert, Jean le Rond, Addition aux recherches sur la courbe que forme une corde tendue mise en vibration (1750), *Mémoires de l'Académie des sciences de Berlin* 6, 1752, pp. 355–360
8. d'Alembert, Jean le Rond, Recherches sur la courbe que forme une corde tendue mise en vibration (1747), *Mémoires de l'Académie des sciences de Berlin* 3, 1749, pp. 214–219
9. P.S. Alexandrov, Poincaré and topology (speech given at the celebration session at the International Congress of Mathematicians in honor of the centenary of Poincaré's birth). Uspekhi Mat. Nauk **27** (1972) **1**(163), 147–158, Russian Math. Surv. **34**(6) (1979), 267–302; **35**(3) (1980), 315–358. With an appendix by V.A. Zorin
10. K.A. Andreiev, Démonstration d'une propriété générale des polyèdres. Société Mathématique de Moscou **6**, 457–466 (1873)
11. Apollonius: Les Coniques, tome 1.1: Livre I, ed. R. Rashed, commentaire historique et mathématique, édition et traduction du texte arabe, de Gruyter, 2008, p. 666
12. Apollonius: Les Coniques, tome 2.2: Livre IV, ed. R. Rashed, commentaire historique et mathématique, édition et traduction du texte arabe, de Gruyter, 2009, p. 319
13. Apollonius: Les Coniques, tome 3: Livre V, ed. R. Rashed, commentaire historique et mathématique, édition et traduction du texte arabe, de Gruyter, 2008, p. 550
14. Apollonius: Les Coniques, tome 4: Livres VI et VII, ed. R. Rashed, commentaire historique et mathématique, édition et traduction du texte arabe, Scientia Graeco-Arabica, vol. 1.4, de Gruyter, 2009, p. 572
15. Apollonius: Les Coniques, tome 2.1: Livres II et III, ed. R. Rashed, commentaire historique et mathématique, édition et traduction du texte arabe, de Gruyter, 2010, p. 682
16. P. Appell, É. Goursat, *Théorie des fonctions algébriques et de leurs intégrales : Étude des fonctions analytiques sur une surface de Riemann* (Gauthier-Villars, Paris, 1895)
17. L.-F.-A. Arbogast, Mémoire sur la nature des fonctions arbitraires qui entrent dans les intégrales des équations aux différentielles partielles, Académie impériale des sciences, Saint-Petersbourg, 1791
18. Aristotle, Categories, in *The Complete Works of Aristotle: The Revised Oxford Translation*, ed. by J. Barnes, vols. 1, 1–27, Translated by J.L. Ackrill, (Princeton University Press, Princeton, 1984)
19. Aristotle, the Metaphysics, in *The Complete Works of Aristotle: The Revised Oxford Translation*, ed. by J. Barnes, vol. 2, pp. 1552–1728, Translated by W. D. Ross, (Princeton University Press, Princeton, 1984)
20. Aristotle, the Physics, in *The Complete Works of Aristotle: The Revised Oxford Translation*, ed. by J. Barnes, vol. 1, pp. 315–446. Translated by R.P. Hardie, R.K. Gaye, (Princeton University Press, Princeton, 1984)
21. Aristotle, On the Heavens, in *The Complete Works of Aristotle: The Revised Oxford Translation*, ed. by J. Barnes, vol. 1, pp. 447–511. Translated by J.L. Stocks, (Princeton University Press, Princeton, 1984)
22. N.V. Bougaïev, *Théorème d'Euler sur les polyèdres; propriété d'un réseau géométrique plan* (Moscou, Société mathématique de Moscou, 1867), pp. 87–92
23. J. Bernoulli, Commercium philosophicum et mathematicum, 1698, I, published Bousquet, by Lausanne et Genève, 1745
24. J. Bernoulli, Remarques sur ce qu'on a donné jusqu'ici de solutions des problèmes sur les isopérimètres, avec une nouvelle méthode courte et facile de les résoudre sans calcul, laquelle s'étend aussi à d'autres problèmes qui ont rapport à ceux-là, Mém. Acad. Royale des Sciences de Paris, 1718, p. 100 (Paris edition), p. 123 (Dutch edition). *Opera*, Lausanne and Geneva, vol. 2, p. 253ff
25. D. Bernoulli, Mémoire sur les vibrations des cordes d'une épaisseur inégale, Histoire de l'Académie Royale des Sciences et des Belles-Lettres de Berlin, vol. 87 (1767), pp. 281–306

26. J.-L.-F. Bertrand, Observations aux remarques de M. Prouhet sur un passage des œuvres inédites de Descartes, Comptes rendus des séances de l'Académie des sciences de Paris **50**, 781 (1860)
27. C. Brianchon, Théorème nouveau sur les polyèdres. J. de l'École Polytech. Paris **15**, 317–319 (1837)
28. C.B. Boyer, *The Foremost Textbook of Modern Times* (International Congress of Mathematicians, Cambridge, MA, 1950). Also published in the Am. Math. Monthly **58**(4), 223–226 (1951)
29. N.L. Biggs, K. Lloyd, R.J. Wilson, *Graph Theory: 1736–1936* (Clarendon Press, Oxford, 1976)
30. P. Carus, *The foundations of mathematics: A contribution to the philosophy of geometry* (Open Court, Chicago, 1908)
31. E. Catalan, Mémoire sur la théorie des polyèdres. J. de l'École Polytech. **24**, 1–71 (1865)
32. A.-L. Cauchy, Recherches sur les polyèdres, Premier mémoire. J. de l'École Polytech. **9**, 68–86 (1813); *Œuvres complètes*, 2e série, tome I, pp. 7–25
33. A.-L. Cauchy, Sur les polygones et les polyèdres, Second mémoire. J. de l'École Polytech. **9**, 87–98 (1813); *Œuvres complètes*, 2e série, tome I, pp. 26–38
34. A.-L. Cauchy, Cours d'analyse de l'École Royale Polytechnique, 1re partie. Analyse algébrique. Imprimerie Royale, Paris, 1821. Œuvres complètes, série 2, tome III
35. A.-L. Cauchy, Résumé des leçons données à l'École Royale Plytechnique sur le calcul infinitésimal, tome I, Paris, 1923. In Œuvres, série 2, tome IV, pp. 5–261
36. A.-L. Cauchy, Mémoire sur les intégrales définies prises entre des limites imaginaires, Imprimerie Royale, Paris, 1825, Œuvres complètes, série 2, tome XV, pp. 41–89
37. A.-L. Cauchy, Mémoire sur les fonctions continues, C. R., t. XVIII, 116–129 (22 janvier 1844), Œuvres complètes, série 1, tome 8, pp. 145–160
38. A.-L. Cauchy, Sur les fonctions de variables imaginaires, Comptes Rendus Acad. Sci. Paris **32**, 207–215 (1851), Œuvres, série I tome XI, pp. 301–304
39. A.-L. Cauchy, Sur les fonctions monotypiques et monogènes, Comptes Rendus Acad. Sci. Paris **32**, 484–487 (1851). Œuvres, série 1, tome XI, pp. 376–380
40. A.-L. Cauchy, Sept leçons de physique générale, Paris, bureau du journal Les Mondes et Gauthier-Villars, 1868
41. A. Cayley, Chapters in the analytic geometry of (n) dimensions. Cambridge Math. J. A, 119–127. Reprinted in Collected mathematical papers of Arthur Cayley, Vol. 1. Univ. Press, Cambridge 1889–1897, 55–62 (1843)
42. P.L. Chebyshev, Sur la fonction qui détermine la totalité des nombres premiers inférieurs à une limite donnée, J. Math. Pures Appl. **17**(1), 341–365 (1852)
43. P.L. Chebyshev, Mémoire sur les nombres premiers, J. Math. Pures Appl. **17**(1), 366–390 (1852)
44. P.L. Chebyshev, Rapport du professeur extraordinaire de l'université de Saint-Pétersbourg Tchebychef sur son voyage à l'étranger. Reprinted in P. L. Tchebycheff, Œuvres [45], vol. 2, pp. vii–xiii
45. P.L. Chebyshev, Œuvres, ed. by A. Markoff, N. Sonin, Imprimerie de l'Académie Impériale des Sciences, Saint Petersburg, 2 vols, 1899–1907
46. W.K. Clifford, On the space-theory of matter. Proc. Cambridge Philos. Soc. **2**, 157–158 (1870). Reprinted in: The Concepts of Space and Time, M. Capek (ed.) Volume 22 of the series Boston Studies in the Philosophy of Science, pp. 295–296
47. R. Descartes, *Exercices pour les éléments des solides : essai en complément d'Euclide; Progymnasmata de solidorum elementis, édition critique avec introduction, traduction, notes et commentaire par Pierre Costabel* (Presses universitaires de France, Paris, 1987)
48. G. Darboux, Review of Fourier's *Œuvres*, Tome I. Bulletin des Sciences Mathématiques, 2e série, t. XII, Mars 1888, pp. 57–59
49. G. Darboux, Notice historique sur le général Meusnier, Mémoires de l'Académie des Sciences de l'Institut de France, lue à la séance annuelle du 20 décembre 1901, pp. I–XXXVIII

50. R. Descartes, Les principes de la philosophie, 1st edn, in Latin, Amsterdam, 1644, French edition, translation by L'abbé Picot, First published in 1647. In: *Œuvres de Descartes*, ed. V. Cousin, Tome III, Paris, Levrault, 1824
51. R. Descartes, *Œuvres complètes*, publiées par C. Adam et P. Tannery, 1897–1913, 11 vol. réimpression : Bibliothèque des textes philosophiques, Vrin, Paris, 1996
52. Diophante : Les Arithmétiques, Livre IV, vol. 3, ed. R. Rashed, Collection des Universités de France, Paris, Les Belles Lettres, 1984, p. 487
53. Diophante : Les Arithmétiques, Livres V, VI, VII, vol 4, ed. R. Rashed, Collection des Universités de France, Paris, Les Belles Lettres, 1984, p. 451
54. Les Arithmétiques de Diophante : Lecture historique et mathématique, ed. by R. Rashed, Ch. Houzel, de Gruyter, 2013, p. 629
55. P.G. Dirichlet (Lejeune), Sur la convergence des séries trigonométriques qui servent à représenter une fonction arbitraire entre des limites données. J. für die reine und Angew. Math. **4**, 157–169 (1829)
56. P.G. Dirichlet (Lejeune), Beweis des Satzes, dass jede unbegrenzte arithmetische Progression, deren erstes Glied und Differenz ganze Zahlen ohne gemeinschaftlichen Factor sind, unendlich viele Primzahlen enthält, Abh. Kgl. Preuss. Akad. Wiss. Berlin (gelesen in der Akad. der Wiss., 45–81, 1839. Also. In: Werke 1 (1889), 313–342 (1837)
57. P. Duhem, Le système du monde; histoire des doctrines cosmologiques de Platon à Copernic, Hermann, Paris, 10 vols., 1913–1959
58. Ch. Dupin, Développements de géométrie, avec des applications à la stabilité des vaisseaux, aux déblais et remblais, au défilement, à l'optique, etc. pour faire suite à la géométrie descriptive et à la géométrie analytique de M. Monge : Théorie. Courcier, Paris, 1813
59. J. Dutka, The early history of the hypergeometric function. Arch. Hist. Exact Sci. **31**(1), 15–34 (1984)
60. G. Enstöm, Bibl. Math. III, 12 (1912)
61. L. Euler, *Introductio in analysin infinitorum*, First edition: Lausannae: Apud Marcum-Michaelem Bousquet & socios., 1748. *Opera omnia*, Series 1, vol. VIII. English translation by J.T. Blanton, 2 vol. (Springer, New York, 1988, 1990)
62. L. Euler, Introduction to Analysis of the Infinite, English translation of [62] by J.T. Blanton, 2 vol. (Springer, New York, 1988)
63. L. Euler, Mechanica, 2 volumes, first edition, 1736 *Opera Omnia*, Series 2, Volumes 1 and 2
64. L. Euler, Rechenkunst, Anmerck ungen uber die zeitungen, 1738. *Opera Omnia*, Series 3, vol. 2, pp. 1–304
65. L. Euler, De progressionibus transcendentibus seu quarum termini generales algebraice dari nequeunt, Commentarii academiae scientiarum Petropolitanae **5**(1738), 36–57. *Opera Omnia*, Series 1, vol. 14, pp. 1–24
66. L. Euler, De summatione innumerabilium progressionum Commentarii academiae scientiarum Petropolitanae **5** (1738), 91–105. *Opera Omnia*, Series 1, vol. 14, pp. 25–41
67. L. Euler, *Dissertatio physica de sono*, E., J.R. Thurneisen Brothers, Basel, 1727. *Opera omnia*, Series 3, vol. 1, pp. 181–196
68. L. Euler, Sur la vibration des cordes (1748), traduit du latin, *Mémoires de l'Académie des sciences de Berlin*, vol. 4, 1750, pp. 69–85. *Opera omnia*, Series 2, vol. 10, pp. 63–77
69. L. Euler, Observationes de theoremate quodam Fermatiano aliisque ad numeros primos spectantibus, Commentarii academiae scientiarum Petropolitanae **6**, 103–107 (1738). *Opera Omnia*, Series 1, vol. 2, pp. 1–5
70. L. Euler, Specimen de constructione aequationum differentialium sine indeterminatarum separatione, Commentarii academiae scientiarum Petropolitanae **6**, 168–174 (1738). *Opera Omnia*, Series 1, vol. 20, pp. 1–7
71. L. Euler, Tentamen novae theoriae musicae ex certissimis harmoniae principiis dilucide expositae, Saint Petersburg Academy, 1739, 263 pages. *Opera Omnia*, Series 3, vol. 1, pp. 197–427
72. L. Euler, De summis serierum reciprocarum, Commentarii academiae scientiarum Petropolitanae **7**, 123–134 (1740). *Opera Omnia*, Series 1, vol. 14, pp. 73–86

73. L. Euler, Methodus inveniendi lineas curvas maximi minimive proprietate gaudentes, sive solutio problematis isoperimetrici lattissimo sensu accepti, Lausanne and Geneva, 1744 *Opera Omnia*, Series 1, vol. 24
74. L. Euler, Solutio problematis ad geometriam situs pertinentis, Commentarii academiae scientiarum Petropolitanae **8**, 128–140 (1741); *Opera omnia*, Series 1, vol. 7, pp. 1–10
75. L. Euler, Solutio trium problematum difficiliorum ad methodum tangentium inversam pertinentium, Mémoires de l'académie des sciences de Saint-Pétersbourg 10, 1826 (16–26), *Opera Omnia*, Series 1, vol. 29, pp. 320–333
76. L. Euler, Variae observationes circa series infinitas, Commentarii academiae scientiarum Petropolitanae **9**, 160–188 (1744). In *Opera Omnia*, Series 1, vol. 14, pp. 217–244
77. L. Euler, Recherches sur la question des inégalités du mouvement de Saturne et de Jupiter, Pièce qui a remporté le prix de l'Académie Royale des Sciences (1748), 1–123 (1749). *Opera Omnia* Series 2, vol. 25, pp. 45–157
78. L. Euler, Recherches sur l'origine des forces, Mémoires de l'académie des sciences de Berlin **6**, 419–447 (1752). *Opera Omnia* Series 2, vol. 5, pp. 109–131
79. L. Euler, Réflexions sur l'espace et le temps, Mémoires de l'Académie des sciences de Berlin **4**, 324–333 (1750). *Opera Omnia*, Series 3, vol. 2, pp. 376–383
80. L. Euler, Découverte d'une loi tout extraordinaire des nombres par rapport à la somme de leurs diviseurs, Bibliothèque impartiale **3**, 10–31 (1751). *Opera Omnia*, Series 1, vol. 2, pp. 241–253
81. L. Euler, Essai d'une démonstration métaphysique du principe général de l'équilibre, Mémoires de l'Académie des sciences de Berlin **7**, 246–254 (1753). *Opera Omnia*, Series 2, vol. 5, pp. 250–256
82. L. Euler, Problema, ad cuius solutionem geometrae invitantur; theorema, ad cuius demonstrationem geometrae invitantur, Nova acta eruditorum, 1754, p. 40
83. L. Euler, Institutiones calculi differentialis cum eius usu in analysi finitorum ac doctrina serierum, in two books, first published in 1755. *Opera Omnia*, Series 1, vol. 10. English translation of Book I, by J.D. Blanton, (Springer, 2000)
84. L. Euler, Elementa doctrinae solidorum, Novi Commentarii academiae scientiarum Petropolitanae **4**, 1758, pp. 109–140 ; *Opera omnia*, Series 1, vol. 26, pp. 71–93
85. L. Euler, Demonstratio nonnullarum insignium proprietatum, quibus solida hedris planis inclusa sunt praedita, Novi Commentarii academiae scientiarum Petropolitanae **4**, 1758, p. 140–160 ; *Opera omnia*, Series 1, vol. 26, pp. 94–108
86. L. Euler, De motu corporum coelestium a viribus quibuscunque perturbato, Novi Commentarii academiae scientiarum Petropolitanae **4**, 161–196 (1758). *Opera Omnia*, Series 2, vol. 25, pp. 175–209. *Opera Omnia*, Series 1, vol. 20, pp. 56–57
87. L. Euler, De integratione aequationis differentialis $\frac{mdx}{\sqrt{1-x^4}} = \frac{ndy}{\sqrt{1-y^4}}$, Novi Commentarii academiae scientiarum Petropolitanae **6**, 37–57 (1761). *Opera Omnia*, Series 1, vol. 20, pp. 58–79
88. L. Euler, Observationes de comparatione arcuum curvarum irrectificibilium, Novi Commentarii academiae scientiarum Petropolitanae **6**, 58–84 (1761). *Opera Omnia*, Series 1, vol. 20, pp. 80–107
89. L. Euler, Specimen novae methodi curvarum quadraturas et rectificationes aliasque quantitates transcendentes inter se comparandi, Novi Commentarii academiae scientiarum Petropolitanae **7**, 83–127 (1761). *Opera Omnia*, Series 1, vol. 20, pp. 108–152
90. L. Euler, Demonstratio theorematis et solutio problematis in actis erud. Lipsiensibus propositorum, Novi Commentarii academiae scientiarum Petropolitanae **7**, 128–162 (1761). *Opera Omnia* Series 1, vol. 20, pp. 201–234
91. L. Euler, Recherches sur la courbure des surfaces. Memoires de l'académie de science de Berlin, **16**, 119–143 (1767)
92. L. Euler, Institutionum calculi integralis, 3 volumes, original versions, 1768, 1769, 1770. *Opera Omnia*, Series 1, vols. 11, 12, 13

93. L. Euler, Integratio aequationis $\frac{dx}{\sqrt{\alpha+\beta x+\gamma x^2+\delta x^3+\epsilon x^4}} = \frac{dy}{\sqrt{\alpha+\beta y+\gamma y^2+\delta y^3+\epsilon y^4}}$, Novi Commentarii academiae scientiarum Petropolitanae **12**, 3–16 (1768). *Opera Omnia*, Series 1, vol. 20, pp. 302–317
94. L. Euler, Remarques sur un beau rapport entre les séries des puissances tant directes que réciproques, Mémoires de l'académie des sciences de Berlin **17**, 83–106 (1768). *Opera Omnia*, Series 1, vol. 15, pp. 70–90
95. L. Euler, De solidis quorum superficiem in planum explicare licet, Novi Commentarii academiae scientiarum Petropolitanae **16**, 3–34 (1772). *Opera Omnia*, Series 1, vol. 28, pp. 161–186
96. L. Euler, Problematis cuiusdam Pappi Alexandrini constructio, Acta Academiae Scientarum Imperialis Petropolitinae **4**, 91–96 (1783). *Opera Omnia*, Series 1, vol. 26, pp. 237–242
97. L. Euler, Methodus succinctior comparationes quantitatum transcendentium in forma $\frac{P\partial z}{\sqrt{A = 2BZ + CZZ + 2DZ^3 + EZ^4}}$, Institutiones calculi integralis **4**, 504–524 (1794). *Opera Omnia*, Series 1, vol. 21, pp. 207–236
98. L. Euler, Anleitung zur Naturlehre. Opera Postuma **2**, 1862, 449–560, and *Opera Omnia* Series 3, vol.1, pp. 16–180
99. L. Euler, De la propagation du son (1759), Mémoires de l'Académie des sciences de Berlin **15**, 185–209 (1766). *Opera omnia*, Series 3, vol. 1, pp. 428–451
100. L. Euler, Supplément aux recherches sur la propagation du son (1759), Mémoires de l'Académie des sciences de Berlin **15**, 210–240 (1766); *Opera Omnia*, Series 3, vol. 1, pp. 452–483
101. L. Euler, Continuation des recherches sur la propagation du son (1759), Mémoires de l'Académie des sciences de Berlin **15**, 241–264 (1766). *Opera Omnia*, Series 3, vol. 1, pp. 484–507
102. L. Euler, Éclaircissemens plus détaillés sur la génération et la propagation du son et sur la formation de l'écho (1765), Mémoires de l'Académie des sciences de Berlin **21**, 335–363 (1767). *Opera omnia*, Series 3, vol. 1, pp. 540–567
103. L. Euler, Remarques sur les mémoires précédents de M. Bernoulli, Mémoires de l'académie des sciences de Berlin **9**, 196–222 (1755). *Opera Omnia*, Series 2, vol. 10, pp. 233–254
104. L. Euler, De repraesentatione superficiei sphaericae super plano, Acta Academiae Scientarum Imperialis Petropolitinae **1777, 1778**, pp. 107–132. *Opera Omnia* Series 1, vol. 28, pp. 248–275
105. L. Euler, De proiectione geographica superficiei sphaericae, Acta Academiae Scientarum Imperialis Petropolitinae **1777, 1778**, pp. 133–142, *Opera Omnia* Series 1, vol. 28, pp. 276–287
106. L. Euler, De proiectione geographica Deslisliana in mappa generali imperii russici usitata, Acta Academiae Scientarum Imperialis Petropolitinae **1777, 1778**, pp. 143–153, *Opera Omnia* Series 1, vol. 28
107. L. Euler, In: Correspondance de Leonhard Euler avec C. Clairaut, J. d'Alembert et J.-L. Lagrange, Euler's *Opera Omnia* Ser. IVA, vol. V, ed. by A. Juskevich, R. Taton, 1980, Birkhäuser
108. L. Euler, J.-L. de Lagrange, *Correspondance de Lagrange avec Euler. Opera omnia*, Series IV, vol.1, Birkhäuser, Boston–Basel, 24 Nov 1759
109. L. Euler, Correspondence of Leonhard Euler with Christian Goldbach, Birkhäuser, ed. by F. Lemmermeyer, M. Mattmüller, L. Euleri *Opera Omnia*, Series IVA, vol. 4 (two parts) Basel, 2015
110. G.C. Fagnano, Teorema da cui si deduce una nuova misura degli Archi Elittici, Iperbolici, e Cicloidali: Giornale de' Letterati d'Italia XXVI (1716), pp. 266–279, in *Produzioni Matematiche*, Gavelli: Pesaro, 1750, t. II, pp. 336–342
111. G.C. Fagnano, Metodo per misurare la lemniscata Giorn. de' Letterati d'Italia, 1718
112. G.C. Fagnano, in *Opere matematiche del marchese G. C. de' Toschi di Fagnano*, ed. by D. Gambioli, G. Loria, V. Volterra, 3 vols. Societá italiana per il progresso delle scienze, 1911–1912

113. M. Feil, Über Euler'sche Polyeder, Sitzungsberichte der Kgl. Akademie der Wissenschaften in Wien **93**, 869–898 (1886)
114. L.-A. Foucher de Careil (dir.), *Œuvres inédites de Descartes*, précédées d'une Introduction sur la Méthode, 2 vol. (Lagrange, Paris, 1860)
115. J. Fourier, Théorie de la propagation de la chaleur dans les solides, Mémoire lu le 21 décembre 1807, présenté à l'Académie des Sciences, 29 octobre, 1809. Edited in [134]
116. J. Fourier, *Théorie analytique de la chaleur*, Paris, Firmin Didot, 1822. In *Œuvres*, ed. G. Darboux, Tome 1, 1888
117. N.I. Fuss, Éloge de Monsieur Léonard Euler, lu à l'Académie Impériale des Sciences dans son assemblée du 23 octobre 1783. St-Pétersbourg, 1783. Reprint available at Kessinger Publishing Legacy's Reprints
118. P.H. Fuss, *Correspondance mathématique et physique de quelques célèbres géomètres du XVI-IIe siècle, Saint-Pétersbourg, 1843* (Johnson Reprint Corp., New York and London, Reeprint, 1968)
119. C.F. Gauss, Demonstratio nova theorematis omnem functionem algebraicam rationalem integram unius variabilis in factores reales primi vel secundi gradus resolvi posse. Quam pro obtinendis summis in philosophia honoribus inclito philosophorum ordini Academiae Iuliae Carolinae, Helmstadii : apud C. G. Fleckeisen, 1799
120. C.F. Gauss, Theoria residuorum biquadraticorum. Commentatio prima et secunda, Göttingenn Dieterich, 1828 and 1832
121. C.F. Gauss, Disquisitiones Arithmeticae, Gehr. Fleischer, Leipzig, *English translation by A* (Yale University Press, A. Clarke, 1801). 1965
122. C.F. Gauss, Disquisitiones generales circa serium $1 + \frac{\alpha\beta}{1.\gamma}x + \frac{\alpha(\alpha+1)\beta(\beta+1)}{1.2\gamma(\gamma+1)}xx + \frac{\alpha(\alpha+1)(\alpha+2)\beta(\beta+1)(\beta+2)}{1.2.3\gamma(\gamma+1)(\gamma+2)}x^3 + \ldots$ etc. pars prior. Königliche Gesellschaft der Wissenschaften, Göttingen, 1812. In: *Werke*, vol. 3, pp. 123–162
123. C.F. Gauss, Determinatio seriei nostrae per aequationem differentialem secundi ordinis. Gauss's *Werke* vol. 3, pp. 207–230
124. C.F. Gauss, General Investigations of Curved Surfaces. Translated from the Latin and German by A. Hiltebeitel and J. Morehead, (Princeton University Library, Princeton, *New edition with an Introduction and Notes* by P. Pesic, Dover, 2005)
125. C.F. Gauss, translation of his notebook: Le Journal mathématique de Gauss: Traduction française annotée par P. Eymard, J.P. Lafon. Revue d'histoire des sciences et de leurs applications, vol. 9, no. 1 (Janvier-Mars 1956), pp. 21–51
126. C.F. Gauss, *Werke* (Königliche Gesellschaft der Wissenschaften, Göttingen, 1900)
127. K. Gödel, A remark about the relationship between relativity theory and idealistic philosophy, In: *Collected works of Kurt Gödel*, vol. II (Oxford University Press, New York, 1990), pp. 202–207
128. L.J. Goldstein, A history of the prime number theorem. Am. Math. Mon. **80**(6), 599–615 (1973)
129. L. Ahrndt Golland, R.W. Golland, Euler's troublesome series: An early example of the use of trigonometric series. Historia Mathematica **20**, 54–67 (1993)
130. J.D. Gergonne, Sur le théorème d'Euler relatif aux polyèdres. Annales de Gergonne **19**, 333 (1828)
131. H. Grassmann, *Die Lineale Ausdehnungslehre ein neuer Zweig der Mathematik: Dargestellt und durch Anwendungen* (O. Wigand, Leipzig, 1844)
132. I. Grattan-Guinness (in collaboration with J. R. Ravetz), Joseph Fourier (1768–1830), A survey of his life and work, based on a critical edition of his monograph on the propagation of heat, presented to the Institut de France, (MIT Press, Massachussets, London, Cambridge, 1972)
133. A. Grothendieck, Récoles et semailles : Réflexions et témoignage sur un passé de mathématicien, manuscript, 1983–1986 (To appear as a book)
134. A. Grothendieck, Esquisse d'un programme (Sketch of a program), unpublished manuscript (1984), English translation by P. Lochak, L. Schneps in *Geometric Galois Actions*, vol. 1,

Around Grothendieck's Esquisse d'un Programme, ed. by L. Schneps, P. Lochak, London Mathematical Society. Lecture Note Series, vol. 242 (Cambridge University Press, Cambridge, 1997), pp. 5–48
135. J.A. Grunert, Einfacher Beweis der von Cauchy und Euler gefundenen Sätze von Figurennetzen und Polyedren, J. für die reine und Angew. Math. Berlin **2**, 367 (1827)
136. J. Hadamard, Notions élémentaires sur la géométrie de situation, Ann. de math., 4e série, (1909) t. 9, pp. 193–235
137. J. Hadamard, The psychology of invention in the mathematical field, 1st edn. (Princeton, University Press, 1954)
138. D. Hilbert, S. et Cohn-Vossen, Anschauliche Geometrie, *English translation: Geometry and the Imagination* (Chelsea, New York, 1932). 1952
139. P. Hilton, J. Pedersen, The Euler Characteristic and Pólya's Dream. Am. Math. Mon. **103**(2), 121–131 (1996)
140. H. Hopf, *Differential Geometry in the Large-1956 Lectures Notes*, vol. 1000, Lectures Notes in Mathematics (Springer-Verlag, Heidelberg-Berlin, 1983)
141. C. Houzel, Riemann's Memoir *Über das Verschwinden der ϑ-Functionen*, this volume, pp. 125–133
142. C. Houzel, The Work of Niels Henrik Abel, In: The legacy of Niels Henrik Abel–The Abel Bicentennial, ed. by O.A. Laudal, R. Piene, (Oslo 2002, Pringer Verlag, 2004)
143. C. Huffman, *Archytas of Tarentum: Pythagorean* (Philosopher and Mathematician King. Cambridge University Press, Cambridge, 2005)
144. Ch. Huygens, *Œuvres complètes*, publiées par la Société hollandaise des sciences, Martinus Nijhoff, La Haye, 1888–1950
145. C.-G.-J. Jacobi. De binis quibus libet functionibus homogeneis secundi ordinis per substitutiones lineares in alias binas transformandis, quae solis quadratis variabilum constant; una cum variis theorematis de transformatione et determinatione integrlium multiplicium. Journal für die reine und angewandte Mathematik. Journal de Crelle. Berlin, **12** (1834), 1–69
146. C.G.J. Jacobi, Gesammelte Werke, ed. by C.W. Borchardt, Herausgegeben auf Veranlassung der königlich preussischen Akademie der Wissenschaften, 8 volumes, 1881–1891. New edition, Cambridge Library Collection
147. P. Jehel, Une lecture moderne d'un mémoire d'Euler: les Éclaircissemens plus détaillés sur la génération et la propagation du son et sur la formation de l'écho. (A modern reading of a memoir of Euler: Detailed clarifications concerning the generation and propagation of sound and the formation of echoes). In: Leonhard Euler, Mathématicien, musicien et théoricien de la musique (X. Hascher and A. Papadopoulos, ed.), 275–300. Collection Sci. Musique Ser. Études, CNRS Éditions, Paris, 2015
148. E.-J.-Ph. Jonquières (Fauque de), Sur un point fondamental de la théorie des polyèdres, Comptes rendus des séances de l'Académie des sciences de Paris, **110**, 1890, 110–115
149. E.-J.-Ph. Jonquières (Fauque de), Note sur le théorème d'Euler dans la théorie des polyèdres, Comptes rendus des séances de l'Académie des sciences de Paris, **110**, 1890, 169–173
150. E.-J.-Ph. Jonquières (Fauque de), Note sur un Mémoire de Descartes longtemps inédit, et sur les titres de son auteur à la priorité d'une découverte dans la théorie des polyèdres ; Comptes rendus des séances de l'Académie des sciences de Paris, **110**, 1890, 261–266
151. E.-J.-Ph. Jonquières (Fauque de), Écrit posthume de Descartes sur les polyèdres, Comptes rendus des séances de l'Académie des sciences de Paris, **110**, 1890, 315–317
152. E.-J.-Ph. Jonquières (Fauque de), Note sur un Mémoire présenté, qui contient, avec le texte complet et revu de l'écrit posthume de Descartes: De solidorum elementis, la traduction et le commentaire de cet Ouvrage, Comptes rendus des séances de l'Académie des sciences de Paris, **110**, 1890, 677–680
153. C. Jordan, Sur la déformation des surfaces, Journal de mathématiques pures et appliquées, ser. 2, t. XI (1866)
154. C. Jordan, Résumé de recherches sur la symétrie des polyèdres non eulériens. J. für die reine und Angew. Math. (Journal de Crelle), **66**, 86–91 (1866)

155. C. Jordan, Recherches sur les polyèdres. J. für die reine und Angew. Math. (Journal de Crelle), **66**, 22–85 (1866)
156. C. Jordan, Recherches sur les polyèdres. Comptes Rendus des Séances de l'Académie des Sciences. Paris **62**, 1339–1341 (1866)
157. C. Jordan, Recherches sur les polyèdres (second Mémoire). J. für die reine und Angew. Math. (Journal de Crelle), **68**, 297–349 (1868)
158. C. Jordan, Note sur la symétrie inverse des polyèdres non eulériens, J. für die reine und Angew. Math. (Journal de Crelle), **68**, 350–353 (1868)
159. I. Kant, *De Mundi Sensibilis atque Intelligibilis Forma et Principiis (Inaugural dissertation, On the form and principles of the sensible and the intelligible world), 1770, translated by William J* (Eckoff, New York, Columbia College, 1894)
160. A.B. Kempe, On a General Method of describing Plane Curves of the nth degree by linkwork. Proc. London Math. Soc. **7**, 213–216 (1876)
161. F. Klein, *Über Riemanns Theorie der algebraischen Funktionen und ihrer Integrale*, Teubner, Leipzig, 1882. English translation : *On Riemann's theory of algebraic functions and their integrals; a supplement to the usual treatises*. Translated from the German by Frances Hardcastle, Macmillan and Bowes, Cambridge, 1893. Reprint : New York, Dover Publications, 1963
162. F. Klein, Vorlesungen über die Entwicklung der Mathematik im 19. 2 volumes, Julius Springer, Berlin, 1926. English translation by M. Ackermann: Development of mathematics in the 19th century, Volume IX of the Series Lie Groups: History, Frontiers and Applications, ed. by R. Hermann, (Math. Sci. Press, Brookline, Brookline, MA, 1979)
163. A.W. Knapp, André Weil: a prologue. Notices. AMS **46**(4), 434–439 (1999)
164. J.-L. de Lagrange Théorie des fonctions analytiques contenant les principes du calcul différentiel dégagés de toute considération d'infiniment petits ou d'évanouissans de limites ou de fluxions, Paris, Imprimerie de la République, 1797
165. J.-L. de Lagrange, Sur la construction des cartes géographiques, Nouveaux mémoires de l'Académie Royale des Sciences et Belles-lettres de Berlin, année 1779, Premier mémoire, *Œuvres complètes*, tome 4, 637-664. Second mémoire *Œuvres complètes*, tome 4, 664-692
166. J.-L. Lagrange, Œuvres, 14 volumes, published by J.-A. Serret (t. I–X and XIII) and G. Darboux, Paris, Gauthier-Villars, 1867–1892
167. J.-L. de Lagrange, J. Le Rond d'Alembert, Correspondance inédite de Lagrange et d'Alembert, *Œuvres de Lagrange*, publiées par J.-A. Serret, t. 13, Gauthier-Villars, Paris, 1882
168. H. Lebesgue, Remarques sur les deux premières démonstrations du théorème d'Euler relatif aux polyèdres. Bulletin de la Société mathématique de France, tome **52**, 315–336 (1924)
169. H. Lebesgue, *Leçons sur l'intégration et la recherche des fonctions primitives* (Gauthier-Villars, Paris, 1904)
170. A.-M. Legendre, Essai sur la théorie des nombres, Duprat, Paris, 1798
171. A.-M. Legendre, Exercices de calcul intégral sur divers ordres de transcendantes et sur les quadratures, 3 volumes, Paris, Courcier, 1811–1816
172. A.-M. Legendre, Traité des fonctions elliptiques et des intégrales eulériennes, avec des tables pour en faciliter le calcul numérique, 3 volumes, Paris, Huzard-Courcier, 1825–1828
173. A.-M. Legendre, Éléments de géométrie, avec des notes, Paris, Firmin Didot, 1794
174. G.W. Leibniz, *Mathematische Schriften, 1 Abt*, vol. II (C. I. Gerhardt, Berlin, 1850)
175. G.W. Leibniz, La caractéristique géométrique, text edited and annotated by J. Acheverría, translation, notes and psotface by M. Parmentier, Coll. Mathesis, Vrin, Paris, 1995
176. G.W. Leibniz, Der Briefwechsel von Gottfried Wilhelm Leibniz mit Mathematikern. Erster Band. Hrsg. von C. I. Gerhardt. Mit Unterstützung der Königl. Preussischen Akademie der Wissenschaften (Mayer & Müller, Berlin, 1899)
177. S.-A.-J. Lhuillier, Démonstrations diverses du théorème d'Euler. Annales de Gergonne **3**, 169–189 (1812)
178. G. Loria, Commemorazione del compianto Socio Prof. Placido Tardy, rend. Acad. Lincei (V) 24 (1° semestre) 1915, pp. 521–521
179. C. MacLaurin, A treatise on fluxions, two volumes London, Baynes, 1801

180. P. Mengoli, Novae quadrature arithmeticae, seu de additione fractionum, Bononiæ, ex Typografia Iacobi Montij, 1650
181. J.-B. Meusnier, Mémoire sur la courbure des surfaces, Mém. Div. savants **10**, 1785, pp. 477–510
182. A. de Moivre, Miscellanea analytica de seriebus et quadraturis. London, J. Tonson & J. Watts, 1730
183. G.H. Moore, The evolution of the concept of homeomorphism. Historia Mathematica **34**(3), 333–343 (2007)
184. G. Monge, Mémoire sur les développées, les rayons de courbure, et les différens genres d'Inflexions des courbes a double courbure, Paris, Imprimerie Royale, 1785
185. G. Monge, Mémoire sur les propriétés de plusieurs genres de surfaces courbes et particulièrement sur celles des surfaces développables avec une application à la théorie générale des ombres et des pénombres, Mém. Div. savants **9**, 382–440 (1780)
186. C. Neumann, Vorlesungen über Riemann's Theorie der Abel'schen Integrale, Leipzig, Teubner, 1865, 2nd revised edn. (1884)
187. I. Newton, Sir Isaac's Newton's mathematical principles of natural philosophy and his system of the world. Translated by A. Motte (1729), Revision by F. Cajori, (Cambridge University Press, 1934)
188. A.B. Niewenglowski, *Exposition de la méthode de Riemann pour la détermination des surfaces minima de contour donné* (Thesis submitted to the Faculté des Sciences, Paris, Gauthier-Villars, 1880)
189. A. Papadopoulos, Euler et les débuts de la topologie. In: Leonhard Euler : Mathématicien, physicien et théoricien de la musique, ed. by X. Hascher, A. Papadopoulos (CNRS Editions, 2015), pp. 321–347
190. A. Papadopoulos, Physics in Riemann's mathematical papers, this volume, pp. 151–199
191. A. Papadopoulos, Cauchy and Puiseux: Two precursors of Riemann, this volume, Chapter 7, pp. 209–232
192. A. Papadopoulos, Riemann surfaces: reception by the French school, this volume, pp. 237–286
193. A. Papadopoulos, Euler and Chebyshev: from the sphere to the plane and backwards. Proc. Cybern. (A volume dedicated to the jubilee of Academician Vladimir Betelin) **2**, 55–69 (2016)
194. A. Papadopoulos, Quasiconformal mappings, from Ptolemy's geography to the work of Teichmüller, to appear, in *Uniformization, Riemann-Hilbert Correspondence, Calabi-Yau Manifolds, and Picard-Fuchs Equations*, ed. by L. Ji, S.-T. Yau (International Press and Higher Education Press, 2017), 65 pp
195. J.F. Pfaff, Disquisitiones analyticae maxime ad calculum integralem et doctrinam serierum pertinentes, Helmstadii, leckeisen, 1797
196. É. Picard, L'œuvre mathematique de É. Galois, Revue générale des sciences pures et appliquées, t. VIII (1897), pp. 339–340
197. É. Picard, Traité d'Analyse, 3 volumes. Paris, Gauthier-Villars, (1891, 1893, 1896)
198. É. Picard, *Sur le Développement de l'Analyse et ses Rapports avec diverses Sciences: Conférences faites en Amérique* (Gauthier-Villars, Paris, 1905)
199. H. Poincaré, Sur la généralisation d'un théorème d'Euler relatif aux polyèdres, *Comptes rendus des séances de l'Académie des sciences de Paris* **117**, 144–145 (1893); *Œuvres*, vol. XI, pp. 6–7
200. H. Poincaré, Analysis situs. J. de l'École Polytech. **1**(2), 1–121 (1895); *Œuvres*, vol. VI, pp. 193–288
201. H. Poincaré, Sciences et méthode, Paris, Flammarion, 1908, several later editions
202. H. Poincaré, Analyse des travaux scientifiques de Henri Poincaré faite par lui-même. Acta Mathematica **38**, 36–135 (1921)
203. L. Poinsot, Note sur la théorie des polyèdres. Comptes rendus des séances de l'Académie des sciences de Paris **46**, 65–79 (1858)
204. L. Poinsot, Mémoire sur les polygones et les polyèdres. J. de l'École Polytech., Paris **4**, 16–49 (1810)

205. S.-D. Poisson, Mémoire sur la courbure des surfaces. J. Reine Angew. Math. **8**, 280–297 (1832)
206. J.-C. Pont, *La topologie algébrique des origines à Poincaré* (Presses Universitaires de France, Paris, 1974)
207. C.A. Possé, Excerpts of a biography of Chebyshev, contained in his *Collected Works*, Edited by A. Markoff and N. Sonin, Vol. II, pp. I–VI
208. P.-M.-E. Prouhet, Sur le nombre des diagonales d'un polyèdre. Nouvelles Annales mathématiques, Paris **2**, 77–79 (1863)
209. P.-M.-E. Prouhet, Remarques sur un passage des Œuvres inédites de Descartes. Comptes rendus des séances de l'Académie des sciences de Paris **50**, 779–781 (1860)
210. R. Rashed, Histoire de l'analyse diophantienne classique : D'Abū Kāmil à Fermat, de Gruyter 349 pp (2013)
211. R. Rashed, Les mathématiques infinitésimales du IXème au XIème siècle. Vol. IV. Ibn al-Haytham. Méthodes géométriques, transformations ponctuelles et philosophie des mathématiques. Al-Furqān Islamic Heritage Foundation Publication. 60. London: Al-Furqān Islamic Heritage Foundation, 1064 pp (2002)
212. R. Rashed, A. Papadopoulos, *Menelaus' Spherics*, Critical edition with historical and mathematical commentaries, (Scientia Graeco-Arabica 21, de gruyter, 2017)
213. R. Remmert, *Theory of complex variables* (Springer, New York, English edition, 1989)
214. B. Riemann, Grundlagen für eine allgemeine Theorie der Functionen einer veränderlichen complexen Grösse, (Göttingen, 1851), [227] pp. 3–48
215. B. Riemann, Über die Darstellbarkeit einer Function durch eine trigonometrische Reihe. Aus dem dreizehnten Bande der Abhandlungen der Königlichen Gesellschaft der Wissenschaften zu Göttingen, 1867
216. B. Riemann, Commentatio mathematica, qua respondere tentatur quaestioni ab Illma Academia Parisiensi propositae: "Trouver quel doit être l'état calorifique d'un corps solide homogène indéfeni pour qu'un système de courbes isothermes, à un instant donné, restent isothermes après un temps quelconque, de telle sorte que la température d'un point puisse s'exprimer en fonction du temps et de deux autres variables indépendantes, in *Bernhard Riemann's Gesammelte Mathematische Werke [227]*, 2nd edn (Teubner, 1892), pp. 391–404
217. B. Riemann, Theorie der Abel'schen Functionen. J. für die Reine und Angew. Math. **54**, 115–155 (1857). Reprinted in his *Gesammelte mathematische Werke* [227], pp. 88–144
218. B. Riemann, Lehrsätze aus der analysis situs für die Theorie der Integrale von zweigliedrigen vollständigen Differentialien. für die Reine und Angew. Math. **54**, 105–110 (1857)
219. B. Riemann, Über die Anzahl der Primzahlen unter einer gegebenen Grösse, Monatsberichte der Berliner Akademie, November 1859, pp. 671–680, *Gesammelte mathematische Werke* [221], pp. 145–153
220. B. Riemann, Über die Fläche vom kleinsten Inhalt bei gegebener Begrenzung, Abh. Königl. d. Wiss. Göttingen, Mathem. Cl., vol. 13 (1867), pp. 3–52
221. B. Riemann, Ein Beitrag zu den Untersuchungen über die Bewegung eines flüssigen gleichartigen Ellipsoides. Abhandlungen der Königlichen Gesellschaft der Wissenschaften zu Göttingen **9**, 3–36 (1861)
222. B. Riemann, Beiträge zur Theorie der durch die Gauss'sche Reihe $F(\alpha, \beta, \gamma, x)$ darstellbaren Functionen. Aus dem siebenten Band der Abhandlungen der Königlichen Gesellschaft der Wissenschaften zu Göttingen (1857), pp. 3–32
223. B. Riemann, Ein Beitrag zur Elektrodynamik, Ann. Phys. **131**, 237–243 (1867). English translation: A contribution to electrodynamics, Phil. Mag., Ser. **4**, 35, 368–372 (1867)
224. B. Riemann, Über das Verschwinden der ϑ-Functionen. J. für die R. und a. Math. **65**, 161–172 (1866). Ges. Math. Werke, pp. 212–224
225. B. Riemann, Gesammelte mathematische Werke, wissenschaftlicher Nachlass und Nachträge. Nach der Ausgabe von Heinrich Weber und Richard Dedekind neu herausgegeben von Raghavan Narasimhan. (Collected mathematical works, scientific posthumous works and supplements. According to the edition by H. Weber and R. Dedekind newly edited by R. Narasimhan). Teubner Verlagsgesellschaft, Leipzig, 1862 (Springer, Berlin, 1990)

226. B. Riemann, Gleichgewicht der Electricität auf Cylindern mit kreisförmigen Querschnitt und parallelen Axen, Conforme Abbildung von durch Kreise bergrenzten Figuren, (Nachlass XXVI) 1857, in *Ges. math. Werke*, pp. 472–476
227. B. Riemann, K. Hattendorff, *Schwere, Elektrizität und Magnetismus, Hannover, Carl Rümpler, 1876* (Nachdruck VDM, Müller, Saarbrücken, 2006)
228. B. Riemann, *Collected papers, English translation* by R. Baker, Ch. Christenson and H. Orde, Kendrick Press, Heber City, UT, 2004
229. B. Riemann, Collected works translated into French: Œuvres mathématiques de Riemann, traduites par L. Laugel, avec une préface de C. Hermite, Paris, Gauthier-Villars, 1898
230. B. Riemann, Über die Hypothesen, welche der Geometrie zu Grunde liegen, published by R. Dedekind, after Riemann's death, in Abhandlungen der Königlichen Gesellschaft der Wissenschaften zu Göttingen, Vol. 13, 1867. French translation in [231] pp. 280–299
231. B. Riemann, On the Hypotheses which lie at the Bases of Geometry. Translation of [232] by W.K. Clifford, Nature **VIII**(183, 184), 14–17, 36, 37
232. B. Riemann, H. Weber, *Die partiellen Differential-Gleichungen der mathematischen Physik nach Riemanns Vorlesungen, 2 vols* (Vieweg, Braunschweig, 1912)
233. The Riemann letters at the Prussian cultural archive, translated by R. Gallagher and M. Weissbach, mimeographed notes, 1981
234. V. de Risi, Geometry and Monadology: Leibniz's Analysis Situs and Philosophy of Space, Birkhäuser, 2007
235. O. Rodrigues, Recherches sur la théorie des lignes et des rayons de courbure des surfaces, et sur la transformation d'une classe d'intégrales doubles qui ont un rapport direct avec les formules de cette théorie. Correspondance de l'Ecole Polytechnique **3**, 162–183 (1815)
236. B. Russell, *Principles of Mathematics* (Cambridge, 1903)
237. M. Schmitz, The life of Gotthold Ferdinand Eisenstein, Res. Lett. Inf. Math. Sci. 6–13 (2004)
238. A. Sossinsky, Configuration spaces of planar linkages, in *Handbook of Teichmüller theory*, ed. by A. Papadopoulos, Vol. VI (European Mathematical Society, Zurich, 2016), pp. 335–373
239. J. Steiner, Leichter Beweis eines stereometrischen Satzes von Euler nebst einem Zusatz X, S. 48, tome 1. J. für die Reine und Angew. Math. 364–367 (1826)
240. J. Tannery, Pensées, ed. É. Boutoux and É. Borel, Revue du mois, 10 mars 1911 and 10 avril 1911. Re-edition: En souvenir de Jules Tannery, brochure, Imprimerie Crété, Corbeil, 1912
241. P. Tannery, Le concept scientifique du continu: Zénon d'Elée et Georg Cantor, Revue philosophique de la France et de l'Étranger, Xe année, t. XX (1885), pp. 385–410
242. B. Taylor, De motu nervi tensi, Philos. Trans. R. Soc. Lond. **28** (1713, published in 1714), 26–32
243. B. Taylor, Methodus incrementorum directa et inversa, Impensis Gulielmi Innys, London, 1715; 2nd edn, 1717
244. J.-M. Thiel, Démostration nouvelle du théorème d'Euler pour des polyèdres convexes, Nieuw Archief voor wiskunde uitgegeven door bet Wiskundig Genootschap. Amsterdam **19**, 98–99 (1892)
245. R. Thiele, The Rise of the Function Concept in Analysis, In *Euler Reconsidered: Tercentenary Essays*, Kendrick Press, Heber City, UT, pp. 422–461
246. R. Thom, Les intuitions topologiques primordiales de l'aristotélisme, *Revue thomiste*, juillet-septembre 1988, XCVI^{e} année **88**(3), 393–409 (1988)
247. R. Thom, Aristote topologue. Revue de synthèse **120**(1), 39–47 (1999)
248. W.P. Thurston, *The Geometry and Topology of Three-Manifolds* (Princeton University, Princeton, 1976)
249. J.-P.-F. Valat, Nouvelles remarques sur l'interprétation d'un théorème de Descartes. Comptes rendus des séances de l'Académie des sciences de Paris **51**, 1031–1033 (1860)
250. V. Volterra, Betti, Brioschi, Casorati : Trois analystes italiens et trois manières d'envisager les questions d'analyse, Comptes Rendus du deuxième congrès international des mathématiciens tenu à Paris. Gauthier-Villars 1902, 43–57 (1900)
251. J. Wallis, Arithemetica infinitorum, 1656. *Opera Mathematica*, vol. 1 (Oxford, 1695)

252. J. Wallis, A treatise on algebra, both historical and practical, shewing, the original, progress, and advancement thereof, from time to time, and by what steps it hath attained to the heighth at which now it is. With some additional treatises, I. Of the cono-cuneus; being a body representing in part a conus, in part a cuneus. II. Of angular sections; and other things relating there-unto, and to trigonometry. III. Of the angle of contact; with other things appertaining to the composition of magnitudes, the inceptives of magnitudes, and the composition of motions, with the Results thereof. IV. Of combinations, alternations, and aliquot parts.London, John Playford for Richard Davis, 1685
253. A. Weil, *Essais historiques sur la théorie des nombres, Monographie NJ. 22 de L'Enseignement Mathématique* (Université de Genève, Genève, Imprimerie Kundig, 1975)
254. A. Weil, *History of Mathematics: Why and How* (Proceedings of the International Congress of Mathematicians, Helsinki, 1978)
255. A. Weil, *Number Theory: An Approach Through History from Hammurapi to Legendre*, 1st edn (Birkhäuser, Boston-Basel-Berlin, 1984)
256. A. Weil, Prehistory of the zeta function, in *Number theory, trace formulas and discrete groups: Symposium in Honor of Atle Selberg, Oslo, Norway, July 14–21, 1987*, ed. by K.E. Aubert, E. Bombieri, D. Goldfeld (Acad. Press, Boston, 1989), pp. 1–9
257. A. Weil, Riemann. Betti and the birth of topology, Archive for History of Exact Sciences **20**(2), 91–96 (1979)
258. H. Weyl, Philosophy of Mathematics and Natural Science, Princeton University Press, New edition, 2009, With a new introduction by Frank Wilczek. Translation of *Philosophie der Mathematik and Naturwissenschaft*, Munich, R. Oldenburg, 1927
259. S. Yamada, *Riemann on Minimal Surfaces*, this volume, Chapter 5, pp. 135–150
260. A.P. Youschkevitch, The concept of function up to the middle of the 19th century. Arch. Hist. Exact Sci. **16**, 37–85 (1976)
261. A.P. Youschkevitch, R. Taton, Introduction to Volume V of Series IV A of Euler's *Opera Omnia* (The correspondence of Euler with Clairaut, d'Alembert and Lagrange)

LEONHARD EULER
J. Darbes pinxit.
J. Kütner sc. Mitau. 1780.

Part I
Mathematics and Physics

Riemann on Geometry, Physics, and Philosophy—Some Remarks

Jeremy Gray

Abstract Riemann's paper 'On the hypotheses that lie at the foundations of geometry' is one of the fundamental papers in the creation of modern geometry. We analyse its content, look at the influence the work of Gauss and Herbart exercised on Riemann, and discuss other of Riemann's papers that shed light on his ideas, in particular on his appreciation of the concept of curvature.

1 Introduction

Riemann's paper 'On the hypotheses that lie at the foundations of geometry' (henceforth, *Hypotheses*) [24] is generally regarded as one of the most important papers ever written in mathematics. As such, it was read by generations of mathematicians, most notably in the Göttingen tradition that reached from him to Hermann Weyl, and its ideas continue to influence mathematics today. Without it, Einstein's general theory of relativity would have been unthinkable.

Unsurprisingly, therefore, it has been worked over by historians of mathematics, historically-minded mathematicians, and philosophers of mathematics looking for its key ideas and a historical and intellectual context into which to put them. The results are intriguingly meagre. The *Hypotheses* is not the last step in a complicated chain of arguments involving Riemann with numerous predecessors, nor is it the response to a perceived crisis. Rather, it is, as it is presented, the next step after the work of Gauss [12] and, partly, as a response to shifting philosophical ideas about the nature of geometry that may have also caught Riemann's attention because of their implications for physics.

J. Gray (✉)
Mathematics Institute, Open University and University of Warwick, Zeeman Building, Coventry CV4 7AL, UK
e-mail: j.j.gray@open.ac.uk

L. Ji et al. (eds.), *From Riemann to Differential Geometry and Relativity*,
DOI 10.1007/978-3-319-60039-0_2

This paper will first look at the *Hypotheses* in detail, and then consider its relation to Gauss's ideas about differential geometry. Then it will turn to the connections to physics and philosophy, and conclude by looking at the other relevant paper Riemann wrote, the *Commentatio* [27] and the the discussions that it provoked.

2 The *Hypotheses*

Riemann's essay [24] was published posthumously in 1867, and is based on the lecture he gave in 1854 as part of the process of obtaining his Habilitation, the necessary and sufficient qualification for obtaining a teaching position at a German university. As such it was given to the Philosophy Faculty at Göttingen, of which Mathematics was a Department, with Gauss as one of the examiners. These circumstances explain the unfortunate absence of formulae that would otherwise have assisted subsequent readers.

Riemann began by remarking that geometry "takes for granted the notion of space" as well as the first principles of constructions in space. The basic concepts have only nominal definitions and the crucial properties are determined from axioms, but this leaves the relationships between the axioms obscure; in fact, said Riemann, it is not even clear a priori if the relationships are possible.

This opening paragraph makes two points. First, that it is not clear what the axioms or postulates of Euclidean geometry are about; second, that it is not even clear that they are consistent.

But Riemann did not take his analysis in the direction of a more refined study of the axioms. It is likely that he saw the work of Legendre, whose name he mentions, as indicative of the poverty of such work, not only because in all the editions of his *Géométrie* [16] Legendre had failed to prove the parallel postulate, but because Riemann thought that the whole axiomatic attempt to give a geometrical account of physical space was misguided. In unpublished notes from the early 1850 s he called such enquiries "extremely unfruitful" (see Scholz ([30] p. 218)), which, as Scholz points out, makes it very unlikely that Riemann had seen any of Lobachevskii's work.

Instead, Riemann began his paper [24] by remarking that the general notion of quantity was multi-dimensional, and "it emerges from this study that a multi-dimensional object is capable of being measured in different ways and that space is only a particular example of a three-dimensional quantity." Moreover, the properties that characterise space among all three-dimensional quantities can only be determined experimentally. From this perspective the axioms of Euclid are only hypotheses, although highly probable outside "the realms of both the immeasurably great and the immeasurably small."

This opening page is one of the first places where a characteristically modern mathematical approach is taken to what many had seen as a straight-forward question. Riemann did not say that Euclidean geometry needs fixing and offer a proposal. He stepped back and asked himself: what are we studying when we study geometry? His answer was quantity, and for him that was a multi-dimensional object—the sort of thing that is described, as we shall see, with coordinates. In Riemann's opinion a geometer, even a practical one, should not concentrate on the passage from space to a mathematical theory of space (still less take one for granted) but first build up a theory of multi-dimensional quantities, and this he turned to do.

He began Part I of the paper by regretting that apart from some ideas of Gauss [14] published in his second memoir on biquadratic residues and some philosophical remarks of Herbart [15] there was little to guide him. He indicated that there are continuous and discrete manifolds depending on how the elements are determined, and here a manifold is a vague term, little more than a collection of elements. Discrete manifolds in this sense are frequently encountered, said Riemann, but continuous ones less so, and he gave examples of the location of material objects and their colours. To make sense of continuous manifolds we rely on measurement, which presupposes a measuring unit that can be freely transported. When this is not available we have only the general concept of a manifold, and this difficulty may be why the work of Lagrange, Pfaff, and Jacobi on many-valued analytic functions has been so unfruitful so far.

Riemann had already rewritten the theory of functions of a complex variable in his doctoral theses [23], published in 1851. It is likely that his ideas about surfaces were what he was alluding to here, although the great work on Abelian functions would not be published until 1857.

How then to determine position in a manifold? Riemann explained that if the manifold has dimension one then position is determined by moving forwards and backwards using some unspecified concept of length. If this one-dimensional manifold is itself then moved forwards and backwards in a different dimension then a two-dimensional manifold is obtained, and so on. The converse also holds: one can break an n-dimensional manifold down into smaller ones along which some function has a constant value, and exceptional cases aside these sets where the function takes a fixed value are $n-1$-dimensional submanifolds.

In Part II of his paper [24] Riemann explained how to introduce metrical relations in a manifold on the assumption that lines have a length independent of their position and every line can be measured by every other line. Here he was happy to acknowledge the work of Gauss [12] on curved surfaces (which we shall look at below).

Riemann now supposed that the position of a point in perhaps some region of an n-manifold is determined by its n coordinates $(x_1, x_2, \ldots, x_n)$. He restricted his attention to continuous systems in which the coordinates can vary by amounts dx and sought an expression for the line element ds in terms of the $dx_1, dx_2, \ldots, dx_n$. He further assumed that the length of a line element is unaltered if all its points undergo the same infinitesimal displacement. If moreover distance increases as points move away from the origin and the first and second derivatives are finite then the first derivative must vanish and the second cannot be negative, so Riemann took it to be positive. He deduced that the line element ds could be "the square root of an everywhere positive quadratic form in the variables dx", as for example we take to be the case for space when we write

$$ds = \sqrt{\sum (dx)^2}.$$

Riemann noted that there are other possibilities. For example, ds could be the fourth root of a fourth power expression, but he did not see many possibilities for geometry there and he set this aside.

The quadratic form, however, did interest him. It contains $n(n+1)/2$ coefficients, of which n can be altered by a change of variables, so it depends essentially on $n(n-1)/2$ coefficients that are determined by the manifold. The example of $\sqrt{\sum (dx)^2}$ is therefore special, and Riemann proposed to call such manifolds flat.

To proceed further, Riemann considered the infinitesimal triangle with one vertex at the origin, one on a geodesic out of the origin to the point $(x_1, x_2, \ldots, x_n)$, and one on a geodesic out of the origin to the point $(dx_1, dx_2, \ldots, dx_n)$. The quotient of $\sqrt{\sum (dx)^2}$ by the area of this triangle measures the departure of this infinitesimal region from flatness, and divided by $-3/4$ is in fact the Gaussian curvature of the surface. So the curvature of an n-manifold can be understood by knowing $n(n-1)/2$ surface curvatures (the sectional curvatures, as we would say).

This led Riemann to explain the difference between intrinsic and extrinsic properties of a surface. He explained that for a sphere, which has an intrinsic geometry different from a plane, the Gaussian curvature multiplied by the area of an infinitesimal geodesic triangle is half the excess of the sum of its angles over π. This allowed him to express the belief that the geometry of an n-dimensional manifold could be understood by understanding its sectional curvatures.

Flat manifolds, he observed, have every sectional curvature zero. They are therefore a special case of the manifolds of constant curvature (that is, having the same sectional curvatures everywhere) and in these manifolds geometric figures can be freely moved around without stretching. To give examples of such manifolds he wrote down the metric

$$ds = \frac{1}{1 + \frac{\alpha}{4} \sum x^2} \sqrt{\sum (dx)^2},$$

where α is the curvature.

It was evident that the surfaces of constant positive curvature are spheres; the sphere of radius r has curvature r^{-2}. Riemann gave a complicated description of how to fit all the surfaces of constant curvature into one family. On this description, the cylinder is the example of a surface of zero curvature, and surfaces of constant negative curvature are locally like the saddle-shaped part of a torus.

In the third and final part of his paper [24] Riemann discussed how his ideas might apply to space. If all the sectional curvatures are zero, the space is Euclidean. But if we assume only that there is free mobility of bodies then space is described as a three-dimensional manifold of constant curvature, which can be determined from the knowledge of the sum of the angles in any triangle. Or, one could assume that length is independent of position but not direction.

As for empirical confirmation, the topological structures available to describe three-dimensional manifolds form a discrete set, so exact statements can be made about them even if one can never be certain of their truth. As for the metrical relations, however, these are necessarily inexact because every measurement is imprecise. This has implications for the immeasurably large and the immeasurably small.

The immeasurably large divided into spaces that are infinite and spaces that are merely unbounded, as for example a sphere. That said, Riemann regarded questions about the immeasurably large as irrelevant to the elucidation of natural phenomena, if only because existing astronomical measurements show that any non-zero sectional curvature of space can only be detected in regions vastly greater than the range of our telescopes. This seems to belong to a Göttingen tradition going back to Gauss and extending at least as far as Schwarzschild, who reported on the implications of measurements of the parallel of stars for the curvature of space in 1899 (see (Epple [7])).

Not so questions about the immeasurably small. Here "the concept of a rigid body, and the concept of a light ray, cease to be meaningful". But, Riemann concluded his lecture, these questions take us into physics "which the nature of today's occasion does not allow us to explore".

3 Influences

3.1 Gauss

It is possible to read Riemann's paper [24] in various ways. A modern mathematician can supply the missing details, or, at the other extreme, regard it as almost incoherent. It is possible for a mathematician to offer a comparably deep vision of new mathematics today, but it would be couched in a language of possible definitions, possible methods, and likely theorems that, conjecturally, resolve outstanding problems. Riemann's paper is more philosophical–in the good sense of challenging one to be clear about what is involved in an enquiry–and more speculative.

As Riemann made clear, among the few antecedents he could acknowledge were two papers by Gauss [11, 12]. The one on differential geometry is easy to appreciate. In the 1810s and 1820s Gauss had re-defined the subject in two memoirs. In his *Disquisitiones generales circa superficies curvas* of 1828 he introduced the concept of the intrinsic curvature of a surface. Gauss began his exposition by taking his readers through three definitions of a surface: in the first a surface is given by an expression of the form $z = f(x, y)$; in the second by an expression of the form $f(x, y, z) = 0$; and in the third in the parameterised form $(x(u, v), y(u, v), z(u, v))$. For each of these approaches he showed what the implications were calculating the curvatures of the principal curves at each point, which Euler [8] had showed are a good way to understand how curved a surface is at each point.

Gauss then introduced the map later known as the 'Gauss map'. At each point P of the surface he supposed there was a vector of unit length and normal to the surface, PP', and he considered the unit vector OQ parallel to PP' that has its base point at the centre of a fixed sphere of unit radius. The image of P' on the surface under the Gauss map is the point Q on the unit sphere.

Gauss then proved that the 'Gauss map' has a simple effect on areas: it multiplies the infinitesimal area around a point by an amount equal to the product of the principal curvatures. This product he proposed to call the curvature of the surface, and he showed that it depends only on E, F, and G and their derivatives with respect to u and v, but not on $x(u, v)$, $y(u, v)$, and $z(u, v)$. It is therefore intrinsic to the surface—a result that surprised him so much he called it the Theorema egregium or exceptional theorem.

One reason Gauss regarded the third form of presenting a surface as not only the most general but the most important, was because it allows u and v to be used as coordinates, and because it allows for a study of maps between one surface and another. In particular, given two surfaces defined by

$$\mathbf{r} = (x(u, v), y(u, v), z(u, v)) \text{ and } \mathbf{r}' = (x'(u', v'), y'(u', v'), z'(u', v'))$$

and a map between them, one can compare the line elements

$$ds^2 = E(u, v)du^2 + 2F(u, v)dudv + G(u, v)dv^2$$

and

$$ds'^2 = E'(u', v')du'^2 + 2F'(u', v')du'dv' + G'(u', v')dv'^2,$$

where $E(u, v) = \mathbf{r}_u.\mathbf{r}_u, \ldots, G'(u', v') = \mathbf{r}'_{v'}.\mathbf{r}'_{v'}$.

For example, the map is conformal or angle preserving if

$$ds^2 = \Phi(u', v')ds'^2,$$

for some function Φ, so in particular a map between a plane and a surface with $ds^2 = E(u, v)du^2 + 2F(u, v)dudv + G(u, v)dv^2$ is conformal if and only if $E = G$ and $F = 0$. Gauss [12] had made a detailed study of maps in connection with the survey of Hannover and South Denmark on which he worked in the 1820s, and had explicitly remarked that a map between planes is conformal if and only if it is given by a complex analytic function; this was only one of several occasions when he hinted at a theory of such functions that he was never to pull together and publish. It is very likely, however, that Riemann knew some of these ideas, but it is often impossible to say if he learned of them in discussions with Gauss or only by reading Gauss's papers [11, 12] after Gauss died. However, he wrote explicitly in his doctoral paper (1851) that the conformal nature of a complex analytic map was something that he learned from Gauss's paper [12] on conformal maps, and he stressed the importance of this geometrical aspect of the maps.

So Riemann took two ideas from Gauss's work on geometry. The conformal nature of a complex analytic map (away from any branch points) surely suggested to Riemann that there was significant geometrical features of a surface as early as 1851. But the idea of the intrinsic curvature of a surface was one Riemann took far beyond what Gauss had done with it.

Gauss [12] had identified the intrinsic feature of the geometry of a surface in $\mathbb{R}^3$, but he continued to think of surfaces as lying in $\mathbb{R}^3$. The idea that a region—a surface—with two coordinates u and v and a metric

$$ds^2 = E(u, v)du^2 + 2F(u, v)dudv + G(u, v)dv^2$$

is a fit subject for geometry already, whether or not there an embedding of it in space given by functions $x(u, v)$, $y(u, v)$, and $z(u, v)$ is due entirely to Riemann. It is almost certainly what Gauss had in mind when he said to his friend Wilhelm Weber after listening to Riemann's lecture that the profundity of the ideas that Riemann had put forward had greatly astonished him (see (Dedekind [6] 581)).

But, as Riemann's paltry citations indicate, there was very little done to extend Gauss's ideas of the intrinsic geometry of surfaces in the three decades that separate Gauss's memoir [12] from Riemann's lecture. One of the few papers written on the subject was by H.F. Minding [21], who investigated surfaces of constant negative curvature in his (1839). Bonnet, and Liouville in (Monge [22]), brought Gauss's theory to France, but Riemann went only once to Paris, in 1860, and it is not clear what he knew of French work before.

Riemann had also mentioned Gauss's second memoir on biquadratic residues (Gauss [14]). This is the work in which Gauss introduced what are today called the Gaussian integers, and explained at some length what he had been less overt in his *Disquisitiones Arithmeticae*, that the complex numbers can be thought of as points in a plane. Here (see § 16) he stressed the highly intuitive character of this representation, and also–which is what Riemann surely picked up on–that this illuminated the true metaphysics of imaginary quantities.[1]

Gauss went on (§20) to stress that one goes beyond the positive numbers only when what is counted has an opposite (a negative) and what is then counted is not a substance (an object thinkable in itself) but a relation between two objects. More generally one creates new objects when one has a relation that admits a concept of opposite. Then (§22) "The mathematician abstracts totally from the nature of the objects and the content of their relations; he is concerned solely with the counting and the comparison of the relations among themselves." Nonetheless, intuitive representations are helpful and once an intuitive meaning for $\sqrt{-1}$ is completely established "one needs nothing further to admit this quantity into the domain of objects of arithmetic."

By 1850 Riemann did not need to be told that complex numbers were admitted into mathematics. Like Gauss, and Cauchy, he knew that the problem was not with complex numbers but with how to define complex functions. But he may well have

[1]See also the English translation, Ewald ([9], I, 312–313).

appreciated the metaphysics, the abstract character of mathematical objects and their relations, and the connection to intuition that, as we shall now see, was also a theme of Herbart's philosophy [15] and Riemann's physics.[2]

3.2 Herbart

The emphasis Riemann placed on Herbart's ideas came from Riemann's interest in philosophy. Herbart had been a philosopher at Göttingen from 1805 until his death in 1841, and his main book, the *Psychology as science newly founded on experience, metaphysics and mathematics* [15], appealed strongly to Riemann. But Riemann was also critical: he wrote in the philosophical passages in his collected works (1990, 539) that he could agree with almost all of Herbart's earliest research, but could not agree with his later speculations at certain essential points to do with his Naturphilosophie and psychology. He also identified himself as a Herbartian in psychology and epistemology, but not in ontology and synechology (a discipline concerned with space, time, and motion, and in particular with intelligible space, regarded as a mental construct that makes the explanation of matter possible).

In Riemann's view, natural science is the attempt to comprehend nature by precise concepts, and if concepts yield inaccurate predictions then the concepts must be modified. As a result, the more of nature we understand the more it sinks below the surface of phenomena. Riemann approved of Herbart's anti-Kantian epistemology, because Herbart [15] had argued that all our concepts arise by modifying earlier ones, and the most primitive concepts originate from attempts to understand what our senses tell us, which is why we have the possibility of forming concepts adequate for natural science. In particular they need not be a priori, as the Kantian ones are.

Herbart was a powerful source for the idea of varying quantities–ultimately, manifolds–although Herbart remained fixed on the idea that geometry was necessarily three-dimensional. But Riemann aimed at constructing coherent systems of concepts that could then be matched against the coherence of the natural world. He did not agree with Herbart's account of how our ideas of space are generated from experience, and went directly to systems of mathematical concepts. The elucidation of fundamental concepts was characteristic of Riemann's work, and it was an approach he shared with Herbart even when he did not use the same concepts himself.

3.3 Physics: Newton and Euler [8]

In a note Riemann [28] made on his work (*Werke*, 539) he wrote

> My main work consists in a new formulation of the known natural laws – expressing them in terms of other fundamental ideas – so as to make possible the use of experimental data

[2]This account draws on Bottazzini and Tazzioli (1995 [2]) and Scholz ([31] 1982b).

> on the exchanges between heat, light, magnetism, and electricity. In researching their interrelationship, I have been guided principally by the study of the work of Newton, Euler and – on the other side – Herbart.

This is a striking assessment; Riemann belongs to a list of brilliant mathematicians whose lasting contributions are more in mathematics than physics, contrary to their hope.

Riemann had no sympathy for action at a distance, and Dedekind [6] in his *Life of Riemann* (Riemann *Werke* [28]) tells us that Riemann was very pleased to discover from Brewster's biography of Newton that Newton too disliked the idea. Instead, Riemann imagined space filled with an ether, whose properties were responsible for the transmission of force and other physical quantities from place to place, and he hoped to unify in this way the theories of gravitation, electromagnetism, heat, and light.

Riemann imagined a substance that flowed between and through atoms, being created in some and vanishing in others. A point-particle is surrounded in this model by something like an elastic medium or ether that is described by a system of curvilinear coordinates centred at the point and varying in time. Deformations in the medium are captured by the equivalent of the strain tensor in elasticity theory, and variations in the metric reduce a force that is propagated through space because the point-particle opposes the deformation.

By 1853 he had brought these very vague ideas to the point where they provided a framework in which to speculate about how heat, light, and gravitation propagate. The mechanism was to be entirely through the action of neighbouring points, and this would involve the point-particle resisting a change in volume and the physical line element associated with the coordinate frame opposing a change in length (see [28], p. 564; [29], p. 511).

> Both classes of phenomena may be explained, if we suppose that the whole of infinite space is filled with a uniform substance, and each particle of substance acts only on its immediate neighbourhood.
>
> The mathematical law according to which this occurs can be considered as divided into
>
> 1) the resistance of a particle of substance to alteration in volume;
>
> 2) the resistance of a physical line element to alteration in length.
>
> Gravitation and electrostatic attraction and repulsion are founded on the first part; propagation of light and heat, and electrodynamic or magnetic attraction and repulsion on the second.

He then investigated "the laws of motion of a substance in empty space". He regarded the motion as the sum $u + v$ of a term u associated with the propagation of gravity and of light respectively. The usual separate equations for each process in a system of equations that Riemann believed gave an account of how the motion of a particle depends only on the particles around it.

As Speiser [32] (1927) was the first to point out, some of these ideas go back to Euler [8], who had attempted to formulate a theory of gravitation, light, electricity and magnetism in terms of an infinite, flowing ether. He had set out this view in his *Letters to a German Princess* in the early 1760s, and succeeded in using it to discuss the propagation of light. Speiser reported that Euler's [8] views were well regarded in his day but have since been largely forgotten.

In 1858 Riemann pushed his ideas further, and came up with a flawed theory of electrodynamics that is nonetheless interesting. The derivation of the equations rested at one point on a faulty exchange of the order of integration of two integrals, which may be why Riemann withdrew it from publication, and his theory involved a retarded potential. In this theory electromagnetism travelled at a speed α, which Riemann related it to the velocity of light, c, by the equation $\alpha^2 = \frac{1}{2}c^2$. In subsequent lectures, although not in the paper itself, he tried to ground his theory in the propagation of light between neighbouring particles.

All this gives weight to the observation that Pearson raised when editing (Clifford 1885, 203) and that Bottazzini and Tazzioli usefully repeat (1995, 32): "whether physicists might not find it simpler to assume that space is capable of varying curvature, and of a resistance to that variation, than to suppose the existence of a subtle medium pervading an invariable homaloidal [Euclidean] space."However, there is no evidence that Riemann took that step.

4 Heat Diffusion and the *Commentatio*

The *Hypotheses* paper [24] was far from being helpful to mathematicians, who might well have preferred more formulae to help them work out Riemann's visionary ideas. They had, in fact, one other paper to refer to, known as the *Commentatio* [27] or Paris memoir, recently and ably discussed in (Cogliati [4] 2014) and (Darrigol [5] 2015). This was an essay, written in Latin, that Riemann submitted, unsuccessfully, for a prize competition on the diffusion of heat in 1861, and which was published with several other of his unpublished papers in the first edition of his collected works [28].[3]

The question asked for conditions on the distribution of heat in an infinite, homogeneous, solid body so that a system of isothermal curves would remain a system of isothermal curves for an indefinite period of time, and moreover the temperature will become a function of time and two other variables.

Riemann viewed the question as concerning a positive definite quadratic form at each point that governed the flow of heat, and because the body is assumed to be homogeneous the coefficients entering the quadratic form are constants. He then looked for the conditions under which a quadratic form with variable coefficients $b_{i,j}$ can be diagonalised. He wrote the quadratic form as a differential form, so the question became one of finding

> conditions under which the expression $\sum_{i,j} b_{i,j} ds_i ds_j$ can be transformed into the form $\sum_{i,j} a_{i,j} dx_i dx_j$, with constant coefficients $a_{i,j}$, by taking the quantities s to be suitable functions of x.

[3]See also Spivak ([33] 1970–1975, Chap. 6, Add. 2).

This turned the question into one of reducing the first form to $\sum_i dx_i^2$, because any positive definite quadratic form with constant coefficients can be so reduced.

A sketchy analysis that was difficult to follow but surely rested on some good unstated reasons led Riemann to claim that the reduction can be carried out provided a very complicated expression in the derivatives of the coefficients $b_{i,j}$, that Riemann abbreviated to (i, i', i'', i''') vanishes, so the question became: what is the meaning of this quantity?

At this point Riemann wrote "In order to understand the structure of these equations better, we form the expression $[X]$". Here, following Darrigol [5] (2015) we have introduced the symbol X for a complicated three-term expression that will be defined below.

A variational argument now gave Riemann a coordinate-free expression for X that involves (i, i', i'', i'''), and at this point he produced a geometrical analogy. He wrote that the expression $\sqrt{\sum_{i,j} b_{i,j} ds_i ds_j}$ can be interpreted as the line element in a general n-dimensional space, and the invariant just obtains appears in this setting as the curvature of the surface at a point. In the case at hand there are three variables, and so six equations that the $b_{i,j}$ must satisfy, of which only three are independent. In short, the reduction of the quadratic form in the heat diffusion problem to a sum of squares with constant coefficients is possible under exactly the same conditions as the reduction of a metric to the Euclidean case: it depends on the curvature vanishing.

When the paper appeared in Riemann's *Werke* [28], Heinrich Weber, one of the editors, supplied a lengthy commentary based on some remarks by Dedekind [6], the other editor. In the second edition he replaced these remarks with some new ones, in which he noticed that several authors had also looked at Riemann's essay: Christoffel, Lipschitz, and Beez among them. In fact, Christoffel [3] and Lipschitz [19] had set themselves the task independently, in attempts to understand the effect of coordinate transformations on quadratic forms in the wake of the publication of Riemann's *Hypotheses* [24] in 1867. Lipschitz returned to the subject in 1876, and Richard Beez's contribution [1] was an attempt write the matter up fully. Thereafter several mathematicians were drawn to *Commentatio* [27], notably Levi-Civita in his paper [17] on parallel displacement (1917).

Much of Weber's commentary consists in very helpfully going through Riemann's calculations more slowly and in more detail, first for a geodesic normal system of coordinates and then by indicating the changes that must be made to deal with a general coordinate system. It was a sensible strategy, but even so Weber made mistakes, and admitted that he had not been able to clear up the paper entirely. And indeed, Riemann had also made a mistake, and attempts to clarify it occupy a fair number of pages in the subsequent literature. Thus it seems that Levi-Civita [17] was led astray in his explanation of Riemann's reasoning, but that Lipschitz [18–20], and Beez [1] before him had understood it better.

Darrigol (2014) gives a thorough account of the developments from Riemann to Levi-Civita, and is particularly interested in on how Riemann came to his final results. It is only too clear that in Riemann's paper [27] we meet for the first time the sheer complexity that we handle today with Christoffel symbols, tensor analysis, Bianchi identities and the like, and Darrigol [5] investigates whether Riemann took a largely algebraic path or one guided by some identifiable geometric intuitions. On the basis of some previously unpublished notes in the Riemann *Nachlaß* [29] he concludes that a geometric insight into how curvature varies suggested some algebraic methods to Riemann.

It is when Riemann turned to the geometric analogy that we have to examine the symbol X. It is defined as

$$X = \delta^2 \sum b_{i,j} ds_i ds_j - 2d\delta \sum b_{i,j} ds_i \delta s_j + d^2 \sum b_{i,j} \delta s_i \delta s_j,$$

and Riemann immediately wrote it as

$$\sum (ij, kl)(ds_i \delta s_j - ds_j \delta s_i)(ds_k \delta s_l - ds_l \delta s_k).$$

The problem here is, as Beez [1] was the first to point out, the deduction is seemingly invalid, but it becomes valid if the term

$$2d\delta \sum b_{i,j} ds_i \delta s_j$$

is replaced by

$$d\delta \sum b_{i,j} ds_i \delta s_j + \delta d \sum b_{i,j} ds_i \delta s_j.$$

Both Darrigol [5] and Cogliati [4] point out that in fact this disparity disappears because the second-order terms are contracted with $d_i ds_j \delta s_k \delta s_l$, but Cogliati adds that Riemann's expression is a natural one to find if Riemann had worked with a normal coordinate system and then appealed to the invariance.

From a historian's point of view, one important point out which Cogliati [4] and Darrigol [5] agree, against some recent historical interpretations, is that Riemann and all his mathematical successors interpreted the expression (i, i', i'', i''') as a curvature and appreciated the use of geometrical reasoning in a problem on heat conduction. The alternative view, that much of this work was a species of tensor calculus without geometrical significance, seems to be an untenable distinction in the period.

References

1. R. Beez, Über das Riemannsche Krümmungsmass höherer Mannigfaltigkeiten. Zeitschrift für. Mathematik und Physik **17**, 65–82 (1879)
2. U. Bottazzini, R. Tazzioli, Naturphilosophie and its role in Riemann's mathematics. Revue d'histoire des mathématiques **1**, 3–38 (1995)
3. E.B. Christoffel, Über die Transformation der homogenen Differentialausdrücke zweiten grades. Journal für Mathematik **70**, 46–70 (1869)
4. A. Cogliati, Riemann's commentatio mathematica: a reassessment. Revue d'histoire des mathématiques **20**, 73–94 (2014)

5. O. Darrigol, The mystery of Riemann's curvature, *Historia Mathematica* **42**, 47–83 (2015)
6. R. Dedekind, *Bernhard Riemann Lebenslauf* (Riemann, 1876), pp. 507–526
7. M. Epple, From quaternions to cosmology: Spaces of constant curvature, ca. 1873–1925. *Proceedings of the International Congress of Mathematicians, Beijing* 3 (2002), pp. 1–3
8. L. Euler, 1768–1744 *Lettres à une princesse d'Allemagne sur divers sujets de physique et de philosophie*, Frankfurt. Angle. transl. H. Hunter vol. **3**, (St.Petersburg, 1975)
9. W.B. Ewald, *From Kant to Hilbert*, 2 volumes (Oxford University Press, 1996)
10. C.F. Gauss, *Disquisitiones arithmeticae*, in *Werke*, ed. by G. Fleischer, Leipzig, 1, (1801)
11. C.F. Gauss, Allgemeine Auflösung der Aufgabe: die Theile einer gegebenen Fläche auf einer andern gegebenen Fläche so abzubilden, dass die Abbildung dem Abgebildeten in den kleinisten Theilen ähnlich wird. *Astronomische Abhandlungen* **3**, (1–30), in *Werke* 4, pp. 189–216 (1825)
12. C.F. Gauss, Disquisitiones generales circa superficies curvas. *Comm. Soc. Göttingen* **6**, 99–146 in *Werke* 4, 217–258, 1828. Rep. with a reprint of the Engl. transl. [1902] in *Astérisque* 62 (1979) 3–81. Engl. transl. as *General investigations of curved surfaces*, ed. by P. Pesic (Dover, New York, 2005)
13. C.F. Gauss, Theoria residuorum biquadraticorum. Commentatio secunda. *Göttingische gelehrte Anzeigen*, Selbstanzeige, 625–638 in *Werke* 2, 169–178, 1831, in *Ewald*, vol. **1**, pp. 306–313 (1996)
14. C.F. Gauss, Theoria residuorum biquadraticorum. Commentatio secunda. *Comm. Soc. Göttingen* **7** (89–148) in *Werke* 2, (93–148), 1832
15. J.F. Herbart, *Psychologie als Wissenschaft neu gegrundet auf Erfahrung, Metaphysik, und Mathematik*, vol. **2**, pp. 1824–1825
16. A.-M. Legendre, *Éléments de Géométrie, avec des notes ...* 4me édition, (Paris, 1817)
17. T. Levi-Civita, Nozione di parallelismo in una varietà qualunque e conseguente specificazione geometrica della curvatura Riemanniana. Rend. Circ. Mat. Palermo **42**, 173–205 (1917)
18. R. Lipschitz, Untersuchungen in Betreff der ganzen homogenen Functionen von n Differentialen. Journal für Mathematik **70**, 71–102 (1869)
19. R. Lipschitz, Entwicklung einiger Eigenschaften der quadratischen Formen von n Differentialen. Journal für Mathematik **71**, 274–287, 288–295 (1870)
20. R. Lipschitz, Fortgesetzte Untersuchungen in Betreff der ganzen homogenen Functionen von n Differentialen. Journal für Mathematik **72**, 1–56 (1870)
21. F. Minding, Wie sich entscheiden lässt, ob zwei gegebene krumme Flächen auf einander abwickelbar sind oder nicht; nebst Bemerkungen über die Flächen von unveränderlichem Krümmungsmaaße. Journal für die reine und angewandte Mathematik **19**, 370–387 (1839)
22. Monge, G., *Application de l'analyse à la géométrie...* 5me édition, revue, corrigée et annotée par M. Liouville. Paris, 1850
23. B. Riemann, *Grundlagen für eine allgemeine Theorie der Functionen einer veränderlichen complexen Grösse.* Inauguraldissertation. Göttingen (1851). In *Werke*, 35–80. Engl. transl. in ([29], 1–42)
24. B. Riemann, 1854. Ueber die Hypothesen welche der Geometrie zu Grunde liegen. *Göttingen Abh.* 13 (1867) 133–152 in *Werke*, 304–319. Engl. transl. in [29], pp. 257–272
25. B. Riemann, 1857. Theorie der Abelschen Functionen. *JfM* **54**, 115–155, in *Werke*, 120–144. Engl. transl. in [29], pp. 79–134
26. B. Riemann, 1861/1876. Commentatio mathematica, qua respondere tentatur quaestioni ab Illma Academia Parisiensi propositae, etc., *Werke*, 2nd ed. Teubner 1892, 391–404), Weber, Anmerkungen 405–423. Weber's remarks, 383–408
27. B. Riemann, *Gesammelte mathematische Werke*, ed. by R. Dedekind, H. Weber. Teubner, Leipzig. 2nd ed. with Nachträge, ed. by M. Noether, W. Wirtinger. Teubner, Leipzig 1892 and 1902
28. G.F.B. Riemann, *Bernhard Riemanns gesammelte mathematische Werke und wissenschaftliche Nachlass*, ed. by R. Narasimhan, (Springer, New York, 1990)
29. G.F.B. Riemann, *Collected Papers*. Transl., Roger Baker, Charles Christenson, and Henry Orde. (Utah: Kendrick Press, Heber City, 2004)

30. E. Scholz, Riemann's frühe notizen zum mannigfaltigkeitsbegriff und zu den Grundlagen der Geometrie. Arch. Hist. Exact Sci. **27**, 213–232 (1982)
31. E. Scholz, Herbarts influence on Bernhard Riemann. Historia Mathematica **9**, 413–440 (1982)
32. A. Speiser, Naturphilosophische untersuchungen von Euler und Riemann. Journal für die reine und angewandte Mathematik **157**, 105–114 (1927)
33. M. Spivak, *A Comprehensive Introduction to Differential Geometry*. (Publish or Perish, Boston, 1970–1975)

Some Remarks on "A Contribution to Electrodynamics" by Bernhard Riemann

Hubert Goenner

Abstract Around 1850, the idea originated that electromagnetic forces between moving charges in circuits are propagated with the velocity of light. After such a speculation by C. F. Gauss in 1845, B. Riemann, in 1858, suggested the inhomogeneous wave equation in 3-dimensional space for the modeling of this propagation. He found a particular solution replacing Coulomb's potential, now called the retarded potential. His attempt failed to derive from this solution Weber's action-at-a-distance potential. Riemann withdrew his pertinent paper before it became printed. After a description of some aspects of research by Gauss, Weber and Riemann, a likely reason for Riemann's withdrawal is specified differing from recent suggestions by historians of mathematics.

1 Introduction

After James Clerk Maxwell's equations for electrodynamics, suggested already in 1864, had been generally accepted, the early contributions to this field by other mathematicians and physicists like Ludvig Lorenz (1829–1891), Franz Neumann (1798–1895), Rudolf Clausius (1822–1888), Hermann von Helmholtz (1821–1894), and Carl Neumann (1832–1925) have been largely forgotten by physicists during the 20th century. It is left to historians of science to maintain the memory of these men and of their achievements (cf. [7], [15]). The reason for this situation is twofold: since Maxwell, field theory with its "near"—interaction has supplanted the previous particle theories with their instantaneous interaction at-a-distance. Secondly, as an invariance group the Poincaré group has replaced the Galilei group ("relativistic" theories).

Surprisingly, within the then reigning view of electromagnetism as a particle theory, we can note a relativistic input, made by the famous mathematician Bernhard Riemann (1826–1866): His introduction of the retarded scalar potential into

H. Goenner (✉)
Institute for Theoretical Physics, University of Göttingen,
Friedrich-Hund-Platz 1, 37073 Göttingen, Germany
e-mail: goenner@theorie.physik.unigoettingen.de

L. Ji et al. (eds.), *From Riemann to Differential Geometry and Relativity*,
DOI 10.1007/978-3-319-60039-0_3

theoretical electrodynamics is still valid, but remains unknown to the overwhelming majority of today's theoretical physicists (Sect. 2). In this chapter, we will try to answer several questions: why Riemann has withdrawn the relevant paper from publication during his lifetime, what brought him to the discovery of the retarded potential, and why did he not further use this potential in his course on electricity and magnetism. Up to now, mathematicians have held accountable a trivial mistake in his paper for the withdrawal by Riemann, i.e., a forbidden interchange of integrations ([1], pp 54–56). Occasionally, it is also claimed that Riemann did make inadmissible approximations in his calculations ([17], p. 265). After recalling ideas of C. F. Gauss and W. Weber concerning a possible propagation of what now is called the electromagnetic field (Sect. 3), we will point to a more serious flaw in Riemann's paper, very likely discovered by the author himself soon after handing in his manuscript to the Royal Academy of Science in Göttingen (Sect. 4). Some helpful concepts of Maxwell's theory, a special relativistic theory of the electromagnetic field, leading to the retarded potential are introduced in Appendix 1.

2 Riemann's New Result of 1858: The Retarded Potential

Riemann's manuscript of 1858 "A contribution to electrodynamics" [25], became published only after his death in 1867 in the journal Annalen der Physik [24], with the same volume also containing the paper by L. Lorenz [18] who, in addition, displayed the retarded *vector* potential. A footnote in the English translation of Riemann's note [23] stated: "This paper was laid before the Royal Academy of Sciences at Göttingen on the 10th of February 1858, by the author [..], but appears, from a remark added to the title by the then Secretary to have been subsequently withdrawn."

The gist of his paper is stated right at its beginning:

> I have found that the electrodynamic actions of galvanic currents may be explained by assuming that the action of one electrical mass on the others is not instantaneous, but is propagated to them with a constant velocity which, within the limits of errors of observation, is equal to that of light.[1]

Moreover, he concluded that "[..] the differential equation for the propagation of the electrical force is the same as that for light and of radiant heat."

His idea was to derive Weber's law for the force between two pointlike electrical charges from a partial differential equation in the same way as Coulomb's potential V had been a consequence of Poisson's equation:

$$\nabla^2 V = \frac{\partial^2 V}{\partial x^2} + \frac{\partial^2 V}{\partial y^2} + \frac{\partial^2 V}{\partial z^2} = 4\pi\rho \,, \tag{1}$$

[1]The translation is taken from [23], p. 368. If not indicated otherwise, translations are made by myself.

where ρ is the electrical charge density. He knew that in order to allow for propagation, the PDE ought to be of the hyperbolic type. As to the type of propagation, in the same lecture course Riemann had also dealt with the parabolic diffusion equation:

$$\alpha\frac{\partial u}{\partial t} + \rho + \beta^2[\frac{\partial^2 u}{\partial x^2} + \frac{\partial^2 u}{\partial y^2} + \frac{\partial^2 u}{\partial z^2}] = 0 . \tag{2}$$

Already in the term 1854/55, in his first course on PDEs and their application to problems of physics, Riemann had studied the 1-dimensional wave equation [28]

$$\frac{\partial^2 u}{\partial t^2} = a^2\frac{\partial^2 u}{\partial x^2}$$

(§43, S. 111), oscillations of a strained string in §74, §75 and solutions by D'Alembert (p. 188).

In §43 the general solution with initial conditions at $t = 0$

$$u = f(x) , \frac{\partial u}{\partial t} = F(x)$$

is written down:

$$u = 1/2[f(x + at) + f(x - at)] + \frac{1}{2a}\int_{x-at}^{x+at} F(\lambda)d\lambda . \tag{3}$$

(His formula (III) on p. 113.) It should not have been a problem for Riemann to generalize the 1-dimensional wave equation to three space-dimensions and to replace the argument $x - at$ by $r - at$ with $r^2 = x^2 + y^2 + z^2$. However, the new physics comes from the combination with d'Alembert's inhomogeneous PDE:

$$\frac{\partial^2 V}{\partial t^2} - \alpha^2(\frac{\partial^2 V}{\partial x^2} + \frac{\partial^2 V}{\partial y^2} + \frac{\partial^2 V}{\partial z^2}) + \alpha^2 4\pi\rho = 0 , \tag{4}$$

and this is exactly the equation he wrote down in his paper for the Royal Academy. Without giving a calculation, he presented as a particular solution of (4) what is now called the "retarded potential":

$$V = \frac{f(t - \frac{r}{\alpha})}{r} , \tag{5}$$

with $r = (x^2 + y^2 + z^2)^{\frac{1}{2}}$, and α a velocity.[2] He thus had chosen the correct physical solution by leaving aside the advanced potential $V = \frac{f(t+\frac{r}{\alpha})}{r}$.

[2]In today's view, he used $\rho = f\delta(r)$, where δ is Dirac's distribution; cf. Appendix 1.

Maxwell's reaction to the retarded potentials of Riemann and Lorenz when they were published in 1867 was entirely negative:

> We are unable to conceive of a propagation in time except either as the flight of a material substance through space or as the propagation of a condition of motion or stress in a medium already existing in space. (Quoted from [22], p. 185.)

For him, the scalar potential was not an observable of the "state" of the electrical field unlike Lorentz's vector potential. Leaving aside the question of observability, there in fact is an epistemological problem when the switch from Weber's theory for point particles to a field theory lying behind the new concept of propagation is to be made: the aether problem.[3]

3 Gauss, Weber, and Riemann on Electrodynamic Interaction

In the first half of the 19th century, from electrostatics and magnetism as generated by electrical currents, electrodynamics developed. For the sources of electricity, the picture of an electrical fluid became replaced by the concept of charged electrical particles. In a first approach, still within theories with action at-a-distance, potentials depending on the velocity of such particles were introduced by C. F. Gauss, W. Weber, F. Neumann, and R. Clausius.[4] At the time, from experiments no convincing conclusion could be drawn as to which of these potentials described the phenomena best. A comparison by help of thought experiments or exemplary calculations was rarely tried; the dissertation by a student of Clausius is an example [30].

3.1 Gauss

Riemann attained the idea that a force between electrical currents need not be instantaneous but propagates from Gauss via Wilhelm Weber. In fact, in his letter of 19 March 1845 to Weber, Gauss wrote:

> Without doubt, I would have given notice of my investigations a long time ago, had I not missed at the time when I stopped them what I considered the real cap stone. To wit: the derivation of the additional forces (supervening the forces of the mutual interaction of electrical parts at rest when they are in motion) from the action which is not instantaneous but propagated in time (similarly as with light) ([9], p. 627–629.)

---

[3]We do not dwell here on Riemann's ideas about the nature of the medium through which the electrical forces are propagated. Cf. [20], p. 529, 532, 534 with the pagination after the 2nd edition of Riemann's collected papers of (1892).

[4]Even before R. Clausius, H. Grassmann had suggested the same potential as Clausius [11], ([19], III, 203–210).

But Gauss had more up his sleeve. In his unpublished notes, we find a remark entitled "Fundamental law for all interactions of galvanic currents (found in July 1835)" ([9], p. 616–617). Let, e, e' be the electric charges, x, y, z and x', y', z' their coordinates, $r^2 = (x'-x)^2 + (y'-y)^2 + (z'-z)^2$. For the mutual action (repulsive force) of the charges in motion, Gauss then gave the expression:

$$\frac{ee'}{r^2}\{1 + k[(\frac{d(x'-x)}{dt})^2 + (\frac{d(y'-y)}{dt})^2 + (\frac{d(z'-z)}{dt})^2 - \frac{3}{2}(\frac{dr}{dt})^2]\}\,, \qquad (6)$$

where "$\sqrt{\frac{1}{k}}$ represents a determined speed". The corresponding potential has been correctly reported in [8] to be:

$$\phi = \frac{Q}{4\pi\epsilon_0 r}[1 + (\frac{\vec{v}_{rel}}{c})^2 - \frac{3}{2c^2}(\vec{v}_{rel}\cdot\frac{\vec{x}}{|\vec{x}|})^2]\,. \qquad (7)$$

Here, Q is the electric charge, c the velocity of light in vacuum, and $\vec{v}_{rel}$ the relative velocity of the two charges. Hence it is to be noted, that a velocity-dependent potential already occurred in the work of Gauss, but remained unpublished during his lifetime.

3.2 *Weber*

As Gauss had done more than a decade earlier, in 1846 Wilhelm Weber derived his law for the absolute value of the force between two charges in relative motion from Ampère's law[5] [37], [35]:

$$\frac{e_1e_2}{r^2}\{1 + \frac{r\ddot{r}}{c^2} - \frac{\dot{r}^2}{2c^2}\}\,, \qquad (8)$$

where $r = |\vec{r}| = |\vec{r_1} - \vec{r_2}|$. In order to do so, assumptions about the distribution and velocity of the charges in the currents had to be made such as: (1) positive and negative charges move with the same speed; (2) In each volume element, always the same amount of positive and negative charges must be present. (8) can be obtained from a Lagrangian:

$$V = \frac{1}{c^2}\frac{e_1e_2}{r}(1 - (\frac{dr}{dt})^2)\,. \qquad (9)$$

Weber's approach was criticized immediately by H. Helmholtz on the false premise that it would violate conservation of energy [13] and, with the same argument, by

[5]Instead of by expression (8), Weber's force also is given in the form resulting from the substitution $c \to \sqrt{2}c$. In Weber's original paper [37] the coefficient of $\dot{r}^2$ had been $\frac{a^2}{16}$. This was changed later into c^2 by Weber, but his c corresponds to $(\sqrt{2})^{-1}\times$velocity of light.

Thomson and Tait in their influential textbook [34][6] until the mistake became obvious. Maxwell rejected Weber's law because it followed from electromagnetism as described by a theory of particles with interaction at-a-distance; he preferred a field theoretic description [6].[7]

Despite his work within a theory of instantaneous action at-a-distance, Weber, besides Kirchhoff, was first in correctly describing the propagation with the velocity of light of oscillations of the electric current in wires of negligible resistance [36], [16]. He also determined the velocity of light in vacuum by electrodynamic measurements with highest precision [2], [4], [3].

3.3 Riemann

B. Riemann joined Weber in his description of an electric current by moving point-like electrical charges and the interaction with other currents as an interaction at-a-distance between two charges (2-body forces). He introduced a further potential ("Riemann's potential") ([28], p. 326) containing only the relative velocity of the particles:

$$V^* = \frac{e_1 e_2}{r}\{(\frac{dx_1}{dt} - \frac{dx_2}{dt})^2 + (\frac{dy_1}{dt} - \frac{dy_2}{dt})^2 + (\frac{dz_1}{dt} - \frac{dz_2}{dt})^2\}\,. \tag{10}$$

As seen above in (6), Gauss already had thought about this term. Yet, in all likelihood, Riemann was unaware of the expression given by Gauss. As is clear from letters to his sister Ida and his brother Wilhelm, he had to guess what the results of Gauss were. Already on 28 December 1853 he wrote to Wilhelm:

> Right after my Habilitationsschrift, I had taken up again my further investigations about the connection of electricity, galvanism, light and gravity. I reached the point that I can publish them in this form without risk. But in the course of this work, I became ever more sure that Gauss works in this field since a couple of years and has told some friends, e.g., A. Weber, of it under the promise to keep it secret [..]. ([27], p. 547.)

Five years later, when he had submitted his results, he still had not learned more on the work by Gauss and let Ida know:

> I have handed over to the Royal Soc. my discovery about the connection between electricity and light. From some utterances which I heard, I must conclude that Gauss, in this context, has set up a theory different from mine. But I am fully convinced that mine is the correct one, [..]. (Letter to Ida early in 1858 [26], p. 585.)

The potential (10) can be found already in Riemann's course on "The mathematical theory of electricity and magnetism" of summer 1858 [33].[8]

---

[6]Translated into German by H. Helmholtz and G. Wertheim [14].

[7]A comparison between Maxwell's and Weber's electrodynamics is presented in [5].

[8]In fact, in the notes by Eduard Sellin, Riemann's second course of summer 1858 on Selected physical problems is also mixed in.

4 Riemann's Paper

In his paper "A contribution to electrodynamics," Riemann set out from "the potential of the forces exerted by [the circuit] S upon S':

$$-\frac{2}{c^2}\int\int\frac{uu'+vv'+ww'}{r}dSdS' \,, \tag{11}$$

this integral being extended over the whole of the elements dS and dS' of the conductors S and S' ." Here, u, v, w; u', v', w' are the components of the specific intensity of the currents. In the particle picture, with charges e, e' and their velocities $\frac{dr}{dt}$, $\frac{d'r}{dt}$ in the conductors S and S', Riemann wrote (11) in the form:

$$V=\frac{1}{c^2}\Sigma\Sigma\frac{ee'}{r}\frac{dd'(r^2)}{dt^2} \,. \tag{12}$$

The summations are taken over the charges e in conductor S and the charges e' in conductor S'. (12) is equal to Neumann's potential [21]. After some manipulations depending on an assumption concerning the motion of electric charges of different sign[9] Riemann arrived at:

$$V=\frac{1}{c^2}\Sigma\Sigma ee'r^2\frac{dd'(\frac{1}{r})}{dt^2} \,. \tag{13}$$

He intended to derive in a different way the expression integrated over time:

$$P=\frac{1}{c^2}\int_0^t d\tau\Sigma\Sigma ee'r^2\frac{dd'(\frac{1}{r})}{d\tau^2} \,. \tag{14}$$

At this point, Riemann's new retarded potential came in. By introducing the function

$$F(t,t')=r(t,t')^{-1} \tag{15}$$

with $r(t,t')=[(x_t-x'_{t'})^2+(y_t-y'_{t'})^2+(z_t-z'_{t'})^2]^{\frac{1}{2}}$, coordinates x_t, y_t, z_t of charge e at time t and $x'_{t'}, y'_{t'}, z'_{t'}$ of charge e' at time t', he went over from (14) to

$$P=\frac{1}{c^2}\int_0^t d\tau\Sigma\Sigma ee'F(\tau-\frac{r}{\alpha},\tau) \,. \tag{16}$$

---

[9]For the motion of the electrical particles I assume that for each part of the conductor the sum of the fundamental actions exerted by the particles with positive and negative electricity is still almost the same during a span of time in which a very large flow passes through. It is known that this assumption is justified as well by experience as by inspection of the electro-motoric forces ([29], Blatt 10).

Expression (16) is interpreted by him as: "the potential of the forces exerted by all masses ϵ [$= e$] of conductor S on the masses ϵ' [$= e'$] of conductor S' during the time 0 to t." ([29], Blatt 14, verso.) On the same page of these handwritten notes, another assumption is formulated: "It is now assumed that the electrical masses cover only a very small distance during the time of the force's propagation; the effect is considered during a time span with regard to which the time of propagation is vanishing."

On the two following pages of his paper of 1858, Riemann replaced $F(\tau - \frac{r}{\alpha}, \tau)$ by $-\int_0^{\frac{r}{\alpha}} d\sigma F(\tau - \sigma, \tau)$, inverted integrations, omited small terms ("it is easily seen...") and then claimed "The value of P from our theory agrees with the experimental one (14), if we assume $\alpha^2 = \frac{1}{2}c^2$."

The flaw in this argument lies in (16): A comparison with (5) shows that Riemann has introduced retarded time also in the distance in the denominator. Thus he has lost his exact solution of the (inhomogeneos) wave equation. It seems that Emil Wiechert (1861-1828) who, independently from Alfred-Marie Liénard, also introduced the retarded potential, has seen this. In his paper of 1900 he wrote: "At first, a conjecture could have been that [..] for a single electron with charge l and velocity v, one could simply set:

$$\phi_{t=t_0} = \frac{1}{r_{t=t_0-\frac{r}{v}}}\,, \Gamma_\nu = l(\frac{1}{r}\frac{v_\nu}{v})_{t=t_0-\frac{r}{v}}\,, \tag{17}$$

and in fact this was assumed at the time by Riemann. Yet this approach leads to contradictions with the fundamental assumptions of our theory." ([38], p. 563.) In (17) ϕ and Γ_ν denote the scalar and vector potentials.

How did Riemann arrive at the expression (15)? This remains unclear even from Riemann's handwritten notes. At some point, he looked at

$$r^2 = a^2 + 2a(x'_{t'} - x_t) + (x'_{t'} - x_t)^2 + (y'_{t'} - y_t)^2 + (z'_{t'} - z_t)^2 \tag{18}$$

and expanded in terms of $\frac{r}{a}$ ([29], Blatt 11, verso). On another page he suggested that Poisson's equation be replaced by the (inhomogeneous) wave equation and in the next line wrote ([29], Blatt 16, recto, Blatt 17 recto):

$$rr = (r^2) = (x'_t - x_{t-\frac{r}{a}})^2 + (y'_t - y_{t-\frac{r}{a}})^2 + (z'_t - z_{t-\frac{r}{a}})^2\,, \tag{19}$$

and added "The assumption concerning the electrostatic effect by arbitrarily distributed electrical masses can be expressed as such." Riemann's fallacy thus can be localized in his notes: When he passed over from Poisson's PDE, with the particular solution $\frac{1}{r}$ written down by him, to the wave equation a particular solution of which he also had found, for reasons of similarity he was intrigued by the idea that the time-independent r in Coulomb's potential must be replaced by (19). Apparently, he did not check whether this also was a solution of the wave equation, and he did not see a contradiction with the form of the retarded potential given in the same paper. Perhaps, he has been in a hurry: some of his calculations were made on sheets intended for letters dated January 28 and 29, 1858, i.e., just two weeks before he handed in his paper to the Academy.

5 Concluding Remarks

What then is the importance of Riemann's paper of 1858? Three main points were made by him :

(1) The "electrical force" is propagated with the velocity of light and this propagation is the same as that for light and of radiant heat;
(2) For moving electrical charges the retarded potential replaces the Coulomb potential;
(3) Weber's potential can be derived by help of the retarded potential.

The first two statements correspond precisely to what we accept today as consequences of Maxwell's theory and are a remarkable anticipation of Maxwell. Only the third point is mistaken; this very likely is the reason why Riemann has withdrawn his paper from publication. In his attempted proof, Riemann started from an expression *different* from the retarded potential and consequently failed to establish a link between the retarded potential and Weber's potential. The inadmissible inversion of two integrals was only a minor additional blemish. In his subsequent course of summer 1861 on "The mathematical theory of gravitation, electricity and magnetism," about which notes by a student are available ([32], pp. 192–199), he changed his previous proof and derived Weber's law with the help of energy conservation in the form of what he called the Lagrange principle—without mentioning the retarded potential at all (cf. also [15], p. 180–181).[10] We do not have the slightest documentary evidence about whether Riemann tried to re-do his calculation with the correct expression for the retarded potential just to conclude that he could not reach Weber's potential in this way.

Another possible reason for the withdrawal might have been that, around the time of the submission of his paper, he had found his additional velocity-dependent (Riemann-)potential.[11] This would have weakened the importance of the suspected connection between the retarded potential and Weber's potential. Some support may be seen in the report by Riemann's colleague, the mathematician and astronomer Ernst Schering (1833–1897), that Riemann had expressed his satisfaction, that [his manuscript] back then had not been printed, because in the meantime he had found a specification of his law as a consequence of which it would satisfy certain general principles like the other fundamental laws for forces [31].

---

[10]By the same approach, Riemann's potential could be derived as well. Thus Riemann had achieved what Gauss had had in mind, i.e., "the derivation of the additional forces [..] from the action".

[11]As mentioned above, he first presented his potential in one of his two summer courses of 1858.

For Riemann, a possible relation between the retarded potential and Weber's potential apparently was more important than the study of the retarded potential for its own sake. Thus he missed the discovery of Lorentz invariance of the wave equation (4) (with $\rho = 0$). Unfortunately, his handwritten notes for the paper withdrawn do not reveal calculations showing how he arrived at (5). Perhaps, with his expertise in the field of PDEs, he had made the calculations already some years earlier; perhaps he had found the particular solution of the wave equation by pure intuition. That he failed to relate it to Weber's potential may have discredited the retarded potential in his eyes. In accord with his idea that the electromagnetic interaction between charges is propagated with the velocity of light, Riemann might have believed that Weber's potential already reflected this propagation. Despite his ingenuity, Riemann thus could not pave the way toward a relativistic electrodynamics for physics. This was left to H. Poincaré and H. A. Lorentz.

Acknowledgements For the invitation to contribute to this volume and for his helpful remarks I am grateful to A. Papadopoulos, Strasbourg.

Appendix 1

Electric field $\vec{E}$ and magnetic field $\vec{B}$ are combined in the the field tensor of the electromagnetic field $F = F_{ik} dx^i \wedge dx^k$ $(i, k = 0, 1, 2, 3)$, which can be expressed by the 4-potential $A = A_i dx^i$ as $F = dA$ with (in components) $A_i \simeq (\phi, -\vec{A})$ where ϕ is the scalar, $\vec{A}$ the vector potential. Thus, $F_{ik} = \partial_i A_k - \partial_k A_i$, with $F_{0k} \simeq \vec{E} = -\vec{\nabla}\phi - \frac{1}{c}\frac{\partial \vec{A}}{\partial t}$, $F_{\mu\nu}$ $(\mu, \nu = 1, 2, 3) \rightarrow \vec{B} = \vec{\nabla} \times \vec{A}$. From the first of Maxwell's equations :

$$\partial_l F^{il} = \frac{4\pi}{c} j^i \ , \quad \partial_l F^{*il} = 0 \tag{20}$$

with $F^{ik} = \eta^{ir}\eta^{ks} F_{rs}$, $F^{*ik} = \frac{1}{2}\epsilon^{iklm} F_{lm}$, the Minkowski metric η_{ik}, and the 4-current $J^i \simeq (c\rho, \vec{j})$, we obtain:

$$\partial^i \partial_l A^l - \partial_l \partial^l A^i = \frac{4\pi}{c} j^i \ . \tag{21}$$

As the vector potential is determined only up to gauge transformations $A \rightarrow A' = A + d\lambda$ with a scalar function λ, a so-called gauge condition may be added. Taking the *Lorenz gauge* $\partial_l A^l = 0$, from (2) the inhomogeneous wave equation follows:

$$\Box A^i = -\frac{4\pi}{c} j^i \tag{22}$$

with $\Box = \partial_s \partial^s = \eta^{rs} \partial_r \partial_s$. The Lorentz gauge condition then leads to $\partial_s j^s = 0$, i.e., to the equation for the conservation of electrical charge. For the scalar potential, then

$$\Box\phi = \frac{1}{c^2}\frac{\partial^2 \phi}{\partial t^2} - \vec{\nabla} \cdot \vec{\nabla}\phi = -4\pi\rho\,. \tag{23}$$

For a static electric field, Poisson's equation follows with the Coulomb potential

$$\phi(x) = \frac{1}{4\pi}\int d^3x' \frac{\rho(x')}{|\vec{x} - \vec{x}'|}\,. \tag{24}$$

The retarded potential is a particular solution of (23):

$$\phi(x) = \frac{1}{4\pi}\int d^3x' \frac{\rho(x', t - \frac{|\vec{x}-\vec{x}'|}{c})}{|\vec{x} - \vec{x}'|} \tag{25}$$

vanishing at spacelike infinity. It replaces Coulomb's potential for an arbitrarily time-dependent charge distribution.

Retarded and advanced solutions are combined in:

$$A^i = \frac{2}{c}\int d^4x' \theta(x' - x)\delta[(x^s - x'^s)(x_s - x'_s)]j^i \tag{26}$$

with Dirac's δ-distribution and the characteristic function $\theta(x' - x) = 0, +1$ or $0, -1$ selecting directions into the future and past lightcone [10]. With the expression for the electrical current

$$j^i = ce\int_{-\infty}^{+\infty} ds\; u^i \delta^4(x - x')\,, \tag{27}$$

where $u^i \simeq \gamma(c, \vec{v})$, $\gamma = (1 - \frac{v^2}{c^2})^{-\frac{1}{2}}$, and e the electrical charge of a point particle, then the so-called Liénard-Wiechert potential results:

$$\phi = \frac{e}{|\vec{x} - \vec{x}'|}(1 - \frac{1}{c}\vec{n}\cdot\vec{v})^{-1}\,,\quad \vec{A} = \frac{e\vec{v}}{|\vec{x} - \vec{x}'|}(1 - \frac{1}{c}\vec{n}\cdot\vec{v})^{-1}\,, \tag{28}$$

with $\vec{v}, \vec{x}'$ taken at the retarded time; $\vec{n} = (\vec{x} - \vec{x}')(|\vec{x} - \vec{x}'|)^{-1}$.

(28) is different from Riemann's Ansatz (16) criticized by Wiechert; cf. (17) in Sect. 4.

References

1. T. Archibald, Physics as a constraint on mathematical research: the case of potential theory and electrodynamics, in *The History of Modern Mathematics II: Institutions and Applications*, ed. by D. E. Rowe, J. McCleary, (Academic Press, Boston, MA 1989), pp. 29–75
2. A. K. T. Assis, On the first electromagnetic measurement of the velocity of light by Wilhelm Weber and Rudolf Kohlrausch, in *Volta and the History of Electricity*, ed. by F. Bevilacqua, E. A. Giannetto, (Ulrico Hoepli, Milano 2003), pp. 267–286
3. A.K.T. Assis, *Weber's Electrodynamics* (Kluwer Academic Publishers, Dordrecht, 1994)
4. A.K.T. Assis, On the propagation of electromagnetic signals in wires and coaxial cables according to Weber's electrodynamics. Found. Phys. **30**, 1107–1121 (2000)
5. A.K.T. Assis, H. Torres, Silva, Comparison between Weber's electrodynamics and classical electrodynamics. Pramana. J. Phys. **55**(3), 393–404 (2000)
6. J. Clerc, Maxwell, A dynamical theory of the electromagnetic field. Philos. Trans. R. Soc. Lond. **166**, 459–512 (1865)
7. O. Darrigol, *Electrodynamics from Ampère to Einstein* (University Press, Oxford, 2000)
8. D. Eberle Spencer and S.Y. Uma, Gauss and the electrodynamic force, in *The mathematical Heritage of C. F. Gauss*, Ed. George M. Rassias, 685–711, (World Scientific, Singapore, 1991)
9. C.F. Gauss, *Gesammelte Werke*, vol. 5, edited by the Royal Society of Göttingen. Dieterich, Göttingen (1876)
10. H. Goenner, *Spezielle Relativitätstheorie und die klassische Feldtheorie* (Elsevier Spektrum Akademischer Verlag, Heidelberg, 2004)
11. H. Grassmann, Zur Elektrodynamik, Crelles J. **64**, 57–64 (1845)
12. K. Hattendorff, *Schwere, Elektrizität und Magnetismus nach den Vorlesungen von Bernhard Riemann bearbeitet von Karl Hattendorff* (Carl Rümpler, Hannover, 1876)
13. H. Helmholtz, Über die Erhaltung der Kraft. Nachdruck der Veröffentlichung von 1847 als *Oswald's Klassiker der exakten Wissenschaften*, (Nr. 1, Engelmann, Leipzig 1889)
14. H. Helmholtz, G. Wertheim, *Handbuch der Theoretischen Physik* (Vieweg, Braunschweig, 1874)
15. C. Jungnickel and R. McCormmach, *Intellectual Mastery of Nature.* Vol. 1: The torch of mathematics, (University of Chicago Press, 1986)
16. G. Kirchhoff, On the motion of electricity in wires. Philoso. Mag. **13**, 393–412 (1857)
17. D. Laugwitz, *Bernhard Riemann 1826–1866* (Birkhäuser, Basel/Boston/Berlin, 1996)
18. L. Lorenz, Ueber die Identität der Schwingungen des Lichts mit den elektrischen Strömen, *(Poggendorff's) Annalen der Physik* **131** (1867), 243–263. Danish original in Oversigt over det K. Danske Vidensk. Selsk. Forhandl. 1867, Nr. 1
19. J. Lüroth and F. Engel eds., *Hermann Grassmann. Gesammelte mathematische und physikalische Werke* Zweiten Bandes zweiter Theil: Die Abhandlungen zur Mechanik und zur Mathematischen Physik, B. G. Teubner, Leipzig (1902)
20. R. Narasimhan (ed.), *B. Riemann: Gesammelte Mathematische Werke, wissenschaftlicher Nachlass und Nachträge*, (Springer, Berlin; Teubner, Leipzig 1990)
21. F. Neumann, Die mathematischen Gesetze der inducirten elektrischen Ströme, Abhandlungen der Preussischen Akademie, 1–87, Allgemeine Gesetze der inducirten elektrischen Ströme. Annalen der Physik **67**, 31–44 (1846)
22. A. O'Rahilly, *Electromagnetics (Green & Co* (London & Cork; University Press, Cork, A Discussion of Fundamentals, Longmans, 1938)
23. B. Riemann, A contribution to electrodynamics, Philosophical Magazine 34. Series **4**, 368–372 (1867)
24. B. Riemann, Ein Beitrag zur Elektrodynamik, (Poggendorff's). Annalen der Physik **131**(6), 289–293 (1867)
25. B. Riemann, Ein Beitrag zur Elektrodynamik. Handschriftenabteilung der Staats- und Universitätsbibliothek Göttingen. Riemann papers. Cod. Ms. B. Riemann **24**, BI. 1–7 (1858)

26. B. Riemann, Letter to Ida Riemann, quoted from curriculum vitae by R. Dedekind in 2nd edition of Riemann's Gesammelte Mathematische Werke, 2. Aufl. 1892. According to [1], this letter is in the Archive of Stiftung Preussischer Kulturbesitz, Berlin, Nachlass Riemann Acc. 17
27. B. Riemann, Letter to Wilhelm Riemann 28. 12. 1853, quoted from curriculum vitae by R. Dedekind in 2nd edition of Riemann's Gesammelte Mathematische Werke, 2. Aufl. 1892
28. B. Riemann, *Partielle Differentialgleichungen und deren Anwendungen auf physikalische Fragen.* Vorlesungen. Herausgegeben von Karl Hattendorff. Braunschweig: Vieweg 1869 (1969). Available in the internet, http://gdz.sub.uni-goettingen.de/dms/load/img/?PPN=PPN234595299&IDDOC=46562
29. B. Riemann, N. Riemann. Handschriftenabteilung der Staats- und Universitätsbibliothek Göttingen. Riemann papers. Cod. Ms. B. Riemann 24 1858
30. F. Schatz, Über das Grundgesetz der Elektrodynamik, Inaugural-Dissertation der phil. Fak. d. Rhein Friedrich-Wilhelms Universität zu Bonn vom 17. 3. 1880. Staats- und Universitätsbibliothek, Göttingen
31. E. Schering, Zum Gedächtnis an B. Riemann in *Gesammelte Mathematische Werke*, ed. By R. Haussner, Karl Schering. (Mayer und Müller, Berlin, 1909)
32. E. Schultze, *B. Riemann "Schwere, Elektricität und Magnetismus."* Mitschrift der Vorlesung von 1861, 220 Seiten, Nachlass Schwarz, Nr. 677. Akademie der Wissenschaften der DDR, Berlin. Digitalized by the University of Göttingen: http://gdz.sub.uni-goettingen.de/dms/load/img/?PPN=PPN631015833&LOGID=LOG_0002
33. E. Sellien, Notes on Riemann's courses "The mathematical theory of electricity and magnetism" and "Selected physical problems" of summer 1858. Nachlass Riemann. Handschriftenabteilung der Staats- und Universitätsbibliothek Göttingen. Riemann papers. Cod. Ms. B. Riemann 45, Blatt 1–23 1858
34. W. Thomson, P. G. Tait, *Treatise on Natural Philosophy* (University Press, Cambridge, 1867)
35. W. Weber, Elektrodynamische Maassbestimmungen (Poggendorffs). Annalen der Physik **73**, 193–240 (1848)
36. W. Weber, Elektrodynamische Maassbestimmungen insbesondere über elektrische Schwingungen. Abhandlungen der Königlich Sächsischen Gesellschaft der Wissenschaften, mathematisch-physische Klasse **6**, 571–716 (1864)
37. W. Weber, Elektrodynamische Maassbestimmungen, *Abhandlungen bei Begründung der Königlich Sächsischen Gesellschaft der Wissenschaften am Tage der zweihundertjährigen Geburtsfeier Leibnizens* (Herausgegeben von der Fürstlich Jablonowskischen Gesellschaft. Weidmann, Leipzig, 1846)
38. E. Wiechert, Elektrodynamische Elementargesetze, Archives Néerlandaises des Sciences Exactes et Naturelles Serie **2**(5), 549–573 (1900)

Riemann's Memoir Über das Verschwinden der ϑ-Functionen

Christian Houzel

Abstract In the second part of his great memoir *Theorie der Abel'schen Functionen* (1857), Riemann gives a solution to Jacobi's problem of inversion for the general case of integrals of algebraic functions, now called *Abelian integrals*. The case of hyperelliptic integrals had been treated, for the genus 2, by Göpel and Rosenhain and, for any genus, by Weierstrass in a series of memoirs between 1848 and 1856. The proof developed by Riemann in his 1857 paper is not complete and the memoir *Über das Verschwinden der ϑ-Functionen* (1865) completes it.

1 Jacobi's Inversion Problem

In the second part of his great memoir *Theorie der Abel'schen Functionen* (1857) [8], Riemann gives a solution to Jacobi's problem of inversion for the general case of integrals of algebraic functions, now called *Abelian integrals*. The case of hyperelliptic integrals had been treated, for the genus 2, by Göpel [1] and Rosenhain [10, 11] and, for any genus, by Weierstrass in a series of memoirs between 1848 and 1856 [12, 13]. The proof developed by Riemann in his 1857 paper is not complete and the memoir *Über das Verschwinden der ϑ-Functionen* (1865) [9] completes it; Riemann had previously exposed this complement in the lectures of 15 to 17 January 1862.[1]

Let us first explain how Riemann formulates his solution in 1857. A class of algebraic functions is represented by a compact Riemann surface T; if z, s are meromorphic functions on T with n and m simple poles respectively, there exists a polynomial $F(s,z)$ of degree n with respect to s and m with respect to z such that $F(s,z) = 0$. In general, F is a power of an irreducible polynomial and, if F itself is irreducible, the covering $z : T \to \mathbb{P}^1(\mathbb{C})$ is ramified as the algebraic function s of z and T is the normalization of the algebraic curve defined by the equation $F(s,z) = 0$. The space

[1] According to the notebooks of Prym and Minnigerode.

C. Houzel (✉)
Laboratoire SPHERE, UMR 7219, Universit Paris Diderot, btiment Condorcet, case 7093, 5 rue Thomas Mann, 75205 Paris cedex 13, France
e-mail: Houzel@vjf.cnrs.fr

L. Ji et al. (eds.), *From Riemann to Differential Geometry and Relativity*,
DOI 10.1007/978-3-319-60039-0_4

of differential forms everywhere regular on T is of dimension p, the genus of T; they are called differential forms of the first kind. Each such form may be written as

$$\frac{\phi dz}{\partial F/\partial s},$$

where ϕ is a polynomial of degree $n-2$ with respect to s and degree $m-2$ with respect to z, which vanishes for every pair (s, z) such that

$$\frac{\partial F}{\partial s}(s, z) = \frac{\partial F}{\partial z}(s, z) = 0.$$

Riemann introduces a system of cuts in order to render T simply connected; it consists of p pairs of closed cuts (a_ν, b_ν) with one origin for each ν, and $p-1$ cross-cuts c_ν linking b_ν to $a_{\nu+1}$. There is a basis $(du_1, du_2, \ldots, du_p)$ of the space of differential forms of the first kind such that the integral $\int_{b_\nu} du_\mu$ is equal to $i\pi$ when $\mu = \nu$ and to 0 when $\mu \neq \nu$; then

$$\int_{a_\nu} du_\mu = a_{\mu\nu}$$

are the elements of a symmetric matrix of which the real part is *negative and not degenerate*. The value of the integral $u_\mu(x)$ of du_μ from a fixed origin x_μ to a point x of $T' = T - \{a_\nu, b_\nu\}$ is defined up to the addition of a period of u_μ, that is a linear combination of πi and the $a_{\mu\nu} (1 \leq \nu \leq p)$ with integral coefficients. Jacobi's inversion problem consists in the determination of $\eta_1, \eta_2, \ldots, \eta_p \in T$ as functions of $e = (e_1, e_2, \ldots, e_p) \in \mathbb{C}^p$ in such a way that

$$\sum_{\nu=1}^{p} u_\mu(\eta_\nu) - e_\mu$$

be a period of u_μ for $1 \leq \mu \leq p$; it is a generalization to the genus p of the inversion of elliptic integrals (genus 1). The problem is symmetrical with respect to $\eta_1, \eta_2, \ldots, \eta_p$ so that only the (elementary) symmetric functions of $(\eta_1, \eta_2, \ldots, \eta_p)$ are expected to be single-valued functions of e; Jacobi [2–4] conjectured that these symmetric functions might be expressed by means of theta functions of the p variables e_μ.

The theta function associated by Riemann to T is defined by the series

$$\vartheta(v_1, v_2, \ldots, v_p) = \sum_{m \in \mathbb{Z}^p} \exp\Big(\sum_{\mu,\mu'} a_{\mu,\mu'} m_\mu m_{\mu'} + 2 \sum_\mu m_\mu \nu_\mu\Big), \tag{1}$$

where $\nu = (\nu_1, \nu_2, \ldots, \nu_p) \in \mathbb{C}^p$; it is convergent because the real part of $(a_{\mu\nu})$ is negative and not degenerate. The theta-function is characterized (up to a constant factor) by the equations

$$\vartheta(\nu) = \vartheta(\nu + \pi i B_\mu) = \exp(2\nu_\mu + a_{\mu\mu})\vartheta(\nu + A_\mu) \quad (1 \leq \mu \leq p) \tag{2}$$

where $\nu = (\nu_1, \nu_2, \ldots, \nu_p)$, $(B_\mu)_\mu$ is the canonical basis of $\mathbb{C}^p$ and $A_\mu = (a_{\nu\mu})_\nu$. The $2p$ vectors $\pi i B_\mu$ and A_μ generate a discrete subgroup Γ of $\mathbb{C}^p$ and the factor group $\mathbb{C}^p/\Gamma = J$ is called the *jacobian* of T; it is a complex torus of dimension p. In modern terms, ϑ may be interpreted as a section of an invertible sheaf on J. Riemann's aim is to prove that the mapping

$$\phi^{(p)} : \eta = \{\eta_1, \eta_2, \ldots, \eta_p\} \mapsto \text{class mod } \Gamma \text{ of } \Big(\textstyle\sum_{\nu=1}^{p} u_\mu(\eta_\nu)\Big)_\mu \tag{3}$$

of the symmetric product $\mathrm{Sym}^p T$ of p copies of T in J is *onto*: for every $e \in \mathbb{C}^p$, there exists an $\eta \in \mathrm{Sym}^p T$ such that $\phi^{(p)}(\eta) \equiv e(\text{mod } \Gamma)$. For this purpose, he observes that, if the function $\vartheta_e : x \mapsto \vartheta(u_1(x) - e_1, u_2(x) - e_2, \ldots, u_p(x) - e_p) = \vartheta(u - e)$ (where $u = (u_\mu)_\mu$) is not identically 0 in T', it has exactly p zeros: indeed the integral of the logarithmic derivative of ϑ_e on each a_μ (resp. b_μ, c_μ) is equal to $2\pi i$ (resp. 0) so that the integral on the boundary of $T \setminus \{a, b, c\}$ is $2\pi i p$.

Let $\eta_1, \eta_2, \ldots, \eta_p$ be these zeros and let ℓ_μ be a cut joining η_μ to the common origin of a_μ and $b_\mu (1 \leq \mu \leq p)$, so that $\log \vartheta_e$ becomes a single-valued function in $T^* = T - \{a_\mu, b_\mu, \ell_\mu\}$. The differences of the values of $\log \vartheta_e$ when crossing the cuts ℓ_ν, a_ν and b_ν are respectively

$$-2\pi i, 2\pi i g_\nu, \text{ and } -2(u_\nu - e_\nu) - 2\pi i h_\nu,$$

where g_ν and h_ν are integers; then $\int_{\partial T^*} \log \vartheta_e du_\mu$ is equal to

$$2\pi i\Big(\sum_{\nu=1}^{p} u_\mu(\eta_\nu) + \sum_{n=1}^{p} g_\nu a_{\nu\mu} - e_\mu + \pi i h_\mu + K_\mu\Big), \tag{4}$$

where the K_μ are constants which depend only on the origins x_μ of the integrals u_μ (and not on the choice of e). A convenient choice of the x_μ reduces K_μ to 0.

As $\log \vartheta_e$ is single-valued in T^*, the expression (4) must be equal to 0 and this proves the fact that, when the ϑ_e is not identically 0, the class of e (mod Γ) belongs to $\phi(p)(\mathrm{Sym}^p T)$:

$$e_\mu \equiv \sum_{\nu=1}^{p} u_\mu(\eta_\nu)(\text{mod } \Gamma). \tag{5}$$

Now the element $\{\eta_1, \eta_2, \ldots, \eta_p\}$ of $\mathrm{Sym}^p T$ such that (5) is satisfied is *unique*; indeed, if $e \equiv \sum_{\nu=1}^{p} u_\mu(\eta'_\nu)(\text{mod } \Gamma)$ for another $\{\eta'_1, \eta'_2, \ldots, \eta'_p\} \in \mathrm{Sym}^p T$ with $\eta'_p \neq \eta_p$, $\vartheta(u - e)$ vanishes in the $p + 1$ points $\eta_1, \eta_2, \ldots, \eta_p$ and η'_p, which is impossible.

The end of the memoir of 1857 is devoted to the proof that the integrals of the algebraic differential forms on T may be expressed as quotients of products of translated theta functions.

2 A Crucial Observation on Theta Functions

Riemann observes that $\vartheta(u(\eta_p) - e) = \vartheta_e(\eta_p) = 0$ implies

$$\vartheta\Big(\sum_{\nu=1}^{p-1} u(\eta_\nu)\Big) = \vartheta\Big(-\sum_{\nu=1}^{p-1} u(\eta_\nu)\Big) = 0 \tag{6}$$

(ϑ is an even function of ν); this is proved for every system of $p-1$ zeros $\eta_1, \eta_2, \ldots, \eta_{p-1}$ of a function $\vartheta(u-e)$ supposed not to be identically zero. Conversely, if $\vartheta(r) = 0$ for an $r \in \mathbb{C}^p$ such that $\vartheta(u - u(\eta_0) + r)$ is not identically 0, with $\eta_0 \in T$ conveniently chosen, the application of what precedes to $e = u(\eta_0) - r$ gives $r \equiv -(\eta_\nu)(\mathrm{mod}\ \Gamma)$, where $\eta_1, \eta_2, \ldots, \eta_{p-1}$ are the zeros of $\vartheta(u - u(\eta_0) + r)$ other than η_0.

Riemann proves in his 1865 memoir [9] that the set $\vartheta^{-1}(0) \subset J$ is exactly the image W_{p-1} of the mapping

$$\phi^{(p-1)} : \eta_1, \eta_2, \ldots, \eta_{p-1} \mapsto \text{ class of} \sum_{\nu=1}^{p-1} u_\mu(\eta_\nu)(\ \mathrm{mod}\ \Gamma)$$

of $\mathrm{Sym}^{\mathrm{p-1}}\mathrm{T}$ into J. Let us admit this result and consider an $e \in \mathbb{C}^p$ such that $\vartheta_e = \vartheta(u-e)$ is identically 0; for each $\eta_p \in T$, there exist $\eta_1, \eta_2, \ldots, \eta_{p-1}$ such that $u(\eta_p) - e \equiv -(\eta_\nu)(\ \mathrm{mod}\ \ \Gamma)$ so that $e \equiv \sum_{\nu=1}^{p} u(\eta_\nu)(\ \mathrm{mod}\ \Gamma)$, but this time not in a unique way, for η_p is arbitrary. If e is fixed and η_p varies, $\sum_{\nu=1}^{p} du_{\eta_\nu} = 0$; now Riemann has proved (in the 1857 memoir [8]) that this implies the existence of a differential form of the first kind which vanishes at $\eta_1, \eta_2, \ldots, \eta_p$. Such a form has $2p-2$ zeros, so there are $p-2$ other zeros $\eta_{p+1}, \ldots, \eta_{2p-2}$ and they are linked by the equation $\sum_{\nu=1}^{2p-2} u(\eta_\nu) = 0$ from which we get

$$e \equiv -\sum_{\nu=p}^{2p-2} u(\eta_\nu)(\ \mathrm{mod}\ \Gamma). \tag{7}$$

Conversely, if e satisfies (7), $\vartheta(u(\eta_p) - e) = \vartheta\Big(\sum_{\nu=p}^{2p-2} u(\eta_\nu)\Big) = 0$ for all $\eta_p \in T$, which means that $\vartheta(u-e)$ is identically 0. In other words, the subset of J above which $\phi^{(p)}$ is not one-to-one is the image of the mapping $\{\eta_1, \eta_2, \ldots, \eta_{p-2}\} \mapsto -\sum_{\nu=1}^{p-2} u(\eta_\nu)$ of $\mathrm{Sym}^{\mathrm{p-2}}\mathrm{T}$ into J.

3 The First Step of Riemann's Proof

The 1865 memoir [9], begins with the proof of the fact that ϑ vanishes on $W_{p-1} = \phi^{(p-1)}(\mathrm{Sym}^{p-1}T)$; this is a consequence of the principle of analytic continuation. As ϑ is not identically 0, there exists a non-empty open part $E \neq \emptyset$ of $\mathbb{C}^p$ such that $\vartheta_e = \vartheta(u-e)$ does not vanishes identically when $e \in E$; the Eq. (7) shows that $E \mod \Gamma$ is contained in $\phi^{(p)}(H)$ where $H \subset \mathrm{Sym}^p T$ is the corresponding set of $\{\eta_1, \eta_2, \dots, \eta_p\}$ (zeros of ϑ_e) and H must be a non-empty open set of $\mathrm{Sym}^p T$. Then ϑ vanishes on $\phi^{(p-1)}(\eta_1, \eta_2, \dots, \eta_{p-1})$ for $\{\eta_1, \eta_2, \dots, \eta_p\} \in H$, so it vanishes everywhere.

When ϑ_e is identically 0, let m be the least integer such that

$$\vartheta\Big(u(\eta_0) + \sum_{\nu=1}^{m} \big(u(\eta_\nu) - u(\varepsilon_\nu)\big) - e\Big) \tag{8}$$

does not vanish for all $\eta_0, \eta_1, \dots, \eta_m, \varepsilon_1, \dots \varepsilon_m \in T$: in the sequel of the paper, Riemann will prove that such an m always exists. The set of $(\eta_\mu, \varepsilon_\nu)$ such that (8) is different from 0 is a non-empty open part of T^{2m+1}. The function

$$x \mapsto \vartheta\Big(u(x) + \sum_{\nu=1}^{m} \big(u(\eta_\nu) - u(\varepsilon_\nu)\big) - e\Big)$$

vanishes at the points $\varepsilon_1, \dots, \varepsilon_m$ so $m \leq p$ and there are $p-m$ other zeroes $\eta_{m+1}, \dots, \eta_p$, which are uniquely determined. Equation (5) applied to $e' = e - \sum_{\nu=1}^{m} \big(u(\eta_\nu) - u(\varepsilon_\nu)\big) (\mathrm{mod}\ \Gamma)$ gives

$$\sum_{\nu=1}^{p-m} u(\eta_{m+\nu}) \equiv e - \sum_{\nu=1}^{m} u(\eta_\nu) \mod \Gamma$$

or $e \equiv \sum_{\nu=1}^{p} u(\eta_\nu)$ (mod Γ), so Eq. (5) is still valid but the set of $\eta \in \mathrm{Sym}^p T$ satisfying it is now of dimension m, for η is determined by $\eta_1, \dots, \eta_m$: it is a case of indetermination for Jacobi's inversion problem.

On the other hand, the function

$$x \mapsto \vartheta\Big(u(\eta_0) + \sum_{\nu=1}^{m} \big(u(\eta_\nu) - u(\varepsilon_\nu)\big) + u(\eta_m) - u(x) - e\Big)$$

vanishes for $x = \eta_0, \eta_1, \dots, \eta_m$ so $m \leq p-1$ and there are $p-m-1$ other zeroes $\varepsilon_{m+1}, \dots, \varepsilon_{p-1}$ determined in a unique manner; one has

$$-\sum_{\nu-1}^{m-1} u(\varepsilon_\nu) - e \equiv \sum_{\nu=1}^{p-m-1} u(\varepsilon_{m+\nu}) (\mathrm{mod}\ \Gamma) \quad \text{or} \quad -e \equiv \sum_{\nu=1}^{p-1} u(\varepsilon_\nu) \ (\mathrm{mod}\ \Gamma),$$

with a set of $\varepsilon \in \mathrm{Sym}^{p-1}T$ of dimension $m-1$. So, for each class $e(\mathrm{mod}\Gamma) \in J$,

$$\dim(\phi^{(p-1)})^{-1}(-e(\mathrm{mod}\ \Gamma)) = \dim(\phi^{(p)})^{-1}(e(\mathrm{mod}\ \Gamma)) - 1$$

4 The Second Step of Riemann's Proof

Consider now an $r = (r_\mu)_\mu \in \mathbb{C}^p$ such that $\vartheta(r) = 0$ and let m be the biggest integer such that $\vartheta\Big(\sum_{\nu=1}^m \big(u(\varepsilon_\nu) - u(\eta_\nu)\big) + r\Big)$ does not vanish for all $\eta_1, \ldots, \eta_m, \varepsilon_1, \ldots, \varepsilon_m$; if $\eta_1, \ldots, \eta_{m+1}, \varepsilon_1, \ldots, \varepsilon_{m+1}$ are given, the function

$$x \mapsto \vartheta\Big(u(x) - u(\eta_{m+1}) + \sum_{\nu=1}^m \big(u(\varepsilon_\nu) - u(\eta_\nu)\big) + r\Big)$$

vanishes for $x = \eta_1, \ldots, \eta_{m+1}$ and it has $p-m-1$ other zeroes $\varepsilon_{m+1}, \ldots, \varepsilon_{p-1}$, uniquely determined. Applying (5) to

$$e = -r - \sum_{\nu=1}^m u(\varepsilon_\nu) + \sum_{\nu=1}^{m+1} u(\eta_\nu),$$

one gets

$$-r - \sum_{\nu=1}^m u(\varepsilon_\nu) \equiv \sum_{\nu=m+1}^{p-1} u(\varepsilon_\nu)\ (\mathrm{mod}\ \Gamma) \text{ or } -r \equiv \sum_{\nu=1}^{p-1} u(\varepsilon_\nu)\ (\mathrm{mod}\ \Gamma).$$

In the same way, the function $\vartheta\Big(u - u(\varepsilon_{m+1}) - \sum_{\nu=1}^m \big(u(\varepsilon_\nu) - u(\eta_\nu)\big) - r\Big)$ vanishes for $x = \varepsilon_1, \ldots, \varepsilon_{m+1}$ and it has $p-m-1$ other zeros $\eta_{m+1}, \ldots, \eta_{p-1}$ uniquely determined; it implies that $r \equiv \sum_{\nu=1}^{p-1} u(\eta_\nu)\ \ (\mathrm{mod}\ \Gamma)$. So the set $(\phi^{(p-1)})^{-1}(r\ \mathrm{mod}\ \Gamma)$ is non-empty and its dimension is equal to m.

5 The Conclusion of the Proof

The end of Riemann's paper is devoted to the proof of the fact that, for each $r \in \mathbb{C}^p$, the integer $m = \dim(\phi^{(p-1)})^{-1}(r\ \mathrm{mod}\ \Gamma)$ is the biggest such that the differential $d^k\vartheta$ vanishes in r for $0 \le k \le m$; as ϑ is not identically 0 this shows that m exists. When $m \ge 1$, the variety $W_{p-1} = \phi^{(p-1)}(\mathrm{Sym}^{p-1}T) = \vartheta^{-1}(0) \subset J$ has a singularity of multiplicity $m+1$ in r.

Assume that

$$\vartheta\Big(\sum_{\nu=1}^m \big(u(\varepsilon_\nu) - u(\eta_\nu)\big) + r\Big) = 0$$

for all $\eta_1, \ldots, \eta_m, \varepsilon_1, \ldots, \varepsilon_m$; if $0 \leq n \leq m$, making $\varepsilon_\nu = \eta_\nu$ for $m - n + 1 \leq \nu \leq m$, we see that

$$\vartheta\Big(\sum_{\nu=1}^{m-n} \big(u(\varepsilon_\nu) - u(\eta_\nu)\big) + r\Big) = 0.$$

By induction on n, Riemann shows that $d^n\vartheta$ vanishes at $r_n = r + \sum_{\nu=1}^{m-n} \big(u(\varepsilon_\nu) - u(\eta_\nu)\big)$; for $n = 0$ there is nothing to prove so we may suppose that $n \geq 1$ and the result is established for $n - 1$. Then

$$0 = d^{n-1}\vartheta_{r_{n-1}}(r_{n-1} + u(\varepsilon_{m-n}) - u(\eta_{m-n})) = d^n\vartheta_{r_n} \circ du_{\eta_{m-n}}(\zeta) + o(\zeta)$$

where ζ is the value at ε_{m-n} of a uniformizing parameter in the neighborhood of η_{m-n}; as the components of u_μ of u are linearly independent, the linear application $du_{\eta_{m-n}} : \mathbb{C}^p \to \mathbb{C}^p$ is invertible and we have $d^n\vartheta_{r_n} = 0$. Making $\varepsilon_\nu = \eta_\nu$ for $1 \leq \nu \leq m - n$, we see that $d^n\vartheta_r = 0$ for $0 \leq n \leq m$.

The proof of the converse is a bit tricky. In a first step, Riemann assumes that

$$\vartheta\Big(\sum_{\nu=1}^{k-1} \big(u(\varepsilon_\nu) - u(\eta_\nu)\big) + r\Big) = 0 \quad \text{for all } \eta_1, \ldots, \eta_{k-1}, \varepsilon_1, \ldots \varepsilon_{k-1}$$

but that

$$\vartheta\Big(\sum_{\nu=1}^{k} \big(u(\varepsilon_\nu) - u(\eta_\nu)\big) + r\Big)$$

is not identically equal to 0; the first part of the proof shows that

$$d^{n-1}\vartheta_{r_n} = 0 \quad \text{for } 1 \leq n \leq k.$$

Let $t = (t_\mu) \in \mathbb{C}^p$ be such that $\vartheta(t) = 0$ but $\theta(u(\varepsilon_1) - u(\eta_1) + t) = 0$ does not vanish for all $\eta_1, \varepsilon_1 \in T$, and consider the expression

$$\Psi = \vartheta\Big(\sum_{\nu=1}^{k} \big(u(\eta_\nu) - u(\varepsilon_\nu)\big) + r\Big)\vartheta\Big(\sum_{\nu=1}^{k} \big(u(\varepsilon_\nu) - u(\eta_\nu)\big) + r\Big) \times \frac{\prod_{\rho \neq \rho'} \vartheta\big(u(\eta_\rho) - u(\eta_{\rho'})\big) + t\big)\vartheta\big(u(\varepsilon_\rho) - u(\varepsilon_{\rho'})\big) + t\big)}{\prod_{\rho,\rho'=1}^{k} \vartheta\big(u(\eta_\rho) - u(\varepsilon_{\rho'})\big) + t\big)\vartheta\big(u(\varepsilon_\rho) - u(\eta_{\rho'})\big) + t\big)}$$

as a function of η_μ; as all its periods are 0, it is an algebraic function on T, more precisely a rational function of (s, z). Such an algebraic function is determined, up to a factor independent of η_μ, by the knowledge of its zeroes and poles; for $\eta_\mu = \varepsilon_\rho$, the numerator and the denominator vanish to order 2, so Ψ is regular at $\varepsilon_1, \ldots, \varepsilon_k$. The numerator or the denominator of Ψ vanishes at other points which do not depend on $\varepsilon_1, \ldots, \varepsilon_k$ but only on r, t and $\eta_\nu, \nu \neq \mu$. Thus we see that Ψ is an algebraic function of $\eta_1, \ldots, \eta_k$; uniquely determined by r and t up to a factor which depends on $\varepsilon_1, \ldots, \varepsilon_k$; exchanging the roles of η and ε, we get

$$\Psi = \psi(\eta_1, \ldots, \eta_m)\psi(\varepsilon_1, \ldots, \varepsilon_m)$$

where ψ is an algebraic function.

When ε is close to η , let us write $\varepsilon_\nu = \eta_\nu + \zeta_\nu$ so that

$$u(\varepsilon_\nu) = u(\eta_\nu) + du_{\eta_\nu}(\zeta_\nu) + o(\zeta_\nu)$$

and, by induction on n,

$$\vartheta\Big(\sum_{\nu=1}^{k} \big(u(\varepsilon_\nu) - u(\eta_\nu)\big) + r\Big)$$

$$= d^n\vartheta_{r_n} \circ (du_{\eta_{k-n+1}} \otimes \cdots \otimes du_{\eta_k})(\zeta_{k-n+1}, \ldots, \zeta_k) + o(\zeta_{k-n+1}, \ldots, \zeta_k)$$

$$= d^k\vartheta_r \circ (du_{\eta_1} \otimes \cdots \otimes du_{\eta_k})(\zeta_1, \ldots, \zeta_k) + o(\zeta_1, \ldots, \zeta_k),$$

and $\vartheta\Big(\sum_{\nu=1}^{k} \big(u(\eta_\nu) - u(\varepsilon_\nu)\big) + r\Big)$ is treated in a similar manner. The factors containing t in the numerator of Ψ differ from the corresponding factors in the denominator by negligible terms and, for $\varepsilon = \eta$, we thus obtain, after computing a square root,

$$\psi(\eta_1, \ldots, \eta_k) = \pm\frac{d^k\vartheta_r \circ (du_{\eta_1} \otimes \cdots \otimes du_{\eta_k})}{\Pi_\rho^k d\vartheta_t \circ du_{\eta_\rho}} = \pm\frac{d^k\vartheta_r(\varphi(\eta_1), \ldots, \varphi(\eta_k))}{\Pi_{\rho=1}^k d\vartheta_t(\varphi(\eta_\rho))},$$

where $\varphi = (\varphi_1, \ldots, \varphi_p)$ and $du_\mu = \frac{\varphi_\mu dz}{\partial F/\partial s}$ for $1 \le \mu \le p$. Now if $d^k\vartheta_r = 0$, $\psi = 0$ and $\Psi = 0$ identically, so $\vartheta\Big(\sum_{\nu=1}^{k} \big(u(\varepsilon_\nu) - u(\eta_\nu)\big) + r\Big) = 0$ for all $\eta_1, \ldots, \eta_k$, $\varepsilon_1, \ldots, \varepsilon_k$, contrary to the assumption.

What precedes shows that, when $\vartheta\Big(\sum_{\nu=1}^{k-1} \big(u(\varepsilon_\nu) - u(\eta_\nu)\big) + r\Big) = 0$ identically and $d^k\vartheta_r = 0$, we have $\vartheta\Big(\sum_{\nu=1}^{k} \big(u(\varepsilon_\nu) - u(\eta_\nu)\big) + r\Big) = 0$ identically. This allows a proof by induction on $k \in [0, m]$ of the fact that, when m is the biggest integer such that $d^k\vartheta_r = 0$ for $0 \le k \le m$, one has

$$\vartheta\Big(\sum_{\nu=1}^{k} \big(u(\varepsilon_\nu) - u(\eta_\nu)\big) + r\Big) = 0 \quad \text{identically for } 0 \le k < m$$

For $k = 0$ there is nothing to prove, so that we may suppose $k \ge 1$ and the statement true for $k - 1$; thus $\vartheta\Big(\sum_{\nu=1}^{k-1} \big(u(\varepsilon_\nu) - u(\eta_\nu)\big) + r\Big) = 0$ identically and, as $d^k\vartheta_r = 0$, we conclude that

$$\vartheta\Big(\sum_{\nu=1}^{k}\big(u(\varepsilon_\nu)-u(\eta_\nu)\big)+r\Big)=0 \quad \text{identically.}$$

For $k = m$, this completes the proof of Riemann's theorem.

6 Later Developments

To sum up, Riemann has demonstrated that $\phi^{(p)} : \mathrm{Sym}^p T \to J$ is onto but not one-to-one when $p \geq 2$; it only *birational*. Riemann gave a characterization of the subset of J above which $\phi^{(p)}$ is not one-to-one and, for each $e \in J$, he gave a rule to know the dimension of $(\phi^{(p)})^{-1}(e)$. The other result is that the image W_{p-1} of $\phi^{(p-1)}$: $\mathrm{Sym}^{p-1}T \to J$ is equal to $\vartheta^{-1}(0)$; moreover, the multiplicity of W_{p-1} in a point r is equal to $\dim(\phi^{p-1})^{-1}(r) + 1$.

Let us mention some more recent results in the line of Riemann's research. Weil [14] exploited the birational character of $\phi(p)$ in order to define the jacobian of an algebraic curve over an abstract commutative field k. The Riemann-Roch theorem enabled him to define a partial group law $U \times U \to \mathrm{Sym}^p T$, where U is a Zariski open set in $\mathrm{Sym}^p T$; then he showed how to obtain an algebraic group J birationally equivalent to $\mathrm{Sym}^p T$, with a group law continuing the partial law of $\mathrm{Sym}^p T$.

In the years 1971–73, G. Kempf [5, 6] completed Riemann's results. He considered, for each $k \in [1, p-1]$, the mapping $\phi^{(k)} : \eta_1, \ldots, \eta_k \mapsto \sum_{\nu-1}^{k} u(\eta_\nu) \pmod{\Gamma}$ of $\mathrm{Sym}^k T$ into J and its image $\phi^{(k)}(\mathrm{Sym}^k T) = W_k$; he proved the following results [7]:

(a) for each $r \in W_k$, $(\phi^{(k)})^{-1}(r)$ is isomorphic to a projective space;

(b) the multiplicity of W_k in r is equal to

$$\binom{p-k+m}{m},$$

where m is the dimension of $(\phi^{(k)})^{-1}(r)$ and the tangent cone to W_k at r is the union of the images by $d\phi^{(k)}$ of the tangent spaces to $\mathrm{Sym}^k T$ at the points $\eta \in (\phi^{(k)})^{-1}(r)$;

(c) W_k is a determinental variety; more precisely, in the neighborhood of each point r, there exists a $(m+1) \times (p-k+m)$-matrix for holomorphic functions such that W_k is the set of zeroes of its $(m+1) \times (m+1)$-minors.

References

1. G.A. Göpel, Theoriae transcendentium Abelianarum primi ordinis adumbratio levis. Journ. für die reine und angewandte Math. **35**, 277–312 (1847)
2. C.G. Jacobi, Considerationes generales de transcendentibus Abelianis. J. für die r. und a. Math. **9**, 394–403, Ges. Werke, t. II, 5–16 (1832)

3. C.G. Jacobi, De functionibus duarum variabilium quadrupliciter periodicis quibus theoria transcendentium Abelianarum innititur. J. für die r. und a. Math. **13**, 55–78, Ges. Werke, t. II, 28–50 (1835)
4. C.G. Jacobi, Note sur les fonctions Abéliennes. J. für die r. und a. Math. **30**, 183–184, Ges. Werke, t. II, 86–87 (1846)
5. G. Kempf, *Schubert Methods With an Application to Algebraic Curves* (Stichting math. Centrom, Amsterdam, 1971)
6. G. Kempf, On the geometry of a theorem of Riemann. Ann. Math. **98** (1973)
7. D. Mumford, *Curves and Their Jacobians*, 2nd edn. (The University of Michigan Press, Ann Arbor, 1975), pp. 1–104. (In the *Red Book of Varieties and Schemes*, Springer L. N. 1358, pp. 225–304 (2010))
8. B. Riemann, Theorie der Abel'schen Functionen. J. für die r. und a. Math. **54**, 115–155 (1857), Ges. Math. Werke, 88–144
9. B. Riemann, Über das Verschwinden der ϑ-Functionen. J. für die r. und a. Math. **65**, 161–172, Ges. Math. Werke, 212–224 (1866)
10. J.G. Rosenhain, Auszug mehrerer Schreiben des Dr. Rosenhain an Herrn Professor Jacobi über die hyperelliptischen Transcendenten. J. für die r. und a. Math. **40**, 319–360 (1850)
11. J.G. Rosenhain, Sur les Fonctions de deux variables et à quatre périodes, qui sont les intégrales ultra-elliptiques de la première classe, Ac. des Sc. de Paris, Mémoires présentés par divers Savants, 2e série **11**, 361–468 (1851)
12. K. Weierstrass, J. der Zur Theorie, Abelschen Functionen, für die r. und a. Math. **47**, 289–306, Math. Werke, t. I, 131–133 (1854)
13. K. Weierstrass, J. Theorie der Abelschen, Functionen, für die r. und a. Math. **52**, 285–379, Math. Werke, t. I, 297–355 (1856)
14. A. Weil, *Variétés abéliennes et courbes algébriques*, 2nd edn (1948). Courbes algébriques et varieties abéliennes, (Hermann, Paris, 1971), pp. 87–249

Riemann's Work on Minimal Surfaces

Sumio Yamada

Abstract Three months before his death in 1866, Riemann left a set of notes to K. Hattendorff, a disciple of his, on minimal surfaces with boundary. Afterwards, Hattendorff supplied the text to the notes mostly consisting of computations, which became the two papers on the subject: "On the surface of least area with a given boundary" and "Examples of surfaces of least area with a given boundary." We will go over the expositions and provide an overview from the modern viewpoint, make some comments on Riemann-Hattendorff's text, and compare the work with that of Weierstrass on the same subject.

Keywords Riemann · Minimal surface · Calculus of variations

2000 Mathematics Subject Classication: 53A10 · 30F45

1 Introduction

We first recall the statement of the Riemann mapping theorem,

Theorem 1.1 *If Ω is a non-empty simply connected open proper subset of the complex plane $\mathbb{C}$, then there exists a bi-holomorphic mapping f from Ω onto the open unit disk $D = \{z \in \mathbb{C} \mid |z| < 1\}$.*

We remark that the resulting plane region Ω is a minimal surface spanned by the boundary $\partial\Omega$ when the complex plane is identified with the xy-plane in the three dimensional ambient space $\mathbb{R}^3$.

It was in Riemann's thesis [8] in 1851 where the cerebrated Riemann Mapping Theorem was first presented. There he utilized the Dirichlet principle in order to obtain a map which is harmonic and conformal from a simply connected region of the complex plane to the unit disc. Riemann's proof was incomplete as the existence of such harmonic functions minimizing the Dirichlet energy functional is not always

S. Yamada (✉)
Department of Mathematics, Gakushuin University, Tokyo 171-8588, Japan
e-mail: yamada@math.gakushuin.ac.jp

L. Ji et al. (eds.), *From Riemann to Differential Geometry and Relativity*,
DOI 10.1007/978-3-319-60039-0_5

guaranteed, a defect first pointed out by Weierstrass in 1859. The subsequent important historical development in analysis and geometry initiated by Riemann's idea of using the Dirichlet principle, which led to the wealth of mathematics around the so-called Plateau problem, is comprehensively described in Courant's book [2].

Hans Lewy, in his introduction [5] to an edition of collected works of Riemann (1953), speculates that the work of Riemann on the subject of minimal surfaces in $\mathbb{R}^3$ [9, 10] may well be an attempt to rectify his own proof of the Riemann Mapping Theorem, as the inverse map $f^{-1} : D \to \Omega \subset \mathbb{R}^3$ of the Riemann mapping provides a harmonic and conformal parameterization of the *minimal* surface $\Omega \subset xy$-plane, where the xy-plane is identified as $\mathbb{C}$. The set of minimal surfaces Riemann succeeded in constructing, however, have boundary sets only of special types; lines, line segments, circles, with which one would not be able to approximate an arbitrary boundary curve $\partial\Omega$ as required by the statement of the Riemann Mapping Theorem.

Having stated this observation about a failed attempt, however, it is important to recognize the true value of what Riemann created in the field of minimal surfaces within the two posthumously published papers [9, 10]. Historically, the study of minimal surfaces and later minimal submanifolds has led to many interesting applications, not only in differential geometry, but also in general relativity, material sciences, industrial design, among others. Riemann was correct to foresee the scientific potential the subject offered. As for the minimal surfaces in $\mathbb{R}^3$, much of the subsequent development on the subject up to the present day is based on the so-called Weierstrass-Enneper representation formula, which is comprehensively presented in [11].

The goal of this chapter is to illustrate that inside the notes Riemann left to Hattendorff, based on the computations Riemann had made over 1860–61, much of the well-known classical results on minimal surfaces including the Weierstrass-Enneper representation, Schwarz's explicit construction of minimal surfaces, as well as the Schwarz-Christoffel transformation, are contained in essence. We will demonstrate this by reading the text of [9] through §1–13 with comments added as appropriate. In particular in the second to the last section, we provide a direct comparison between the Weierstrass-Enneper representation and the much less known Riemann representation of minimal surface.

2 On the Surface of Least Area with a Given Boundary

Regarding the treatment of surfaces in $\mathbb{R}^3$ in this article, it is safe to assume that Riemann was fully informed of the surface theory of Gauss [4] where every surface is locally parameterized by two independent real variables u, v.

In §1 and §2, a disc-type surface Ω is parameterized by two parameters p, q which are effectively the polar coordinates in the two-dimensional disc D_r; $p \in [0, r]$ on the radial set $\{q = \text{const.}\}$ and $q \in [0, 2\pi\rho)$ on the circle $\{p = \rho\}$. Then a change of variable formula is given so that for a new set of parameters $\phi(u, v) := (f(u, v), g(u, v))$, the area functional on $\phi(D_r) := \Omega$ with respect to the coordinates (f, g) is written as

$$\iint_{\Omega} df \wedge dg = \int\int_{D} \left(\frac{\partial f}{\partial p}\frac{\partial g}{\partial q} - \frac{\partial f}{\partial q}\frac{\partial g}{\partial p}\right) dp \wedge dq$$

which then, is shown to be equal to

$$\int_{\partial\Omega} f dg = -\int_{\partial\Omega} g df$$

by using the fact that

$$\frac{\partial f}{\partial p}\frac{\partial g}{\partial q} - \frac{\partial f}{\partial q}\frac{\partial g}{\partial p} = \frac{\partial\left(f\frac{\partial g}{\partial q}\right)}{\partial p} - \frac{\left(f\frac{\partial g}{\partial p}\right)}{\partial q},$$

that f and g are periodic in q, and that $\frac{\partial g}{\partial q} = 0$ on $\{p = 0\}$, as well as the fundamental theorem of calculus.

In §3, the surface Ω in $\mathbb{R}^3$ parameterized by (p, q) is again parameterized by a different pair of parameters, namely the image of the Gauss map $\nu : \Omega \to S^2$, or rather, as introduced in §5, the stereographic projection P_S of the image of the Gauss map ν from the south pole $S = (-1, 0, 0)$ of S^2:

$$\eta := P_S \circ \nu : \Omega \to yz\text{-plane}.$$

This new complex variable η becomes useful and indeed central in Riemann's thinking of the conformal geometry of minimal surfaces, as we will see below.

In §4, the Euler-Lagrange equation for the minimal surface is obtained. For each point $P = (x, y, z)$ of the surface Ω, let $\nu(P)$ be the unit normal vector to Ω at P. The unit normal ν in S^2 has the spherical coordinates $\nu(p, q) = (r, \phi) \in [0, \pi] \times [0, 2\pi)$ where $\{r = 0\}$ stands for the point $(1, 0, 0)$ and $(r, \phi) = (\pi/2, 0)$ stands for $(0, 1, 0)$. Further assume that near P, locally the surface is a graph $\{(x, y, z) \mid x = x(y, z)\}$ over a region $\tilde{\Omega}$ over the yz-plane. Then the tangent plane at P is $dx = (\partial x/\partial y)dy + (\partial x/\partial z)dz$, from which we deduce the following set of equalities

$$\cos r dx + \sin r \cos\phi\, dy + \sin r \sin\phi\, dz = 0, \tag{2.1}$$

as well as

$$\cos r = \pm\frac{1}{\sqrt{1 + \left(\frac{\partial x}{\partial y}\right)^2 + \left(\frac{\partial x}{\partial z}\right)^2}}, \tag{2.2}$$

$$\sin r \cos\phi = \mp \frac{\frac{\partial x}{\partial y}}{\sqrt{1+\left(\frac{\partial x}{\partial y}\right)^2+\left(\frac{\partial x}{\partial z}\right)^2}}, \tag{2.3}$$

$$\sin r \sin\phi = \mp \frac{\frac{\partial x}{\partial z}}{\sqrt{1+\left(\frac{\partial x}{\partial y}\right)^2+\left(\frac{\partial x}{\partial z}\right)^2}}, \tag{2.4}$$

where the double $\pm$ signs correspond to each other. Now the point P on the surface is assigned to the spherical coordinates $\nu(P(p,q)) = (r(p,q), \phi(p,q))$ and one calculates the area of the surface Ω by integrating over $\tilde{\Omega}$

$$S = \iint_{\Omega} \frac{1}{\cos r} dydz = \iint_{\nu(\Omega)} \pm\sqrt{1+\left(\frac{\partial x}{\partial y}\right)^2+\left(\frac{\partial x}{\partial z}\right)^2}\, dydz$$

where $\frac{1}{\cos r} dydz$ is the area element of the graph $\{x = x(y,z)\}$. Note that the original parameters p and q are suppressed. As for the double signs, we remark that when $r > \pi/2$, the 2-form $dydz$ is of the form $dz \wedge dy$.

The first variation of the area functional under a variation vector field δx which is compactly supported on Ω is then given by

$$\delta S = \iint_{\nu(\Omega)} \left[\frac{\partial(\delta x)}{\partial y} \frac{\frac{\partial x}{\partial y}}{\sqrt{1+\left(\frac{\partial x}{\partial y}\right)^2+\left(\frac{\partial x}{\partial z}\right)^2}} + \frac{\partial(\delta x)}{\partial z} \frac{\frac{\partial x}{\partial z}}{\sqrt{1+\left(\frac{\partial x}{\partial y}\right)^2+\left(\frac{\partial x}{\partial z}\right)^2}} \right] dydz$$

so that the Euler-Lagrange equation for the area functional is

$$\frac{\partial}{\partial y}\left(\frac{\frac{\partial x}{\partial y}}{\sqrt{1+\left(\frac{\partial x}{\partial y}\right)^2+\left(\frac{\partial x}{\partial z}\right)^2}} \right) + \frac{\partial}{\partial z}\left(\frac{\frac{\partial x}{\partial z}}{\sqrt{1+\left(\frac{\partial x}{\partial y}\right)^2+\left(\frac{\partial x}{\partial z}\right)^2}} \right) = 0 \tag{2.5}$$

which, using the Eqs. (2.3) and (2.4), can be rewritten in (r, ϕ) coordinates as

$$\frac{\partial \sin r \cos\phi}{\partial y} + \frac{\partial \sin r \sin\phi}{\partial z} = 0. \tag{2.6}$$

Equation (2.5), though it does not appear explicitly in Riemann's manuscript (Eq. (2.6) does), is the minimal surface equation for the graph $x = x(y,z)$, first obtained by Lagrange in "Essai d'une nouvelle méthode pour déterminer les maxima et les minima des formules intégrales indéfinies." (1760–61.) We also note that the expression on the left hand side is the mean curvature of the graphical surface. Riemann was under a direct influence of Dirichlet in his Berlin years, and he was most likely well informed of the development of the calculus of variations by pioneers such as Euler and Lagrange.

Equation (2.6) is now regarded as an integrability condition: Namely the 1-form $-\sin r \sin \phi dy + \sin r \cos \phi dz$ is exact, and there is a potential function $\mathfrak{x}$ so that

$$d\mathfrak{x} = -\sin r \sin \phi dy + \sin r \cos \phi dz. \tag{2.7}$$

On the other hand, Eq. (2.1) is an integrability condition in the sense that x is the potential for the 1-form:

$$dx = -\tan r \cos \phi dy - \tan r \sin \phi dz. \tag{2.8}$$

We note that the stereographic projection from the south pole of the image of the Gauss map $\eta = P_S \circ \nu$ is, in terms of $(r, \phi) \in S^2$,

$$\eta = \tan \frac{r}{2} e^{\phi i}.$$

Denoting the complex conjugate of η and $s := y + iz$ respectively by

$$\eta' = \tan \frac{r}{2} e^{-\phi i}, \qquad s' = y - iz,$$

the pair of integrability conditions (2.7) and (2.8) become

$$(1 - \eta\eta')dx + \eta' ds + \eta ds' = 0$$

and

$$i(1 + \eta\eta')d\mathfrak{x} - \eta' ds + \eta ds' = 0.$$

By introducing a complex coordinate X for the surface Ω, and its conjugate X' by

$$x + i\mathfrak{x} =: 2X \qquad x - i\mathfrak{x} =: 2X',$$

the pair of the integrability conditions becomes

$$ds = \eta dX - \frac{1}{\eta'} dX' \tag{2.9}$$

and

$$ds' = -\frac{1}{\eta} dX + \eta' dX'. \tag{2.10}$$

As ds and ds' are both exact differentials, this would imply

$$\frac{\partial \eta}{\partial X'} = 0, \qquad \frac{\partial \eta'}{\partial X} = 0,$$

namely the complex analyticity of η as a function of X.

Conversely, provided that the function η is univalent, $X = \frac{1}{2}(x + i\mathfrak{x})$ is a complex analytic function of η. In particular, x is a harmonic function, and the new real variable $\mathfrak{x}$ stands for the harmonic conjugate of x on Ω. Clearly the argument above can be repeated for y and z to conclude that the coordinate functions (x, y, z) as well as $(\mathfrak{x}, \mathfrak{y}, \mathfrak{z})$ are all harmonic.

In modern textbooks on minimal surface (for example [12]) this statement is expressed as follows.

Theorem 2.1 *A surface Ω is minimal if and only if its coordinate functions (x, y, z) are harmonic on the surface.*

In the last paragraph of §5, it is mentioned that by integrating the 1-forms ds and ds', the variables s and s' are expressed as functions of X, X' and η. Once η can be expressed as a function of X, as well as η' as a function of X' by taking the conjugate, $s = y + iz$ and $s' = y - iz$ are represented as functions of X and X' only. By eliminating the imaginary part $\mathfrak{x}$, one obtains an equation between x, y and z, an implicit representation for the minimal surface Ω.

In §6 and §7, the area functional and the first fundamental form of the minimal surface are written respectively as an integral and a tensor with respect to the local coordinate η.

In order to do so, we also introduce the new complex variables

$$Y = \int \frac{\partial y}{\partial \eta} d\eta \quad \text{and} \quad Y' = \int \frac{\partial y}{\partial \eta'} d\eta'$$

and

$$Z = \int \frac{\partial z}{\partial \eta} d\eta \quad \text{and} \quad Z' = \int \frac{\partial z}{\partial \eta'} d\eta'$$

where $y = Y + Y', \mathfrak{y} = Y - Y'$ and $z = Z + Z', \mathfrak{z} = Z - Z'$, where $\mathfrak{y}$ and $\mathfrak{z}$ are defined as the corresponding potential functions of the exact one forms induced from the Euler-Lagrange equations of the area functional. By analogous arguments to the one for X, the complex variables Y and Z are also shown to be complex analytic in η.

The ingredients for the calculations are the two integrability conditions (2.9) and (2.10) as well as the holomorphicity of X and the anti-holomorphicity of X' with respect to η:

$$dX = \frac{\partial X}{\partial \eta} d\eta = \frac{\partial x}{\partial \eta} d\eta \quad \text{and} \quad dX' = \frac{\partial X}{\partial \eta} d\eta = \frac{\partial x}{\partial \eta} d\eta,$$

and the corresponding relations for Y, Y', Z, Z' and η.

The resulting formula for the area functional is

$$S = \iint \frac{1}{\cos r} dy dz = -i \iint \Big(\frac{\partial x}{\partial \eta}\frac{\partial x}{\partial \eta'} + \frac{\partial y}{\partial \eta}\frac{\partial y}{\partial \eta'} + \frac{\partial z}{\partial \eta}\frac{\partial z}{\partial \eta'}\Big) d\eta \wedge d\eta'.$$

It is this expression that made Hans Lewy [5] speculate on Riemann's motivation for investigating minimal surfaces with boundary. Namely through the isothemal coordinate η, the area S is expressed as the Dirichlet energy of the map $\eta \mapsto (x(\eta), y(\eta), z(\eta))$, and the minimal surface is nothing but the energy minimizing map which was at stake in Riemann's thesis, especially the argument in showing the existence of the Riemann mapping.

The area S can also be expressed as

$$\begin{aligned} S &= -i \iint dX \wedge dX' + dY \wedge dY' + dZ \wedge dZ' \\ &= \frac{1}{2} \iint dx \wedge d\mathfrak{x} + dy \wedge d\mathfrak{y} + dz \wedge d\mathfrak{z}. \end{aligned}$$

The first fundamental form of the minimal surface is given as

$$dx \otimes dx + dy \otimes dy + dz \otimes dz = 2\Big(\frac{\partial x}{\partial \eta}\frac{\partial x}{\partial \eta'} + \frac{\partial y}{\partial \eta}\frac{\partial y}{\partial \eta'} + \frac{\partial z}{\partial \eta}\frac{\partial z}{\partial \eta'}\Big) d\eta \otimes d\eta'. \tag{2.11}$$

This representation of the first fundamental form, or equivalently, the induced metric, involves some calculations, which include verifying the equalities

$$dX \otimes dX + dY \otimes dY + dZ \otimes dZ = 0 \tag{2.12}$$

and

$$dX' \otimes dX' + dY' \otimes dY' + dZ' \otimes dZ' = 0, \tag{2.13}$$

which we will come back to, in reference to the Weierstrass representation.

In particular, it follows from the representation Eq. (2.11) that the η coordinate induces an isothermal coordinate to the minimal surface Ω. In modern differential geometry textbooks, this fact corresponds to the following statement (see [12] for example):

Theorem 2.2 *The Gauss map is conformal at $P \in \Omega$ either if the surface is umbilical at P or if the mean curvature vanishes. If restricted to surfaces of Gauss curvature $K \leq 0$ then the surface is minimal if and only if its Gauss map is conformal.*

We recall here that the stereographic projection is conformal.

Furthermore, when X, Y and Z are considered as holomorphic functions in η, Riemann writes down the pull-back metrics of the Euclidean metric on the respective complex plane as

$$dX \otimes dX' = \frac{\partial x}{\partial \eta}\frac{\partial x}{\partial \eta'} d\eta \otimes d\eta',$$

$$dY \otimes dY' = \frac{\partial y}{\partial \eta}\frac{\partial y}{\partial \eta'} d\eta \otimes d\eta',$$

and

$$dZ \otimes dZ' = \frac{\partial z}{\partial \eta}\frac{\partial z}{\partial \eta'} d\eta \otimes d\eta',$$

all of which, induce isothermal coordinates.

In §8 and §9, Riemann-Hattendorff is concerned with explicit parametrizations of the minimal surface, not in terms of η which was the Gauss map post-composed with the stereographic projection, but a new variable u which takes away the gauge freedom SO(3) arising from the position of the pole of the stereographic projection, which corresponds to the choices of oriented hyperplanes in $\mathbb{R}^3$.

Let α be a point in the yz-plane (or equivalently the η-plane) which is the image by the stereographic projection $P_S : S^2 \setminus \{S\} \to yz$-plane of a point $P_S^{-1}(\alpha) \in S^2 \setminus \{S\}$. (Recall $S = (-1, 0, 0)$.) We introduce a change of variables $\eta_\alpha := e^{i\theta}\frac{\eta-\alpha}{1+\alpha'\eta}$ which represents the transformation law which relates the image η of the stereographic projection of $S^2 \setminus \{S\}$ onto the yz-plane to the image η_α of the stereographic projection of $S^2 \setminus P_S^{-1}\{-\alpha\}$ onto the hyperplane Π_α whose unit normal vector is α. Note that the map $\eta \mapsto \eta_\alpha$ is conformal. For $P \in \Omega$, we then define $x_\alpha(P)$ to be the height of a point P over the hyperplane Π_α.

It is shown then that for $\eta_\alpha := e^{i\theta}\frac{\eta-\alpha}{1+\alpha'\eta}$,

$$(d\log\eta_\alpha)^2 \frac{\partial x_\alpha}{\partial \log \eta_\alpha} = (d\log\eta)^2 \frac{\partial x}{\partial \log \eta},$$

resulting in defining a new variable u

$$u = \int \sqrt{i\frac{\partial x}{\partial \log \eta}} d\log\eta, \tag{2.14}$$

which is independent of the gauge α.

This new variable u, a function in η, effectively contains all the information about the minimal surface Ω. It can be used to recover the three coordinate functions x, y and z as follows. By rewriting x as a function of u in Eq. (2.14) we obtain

$$x = -i \int \Big(\frac{du}{d\log\eta}\Big)^2 d\log\eta + i \int \Big(\frac{du'}{d\log\eta'}\Big)^2 d\log\eta'.$$

Recall that $\eta = \eta_0$ is the stereographic projection of $S^2 \setminus \{S\}$ from the south pole $S = (-1, 0, 0)$ so that the north pole $(1, 0, 0)$ is sent to $\alpha = (0, 0)$ in the yz-plane, or the η-plane. And x_0 is the height of the point P in Ω measured from the yz-plane. Analogously, by choosing $\alpha = (1, 0) = 1 + 0i$ in yz-plane, x_α represents the height of P measured from the xz-plane. Namely by substituting

$$\eta_\alpha = \frac{\eta - 1}{1 + \eta}$$

we obtain the representation

$$y = -\frac{i}{2} \int \Big(\frac{du}{d\log\eta}\Big)^2 \Big(\eta - \frac{1}{\eta}\Big) d\log\eta + \frac{i}{2} \int \Big(\frac{du'}{d\log\eta'}\Big)^2 \Big(\eta' - \frac{1}{\eta'}\Big) d\log\eta'.$$

Similarly, for $\alpha = (0, 1) = 0 + 1i$,

$$\eta_\alpha = \frac{\eta - i}{1 - i\eta}$$

we have

$$z = -\frac{1}{2} \int \Big(\frac{du}{d\log\eta}\Big)^2 \Big(\eta + \frac{1}{\eta}\Big) d\log\eta - \frac{1}{2} \int \Big(\frac{du'}{d\log\eta'}\Big)^2 \Big(\eta' + \frac{1}{\eta'}\Big) d\log\eta'$$

the height of (x, y, z) over the η_α-plane, which is the xy-plane.

In §10, §11 and §12, the three complex-valued functions X, Y and Z are regarded as holomorphic functions of η, and also of u, and they are locally analytically extended near a point, either in the interior, or on the boundary set of the minimal surface. As for the boundary set, it is bounded by Euclidean lines, line segments, or line segments intersecting at a rational angle $q\pi$ with $q \in \mathbb{Q}$.

In particular, Riemann makes an important geometric observation in §10, where a point P in Ω is regarded as the origin of $\mathbb{R}^3$, and the tangent plane $T_P\Omega$ is identified with the yz-plane. Then Ω is locally the graph of a function $x = x(y, z)$ with

$$x(P) = 0, \qquad \frac{\partial x}{\partial y}(P) = \frac{\partial x}{\partial z}(P) = 0$$

and the scalar function x is locally approximated by a harmonic function over the yz-plane,

$$\frac{\partial^2 x}{\partial y^2} + \frac{\partial^2 x}{\partial z^2} = 0.$$

When the second order Taylor expansion terms are nontrivial, the graph at P has nontrivial curvature, and the eigenvalues of the Hessian of $x(y, z)$ are the principal curvatures of the surface at P, of opposite signs and of the same magnitude, namely the mean curvature vanishes. When all the second order Taylor expansion terms vanish, the complex-valued function X has zeros of higher order.

Namely by supplying the harmonic conjugates $\mathfrak{x}$, $\mathfrak{y}$, $\mathfrak{z}$ to the real coordinates x, y and z, one can consider the relation among the complex valued functions X, Y and Z in η. As η can be seen as a holomorphic function in the previously introduced complex variable u, X, Y and Z are in turn holomorphic functions in u. In the consecutive sections §10, §11 and §12, functional equations of these variables are written down, at interior points, interior branch points, unramified as well as ramified boundary points, and at a boundary points where two line segments meet at a rational angle.

In §12, also considered is the situation where the minimal surface is simply connected, unbounded in $\mathbb{R}^3$ and bounded by two non-intersecting infinite lines. In commenting the setting, another geometric observation is made. When the minimal surface is bounded by a line ℓ, choose the coordinates in $\mathbb{R}^3$ so that the x-axis coincides with the line ℓ. Then the unit normal $\nu(P)$ to Ω, written as a vector in $\mathbb{R}^3$, has no x component. When the unit sphere is centered at the origin, and the north pole is on the positive x-axis, the image of the Gauss map $\nu : \ell \to S^2$ lies on the equator $\{(0, y, z) \mid y^2 + z^2 = 1\} \subset \mathbb{R}^3$, and consequently the image η of the stereographic projection $P_S : S^2 \setminus \{S\} \to yz$-plane is the equator $\{|\eta| = 1\}$ itself. Hence $\log \eta = \log |\eta| + i \arg \eta$ is purely imaginary along the boundary line of the x-axis. Then we conclude when x is seen as a function in η instead,

$$i \frac{\partial x}{\partial \log \eta}$$

is real along the boundary line. Similarly if we designate another boundary line as the $\hat{x}$-axis of another coordinate system $(\hat{x}, \hat{y}, \hat{z})$, then

$$i \frac{\partial \hat{x}}{\partial \log \eta}$$

is real along that boundary line. Now recall that the quantity

$$u = \int \sqrt{i \frac{\partial x}{\partial \log \eta}} d \log \eta$$

is independent of the coordinate system on $\mathbb{R}^3$, and hence is independent of the procedure of identifying the x-axis with the boundary line. Namely

$$du = \sqrt{i \frac{\partial x}{\partial \log \eta} d \log \eta}$$

is either real or purely imaginary, depending on the sign of the real number $i \frac{\partial x}{\partial \log \eta}$.

We recollect the construction here: when the minimal surface is bounded by a line segment, either bounded or unbounded, the neighborhood of a boundary point is realized by a conformal map from a region in the unit sphere bounded by a great circle. This fact is reflected in the observation that the function $du/d \log \eta$ is either real or purely imaginary.

Up to this point, several choices of complex parameters are presented; η, u, X, Y, Z, each of which can be used to parameterize the minimal surface Ω locally. In §13, the issue of transcribing one by another among those numbers is treated. The method is effectively what is known as the Schwarz-Christoffel mapping, where in Eq. (2.11) an explicit map $u(t)$ from the upper half t-plane $\mathbb{H}$ to the region in the u-plane, or to be exact, the region that is a ramified cover over the u-plane bounded by totally real or totally imaginary lines, is constructed. The coordinate t of the upper half plane corresponds to u by

$$t = \frac{\text{const.}}{u - b} + \text{terms holomophic in } u$$

where b is a value of u corresponding to a boundary point, and where the constant coefficient of the simple pole $u = b$ is determined by the conditions the imaginary part of t is (1) zero along the boundary, and (2) positive in the interior of the surface. By this local information, u is globally defined by the expression of Eq. (2.11) of the original paper:

$$u = \text{const.} + \text{const.} \int \sqrt{\frac{\Pi(t-a)\Pi(t-a')\Pi(t-b)}{\Pi(t-c)}} \frac{dt}{\Pi(t-e)}$$

is obtained, where $a = \cup a_i$ denotes the set of branch points in the t-plane, $\Pi(t - a)$ denotes the product of $(t - a_i)$ over i, a' the complex conjugates of a, b the branch points on the real axis, c the pre-images of the vertices of the boundary where two line segments meet, and e stands for vertices in the unbounded sectors spanned by a pair of line segments, where each pair is designated with an angle $\alpha\pi$ between the projections of the two lines to the plane perpendicular to both lines, as well as the length of the shortest connecting line segment between the two lines. In the expression in Eq. (2.11), there appear only square roots due to the fact that Riemann considers only simple branch points, unlike the general formula of Schwarz-Christoffel's (cf [1]) where the power of $(t - c)$ could be any number in $[0, 1]$ representing the exterior angle at a vertex of the polygon.

Note that the function u above is defined over the entire t-plane. This is explained in the original article as

> In order to form the expression for $\frac{du}{dt}$ we must observe that dt is always real along the boundary, and du is either real or pure imaginary. Hence $(du/dt)^2$ is real when t is real. This function can be continuously extended over the line of real values of t by the condition that, for conjugate values t and t' of the variable, the function will also have conjugate values. Then $(du/dt)^2$ is determined for the whole t-plane and turns out to be single-valued.

Thus it is justified that du/dt is identified with the integrand of the complex line integral of Eq. (2.11). In this manner, the function u is represented by the complex variable t on the upper half-space, which then is used to reproduce the minimal surface $\Omega = \{(x(t), y(t), z(t))\}$. The by-product of the construction is the extension of the domain of u over the entire complex plane. In modern textbooks, this fact is expressed in the following important statement, often attributed to Schwarz.

Theorem 2.3 (Reflection Principle) *Let $U(t)$ be a minimal surface in isothermal parameters defined in a semi-disk $D_+ = \{|t| < \varepsilon \mid \text{Im}(t) > 0\}$. Suppose there exists a line L in $\mathbb{R}^3$ such that $U(t) \to L$ when $\text{Im}(t) \to 0$. Then $U(t)$ can be extended to a generalized minimal surface defined in the full disk $D = \{|t| < \varepsilon\}$. Furthermore this extended surface is symmetric in L.*

By utilizing this *universal* variable t, Riemann-Hattendorff (see also the paper [10], in addition to [9]) sets up an ansatz in the following geometric settings for the boundary of the minimal surface Ω:

- two non-intersecting non-parallel infinite lines, which would produce a part of the helicoid bounded by a pair of ruling lines;
- two lines which intersect at a point and a line on the plane whose normal vector is perpendicular to the first two lines;
- four intersecting line segments obtained by removing two edges that do not touch each other from the one skeleton of an arbitrary tetrahedron. This includes the so-called Schwarz surface under the additional symmetry;
- three infinite lines mutually skewed and nonintersecting, closely investigated in the recent work of B. Daniel [3];
- two circles on a pair of parallel planes, which would produce the celebrated "Riemann example" which is a minimal surface bounded by two parallel infinite lines, and foliated by circles. The example is explained in detail by Meeks-Pérez [6];
- two convex polygons on a pair of parallel planes, with and without symmetries, which include the so-called Schwarz P-surface and Schwarz H-surface.

A copy of the surface bounded by two circles on a pair of parallel planes can be extended across the boundary lines via the reflection, which would result in a complete periodic surface of genus zero. This example has recently received renewed attention in minimal surface theory ([6]), as the following classification [7] was demonstrated:

Theorem 2.4 *The plane, the helicoid, the catenoid and the Riemann minimal examples are the only properly embedded minimal surfaces in $\mathbb{R}^3$ with the topology of a planar domain.*

The Schwarz-Christoffel mapping is closely related to the theory of elliptic functions and hypergeometric functions, and those examples Riemann calculated fully make use of these theories. In this article, we will not go into the discussion of the papers [9, 10]. Interested readers are directed to Nitsche's book [11] on the subject, which records all the post-Riemannian development of the subject.

3 Representation Formulas by Riemann and Weierstrass-Enneper

In Osserman's book [12], a *generalized* minimal surface is defined as follows

Definition 3.1 A generalized minimal surface Ω in $\mathbb{R}^n$ is a non-constant map $x(p) : M \to \mathbb{R}^n$, where M is a 2-manifold with a conformal structure defined by an atlas, such that each coordinate function $x_k(p)$ is harmonic on M and furthermore

$$\sum_{k=1}^{n} \phi_k^2(\zeta) = 0 \tag{3.1}$$

where $\zeta = \xi_1 + i\xi_2$ is a complex-valued local coordinae of M, and the embedding $x(p)$ is given by

$$\phi_k(\zeta) = \frac{\partial x_k(\zeta)}{\partial \xi_1} - i\frac{\partial x_k}{\partial \xi_2}.$$

In this article, we are only interested in the case $n = 3$.

As the map $x(p)$ is non-constant, at least one of the functions x_k $(k = 1, 2, 3)$ is non-constant, which then implies that the corresponding holomorphic function $\phi_k(\zeta)$ can have at most isolated zeros. Thus the singular points of the map x satisfying

$$\sum_{k=1}^{n} |\phi_k(\zeta)|^2 = 0 \tag{3.2}$$

can exists at most at isolated points on M.

Now the following procedure to specify generalized minimal surfaces is called the Weierstrass-Enneper representation, first devised by A. Enneper (1864) and Weierstrass (1866). First we specify the triplets of holomorphic data:

Lemma 3.2 (cf. [12]) *Let D be a domain in the complex ζ plane, $\eta(\zeta)$ an arbitrary meromorphic function in D and $f(\zeta)$ an analytic function in D having the property that at each point where $\eta(\zeta)$ has a pole of order m, $f(\zeta)$ has a zero of order at least $2m$. Then the functions*

$$\phi_1 = f\eta, \quad \phi_2 = \frac{1}{2}f(1-\eta^2), \quad \phi_3 = \frac{i}{2}f(1+\eta^2) \tag{3.3}$$

are analytic in D and satisfy Eq. (3.1). Conversely every triple of analytic functions in D satisfying Eq. (3.1) may be represented in the form Eq. (3.3), except for the case of $\phi_3 = 0$, which would imply $\phi_1 = i\phi_2$.

Then every simply connected minimal surface in $\mathbb{R}^3$ is represented in the form

$$x_k(\zeta) = c_k + \Re\Big(\int_0^\zeta \phi_k(z)\,dz\Big) \qquad k = 1, 2, 3$$

where ϕ_k are defined by Eq. (3.3). The pair (f, η) is called the Weierstrass data of x.

We write out for the sake of comparison the Weierstrass-Enneper representation explicitly:

$$\begin{aligned} x &= c_1 + \Re\int f\eta\, d\zeta, \\ y &= c_2 + \Re\int \frac{1}{2} f(1-\eta^2)\, d\zeta, \\ z &= c_3 + \Re\int \frac{i}{2} f(1+\eta^2)\, d\zeta. \end{aligned}$$

Let us recall the Riemann representation formula

$$\begin{aligned} x &= c_1 - 2\Re\Big(i\int \Big(\frac{du}{d\log\eta}\Big)^2 d\log\eta\Big), \\ y &= c_2 - \Re\Big(\frac{i}{2}\int \Big(\frac{du}{d\log\eta}\Big)^2 \Big(\eta - \frac{1}{\eta}\Big) d\log\eta\Big), \\ z &= c_3 - \Re\Big(\int \Big(\frac{du}{d\log\eta}\Big)^2 \Big(\eta + \frac{1}{\eta}\Big) d\log\eta\Big). \end{aligned}$$

We now relate the two representations, Riemann's and Weierstrass-Enneper's, to each other. Calculating the Gauss map of (x, y, z), one verifies that

$$\nu = \Big(\frac{-2\Re\eta}{1+|\eta|^2}, \frac{2\Im\eta}{1+|\eta|^2}, \frac{1-|\eta|^2}{1+|\eta|^2}\Big)$$

which is the inverse map of the stereographic projection $P_S : S^2 \setminus \{(-1, 0, 0)\} \to$ yz-plane with $P_S = \eta$. Hence η of the Weierstrass data (f, η) is the same as the η of Riemann. Furthermore, we have seen the equation $\phi_1^2 + \phi_2^2 + \phi_3^2 = 0$, in the form of Eq. (2.12)

$$dX \otimes dX + dY \otimes dY + dZ \otimes dZ = 0$$

which would suggest the following correspondence:

$$dX = \phi_1 d\eta, \quad dY = \phi_2 d\eta, \quad dZ = \phi_3 d\eta$$

after identifying the isothermal complex variable ζ with Riemann's η. In other words,

$$\phi_1(\eta) = \frac{\partial X}{\partial \eta} = \frac{\partial(x + i\mathfrak{x})}{\partial \eta} = \frac{\partial x}{\partial \eta} = \frac{\partial x}{\partial \eta_1} - i\frac{\partial x}{\partial \eta_2},$$

for $\eta = \eta_1 + i\eta_2$. In this way, we recover the definition of the generalized minimal surface we have seen above.

Lastly, by comparing the two sets of representation formulas, the function f of the Weierstrass data (f, η) can be identified with Riemann's u as follows;

$$f = i\left(\frac{du}{d\eta}\right)^2.$$

4 Closing Remarks

In the latter half of the 19th century (cf. [11]), the theory of minimal surface developed rapidly with the Weierstrass-Enneper representation formula, together with the theory of elliptic integrals, and hypergeometric functions. Consequently Riemann's representation was mostly pushed aside and left unacknowledged. In hindsight, the Weierstrass-Enneper representation is simpler, and though both representations are local in nature, the Weierstrass-Enneper formula can be modified so that one obtains a global formula for the minimal surface. Furthermore the Weierstrass holomorphic data (f, η) is purely algebraic and formal, while Riemann's is dependent on the particular complex coordinate η, which is the value of the Gauss map.

Geometrically both representations rely on the complex analytic method, which is congenial to minimal surfaces with its boundary being lines and circles, including the constructions of the Enneper surface, the catenoid, the helicoid and the Schwarz surfaces. For general boundary sets, however, the procedure is not very useful, and consequently the complete solution of the Plateau problem by Douglas and Rado was not obtained until the 1930s (cf. [2].)

However, it is clear, judging from the content of the manuscript Riemann had left behind, that he had, as of 1860–61, ahead of Weierstrass, Schwarz and Enneper, captured the theoretical essence of the Weistrass-Enneper formula, the Schwarz reflection and the Schwarz-Christoffel transformation as well as the wealth of examples that are much prized in the following years.

References

1. L. Ahlfors, *Complex Analysis*, International Series in Pure & Applied Mathematics (McGraw-Hill, 1979)
2. R. Courant, *Dirichlet's Principle, Conformal Mapping and Minimal Surafaces*, Pure and applied mathematics, **3**, (Interscience, 1950)

3. B. Daniel, Minimal disks bounded by three straight lines in Euclidean space and trinoids in hyperbolic space. J. Differ. Geom. **72**, 467–508 (2006)
4. C.F. Gauss, General Investigations of Curved Surfaces (1825) and (1827). Princeton University Library, also available as *General Investigations of Curved Surfaces* (Dover, 2005)
5. H. Lewy, Introduction in *Collected Works of Bernhard Riemann* (Dover, 1953)
6. W.H. Meeks III, J. Pérez, The Riemann minimal examples, in *The Legacy of Bernhard Riemann After One Hundred and Fifty Years* ALM vol. 35, pp. 417–457
7. W.H. Meeks III, J. Pérez, A. Ros, Properly embedded minimal planar domains. Ann. Math. **181**, 473–546 (2015)
8. B. Riemann, Grundlagen für eine allgemeine Theorie der Functionen einer veränderlichen complexen Grösse, dissertation, Göttingen (1851)
9. B. Riemann, Über die Fläche vom kleinsten Inhalt bei gegebener Begrenzung (On the surface of least area with a given boundary) Abh. Königl, d. Wiss. Göttingen Mathem. Cl. **13**, 3–52 (1867)
10. B. Riemann, *Beispiele von Flächen kleinsten Inhalts bei gegebener Begrenzung (Examples of surfaces of least area with a given boundary), Chap. XXVII in Gesammelte mathematishe Werke* 2nd edn. (B.G. Teubner, Leibzig, 1892)
11. J.C.C. Nitsche, *Lectures on Minimal Surfaces: Volume 1, Introduction, Fundamentals, Geometry and Basic Boundary Value Problems* (Cambridge University Press, 2011)
12. R. Osserman, *A Survey of Minimal Surfaces* (Dover, 1986)

Physics in Riemann's Mathematical Papers

Athanase Papadopoulos

Abstract Riemann's mathematical papers contain many ideas that arise in physics, and some of them are motivated by problems from physics. In fact, it is not easy to separate Riemann's ideas in mathematics from there in physics. Furthermore, Riemann's philosophical ideas are often in the background of his work on science. The aim of this chapter is to give an overview of Riemann's mathematical results based on physical reasoning or motivated by physics. We also elaborate on the relation with philosophy. While we discuss some of Riemann's philosophical points of view, we review some ideas on the same subjects emitted by Riemann's predecessors, and in particular Greek philosophers, mainly the pre-socratics and Aristotle.

Keywords Bernhard Riemann · Space · Riemannian geometry · Riemann surface · Trigonometric series · Electricity · Physics

AMS Mathematics Subject Classification: 01-02 · 01A55 · 01A67

1 Introduction

Bernhard Riemann is one of these pre-eminent scientists who considered mathematics, physics and philosophy as a single subject, whose objective is part of a continuous quest for understanding the world. His writings not only are the basis of some of the most fundamental mathematical theories that continue to grow today, but they also effected a profound transformation of our knowledge of nature, in particular through the physical developments to which they gave rise, in mechanics, electromagnetism, heat, electricity, acoustics, and other topics. In Riemann's writings, geometry is at the center of physics, and physical reasoning is part of geometry. His ideas on space

---

A. Papadopoulos (✉)
Institut de Recherche Mathématique Avancée, Université de Strasbourg and CNRS,
7 Rue René Descartes, 67084 Strasbourg Cedex, France
e-mail: athanase.papadopoulos@math.unistra.fr

A. Papadopoulos
Mathematics Department, Brown University, 151 Thayer Street, Providence, RI 02912, USA

L. Ji et al. (eds.), *From Riemann to Differential Geometry and Relativity*,
DOI 10.1007/978-3-319-60039-0_6

and time affected our knowledge in a profound way. They were at the basis of several elaborate theories by mathematicians and physicists, and one can mention here the names of Hermann Weyl and Albert Einstein. Likewise, Riemann's speculations on the infinitely small and the infinitely large go beyond the mathematical and physical setting, and they had a non-negligible impact on philosophy.

In the present chapter, we survey some of Riemann's ideas from physics that are contained in his mathematical works. It is not easy to separate Riemann's ideas on physics from those on mathematics. It is also a fact that one cannot consider the fundamental questions that Riemann addressed on physics without mentioning his philosophical background. This is why our survey involves philosophy, besides physics and mathematics. A certain number of papers and fragments by Riemann on philosophy, psychology, metaphysics and gnosiology were collected by Heinrich Weber and published in his edition of Riemann's Collected Works (p. 507–538). We also mention the name of Gilles Deleuze (1925–1995), a twentieth-century French philosopher who was influenced by Riemann. The name of Deleuze is not commonly known to mathematicians. The relation of his work with Riemann's ideas is highlighted in two chapters of the present volume (see [80, 119]).

As a mathematician, physicist and philosopher, Riemann belongs to a long tradition of thinkers which can be traced back to ancient Greece. One of the main outcomes of his Habilitation lecture *Über die Hypothesen, welche der Geometrie zu Grunde liegen* (On the hypotheses that lie at the bases of geometry) [146] (1854), which we discuss more thoroughly in Sect. 4 of the present chapter, is the merging of philosophy, geometry and physics. The fundamental questions that he addresses explicitly in this work, on space, form, dimension, magnitude, the infinite and the infinitesimal, the discrete and the continuous, are precisely the questions that obsessed the Greek philosophers, starting with the Milesians and the Pythagoreans, and passing through Plato, Aristotle, Archimedes and several others. One important fact to recall is that the Greeks had a *name* for infinity, *apeiron*. The name was used by Anaximander, in the sixth century B.C. There is an extensive literature on the word *apeiron*, whether it denoted an unlimited extent, or a boundless shape, whether it applies to quantity or to shape, etc. The Greeks thoroughly considered the question of infinity, both mathematically and philosophically, and it is often difficult to make the distinction between the two points of view. A mathematical method of dealing with infinitely small quantities called "method of exhaustion," which is very close to what we use today in infinitesimal calculus, was developed in the fourth century B.C. by Eudoxus of Cnidus, a student of Plato. This method is used by Euclid in the proofs of several propositions of the *Elements*. Dedekind was inspired by this method when he introduced the so-called Dedekind cuts. It is also well known that the philosophical reflections on the infinitely small are not foreign to Leibniz's and Newton's work on the foundations of infinitesimal calculus.

A certain number of these thinkers wondered about the smallest particles of matter, for which they invented a name: atoms, they speculated about their shape and their arrangement and how they fit in an ambient space, they meditated on characters of these atoms: cold, waterly, etc. The thinkers belonging to the "atomist" tradition believed that the universe is a mixture of such atoms, that is, uncuttable, or indivisible matter, and void. Riemann had his own ideas about matter and void. Klein, in his *Development of mathematics in the 19th century* [87] (p. 235), reporting on some of Riemann's ideas from his *Nachlass* (the collection of manuscripts, notes and correspondence that he left), writes:

> Riemann thinks of space as being filled with continuous matter *[Stoff]*, which transmits the effect of gravity, light, and electricity. He has throughout the idea of a temporal extension of process. A remark on this topic is found in a personal letter from Gauss to Weber – with an express request for complete secrecy. And now I again ask, how did these things come to Riemann? It is just mystical influence, which cannot be defined and yet cannot be clearly grasped, of the general atmosphere of a receptive spirit.

Long before Riemann, the pre-socratics Parmenides and Zeno (sixth century B.C.), and then Leucippius and Democritus and other thinkers of the fifth century B.C., thoroughly gazed at the notions of atom and indivisible matter. Their opinions are reported on by Aristotle, who made a systematic study of this matter in several texts (*Metaphysics* V, *Physics* V and VI, *Categories* IV, etc.). Other Greek thinkers considered that matter is continuous, rather than atomic, asserting that the atomic structure requires the existence of a void, and claiming that the existence of a void contradicts several laws of physics. They stressed instead the geometric structure of the universe. A theory of chaos, in the sense of unformed matter arising from the void had also its supporters—Chaos is an important notion in Greek mythology—but in general, the Greek philosophers considered that nature is governed by natural laws which they tried to understand. Concerning these thinkers, let us quote Hermann Weyl, one of the best representatives of Riemann's tradition of thought, from the beginning of his book *Philosophy of mathematics and natural science* [166] (p. 3):

> To the Greeks we owe the insight that the structure of space, which manifests itself in the relations between spatial configurations and their mutual lawful dependences, is something entirely rational.

Thus, talking about the origin of Riemann's ideas, we shall often mention his Greek predecessors.

One should also recall that the exceptional rise of Greek science that started in the sixth century B.C., in the form of precise questions whose aim was to understand the universe, was accompanied by a profound philosophical reflection on the nature and the goal of sciences, and in particular mathematics. Aristotle, who is probably the best representative of the Greeks thinkers of the culminating era, in Book VI of his *Metaphysics* [11], states that among the sciences, three have the status of being *theoretical*: mathematics, physics and theology, the latter, for him, being close to what we now understand as philosophy.[1] Let us note right away that these are precisely

---

[1]Cf. [11] p. 1619: "And since natural science, like other sciences, confines itself to one class of being, i.e. to that sort of substance which has the principle of its movement and rest present in itself, evidently it is neither practical nor productive. For the principle of production is in the producer—it is either reason or art or some capacity, while the principle of action is in the doer—viz. choice, for that which is done and that which is chosen are the same. Therefore, if all thought is either practical or productive or theoretical, natural science must be a theoretical, but it will theorize about such being as admits of being moved, and about substance which in respect of its formula is from the

the three branches of knowledge that constitute the background of Riemann (who, by the way, was also trained in theology). We also note that although Pythagoras is supposed to have coined the term $\phi\iota\lambda o\sigma o\phi \acute{\iota}\alpha$, *communis opinio* now seems that its current meaning (striving for knowledge) goes back to Plato.

In the same work and in others, Aristotle discusses at length the role of each of these three sciences and the relations among them. He also addresses thoroughly the question of whether mathematics has a purely ideal character or whether it reflects the real world. Such interrogations lead directly to the most fundamental questions that Riemann addressed in his Habilitationsvortrag and in his other writings. We shall say more on the lineage of Riemann's ideas to Greek philosophy in Sect. 4 where we discuss this work.

Riemann's interest in physics was constant during his lifetime. Since his early twenties, he tried to develop a theory that would unify electricity, magnetism, light and gravitation—the same quest that Poincaré, Lorentz and Einstein had after him, culminating in the theory of general relativity. One of Riemann's manuscripts, *Ein Beitrag zur Electrodynamik* (A contribution to electrodynamics) [140], whose subject is electrodynamics and which is related to his search for the unification of the various forces of nature, published posthumously, is discussed in the chapter [68] by Hubert Goenner in the present volume. In this paper, Riemann develops a theory of electromagnetism which is based on the assumption that electric current travels at the velocity of light. Furthermore, he considers that the differential equation that describes the propagation of electric force is the same as that of heat and light propagation. Goenner, in his paper, mentions the works of other physicists of the same period, including Maxwell, Lorenz, Helmholtz, Carl Neumann and Franz Neumann. Being himself a physicist, Goenner writes:

> Surprisingly, within the then reigning view of electromagnetism as a particle theory, we can note a relativistic input, made by the famous mathematician Bernhard Riemann. His introduction of the retarded scalar potential into theoretical electrodynamics is still valid, but remains unknown to the overwhelming majority of today's theoretical physicists.

Several other commentaries on Riemann's manuscript exist, and some of them are mentioned in the bibliography of [68]. Enrico Betti, who was Riemann's friend and who translated into Italian several of his works and wrote commentaries on them, had already commented on that paper in 1868, see [18].

It is known that, as a student, in Göttingen and Berlin, Riemann attended more courses and seminars on (theoretical and experimental) physics than on mathematics. One may also mention here an essay in Riemann's *Nachlass*, entitled *Gravitation*

(Footnote 1 continued)
most part not separable from matter. Now, we must not fail to notice the nature of the essence of its formula, for, without this, inquiry is but idle. [...] That natural science, then, is theoretical, is plain from these considerations. Mathematics also, however, is theoretical; but whether its objects are immovable and separable from matter, is not at present clear; it is clear, however, that it considers some mathematical objects *qua* immovable and *qua* separable from matter. But if there is something which is eternal and immovable and separable, clearly the knowledge of it belongs to a theoretical science,—not, however, to natural science (for natural science deals with certain movable things) nor to mathematics, but to a science prior to both."

und Licht (Gravitation and light) [141] p. 532–538, whose subject is the theoretical connection between gravitation and light. Betti also wrote a paper, entitled *Sopra una estensione dei principii generali della dinamica* (On the extension of the general principles of dynamics) [19], in which he announces several results which are based on ideas contained in Riemann's lectures *Schwere, Electricität und Magnetismus*, edited by the latter's student Karl Hattendorf (Hannover, 1880) [143]. Riemann establishes in these lectures necessary and sufficient conditions under which Hamilton's principle on the motion of a free system subject to time-independent forces that depend on the position and the motion of the system is satisfied. Chapter VII of Picard's famous *Traité d'analyse* [115] contains a chapter called *Attraction and potential*. The author declares there (p. 167) that he uses a transformation from Riemann's posthumous memoir *Schwere, Electricität und Magnetismus*. Finally, we note that Maxwell discussed Riemann's theory of electrodynamics in his *Note on the electromagnetic theory of light*, an appendix to his paper [96].

In talking about Riemann's background in physics, we take this opportunity to recall a few facts about Riemann's studies.

In a letter to his father on April 30, 1845, while he was still in high-school, Riemann informs the latter that he starts being more and more attracted by mathematics. He also tells him in the same letter that he plans to enroll the University of Göttingen to study theology, but that in reality he must decide for himself what he shall do, since otherwise he will not bring anything good to a subject. Cf. [149]. Riemann entered the University of Göttingen in 1846, as a student in theology. He stayed there for one year and then moved to the University of Berlin where he spent two years, attending lectures by Jacobi and Dirichlet. In a letter to his father, dated May 30, 1849, Riemann writes ([149])[2]: "I had come just in time for the lectures of Dirichlet and Jacobi. Jacobi has just begun a series of lectures in which he lead off once again with the entire system of the theory of elliptical functions in the most advanced, but elementary way." Jacobi was highly interested in mechanics, and it would not be surprising if his interest in elliptic functions was motivated by their applications to mechanics. In another letter, written to his brother, dated November 29, 1847, Riemann writes:

> When I arrived, I found to my great joy that Jacobi, who had announced no course in the catalog, had changed his mind. He plans to lecture on mechanics. I would, if possible, stay here for another semester just to attend it. Nothing could be more satisfying to me than this. [...] The next day I went to see Jacobi in order to enroll in his course. He was very polite and friendly, because in his previous lecture he had dealt with a subject related to the problem I had just solved, I brought it up and told him of my work. He said if it was a nice job, he would send it to Crelle's Journal as soon as possible. Unfortunately my time will be somewhat tight for writing it up. Also I don't know whether the complete solution of the problem will take yet more time.[3]

[2]We are using the translation by Gallagher and Weissbach.

[3]It is not clear to the author of the present article what the work that Riemann is talking about is. It might be that there is a mistake in the date of the letter. Nevertheless, the content is interesting for us here regardless of the date.

Dirichlet's lectures in Berlin, at that epoch, were centered mostly on theoretical physics (partial differential equations). It is from these lectures that Riemann became familiar with potential theory, a topic which was about to play an important role in his later work. Klein, in his *Development of mathematics in the 19th century* ([87], p. 234–235), writes:

> Dirichlet loved to make things clear to himself in an intuitive substrate; along with this he would give acute, logical analyses of foundational questions and would avoid long computations as much as possible. His manner suited Riemann, who adopted it and worked according to Dirichlet's methods.

Riemann's admiration for Dirichlet is expressed at several places of his writings, for instance in the third section of the historical part of his habilitation dissertation on trigonometric series which we shall analyse in Sect. 3 below.[4]

During his stay at Berlin, Riemann also attended lectures by Dove on optics, which he found very interesting, and by Enke on astronomy (letter without date, quoted in [149]). About the latter, Riemann says that "his presentation for the most part is rather dry and boring, however the time that we spend at the observatory once a week, from 6pm to 8pm, is useful and instructive."

After the two years spent in Berlin (1847–1849), Riemann returned to Göttingen, where he attended the lectures and seminars of the newly hired physics professor Wilhelm Eduard Weber (1804–1891),[5] who was also Gauss's collaborator and close friend. Klein writes, in his *Development of mathematics in the 19th century*, ([87], p. 235): "In Weber, Riemann found a patron and a fatherly friend. Weber recognized Riemann's genius and drew the shy student to him. [...] Riemann's interest in the mathematical treatment of nature was awakened by Weber, and Riemann was strongly influenced by Weber's questions."

From Riemann's posthumous papers, we read[6]:

---

[4]It is also true that Riemann, with his extreme sensibility, was at some point disappointed of Dirichlet being less amicable with him. In a letter to his brother, dated April 25, 1857, he writes: [...] Also Dirichlet appeared, if still very polite, yet not so well-disposed towards me as before. This also was agony for me.

[5]This was the second time that Weber was hired in Göttingen. The first time was in 1831, at Gauss's recommendation (Weber was 27). Weber developed a theory of electromagnetism which was eventually superseded by Maxwell's. Weber and Gauss published joint results and they constructed the first electromagnetic telegraph (1833), which operated between the astronomical observatory and the physics laboratory of the University of Göttingen (the locations were 3 km apart). In 1837, as the result of a repression, led by the new King of Hannover Ernest Augustus (who reigned between 1837 and 1851) and caused by political events, Weber, together with six other leading professors (including the two brothers Grimm), was dismissed from his position at the University of Göttingen. He came back to this university in 1849 and served intermittently as the administrator of the Astronomical Observatory. The position of director had been occupied by Gauss since its foundation in 1816. Gauss was more than seventy at the time Weber returned to Göttingen.

[6]Cf. Bernhard Riemann's *Gesammelte mathematische Werke und wissenschaftlicher Nachlass*, ed. H. Weber and R. Dedekind, [141] 2nd edition, Leipzig, 1892, p. 507. The translation of this passage is borrowed from the English translation of Klein's *Development of mathematics in the 19th century*, ([87], p. 233).

> My main work concerns a new conception of the known works of nature – their expression by means of other basic concepts – whereby it became possible to use the experimental data on the reciprocal actions between heat, light, magnetism and electricity to investigate their connections with each other. I was led to this mainly through studying the works of Newton, Euler, and – from another aspect – Herbart.

The name of Herbart, which appears at the end of this quote, will be mentioned several times in the present chapter, and it is perhaps useful to say right away a few words on him.

Johann Friedrich Herbart (1776–1841) started his studies in philosophy in Iena under Fichte but he soon disagreed with his ideas and went to Göttingen where he received his doctorate and habilitation, and after that he taught there pedagogy and philosophy. In 1809, he was offered the chair formerly held by Kant in Königsberg. His philosophy relies on Leibniz's theory of monads. Herbart was conservative and anti-democratic. He was an advocate of the view that the state higher officials should be appointed among those who have a strong cultural education. He wrote in 1824 a treatise entitled *Psychology as a science newly founded on experience, metaphysics and mathematics* [75]. In his research on psychology, he made use of infinitesimal calculus, and he was probably the first to do this. For him, psychology is a science which is based at the same time on experimentation, mathematics and metaphysics, and he made a parallel between this new field and the field of physics in the way Newton conceived it. Sygmund Freud was profoundly influenced by Herbart.[7] Herbart returned to Göttingen in 1833 where he taught philosophy and pedagogy until his death. Riemann was 15 and it is unlikely that he followed any course of Herbart. Erhard Scholz, in his paper [152], reports on several sets of notes written by Riemann and preserved in the Riemann archives in Göttingen, which concern the philosophy of Herbart. These notes show that Riemann was indeed influenced by the philosopher, for what concerns epistemology and the philosophy of science, and in particular for his ideas on space. After analyzing some of Riemann's fragments,

---

[7]From the article on Herbard in the *Freud encyclopedia: Theory, therapy and culture* ([58] p. 254), we read: "The ghost of the philosopher Johann Friedrich Herbart hovers over all of Freud's works, an inseparable albeit unacknowledged presence. Herbart, the successor of Kant in Königsberg, arguably exercised a more profound, more persuasive influence on Freud than either Schopenhauer or Nietzsche, whom many scholars regard as sources for some of his major concepts. From Herbart, Freud derived such ideas as the mental activity can be conscious, preconscious, or unconscious, that unconscious mental activity is a continuous determinant of conscious activity, and that the present is unceasingly shaped by the past, whether remembered or forgotten. From Herbart, he also borrowed some essentials of his model, the idea of conflicting conscious and unconscious psychic forces, the censorship-exercising ego, the threshold of consciousness, 'resistance,' 'repression' and much else. [...] It was Herbart's ambition to contribute to the establishment of 'a research of the mind which will be the equal of natural science, insofar as this science everywhere presupposes the absolutely regular connection between appearances.' He compared the situation in psychology with that of astronomy: in the pre-Copernician era, the motions of the planets had seemed irrational; every so often these heavenly bodies inexplicably seemed to reverse their course; for this reason they were known as the 'wanderers.' These peculiar paths, however, were recognized as entirely lawful as soon as the heliocentric theory was introduced. The hypothesis of unconscious thought performed the same service to the mind, Herbart maintained." We encourage the interested reader to go through this entire article by R. Sand.

Scholz writes: "Riemann's views on mathematics seem to have been deepened and clarified by his extensive studies of Herbart's philosophy. Moreover, without this orientation, Riemann might have never formulated his profound and innovative concept of manifold. This represents an indirect but nevertheless effective influence of Herbart on Riemann's mathematical and (in particular) his geometrical thinking." The last paragraph we quoted from Riemann's *Gesammelte mathematische Werke* continues as follows:

> As for the latter, I could almost completely agree with Herbart's earliest investigations, whose results are giving in his graduating and qualifying theses *[Promotions – und Habilitationsthesen]* (of October 22 and 23, 1802), but I had to veer away from the later course of his speculations at an essential point, thus determining a difference with respect to his philosophy of nature and those propositions of psychology which concern its connections with the philosophy of nature.

When Riemann became, in 1854, *Privatdozent* at the University of Göttingen, the subject of the first lessons he gave was differential equations with applications to physics. These lectures became well known even outside Germany. In a letter to Houël sent in 1869 (see [32] p. 90), Darboux writes:[8]

> I wonder if you know a volume by Riemann on mathematical physics entitled *On partial differential equations*. I was very much interested in this small volume, it is clear and could be put in the hands of the students of our universities. I think that you will appreciate it; if you are a little bit concerned with mathematical physics it will be of interest to you. Above all, hydrodynamics seems to me very well treated.[9]

Three editions of Riemann's notes from his lectures on differential equations applied to physics appeared in print, the last one in 1882, edited by his student Karl Hattendorf. A work in two volumes entitled *Die partiellen Differential-Gleichungen der mathematischen Physik* (The partial differential equations of mathematical physics) [148] appeared in 1912, written by Heinrich Weber.[10] This work is considered as a revised version of Riemann's lectures on this subject. The applications

[8]In the present chapter, the translations from the French are mine, except if the contrary is indicated.

[9]Je ne sais si vous connaissez un volume de Riemann sur la physique mathématique intitulé *Sur les équations aux dérivées partielles*. Ce petit volume m'a beaucoup intéressé, il est clair et pourrait être mis avec avantage entre les mains des auditeurs de nos facultés. Je crois que vous en serez content ; si vous vous occupez un peu de physique mathématique il vous intéressera, l'hydrodynamique surtout m'y a paru très bien traitée.

[10]Heinrich Weber (1842–1913) taught mathematics at the University of Strasbourg—a city which at that time belonged to Germany—from 1895 till 1913. He is the co-editor, with Dedekind, of Riemann's collected works. Dedekind followed Riemann's lectures in 1855–1856, and they became friends. Regarding the relation between the two mathematicians, let us mention the following. In 1858, a position of professor of mathematics was open at the Zurich Eidgenössische Technische Hochschule. Both Riemann and Dedekind applied, and Dedekind was preferred, probably because he had more experience in teaching elementary courses. Indeed, a Swiss delegation visited Göttingen to examine the candidates, and considered that Riemann was "too introverted to teach future engineers." After he left for Zurich, Dedekind remained faithful to Riemann. Klein, in his Lectures on the history of nineteenth century mathematics, characterized Dedekind as a major representative of the Riemann tradition. At Riemann's death in 1866, Dedekind was given the heavy load of editing Riemann's works. This is where he asked the help of his friend Heinrich Weber. The first edition of the Collected works was published in 1876, and it included a biography of Riemann written

to physics include heat conduction, elasticity and hydrodynamics. The work was used for many years as a textbook in various universities.[11] In a biography of Neugebauer [151] (p. 16), the author reports that in the 1920s, Courant was using these books in his teaching:

> Courant was hardly a brilliant lecturer, but he did have the ability to spark interest in students by escorting them on the frontiers of research in analysis. In his course on partial differential equations, he stressed two sharply opposed types of literature: works that expound general theory, on the one hand, and those that pursue special problems and methods, on the other [...] For the second type of literature, Courant's background and personal preferences came to the fore. Here he recommended the *Ausarbeitung* of Riemann's lectures on partial differential equations prepared by Karl Hattendorf together with Heinrich Weber's subsequent volume on Riemann's theory of PDEs in mathematical physics.

Concerning physics in Riemann's work, let us also quote Klein from his address delivered at the 1894 general session of the *Versammlung Deutscher Naturforscher und Ärtzte* (Meeting of the German naturalists and physicians; Vienna, September 27, 1894), cf. [86]:

> I must mention, first of all, that Riemann devoted much time and thought to physical considerations. Grown up under the tradition which is represented by the combinations of the names of Gauss and Wilhelm Weber, influenced on the other hand by Herbart's philosophy, he endeavored again and again to find a general mathematical formulation for the laws underlying all natural phenomena. [...] The point to which I wish to call your attention is that *these physical views are the mainspring of Riemann's purely mathematical investigations.*

Riemann wrote papers on physics (a few, but may be as many as his papers on mathematics), and they will only be briefly mentioned in the present chapter. Our main subject is *not* Riemann's work on physics, but his ideas concerning physics that are present in his mathematical papers. In presenting these ideas, we shall comment on his motivation. Jeremy Gray, the author of the chapter of the present volume entitled *Riemann on geometry, physics, and philosophy* [69], writes in that chapter that "Riemann belongs to a list of brilliant mathematicians whose lasting contributions are more in mathematics than physics, contrary to their hope." We shall mention

---

(Footnote 10 continued)
by Dedekind. Before that, Dedekind had edited, in 1868, the two habilitation works of Riemann, *Über die Darstellbarkeit einer Function durch eine trigonometrische Reihe* (On the representability of a function by a trigonometric series) [131] and *Über die Hypothesen, welche der Geometrie zu Grunde liegen* [146] which we already mentioned. Klein says in his lectures on the history of 19th century mathematics [87] that at Riemann's death, the latter's heirs entrusted Dedekind with the edition of the *Nachlass*, that Dedekind started working on that and wrote illuminating comments, but that he was not able to continue that work alone. In 1871 he asked the help of Clebsch, but the latter died soon after (in 1872). He then asked Weber to help him completing the work. In 1882, Dedekind and Weber published a paper entitled *Theorie der algebraischen Functionen einer Veranderlichen* (The theory of algebraic functions in one variable) [34] in which they developed in a more accessible manner Riemann's difficult ideas on the subject. An analysis of this paper, including a report on its central place in the history of mathematics, is contained in Dieudonné [35] p. 29–35. Finally let us mention that Weber is also a co-editor of the famous Klein *Encyklopädie der Mathematischen Wissenschaften.*

[11] See [8] for a review of this book.

Riemann's impact of some of his mathematical work on the later development of physics. At the same time, we shall give an overview of some important mathematical works of Riemann, in particular on the following topics:

(1) *Riemann surfaces and functions of a complex variable*. Riemann approached complex analysis from the point of view of potential theory, that is, based on the theory of the Laplace equation

$$\frac{\partial^2 u}{\partial x^2} + \frac{\partial^2 u}{\partial y^2} = 0.$$

Here, the function $u = u(x, y)$ represents the potential function that gives rise to a streaming for an incompressible flow contained between two planes parallel to the x,y plane (the flow may be an electric field, in which case u is the electrostatic potential). The Laplace equation expresses the fact that there is as much fluid that flows into an element of area per unit time than fluid that flows out. The bases of the theory of Riemann surfaces are contained in Riemann's doctoral dissertation [130] and his paper on Abelian functions [133]. We shall review the role of physical reasoning in these works. Conversely, Riemann made the theory of functions of a complex variable, based on his approach using partial differential equations in the real domain, a basic tool in mathematical physics. We shall mention below some of the tremendous impact of the theory of Riemann surfaces in modern theoretical physics.

(2) *Trigonometric series*: Riemann's work on trigonometric series is contained in his habilitation dissertation (*Habilitationsschrift*) [131]. His motivation, as Riemann himself writes, comes from the theory of sound. The origin of the questions he tackled lies in seventeenth-century physics and mathematics, and they led then to a harsh debate that involved several scientists including Euler, Lagrange, d'Alembert and Daniel Bernoulli (to mention only the most famous ones). From the mathematical point of view, the main issue was the nature of the functions that were admitted as solutions of the wave equation. Riemann eventually concluded the debate, showing the generality of those functions that have to be included as solutions of these equations. In the same paper, Riemann laid the foundations of what became known later on as the theory of the Riemann integral. This came from his effort to clarify the nature of the coefficients of a trigonometric series associated with a function. These coefficients are indeed given in the form of integrals.

(3) *Riemannian geometry*. This is contained in Riemann's habilitation lecture [146] and his later paper, the *Commentatio* [132]. In the development of this theory, Riemann was motivated in part by physics, and in part by philosophy. In his habilitation lecture, Riemann's bond of filiation with Greek philosophy, and in particular with Aristotle, is clear. We shall comment on this and we shall also recall the huge impact of these two works of Riemann on the later physical theories.

(4) *Other works*. In the last section of this chapter, we shall analyze more briefly some other papers of Riemann related to our subject.

To close this introduction, we mention that the fact that Riemann, in his mathematical work, was motivated by physics was also common to other mathematicians of the eighteenth and the nineteenth centuries. One may recall that Gauss, who was Riemann's mentor, considered himself more as a physicist than a mathematician. We refer to [59] for a review of Gauss's contribution to geomagnetism. Gauss was in charge of the practical task of surveying geodetically the German kingdom of Hannover. In the preface of his paper [61] which we already mentioned, published in 1825, about the same time he wrote his famous *Disquisitiones generales circa superficies curvas* (General investigation of curved surfaces) (1825 and 1827) [66], Gauss writes that his aim is only to construct geographical maps and to study the general principles of geodesy for the task of land surveying. Surveying the kingdom of Hannover took nearly two decades to be completed. It led Gauss gradually to the investigation of triangulations, to the use of the method of least squares in geodesy,[12] and then to his *Disquisitiones generales circa superficies curvas*. In the latter, we can read, for instance, in §27 (p. 43 of the English translation [66]): "Thus, e.g., in the greatest of the triangles which we have measured in recent years, namely that between the points Hohenhagen, Brocken, Inselberg, where the excess of the sum of the angles was 14."85348, the calculation gave the following reductions to be applied to angles: Hohehagen: 4."95113; Brocken: 4."95104; Inselberg: 4."95131."

It is also interesting to know that Jacobi, after Gauss, studied similar problems of geodesy, using elliptic functions. In a paper entitled *Solution nouvelle d'un problème fondamental de géodésie* (A new solution of a fundamental problem in geodesy) [79], he considers, on an ellipsoid having the shape of the earth, a geodesic arc whose length, the latitude of its origin and its azimuth angle at that point are known. The question is then to find the latitude, the azimuth angle of the extremity of this arc, as well as the difference in longitudes between the origin and the extremity. He then declares: "The problem of which I just gave a new solution has been recently the subject of a particular care from Mr. Gauss, who treated it in various memoirs and gave different solutions of it."[13]

Riemann was profoundly influenced by Gauss. We emphasize this fact because it is written here and there that Riemann did not learn a lot from Gauss, since when Riemann started his studies, Gauss was already old, and that in any case, Gauss was never interested in teaching. Klein writes in his *Development of mathematics in the 19th century* ([87], p. 234 of the English translation):

> Gauss taught unwillingly, had little interest in most of his auditors, and was otherwise quite inaccessible. Nevertheless, we call Riemann a pupil of Gauss; indeed he is Gauss's only true pupil, entering into his inner ideas, as we now are coming to see in outline from the *Nachlass*.

[12] Gauss first published his method of least squares in an important treatise in two volumes calculating the orbits of celestial bodies in 1809 [60], but in that work he claims that he knew the method since 1795. This led to a priority controversy between Gauss and Legendre, who published the first account of that method in 1805 [92].

[13] Le problème dont je viens de donner une solution nouvelle a été dans ces derniers temps l'objet de soins particuliers de la part de M. Gauss, qui en a traité dans différents mémoires et en a donné plusieurs solutions.

Before Gauss, Euler, whose work was also a source of inspiration for Riemann, was likewise thoroughly involved in physics. His work on partial differential equations was motivated by problems from physics. In fact, Euler tried to systematically reduce every problem in physics to the study of a differential equation. Euler was also very much involved in acoustics. The initial attraction by Euler to number theory arose in his work on music theory; cf. [21] where this question is thoroughly discussed. Later in this chapter, we shall have the occasion to talk about Euler's and Riemann's works related to acoustics. The influence of Euler and Gauss on Riemann is thoroughly reviewed in Chap. 1 of the present volume [104].

2 Function Theory and Riemann Surfaces

Riemann's work on the theory of functions of a complex variable is developed in his two memoirs *Grundlagen für eine allgemeine Theorie der Functionen einer veränderlichen complexen Grösse* (Foundations of a general theory of functions of a variable complex magnitude) [130] (1851) and *Theorie der Abel'schen Functionen* (Theory of Abelian functions) [133] (1857). The first of these memoirs is Riemann's doctoral dissertation. The text of this dissertation was submitted to the University of Göttingen on November 14, 1851, and the defense took place on December 16 of the same year. Several ideas introduced in these two papers are further developed in subsequent works of Riemann. We mention in particular a fragment on the theory of Abelian functions published posthumously and which is part of Riemann's collected papers editions [141, 145]. There are also other works of Riemann that involve in an essential way functions of a complex variable; a famous example is his extension of the real zeta function[14] to the complex domain, which turned out to be a huge step in the study of this function. Finally, we mention that there are lecture notes of Riemann on functions of a complex variable available at Göttingen's library, and there is an outline of these lectures in Narasimhan's article [100]. It is not our intention here to comment on Riemann's fundamental work on functions of a complex variable and its importance for later mathematics; we shall only concentrate on its relation to physics. However, we start with a few comments on the theory of functions of a complex variable, before Riemann started working on it, because this will help including Riemann's work in its proper context.

The notion of function of a complex variable can be traced back to the beginning of the notion of function, which, in the form which is familiar to us today, is usually attributed to Johann Bernoulli and Euler.[15] A precise definition of a function, based on a careful description of the notion of variable, is contained in Euler's treatise

[14]The real zeta function was already considered by Euler; cf. in particular his papers [38, 39]. Note that Euler did not use the notation ζ. In his book, *Introductio in analysin infinitorum* (Introduction to the analysis of the infinite) [49], where he considers this function for integer values of the variable, he denotes it by P.

[15]One should emphasize that the seventeenth-century infinitesimal calculus of Leibniz and Newton, which was developed before Euler, dealt with curves, and not with functions.

Introductio in analysin infinitorum [49] (1748). This book was published one year after the appearance of the famous memoir of d'Alembert [4] in which the latter gave the wave equation. We mention this fact because the main mathematical question that was motivated by d'Alembert's memoir turned out to be the question of the nature of functions that are solutions of the wave equation. Hence, the general question was addressed: *What is a function?* Furthermore, this memoir of d'Alembert was the original motivation for the study of trigonometric series, which was the subject of Riemann's *Habilitationsschrift* which we discuss in Sect. 3. The *Introductio* consists of two volumes. In Chap. 1 of the first volume, after he defines functions, Euler writes: "[...] Even zero and complex numbers are not excluded from the signification of a variable quantity." Thus, complex functions were considered by Euler from the very outset of his work on general functions. We shall come back in Sect. 3 to Euler's definition of a function. Two years after his *Introductio*, Euler published his famous memoir *Sur la vibration des cordes* (On string vibration) [42] which we shall also discuss below.

After Euler, one has to mention Cauchy, who made a thorough and profound contribution to the theory of analytic functions of a complex variable, during the three decades that preceded Riemann's work on the subject. In a series of *Comptes Rendus* Notes and in other publications, including his *Cours d'analyse de l'École Royale Polytechnique* (A course of analysis of the École Royale Polytechnique) [24] (1821) and his *Mémoire sur les intégrales définies prises entre des limites imaginaires* (Memoir on the definite integrals taken between two imaginary limits) [25] (1825), Cauchy introduced several fundamental notions, such as the disc of convergence of a power series, and path integrals between two points in the complex plane, with the study of the dependence on the path.[16] He dealt with functions which may take the value infinity at some points, and he invented the calculus of residues and the characterization of complex analyticity by the partial differential equations satisfied by the real and imaginary parts of the function, which were called later the Cauchy–Riemann equations.

Besides the work of Cauchy, we mention that of his student Puiseux who further developed some of his master's ideas and brought new ones, essentially in two papers [126, 127]. In the 177 pages paper [126], Puiseux uses the methods introduced by Cauchy on path integration in the study of the problem of uniformization of an algebraic function $u(z)$. This is a function defined implicitly by an equation of the form $P(u, z) = 0$ where P is a two-variable polynomial. The uniformization problem, in this setting, is to get around the fact that such a function u is multi-valued and to make it univalued (uniform). In doing this, Puiseux also developed the theory of functions of a complex variable which are of the form $\int u dz$, where u is as above. He highlighted the role of the critical points of the function u in this line integral, and the fact that integrating along the loops that contain one such point one gets different values for the function. Using this fact, he gave an explanation

[16]One should mention that the idea of integration along paths was present in the works of Gauss [65] and Poisson [123]. They both considered line integrals in the complex plane and they noticed that these integrals depend on the choice of a path.

for the periodicity of the complex circular functions, of elliptic functions, and of the functions defined by integrals introduced by Jacobi. He showed that for a given z, the various solutions $u(z)$ of the equation $f(u, z) = 0$ constitute a certain number of "circular systems," and he gave a method to collect them into groups. In doing this, he developed a geometric Galois theory, discussing the "substitutions" which act on the solutions of the algebraic equation. He also gave a method to find expressions for these solutions as power series with fractional exponents. There is a profound relation between the results of Puiseux on algebraic functions and Riemann surfaces. This is also surveyed in the chapter [105] in the present volume.

Since we talked about elliptic functions, whose study was one of Riemann's main subjects of interest, let us mention that these functions were also used in physics, and that this was certainly one of the reasons why Riemann was interested in them. Already Euler, in his numerous memoirs on elliptic integrals, studied their applications to the oscillations of the pendulum with large amplitudes, to the measurement of the earth, and to the three-body problem. In the preface of the treatise *Théorie des fonctions doublement périodiques et, en particulier, des fonctions elliptiques* (Theory of doubly periodic functions, and in particular elliptic functions) by Briot and Bouquet which we review in another chapter of the present volume [106] in relation with Riemann's work, the authors write:

> One encounters frequently elliptic functions in questions of geometry, mechanics, or mathematical physics. We quote, as examples, the ordinary pendulum, the conical pendulum, the ellipsoid attraction, the motion of a solid body around a fixed point, etc. Mr. Lamé published last year a very interesting work, where he shows that that elliptic functions enter into questions relative to heat distribution and of isothermal surfaces.[17]

Regarding the same subject, we note that the second volume of Halphen's *Traité des fonctions elliptiques et de leurs applications* (Treatise on elliptic functions and their applications) [73] carries the subtitle: *Applications à la mécanique, à la physique, à la géodésie, à la géométrie et au calcul intégral* (Applications to mechanics, physics, geodesy, geometry and integral calculus).

Riemann adopted a physical approach to functions of a complex variable. This point of view was new, compared to that of Cauchy, although both men reached simultaneously the characterization of conformal mappings in terms of the partial differential equations which are called the Cauchy–Riemann equations.[18] Riemann, just after establishing these equations, notes that the real and imaginary parts of such a function satisfy the Laplace equation. Ahlfors writes, in [2] p. 4: "Riemann virtually puts equality signs between two-dimensional potential theory and complex function theory." We shall say more about Riemann's use of the Dirichlet principle,

[17]On rencontre fréquemment les fonctions elliptiques dans les questions de géométrie, de mécanique ou de physique mathématique. Nous citerons, comme exemples, le pendule ordinaire, le pendule conique, l'attraction des ellipsoïdes, le mouvement d'un corps solide autour d'un point fixe, etc. M. Lamé a publié l'année dernière un ouvrage très-intéressant, où il montre que les fonctions elliptiques s'introduisent dans les questions relatives à la distribution de la chaleur et aux surfaces isothermes.

[18]Riemann gives this characterization at the beginning of his doctoral dissertation [130], defended in 1851, and Cauchy in his papers [27, 28], published the same year.

in particular in his paper on Abelian functions [133], where he solves the question of the determination of a function of a complex variable by given conditions on the boundary and the discontinuity points. Klein, in his article on *Riemann and his significance for the development of modern mathematics* (1895) [86], recalls the importance of potential theory and the influence of Dirichlet on Riemann. He writes:

> It should also be observed that the theory of the potential, which in our day, owing to its importance in the theory of electricity and in other branches of physics, is quite universally known and used as an indispensable instrument of research, was at that time in its infancy. It is true that Green had written his fundamental memoir as early as 1828; but this paper remained for a long time almost unnoticed. In 1839, Gauss followed with his researches. As far as Germany is concerned, it is mainly due to the lectures of Dirichlet that the theory was farther developed and became known more generally; and this is where Riemann finds his base of operations.

Among the works in potential theory that had a great impact later on, that of George Green,[19] mentioned by Klein in the last quote, is worth singling out because his author did it in isolation and never obtained, during his lifetime, the credit he deserves. Green, in 1828, gave the famous *Green formula*, in his paper entitled *An essay on the application of mathematical analysis to the theory of electricity and magnetism* [70]. This article contains the basis of what we call now Green's functions and Green's potential. The introductory part of the essay emphasizes the role of a potential function, and this notion is then used in the setting of electricity and magnetism. The work also contains an early form of Green's Theorem (p. 11–12) which connects a line integral along a simple closed curve and the surface integral over the region bounded by that curve.[20] There is an analogous theorem which relates volume and surface integrals contained in Riemann's 1851 inaugural dissertation. It might be noted that the result is stated (without proof) in an 1846 paper by Cauchy [26]. Cauchy was

[19]George Green (1793–1841) was a British mathematician and physicist who was completely self-taught. His father, also called George, was a baker, and the young George began working to earn his living at the age of five. He went to school for only one year, between the ages of 8 and 9. While he was working full-time in his father's mill, Green used the small amount of time that was left to him to study mathematics without the help of anybody else. On his own, Green became one of the main founders of potential theory. The word "potential" was coined by him, although the notion existed before, e.g. in the works of Laplace and Poisson on hydrostatics. Besides, Green developed mathematical theories of magnetism and electricity that later on inspired the works of Maxwell and William Thomson (later known as Lord Kelvin). Green's father died one year after the publication of the his son's *Essay on the Application of Mathematical Analysis to the Theories of Electricity and Magnetism*, printed at the author's own expense. In the meantime, George Sr. Green had gained some wealth, and what he left was sufficient for his son to put an end to his activities in the mill and to dedicate himself to mathematics. At the recommendation of some influential acquaintance, Green was admitted at the University of Cambridge, as an undergraduate, in 1833, at the age of forty. The difficulties he encountered were not in catching up in the sciences, but in Greek and Latin. Green sat for the bachelor examination five years later, the same year as Sylvester, who was 21 years younger than him. During his relatively short career, Green wrote, besides the paper we mentioned above, several others, on optics, hydrodynamics, gravitation, and the theory of sound. He spent the last part of his life in Cambridge in isolation, addicted to alcohol. His work was rediscovered after his death by Thomson and his ideas blossomed in physics and mathematics.

[20]Basically, Green proved Stokes' theorem for surfaces embedded in 3-space.

also a physicist. We owe him important works on hydrodynamics, elasticity, celestial mechanics and several other topics. But unlike Riemann, Cauchy's mathematical papers do not contain references to physics. Cauchy made a clear distinction between the methods of the two subjects. In the introduction to the first volume of his famous *Cours d'analyse de l'École Polytechnique* [24], we can read:

> Without any doubt, in the sciences which we call natural, the only method which is worth using with success consists in observing the facts and submitting later on the observations to calculus. But it would be a big mistake to think that we can only find certainty in geometric proofs or in the evidence of senses. [...] Let us cultivate with hard work the mathematical sciences, without intending to extend them beyond their domain; and let us not imagine that one can address history with formulae, neither giving as sanctions for morals, theorems from algebra or integral calculus.[21]

Turning back to the Ancients, we quote a related phrase from Book I of the *Nicomachean Ethics*; cf. [9] 1094b[22]:

> [...] for it is the mark of an educated mind to expect that amount of exactness in each kind which the nature of the particular subject admits. It is equally unreasonable to accept merely probable conclusions from a mathematician and to demand strict demonstration from an orator.

At the beginning of his dissertation, Riemann introduces the definition of conformality in terms of the existence of a complex derivative. From this point of view, a function w of a complex variable z is conformal if the derivative $\frac{dw}{dz}$ exists and is independent of the direction. This is equivalent to the infinitesimal notion of angle-preservation. As a matter of fact, conformality of maps in the sense of angle-preservation was already rooted in physics before Riemann. It is important to remember that the question of representing conformally the surface of a sphere onto the plane was already addressed by Hipparchus (second century B.C.), Ptolemy (first century A.D.), and certainly other Greek geometers and astronomers in their work on spherical geometry and cartography, see [113] p. 405ff. We refer the reader to the recent surveys [108, 109] regarding the relation between geography and conformal and quasiconformal (in the sense of close-to-conformal) mappings. One may mention in particular Euler who studied general conformal maps from the sphere to the plane in his memoirs [51–53] which he wrote in relation with his work as a cartographer.[23] In these memoirs, Euler expressed the conformality of projection maps from the sphere onto a Euclidean plane in terms of partial differential equations. Lambert,

---

[21]Sans doute, dans les sciences qu'on nomme naturelles, la seule méthode qu'on puisse employer avec succès consiste à observer les faits et à soumettre ensuite les observations au calcul. Mais ce serait une erreur grave de penser qu'on ne trouve la certitude que dans les démonstrations géométriques, ou dans le témoignage des sens [...] Cultivons avec ardeur les sciences mathématiques, sans vouloir les étendre au-delà de leur domaine; et n'allons pas nous imaginer qu'on puisse attaquer l'histoire avec des formules, ni donner pour sanction à la morale des théorèmes d'algèbre ou de calcul intégral.

[22]I thank M. Karbe for this reference.

[23]At the Academy of Sciences of Saint Petersburg, Euler, among his various duties, had the official charge of cartographer and participated in the huge project of drawing maps of the Russian Empire.

in his paper [91], also formulated problems concerning the projection of subsets of the sphere onto the Euclidean plane in terms of partial differential equations. Likewise, Lagrange used the notion of conformal map in his papers on cartography [88], and the same notion is inherent in Gauss's work. The terminology "isothermal coordinates" which the latter introduced, referring to a locally conformal map between a subdomain of the plane and a subdomain of the surface, indicates the relation with physics. Riemann, in his dissertation, refers to an 1822 paper by Gauss, published in the *Astronomische Abhandlungen* in 1825 (but written several years before).[24] The title of that paper is *Allgemeine Auflösung der Aufgabe: die Theile einer gegebnen Fläche auf einer andern gegebnen Fläche so abzubilden, daß die Abbildung dem Abgebildeten in den kleinsten Theilen ähnlich wird* (General solution of the problem: to represent the parts of a given surface on another so that the smallest parts of the representation shall be similar to the corresponding parts of the surface represented), a paper presented to a prize question proposed by the Royal Society of Sciences at Copenhagen, [61]. In this paper, Gauss shows that every sufficiently small neighborhood of a point in an arbitrary real-analytic surface can be mapped conformally onto a subset of the plane.[25]

After recalling the definition of a conformal map, Riemann passes to the equivalent condition expressed in terms of partial differential equations. Here, we are given a function f of a complex variable which is composed of two functions u and v of two real variables x and y:

$$f(x + iy) = u + iv.$$

The functions u and v are differentiable and satisfy the Cauchy–Riemann equations. They appear as *potentials* in the space of the two variables x and y. Klein, in his article on *Riemann and his significance for the development of modern mathematics* writes ([86] p. 168):

> Riemann's method can be briefly characterized by saying that *he applies to these parts u and v the principles of the theory of the potential.* In other words, his starting point lies in the domain of mathematical physics.

In the same article (p. 170), after explaining some of Riemann's tools, Klein adds:

> All these new tools and methods, created by Riemann for the purpose of pure mathematics out of the physical intuition, have again proved of the greatest value for mathematical physics. Thus, for instance, we now always make use of Riemann's methods in treating the *stationary* flow of a fluid within a two-dimensional region. A whole series of most interesting problems, formerly regarded as insolvable, had thus been solved completely. One of the best known problems of this kind is Helmholtz's determination of the shape of a free liquid jet.

---

[24]The paper won a prize for a question proposed by the Copenhagen Royal Society of Sciences in 1822. The subject of the competition was: "To represent the parts of a given surface onto another surface in such a way that the representation is similar to the original in its infinitesimal parts." A letter from Gauss to Schumacher dated July 5, 1816 shows that the solution was already known to Gauss at that time; cf. Gauss's *Werke* Vol. 8, p. 371.

[25]Gauss did not solve the problem of mapping conformally an arbitrary finite portion of the surface; this was one of the questions considered by Riemann. An English translation of Gauss's paper is published in [153], Volume 3.

Klein, who spent a significant part of his time advertising and explaining Riemann's ideas, completely adhered to his physical point of view. In 1882, he wrote a booklet entitled *Über Riemanns Theorie der algebraischen Funktionen und ihrer Integrale* (On Riemann's theory of algebraic functions and their integrals: A supplement to the usual treatises) [85] in which he explains the main ideas in Riemann's 1857 article on Abelian functions. This booklet is a redaction of part of a course that Klein gave in 1881 at the University of Leipzig, and it had a great influence in making Riemann's ideas known.[26] The excerpts we present here and later in this chapter are from the English translation [85], published a few years after the German original.

According to Klein, the point of view on analytic functions based on the Cauchy–Riemann equations is supported by physics. The first paragraph of his exposition [85] (p. 1) is entitled *Steady streaming in the plane as an interpretation of the functions of* $x + iy$. He writes there: "The physical interpretation of those functions of $x + iy$ which are dealt with in the following pages is well known." He refers to Maxwell's *Treatise of electricity and magnetism* (1873), and he adds: "So far as the intuitive treatment of the subject is concerned, his point of view is exactly that adopted in the text." Maxwell, at the beginning of the 1860s, developed a theory of electricity and magnetism and established the partial differential equations that carry his name, which describe the generation of electric and magnetic fields and the relation between them. In some sense, Maxwell's equations are a generalization of the Cauchy–Riemann equations. Klein is among the first mathematicians who stressed this point. After he states the Cauchy–Riemann equations, Klein continues:

> In these equations we take u to be the *velocity-potential*, so that $\frac{\partial v}{\partial y}$, $\frac{\partial u}{\partial x}$ are the components of the velocity of a fluid moving parallel to the xy plane [...] For the purposes of this interpretation it is of course indifferent of what nature we may imagine the fluid to be, but for many reasons it will be convenient to identify it here with the *electric fluid*: u is then proportional to the electrostatic potential which gives rise to the streaming, and the apparatus of experimental physics provide sufficient means for the production of many interesting systems of streaming.

Later in the text, in dealing with residues, Klein writes:

> The reason that the residue of z_0 must be equal and opposite to that of z_1 is now at once evident: the streaming is to be steady, hence the amount of electricity flowing at one point must be equal to that flowing out at the other.

In another report on Riemann's work, [86] p. 175, Klein states:

> Riemann's treatment of the theory of function of complex variables, founded on the partial differential equation of the potential, was intended by him to serve merely as an *example* of the analogous treatment of all other physical problems that lead to partial differential equations, or to differential equations in general. [...] The execution of this programme which has since been considerably advanced in various directions, and which has in recent years been taken up with particular success by French geometers, amounts to nothing short of a *systematic reconstruction of the methods of integration required in mechanics and in mathematical physics.*

---

[26]Constance Reid reports on p. 178 of her biography of Hilbert [129] that at a meeting of the Göttingen Scientific Society dedicated to the memory of Klein, held a few months after his death, Courant declared: "If today we are able to build on the work of Riemann, it is thanks to Klein."

It is also interesting to recall that according to Klein [87], Riemann started studying Abelian functions because of their use in his research on galvanic currents.

An important element in Riemann's theory of functions of a complex variable is the so-called Dirichlet problem.[27] Stated with a minimal amount of hypotheses, the problem, from the mathematical point of view, asks for the following: Given an open subset Ω of $\mathbb{R}^n$ and a continuous function f defined on the boundary $\partial\Omega$ of Ω, to find an extension of f to Ω which is harmonic and continuous on the union $\Omega \cup \partial\Omega$. The problem has more than one facet and there are several ways of dealing with it. Physicists consider that the problem has obviously a positive solution under very mild conditions, and that this solution is unique. In this setting, one thinks of the function f on $\partial\Omega$ as a time-independent potential (electric, gravitational, etc.). Letting the system evolve, it will attain an equilibrium state, and the solution will necessarily satisfy a mean value property, that is, it will be harmonic. The harmonicity property is also formulated in terms of realizing the minimum of the energy functional

$$\int\int\left((\frac{\partial u}{\partial x})^2 + (\frac{\partial u}{\partial y})^2\right) dxdy. \tag{1}$$

All these ideas were known to eighteenth century physicists and in fact, most of them can be traced back to Newton.

The "Dirichlet principle" is a method for solving the Dirichlet problem. It is Riemann who coined the term. The principle is based on an assertion he took for granted, namely, that an infimum of the energy functional is attained. This infimum is necessarily harmonic and forms a harmonic function u. Riemann used the Dirichlet principle to construct analytic functions, not only on the disc, but on an arbitrary Riemann surface, after cutting it along a system of arcs so that it becomes simply-connected. Riemann also used the Dirichlet principle at other places in his doctoral dissertation and in his paper on Abelian functions.[28] At the time Riemann appealed to the Dirichlet problem, several other eminent mathematicians used an analogous

---

[27] Concerning the terminology, Klein writes in his *Development of mathematics in the 19th century* ([87], p. 242 of the English translation): "This is the first boundary value problem, which the French, unhistorical as they are, call the "Dirichlet problem": to determine a function u if its boundary values and definite physically possible discontinuities are given—there will be one and only one solution.

[28] The name "Dirichlet principle" is used in the paper [133] on Abelian functions (§III, IV, Preliminaries), but not in the doctoral dissertation [130]. Riemann, in his existence proof of meromorphic functions on general Riemann surfaces, defined these functions by their real parts, which are harmonic functions, using this principle. He also used it in his proof of the Riemann Mapping Theorem. In fact, it is well known today that the Riemann Mapping Theorem, the existence of meromorphic functions, and the Dirichlet problem, are all equivalent. Riemann writes in §III of the Preliminary section of his paper on Abelian functions [133] that in the study of integrals of algebraic functions and their inverses, one can use a principle which Dirichlet used several years before in his lectures on the forces that act by the inverse of the square of the distance, to solve of a problem related to a function of three variables satisfying the Laplace equation. He adds that Dirichlet was probably inspired by an analogous idea of Gauss. In fact, Gauss used such a principle in his 1839 paper [62]. He assumed there without proof that for a given constant potential distribution, an equilibrium state is attained and is unique and corresponds to the minimum of the energy. It is possible that Riemann chose to call this principle the "Dirichlet principle" out of faithfulness to the mathematician from

principle, in physics and in mathematics. This includes Laplace, Fourier and Poisson in France, Green, Thomson and Stokes in England, and Gauss in Germany. Helmholtz used this principle in his work on acoustics [74]. Riemann's use of the Dirichlet principle was criticized by Weierstrass [162]. Klein writes in his *Development of mathematics in the 19th century* ([87], p. 248 of the English translation):

> With this attack by Weierstrass on Dirichlet's principle, the evidence to which Dirichlet, and after him, Riemann, had appealed, became fragile [...] The majority of mathematicians turned away from Riemann; they had no confidence in the existence theorems, which Weierstrass's critique had robbed of their mathematical supports.
>
> The physicists took yet another position: they rejected Weierstrass's critique. Helmholtz, whom I once asked about this, told me: "For us physicists the Dirichlet principle remains a proof." Thus he evidently distinguished between proofs for mathematicians and physicists; in any case, it is a general fact that physicists are little troubled by the fine points of mathematics – for them the "evidence" is sufficient.

The mathematicians' doubts concerning Riemann's use of the Dirichlet principle were removed only several years later. We refer the reader to [99] for the details of this interesting story.

In the preface to his booklet [85], (p. IX), Klein writes:

> [...] there are certain physical considerations which have been lately developed, although restricted to simpler cases, from various points of view.[29] I have not hesitated to take these physical conceptions as the starting point of my presentation. Riemann, as we know, used Dirichlet's Principle in their place in his writings. But I have no doubt that he started from precisely those physical problems, and then, in order to give what was physically evident the support of mathematical reasoning, he afterwards substituted Dirichlet's Principle.

Klein adds:

> Anyone who clearly understands the conditions under which Riemann worked in Göttingen, anyone who has followed Riemann's speculations as they have come down to us, partly in fragments, will, I think, share my opinion. However that may be, the physical method seemed the true one for my purpose. For it is well known that Dirichlet's principle is not sufficient for the actual foundation of the theorems to be established; moreover, the heuristic element, which to me was all-important, is brought out far more prominently by the physical method. Hence the constant introduction of intuitive considerations, where a proof by analysis would not have been difficult and might have been simpler, hence also the repeated illustration of general results by examples and figures.

Poincaré was, like Riemann, a pre-eminent representative of a philosophical tradition of thought in geometry and physics which was invoked at the outset, a tradition combining mathematical, physical, and philosophical thinking. In his booklet

(Footnote 28 continued)
whom he learned most. Klein writes in [87]: "Riemann was bound to Dirichlet by the strong inner sympathy of a like mode of thought. Dirichlet loved to make things clear to himself in an intuitive substrate; along with this he would give acute, logical analyses of foundational questions and would avoid long computations as much as possible. His manner suited Riemann, who adopted it and worked according to Dirichlet's methods."

[29][Klein's footnote] Cf. C. Neumann, Math. Ann. t. X, pp. 569–771. Kirchoff, Berl. Monatsber., 1875, pp. 487–497. Töpler, Pogg. Ann. t. CLX., pp. 375–388.

La valeur de la science (The value of science) [121] (1905), commenting on Klein's method, Poincaré writes:

> [...] On the contrary, look at Mr. Klein: He is studying one of the most abstract questions in the theory of functions; namely, to know whether on a given Riemann surface there always exists a function admitting given singularities: for instance, two logarithmic singular points with equal residues of opposite signs. What does the famous German geometer do? He replaces his Riemann surface by a metal surface whose electric conductivity varies according to certain rules. He puts the two logarithmic points in contact with the two poles of a battery. The electric current must necessarily pass, and the way this current is distributed on the surface defines a function whose singularities are the ones prescribed by the statement.[30]

To end this section, let us mention some of the numerous applications of Riemann surfaces in modern mathematical physics.

One of the major applications of the theory of Riemann surfaces in physics is the Atiyah-Singer index theorem. This theorem, obtained in 1963, gives an information on the dimension of the space of solutions of a differential operator (the *analytical index*) in terms of topology (the *topological degree*). The theorem is used in the theory of the Einstein equation, the instanton equation, the Dirac operator, etc. It is considered as a vast generalization of the classical version of the theorem of Riemann–Roch, which is an equality, half of which contained in Riemann's paper on Abelian functions [133], and the other half in the dissertation of his student Roch [150]. (See [1] in this volume for a review of this theorem.)

We also mention string theory, in which (0-dimensional) particles of physics are replaced by (1-dimensional) strings. This theory was developed as a framework that would hopefully solve some problems that cannot be handled by the theory of relativity. At some point (and it still is, albeit with a more skepticism on physical and mathematical grounds) string theory was considered as a possible theory for the unification of the fundamental forces in nature: gravitation and quantum theory, including electromagnetism—another attempt to realize Riemann's long-life insight. In this theory, one follows the history of a closed string, that is, a closed loop in 3-space. While it propagates, such a loop sweeps out a surface. For reasons that have to do with the consistency of the theory, the surface turns out to be equipped with a 1-dimensional complex structure, that is, it is a Riemann surface. If the string does not interact with anything else, then the swept-out surface is a cylinder, but in general, the string, under some interaction, splits into two other strings, which join again, etc. creating a Riemann surface of higher connectivity (Figs. 1 and 2). Seen from very large distances, strings look like ordinary particles, they have mass and charge, but they can also vibrate. This vibration leads to a hypothetical quantum mechanical

[30] [...] Voyez au contraire M. Klein: il étudie une des questions les plus abstraites de la théorie des fonctions; il s'agit de savoir si sur une surface de Riemann donnée, il existe toujours une fonction admettant des singularités données: par exemple, deux points singuliers logarithmiques avec des résidus égaux et de signe contraire. Que fait le célèbre géomètre allemand? Il remplace sa surface de Riemann par une surface métallique dont la conductibilité électrique varie suivant certaines lois. Il met les deux points logarithmiques en communication avec les deux pôles d'une pile. Il faudra bien que le courant passe, et la façon dont ce courant sera distribué sur la surface définira une fonction dont les singularités seront précisément celles qui sont prévues par l'énoncé.

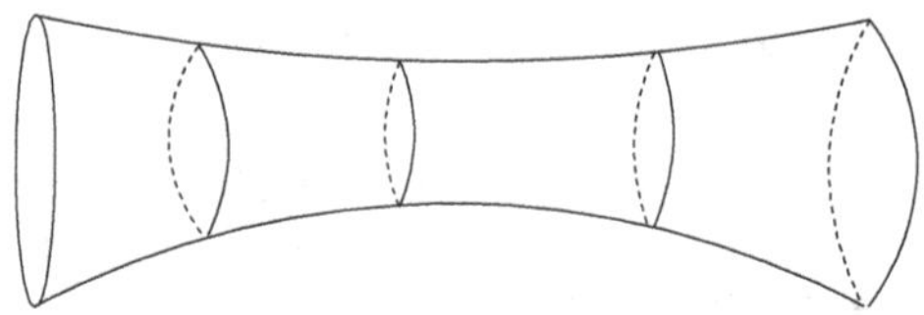

Fig. 1 A moving string, sweeping out a cylinder

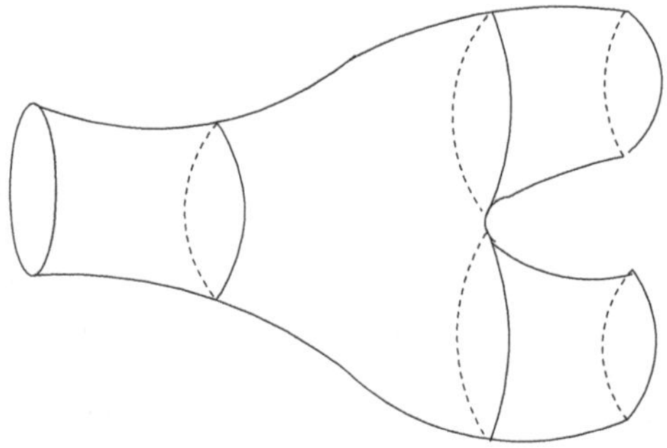

Fig. 2 A string splitting up into two pieces sweeping out a surface of higher connectivity

particle called graviton, which is supposedly responsible for the gravitational force. It is in this sense that string theory is a theory of quantum gravity.

One may also talk about Polyakov's perturbative quantum string theory, a physical theory which involves summations over all Riemann surfaces of arbitrary genus, see [124, 125].

Riemann surfaces are at the basis of conformal field theories (CFT), in which one associates to a marked Riemann surfaces a vector space satisfying certain natural axioms. These surfaces also appear as a major ingredient in the topological quantum field theories (TQFT) developed by Witten and others, which are based on sets of axioms that provide functors from a certain category of cobordisms to the category of vector spaces (Segal and Atiyah gave such sets of axioms). TQFTs lead to results in physics (relativity, quantum gravity, etc.) and at the same time to results in mathematics, where they provide quantum invariants of 3-manifolds. They have applications in symplectic geometry, representation theory of Lie groups and algebraic geometry, in particular in the study of moduli spaces of holomorphic vector bundles over Riemann surfaces. One may also mention that the famous geometric Langlands correspondence is based on the theory of Riemann surfaces. Stated loosely, in the geometric Langlands correspondence one assigns to each rank n holomorphic vector bundle with a holomorphic connection on a complex algebraic curve, a Hecke eigensheaf on the moduli space of rank n holomorphic vector bundles on that curve, cf. [57].

Riemann surfaces are also the main ingredients in the theory of Higgs bundles. These objects arose in the study made by Nigel Hitchin of the self-duality equation on a Riemann surface. From the physical point of view, Higgs bundles describe particles like the Higgs boson. Conversely, the physical methods of Higgs bundle theory are used in the study of moduli spaces of representation of surface groups. Hitchin's motivation arose from his work done in the 1970s with Atiyah, Drinfield and Manin on the so-called instanton equation, another theory combining in an essential way mathematics and physics [13].

Finally, let us mention that Riemann surfaces are used in biology, cf. the recent survey [114].

3 Riemann's Memoir on Trigonometric Series

The habilitation degree, which was required in Germany in order to hold a university teaching position, involved two presentations: the *Habilitationsschrift*, a written original work on a specialized subject, and the *Habilitationsvortrag*, a lecture on a subject chosen by the university council. The present section is devoted to Riemann's *Habilitationsschrift* [131]. We shall discuss his Habilitationsvortrag in the next one.

Riemann's Habilitationsschrift is entitled *Über die Darstellbarkeit einer Function durch eine trigonometrische Reihe* (On the representability of a function by a trigonometric series). It is generally considered that Riemann worked on it during thirty months. He presented it to the university in December 1853. About this work, in a letter to Houël, dated March 18, 1873, and quoted in [33], Darboux writes: "This memoir of Riemann is a masterpiece which is similar to these old paintings of which some small parts in full light make you regret what time has destroyed or what the author has neglected."[31]

This theory of trigonometric series finds its origin in eighteenth century physics, more precisely, in the introduction by d'Alembert, in 1747, of the vibrating string equation (also called the wave equation). To understand the context of Riemann's contribution, it might be useful to recall a few key events in the history of the subject. This theory expanded very slowly, and it was eventually put on firm bases in the nineteenth century, mainly by Joseph Fourier, while he was working on another problem arising from physics, namely, heat diffusion. In the meantime, many preeminent mathematicians and physicists worked on trigonometric series, and we shall mention a few of them. Furthermore, the work done during the first decades after the introduction to the vibrating string equation gave rise to one of the most passionate controversies in the history of mathematics and physics whose scope was larger than the subject of trigonometric series, and we shall say a few words about it. The controversy involved Euler, d'Alembert, Lagrange, Daniel Bernoulli and other major scientists. In particular, a quarrel between Euler and d'Alembert lasted from 1748 until 1783 (the year both of them died). Later on, a dispute concerning the same subject broke out between Fourier and Poisson. The question was about the "continuity" of the functions representing the solutions. This dispute is thoroughly discussed in the introductory part of volume IV of Series A of Euler's *Opera omnia* [48], a volume containing the correspondence between Euler and d'Alembert.

[31]Ce mémoire de Riemann est un chef-d'œuvre semblable à ces vieux tableaux dont quelques parties en pleine lumière vous font regretter ce que le temps a détruit ou ce que l'auteur a négligé.

A comprehensive survey of this controversy is also made in [81] and in Chap. 1 of the present volume [104].

When d'Alembert discovered the vibrating string equation, Euler immediately became interested. He had already been dealing with partial differential equations for several years. In fact, he started working, around the year 1735, on partial differential equations and their applications in geometry and physics. Furthermore, the theory of sound was one of his favorite subjects.[32] This subject was not new, and, in fact, it is worth recalling that the physics of vibrating strings was one of the main problems studied by the Pythagoreans, back in the sixth century B.C. Indeed, most of the ancient biographers of Pythagoras describe his experiments on pitch production, cf. [78]. For a recent scholarship on Pythagoras and the early Pythagoreans, the reader may consult [168].

The heart of the controversy on the vibrating string lies in the question of the clarification of the notion of function, more precisely, the nature of the functions that are solutions of the partial differential equation representing the vibration of a string.

We discuss this matter in Chap. 1 of the present volume [104]. Instead, we make here an excursion to the origin of the theory of sound production, in order to make fully clear that Riemann's investigations on trigonometric functions and integration theory originate in physics.

The first part of Riemann's *Habilitationsschrift* is a historical report on the representation of a function by a trigonometric series, and in fact, it is motivated by the theory of theory of the vibrating string. In a letter to his father, written in the autumn of 1852, Riemann says that he learned the historical details from Dirichlet, who explained them to him in a two-hour session. (The letter is reproduced in Riemann's Collected Works, [141] p. 578.) Riemann starts his historical survey by recalling that this subject is important for physics:

> Trigonometric series, which are given this name by Fourier, that is, series of the form
>
> $$a_1 \sin x + a_2 \sin 2x + a_3 \sin 3x + \dots$$
>
> $$+\frac{1}{2}b_0 + b_1 \cos x + b_2 \cos 2x + b_3 \cos 3x + \dots$$
>
> play a substantial role in the part of mathematics where we encounter functions which are completely arbitrary. We also have reasons to say that the progress of this part of mathematics, which is so important for physics, has been subject to a more precise knowledge of the nature of these series.

Riemann's excursion in history is divided into three periods, and we shall say a few words about each period.

The first period is concerned with the controversies that arose concerning the notion of function which led to the question of representing arbitrary functions by a

[32]Euler writes in his memoir [47] that "the most sublime research that scientists successfully undertook these days is in all respects without question that of propagation of sound." [La plus sublime recherche que les géomètres aient entreprise de nos jours avec succès est sans contredit à tous égards celle de la propagation du son.] We also recall that the subject of Euler's first published memoir is the theory of sound [40].

trigonometric series. D'Alembert, in 1747, wrote two papers, which were published in the Memoirs of the Berlin Academy and under the titles *Recherches sur la courbe que forme une corde tendue mise en vibration* (Researches on the curve that is formed by a stretched vibrating string) [4] and *Suite des recherches sur la courbe que forme une corde tendue mise en vibration* (Sequel to the researches on the curve that is formed by a stretched vibrating string) [5]. In these memoirs, d'Alembert, relying on the fundamental principle of dynamics, gave the partial differential equation that represents the motion of a point on a vibrating string subject to small vibrations:

$$\frac{\partial^2 y}{\partial t^2} = \alpha^2 \frac{\partial^2 y}{\partial x^2}. \tag{2}$$

Here, α is a constant and y is the oscillation of the string, a function of time, t, and distance along the string, x. In the same memoir, d'Alembert wrote the first general solution to the problem, with the given boundary conditions, in the form

$$y(x,t) = \frac{1}{2}\left(\phi(x+\alpha t) + \phi(x-\alpha t)\right),$$

where ϕ is an "arbitrary" periodic function whose period is the double of the length of the string. D'Alembert used a method he attributes to Euler for the integration of partial differential equations. The problem was to give a meaning to the adjective "arbitrary," and this is where the more basic question of *What is a function*? was raised.

It is natural to assume that the solution of d'Alembert's vibration equation should be (twice) differentiable, since the equation involves second partial derivatives, and this is what d'Alembert did. Euler was not of the same opinion. The reason he gave is physical, namely, that one can give a non-smooth initial form to the string (for example a curve with corners) which is being pinched, therefore the function that represents the shape of the string could be quite arbitrary. This implies that the solution may be arbitrary. Euler published his remarks in his memoir *Sur la vibration des cordes* (On the vibration of strings)[33] [42] in which he reviews d'Alembert's work on the wave equation. These remarks introduced some doubts concerning the work of d'Alembert, who wrote a new memoir on the same subject, in which he confirms his ideas, *Addition aux recherches sur la courbe que forme une corde tendue mise en vibration* (Addition to the researches on the curve formed by a stretched vibrating string) [6].

Another pre-eminent scientist who became involved in these questions was Daniel Bernoulli, who was primarily a physicist. Before talking about his contribution to the subject, one should recall that Brook Taylor, in his memoir *De motu nervi tensi* (On the motion of a tense string) [156] (1713) and later in his treatise *Methodus incrementorum directa et inversa* (Direct and Indirect Methods of Incrementation) [157]; first edition 1715, noted that a trigonometric function like $f(x) = \sin x$ represents a

[33] Euler wrote two versions, one in Latin and one in French, the French version bearing the mention "Translated from the Latin."

periodic phenomenon, a wave. In his work on the subject, Taylor was motivated by music theory. Bernoulli came out with a formula of the form

$$\sum_{n=1}^{\infty} \sin \frac{n\pi x}{l} \cos \frac{n\pi \alpha t}{l}. \tag{3}$$

Like Taylor, he was motivated by music. In fact, among all the scientists of the Bernoulli family, Daniel was the most inclined towards physics. His intuition concerning Formula (3) originates in the fact known to all music theorists that the string vibration produces, together with the fundamental pitch, an infinite sequence of harmonics. It is interesting to read some excerpts of Bernoulli's writings on this subject. In his memoir *Mémoire sur les vibrations des cordes d'une épaisseur inégale* (Memoir on the vibrations of a string of uneven width), he writes [16] (p. 173):

> I showed furthermore in the Berlin Memoirs that the vibrations of various orders, however one takes them, may coexist in one and the same string, without disturbing each other in any way, these various kinds of coexisting vibration being absolutely independent of each other. Hence this multiplicity of harmonic sounds which we hear at the same time and with one and the same string. If all the modes of vibration start at the same instant, it may happen that the first vibration of first order, the second vibration of second order, the third vibration of third order, etc. terminate at the same moment. This is in some sense an apparent synchronism which is nothing less than general, since there are infinitely many vibrations which do not terminate at the same instant.[34]

In his *Réflexions et éclaircissements sur les nouvelles vibrations des cordes* (Reflections and clarifications on the new vibrations of strings), Bernoulli writes ([14] p. 152–153):

> Indeed, all musicians agree that a long pinched string gives at the same time, besides its fundamental tone, other tones which are much more acute; most of all they will notice the mixture of the twelfth and the minor sixteenth: in case they don't notice as much distinctly the octave and the double octave, it is only because of the very big resemblance of these two tones with the fundamental. This is an evident proof that there could occur in one and the same string a mixture of several sorts of Taylorian vibrations at the same time. In the same manner, we hear in the sound of large bells a mixture of different tones. If we hold by the middle a steel stick, and if we hit it, we hear at the same time a confused mixture of several tones, which, when appreciated by a skilled musician, turn out to be extremely inharmonious, in such a way that a combination of vibrations is formed, which never start and finish at the same moment, except by a happenstance: hence we see that the harmony of sounds, which we hear at the same time in one sonorous body, is not essential to that material, and should not serve as a principle for systems in music. Air is not free of this multiplicity of coexisting sounds: it often happens that one extracts two different sounds

[34] J'ai démontré de plus dans les Mémoires de Berlin, que les vibrations de différents ordres, quels qu'on les prenne, peuvent coexister dans une seule et même corde, sans se troubler en aucune façon, ces différentes espèces de vibration coexistantes étant absolument indépendantes les unes des autres. De là cette pluralité de sons harmoniques qu'on entend à la fois d'une seule et même corde. Si toutes espèces de vibration commencent au même instant, il arrivera que la première vibration du premier ordre, la seconde vibration du second ordre, la troisième vibration du troisième ordre etc. finiront au même instant. C'est là un synchronisme apparent dans un certain sens, et qui n'est rien moins que général, puisqu'il y a une infinité d'autres vibrations qui ne finissent pas au même instant.

> from a pipe; but the best proof of how much the various air waves may prevent each other is that we hear distinctly every part of a concert, and that all the waves due to these different parts are formed from the same mass of air without disturbing each other, very much like light rays entering in a dark room from a small hole do not disturb each other.[35]

In the same memoir, Bernoulli writes ([14] p. 151):

> My conclusion is that every sonorous body contains essentially an infinity of sounds, and infinitely corresponding ways of performing their regular vibrations; finally, that in each different way of vibrating the variations in the parts of the sonorous body are formed in a different way.[36]

In the same memoir, Bernoulli writes ([14] p. 148):

> [...] without any less esteem for the calculations of Messrs d'Alembert and Euler, which certainly include the most profound and exquisite things that analysis contains; but which show at the same time that an abstract analysis, which we follow without any synthetic examination of the proposed question, may be surprising rather than enlightening for us. It seems to me that one had only to be attentive to the nature of simple vibrations of a string in order to foresee without any calculation everything these two geometers found by the most tricky and abstract calculations with which an analytical mind has been instructed.[37]

It may seem surprising that Euler, who was as much involved in music theory than Bernoulli—he had even corresponded with Rameau on overtones back in 1752 (see

---

[35]Effectivement tous les Musiciens conviennent, qu'une longue corde pincée donne en même temps, outre son ton fondamental, d'autres tons beaucoup plus aigus; ils remarquent surtout le mélange de la douzième et de la dix-septième majeure: s'ils ne remarquent pas aussi distinctement l'octave et la double octave, ce n'est qu'à cause de la trop grande ressemblance de ces deux tons avec le ton fondamental. Voilà une preuve évidente, qu'il peut se faire dans une seule et même corde un mélange de plusieurs sortes de vibrations Tayloriennes à la fois. On entend pareillement dans le son des grosses cloches un mélange de tons différents. Si l'on tient par le milieu une verge d'acier, et qu'on la frappe, on entend à la fois un mélange confus de plusieurs tons, lesquels étant appréciés par un habile Musicien se trouvent extrêmement désharmonieux, de sorte qu'il se forme un concours de vibrations, qui ne commencent et ne finissent jamais dans un même instant, sinon par un grand hazard: d'où l'on voit que l'harmonie des sons, qu'on entend dans une même corps sonore à la fois, n'est pas essentielle à cette matière, et ne doit pas servir de principe pour les systèmes de Musique. L'air n'est pas exempt de cette multiplicité de sons coexistants : il arrive souvent qu'on tire deux sons différents à la fois d'un tuyau; mais, ce qui prouve le mieux, combien peu les différentes ondulations de l'air s'entre-empêchent, est qu'on entend distinctement toutes les parties d'un concert, et que toutes les ondulations causées par ces différentes parties se forment dans la même masse d'air sans se troubler mutuellement, tout comme les rayons de la lumière, qui entrent dans une chambre obscure à travers une petite ouverture, ne se troublent point.

[36]Ma conclusion est, que tous les corps sonores renferment en puissance une infinité de sons, et une infinité de manières correspondantes de faire leurs vibrations régulières; enfin, que dans chaque différentes espèce de vibrations les inflexions des parties du corps sonore se font d'une manière différente.

[37][...] je n'en estime pas moins les calculs de Mrs. d'Alembert et Euler, qui renferment certainement tout ce que l'Analyse peut avoir de plus profond et de plus sublime ; mais qui montrent en même temps, qu'une analyse abstraite, qu'on écoute sans aucun examen synthétique de la question proposée, est sujette à nous surprendre plutôt qu'à nous éclairer. Il me semble à moi, qu'il n'y avait qu'à faire attention à la nature des vibrations simples des cordes, pour prévoir sans aucun calcul tout ce que ces deux grands géomètres ont trouvé par les calculs les plus épineux et les plus abstraits, dont l'esprit analytique se soit encore avisé.

[21], Vol. II)—did not state this idea before, especially that Euler had already heavily manipulated infinite series. It is also a fact that the techniques of trigonometric series are quite different from those of power series.

When Daniel Bernoulli suggested that an "arbitrary" function defined on a finite interval can be expanded as a convergent trigonometric series, several basic questions appeared at the forefront of research:

(1) What is the meaning of such an infinite sum, that is, in what sense does it converge?
(2) In what sense functions possess trigonometric series expansions, and how can such a result be proved.
(3) What is an "arbitrary" function (a question that had been thoroughly investigated without reaching any definite conclusion), and more precisely, is there a definition of an arbitrary function such that it coincides with functions expressible by such an infinite sum?

The second period of Riemann's historical report is dominated by Joseph Fourier (1768–1830) who, in his *Théorie analytique de la chaleur* (Analytic theory of heat) [56] (1822), developed the theory of trigonometric series, while he was studying the heat equation. Let us say a few words about Fourier's treatise.

The introduction (Discours préliminaire) of this treatise is interesting. It concerns the importance of heat in our universe. Fourier writes, at the beginning of that introduction (p. i):

> Heat, like gravity, penetrates all substances of the universe, its rays occupies all parts of space. The goal of our work is to present the mathematical laws that govern this element. From now on, this theory will constitute one of the most important branches of general physics.[38]

Fourier then mentions the works of Archimedes, Galileo and Newton, and he comments on the importance of the effect of sun rays on every element of the living world. It is interesting to note that Archimedes, Galileo and Newton are again mentioned, together, in the introduction to Riemann's habilitation lecture 1854, which we consider in Sect. 4.

After that, he arrives at the mathematical principles of that theory. The problem, which turned out to be very difficult to solve, is stated very clearly ([56] p. 2):

> When heat is unevenly distributed between the various points of a solid mass, it tends to an equilibrium position, and it slowly passes from the overheated parts to the ones which are less heated. At the same time, it dissipates through the surface, and it gets lost in the ambient space or in void. This tendency towards a uniform distribution, and this spontaneous emission which takes place at the surface of bodies, causes a continuous change in the temperature at the various points. The question of the propagation of heat consists in determining what is

[38]La chaleur pénètre, comme la gravité, toutes les substances de l'univers, ses rayons occupent toutes les parties de l'espace. Le but de notre ouvrage est d'exposer les lois mathématiques que suit cet élément. Cette théorie formera désormais une des branches les plus importantes de la physique générale.

the temperature at each point of a body at a given time, assuming that the initial temperatures are known.[39]

In Chap. 2 of his treatise, Fourier establishes (p. 136) the equation which is known nowadays as the "heat equation":

$$\frac{\partial u}{\partial t} = k^2 \left(\frac{\partial^2 u}{\partial x^2} + \frac{\partial^2 u}{\partial y^2} + \frac{\partial^2 u}{\partial z^2} \right).$$

In Chap. 3, he gives the solution of this equation in the form of a trigonometric series. On page 243, he writes: "The preceding analysis gave us the way to develop an arbitrary function as a series of sines and cosines of multiple arcs." Then he announces that he will apply these results to some particular cases which show up in physics, as solutions of partial differential equations. On page 249, he considers the problem of the vibrating string, and he declares that the principles he established solve the difficulties that are inherent in the analysis done by Daniel Bernoulli. He recalls that the latter gave a solution that assumes that an arbitrary function can be developed as a trigonometric series, but that of all the proofs that were proposed of this fact, the most complete is the one where we can determine the coefficients of such a function. This is precisely what Fourier does. For a given trigonometric series,

$$f(x) = a_1 \sin x + a_2 \sin 2x + \ldots - \frac{1}{2} b_0 + b_1 \cos x + b_2 \cos 2x + \ldots,$$

Fourier provides the (now well-known) integral formula for the coefficients:

$$a_n = \frac{1}{\pi} \int_{-\pi}^{\pi} f(x) \sin nx dx$$

and

$$b_n = \frac{1}{\pi} \int_{-\pi}^{\pi} f(x) \cos nx dx.$$

Picard, in the series of three lectures on the history of analysis that he gave in America [117], says (p. 7) that these integral formulae were known to Euler, who mentioned them incidentally. In the same lectures, Picard insists on the fact that Fourier had an audacious method which involved the solution of an infinite number of first-order equations, with an infinite number of unknowns.

[39] Lorsque la chaleur est inégalement distribuée entre les différents points d'une masse solide, elle tend à se mettre en équilibre, et passe lentement des parties les plus échauffées dans celles qui le sont moins ; en même temps elle se dissipe par la surface, et se perd dans le milieu ou dans le vide. Cette tendance à une distribution uniforme, et cette émission spontanée qui s'opère à la surface des corps, changent continuellement la température des différents points. La question de la propagation de la chaleur consiste à déterminer quelle est la température de chaque point d'un corps à un instant donné, en supposant que les températures initiales sont connues.

Fourier also showed that his theory can be applied to functions which may have discontinuities.[40] It is also good to recall that Fourier stated, back in 1807 (the paper was published in 1808, [55]), the fact that a function, given graphically in an arbitrary manner, may be expressed by a trigonometric series.

Riemann reports in his memoir on trigonometric functions that at that time Lagrange vigorously rejected Fourier's assertion. He also recalls the rivalry between Fourier and Poisson, and that the latter took the defense of Lagrange. Riemann analyzes some passages from Lagrange's work, and he concludes by repeating that Fourier was the first to understand exactly and completely the nature of trigonometric series. He adds that after Fourier's work, these series appear in several ways in mathematical physics, as representations of arbitrary functions. He declares that in each particular case one was able to prove that the Fourier series indeed converges to the value of the function, but that it took a long time before such an important theorem was proved in full generality. He recalls that in 1826 Cauchy attempted a proof of that result using complex numbers, in a memoir of the Academy of Sciences (t. VI, p. 603), but that this proof is incomplete, as was shown by Dirichlet.[41] Riemann declares that he completed Cauchy's proof in his inaugural dissertation.

The third section of the historical part of Riemann's *Habilitationsschrift* concerns the work of Dirichlet. The latter, whom we already mentioned several times and who had been one of Riemann's teachers in Berlin, gave a necessary and sufficient condition under which a periodic function can be expanded as a trigonometric series ([94], 1829). In the same memoir, Dirichlet obtained the general theorem concerning the convergence of Fourier series after he pointed out some errors in Cauchy's proof of that result. Riemann, in his *Habilitationsschrift*, considers that it is Dirichlet who closed the controversy. He declares that the latter, in a publication which appeared in 1829 in Crelle's journal (t. IV),[42] gave a "very rigorous" theory of representation by trigonometric series of general functions under the hypothesis that they are integrable, that they do not have infinitely many maxima or minima, and that at the points of discontinuity, the value of the function is the arithmetic mean of its left and right limits. Dirichlet left open the converse: given a function that does not satisfy the first two conditions (the third one must obviously be satisfied), under what conditions can it be represented by a trigonometric series? This is one of the questions that Riemann solved in his habilitation memoir. For more details, the interested reader is referred to the exposition in Chap. 1 of the present volume [104].

In conclusion, the question of the meaning of a function started with physics: the vibration equation discovered by d'Alembert, and it ended again with physics: the study of heat, by Fourier, and the discovery of Fourier series, which are extensively used in mathematical physics. Picard writes in his historical survey [117] that the development of a function as a series is a remarkable example of the intimate solidarity that unites at certain points pure analysis and applied mathematics.

---

[40]The word "discontinuity" is understood here in the modern sense of the word, and not in the sense of Euler. Cf. the explanation in Chap. 1 of the present volume [104].

[41]Riemann refers to [94].

[42]This is Dirichlet's article [36].

4 Riemann's Habilitationsvortrag 1854—Space and Matter

Riemann's public lecture, his Habilitationsvortrag, *Über die Hypothesen, welche der Geometrie zu Grunde liegen*, which was the final requirement before he was allowed to teach at the university level, was delivered on June 10, 1854. This lecture marks the birth of modern differential geometry. It is a difficult text, involving—like for other writings of Riemann—mathematics, physics and philosophy. It was commented on by many philosophers and scientists, and translated several times into other languages.[43] The earliest translation is probably the one that Clifford made in 1873 for the journal *Nature* [30]. This translation is generally considered as too literal. It is nevertheless interesting because Clifford was at the same time a mathematician, a physicist and a philosopher. He was knowledgeable in the philosophical issues raised by Riemann and he was familiar with the specialized language of philosophy that the latter used.

In a broad sense, the subject of the investigation is geometry, space and the relation between them. Thc discussion takes place at several levels, starting from the foundations of geometry: Riemann mentions the axioms at the beginning of his essay. He introduces several kinds of spaces and the notion of "manifoldness" which we shall discuss below. He mentions in particular discrete and continuous manifoldnesses, infinitesimal and large-scale properties, the ambient physical space, mathematical n-dimensional spaces and n-tuply extended magnitudes. He declares that the propositions of geometry cannot be derived from the general notion of magnitude (the word is taken in the Aristotelean sense), and that the properties which distinguish (physical) space from other conceivable triply extended magnitudes are only to be deduced from experience.

Several authors commented on Riemann's dissertation, and we shall make a few remarks on them below. Riemann's lecture has three parts (We use a slight modification of Clifford's original headlines in *Nature*):

(1) The notion of an n-tuply extended magnitude.
(2) Measure-relations of which a manifoldness of n dimensions is susceptible, on the assumption that lines have a length independent of position, and consequently that every line may be measured by every other.
(3) Applications to space.

Roughly speaking, the first part is philosophical, the second one is mathematical, and the third one deals with applications to physics. But to some extent philosophy and physics are present in the three parts. A detailed explanation of all these notions would take us too far, and it is also known that several points in this essay are very cryptic. Heinrich Weber, Hermann Weyl and many other pre-eminent mathematicians tried to uncover their meaning. Weyl, who had a great devotion for Riemann, edited

[43]There are English translations by M. Spivak in his *Comprehensive Introduction to Differential Geometry* ([154], Volume 2, pp. 132–153) and by H. S. White in Smith's *Source book mathematics* ([153], p. 411–425), and probably others. J. Jost's edition [147] contains Cayley's translation. Italians translations were made by E. Betti, and G. Gabella, and a French one by J. Hoüel.

the Habilitationsvortrag in 1919, [164], together with a commentary, making the link with relativity theory. One of the main features of the local geometry conceived by Riemann is that it is well suited to the study of gravity and more general fields in physics. Relativity theory, which encompasses the largest part of modern physics, relies in a crucial way on the notions introduced by Riemann.

From the purely mathematical point of view, the most important contribution of the Habilitationsvortrag is that it sets the bases of what we call today Riemannian geometry, with the introduction of the curvature tensor and its consequences, including several results such as the fact that the homogeneity (with the inherent notion of transformation group) corresponds to constant curvature. This new geometry can be considered as a far-reaching generalization of Gauss's work on the intrinsic geometry of surfaces, and at the same time it is a generalization of the non-Euclidean geometry (of constant curvature) discovered by Lobachevsky, Bolyai and Gauss, in the few decades that preceded Riemann. Furthermore, Riemann sets in this memoir the bases of several developments made in several directions by Clifford, Christoffel, Bianchi, Ricci, Beltrami, Levi-Civita, Élie Cartan, Einstein and many others.

It is known that the full importance of the Habilitationsvortrag was not recognized in the first years after it was delivered. A report by Dedekind on the mathematical content of the memoir was published only in 1868.[44] But it is also known that Gauss, who, as Riemann's mentor, was present at the lecture, expressed his complete satisfaction with it. This reaction to Riemann's Habilitationsvortrag is described in Dedekind's biography of Riemann published in the Collected Works [141]. Gauss's praise was certainly a cause for Riemann's own contentedness, because Gauss was known to be sparing with compliments.

In the introduction to his dissertation, Riemann declares that he is unexperienced in philosophical questions, and that in preparing the lecture he could rely only on some remarks that Gauss made in his second paper on biquadratic residues and in his Jubilee-book, and some philosophical researches of Herbart. Riemann discusses the question of space and that of manifoldness and its specialization. This specialization may be either continuous or discrete. In the chapter [119] contained in the present volume, Plotnitsly emphasizes that Riemann speaks of discrete manifolds, and then says that, rather than space itself, it is "the reality underlying space" that may be discrete.[45] "Manifolds" as we intend them today are particular cases of manifoldnesses. Riemann writes:

> Manifoldnesses in which, as in the plane and in space, the line-element may be reduced to the form $\sqrt{\sum dx^2}$, are therefore only a particular case of the manifoldnesses to be here investigated; they require a special name, and therefore these manifoldnesses in which the square of the line-element may be expressed as the sum of the squares of complete differentials I will call *flat*.

[44] *Abhandlungen der Königlichen Gesellschaft der Wissenschaften zu Göttingen, 13.*

[45] It may be useful to note that in modern physics, spacetime is studied in its both characters, discrete (e.g., in lattice gauge theories, which are often considered as mathematical discrete approximations) or as continuous (e.g., in general relativity).

Riemann declares that notions whose specializations form a continuous manifoldness are the positions of perceived objects (die Orte der Sinnengegenstände) and colors. It is conceivable that Riemann, in his mention of colors, refers to the fact that one can continuously move from a color to another one, a color being characterized by the proportions of red, green and violet it contains. This makes color a three-dimensional manifoldness. Weyl writes, in his *Space, time, matter*, that Riemann's reference to color "is confirmed by Maxwell's familiar construction of the color triangle" ([165], p. 84 of the English translation). There are also writings of Helmholtz and Thomas Young on this matter. There are particular portions of a manifoldness called *quanta*,[46] whose nature is different from that which is characterized by the discrete and the continuous. "Their comparison with regard to quantity is accomplished in the case of discrete magnitudes by counting, in the case of continuous by measuring." The notions of measurement and of dimension are discussed in the Aristotelean style. The relation between measurement and the axioms of geometry is said to be fundamental, and in some sense this question concerns the relation between the axioms of geometry and the reality of space, that is, between mathematical and empirical truth. Riemann writes:

> Either therefore the reality which underlies space must form a discrete manifoldness, or we must seek the ground of its metric relations outside it, in binding forces which act upon it.[47]

It is important to recall Riemann's words. He declares that geometry depends at the same time on axioms and on observational and experimental physics. He considers that classical geometry, with the first principles and axioms that it assumes and the connections between them, does not lead anywhere, because "we do not perceive the necessity of these connections." What is missing is a notion of "multiply-extended magnitude" (mehrfach ausgedehnte Grösse), a notion which makes space a particular "triply-extended magnitude." He proposes that the properties that distinguish space from other conceivable triply-extended magnitudes be deduced from experience. In particular, the space that Riemann talks about, although built from undefined notions and axioms connecting them, is not the space of traditional geometry. He suggests that this space should reflect the material world around us. He formulates the problem of finding the "simplest matters of fact from which the metric relations (Massverhältnisse) of space may be determined." He declares that these matters of fact are "not of necessity, but only of empirical certainty." They are the "hypotheses" that are referred to in the title of the dissertation. He says that he will investigate the "probability" of these matters of fact, "within the limits of observation," and see whether they may be extended "beyond the limits of observation, both on the side of the infinitely great and of the infinitely small." The most important among these matters of fact is related to the work of Euclid. The geometry that Riemann will

[46]In Aristotle's language, the latin word *quantum* is used as the translation of the word $\pi o \sigma \acute{o} \nu$ (quantity, or "quantified thing"), which is one of Aristotle's categories. In his *Metaphysics*, Aristotle mentions four types of change: of substance, *quale*, *quantum*, or place. (Metaphysics, 1069b9–13).

[47]Es muss also entweder das dem Raume zu Grunde liegende Wirkliche eine discrete Mannigfaltigkeit bilden, oder der Grund der Massverhältnisse ausserhalb, in darauf wirkenden bindenden Kräften, gesucht werden.

construct will be Euclidean at the infinitesimal level.[48] We already noted by the way that the notion of "infinitely small" is treated in several works of the Ancient Greeks. The same notion is thoroughly discussed by Galileo Galilei in his *Discorsi; First day*, to whom Riemann refers in his habilitation, though in a different context.

The rest of the Habilitationsvortrag is a development of the ideas expressed in the introduction. The first part concerns the notion of n-dimensional Mannigfaltigkeit. This term is sometimes translated into English by "manifoldness." Riemann also talks about a "multiply extended magnitude." This is an ancestor to the mathematical notion of manifold. But the meaning of Mannigfaltigkeit, in Riemann's terminology, and that of the mathematical notion of manifold, as it is used today, do not coincide, even though in German the word Mannigfaltigkeit is used for "manifold."[49] There are discrete and continuous manifoldnesses, and there are manifoldnesses which are not mathematical. Riemann says that notions with specializations to discrete manifoldness are very common, but that, by contrast, there are very few notions whose specialization form a continuous manifoldness. We note incidentally that Poincaré pondered this terminology. In a letter to the mathematician and historian of mathematics Gustav Eneström, dated November 19, 1883 (cf. [120] p. 143), he writes:

> I prefer the translation of *Mannigfaltigkeit* by multiplicity, because the two words have the same etymological meaning. The word set is more adapted to the *Mannigfaltigkeiten* considered by Mr. Cantor and which are discrete. It would be less adapted to those which I consider and which are discontinuous. What is the opinion of Mr. Mittag on this matter?[50]

Eneström responds, on November 23:

> Mr. Mittag-Leffler thinks that you may be right, and, consequently, one should prefer the word multiplicity.[51]

In fact, Poincaré used the word multiplicity in its French form ("multiplicité") to denote Riemann's moduli space.

In the treatise *Théorie des fonctions algébriques de deux variables indépendantes* (Theory of algebraic functions of two independent variables) by Picard and Simart

[48]It may be worth recalling that Lobachevsky, in his various works on non-Euclidean geometry that he started in the late 1820s, systematically checked that the formulae that obtained in his new geometry give, at the infinitesimal level, the Euclidean formulae. See e.g. [93] p. 31.

[49]Jost, in [147], tries to sort out this complex terminology. He writes on p. 29: "The English of Clifford may appear somewhat old-fashioned for a modern reader. For instance, he writes 'manifoldness' instead of the simpler modern translation 'manifold' of Riemann's term 'Mannigfaltigkeit.' But Riemann's German sounds likewise somewhat old-fashioned, and for that matter, 'manifoldness' is the more accurate translation of Riemann's term. In any case, for historical reasons, I have selected that translation here."

[50]Je préfère la traduction de *Mannigfaltigkeit* par multiplicité, car les deux mots ont même sens étymologique. Le mot ensemble convient bien aux *Mannigfaltigkeiten* envisagés par M. Cantor et qui sont discrètes; il conviendrait moins à celles que je considère et qui sont discontinues. Qu'en pense M. Mittag à ce sujet?

[51]M. Mittag-Leffler pense que vous pouvez avoir raison, et que, par conséquent, il faut préférer le mot multiplicité.

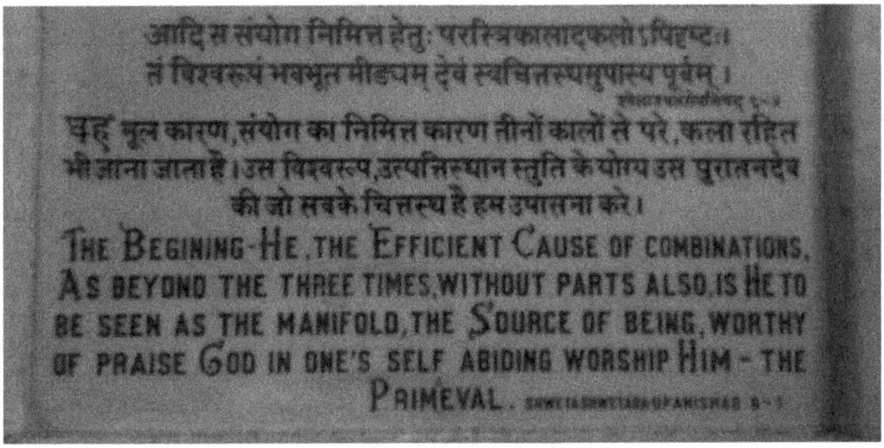

Fig. 3 From the Birla Temple of Varanasi, devoted to Shiva. The word Manifold is used as an attribute of God. The languages are Sanskrit at the top, and Hindi in the middle. The word used in these texts is *viśvarūpa*, which is the composite of viśva which stands for universe and rūpa which corresponds to something like "form." The word conveys God bearing the form of the universe itself. The English word "manifold" seems to have been used for want of a better choice for the context. I thank S. G. Dani for his help in this explanation *(Photo A. Papadopoulos)*

[116] on which we report in Chap. 8 of the present volume [106], the authors use interchangeably the words "variété," "multiplicité" and "continuum" to denote "a certain *continuous set* of points depending on a number of parameters which is equal to the dimension of this variety or continuum." (p. 20) It is interesting to note that Grothendieck, in his *Esquisse de programme* (A sketch of a program) [71], uses the same word. We refer the reader to the chapters [103] by Ohshika and [119] by Plotnitsky in the present volume for further discussion of the notion of Mannigfaltigkeit in relation with manifolds. We also note that the word "manifold" itself is also used in Hinduism, see Fig. 3.

The second part of the habilitation lecture, which concerns metric relations (Maassverhältnisse), is more mathematically-oriented. It contains, condensed in six or seven pages, the foundations of Riemannian geometry. The exposition contains a minimum amount of formulae. In fact, there are essentially two formulae. The first one gives the line element in (Euclidean) "space" as a square root of squares of differentials of the coordinates:

$$ds^2 = \sum dx_i^2. \tag{4}$$

This formula is an "infinitesimal Euclidean Pythagorean theorem." It expresses the fact that at the infinitesimal level the metric is Euclidean. The second formula gives the line element in a curved space:

$$\frac{1}{1+\frac{\alpha}{4}\sum x^2}\sqrt{\sum dx^2}$$

where α denotes, in Riemann's notation, the curvature. As is well known, this formula gives the one of the Poincaré metric of the disc in the case where the curvature constant is negative.

After Riemann gives the general expression of the infinitesimal line element as the square root of a quadratic form, and the curvature representing a deviation from flatness, he states that to know curvature at a point in a manifoldness of dimension n, it is sufficient to know it in $n(n-1)/2$ surface directions. He notes that if the length of a line element is independent from its position (that is, the group of motions acts transitively m line elements), then the space must have constant curvature.

The third part, called "Applications to space," contains in particular Riemann's famous discussion of the difference between *unbounded* and *infinite extent*. We quote him again:

> In the extension of space-construction to the infinitely great, we must distinguish between *unboundedness* and *infinite extent*, the former belongs to the extent relations, the latter to the measure-relations. That space is an unbounded three-fold manifoldness, is an assumption which is developed by every conception of the outer world; according to which every instant the region of real perception is completed and the possible positions of a sought object are constructed, and which by these applications is for ever confirming itself. The unboundedness of space possesses in this way a greater empirical certainty than any external experience. But its infinite extent by no means follows from this; on the other hand if we assume independence of bodies from position, and therefore ascribe to space constant curvature, it must necessarily be finite provided this curvature has ever so small a positive value. If we prolong all the geodesics starting in a given surface-element, we should obtain an unbounded surface of constant curvature, i.e., a surface which in a *flat* manifoldness of three dimensions would take the form of a sphere, and consequently be finite.

This is a famous passage for which Riemann's name is associated with the geometry of the sphere (constant positive curvature). It has been commented on by mathematicians, and also by philosophers, and it is related to the question of whether the universe has a spherical shape or not. Again, we can quote related texts from Greek antiquity, e.g. from Empedocles concerning the universe as a round "boundless" sphere, of which only a few fragments remain [20]:

> The Sphere on every side the boundless same,
> Exultant in surrounding solitude.

One may also quote Plato, who considers, for philosophical reasons, in the *Timaeus* ([112], 33b), that the universe is spherical, and hence, bounded:

> He wrought it into a round, in the shape of a sphere, equidistant in all directions from the center to the extremities, which of all shapes is the most perfect and the most self-similar, since he deemed that the similar is infinitely fairer than the dissimilar. And on the outside round about, it was all made smooth with great exactness, and that for many reasons.

Other Greek philosophers considered that the universe is infinite. This is an endless discussion.

Riemann wanted his (Riemannian) geometry to represent at the same time the large-scale and, most of all, the small-scale geometries of space. The progress made in mechanics in the preceding centuries, he says, is due to the invention of the infinitesimal calculus and to the simple principles discovered by Archimedes, Galileo and Newton. The natural sciences, are still in want of *simple* principles. Let us quote Riemann again:

> The questions about the infinitely great are for the interpretation of nature useless questions. But this is not the case with the questions about the infinitely small. It is upon the exactness with which we follow phenomena into the infinitely small that our knowledge of their causal relations essentially depends. The progress of recent centuries in the knowledge of mechanics depends almost entirely on the exactness of the construction which has become possible through the invention of the infinitesimal calculus, and through the simple principles discovered by Archimedes, Galileo, and Newton, and used by modern physics.

The synopsis of the Habilitationsvortrag (Clifford's translation [30]) ends with two questions.

(1) How far is the validity of these empirical determinations probable beyond the limits of observations towards the infinitely great?
(2) How far towards the infinitely small? Connection of this question with the interpretation of nature.

It is interesting to put again in parallel Riemann's writings with some texts of Aristotle on related matters, and there are many of them. We choose an excerpt from the beginning of the treatise *On the Heavens* [12]. It concerns at the same time magnitude, continuum, divisibility, infinity, dimension, infinitesimals, and the importance of these questions for understanding nature:

> The science which has to do with nature clearly concerns itself for the most part with bodies and magnitudes and their properties and movements, but also with the principles of this sort of substance, as many as they may be. For of things constituted by nature some are bodies and magnitudes, some possess body and magnitude, and some are principles of things which possess these. Now a continuum is that which is divisible into parts always capable of subdivision, and a body is that which is every way divisible. A magnitude if divisible one way is a line, if two ways a surface, and if three a body. [...] All magnitudes, then, which are divisible are also continuous. Whether we can also say that whatever is continuous is divisible does not yet, on our present grounds, appear. [...] The question as to the nature of the whole, whether it is infinite in size or limited in its total mass, is a matter for subsequent inquiry. [...] This being clear, we must go on to consider the questions which remain. First, is there an infinite body, as the majority of the ancient philosophers thought, or is this an impossibility? The decision of this question, either way, is not unimportant, but rather all-important, to our search for the truth. It is this problem which has practically always been the source of the differences of those who have written about nature as a whole. So it has been and so it must be; since the least initial deviation from the truth is multiplied later a thousandfold. Admit, for instance, the existence of a minimum magnitude, and you will find that the minimum which you have introduced, small as it is, causes the greatest truths of mathematics to totter.

Concerning the particular notion of infinite, we choose two texts from Aristotle's *Physics* [10]. We stress on the fact that even though the Greek philosophers, represented by Aristotle, did not formulate an axiomatic (in a mathematical sense) notion of infinity as we do it today, one should not underestimate the importance of the fact that they considered this notion as a fundamental philosophical notion, they asked many questions around it and about its role, and they also regarded it as central in physics and in mathematics.

The first text we choose is from Book III of Aristotle's *Physics*:

> [...] But on the other hand to suppose that the infinite does not exist in any way leads obviously to many impossible consequences: there will be a beginning and an end of time, a magnitude will not be divisible into magnitudes, number will not be infinite. If, then, in view of the above considerations, neither alternative seems possible, an arbiter must be called in; and clearly there is a sense in which the infinite exists and another in which it does not. We must keep in mind that the word "is" means either what potentially is or what fully is. Further, a thing is infinite either by addition or by division. Now, as we have seen, magnitude is not actually infinite. But by division it is infinite. (There is no difficulty in refuting the theory of indivisible lines.) The alternative then remains that the infinite has a potential existence.

The second text is from Book V of the *Physics*:

> Now it is impossible that the infinite should be a thing which is in itself infinite, separable from sensible objects. If the infinite is neither a magnitude nor an aggregate, but is itself a substance and not an accident, it will be indivisible; for the divisible must be either a magnitude or an aggregate. But if indivisible, then not infinite, except in the way in which the voice is invisible. But this is not the way in which it is used by those who say that the infinite exists, nor that in which we are investigating it, namely as that which cannot be gone through. But if the infinite is accidental, it would not be, qua infinite, an element in things, any more than the invisible would be an element of speech, though the voice is invisible.
>
> Further, how can the infinite be itself something, unless both number and magnitude, of which it is an essential attribute, exist in that way? If they are not substances, a fortiori the infinite is not.
>
> It is plain, too, that the infinite cannot be an actual thing and a substance and principle. For any part of it that is taken will be infinite, if it has parts; for to be infinite and the infinite are the same, if it is a substance and not predicated of a subject. Hence it will be either indivisible or divisible into infinites. But the same thing cannot be many infinites. (Yet just as part of air is air, so a part of the infinite would be infinite, if it is supposed to be a substance and principle.) Therefore the infinite must be without parts and indivisible. But this cannot be true of what is infinite in fulfillment; for it must be a definite quantity.
>
> Belief in the existence of the infinite comes mainly from five considerations: From the nature of time—for it is infinite; From the division of magnitudes—for the mathematicians also use the infinite [...]

Finally, let us note that in Book IV of the *Physics* [10] there is a long discussion about time, with its relation to measure and change:

> As to what time is or what is its nature, the traditional accounts give us as little light. [...] It is evident, then, that time is neither movement nor independent of movement. We must take this as our starting-point and try to discover—since we wish to know what time is—what exactly it has to do with movement.

Closer to us, another mathematician-philosopher who was fascinated by infinity is Blaise Pascal. He wrote on this theme, in his mathematical and philosophical writings. From his *Pensées* [111], we read:

> Unity added to infinity adds nothing to it, any more than does one foot added to infinite length. The finite is annihilated in presence of the infinite, and becomes pure nothingness.[52]

> Our soul has been cast into the body, where it finds number, time and dimension. It reasons thereupon, and calls it nature, necessity, and can believe nothing else.[53]

> The eternal silence of these infinite spaces terrifies me.[54]

One should also talk about modern physics, where the same kind of questions are still the basic ones: What is space? What is time? What physical theories describe at the same time the macroscopic and the microscopic worlds? What are the relations between these worlds? How do we pass between the discrete and the continuous?

Riemann's last sentence in the Habilitationsvortrag shows that he modestly considered that in his work, he did not make any significant advance in the direction of physics:

> This leads us into the domain of another science, of physics, into which the object of this work does not allow us to go today.

Higher-dimensional spaces, from the mathematical point of view were surely considered before Riemann. But for the first time, Riemann's major achievement was to introduce on these spaces a geometry that was necessary for the development of modern physics. The physical theories of superstrings and supergravity need ten or eleven dimensions. The spacetime of special relativity—Minkowski's spacetime—is a four-dimensional manifold equipped with a structure that generalizes the one that Riemann considered. In an address to the 80th Assembly of German Natural Scientists and Physicians, (Sep 21, 1908), Minkowski declares (cf. [97]):

> The views of space and time which I wish to lay before you have sprung from the soil of experimental physics, and therein lies their strength. They are radical. Henceforth, space by itself, and time by itself, are doomed to fade away into mere shadows, and only a kind of union of the two will preserve an independent reality.

In this setting, Riemann's formula (4) is replaced by the formula

$$ds^2 = c^2dt^2 - dx^2 - dy^2 - dz^2 \tag{5}$$

where t is the time component, (x, y, z) the space components and c the velocity of light. The geometry of Minkowski spacetime is included in the setting of semi-Riemannian geometry, a geometry in which the metric tensor is not necessarily positive-definite. This incorporates in the theory the fact that particles cannot move at a speed which is larger than the speed of light. But the basic features that Riemann conceived are there. In general relativity, the metric tensor is an expression of the gravitational potential, in the trend of Riemann's ideas.

[52]L'unité jointe à l'infini ne l'augmente de rien, non plus qu'un pied à une mesure infinie, le fini s'anéantit en présence de l'infini et devient un pur néant.

[53]Notre âme est jetée dans le corps où elle trouve nombre, temps, dimensions, elle raisonne là-dessus et appelle cela nature, nécessité, et ne peut croire autre chose.

[54]Le silence éternel de ces espaces infinis m'effraye.

Let us now mention some comments by various authors (especially physicists) on the Habilitationsvortrag. We quote Clifford and Weyl whom we already mentioned.

On February 21, 1870, Clifford presented a paper to the Cambridge Philosophical Society whose title is *On the space theory of matter* [31], in which he stressed the relation of the new geometry with physics. It is interesting to read the abstract of that paper, for it gives quite a good idea of its physical background. Clifford writes:

> Riemann has shown that as there are different kinds of lines and surfaces, so there are different kinds of spaces of three dimensions; and that we can only find out by experience to which of these kinds the space in which we live belongs. In particular, the axioms of plane geometry are true within the limits of experiment on the surface of a sheet of paper, and yet we know that the sheet is really covered with a number of small ridges and furrows, upon which (the total curvature not being zero) these axioms are not true. Similarly, he says that although the axioms of solid geometry are true within the limits of experiment for finite portions of our space, yet we have no reason to conclude that they are true for very small portions; and if any help can be got thereby for the explanation of physical phenomena, we may have reason to conclude that they are not true for very small portions of space.
>
> I wish here to indicate a manner in which these speculations may be applied to the investigation of physical phenomena. I hold in fact
>
> (1) That small portions of space *are* in fact of a nature analogous to little hills on a surface which is on the average flat; namely, that the ordinary laws of geometry are not valid in them.
>
> (2) That this property of being curved or distorted is continually being passed on from one portion of space to another after the manner of a wave.
>
> (3) That this variation of the curvature of space is what really happens in that phenomenon which we call the *motion of matter*, whether ponderable or etherial.
>
> (4) That in the physical world nothing else takes place but this variation, subject (possibly) to the law of continuity.
>
> I am endeavoring in a general way to explain the laws of double refraction on this hypothesis, but have not yet arrived at any results sufficiently decisive to be communicated.

It is not superfluous to recall that Gauss, who was Riemann's mentor, was also interested in the philosophical implications of the new discoveries of geometry. In a letter dated March 6, 1832 (see [155] and Gauss's Collected Works Vol. VI [67]), Gauss writes to his friend Wolfgang Bolyai that Kant was wrong in declaring that space is *only the form*[55] of our intuition. These remarks are made amid a discussion on non-Eucldiean geometry. In the same letter, Gauss refers to an article on the subject that he published in the Göttingische Gelehre Anzeigen, in 1831. This article is reproduced in Volume II of Gauss's *Werke*. Gauss criticizes an argument, which is independent of non-Euclidean geometry, which Kant gave in support of his assumption (and his proof) that space is only a form of our exterior intuition. The argument is in Kant's *Prolegomena zu einer jeden künftigen Metaphysik, die als Wissenschaft wird auftreten können* (Prolegomena to any future metaphysics that will be able to present itself as a science) [83] §13, and it is based on the existence of symmetries. Gauss's position was that, on the contrary, space has a real significance, independent of our mode of intuition. An excerpt on space of Kant's inaugural dissertation—

[55]Gauss's emphasis.

in fact an excerpt concerned by Gauss's critic—is quoted in Chap. 1 of the present volume (in the section concerning space).

The questions of space and of time remained among the major preoccupations of Kant. They are developed in particular in his habilitation [82] and in his *Critik der reinen Vernunft* (Critique of pure reason) [84] (1781), which is one of the most influential philosophical works ever written. In this work, like in his inaugural dissertation, Kant addresses the fundamental questions that were addressed before him by Leibniz, Newton and others, namely, *What is space? What is time? What is the relation between space, time and the mind? Is this relation real or ideal? Do space and time have subjective existence, beyond our intuition of them? Are they empirical concepts? Are they substances or the product of our mind? Do they exist independently of objects and their relation? Are they necessary tools for our understanding?* Kant also analyses our representation of space and its relation to geometry. Elaborating on these most difficult questions is the subject of the fundamental contribution of Kant to philosophy.

Weyl's book *Space, time, matter*,[56] (first edition 1918),[57] is an introduction to the theory of relativity, based on lectures he gave at Zurich's ETH. This work of Weyl is a celebration of the idea that Einstein's theory of relativity is an accomplishment of Riemann's geometry. In the introduction, Weyl writes:

> It was my wish to present this great subject as an illustration of the intermingling of philosophical, mathematical, and physical thought, a study which is dear to my heart. This could be done only by building up the theory systematically from the foundations, and by restricting attention throughout the principles. But I have not been able to satisfy these self-imposed requirements: the mathematician predominates at the expense of the philosopher.

The mathematician's role is played essentially by Riemann. In Riemannian geometry, the space (a manifold) is equipped at each tangent space with a quadratic form defining a geometry which is Euclidean. Weyl comments on this fact and on its relation with physics. He writes ([165] p. 91):

> The transition from Euclidean geometry to that of Riemann is founded in principle on the same idea as that which led from physics based on action at a distance to physics based on infinitely close action. We find by observation, for example, that the current flowing along a conduction wire is proportional to the difference of potential between the ends of the wire (Ohm's Law). But we are firmly convinced that this result of measurement applied to a long wire does not represent a physical law in its most general form; we accordingly deduce this law by reducing the measurements obtained to an infinitely small portion of wire. But this means we arrive at the expression on which Maxwell's theory is founded. Proceeding in the reverse direction, we derive from this differential law by mathematical processes the integral law, which we observe directly, on the supposition that conditions are everywhere similar (homogeneity). We have the same circumstance here. The fundamental fact of Euclidean geometry is that the square of the distance between two points is a quadratic form of the relative co-ordinates of the two points (*Pythagoras Theorem.*) *But if we look upon*

[56]We already recalled that the triad Matter, Space and Time is *par excellence* an Aristotelean theme. There are numerous references regarding this subject, and the best way for the reader to get into this is to skim Aristotle's works. Some of these works are listed in the bibliography, but there are many others.

[57]The book, under the German title *Raum, Zeit, Materie*, appeared in English translation in 1922.

> *this law as being strictly valid only for the case when these two points are infinitely near, we enter the domain of Riemann's geometry.* [...] We pass from Euclidean "finite" geometry to Riemann's "infinitesimal" geometry in a manner exactly analogous to that by which we pass from "finite" physics to "infinitesimal" (or "contact") physics.

Weyl continues ([165] p. 92):

> The principle of gaining knowledge of the external world from the behavior of its infinitesimal parts is the mainspring of the theory of knowledge in infinitesimal physics as in Riemann's geometry, and, indeed, the mainspring of all the eminent work of Riemann, in particular, that dealing with the theory of complex functions.

In the same book, Weyl writes ([165] p. 98):

> Riemann rejects the opinion that had prevailed up to his own time, namely, that the metrical structure of space is fixed and inherently independent of the physical phenomena for which it serves as a background, and that the real content takes possession of it as of residential flats. *He asserts, on the contrary, that space in itself is nothing more than a three-dimensional manifold devoid of all form; it acquires a definite form only through the advent of the material content filling it and determining its metric relations.*

And then ([165] p. 102):

> Riemann, in the last words of the above quotation, clearly left the real development of his ideas in the hands of some subsequent scientist whose genius as a physicist could rise to equal flights with his own as a mathematician. After a lapse of seventy years this mission has been fulfilled by Einstein.

Relativity theory is based on the fact that space and time cannot be separated and form a four-dimensional continuum in one of the senses that Riemann intuited. Einstein made a profound relation between Riemannian geometry and physics, in particular in his discovery that gravity is the cause of curvature of physical space. Einstein's equation, published for the first time in 1915, which is the main partial differential equation of general relativity, expresses a relation between energy, gravitation and the curvature of spacetime. In this setting, the Lorentzian metric encodes the gravitational effects, and the notion of curvature plays a central role. At several places, Einstein expressed his debt to Riemann. Let us quote him from [37] (p. 281):

> But the physicists were still far removed from such a way of thinking; space was still, for them, a rigid, homogeneous something, incapable of changing or assuming various states. Only the genius of Riemann, solitary and uncomprehended, had already won its way to a new conception of space, in which space was deprived of its rigidity, and the possibility of its partaking in physical events was recognized. This intellectual achievement commands our admiration all the more for having preceded Faraday's and Maxwell's field theory of electricity.

We end this section by quoting Riemann, and his concerns about physics. In a letter to his father, written February 5, 1852 [149], right after he submitted his Habilitationsschrift, Riemann writes:

> Right after the submission of my Habilitationsschrift I resumed my investigations into the coherence of the laws of Nature and got so involved in it that I could not tear myself loose. The continuing preoccupation with it has become bad for my health, in fact, right after

> New Year's my usual affliction set in which such persistence, that I could only obtain relief through the strongest remedies. As a result I felt very ill, felt unable to work, and sought to again put my health in order through long walks.

On June 26 of the next year, he writes to his brother on the same subject:

> I had completed my habilitation paper at the beginning of December, submitted it to the dean, and soon after once again turned to my investigation on the coherence of the fundamental laws in physics; also that I so immersed myself in it that when the theme for my examination lecture was posted at the colloquium, I could not immediately tear myself away. Rightly after I came down sick, partly, of course, as a result of too much brooding, and partly as a result of sitting a lot in my room during bad weather.

5 The *Commentatio* and the *Gleichgewicht der Electricität*

Riemann developed some of his mathematical ideas introduced in his Habilitationsvortrag in a paper, written in Latin, whose extended title is *Commentatio Mathematica, qua respondere tentatur quaestioni ab Illma Academia Parisiensi propositae: Trouver quel doit être l'état calorifique d'un corps solide homogène indéfini pour qu'un système de courbes isothermes, à un instant donné, restent isothermes après un temps quelconque, de telle sorte que la température d'un point puisse s'exprimer en fonction du temps et de deux autres variables indépendantes* (A mathematical treatise in which an attempt is made to answer the question proposed by the most illustrious Academy of Paris: To find what must be the thermal state of an indefinite homogeneous solid body so that a system of isothermal curves, at a given instant, remain isothermal after an arbitrary time, in such a way that the temperature at a point can be expressed in terms of time and of two other independent variables). The memoir, as the name indicates, was presented as a contribution to a problem which was proposed for competition by the Paris Academy of Sciences in 1861. Part of the *Commentatio* is translated and commented by Spivak in Chap. 4 of Volume II of his *Comprehensive introduction to differential geometry* [154].

The problem concerns heat conduction, more precisely, the determination of the temperature of a body endowed with a set of given conductivity coefficients. From the mathematical point of view, it amounts to finding the solution of a partial differential equation—an evolution equation. The "solid body" that is referred to in the statement of the problem becomes, in Riemann's context, a Riemannian manifold. At the same time the terms used have a physical significance. It is not surprising that Riemann got interested in that problem, which combines geometry and potential theory, two of his favorite subjects. The word "isothermal" is also reminiscent of the work done by his mentor, Gauss.

While the Habilitationsvortrag is practically devoid of any mathematical formulae, the *Commentatio* is full of them. In fact, it is in the style of the later papers on Riemannian geometry, and in particular those on general relativity, with their debauchery of indices.[58] Riemann's *Commentatio* also contains new tools that are essential to differential geometry. It is in this paper that Riemann introduced his 4-entry curvature tensor. The authors of [54] consider this paper as a "contribution to the development of what later became known as tensor analysis." As is well known, this topic became an important tool in general relativity. There is a general agreement now that Riemann's paper contains several results that are usually attributed to Christoffel, cf. [54, 169], and also the idea of Finsler geometry.

Let us quote from this paper. Riemann starts his paper by a summary, in which he declares that he will first solve a more general problem:

> We shall approach the question proposed by the Academy in such a way that we shall first solve a more general question: the properties of a body which determine the conduction of heat and the distribution of heat within it such that there exists a system of lines which remain isothermal.
> From the general solution of this problem we shall distinguish those cases in which the properties vary from those in which the properties remain constant, that is where the body is homogeneous.

We recall that in the Habilitationsvortrag, the homogeneity property was proved to be equivalent to the curvature being constant.

The second part of the paper is concerned with the question under equivalence of passing from one quadratic form to another. This is essential in the theory of the transformations that make tensors coordinate-free forms. The reader can find mathematical commentaries on Riemann's memoir in the paper [54].

The *Commentatio*, like Riemann's Habilitationsvortrag is difficult to read, but this time because of the density of its mathematical content. Riemann's article did not win the prize, probably because some details in the proofs were missing. (In fact, the prize was not awarded.) The authors of [54] present a certain number of different and conflicting interpretations of the *Commentatio*, a fact which is uncommon for a mathematical paper. This is another indication of how much Riemann's writing are special and cryptic (even today).

The subject of the unfinished paper *Gleichgewicht der Electricität auf Cylindern mit kreisförmigen Querschnitt und parallelen Axen* (On the equilibrium of electricity on cylinders with circular transverse section and whose axes are parallel) (1857) [142] by Riemann, published posthumously in the second edition of his Collected works, is related to the one of the *Commentatio*. It concerns the distribution of electricity or temperature on infinite cylindrical conductors with parallel generatrices. Riemann gives in this paper a solution of the Dirichlet boundary value problem for plane domains. He declares, at the beginning of the paper, that the physical question considered will be solved if the following mathematical question is solved: On a planar connected surface which simply covers the plane and whose boundary may be arbitrary, to determine a function u of the rectangular coordinates x, y satisfying the equation

---

[58]The expression is due to Élie Cartan, from [23], p. VII: "The distinguished favor that the absolute differential calculus of Ricci and Levi-Civita did for us, and will continue to do, should not prevent us of avoiding the over-exclusively formal computations, where the debauchery of indices hides a geometric reality which is often simple. It is this reality which I tried to bring out."

$$\frac{\partial^2 u}{\partial x^2} + \frac{\partial^2 u}{\partial y^2} = 0$$

and taking arbitrary values on the boundary. Riemann's solution makes use of Green's theorem and of Abelian integrals. This work is another illustration of the fact that Riemann equates potential theory with the theory of Riemann surfaces.

6 Riemann's Other Papers

We discuss briefly some other papers of Riemann related to our subject. Needless to say, the fact that we pass rapidly through these papers does not mean they are less important than those which we discussed more thoroughly in the previous sections.

Darboux, in his famous *Leçons sur la théorie générale des surfaces et les applications géométriques du calcul infinitésimal* (A course on the general theory of surfaces and the geometric applications of infinitesimal calculus), 1896, § 358, discussing the notion of the adjoint equation of a given linear equation, says that the origin of this notion is contained in Riemann's memoir *Über die Fortpflanzung ebener Luftwellen von endlicher Schwingungsweite* (On the propagation of planar air waves that have finite vibration amplitude), [136] 1860; cf. [145] p. 177. He declares that P. du Bois-Reymond, in his work on partial differential equations as well as in a short article he published in Tübingen, called the attention of geometers on that memoir by Riemann. He then presents the work. The content is mathematical, with applications to experimental physics. In the introduction, Riemann writes that his research on this subject is in the lineage of the recent work of Helmholtz on acoustics. He says that his results, besides their theoretical interest in the theory of the nonlinear partial differential equations which determine the motion of gases, should give the bases for experimental research on the subject.[59] He starts in his paper by recalling the physical laws of Boyle, Gay-Lussac and the recent experiments of Regnault, Joule, Thomson and others. About a hundred years later, commenting on the same memoir in the new edition of Riemann's Collected Papers (1990), Peter Lax writes ([141] p. 807): "In this paper, Riemann lays the foundations of the theory of propagation of nonlinear and linear waves governed by hyperbolic equations. The concepts introduced here—Riemann invariants, the Riemann initial value problem, jump conditions for nonlinear equations, the Riemann function for linear equations—are still the basic building blocks of the theory today." Riemann states in the announcement of the paper (cf. Footnote 59) that the solution of that problem would help clarifying a perennial debate that involved the mathematicians Challis, Airy, Stokes, Pretzval,

[59] We note however that in the announcement of this paper, published in the Göttinger Nachrichten, No. 19 (1859), Riemann begins by stating that he does not claim to give any results that are useful in experimental research. At the end of that announcement, he mentions connections with acoustics but he says that their verification seems to be very hard, the reason being that they either involve very small tone differences, which are not noticeable, or large variations which involve many parameters, therefore the causes cannot be separated. He also talks about applications to meteorology.

Doppler and Ettinghausen. Betti wrote an extensive technical report on that paper, [17].

We now briefly review some other papers.

Riemann declares in the introduction to the paper *Ein Beitrag zu den Untersuchungen über die Bewegung eines flüssigen gleichartigen Ellipsoides* (A contribution to the investigation of the movement of a uniform fluid ellipsoid) [138] that he is continuing the last work of Dirichlet, that this work is surprising and that it opens up a new path for mathematicians which is independent of the original motivation of Dirichlet, which originates in a question on the heavenly bodies. This paper is also discussed in a supplement in the new edition of his Collected Works [29, 141]. Riemann's motivation originates in a writing of Newton, more precisely in his proof of the fact that the spheroidal (rather than spherical)[60] form of the earth is due to its rotation. Newton gave the following formula for a homogeneous body in gravitational equilibrium and small rotation:

$$m = \frac{5}{4}\epsilon$$

where ϵ is the ellipticity coefficient, equal to the equatorial radius—polar radius/mean radius, and m the centrifugal acceleration/mean gravitational acceleration on the surface. The formula was generalized by MacLaurin, who removed the restriction to small rotations. Later works and clarifications are due to Lagrange and Jacobi. Dirichlet investigated these problems in his 1856/57 lectures on partial differen-

[60]Newton, in his *Principia* (1687), expected a flattening of the earth at the poles, of the order of 1/230. The real shape of the earth was another major controversial issue in the seventeenth and eighteenth centuries, and it opposed the English scientists, represented by Newton, to the French, who considered themselves as the heirs of Descartes, and who were represented by the astronomer Jacques Cassini (1677–1756), the physicist Jean-Jacques d'Ortous de Mairan (1678–1771) and others who pretended on the contrary that the earth was stretched at the poles. Huygens was on the side of Newton. Maupertuis tried to convince the French Academy of Sciences that the theory of Newton concerning the shape of the earth was sound, and he led an expedition to Lapland, whose aim was to measure the length of a meridian. The expedition, which lasted sixteen months, was successful, and it confirmed Newton's ideas. The mathematician Alexis-Claude Clairaut (1713–1765) and the Swedish astronomer Anders Celsius (1701–1744) were part of the expedition. The controversy on the form of the earth gave rise to an extensive literature, in the seventeenth and eighteenth centuries. In the *Discours préliminaire* (Preliminary discourse) of the *Encyclopédie* (1751), d'Alembert praises Maupertuis who dared to take side for the English. He writes: "The first among us who dared to declare openly that he was Newtonian is the author of the *Discours sur la figure des astres* [...]. Maupertuis thought that one could be a good citizen without blindly adopting the physics of one's country; to attack this physics, he needed a courage for which we have to be grateful to him." Voltaire, who contributed in making Newton's ideas known in France, was among the few major figures on the continent who stood up for the English. He presents these polemics in his famous *Lettres philosophiques* [161] (1734) (No. XIV): "A Frenchman who arrives in London, will find philosophy, like everything else, very much changed there. He had left the world a plenum, and he now finds it a vacuum. At Paris, the universe is seen composed of vortices of subtile matter; but nothing like it is seen in London. In France, it is the pressure of the moon that causes the tides; but in England it is the sea that gravitates towards the moon; so that when you think that the moon should make it flood with us, those gentlemen fancy it should be ebb, which very unluckily cannot be proved. [...] At Paris you imagine that the earth is shaped like a melon, or of an oblique figure; at London it has an oblate one."

tial equations, which were edited in part by Dedekind in 1860. Chandrasekhar and Lebowitz, in a commentary on Riemann's paper [138] which is published in Riemann's Collected Works edition [29, 141], quote Riemann saying:

> In his posthumous paper, edited for publication by Dedekind, Dirichlet has opened up, in a most remarkable way, an entirely new avenue for investigations on the motion of a self-gravitating homogeneous ellipsoid. The further development of this beautiful discovery has a particular interest to the mathematician even apart from its relevance to the forms of heavenly bodies which initially instigated these investigations.

We refer the reader to the analysis of Riemann's paper contained in [141] p. 811–820, where the authors consider this paper to be "remarkable for the wealth of new results it contains and for the breadth of its comprehension of the entire range of problems. [...] This much neglected paper [...] deserves to be included among the other great papers of Riemann that are well known." In their conclusion, they write: "A variety of further developments in astronomy and physics have been made possible by the existence of Riemann's work on ellipsoidal figures. [...] The foregoing brief account of developments in the theory of the classical ellipsoids show how Riemann's investigations, after a lapse of some one hundred years, occupy a central place in theoretical astrophysics today."

Let us now say a few words about Riemann's paper *Beiträge zur Theorie der durch die Gauss'sche Reihe $F(\alpha, \beta, \gamma, x)$ darstellbaren Functionen* (Contribution to the theory of functions representable by Gauss's series $F(\alpha, \beta, \gamma, x)$) [139]. In the introductory part, Riemann announces that in this paper, he investigates the functions representable by Gauss's series using a new method which essentially applies to any function satisfying a linear differential equation with algebraic coefficients. He also says that the main reasons for his investigations are the many applications of this function in physics and astronomy. In the announcement of that memoir, published in the *Göttinger Nachrichten*, No. 1, 1857, Riemann recalls that Gauss, in studying these functions, was motivated by astronomy. Riemann's announcement starts with the words: "This memoir treats a class of functions which are useful to solve various problems in mathematical physics." These functions are still commonly used today in mathematical physics.

Finally, we say a few words on Riemann's paper on minimal surfaces [137],[61] *Über die Fläche vom kleinsten Inhalt bei gegebener Begrenzung* (On surfaces of minimal area, with a given contour). The problem is to find surfaces with minimal area and with fixed boundary. This problem is also related to physics. Again, the mathematical field to which this question belongs can be traced back to the Greeks, namely to works of Archimedes on isoperimetry and isoepiphany. The specific question of minimal surfaces belongs to the calculus of variations, more precisely the so-called multi-dimensional calculus of variations. In dimension two, one minimizes area or the Dirichlet functional over spaces of surfaces with a given boundary (whereas in the problems of the classical, one-dimensional calculus of variations, one typically minimizes the length, energy, etc. functional on a space of curves joining two given

---

[61] The paper was published posthumously in 1867, and according to Hattendorf, quoted in [145] p. 306, it was written around 1860–1861.

points). It was probably Euler, in 1744, who discovered the first minimal surface, the catenoid, the surface of least area whose boundary consists of two parallel circles in space [50]. (The name comes from the fact that this is the surface obtained by rotating a catenary around a line.) One year after, Lagrange, who was 19 years old, studied the question of finding the graph of a surface in space with least area with prescribed boundary in the plane. He found a partial differential equation satisfied by such a surface. This was the birth of the so-called Euler–Lagrange equation. In 1776, Meusnier[62] interpreted Lagrange's equation as the vanishing of the mean curvature. Monge also made substantial contributions to the subject of minimal surfaces. Riemann's contribution (1860–1861) concerns the solution for some given boundary curves. Riemann gave a one-parameter family of examples of minimal surfaces. It was proved recently that the plane, the helicoid, the catenoid and the one-parameter family discovered by Riemann form exactly the set of complete properly embedded, minimal planar domains in $\mathbb{R}^3$, see [98]. Weierstrass made the relation between the Euler–Lagrange and the Cauchy–Riemann equations. Schwarz obtained results on the same question. The extensive study of minimal surfaces based on soap films was conducted by Plateau[63] around the year 1873. Besides the relation with soap films, minimal surfaces appear in physics, in particular in hydrodynamics. We again cite Klein, from his article on *Riemann and his significance for the development of modern mathematics* (1895) [86]:

> Perhaps less attention has been paid to another physical application in which Riemann's ways of looking at things are laid under contribution in a most attractive manner. I have in mind the theory of *minimum surfaces* [...] the problem is to determine the shape of the least surface that can be spread out in a rigid frame, – let us say, the form of equilibrium of a fluid lamina that fits in a given contour. It is noteworthy that, with the aid of Riemann's methods, the known functions of analysis are just sufficient to dispose of the more simple cases.

This paper on minimal surfaces is analyzed by Yamada in the present volume, [167].

One could also talk about Riemann's paper on the zeta function, *Über die Anzahl der Primzahlen unter einer gegebenen Grösse* (on the number of prime numbers less than a given quantity) [135], recalling that the apparent chaotic distribution of primes has been shown to match the classical random models which describe physical phenomena.

---

[62]Jean-Baptiste Marie Charles Meusnier de la Place (1754–1793) was a Revolution general, a geometer and an engineer. Together with the mathematicians Gaspard Monge and Alexandre-Théophile Vandermonde, he belonged to the *société patriotique du Luxembourg*, a patriotic-revolutionary movement. In mathematics, he is known for the Meusnier Theorem on the curvature of surfaces, and for the discovery of the helicoid, a ruled minimal surface.

[63]Joseph Plateau (1801–1883) obtained a doctorate in mathematics and then became a physicist. By his experiments on the retina, and for several machines he invented, Plateau is among the first scientists who contributed to the bases of moving images (cinema).

7 Conclusion

Beyond Riemann's work which is the subject of the present chapter, one may wonder about the interrelation between mathematics and physics. This subject is complex, much on it has been said, and adding something new is not a trivial task. Instead, we quote a text by Picard, from his opening address at the 1920 International Congress of Mathematicians. Picard is one of the main advocates of the theory of functions of one complex variable, a subject that was dear to Riemann.[64] He writes in [118]:

> Any physical theory which is sufficiently elaborate takes necessarily a mathematical form; it seems that the actions and reactions between spirit and objects gradually brought the formation of moulds where reality could fit, at least in part. For sure, many concepts created by mathematicians did not find yet any application in the study of physical phenomena, but history of science shows that it was reckless to assert that such or such notion would never be used one day. Geometers like to recall the word of the great mathematician Lagrange who, one day, comparing mathematics to an animal of which every part can be eaten, said: "Mathematics is like a pig, everything in it is good."[65]

Acknowledgements The author is grateful to Vincent Alberge, Hubert Goenner, Marie Pascale Hautefeuille, Manfred Karbe, Victor Pambuccian and Arkady Plotnitsky who read preliminaries versions of this chapter, made comments and suggested corrections.

References

1. N. A'Campo, V. Alberge, E. Frenkel, The theorem of Riemann–Roch. (This volume, pp. 389–410)
2. L.V. Ahlfors, Development of the theory of conformal mapping and Riemann surfaces through a century. Ann. Math. Stud. **30**, 3–13 (1953)
3. J. (le Rond) d'Alembert, Recherches sur les vibrations des cordes sonores & Supplément, in *Opuscules mathématiques*, Paris, tome 1 (1761), pp. 1–73
4. J. (le Rond) d'Alembert, Recherches sur la courbe que forme une corde tendue mise en vibration (1747), in *Mémoires de l'Académie des sciences de Berlin*, vol. 3 (1749), pp. 214–219
5. J. (le Rond) d'Alembert, Suite des recherches sur la courbe que forme une corde tendue mise en vibration (1747), in *Mémoires de l'Académie des sciences de Berlin*, vol. 3 (1749), pp. 220–249

---

[64] The congress took place at Strasbourg, the place where the idea of the present book came to the editors, at the occasion of two conferences in 2014, the 93th and 94th Encounters between mathematicians and theoretical physicists. The theme of the 93th encounter was "Riemann, topology and physics", and that of the 94th was "Riemann, Einstein and geometry.".

[65] Toute théorie physique, suffisamment élaborée, prend nécessairement une forme mathématique; il semble que les actions et réactions entre l'esprit et les choses ont amené peu à peu à former des moules où peut, partiellement au moins, s'insérer le réel. Certes, beaucoup de concepts créés par les mathématiciens n'ont pas trouvé encore d'applications dans l'étude des phénomènes physiques, mais l'histoire de la science montre qu'il était téméraire d'affirmer que telle ou telle notion ne sera pas un jour utilisée. Les géomètres aiment à rappeler le mot du grand mathématicien Lagrange qui, comparant un jour les mathématiques à un animal dont on mange toutes les parties, disait: "Les mathématiques sont comme le porc, tout est bon."

6. J. (le Rond) d'Alembert, Addition aux recherches sur la courbe que forme une corde tendue mise en vibration (1750), in *Mémoires de l'Académie des sciences de Berlin*, vol. 6 (1752), pp. 355–360
7. J. (le Rond) d'Alembert, Opuscules mathématiques, Paris, Claude-Antoine Jombert (1780)
8. J.S. Ames, Book review of *Die Partiellen Differentialgleichungen der Mathematischen Physik* (4th edition), vol. 1 (1900) and vol. 2 (1901), ed. by Heinrich Weber. Bull. Am. Math. Soc. **8**(2), 81–85 (1901)
9. Aristotle, Nicomachean Ethics, ed. H. Rackham (Harvard University Press, Cambridge, MA; William Heinemann Ltd., London, 1934)
10. Aristotle, The physics, in *The Complete Works of Aristotle: The Revised Oxford Translation*, vol. 1, ed. by J. Barnes (Princeton University Press, Princeton, 1984), pp. 315–446 (Translated by R.P. Hardie, R. K. Gaye)
11. Aristotle, The metaphysics, in *The Complete Works of Aristotle: The Revised Oxford Translation*, vol. 2, ed. by J. Barnes (Princeton University Press, Princeton, 1984), pp. 1552–1728 (Translated by W. D. Ross)
12. Aristotle, On the Heavens, in *The Complete Works of Aristotle: The Revised Oxford Translation*, vol. 1, ed. by J. Barnes (Princeton University Press, Princeton, 1984), pp. 447–511 (Translated by J. L. Stocks)
13. M.F. Atiyah, V.G. Drinfeld, N.J. Hitchin, Y.I. Manin, Construction of instantons. Phys. Lett. A **65**(3), 185–187 (1978)
14. D. Bernoulli, Réflexions et éclaircissements sur les nouvelles vibrations des cordes exposées dans les mémoires de l'académie de 1747–1748, *Histoire de l'Académie Royale des Sciences et des Belles-Lettres de Berlin*, vol. 9 (1755), pp. 147–172
15. D. Bernoulli, Sur le mélange de plusieurs espèces de vibrations simples isochrones, qui peuvent coexister dans un même système de corps, *Histoire de l'Académie Royale des Sciences et des Belles-Lettres de Berlin* vol. 9 (1755), pp. 173–195
16. D. Bernoulli, Mémoire sur les vibrations des cordes d'une épaisseur inégale, *Histoire de l'Académie Royale des Sciences et des Belles-Lettres de Berlin*, vol. 87 (1767), pp. 281–306
17. E. Betti, Sopra la propagazione delle onde piane di un gaz. Annali di matematica pura ed applicata, serie I, t. III, 232–241 (1860)
18. E. Betti, Sopra la elettrodinamica, Nuovo Cimento, ser. I, t. XXVII (1868), pp. 401–407
19. E. Betti, Sopra una estensione dei principii generali della dinamica, Atti della Reale Accademia dei Lincei, ser. III, t. II (1877–1878), pp. 32–34
20. W.E. Leonard, Empedocles of Agrigentum, *The Fragments of Empedocles* (The Open Court Publishing Company, Chicago, 1908)
21. R. Caddeo, X. Hascher, P. Jehel, A. Papadopoulos, H. Papadopoulos, L. Euler, Écrits sur la musique (a critical edition of the complete works and correspondence of Euler on music), 2 volumes, Hermann, Paris (2015)
22. R. Calinger, Euler's "Letters to a Princess of Germany" as an Expression of his Mature Outlook. Arch. Hist. Exact Sci. **15**(3), 211–233 (1976)
23. É. Cartan, Leçons sur la théorie des espaces de Riemann, Paris, Gauthier-Villars, 1928, 2nd edn. (1946)
24. A.-L. Cauchy, Cours d'analyse de l'École Royale Polytechnique, 1re partie. Analyse algébrique. Imprimerie royale, Paris. Œuvres complètes, série 2, tome 3 (1821)
25. A.-L. Cauchy, Mémoire sur les intégrales définies prises entre des limites imaginaires, De Bure Frères, Paris, Œuvres complètes, série 2, tome 2 (1825), p. 57–65
26. A.-L. Cauchy, Sur les intégrales qui s'étendent à tous les points d'une courbe fermée. C. R. Acad. Sci. Paris **23**, 251–255 (1846)
27. A.-L. Cauchy, Sur les fonctions de variables imaginaires. C. R. Acad. Sci. Paris **32**, 207–215 (1851) (Œuvres, série I, tome XI, 301–304)
28. A.-L. Cauchy, Sur les fonctions monotypiques et monogènes, C. R. Acad. Sci. Paris **32**, 484–487 (1851) (Œuvres, série 1, tome XI, 376–380)
29. S. Chandrasekhar, L. Lebovitz, On Riemann's paper: Ein Beitrag zu den Untersuchungen über die Bewegung eines flüssigen gleichartigen Ellipsoides, in [141], pp. 811–820

30. W.K. Clifford, Translation of Riemann's Habilitationsvortrag [146]: On the hypotheses which lie at the bases of geometry. Nature, **8**(183, 184), 14–17, 36, 37 (1873) (Reprinted in Clifford's *Mathematical papers* pp. 55–69, followed by a synopsis of the paper)
31. W.K. Clifford, *On the Space Theory of Matter. Proceedings of the Cambridge Philosophical Society*, vol. II (1876), pp. 157–158 (Clifford's *Mathematical papers*, pp. 21–22)
32. G. Darboux, J. Houël, La correspondance de G. Darboux avec J. Houël. Chronique d'un rédacteur (déc. 1869–nov. 1871), ed. by H. Gispert, *Cahiers du séminaire d'histoire des mathématiques*, 8(1987), pp. 67–202
33. P. Dugac, Éléments d'analyse de Karl Weierstrass. Archive for History of Exat Sciences **10**, 41–167 (1973)
34. R. Dedekind, H. Weber, Theorie der algebraischen Functionen einer Veranderlichen. Journal für die reine und angewandte Mathematik **92**, 181–290 (1882)
35. J. Dieudonné, History of algebraic geometry: An Outline of the History and Development of Algebraic Geometry, English translation by J. Sally, Wadsworth, Monterey, CA, *Original French: Cours de géométrie algébrique* (PUF, Paris, 1985). 1974
36. G.L. Dirichlet, Sur la convergence des séries trigonométriques qui servent à représenter une fonction arbitraire entre des limites données. Journal für die reine und angewandte Mathematik **4**, 157–169 (1829)
37. A. Einstein, Ideas and opinions, based on *Mein Weltbild*, ed. C. Seeling, and other sources. New translations and revisions by S. Bargmann (Crown Publishers, Inc. New York, 1954)
38. L. Euler, Variae observationes circa series infinitas, *Commentarii academiae scientiarum Petropolitanae* 9, 1744, pp. 160–188. (*Opera Omnia*, Series 1, vol. 14, pp. 217–244)
39. L. Euler, Remarques sur un beau rapport entre les series des puissances tant directes que réciproques, Mémoires de l'Académie des Sciences de Berlin 17 (1768), 83–106 (*Opera Omnia*, Series 1, Volume 15, pp. 70–90)
40. L. Euler, *Dissertatio physica de sono*, E. and J. R. Thurneisen Brothers, Basel, 1727 (*Opera omnia*, Series 3, vol. 1, 181–196)
41. L. Euler, *Tentamen novae theoriae musicae ex certissimis harmoniae principiis dilucide expositae* (1731), Saint-Pétersbourg, 1739. (*Opera omnia*, Series 3, vol. 1)
42. L. Euler, Sur la vibration des cordes (1748), traduit du latin, *Mémoires de l'Académie des Sciences de Berlin* 4, 1750, pp. 69–85 (*Opera omnia*, Series 2, vol. 10, pp. 63–77)
43. L. Euler, Découverte d'un nouveau principe de mécanique (1750), *Mémoires de l'Académie des sciences de Berlin* 6, 1752, pp. 185–217 (*Opera omnia*, Series 2, vol. 5)
44. L. Euler, De la propagation du son (1759), *Mémoires de l'Académie des sciences de Berlin* 15, pp. 185–209, 1766 (*Opera omnia*, Series 3, vol. 1, p. 428–451)
45. L. Euler, Supplément aux recherches sur la propagation du son (1759), *Mémoires de l'Académie des sciences de Berlin* 15, pp. 210–240, 1766 (*Opera omnia*, Series 3, vol. 1, pp. 452–483)
46. L. Euler, Continuation des recherches sur la propagation du son (1759), *Mémoires de l'Académie des sciences de Berlin* 15, pp. 241–264, 1766 (*Opera omnia*, Series 3, vol. 1, pp. 484–507)
47. L. Euler, Éclaircissemens plus détaillés sur la génération et la propagation du son et sur la formation de l'écho (1765), *Mémoires de l'Académie des sciences de Berlin*, 21, pp. 335–363, 1767 (*Opera omnia*, Series 3, vol. 1, pp. 540–567)
48. L. Euler, A.-C. Clairaut, J. Le Rond d'Alembert and L.-L. Lagrange, Correspondance de Leonhard Euler avec A. C. Clairaut, J. d'Alembert et J. L. Lagrange. *Opera omnia*, series IV A, volume 5, Birkhäuser, Bâle, 1980
49. L. Euler, *Introductio in analysin infinitorum*, première édition : 1748. *Opera omnia*, Series 1, vol. VIII
50. L. Euler, Methodus inveniendi lineas curvas maximi minimive proprietate gaudentes, sive solutio problematis isoperimetrici lattissimo sensu accepti (A method for finding curved lines enjoying properties of maximum or minimum, or solution of isoperimetric problems in the broadest accepted sense), book in 1744. Opera Omnia: Series 1, Volume 24, Fiisli Turici, Lausanne, 1952

51. L. Euler, De repraesentatione superficiei sphaericae super plano (On the mapping of spherical surfaces onto the plane), Acta Academiae Scientarum Imperialis Petropolitinae 1777, 1778, pp. 107–132 (*Opera Omnia* Series 1, Volume 28, pp. 248–275)
52. L. Euler, De proiectione geographica superficiei sphaericae (On a geographic projection of the surface of the sphere), Acta Academiae Scientarum Imperialis Petropolitinae 1777, 1778, pp. 133–142 (*Opera Omnia* Series 1, Volume 28, pp. 276–287)
53. L. Euler, De proiectione geographica Deslisliana in mappa generali imperii russici usitata (On Delisle's geographic projection and its use in the general map of the Russian empire), Acta Academiae Scientarum Imperialis Petropolitinae 1777, 1778, pp. 143–153 (*Opera Omnia* Series 1, Volume 28)
54. R. Farwell, C. Knee, The missing link: Riemann's "Commentatio," differential geometry and tensor analysis. Hist. Math. **17**(3), 223–255 (1990)
55. J. Fourier, Mémoire sur la propagation de la Chaleur dans les corps solides, Nouveau Bulletin des sciences par la Société philomathique de Paris. No. 6, Mars 1808, pp. 112–116
56. J. Fourier, *Théorie analytique de la chaleur* (Firmin Didot, Paris, 1822)
57. E. Frenkel, Lectures on the Langlands program and conformal field theory, in *Frontiers in Number Theory, Physics, And Geometry II. On Conformal Field Theories, Discrete Groups And Renormalization*, eds. by P. Cartier, et al., (Springer, Berlin, 2007) (Papers from the meeting, Les Houches, France, March 9–21, 2003)
58. E. Erwin (ed.), *Freud Encyclopedia: Theory, Therapy and Culture*, (Routledge, New York and London, 2002)
59. G.D. Garland, The contributions of Carl Friedrich Gauss to geomagnetism. Hist. Math. **6**, 5–29 (1979)
60. C.F. Gauss, *Theoria Motus Corporum Coelestium in Sectionibus Conicis Solem Ambientium, Hamburgi : sumtibus Frid* (Perthes et I. H, Besser, 1809)
61. C.F. Gauss, Allgemeine Auflösung der Aufgabe: die Theile einer gegebnen Fläche auf einer andern gegebnen Fläche so abzubilden, daß die Abbildung dem Abgebildeten in den kleinsten Theilen ähnlich wird, Astronom. Abh. (Schumacher ed.), 3 (1825), 1–30 (Also in Gauss's *Werke*, vol. IV, 189–216)
62. C.F. Gauss, Allgemeine Lehrsätze in Beziehung auf die im verkehrten Verhältnis des Quadrats der Entfernung wirkenden Anziehungs- und Abstossungs-Kräfte. In der Weidmannschen Buchhandlung, Gauss's Werke, vol. V, 243–276 (1840)
63. C.F. Gauss, Allgemeine Lehrsätze in Beziehung auf die im Verkehrten Verhältnisse des Quadrals der Entfernung wirkenden Anziehungs und Abstossungskräfte, Gauss Werke, t. 5, pp. 197–242
64. C.F. Gauss, Allgemeine Lehrsätze in Beziehung auf die im verkehrten Verhältnisse des Quadrats der Entfernung wirkenden Anziehungs- und Abstossungs-Kräfte, Resultate aus den Beobachtungen des magnetischen Vereins im Jahre 1839, or Carl F. Gauss, Werke, Königliche Gesellschaft der Wissenschaften zu Göttingen, Leipzig: Teubner, 1880, 2nd printing, 5:195–242; reprinted as Ostwald's Klassiker der exacten Wissenschaften, Vol. 2, ed. Albert Wangerin, Leipzig: Wilhelm Engelmann, 1889, Vol. 2
65. C.F. Gauss, Über das Wesen und die Definition der Funktionen (Gauss and Bessel, 1811) (See also Werke, Vol. 8, 90–92)
66. C.F. Gauss, *Disquisitiones generales circa superficies curvas*, English translation: General investigations of curved surfaces of 1925 and 1927. Translated with notes and a bibliography by J.C. Morehead and A.M. Hilterbeitel (Princeton, 1902)
67. C.F. Gauss, *Werke* (Königliche Gesellschaft der Wissenschaften, Göttingen, 1900)
68. H. Goenner, Some remarks on a contribution to electrodynamics by Bernhard Riemann. This volume, pp. 111–122
69. J. Gray, Riemann on geometry, physics, and philosophy–some remarks. This volume, pp. 97–108
70. G. Green, An essay on the application of mathematical analysis to the theory of electricity and magnetism, Nottingham, 1828, printed at the author's expenses

71. A. Grothendieck, Esquisse d'un programme (Sketch of a program), unpublished manuscript (1984), English translation by P. Lochak and L. Schneps in *Geometric Galois actions, vol. 1, "Around Grothendieck's Esquisse d'un Programme"* (L. Schneps and P. Lochak, ed.) London Math. Soc. Lecture Note Ser. vol. 242, pp. 5–48, (Cambridge University Press, Cambridge, 1997)
72. A. Grothendieck, Récoles et semailles : Réflexions et témoignage sur un passé de mathématicien (Harvesting and Sowing: Reflections and testimony on a mathematician's past), unpublished manuscript, 1985–1986
73. G.-H. Halphen, Traité des fonctions elliptiques et de leurs applications, 3 volumes. Paris, Gathier-Villars, 1886, 1888 (1891)
74. H. von Helmholtz, Theorie der Luftsschwingungen in Röhren mit offenen Enden. Crelle **57**, 1–72 (1860)
75. J.F. Herbart, Psychology as a science newly founded on experience, metaphysics and mathematics, 1824, Reprinted, in *Classics in Psychology*, ed. by T. Shipley (Philosophical library, New York, 1961), pp. 22–50
76. D. Hilbert, Über das Dirichletsche Prinzip, Jahresber. Deutsch. Math.-Verein. 8, 184–188. Gesammelte Abh. **3**, 10–14 (1900)
77. D. Hilbert, Über das Dirichletsche Prinzip, Jahresber. Math. Ann. 59, 161–186. Gesammelte Abh. **3**, 15–37 (1904)
78. Iamblichus, The Life of Pythagoras, or Pythagoric life, English translation from the Greek by Thomas Taylor, John M. Watkins, London, 1818, New printing, 1965
79. C.G.J. Jacobi, Solution nouvelle d'un problème fondamental de géodésie, Journal für die reine und angewandte Mathematik, 53, 335–341. Gesammelte Werke, t. II, 417–424 (1857)
80. F. Jederzejewski, Deleuze et la goémétrie riemannienne : une topologie des multiplicités. This volume, pp. 311–327
81. P. Jehel, Une lecture moderne d'un mémoire d'Euler: les Éclaircissemens plus détaillés sur la génération et la propagation du son et sur la formation de l'écho. (A modern reading of a memoir of Euler: Detailed clarifications concerning the generation and propagation of sound and the formation of echoes). In: Leonhard Euler, Mathématicien, musicien et théoricien de la musique (X. Hascher and A. Papadopoulos, ed.), 275–300. Collection Sci. Musique Ser. Études, CNRS Éditions, Paris, 2015
82. I. Kant, *De Mundi Sensibilis atque Intelligibilis Forma et Principiis (Inaugural dissertation, On the form and principles of the sensible and the intelligible world), 1770, translated by William J* (Eckoff, New York, Columbia College, 1894)
83. I. Kant, Prolegomena zu einer jeden künftigen Metaphysik, die als Wissenschaft wird auftreten können, Riga, bey Johann Friedrich Hartknoch. 1783
84. I. Kant, Critik der reinen Vernunft, 1st edn. (Riga 1781)
85. F. Klein, *Über Riemanns Theorie der algebraischen Funktionen und ihrer Integrale*, Teubner, Leipzig, 1882. English translation: *On Riemann's theory of algebraic functions and their integrals; a supplement to the usual treatises*. Translated from the German by Frances Hardcastle (Macmillan and Bowes, Cambridge, 1893) (Reprint: New York, Dover Publications, 19630
86. F. Klein, Riemann und seine Bedeutung für die Entwicklung der modernen Mathematik, in F. Klein, Gesammelte mathematische Abhandlungen, Berlin: Springer-Verlag, 1923 (reprinted 1973), 3: 482–497. From an address delivered at the general session of the Versammlung Deutscher Naturforscher und Artzte (Vienna, September 27, 1894). English translation: Riemann and his significance for the development of modern mathematics. Bull. Amer. Math. Soc. 1 (1895) No. 5, 165–180
87. F. Klein, Vorlesungen über die Entwicklung der Mathematik im 19. 2 volumes, Julius Springer, Berlin, 1926. English translation by M. Ackermann: Development of mathematics in the 19th century, Math. Sci. Press, Brookline, Mass., 1979
88. J.-L. de Lagrange, Sur la construction des cartes géographiques, Nouveaux mémoires de l'Académie royale des sciences et belles-lettres de Berlin, année 1779, Premier mémoire 637–664, Second mémoire 664–692. *Œuvres complètes*, tome 4, 637–692

89. J.-L. de Lagrange, Recherches sur la nature et la propagation du son, *Miscellanea philosophico-mathematica*, t. 1, Turin, 1759
90. J.-L. de Lagrange and J. Le Rond d'Alembert, Correspondance inédite de Lagrange et d'Alembert, *Œuvres de Lagrange*, publiées par J.-A. Serret, t. 13, Gauthier-Villars, Paris, 1882
91. J.H. Lambert, Anmerkungen und Zusätze zur Entwerfung der Land- und Himmelscharten (German: Notes and additions for creating ground and aerial charts). In Beyträge zum Gebrauche der Mathematik und deren Anwendung durch J. H. Lambert, Dritter Teil (Berlin, Realschule, 1772)
92. A.-M. Legendre, *Nouvelles méthodes pour la détermination des orbites des comètes* (F. Didot, Paris, 1805)
93. N.I. Lobachevsky, Pangeometry, translation, notes and Commentary by A. Papadopoulos, Heritage of European Mathematics, vol. 4 (European Mathematics Publishing House, 2010)
94. J.P.G. Lejeune Dirichlet, Sur la convergence des séries trigonométriques qui servent à représenter une fonction arbitraire entre des limites données. J. Reine Angew. Math. **4**, 157–169 (1828)
95. J.P.G. Lejeune Dirichlet, Vorlesungen über die im umgekehrten Verhältniss des Quadrats der Entfernung wirkenden Kräfte, ed. F. Grube (Leipzig: Teubner, 1876)
96. J.C. Maxwell, On a method of making a direct comparison of electrostatic with electromagnetic force. Phil. Trans. **158**, 643–657 (1868)
97. H. Minkowski, The Union of Space and Time. In: Space and Time, in The Principle of Relativity (A. Sommerfeld, ed.), Eng. tr. W. Perrett and G. B. Jeffery, London 1923 and Dover, N.Y. n.d., pp. 75–80. Reprinted in: The Concepts of Space and Time, Their Structure and Their Development (M. Capek, ed.), Boston Studies in the Philosophy of Science Series, Volume 22, Springer Science and Business Media, Dordrecht, 1976
98. W.H. Meeks III, J. Pérez, The classical theory of minimal surfaces. Bull. Am. Math. Soc. **48**, 325–407 (2011)
99. A.F. Monna, Dirichlet's principle. A mathematical comedy of errors and its influence on the development of analysis (Oosthoek, Scheltema and Holkema, Utrecht, 1975)
100. R. Narasimhan, Bernhard Riemann. Remarks on his Life and Work. Milan J. Math. **78**, 3–10 (2010)
101. E. Neuenschwander, Über die Wechselwirkungen zwischen der französischen Schule. Riemann und Weierstrass: Eine Übersicht mit zwei Quellenstudien, Archive for History of Exact Sciences **24**, 221–255 (1981)
102. C. Neumann, *Vorelsungen über Riemann's Theorie der Abel'schen Integrale, Leipzig* (Teubner, B. G, 1865)
103. K. Ohshika, The origin of the notion of manifold: from Riemann's Habilitationsvortrag onward. This volume, pp. 295–308
104. A. Papadopoulos, Looking backward: Euler and Riemann. This volume, pp. 1–81
105. A. Papadopoulos, Cauchy and Puiseux: Two precursors of Riemann. This volume, pp. 209–232
106. A. Papadopoulos, Riemann surfaces: reception by the French school. This volume, pp. 237–286
107. A. Papadopoulos, On the origin of topology, from Aristotle to Riemann, to appear in *Geometry in history*, ed. by S.G. Dani, A. Papadopoulos (Springer Verlag)
108. A. Papadopoulos, Nicolas Auguste Tissot: A link between cartography and quasiconformal theory, Archive for History of Exact Sciences (to appear in 2017)
109. A. Papadopoulos, Quasiconformal mappings, from Ptolemy's geography to the work of Teichmüller, to appear in: *Uniformization, Riemann-Hilbert Correspondence, Calabi-Yau Manifolds, and Picard-Fuchs Equations*, ed. by L. Ji, S.-T. Yau (International Press and Higher Education Press. To appear in 2017)
110. Pappus of Alexandria, *La collection mathématique; œuvre traduite pour la premiere fois du grec en français par Paul Ver Eecke* (Bruges, Desclée de Brouwer, Paris, 1933)

111. B. Pascal, Pensées (Everyman's Library No. 874, London, 1932) (Translated by John Warrington)
112. Plato, *Plato in Twelve Volumes*, vol. 9 (Harvard University Press, Cambridge, MA; William Heinemann Ltd., London, 1925) (Translated by W.R.M. Lamb)
113. O. Pedersen, *A Survey of the Almagest* (Odense University Press, 1974) (Reprinted with annotation and a new commentary by A. Jones, Springer Verlag, New York, 2010)
114. R.C. Penner, Moduli spaces and macromolecules. Bull. Amer. Math. Soc. **53**, 217–268 (2016)
115. É. Picard, *Traité d'analyse*, 2 vol. (Gauthier-Villars, Paris, 1891)
116. É. Picard, G. Simart, *Théorie des fonctions algébriques de deux variables indépendantes*, 2 vol. (Gauthier-Villars, Paris, 1897–1906) (Reprinted in one volume, Chelsea Publishing Co., Bronx, N.Y., 1971)
117. É. Picard, *Sur le Développement de l'Analyse et ses Rapports avec diverses Sciences: Conférences faites en Amérique* (Gauthier-Villars, Paris, 1905)
118. É. Picard, Allocution, Comptes rendus du Congrès International des Mathématiciens, Strasbourg, 22–30 septembre 1920, éd. par Henri Villat, Imprimerie et Librairie Éd. Privat, Toulouse, 1921
119. A. Plotnitsky, Comprehending the connections of things: Bernhard Riemann and the architecture of mathematical concepts (This volume, p. 329)
120. H. Poincaré, La correspondance d'Henri Poincaré avec des mathématiciens de A à H. Cahiers du séminaire d'histoire des mathématiques, tome 7, 59–219 (1986)
121. H. Poincaré, La valeur de la science (Paris, Flammarion, 1905) (several later editions)
122. H. Poincaré, Sciences et méthode (Paris, Flammarion, 1908) (several later editions)
123. S.D. Poisson, Suite du mémoire sur les intégrales definies. J. École Polytech. **11**, 295–341 (1820)
124. A.M. Polyakov, Quantum geometry of bosonic strings. Phys. Lett. **103B**(3), 207–210 (1981)
125. A.M. Polyakov, Quantum geometry of fermionic strings. Phys. Lett. **103B**(3), 211–213 (1981)
126. V.-A. Puiseux, Recherches sur les fonctions algébriques. Journal de Mathématiques Pures et Appliquées **15**, 365–480 (1850)
127. V.-A. Puiseux, Nouvelles recherches sur les fonctions algébriques. Journal de Mathématiques Pures et Appliquées **16**, 228–240 (1851)
128. R. Rashed, Les mathématiques infinitésimales du IXème au XIème siècle. Vol. IV. Ibn al-Haytham. Méthodes géométriques, transformations ponctuelles et philosophie des mathématiques. Al-Furqān Islamic Heritage Foundation Publication. 60. London: Al-Furqān Islamic Heritage Foundation, 1064 p. (2002)
129. C. Reid, *Hilbert* (Springer-Verlag, New York, 1996)
130. B. Riemann, Grundlagen für eine allgemeine Theorie der Functionen einer veränderlichen complexen Grösse, (Göttingen, 1851), [141] pp. 3–48
131. B. Riemann, Über die Darstellbarkeit einer Function durch eine trigonometrische Reihe (On the representability of a function by a trigonometric series). Aus dem dreizehnten Bande der Abhandlungen der Königlichen Gesellschaft der Wissenschaften zu Göttingen, 1867
132. B. Riemann, Commentatio mathematica, qua respondere tentatur quaestioni ab Illma Academia Parisiensi propositae: "Trouver quel doit être l'état calorifique d'un corps solide homogène indéfeni pour qu'un système de courbes isothermes, à un instant donné, restent isothermes après un temps quelconque, de telle sorte que la température d'un point puisse s'exprimer en fonction du temps et de deux autres variables indépendantes." In: Bernhard Riemann's Gesammelte Matheamtische WerkeRiemann-Gesammelte, 2nd Edition, Teubner 1892, pp. 391–404
133. B. Riemann, Theorie der Abel'schen Functionen. Journal für die reine und angewandte Mathematik, **54**, 115–155 (1857). (Reprinted in his *Gesammelte mathematische Werke* [141], pp. 88–144)
134. B. Riemann, Lehrsätze aus der analysis situs für die Theorie der Integrale von zweigliedrigen vollständigen Differentialien. Journal für die reine und angewandte Mathematik **54**, 105–110 (1857)

135. B. Riemann, Über die Anzahl der Primzahlen unter einer gegebenen Grösse, Monatsberichte der Berliner Akademie, November 1859, 671–680, *Gesammelte mathematische Werke* [141], pp. 145–153
136. B. Riemann, Über die Fortpflanzung ebener Luftwellen von endlicher Schwingungsweite. Abhandlungen der Königlichen Gesellschaft der Wissenschaften in Göttingen **8**, 43–66 (1860)
137. B. Riemann, Über die Fläche vom kleinsten Inhalt bei gegebener Begrenzung, Abh. Königl. d. Wiss. Göttingen, Mathem. Cl., **13**, 3–52 (1867)
138. B. Riemann, Ein Beitrag zu den Untersuchungen über die Bewegung eines flüssigen gleichartigen Ellipsoides. Abhandlungen der Königlichen Gesellschaft der Wissenschaften zu Göttingen **9**, 3–36 (1861)
139. B. Riemann, Beiträge zur Theorie der durch die Gauss'sche Reihe $F(\alpha, \beta, \gamma, x)$ darstellbaren Functionen. Aus dem siebenten Band der Abhandlungen der Königlichen Gesellschaft der Wissenschaften zu Göttingen pp. 3–32 (1857)
140. B. Riemann, Ein Beitrag zur Elektrodynamik, Ann. Phys. **131**, 237–43 (1867). (English translation: A contribution to electrodynamics, Phil. Mag., ser. 4, 35 (1867), 368–372)
141. B. Riemann, Gesammelte mathematische Werke, wissenschaftlicher Nachlass und Nachträge. Nach der Ausgabe von Heinrich Weber und Richard Dedekind neu herausgegeben von Raghavan Narasimhan. (Collected mathematical works, scientific posthumous works and supplements. According to the edition by H. Weber and R. Dedekind newly edited by R. Narasimhan). Teubner Verlagsgesellschaft, Leipzig, 1862; Springer-Verlag, Berlin (1990)
142. B. Riemann, Gleichgewicht der Electricität auf Cylindern mit kreisförmigen Querschnitt und parallelen Axen, Conforme Abbildung von durch Kreise begrenzten Figuren, (Nachlass XXVI) 1857, In: Ges. math. Werke, p. 472–476
143. B. Riemann, K. Hattendorff, *Schwere, Elektrizität und Magnetismus, Hannover, Carl Rümpler, 1876* (Nachdruck VDM, Müller, Saarbrücken, 2006)
144. B. Riemann, *Collected Papers* (Kendrick Press, Heber City, UT, Baker, 2004) (English translation by R Ch. Christenson and H. Orde)
145. B. Riemann, Collected works translated into French: Oeuvres mathématiques de Riemann, traduites par L. Laugel, avec une préface de C. Hermite (Paris, Gauthier-Villars, 1898)
146. B. Riemann, Über die Hypothesen, welche der Geometrie zu Grunde liegen, published by R. Dedekind, after Riemann's death, in Abhandlungen der Königlichen Gesellschaft der Wissenschaften zu Göttingen, Vol. 13, 1867. French translation in [145] pp. 280–299
147. B. Riemann, On the hypotheses which lie at the bases of geometry. (J. Jost, ed.) Birkhäuser, Springer International Publishing, German edition, 2013; English edition, 2016
148. B. Riemann, H. Weber, *Die partiellen Differential-Gleichungen der mathematischen Physik nach Riemanns Vorlesungen, 2 vols* (Vieweg, Braunschweig, 1912)
149. The Riemann letters at the Prussian cultural archive, translated by R. Gallagher and M. Weissbach, mimeographed notes, 1981
150. G. Roch, Über die Anzahl der willkürlichen Constanten in algebraischen Functionen. J. Reine Angew. Math. **64**, 372–376 (1865)
151. D. E. Rowe, From Graz to Göttingen: Neugebauer's early intellectual journey, in *A Mathematician's Journeys. Otto Neugebauer and Modern Transformations of Ancient Science*, eds. by A. Jones et al., Archimedes 45, (Springer, 2016), pp. 1–59
152. E. Scholz, Herbart's influence on Bernhard Riemann. Historia Mathematica **9**, 413–440 (1982)
153. D.E. Smith, *A Source Book in Mathematics*, vol 3 (McGraw-Hill, 1929) (Mineola, Dover, N.Y., 1959)
154. M. Spivak, *A Comprehensive Introduction to Differential Geometry*, vol. 1, 1st edn. (Publish or Perish, Wilmington, Del., 1970) (vol. V, 1975)
155. P. Stäckel, F. Engel, Gauss, die beiden Bolyai und die nichteuklidische Geometrie. Math. Annalen **2**, 149–167 (1897)
156. B. Taylor, De motu nervi tensi. Philos. Trans. R. Soc. Lond. **28** (1713, published in 1714) 26–32
157. B. Taylor, Methodus incrementorum directa et inversa, Impensis Gulielmi Innys, London, 1715; 2nd ed. 1717

158. R. Thom, Les intuitions topologiques primordiales de l'aristotélisme, *Revue thomiste*, juillet-septembre 1988, XCVI[e] année, 1988 (3), pp. 393–409 (1988)
159. R. Thiele, The rise of the function concept in analysis, in *Euler Reconsidered: Tercentenary Essays* (Kendrick Press, Heber City, UT. 422–461)
160. C.A. Truesdell, The Controversy Over Small Plane Vibrations of a String of Uniform Thickness, 1746–1788, Leonhardi Euleri Opera omnia, series II, volumes 11/2 (Birkäuser, Basel-Boston-Stuttgart, 1960)
161. Voltaire, Lettres philosophiques (Lettres écrites de Londres sur les anglais et leurs sujets), Basel, 1734. English version: Letters on the English. Several new editions
162. K. Weierstrass, Über das sogenannte Dirichlet'sche Princip, Gelesen in der Königl. Akademie der Wissenschaften am 14. Juli 1870, Ausgewählte Kapitel aus der Funktionenlehre, Teubner-Archiv zur Mathematik Volume 9, 1988, pp. 184-189; published in Weierstrass' *Mathematische Werke*, 2, Mayer and Müller, Berlin (1895), pp. 49–54
163. A. Weil, The Apprenticeship of a Mathematician, Springer, Basel, 1991 (Souvenirs d'apprentissage, Basel, Birkhäuser, Translated from the French, 1991)
164. H. Weyl, *Kommentar zu Riemanns "Über die Hypothesen, welche der Geometrie zu Grunde liegen,"* (Springer, Berlin, 1919)
165. H. Weyl, Raum Zeit Materie, J. Springer, Berlin, 1918, several revised editions. English translation: Space-Time-Matter, First edition 1922 (from the 4th German edition), Methuen, London. New Edition, 1950, reprinted, Dover, 1952
166. H. Weyl, Philosophy of Mathematics and Natural Science, Princeton Univesity Press, New edition, 2009, With a new introduction by Frank Wilczek. Translation of *Philosophie der Mathematik and Naturwissenschaft*, Munich, R. Oldenburg, 1927
167. S. Yamada, Riemann on minimal surfaces. This volume, pp. 135–150
168. L. Zhumd, *Pythagoras and the Early Pythagoreans* (Oxford University Press, Oxford, 2012) (Translated from the Russian by K. Windle and R. Ireland)
169. J.D. Zund, Some comments on Riemann's contributions to differential geometry. Historia Mathematica **10**, 84–89 (1983)

Cauchy and Puiseux: Two Precursors of Riemann

Athanase Papadopoulos

Abstract In this chapter, we review the works of Cauchy and Puiseux on the theory of functions of a complex variable that preceded Riemann's introduction of what soon became known as Riemann surfaces. The work of the two French mathematicians (especially that of Puiseux) inaugurates a group-theoretic point of view which complements the topological one discovered by Riemann.

Keywords Riemann surface · Algebraic function · Multi-valued function · Uniformization · Monodromy

AMS Mathematics Subject Classification: 30F10, 30F20, 01A55

1 Introduction

Riemann surfaces were introduced unexpectedly by Riemann in his doctoral thesis, defended on December 16, 1851. I said "unexpectedly" because it was something completely new, difficult to apprehend by Riemann's contemporaries, and it is not clear whether somebody else would have invented this notion even fifty years after Riemann, had he failed to do it. Riemann introduced these surfaces as ground spaces on which holomorphic (or meromorphic) functions are naturally defined. We recall Klein's sentence from his monograph [40] in which he surveys Riemann's ideas (p. 77): "The Riemann surface not only provides an intuitive illustration of the functions in question, but it actually defines them." In particular, a multi-valued function given as the solution of an algebraic equation acquires a new domain of definition, its associated Riemann surfaces, on which it becomes uniform (single-valued). This

A. Papadopoulos (✉)
Institut de Recherche Mathématique Avancée, CNRS et Université de Strasbourg,
7 rue René Descartes, 67084 Strasbourg Cedex, France
e-mail: athanase.papadopoulos@math.unistra.fr

A. Papadopoulos
Mathematics Department, Brown University, 151 Thayer Street,
Providence, RI 02912, USA

L. Ji et al. (eds.), *From Riemann to Differential Geometry and Relativity*,
DOI 10.1007/978-3-319-60039-0_7

idea of working with general surfaces equipped with complex structures, instead of only the sphere or the complex plane, or subsets of them, had a tremendous influence on the development of complex analysis, but also on geometry and topology.

The notion of Riemann surface, as all mathematical notions, has a history. Its discovery was motivated by several questions on which many mathematicians spent their lives. One of the main problems that led to this notion was addressed by the uniformization of algebraic functions. In fact, the notion of "algebraic function" is complicated, because such a "function" is generally not a function in the usual sense: it is multi-valued. In this sense, the uniformization problem asks for a way of getting around this complication. It is in trying to solve this "uniformization problem" that the theory of Riemann surfaces was born.

In this chapter, I will explain how the problem of uniformization of algebraic functions led to results by Puiseux—who was strongly motivated by the work of Cauchy—which, interpreted in the right perspective, are the algebraic counterpart of Riemann surfaces. The work of Puiseux, rather than the one of Riemann, was discussed in the French treatises on analysis during the first decades that followed these works. We discuss this fact in Chap. 8 of the present volume [49].

The outline of the rest of this chapter is the following.

In Sect. 2 we recall the notion of algebraic function and the problem of uniformization of such functions.

Section 3 is the heart of the chapter. We explain there how Puiseux, in his two articles published in 1850 and 1851, using Cauchy's theory of path integration, developed a notion which is a combinatorial analogue of the notion of Riemann surface. Puiseux's theory was also rooted in the theory of the uniformization of algebraic functions, and it makes connections with group theory, in particular with Galois theory. This section also contains information on the life of Puiseux.

In Sect. 4, we give a summary of the important work of Cauchy that was used by Puiseux when he developed his theory. This work was also available to Riemann when he introduced Riemann surfaces.

In Sect. 5, we review Hilbert's 22nd problem which concerns uniformization. In this statement, the word uniformization is slightly different from the one we use in the previous sections, but the two notions are closely related, and the origin of the two words is the same. Our goal in this concluding section is to indicate the development of the theory whose bases were laid down by Puiseux and Riemann.

2 Algebraic Functions and Uniformization

An algebraic function u of the complex variable z is defined by an equation

$$f(u, z) = 0$$

where f is a polynomial in the two variables u and z. Here, u is considered as an implicitly defined function of z. For each value of z, there are generally more than one value of u. (If the equation is irreducible, then the number of values is the highest degree of u in the equation.) The first question that arises is: Can one make a choice between these values so as to get a bona fide function $u(z)$? The obvious answer is clearly "no," but one wishes to understand more. The first approach to the question is a case by case analysis. Consider for example the case where f is the polynomial $f(u, z) = u^2 - z$. Then, for each value of z, we have, except if $z = 0$, two different values of u. Setting $z = re^{i\theta}$ with $r > 0$, the two values are $u_1 = \sqrt{r}e^{i\theta/2}$ and $u_2 = -\sqrt{r}e^{i\theta/2}$. If for some value of z we choose one of the values u_1 or u_2 and assign it to $u(z)$, and if we try to extend u as a function defined on the whole complex plane, we obtain a discontinuous function, which is not desirable. We are then led to define the function $u(z)$ on a subset of the plane, but there is no natural choice of such a subset.

Uniformization originates with this problem, that is, the fact that algebraic equations generally have more than one complex solution. The word "uniformization", in the sense of Riemann, refers to the fact that one would like to have a way of making such a multi-valued function single-valued, or "uniform." The German adjectives *einwerthig* and *mehrwerthig* used by Riemann are translated, in the 1898 French version [64], by *uniforme* and *multiforme*. In the recent English translation by Jason Ross, the same words are used: *uniform* and *multiform*. Riemann utilises the word *mehrwerthig* (multiform) for a function which may assign to a value of the variable more than one value, and *einwerthig* (uniform) otherwise. In the preliminary part (§ 1) of his memoir on Abelian functions, he addresses the problem of extending a holomorphic function defined on a piece of the plane. He writes: "From the nature of the function we wish to extend, this function either will always take, or will not take, the same value for a given value of z regardless of the path along which we are extending. In the first case, I will call it *uniform*: it is then a perfectly determined function for any value of z, and it will never be discontinuous along a line. In the second case, where we shall say it is *multiform*, we first have, in order to understand the motion of this function, to concentrate our attention on certain points of the z-plane, around which the function is extended into another function..." A little bit later, he uses, as an alternative for the word "uniform," the word "monodromic" (see our Footnote 10). Likewise, in his memoir [62], Riemann uses the word *monodromic* as a synonym for *uniform*. Weyl, in [70] (p. 2), also uses the word "uniform." We shall use the word "uniformization" in this original sense of Riemann. There are other meanings for the word uniformization; see Sect. 5 of the present chapter.

The modern definition of function assumes that a function is single-valued, or "uniform," that is, to each value of the variable, the function associates a single value. But this was not the case at the epoch of Riemann or Cauchy or before them. In fact, Euler, to whom the first abstract definition of a function is attributed, considered the possibility that a function is multi-valued. Riemann was a devoted reader of Euler. He mentions his name at several occasions, when he informs his reader about the origin of his own ideas, for instance in his doctoral dissertation, in his Habilitation dissertation [59], in his memoir on Gauss's hypergeometric series [62], and there are several references to Euler in Riemann's memoirs and posthumous papers. The interested reader may find a thorough report on Riemann's debt to Euler in Chap. 1 of the present volume [48]. Among other things, we discuss in that chapter the way the notions of algebraic and multivalued functions appear in Euler's treatise *Introductio in analysin infinitorum* (Introduction to the analysis of the infinite) (1784) [29].

3 Puiseux and Uniformization

Victor Puiseux (1820–1883) defended a doctoral dissertation on astronomy in 1841. He attended Cauchy's courses on analysis and soon became interested in this topic. At the same time, he became Cauchy's closest follower and friend, and he always expressed his respect and admiration for his teacher's work. Puiseux spent a large part of his time developing, correcting and refining results of Cauchy who used to publish very rapidly his ideas, sometimes in rough form. Cauchy's influence on Puiseux was enormous.

Motivated by Cauchy's work, Puiseux wrote two remarkable memoirs, *Recherches sur les fonctions algébriques* (Researches on algebraic functions, 1850) [55] and *Nouvelles recherches sur les fonctions algébriques* (New researches on algebraic functions, 1851) [56]. The second memoir appeared in the year Riemann defended his doctoral dissertation, consisting of his first memoir on the theory of functions of a complex variable.[1] After the publication of Riemann's dissertation, Puiseux practically stopped working on this topic.

Below, we shall give a quick review of the content of the two memoirs of Puiseux.

Puiseux taught mathematics at the École Normale Supérieure. From 1855 to 1859, he worked at the Observatory of Paris, and in 1859 he became a member of the Bureau des longitudes.[2] In 1857, he became the successor of Cauchy on the chair of astronomy at the University of Paris, and he taught there until his health became critical, in 1882, one year before his death. His works included, besides complex analysis, mechanics, observational astronomy and botanics. Puiseux also made important contributions to celestial mechanics. In this field, he solved several difficult questions which had also been addressed by Cauchy. One of his most influential memoirs in this domain is the *Mémoire sur l'accélération séculaire du mouvement de la lune* (Memoir on the secular acceleration of the motion of the moon) [57] 1873, in which he contributed to the difficult problem of explaining the acceleration of the mean motion of the moon. Puiseux is also a precursor of French Alpinism, and a pick (3946 m) in the Alps, which he climbed in 1848, carries his name.

We shall give in the next section several biographical details on Cauchy. Let us mention that like Cauchy, Puiseux was involved in social issues, that he founded several charities, including one for the help of the poor at their home. During his lifetime, Puiseux kept secret most of his philanthropical activity, which was

---

[1] It may be useful to recall that Dedekind, in his notes on Riemann's life published in the *Collected works* edition [63], states that Riemann probably conceived his ideas on Riemann surfaces in 1847.

[2] The Bureau des longitudes is a French institution in charge of geodesy, standardisation of time-keeping, and astronomical measurements. The names of famous members of the Bureau include Lagrange, Laplace and Poincaré.

discovered by his family only after his death [68]. Picard writes about Puiseux as a teacher ([51] p. v)[3]:

> Victor Puiseux's modesty was intimidating, and his patience and politeness were admirable. To a student blundering at some test, he just used to say, with a very sweet tone: "I don't know whether I heard well or whether I am mistaken, but it seems to me that what you said is not completely true."[4]

In an article [55] published in 1850, that is, one year before Riemann defended his doctoral thesis, Puiseux addressed the problem of uniformization. As a result, he did not introduce Riemann surfaces, but he discovered a notion which is close to it. We now give a short description of his work on the subject, using the notation of Sect. 2 above for an algebraic function, which is Puiseux's notation.

Puiseux writes, after introducing the discontinuity problem posed by multi-valued functions which we recalled in the introduction ([55] p. 366):

> We shall avoid this discontinuity by defining in a different way the function u. Let us consider again the equation
>
> $$f(u, z) = 0,$$
>
> of which we may assume the first side to be integral in u and z; let us give to z an arbitrary initial value c, and, for the initial value b of u, let us choose any one of the solutions of the equation
>
> $$f(u, c) = 0.$$
>
> Let us now conceive that z varies in an arbitrary manner starting from the value c, and reaches another value k. Mr. Cauchy showed (*Nouveaux Exercices de Mathématiques*, tome II, p. 109) that the different values of u vary simultaneously in a continuous manner. Thus, there will be one which is first equal to b, which will pass by infinitely small steps to a determinate value h which it will attain for $z = k$. For us, this value of u will be a function of z, and, as we can see, it will be a continuous function. But its determination, for a particular value of z, will depend at the same time on this same value and on the series of values by which e passed starting from its initial value.

Thus, Puiseux solves the continuity problem by declaring that the function $u(z)$ not only depends on the variable z but also on a path that we choose from a basepoint to the point z. Concerning the choice of the path, he writes:

> Let us observe however that the function will stop being determined if, when passing from the value c to the value k, z takes a value for which the equation
>
> $$f(u, z) = 0$$
>
> have equal solutions. But the number of these values being finite, it will always be possible to avoid this circumstance, for any values c and k.

Thus, the chosen path between the basepoint and this point z avoids a certain number of *singular points*. Puiseux investigates in detail the dependence of $u(z)$ on

[3]In this chapter, all the translations from the French are mine.

[4]Victor Puiseux était d'une modestie intimidante, d'une patience et d'une politesse admirables. Quand un élève avait, dans une interrogation, énoncé quelque énormité, il se contentait de lui dire d'un ton très doux: "Je ne sais pas si j'ai bien entendu ou si je me trompe, mais il me semble que ce que vous avez dit n'est pas tout à fait exact."

the path, highlighting the roles of the singular points, which, he says, are of two types: points where the function u becomes infinite, and points which correspond to multiple solutions of the algebraic equation. Cauchy, in his previous works, misunderstood the nature of the singularities, since he considered that the singular points are only those where the algebraic function u becomes infinite. Puiseux presents clearly the invariance of the value of the function at the point z under homotopy of paths.

To show the close relation with Riemann's work, we need to recall Riemann's idea of a Riemann surface, and we consider again the example of the algebraic function $w^2 - z = 0$. In this case, w is the "function" $\sqrt{z}$. We take a basepoint $z_0 \neq 0$ in the complex plane. The function $\sqrt{z}$ is multi-valued at such a point. We take some determination of this function in some neighborhood of z_0. We continue the definition of this function along paths starting at z_0 (we use analytic continuation). We allow ourselves the use of modern terminology. A problem arises when the path comes back at z_0 and encloses the origin 0 of the complex plane. If such a path is not simple (that is, injective), the fact of enclosing the origin means that it has non-zero winding number with respect to 0. If such a path has odd winding number with respect to the origin, the value we get at z_0 is different from the initial one. At this point, Riemann introduced the idea that in this case the endpoint of such a path should not be considered as the basepoint z_0, but a point on a different *sheet* of a new surface on which the function $\sqrt{z}$ should be defined. This is the Riemann surface associated with the function. At the same time, Riemann introduced the notion of covering space. In the example considered, the surface obtained is a two-sheeted branched cover of the complex plane (or of the sphere), and the branching locus is the origin. This construction is very general, that is, it associates to an arbitrary multi-valued function defined by an algebraic equation a Riemann surface which is a branched cover of the sphere and on which the function is defined and becomes single-valued. The degree of the covering is the number of values of w associated with a generic value of z. This construction is described for the first time in Riemann's dissertation [58] and is further developed in the section on preliminaries in his 1857 paper [60].

It is not hard for a topologist to see that Puiseux's description of u as a function, not of z alone, but of a pair (z, γ), where γ a homotopy class of paths joining a basepoint to the variable point z, the homotopy being relative to some finite set of points on the surface (namely, the set of singular points of the algebraic equation $f(z) = 0$), is equivalent to considering that the function is defined on a Riemann surface which is a covering of the complex plane, ramified over this set of singular points. In fact, the usual modern construction of a covering of a surface defines it as a set of equivalence classes of homotopy classes of paths in the base surface subject to certain conditions which can be expressed in terms of group theory. (With no condition on the homotopy classes of paths, we get the universal covering of the base surface.) Let us emphasize though that we know the relation between the two definitions, the one of Puiseux and the one using Riemann surfaces and their coverings, initiated by Riemann, because we are familiar with the theory of surfaces. Thus, we are not claiming that Puiseux discovered Riemann surfaces. But he came very close to them. In fact, the work of Puiseux is group-theoretic, before the formal introduction of groups in the theory of Riemann surfaces. There is a famous result due to Riemann, which he gives in his paper on Abelian functions [60]. The result, stated in modern terms, says that given a finite set of points on the Riemann sphere and a representation of the fundamental group of the complement of these points into a permutation group, there exists a Riemann surface which is a branched covering of the sphere having the given points as branch points and whose monodromy is the given representation. This is one form of the so-called Riemann existence theorem (there are several other forms). The general form of the theorem deals with branched covers of surfaces that are more general than the Riemann sphere. The theorem establishes relations between topology, group theory and function theory. The permutation representation is that which Puiseux studies.

It was natural that Puiseux, in considering functions defined using paths from a basepoint to the variable point, studies line (or path) integrals, especially that the theory of such integrals was part of Cauchy's courses he followed. Starting from § 8 of his memoir [55], Puiseux considers line integrals of the form $\int_c^k u dz$ where c and k are points in the complex plane. In § 9 (p. 373), he proves the following theorem:

> The value of the integral $\int_c^k u dz$, taken along the line CMK, will not change if, the points c and k remaining fixed, this line is deformed without crossing any point for which the function u_1 becomes infinite or equal to another solution of the equation $f(u, z) = 0$.[5]

Puiseux attributes this theorem to Cauchy, like a few others he proves in § 9 to 11 of his memoir, as preliminaries for his main results. However, he brings important complements to Cauchy's results. He states ([55] Note p. 375):

> The theorems in § 9, 10, 11 were given by Mr. Cauchy in the *Comptes Rendus des séances de l'Académie des Sciences*, year 1846. But the famous geometer [Cauchy] characterizes the points which must be avoided by the path that is travelled by the fact that at these points the function becomes discontinuous; but since I restrict here to algebraic functions, I thought I would give more precision to the statements and the proofs by saying that the points considered are those for which the function u either becomes infinite or is a multiple solution of the equation $f(u, z) = 0$.[6]

Here, u_1 is a fonction obtained by starting with one of the branches of the function u defined by the equation $f(u, z) = 0$.

Later in the same paper (§ 53 to 55), Puiseux considers elliptic integrals and their dependence on the integration path, and he makes explicit the periods of the inverse functions. We note incidentally that the study of these integrals was one of the main subjects of interest of Riemann. We shall consider this question again below.

---

[5]L'intégrale $\int_c^k u dz$, prise le long de la ligne CMK, ne changera pas de valeur, si, les points C et K restant fixes, cette ligne vient à se déformer, sans franchir toutefois aucun point pour lequel la fonction u_1 devient infinie ou égale à une autre racine de l'équation $f(u, z) = 0$.

[6]Les théorèmes de nos 9, 10, 11 ont été donnés par M. Cauchy dans les *Comptes Rendus des séances de l'Académie des Sciences*, année 1846. Seulement l'illustre géomètre caractérise les points que le chemin parcouru ne doit pas franchir en disant que, pour ces points, la fonction devient discontinue: comme je me borne ici aux fonctions algébriques, j'ai cru donner plus de précision aux énoncés et aux démonstrations en disant que les points dont il s'agit sont ceux pour lesquels la fonction u devient infinie ou une racine multiple de l'équation $f(u, z) = 0$.

In the second part of his memoir [55] (starting p. 384), Puiseux studies the passage from one value of u to another one corresponding to the same z. This involves a detailed analysis of how the various values $u_1, \ldots, u_p$ corresponding to a given z are interchanged when the point z, seen as a geometric point in the complex plane, describes a small loop. The result will depend on the behavior of the function at the singular points enclosed by the loop.

Puiseux discovered the fact that the solutions of an algebraic equation are grouped into cycles which he called *circular systems* (systèmes circulaires) and he gave a method to perform this grouping. This decomposition into circular systems is related to the fact that the solutions are permuted by following the points geometrically along closed paths, and that an arbitrary permutation may be decomposed into circular permutations, a fact already proved by Cauchy in his paper [16], precisely in the setting of solutions of algebraic equations. In p. 479 of his memoir [55], he writes that the possibility of grouping into circular systems the various solutions $u_1, u_2, \ldots$ and of seeing that these values are interchanged around the points where the function u has multiplicity or takes the value infinity may be deduced from a theorem on substitutions[7] by Cauchy (Journal de l'École Polytechnique, tome X). But he adds that the method that he gives for this grouping is new.

The third part of the memoir [55] concerns applications of the theory to periods of integrals. Again, Puiseux refers to Cauchy's work, declaring that it leads to the existence of periods, that Cauchy recovered in this way the periods of elliptic integrals, but that Cauchy's method does not allow one to recover periods of general integrals. With his results on periods, Puiseux gave an explanation of the periodicity in the determinations of the complex circular functions, of elliptic functions and of other functions defined by integrals (in particular those introduced by Jacobi).

On p. 428 of his memoir, Puiseux says that the propositions he established are also applicable to the case where the function u of the variable z, which was taken to satisfy an algebraic equation $f(u, z) = 0$, is transcendental instead of being algebraic. He declares that the only property that is used is the continuity of u in terms of z, and he says that this question was treated by Cauchy in his *Nouveaux Exercices de Mathématiques*, tome II, p. 109.

In the following year, Puiseux published a second paper [56] in which he gave a method for characterizing periods of integrals in the case where the function f in the equation $f(u, z) = 0$ is an irreducible polynomial.

Puiseux's paper [55] also contains the so-called "Newton–Puiseux polygon," a method for evaluating the value of an algebraic function near a branch point, using so-called Pusieux series. These are a generalization of power series where the exponent may be fractional or negative. In fact, Puiseux did not discover these series, he rather rediscovered them ([55] p. 399), since they were introduced before him by Newton, in 1676.[8] Puiseux came up with these series in the context of his work on separating the various branches of functions defined by algebraic equations. He gave

[7] In this context, a "substitution" means a permutation of letters. This word substitution is used e.g. in Jordan's *Traité des substitutions et des équations algébriques* [39].

[8] Isaac Newton, Letter to Oldenburg, October 24, 1676.

an expansion of each of these branches in such a convergent series. The so-called Newton–Puiseux theorem states that an algebraic equation $f(u, z) = 0$, the variable u, seen as a function of z, may be expanded as a series (called now Puiseux series) that converges in some neighborhood of the origin. Stated differently, the result says that any branch of an algebraic curve can be represented as a Puiseux series. The Newton–Puiseux series has a wide generalization to the study of polynomials over local fields (the classical case being the one where the local field is the field of Laurent polynomials).

The work of Puiseux on solutions of algebraic equations was a forerunner of works of several mathematicians. It was interpreted and generalized in the setting of groups by Hermite and others. One should mention here that group theory was still unborn, or at best, was only in its infancy.[9]

Hermite presented a paper, in 1849, entitled *Sur la théorie des fonctions elliptiques* (On the theory of elliptic functions) [35] where he studies periods of elliptic integrals, and in which he acknowledges Cauchy's influence. We shall soon talk about the work of Hermite in relation with that of Puiseux.

We showed that Puiseux studied how the fact that the roots of an algebraic equation are interchanged when the variable z describes some loops in the plane leads to a group factorized into permutations. He used in this context the word "monodromic," which was already introduced by Cauchy. Hermite continued using this word in [36] (1851).[10] This led eventually to the notion of *monodromy group*, which we still use today. Jordan, in his *Traité des substitutions* [39] (1870), defined a group he called the *algebraic group*, which contains the monodromy group as a normal subgroup. The paper [36] by Hermite in which he studies the solvability of equations by radicals makes the relation between the work of Puiseux and the Galois group of an algebraic equation. Hermite's paper starts as follows:

> It seems to me that the propositions given by Mr. Puiseux, on the roots of algebraic equations considered as functions of a variable z which enters rationally in their first member, open up a wide research field which is intended to shed light on the analytic nature of this kind of quantities. I propose to give here the principle of these researches, and to show how they lead to the knowledge of whether an arbitrary equation
>
> $$F(u, z) = 0$$

[9]It is usually considered that the first abstract definition of a group is contained in the 1854 paper by Arthur Cayley [24]. But the notion of group appears in essence, as a group of permutations of the roots of an algebraic equation, in works of various people on the solutions of polynomial equations of degree ≥ 4, in particular the work of Galois. Klein writes, in his *Development of mathematics in the 19th century* ([42] p. 316 of the English translation), that "group theory first developed in the theory of algebraic equations [...] the central significance of group theory for algebraic equations first appeared in the work of Galois in 1831 (from whom the term 'group' also stems)."

[10]Let us note that Riemann used the word "monodromic" in his memoir on Abelian functions [60] for a function which is uniform, or single-valued. He writes (§ 1): "To simplify the designation of these relations, we shall call the various extensions of *one* function, for some fixed portion of the plane of the z, the *branches* of this function, and a point around which a branch of the function extends in another one a *ramification point* of the function. Everywhere where there is no ramification, the function will be *monodromic* or *uniform*."

> is algebraically solvable, that is, whether the unknown u can be expressed by a function of the variable z, containing only this variable under root extraction signs of integer degree. The theorems to which we will be led in this way will give, from a completely new point of view, the beautiful result obtained by Abel on the possibility of expressing algebraically[11] $\sin \mathrm{am}\left(\frac{x}{n}\right)$ by $\sin \mathrm{am}(x)$. I restrict myself here to the question of the resolution by radicals. Later, I will show how the theorems of Mr. Puiseux lead to a lowering of these equations in the cases announced by Galois, whose principles will serve as a basis for everything we shall say.[12]

Cauchy published two reports [20, 21] on the two memoirs of Puiseux. In the first report, he reviews in detail his own work on the subject, and then presents Puiseux's contribution. He writes in his conclusion:

> Not only Mr. Puiseux added new developments and new improvements to the theory of curvilinear integrals of algebraic functions, but, furthermore, he highlighted, with a lot of wisdom, the rules according to which the various values of an algebraic function are interchanged when the curve which conducts the integration winds around one of the points he calls *principal points*. Finally, he was able to determine in general the number of distinct values and the periods of certain curvilinear integrals which are relative to a very large class of algebraic functions and which contain as particular cases elliptic and Abelian integrals.[13]

In the report on the second memoir, Cauchy mentions Puiseux's new results on the periods of curvilinear integrals and the use that Hermite made of Puiseux's results in his research on the solvability of equations by radicals.

---

[11]The notation am is used in the theory of elliptic functions. It denotes the Jacobi amplitude.

[12]Les propositions données par Mr. Puiseux, sur les racines des équations algébriques considérées comme fonctions d'une variable z, qui entre rationnellement dans leur premier membre, me semblent ouvrir un vaste champ de recherches destinées à jeter un grand jour sur la nature analytique de ce genre de quantités. Je me propose de donner ici le principe de ces recherches, et de faire voir comment elles conduisent à reconnaître si une équation quelconque

$$F(u, z) = 0$$

est résoluble algébriquement, c'est-à-dire si l'inconnue u peut être exprimée par une fonction de la variable z, ne contenant cette variable que sous les signes d'extraction de racines de degré entier. Les théorèmes auxquels nous serons ainsi amenés donneront, et sous un point de vue entièrement nouveau, le beau résultat obtenu par Abel sur la possibilité d'exprimer algébriquement $\sin \mathrm{am}\left(\frac{x}{n}\right)$ par $\sin \mathrm{am}(x)$. Je me borne ici à la question de la résolution par radicaux ; plus tard je ferai, au même point de vue, l'étude des équations modulaires, et je montrerai comment les théorèmes de Mr. Puiseux conduisent à effectuer l'abaissement de ces équations dans les cas annotés par Galois, dont les principes serviront d'ailleurs de base à tout ce que nous allons dire.

[13]Mr. Puiseux a non seulement ajouté de nouveaux développements et des perfectionnements nouveaux à la théorie des intégrales curvilignes des fonctions algébriques, mais, de plus, il a mis en évidence, avec beaucoup de sagacité, les lois suivant lesquelles les diverses valeurs d'une fonction algébrique se trouvent échangées entre elles quand la courbe qui dirige l'intégration tourne autour de l'un des points qu'il nomme *points principaux*; enfin, il est parvenu à déterminer généralement le nombre de valeurs distinctes et le nombre de périodes de certaines intégrales curvilignes, qui sont relatives à une classe très étendue de fonctions algébriques, et qui comprennent comme cas particuliers les intégrales elliptiques et abéliennes.

The work of Puiseux was acknowledged as important by many mathematicians. Bertrand,[14] in his eulogy of Puiseux [6], writes the following:

> Ch. Sturm,[15] our benevolent master of all, but above all proud of his pupil of Collège Rollin, accosted me one day with this question which nobody before Puiseux had addressed: "If you follow along a closed loop the root of an equation whose parameter represents a point of the contour, what do you obtain when you come back to the starting point?" – I responded without hesitation: "I will recover my root." – "Well, no! you will not recover it: Puiseux proves this. He did a beautiful memoir!"[16]

The relation of the work of Puiseux with the notion of Riemann surface has not been sufficiently emphasized. Riemann defined these surfaces as ramified coverings of the plane (more precisely, of the Riemann sphere). The work of Puiseux on algebraic functions, interpreted from a topological point of view, contains in essence the combinatorics of such a surface, giving a description of how its sheets are permuted above a ramification point, and establishing the precise relation between this sheet permutation and the nature of the singularities of the algebraic equation. At the same time, Puiseux's work makes the relation with group theory. At the expense of being anachronical, let us mention that the theory of Puiseux expresses the so-called *monodromy homomorphism* from the fundamental group of the Riemann sphere with a finite set deleted (the singular set of the algebraic equation) into the permutation group on d symbols. The books on the history of nineteenth-century complex analysis hardly mention Puiseux. Gray writes in [33] p. 193: "although we know from Laugwitz [44] that Riemann had read Cauchy's report on Puiseux's memoir by December 1851 it seems unlikely that Riemann had anything to learn from Puiseux by the time he was writing his doctoral thesis."

The work of Puiseux was thoroughly used in several French treatises and dissertations on complex analysis and Riemann surfaces in a period that lasted more than 50 years after the publication of this work. We refer the reader to Chap. 8 of the present volume [49].

---

[14] Joseph Bertrand (1822–1900) taught mathematics and physics at Lycée Saint-Louis, École Polytechnique, École Normale Supérieure and then Collège de France. His name is attached to the "Bertrand series" in analysis and to the "Bertrand postulate" in number theory. He became member of the Académie des Sciences, in 1856, as the successor of Charles Sturm. He was the secretary ("secrétaire perpétuel") of the mathematical section of the Academy from 1874 until his death, after which Darboux became the secretary. This explains the fact that Bertrand wrote several eulogies. Bertrand was also the brother-in-law of Hermite. Paul Appell's wife was a niece of Bertrand and of Hermite and a cousin of Émile Picard.

[15] Charles-François Sturm (1803–1855) whose name is associated with the Sturm-Liouville principle on linear order-two differential equations with a parameter, was one of Puiseux's teachers at the Collège Rollin in Paris, which Puiseux enrolled in 1834.

[16] Ch. Sturm, notre maître bienveillant à tous, mais fier surtout de son élève du collège Rollin, m'aborda un jour par cette question que personne avant Puiseux ne s'était proposée: "Si vous suivez le long d'un contour fermé la racine d'une équation dont un paramètre représente un point du contour, qu'obtiendrez-vous en revenant au point de départ ?"—"Je retrouverai ma racine, répondis-je sans hésiter."—"Eh bien, non ! vous ne la retrouverez pas: ce Puiseux le démontre. Il a fait un bien beau Mémoire !".

In the next section, we give a summary of some of the tools introduced by Cauchy that were available to Puiseux. Riemann had the same tools at his disposal.

4 Cauchy and His Work on Functions of a Complex Variable

When Riemann started working on his doctoral dissertation, functions of a complex variable were already studied by various authors. In particular, such functions were considered by Euler in his 1748 treatise *Introductio in analysin infinitorum* [29]. In 1777, Euler, in a memoir on geographical maps [30], uses complex numbers in his study of maps from the sphere to the complex plane. See also [25] for a commentary on that memoir. More importantly, by the time Riemann started his study of such functions, Cauchy had introduced several of the tools that were needed for the development of the theory of Riemann surfaces. In particular, in a series of articles he published in the 1830s and the 1840s, Cauchy studied line integrals in the complex domain and their dependence of homotopy classes of paths. This inaugurated the use of topological methods in the study of functions of a complex variable.[17] Riemann, who knew the importance of Cauchy's work, was certainly following his publications. Klein, who was probably the most enthusiastic representative of Riemann, in his essay *Riemann and his significance for the development of modern mathematics* [41] (1895), recalls that the foundations of the theory of functions of a complex variable are due to Cauchy. He writes (p. 168):

> The founder of this theory is the great French mathematician Cauchy; but only later, in Germany, did this theory assume its modern aspect which has made it the central point of our present views of mathematics. This was the result of the simultaneous efforts of two mathematicians whom we shall have to name together repeatedly, – of Riemann and Weierstrass.[18]

Weierstrass, who is mentioned in this passage, based on Cauchy's theory, developed the theory of functions of a complex variable in a way different from that of Riemann. He is known for a multitude of interesting works related to the theory of functions. To him is attributed the definition of an analytic function of a com-

[17] One should remember though that the topological notions that appear in Cauchy's work (paths, homotopy, etc.) were still not rigorously defined, and that part of this theory was based on intuitive grounds. One of the earliest rigorous definitions of a path is contained in the much later Jordan's *Cours d'Analyse de l'École Polytechnique*, in three volumes, written between 1882 and 1887 (cf. [38], 2nd. edition, vol. 1, p. 90).

[18] Klein writes in a footnote: "In the text I refrained from mentioning Gauss, who being in advance of his time in this and in other fields, anticipated many discoveries without publishing what he had found. It is very remarkable that in the papers of Gauss we find occasional glimpses of methods in the theory of functions which are completely in line with the later methods of Riemann, as if unconsciously a transfer of leading ideas has taken place from the older to the younger mathematician."

plex variable using convergent power series,[19] which he developed around the year 1841 in a work which was essentially unpublished.[20] This led him to a concept of Riemann surface using the principle of analytic continuation.

One of the facts that emanates from an analysis of Cauchy's work is that although he had most of his ideas early in his career, the fact that his results became precise and rigorous was progressive. Before our exposition of Cauchy's work, we shall say a few words on his life.

Cauchy was born in the year of the French revolution. He belonged to a family who escaped Paris during the revolution and had to remain discreet during the so-called Terror regime which followed it. Later, and due to a sequence of political events, Cauchy had to leave his country several times.

Like Euler and Riemann, Cauchy received his education at home, from his father.[21] Laplace and Lagrange were family friends, and they encouraged Cauchy's father in the education of his son. Like Euler and Riemann, Cauchy was a devout Christian, and this had some effect on his relation with others, in particular, with Puiseux and Hermite who shared the same faith and with whom he had excellent relations, but others considered Cauchy's extreme religiousness problematic. Cauchy founded several charities, in particular the famous *Œuvre d'Orient*, which still operates today. Valson, in his *Vie et travaux du Baron Cauchy* [69], writes that "Cauchy was *par excellence* a man of charities. For them he never bargained his time and effort." The list of mathematicians who were openly hostile to him includes Poinsot, Abel, Poisson, Fourier and there are others. We also learn from his biographers that Cauchy was often sick and had a depressive character. In a letter to Holmboë, dated October 24, 1826, the young Abel, who was visiting Paris, writes: "Cauchy is crazy and it is impossible to deal with him." In the same letter, Abel writes about Cauchy that he is extremely Catholic and bigoted, which Abel finds strange for a mathematician. He adds about him: "he is the only one actively working on pure mathematics. Poisson, Fourier, Ampère, etc. work exclusively on magnetism and other parts of physics."[22] (The text of the letter is contained in [1] p. 45–49.)

Between 1816 and 1830, Cauchy lectured regularly on analysis in Paris, at the École Polytechnique, at the Collège de France and at the Faculté des Sciences. Like

[19] Lagrange defined complex functions using power series, but for him the notion of convergence was a secondary issue.

[20] Weierstrass, at that time was working in isolation, as a high-school teacher.

[21] In one of his writings, quoted by Bertrand [4] p. 187, Cauchy says: "If I know something, it is only through the care of my father." [Si je sais quelque chose, c'est uniquement à cause des soins que mon père a pris de moi.

[22] Picard, in his historical survey [52] (p. 15) describes this epoch, saying that one must not profess opinions which are too much systematic, on this parallel between pure theory and applications, like, he says, Laplace, Fourier, Poisson and the brilliant French school of mathematical physics of the beginning of the nineteenth century. "For them, he says, pure analysis was only the instrument, and Fourier, when he announced to the Academy of sciences the works of Jacobi, said that natural philosophy must be the main object of meditation of geometers." Picard says that such an exclusiveness would mean ignoring the philosophical and artistic value of mathematics.

Euler, Cauchy had a very close relation with his students. Valson, in [69] p. 253 of Vol. I, describes this relation:

> His position of professor did not offer only the satisfaction of that feeling of generous expansion which led him to be in intimate connection with the young men of the schools he liked, whom he admitted into his study like in a lounge, with whom he was used to converse informally as a friend rather than as a master.[23]

Unlike Riemann and Puiseux, Cauchy was very prolific in terms of volume of writings. In this respect, he was also close to Euler. His list of publications includes more than 800 articles, and his collected works edition consists of 28 large volumes, whose publication took almost a century (1882–1974). Cauchy used to publish quickly, and it is rather common knowledge that he made mistakes which for us today seem trivial. For instance, it is considered that he thought he proved that a function of several variables is continuous provided it is continuous separately in each variable (*Cours d'analyse* (1821), [10] p. 37–38; Œuvres, Série II, 3, p. 45–47).[24] Cauchy also "proved" that a convergent series of continuous functions can be integrated term by term ([12] p. 157; Œuvres, Série 2, t. 4, p. 237–238). Chebyshev, who had a lot of respect for French mathematicians, and in particular for Cauchy, pointed out some mistakes of the latter. In one of his first papers, written in 1844, whose title is *Note sur la convergence de la série de Taylor* [26], he writes, after he proves a theorem concerning Taylor expansions of functions: "This theorem is only a simple conclusion of the remarkable discoveries of Mr. Cauchy; but in part, it is contrary to the rule for convergence of series that was given by this famous geometer," and he states the rule[25]:

> If x denotes a real or imaginary variable, a real or complex function of x can be expanded into increasing powers of x provided the value of the modulus of x stays less than the smallest value for which the function or its derivative stop being finite and continuous.

Chebyshev declares that it seems that the inadequacy of this rule comes from the fact that Cauchy assumed that a definite integral may be expanded as a convergent series when the differential between the two limits of integration may be expanded as a convergent sequence. Chebyshev says that "this happens only in particular cases."

Some mathematicians argued however that Cauchy's so-called errors are in fact correct theorems when interpreted in the right setting, using his own concepts. For instance, and especially in the first period of his mathematical works, when Cauchy considers a functions, he means analytical expressions in the sense of Euler where the existence of a derivative follows from the assumptions; see e.g. [32, 45]. We recall

---

[23]Les fonctions de professeur ne lui offraient pas seulement la satisfaction de ce sentiment d'expansion généreuse qui le portait à se mettre en communication intime avec les jeunes gens des écoles qu'il aimait, qu'il admettait dans son cabinet de travail comme dans son salon, avec lesquels il s'entretenait familièrement en ami plutôt qu'en maître.

[24]The first definition of a continuous function of two variables, in the sense we intend it today, using a Euclidean norm on the plane, was given by Darboux in 1872, [28].

[25]Cauchy's *Exercices d'Analyse et de Physique Mathématique*, Tome I, p. 29.

incidentally that there are also gaps and mistakes in some of Riemann's works,[26] and there are also gaps and inconsistencies in the works of several other great mathematicians. Fortunately, mathematicians are not evaluated by their mistakes, but only by their achievements. Mentioning the mistakes does not undervalue their work. Sometimes, on the contrary, it shows how subtle was the new material they were working with, even though today their mistakes seem obvious. Picard, in one of his famous historical talks that he gave in the United States [52], says (p. 5) that "error is sometimes useful, and in epochs of real creativity, an incomplete or approximate truth may be more productive than the same truth accompanied by the necessary restrictions."[27] He gives the examples of Newton and Leibniz, saying that if they knew that there exist continuous function with no derivative, differential calculus would not have been born. Likewise, he says, the false ideas of Lagrange onTaylor expansions were extremely useful. One can find many examples in mathematics where gaps and mistakes led to important developments. Talking about Riemann, we mention that in 1892, Hadamard obtained the Grand Prix of the *Académie des Sciences* for an article on Riemann's zeta function [34], and that the subject of that contest was to fill in a gap in Riemann's work on that function.[28]

Cauchy submitted his first paper on definite integrals of a complex variable, the *Mémoire sur les intégrales définies*, in 1814 [8]. The paper is 188 pages long. Cauchy, at the time he wrote this paper, was 25, the same age at which Riemann submitted his doctoral dissertation, thirty-six years later. This was not Cauchy's first result. Cauchy found in 1805 (he was 16) a solution to a problem of Apollonius concerning a circle tangent to three circles. In 1811, he wrote two articles on polyhedra, generalizing Euler's formula, solving a rigidity problem that Lagrange asked him. Cauchy's name is now attached to this rigidity result. In Chap. 1 of the present volume [48], we comment on the work of Cauchy on polyhedra in relation with Euler's work. Legendre, in a later edition of his *Éléments de géometrie*, included the new proofs and the results of Cauchy on polyhedra. In 1812, Cauchy submitted a memoir on symmetric functions. In 1816, he won a prize for a contest set by the Paris Academy of Sciences concerning the propagation of water waves. The paper he presented for

[26]For instance, Riemann "proved" in a course he gave on complex variables that if a series of functions is convergent, then one can integrate it term by term; cf. [27] p. 13, where Riemann's proof is analyzed.

[27]On peut dire que l'erreur est quelquefois utile, et que, dans les époques vraiment créatrices, une vérité incomplète ou approchée peut être plus féconde que la vérité même accompagnée des restrictions nécessaires.

[28]The subject of the competition was: "The determination of the number of primes smaller than a given quantity" (which is the title of Riemann's article [61]), but in the comments following the problem, it was asked to fill the gaps in Riemann's work on the zeta function. The subject of the contest was chosen by Hermite, with his friend Stieltjes in mind, who had announced in 1885 a proof of the Riemann hypothesis. In the meantime Stieltjes withdrew his "proof," and the prize went to Hadamard [34]. See the details of this story in [46], and also in Chap. 8 of the present volume [49]. Hadamard's contribution followed from the work he did in his doctoral thesis, *Essai sur l'étude des fonctions données par leur développement de Taylor* (Essay on the study of the functions given by their Taylor expansion), devoted to complex function theory and written under Émile Picard and Jules Tannery.

that competition is 300 pages long. Cauchy did these works while he was working as an engineer, at the construction site of the port of Cherbourg (between 1810 and 1813). Bertrand [4] writes that during these three years where he worked as an engineer, "Cauchy reserved several hours every day to the study of Lagrange and Laplace, but original and new ideas were perturbing him at every moment. After they stole from him his sleep, formulae were haunting him on the construction site."[29] After these three years, Cauchy decided to stop working as an engineer and to come back to university.

The 1814 paper [8] of Cauchy is considered as one of his most important. It inaugurated a long series of papers on the theory of definite integrals and on complex functions, two subjects that accompanied Cauchy for the rest of his life. In this paper, Cauchy studies definite integrals in which the limits of integration are real numbers, but where the function that is integrated may be real or complex. Using the standard terminology of his epoch, Cauchy calls such a function "imaginary." Furthermore the function is allowed to become infinite at some points between the limits of integration. This led him to develop a notion of integrals he called *singular*. The 1814 memoir contains in an embryonic form the theory of line integrals in the complex plane which he developed later.

In two memoirs written in 1825, [13, 14], Cauchy initiated the theory of definite integrals taken between complex values. He proved that such an integral can take more than one value, depending on the choice of a path between the (now complex) numbers x_0 and X. Again, this happens in particular when the function f takes the value infinity at some points. Cauchy also gave a method of calculating the difference between two such values in terms of a finite number of "singular integrals." It might be important to note that one of the main reasons for which Cauchy studied integrals of functions of complex variables is that he knew that passing to complex values of the variable and using his residue calculus will also lead to results on definite integrals of functions of a real variable; see e.g. [17]. In fact, getting formulae for definite integrals was a fashionable subject at that time.

It is not possible to mention here the totality of Cauchy's later papers and books on functions of a complex variable (there are too many), and we shall say only a few words on some of them. For a comprehensive exposition of Cauchy's work on functions of a complex variable, the reader may refer to [7, 33]. We give though a list of a few important concepts in the theory of functions that we owe to Cauchy. Our list is very far from being exhaustive, but some of the concepts we present here were crucial in the work of Riemann.

(1) The notion of *path* ("chemin"), in relation with functions of a complex variable, and the notion of path integral.

[29] Cauchy à Cherbourg réservait des heures réglées pour l'étude de Lagrange et de Laplace; mais les idées originales et nouvelles le troublaient à toute heure. Après avoir usurpé sur son sommeil, les formules le poursuivaient sur les chantiers.

(2) Rigorous definitions of limits,[30] of integrals (as limits of sums) and of convergence of series. In the introduction to his *Cours d'analyse de l'École Royale Polytechnique* [10], written in 1821, Cauchy writes:

> [...] Thus, before carrying out the summation of any series, I was led to examine in what cases these series may be summed, or, in other words, what are the conditions of their convergence. And in this respect, I established general rules which I think are worth of some attention.[31]

In particular we owe to Cauchy the espsilon-delta and the epsilon-N definitions of limits and convergence[32] as well as the notion of Cauchy sequence. In this context, Cauchy is considered as one of the main founders of the rigorous methods in analysis as we conceive them today, for what concerns convergence, infinite series, integration, etc.

(3) The notion of circle of convergence of a power series. One might note also that power series were studied by Euler and Lagrange long before Cauchy (and, in fact, the notion of power series, in Lagrange's sense was part of his definition of a function), but that it was Cauchy who considered that a power series makes sense only if it is convergent.

(4) A theorem for local existence results for differential equations (known today as Cauchy's theorem).

(5) The definition of a holomorphic function through the partial differential equations which became known as the Cauchy–Riemann equations.

In 1851, Cauchy discovered the notion of a derivative independent of direction and he called a function with such a property "fonction monogène."[33] He showed that the real and imaginary parts of such a function must satisfy the Cauchy–Riemann equations; cf. his papers [19, 22].[34] This was the same year (1851) that Riemann defined analytic functions using the Cauchy–Riemann equations. In fact, starting from 1831 (see [15]), Cauchy was interested in the question of when a function can be developed as a convergent power series. He introduced, rather unsuccessfully, several conditions, including the fact that the function has

---

[30]Cauchy had rigorous definitions of limit and continuity, although, in some sense, it is difficult to have such rigorous definitions without a rigorous development of the notion of real number, which was done much later.

[31][...] Ainsi, avant d'effectuer la sommation d'acucune série, j'ai dû examiner dans quels cas les séries peuvent être sommées, ou, en d'autres termes, quelles sont les conditions de leur convergence; et j'ai, à ce sujet, établi des règles générales qui me paraissent mériter quelque attention.

[32]On this subject, besides Cauchy, one has to mention the work of Bolzano, done around the same period.

[33]The Greek roots of the French word "monogène" used by Cauchy reflect the fact that this function has a unique derivative. The Greek word "monogenes" has a theological connotation. It is used in the Septuagint translation of the Bible (Hebrews 11–17), for Isaac as Abraham's "only begotten son" and in the Gospel of John (20–31) for Jesus as the "only begotten son" of God.

[34]It is interesting to note that in his doctoral dissertation, Riemann includes in the *definition* of a function of a complex variable the fact of having a derivative independent of direction. The fact that every complex function satisfies the Cauchy–Riemann equations becomes a theorem. Cf. §4 of Riemann's dissertation.

a continuous derivative. It was only at the beginning of the 1850s that he came up with the condition saying that the function has a (unique) complex derivative, which is equivalent to conformality. These hesitations of Cauchy are analyzed in the thesis [50].

It is important to emphasize that even though the Cauchy–Riemann equations were known before Cauchy and Riemann,[35] it is thanks to these two authors that these equations became at the forefront of the theory of functions of one complex variable, and at the same time made the connection between analysis and mathematical physics.

(6) The notion of period of a definite integral [18].
(7) The Cauchy integral formula and the calculus of residues which became known as the Cauchy formula. (An early version appears in his paper [9]).
(8) The notion of *monodromy* associated with a function on a given domain which attains the same value independently of the path chosen in that domain. Cauchy made the relation between this notion and that of being *monogenic* (having a derivative independent of direction) [23].

Cauchy, like Riemann, was also a physicist. He made important contributions to hydrodynamics, elasticity and astronomy. His name is also attached to a hypersurface in spacetime which intersects every inextensible causal curve exactly once. We mention this fact because it is related to relativity theory, a field on which the ideas of Riemann have a large impact and which is the subject of the last three chapters of the present volume.

We review now a major treatise of Cauchy on analysis, his *Cours d'analyse de l'École Royale Polytechnique* [10], written in 1821. This treatise was conceived as a textbook for the first-year students of the École Polytechnique, accompanying Cauchy's lectures whose aim was to present the bases of analysis in the most possible rigorous way.[36] An English translation of Cauchy's *Cours* is available (see [11]). In

---

[35]The Cauchy–Riemann equations are, in themselves, much older than Cauchy and Riemann. They already occur in d'Alembert's works on fluid dynamics, *Essai d'une nouvelle théorie de la résistance des fluides*, Paris, 1752. Klein, in his *Development of mathematics in the 19th century* ([42] p. 239) writes that "perhaps they occur even earlier."

[36]There is a long French tradition of *Cours d'Analyse* for the students of the École Polytechnique. One may mention Lagrange's *Cours* whose complete title is *Théorie des fonctions analytiques, contenant les principes du calcul différentiel, dégagés de toute considération d'infiniment petits ou d'évanouissans, de limites ou de fluxions, et réduits à l'analyse algébrique des quantités finies* (Theory of analytic functions containing the principles of differential calculus, without any consideration of infinitesimal or vanishing quantities, of limits or of fluxions, and reduced to the algebraic analysis of finite quantities), written in 1797, three years after the foundation of the *École*. Cauchy started to teach his course two years after Lagrange's death. One should also mention the *Résumé des leçons données à l'École Royale Polytechnique sur le calcul infinitésimal* (Summary of lectures on infinitesimal calculus given at the *École Royale Polytechnique*) (1823), a treatise which Cauchy wrote for the use of his students, after he modified his lectures because of a change in the official program. One may also mention the *Résumé des cours d'analyse* by Charles Hermite, in two parts (1867–1868 and 1868–1869), the *Cours d'analyse de l'École Polytechnique* by Charles Sturm, the *Cours d'Analyse* by Jacques Hadamard, the more recent *Cours d'analyse* by Laurent Schwartz (1967), and there are several others.

fact, Cauchy published only the first part of his *Cours*, to which he gave the name *Analyse algébrique*. It is conceivable that the sequel of this treatise never appeared because of a change in the curriculum, after which Cauchy published his *Résumé des leçons* [12].

In his *Cours*, Cauchy starts with the notion of variable and constant, then he considers infinitely small quantities, the various kinds of functions of a real or complex variable, logarithms, powers, trigonometric functions, limits of functions and of sequences, convergent and divergent series, methods of solving equations, decomposition of rational functions, continuity, convergence and divergence criteria and many other items that are still taught to students today. In Chap. 1, Sect. 3, Cauchy discusses functions which assign to a given value of the variable more than one value. An example is when the function is defined by a limiting procedure, and the limits are not unique. He calls such values *singular values* of the function. He says that such values can be obtained when the variable takes the value infinity. He writes that "the search of the singular values of functions is one of the most important and delicate questions in analysis."

Between the years 1826 and 1830, Cauchy published, on a monthly basis, a series of papers in volumes which he called *Exercices de Mathématiques*. Between 1840 and 1847, he published another set of four volumes, which he called *Exercices d'analyse et de physique mathématique*. The *Exercices* appeared in the form of a periodical of which Cauchy was the unique author. In several papers published in the *Exercices*, Cauchy rewrites, corrects, improves previous results.[37] In a report that Bertrand wrote on Cauchy's biography by Valson [5], he says (p. 110):

> The genius of Cauchy is worthy of all our respect. But why should we refrain from recalling that the great profusion of his works, which often reduces their precision, has more than one time hidden their force? The dangerous easiness of an immediate publicity was for Cauchy a compelling temptation, and often, a pitfall. His sprit, always in motion, used to bring each week to the Academy works that were barely sketched, projects of memoirs and attempts

[37]In his Éloge of Cauchy, Bertrand writes ([4] p. 114) about the *Exercices*: No mathematical publication, with whatever excellency and number of collaborators, may compete with the eight volumes of the *Exercices*. Avidly expected in their novelty, they are nowadays classical among the masters. No page of the *Exercices* is unknown to any geometer. When Cauchy had to refer to himself, he gladly referred to himself as the author of the *Exercices*. This title was sufficient. If some geometer today dared to publish an *Exercices de mathématiques*, we would be surprised by such a boldness, in the same way, and I am not exaggerating at all, as if a poet, whose name is not Lamartine or Victor Hugo, had dared to publish some *Orientales* or *Méditations poétiques* [Aucune publication mathématique, quelle que fût l'excellence et le nombre de ses collaborateurs, ne pourrait rivaliser avec les huit volumes des *Excercices*. Avidement attendus dans leur nouveauté, ils sont aujourd'hui classiques parmi les maîtres ; aucune page des *Excercices* n'est inconnue à aucun géomètre. Lorsque Cauchy avait à se citer lui-même, il se nommait volontiers: l'auteur des *Exercices*. Ce titre suffisait. Si un géomètre osait aujourd'hui publier des *Exercices de mathématiques*, on s'étonnerait d'une telle audace, tout autant, je n'exagère rien, que si un poète, sans se nommer Lamartine ou Victor Hugo, osait publier des *Orientales* ou des *Méditations poétiques*]. We note that the name *Exercices* for a publication was already used by Legendre, who published a famous multi-volume *Exercices de calcul intégral* (1811–1817) [43], a treatise whose main subject is elliptic integrals and their applications to geometry and analysis, which incidentally was one of the favorite subjects of research of Riemann.

> which were sometimes unsuccessful. Even when a brilliant discovery came to crown his efforts, he used to force his reader to follow him in ways that were often infertile and which were tested and abandoned alternately without any prior notice. Let us take as an example the theory of substitutions and the number of values of a function. To whom does it owe its greatest advances? To Cauchy, without any doubt, and it is true that his name, in the history of this beautiful question, rises to a great hight above all the others. But on that theory, which owes him a lot, Cauchy composed more than twenty memoirs. Two among them are masterpieces. What can we say of the eighteen others? Nothing, except for the fact that their author is searching a new way, follows it for some time, catches a glimpse of light, tries hard pointlessly to attain it, and at the end quits, without showing any embarrassment, the avenues of the edifice which he renounces to build.[38]

We quote now Bertrand, from his *Éloge* of Cauchy ([4] p. 101):

> He was exploring new regions, whose hights were known, but nobody was able to guess the extent, the consistency, and their inexhaustible fertility.[39]

An *Éloge funèbre* is an homily in which the departed person is praised for his life and achievements, and it is natural to find in Bertrand's Éloge such laudatory words. Other people, historians of mathematics, made also very laudatory statements. We quote Bruno Belhoste, from the end of his exquisite biography of Cauchy [3]:

> Thus ended the life of the greatest French mathematician of his times scarcely two years had passed since Gauss had died in Germany. A new age was now opening in the long history of mathematics, an age in which the leading figures in the mathematical sciences would be Germans. Between 1854 and 1859, Riemann, Weierstrass, and Kronecker came onto the scene on the other side of the Rhine. Meanwhile, however, in France, there was a blossoming of works on Cauchy's theory.

Laugwitz notes in his article [44] p. 80 that Cauchy's *Cours d'analyse*, remained for a long time the only treatise containing a complete theory of real and complex power series. He also reports that according to the Göttingen library borrowing list, Riemann, during the years 1846/47, while he was a student, borrowed this book, together with the *Exercices de mathématiques* and other works of Cauchy. Furthermore, in the draft for the defense of his doctoral thesis, Riemann refers to the works

[38] Le génie de Cauchy est digne de tous nos respects; mais pourquoi d'abstenir de rappeler que la trop grande abondance de ses travaux, en diminuant souvent leur précision, en a plus d'une fois caché la force ? La dangereuse facilité d'une publicité immédiate a été pour Cauchy une tentation irrésistible et souvent un écueil. Son esprit, toujours en mouvement, apportait chaque semaine à l'Académie des travaux à peine ébauchés, des projets de Mémoire et des tentatives parfois infructueuses, et lors même qu'une brillante découverte devrait couronner ses efforts, il forçait le lecteur à le suivre dans les voies souvent stériles essayées et abandonnées tour à tour sans que rien vint l'en avertir. Prenons pour exemple la théorie des substitutions et du nombre de valeurs d'une fonction. À qui doit-elle ses plus grands progrès ? à Cauchy sans aucun doute, et il est véritable que son nom, dans l'histoire de la belle question, s'élève à une grande hauteur au-dessus de tous les autres. Mais, sur cette théorie qui lui doit tant, Cauchy a composé plus de vingt mémoires. Deux d'entre eux sont des chefs d'œuvre. Que dire des dix-huit autres ? rien, sinon que le lecteur y cherche une voie nouvelle, la suit quelque temps, entrevoit la lumière, s'efforce inutilement de l'atteindre et quitte enfin, sans marquer aucun embarras, les avenues de l'édifice qu'il renonce à construire.

[39] Il explorait des régions nouvelles, on savait à quelle hauteur : nul n'en pouvait deviner l'étendue, la consistance et l'inépuisable fécondité.

of Cauchy concerning the definition of an analytic function [47]. Neuenschwander adds the following:

> Riemann was suitable, as no other German mathematician was, to effect the first synthesis of the "French" and the "German" approaches in function theory. In his introductory lectures on complex function theory (cf. [65–67]; 1861), Riemann dealt with the Cauchy Integral Formulae, the operations on infinite series, the power series expansion, the Laurent series, the analytic continuation by power series, the argument principle, the product representation of an entire function with arbitrarily prescribed zeros, the evaluation of definite integrals by residues, etc., besides the subjects known from his published papers.

Riemann does not mention Cauchy in his doctoral dissertation [58]. It is not sure that Riemann, even though he borrowed from the library Cauchy's work, really read them. It is possible that he only skimmed them and reconstructed the theory on his own. Riemann however mentions Cauchy's name twice in his paper on Abelian functions [60], at the end of § 2 and in § 6, for a result on the expansion of a function in power series, but he adds, both times, that the result may also be proved using Fourier series. Riemann also mentions Cauchy's work three times in his Habilitation memoir on trigonometric functions [59]. The first time is in § 2, in the historical part of his paper, where Riemann quotes a result where Cauchy was mistaken, and which he says, can be proved using Fourier series.[40] The second time is in § 3 of this memoir, where Riemann says that Cauchy's attempt to prove the convergence of a certain series is unsuccessful. The third time is in § 4, where Riemann introduces his famous theory of integration. He criticizes again Cauchy's attempts to develop a general concept of definite integral. It is possible that Riemann was disturbed by Cauchy's mistakes and for this reason he was not so much inclined to quote him. Cauchy was also hardly quoted by the Germans during the same period. On the contrary, Cauchy was very generous in quoting others. Freudenthal writes his biography [31]:

> Of all the mathematicians of his period he is the most careful in quoting others. His reports on his own discoveries have a remarkably naïve freshness because he never forgot to sum up what he owed to others. If Cauchy were found in error, he candidly admitted his mistake.

We elaborate on the relation between the way Riemann's work on Riemann surfaces was received by the French school in Chap. 8 of the present volume [49]. Hermite, in the introduction to the treatise *Théorie des fonctions algébriques et de leurs intégrales* (Theory of algebraic functions and their integrals) by Appell and Goursat [2], published in 1895, makes a summary of the influence of the ideas of Puiseux, 44 years after their appearence. He writes the following:

> The Memoir on algebraic functions by Puiseux, published in 1854,[41] opened the research ground which led to the great mathematical discoveries of our epoch. These discoveries gave the science of calculus necessary and fruitful principles which were missing until that time. They replaced the notion of function, which was still obscure and incomplete, by a precise conception which transformed analysis by giving it a new basis. Puiseux is the first who shed complete light onto the insufficiency and the defect of the point of view where we represent, in the same way as polynomials and rational fractions, the algebraic

[40] Riemann adds that it was Dirichlet who showed Cauchy's mistake.

[41] The year should be 1851.

> irrationals and all the quantities in infinite number which have their origin in integral calculus. Following the path of Cauchy, considering the succession of imaginary values, the paths described simultaneously by the variable and the roots of an equation, the eminent geometer highlighted, in its essential character, their analytic nature. He discovered the role of critical points, and the circumstances of the exchange of the initial values of the roots, when the variable returns to its starting point, describing a closed loop containing one or several of these roots. He resumed the consequences of these results in the study of the integrals of algebraic differentials. He noticed that the various integration paths give rise to multiple determinations, which led him to the origin – which till then was completely hidden – of the periodicity of circular functions, of elliptic functions, and of the multi-variable transcendants defined by Jacobi as inverse functions of hyperelliptic integrals.[42]

5 Uniformization Again

In the previous sections, we used the word "uniformization" in the original sense intended by Riemann, as finding a ground space on which a multi-valued function defined by an algebraic equation becomes uniform (that is, single-valued). We showed that the question of uniformization, in this sense, was a major factor in the development of the theory of Riemann surfaces. Later on, the word uniformization acquired several new meanings, albeit variations on the original one. One of the alternative formulations of the uniformization problem is the following: Given an algebraic equation $f(z, w) = 0$ as in Sect. 2 above, to find two single-valued functions $z(t)$ and $w(t)$ of one variable t such that the equation $f(z(t), w(t)) = 0$ is satisfied. This is the form in which Poincaré used this word. Besides his formulation of the problem, Poincaré introduced automorphic functions in the study of uniformization. In an 1882 Comptes Rendus note [53], he announces a result saying that for any algebraic curve of genus ≥ 2 defined by an algebraic equation $f(z, w) = 0$ there exists two Fuchsian functions $F(u)$ and $G(u)$ satisfying $f(F(u), G(u)) = 0$. One

[42]Le mémoire de Puiseux sur les fonctions algébriques, publié en 1854, a ouvert le champ de recherches qui a conduit aux grandes découvertes mathématiques de notre époque. Ces découvertes ont donné à la science du calcul des principes nécessaires et féconds qui, jusqu'alors, lui avaient manqué; elles ont remplacé la notion de fonction, restée obscure et incomplète, par une conception précise qui a transformé l'analyse en lui donnant de nouvelles bases. Puiseux a le premier mis en lumière l'insuffisance et le défaut de ce point de vue où l'on se représente, à l'image des polynômes et des fractions rationnelles, les irrationnelles algébriques et toutes les quantités en nombre infini qui ont leur origine dans le calcul intégral. En suivant la voie de Cauchy, en considérant la succession des valeurs imaginaires, les chemins décrits simultanément par la variable et les racines d'une équation, l'éminent géomètre a fait connaître, dans ses caractères essentiels, leur nature analytique. Il a découvert le rôle des points critiques, et les circonstances de l'échange des valeurs initiales des racines, lorsque la variable revient à son point de départ, en décrivant un contour fermé comprenant un ou plusieurs de ces points. Il a poursuivi les conséquences de ces résultats dans l'étude des intégrales de différentielles algébriques. Il a reconnu que les divers chemins d'intégration donnent naissance à des déterminations multiples, ce qui l'a conduit à l'origine, jusqu'alors restée entièrement cachée, de la périodicité des fonctions circulaires, des fonctions elliptiques, des transcendantes à plusieurs variables définies par Jacobi comme fonctions inverses des intégrales hyperelliptiques.

year later, in his paper [54], he stated a general uniformization theorem, in which the reference to algebraic functions disappeared:

> Let y be an analytic function of x, which is non-uniform. We can always find a variable z such that x and y are uniform functions of z.[43]

This is the general form of the uniformization problem. It took several years for Poincaré to provide a proof of this theorem. The attempts to prove this general statement made the subject of uniformization, for several decades, a vast subject of research. Whereas from the French side only one name comes to the forefront: Poincaré, on the German side, a multitude of prominent mathematicians were involved in this uniformization program (Christoffel, Hilbert, Klein, Koebe, Osgood, Schwarz, and there are others). It is not our aim here to enter into this immense research ground, but we would like to recall Hilbert's Problem 22, a problem concerning specifically this general uniformization. This is one of the problems that Hilbert presented in his lecture, delivered on August 1900, at the Second International Congress of Mathematicians held in Paris. The lecture is entitled *The future problems of mathematics*, and the problems he presented became a guide for a substantial part of the mathematical research that was conducted in the twentieth century. Several slightly different versions of Hilbert's problems were published by Hilbert after that lecture, in various journals and in several languages. Moreover, the number of problems is not the same in all these versions. The paper published in the *Bulletin of the American Mathematical Society* ([37], 1901) contains a commented set of twenty-three problems. Problem 22 is entitled *Uniformization of analytic relations by means of automorphic functions.* Hilbert presents the problem completely in the tradition of Poincaré, as the one of reducing a two-variable relation to a one-variable one, by introducing automorphic forms. In his statement of and his comments on the problem, Hilbert mentions several times Poincaré and no other mathematician, except for Picard, whom he mentions at the end of his text, when he suggests a more general uniformization problem, involving algebraic (and, more generally, analytic) equations of three or more variables. Let us review precisely Hilbert's statement:

> As Poincaré was the first to prove, it is always possible to reduce any algebraic relation between two variables by the use of automorphic functions of one variable. That is, if any algebraic equation in two variables be given there can always be found for these variables two such single valued automorphic functions of a single variable that their substitution renders the given algebraic equation an identity. The generalization of this fundamental theorem to any analytic non-algebraic relations whatever between two variables has likewise been attempted with success by Poincaré,[44] though by a way entirely different from that which served him in the special problem first mentioned. From Poincaré's proof of the possibility of reducing to uniformity an arbitrary analytic relation between two variables, however, it does not become apparent whether the resolving functions can be determined to meet certain additional conditions. Namely, it is not shown whether the two single valued functions of the one new variable can be so chosen that, while this variable traverses the *regular* domain of these functions, the totality of all regular points of the given analytic field are actually reached

[43]Soit y une fonction analytique de x, non uniforme. On peut toujours trouver une variable z telle que x et y soient fonctions uniformes de z.

[44][Hilbert's footnote:] Bull. Soc. Math. de France, vol. 11 (1883).

> and represented. On the contrary it seems to be the case, from Poincaré's investigations, that there are beside the branch points certain others, in general infinitely many other discrete exceptional points of the analytic fields, that can be reached only by making the new variable approach certain limiting points of the function. *In view of the fundamental importance of Poincaré's formulation of the question it seems to me that an elucidation and resolution of this difficulty is extremely desirable.*
>
> In conjunction with this problem comes up the problem of reducing to uniformity an algebraic or any other analytic relation among three or more complex variables – a problem which is known to be solvable in many particular cases. Toward the solution of this the recent investigations of Picard on algebraic functions of two variables are to be regarded as welcome and important preliminary studies.

The uniformization problem in its general form was solved eventually by Poincaré and Koebe. There are several modern books and articles that report on this problem and its solution. The interested reader should go through the original papers, guided by the modern reports.

Acknowledgements I would like to thank Vincent Alberge, Ken'ichi Ohshika and François Laudenbach for their comments on a preliminary version of this chapter.

References

1. N.-H. Abel, Niels-Henrik Abel : Mémorial publié à l'occasion du centenaire de sa naissance, ed. E. Holst, C, Størmer, L. Sylow (Kristiania, Brøgger, 1902)
2. P. Appell, É. Goursat, *Théorie des fonctions algébriques et de leurs intégrales : Étude des fonctions analytiques sur une surface de Riemann* (Gauthier-Villars, Paris, 1895)
3. B. Belhoste, Cauchy, Un mathématicien légitimiste du XIXe siècle. Belin, Paris. English translation: *Augustin-Louis Cauchy: A Biography* (Springer, New York, 1991), pp. 1789–1867
4. J. Bertrand, Éloge de Augustin Cauchy, Académie des Sciences de Paris, lu à la séance publique du lundi 10 janvier (1898)
5. J. Bertrand, Rapport sur "La vie et les travaux du Baron Cauchy par C.-A. Valson," Bulletin des sciences mathématiques et astronomiques **1**, 105–117 (1870)
6. J. Bertrand, Éloge de M. Victor Puiseux, lu dans la séance publique annuelle de l'Académie des sciences du 5 mai 1884, Bulletin des sciences mathématiques et astronomiques 2e série, tome **8**(1), 227–234 (1884)
7. U. Bottazini, J. Gray, *Hidden Harmony—Geometric Fantasies: The Rise of Complex Function Theory* (Springer, New York, Heidelberg, Dordrecht, London, 2013)
8. A.-L. Cauchy, Mémoire sur les intégrales définies (1814), Mémoires présentés par divers savants à l'Académie royale des sciences de l'Institut de France 1, 611–799 (1827). Œuvres Complètes, série 1, tome I, 319–506
9. A.-L. Cauchy, Mémoire sur l'intégration des équations linéaires aux différentielles partielles et à coefficients constants. J. Ecole Polytechnique, 12, 511–592 (1823). Œuvres Complètes, série 2, tome I, 275–357
10. A.-L. Cauchy, Cours d'analyse de l'École Royale Polytechnique, 1re partie. Analyse algébrique (Imprimerie royale, Paris, 1821). Œuvres complètes, série 2, tome III
11. A.-L. Cauchy, *Cauchy's Cours d'analyse: An Annotated Translation of [10]*, ed. By R.E. Bradley, C.E. Sandifer Sources and Studies in the History of Mathematics and Physical Sciences (Springer, Berlin, 2009)
12. A.-L. Cauchy, Résumé des leçons données à l'école Royale Polytechnique sur le calcul infinitésimal, tome I, Paris, 1923. Œuvres Complètes, série 2, tome IV, 5–261

13. A.-L. Cauchy, Mémoire sur les intégrales définies prises entre des limites imaginaires (Imprimerie Royale, Paris, 1825). Œuvres complètes, série 2, tome XV, pp. 41–89
14. A.-L. Cauchy, Mémoire sur les intégrales définies prises entre des limites imaginaires, Bulletin des sciences mathématiques, astronomiques, physiques et chimiques (Bulletin de Férussac), tome III (1825), pp. 214–221. Œuvres complètes, série 2, tome II, pp. 57–65
15. A.-L. Cauchy, *Sur la mécanique céleste et sur un nouveau calcul qui s'applique à un grand nombre de questions diverses, Mémoire presented to the Academy of Turin.* Œuvres Complètes, série 2, tome II, pp. 158–168 (1831)
16. A.-L. Cauchy, Sur le nombre de valeurs égales ou inégales que peut acquérir une fonction de n variables indépendantes, quand on permute ces variables entre elles d'une manière quelconque, Comptes Rendus Acad. Sc. Paris, tome XVIII (15 septembre 1845), 593–607. Œuvres complètes, série 1, tome IX, pp. 277–293
17. A.-L. Cauchy, Mémoire sur les intégrales dans lesquelles la fonction sous le signe $\int$ change brusquement de valeur, Comptes Rendus Acad. Sc. Paris, tome XXIII (1846), pp. 557–563. série 1, tome X, pp. 135–143
18. A.-L. Cauchy, Considérations nouvelles sur les intégrales définies qui s'étendent à tous les points d'une courbe fermée, et sur celles qui sont prises entre des limites imaginaires. Comptes Rendus Acad. Sc. Paris **23**, 689–704 (1846), Œuvres complètes, série 1, tome X, pp. 153–168
19. A.-L. Cauchy, Sur les fonctions de variables imaginaires, Comptes Rendus Acad. Sc. Paris **32**, 207–215 (1851). Œuvres Complètes, série I tome XI, pp. 301–304
20. A.-L. Cauchy, Rapport sur un Mémoire présenté à l'Académie par M. Puiseux et intitulé: Recherches sur les fonctions algébriques. C. R., t. XXXII (25 février 1851), p. 276. Œuvres complètes, série 1, tome XI, pp. 325–335
21. A.-L. Cauchy, Rapport sur un Mémoire présenté à l'Académie par M. Puiseux, et intitulé: Nouvelles recherches sur les fonctions algébriques. C. R., t. XXXII (7 avril 1851), p. 493. Œuvres complètes, série 1, tome XI, pp. 380–382
22. A.-L. Cauchy, Sur les fonctions monotypiques et monogènes, Comptes Rendus Acad. Sc. Paris **32**, 484–487 (1851). Œuvres Complètes, série 1, tome XI, 376–380
23. A.-L. Cauchy, Mémoire sur l'application du calcul infinitésimal à la détermination des fonctions implicites. Comptes Rendus Acad. Sc. Paris **34**, 265–273 (1852). Œuvres Complètes, série 1, tome XI, 406–415
24. A. Cayley, On the theory of groups, as depending on the symbolic equation $\theta^n = 1$, Philosophical Magazine 7, 40–47 (1854). The Collected Mathematical Papers of Arthur Cayley, Cambridge University Press, Cambridge **2**, 123–130 (1889)
25. C. Charitos, I. Papadoperakis, On a theorem of Euler on mappings from the sphere to the plane. in *Essays in non-Euclidean geometry* ed. by V. Alberge and A. Papadopoulos (European Mathematical Society, Zurich, to appear in 2017)
26. P.L. Chebyshev, Note sur la convergence de la série de Taylor, Jour. Für die Reine and ungewandte Mathematik, B. **28**, 279–283 (1844). Reprinted in Œuvres de P. L. Tchebycheff, ed. A. Markoff and A. Sonin, Tome I, Saint-Pétersbourg (1899), pp. 9–14
27. P. Dugac, Problèmes de l'histoire de l'analyse mathématique au XIXème siècle. Cas de Karl Weierstrass et de Richard Dedekind, Historia Mathematica **3**(1), 5–19 (1976)
28. G. Darboux, Sur un théorème relatif à la continuité des fonctions. Bull. Sci. math. **3**, 307–313 (1872)
29. L. Euler, *Introductio in analysin infinitorum*, first edition 1748. *Opera omnia*, Series 1, vol. VIII
30. L. Euler, De repraesentatione superficiei sphaericae super plano, Acta Academiae Scientarum Imperialis Petropolitinae 1777, 1778, pp. 107–132, Opera Omnia: Series 1, Vol. 28, pp. 248–275
31. H. Freudenthal, *Article "Cauchy," Dictionary of Scientific Biography*, vol. III (Scribner, New York, 1971), pp. 131–148
32. E. Giusti, Gli "errori" di Cauchy e i fondamenti dell'analisi. Boll. Stor. Sci. Mat. **4**(2), 24–54 (1984)

33. J. Gray, *The Real and the Complex: A History of Analysis in the 19th Century* (Springer, Cham, Heidelberg, New York, Dordrecht, London, 2015)
34. J. Hadamard, Étude sur les propriétés des fonctions entières et en particulier d'une fonction considérée par Riemann. Mémoire couronné en 1892 par l'Académie : Grand Prix des Sciences Mathématiques, J. Math. Pures Appl. Série 4, **9**, 171–215 (1893)
35. Ch. Hermite, Sur la théorie des fonctions elliptiques, Comptes Rendus Acad. Sc. Paris **29**, 594 (1849). Œuvres, t. I, p. 74
36. Ch. Hermite, Sur les fonctions algébriques, Comptes Rendus Acad. Sc. Paris, tome XXXII, 458–461 (1851)
37. D. Hilbert, Mathematische Probleme, *Göttinger Nachrichten*, 1900, pp. 253–297, reprinted in *Archiv der Mathematik und Physik*, 3d. ser., vol. 1 (1901) pp. 44–63 and 213–237. English version, "Mathematical problems," reprinted also in the Bull. Amer. Math. Soc. (N.S.) **37**(4), 407–436 (2000)
38. C. Jordan, Cours d'analyse de l'École Polytechnique (Gauthier-Villars, Paris, 2d ed. 1893, 3d. ed. 1909)
39. C. Jordan, Traité des substitutions et des équations algébriques (Gauthier-Villars, Paris, 1878). New edition Gauthier-Villars, Paris 1957. New edition, Gabay, Paris 1989
40. F. Klein, *Über Riemanns Theorie der algebraischen Funktionen und ihrer Integrale*, Teubner, Leipzig, 1882. English translation: *On Riemann's theory of algebraic functions and their integrals; a supplement to the usual treatises*. Translated from the German by Frances Hardcastle, Macmillan and Bowes, Cambridge, 1893. Reprint: New York, Dover Publications, 1963
41. F. Klein, Riemann und seine Bedeutung für die Entwicklung der modernen Mathematik, in F. Klein, Gesammelte mathematische Abhandlungen, Berlin: Springer-Verlag, 1923 (reprinted 1973), 3: 482–497. English translation: Riemann and his significance for the development of modern mathematics. Bull. AMS **1**(5), 165–180 (1895)
42. F. Klein, Vorlesungen über die Entwicklung der Mathematik im 19. 2 volumes, Julius Springer, Berlin, 1926. English translation by M. Ackermann: Development of mathematics in the 19th century (Math. Sci. Press, Brookline, Mass.,1979)
43. A.-M. Legendre, Exercices de calcul intégral sue divers ordres de transcendantes et sur les quadratures, 3 volumes (Courcier, Paris), pp. 1811–1816
44. D. Laugwitz, Bernhard Riemann (1826–1866), Turning points in the conception of mathematics. translated By A. Schenitzer (Birkhäuser, 1999). Reprint, 2008
45. D. Laugwitz, Infinitely small quantities in Cauchy's textbooks. Hist. Math. **14**(3), 258–274 (1987)
46. V. Maz'ya, T. Shpaposhnikova, Jacques Hadamard, A universal mathematician, History of Mathematics 14, AMS & LMS, 1998
47. E. Neuenschwander, Studies in the history of complex function theory. II: Interactions among the French school, Riemann, and Weierstrass. Bull. Am. Math. Soc. New Ser. **5**, 87–105 (1981)
48. A. Papadopoulos, Looking backward: Euler and Riemann, this volume, pp. 1–81
49. A. Papadopoulos, Riemann surfaces: reception by the French school, this volume, Chap. 8, pp. 237–286
50. J. Pfeiffer, *Les premiers exposés globaux de la théorie des fonctions de Cauchy, thèse de doctorat de 3e cycle* (EHESS, Paris, 1978)
51. E. Picard, La vie et l'œuvre de Jules Tannery, Académie des Sciences, lecture faite dans la séance publique annuelle du 14 décembre 1925
52. É. Picard, *Sur le Développement de l'Analyse et ses Rapports avec diverses Sciences: Conférences faites en Amérique* (Gauthier-Villars, Paris, 1905)
53. H. Poincaré, Sur les fonctions fuchsiennes, C. R. Acad. Sci. Paris, 94, 1038–1040. Œuvres II, 41–43 (1882)
54. H. Poincaré, Sur un théorème de la théorie générale des fonctions, Bull. Soc. Math. Fr. **11**, 112–125 (1883). Œuvres IV, pp. 57–69
55. V.-A. Puiseux, Recherches sur les fonctions algébriques. Journal de Mathématiques Pures et Appliquées **15**, 365–480 (1850)

56. V.-A. Puiseux, Nouvelles recherches sur les fonctions algébriques. Journal de Mathématiques Pures et Appliquées **16**, 228–240 (1851)
57. V.-A. Puiseux, Mémoire sur l'accélération séculaire du mouvement de la lune Paris, Imprimerie Nationale, 1873, 129 p
58. B. Riemann, Grundlagen für eine allgemeine Theorie der Functionen einer veränderlichen complexen Grösse, (Göttingen, 1851), [63] pp. 3–48
59. B. Riemann, Über die Darstellbarkeit einer Function durch eine trigonometrische Reihe. Göttingen, Dieterich, 1867
60. B. Riemann, Theorie der Abel'schen Functionen. Journal für die reine und angewandte Mathematik, 54 (1857), 115–155. Reprinted in his *Gesammelte mathematische Werke* [66], pp. 88–144
61. B. Riemann, Über die Anzahl der Primzahlen unter einer gegebenen Grösse, Monatsberichte der Berliner Akademie, November 1859, 671–680, *Gesammelte mathematische Werke* [63], pp. 145–153
62. B. Riemann, Beiträge zur Theorie der durch die Gauss'sche Reihe $F(\alpha, \beta, \gamma, x)$ darstellbaren Functionen. Aus dem siebenten Band der Abhandlungen der Königlichen Gesellschaft der Wissenschaften zu Göttingen (1857), pp. 3–32
63. B. Riemann, Gesammelte mathematische Werke, wissenschaftlicher Nachlass und Nachträge. Nach der Ausgabe von Heinrich Weber und Richard Dedekind neu hrsgg. von Raghavan Narasimhan. (Collected mathematical works, scientific posthumous works and supplements. According to the edition by H. Weber and R. Dedekind newly edited by R. Narasimhan). Teubner Verlagsgesellschaft, Leipzig, 1862 (Springer, Berlin, 1990)
64. B. Riemann, Collected works translated into French: Œuvres mathématiques de Riemann, traduites par L. Laugel, avec une préface de M. Hermite et un discours de Félix Klein (Gauthier-Villars, Paris, 1898)
65. B. Riemann, Theorie der Functionen complexer Variabeln. Vorlesung des Prof. Riemann. Göttingen, Sommersemester 1861; lecture notes taken by E. Abbe formerly in the possession of G. Thieme, Cod. MS: Riemann 32c, Niedersächsische Staats- und Universitätsbibliothek Göttingen
66. B. Riemann, Theorie complexer Functionen. Im Sommersemester 1861, 4stündig vorgetragen von B. Riemann; lecture notes taken by H. Hankel, formerly in the possession of G. Thieme, Cod. MS: Riemann 32g, Niedersächsische Staats- und Universitätsbibliothek Göttingen
67. B. Riemann, Theorie der Functionen einer complexen Größe. Vorgetragen von Prof. Riemann. Göttingen, Sommer 1861. Winter 1861/62. Friedrich Prym, Doctor phil.; lithographie: Friedr. Umlauft. Vind. (1865)
68. F. Tisserand, Notice sur Victor-Alexandre Puiseux. Bulletin des sciences mathématiques et astronomiques, Sér. 2 **8**(1), 234–245 (1884)
69. C.-A. Valson, La vie et les travaux du Baron Cauchy, vol. 2 (Gauthier-Villars, Paris, 1868)
70. H. Weyl, Die Idee, der Riemannschen Fläche. B.G. Teubner, Leipzig, , 1913. English translation by G.R. MacLane, *The concept of a Riemann surface* (Addison-Wesley, 1955)

Riemann Surfaces: Reception by the French School

Athanase Papadopoulos

La vérité réside dans l'imaginaire

Eugene Ionesco
(Lectures at Brown University, Sept. 1984)

Abstract Riemann introduced in his doctoral dissertation (1851) the concept of Riemann surface as a new ground space for meromorphic functions and in particular as a domain for a multi-valued function defined by an algebraic equation such that this function becomes single-valued when its is defined on its associated Riemann surface. It took several years to the mathematical community to understand the concept of Riemann surface and the related major results that Riemann proved, like the so-called *Riemann existence theorem* stating that on any Riemann surface—considered as a complex one-dimensional manifold—there exists a non-constant meromorphic function. In this chapter, we discuss how the concept of Riemann surface was apprehended by the French school of analysis and the way it was presented in the major French treatises on the theory of functions of a complex variable, in the few decades that followed Riemann's work. Several generations of outstanding French mathematicians were trained using these treatises. At the same time, this will allow us to talk about the remarkable French school that started with Cauchy and expanded in the second half of the nineteenth century. We also comment on the relations between the French and the German mathematicians during that period.

Keywords Riemann surface · Nineteenth century mathematics · Elliptic integral · Algebraic equation

AMS Mathematics Subject Classification: 30F10 · 01A55

A. Papadopoulos (✉)
Institut de Recherche Mathématique Avancée, CNRS et Université de Strasbourg, 7 rue René Descartes, 67084 Strasbourg Cedex, France
e-mail: athanase.papadopoulos@math.unistra.fr

A. Papadopoulos
Mathematics Department, Brown University, 151 Thayer Street, Providence, RI 02912, USA

L. Ji et al. (eds.), *From Riemann to Differential Geometry and Relativity*,
DOI 10.1007/978-3-319-60039-0_8

1 Introduction

The notion of Riemann surface, discovered by Riemann and introduced in his doctoral dissertation (1851), is the culmination of a series of investigations done before him, by Cauchy and others, on the theory of functions of a complex variable. With this discovery, Riemann made a complete transformation of the field of complex analysis, merging it with topology and algebraic geometry. He also paved the way to the methods of hyperbolic geometry combined with group theory that gave rise to automorphic forms, developed by Poincaré, Klein and others, and to many other developments.

In Chap. 7 of the present volume (cf. [77]), we discussed the results of Cauchy and Puiseux on line integrals and their dependence on homotopy classes of paths, and we also mentioned other related results that were available to Riemann when he wrote his doctoral dissertation. Although the problems he addressed were in the continuity of the works of his predecessors, the complete novelty of his ideas, with proofs that rely largely on geometric intuition, sometimes with arguments from physics, led to the fact that these ideas were sometimes poorly understood by Riemann's contemporaries and immediate successors. In particular, this led Klein to spend a substantial part of his life explaining Riemann's work and trying to make it more accessible. He did this in numerous lectures and books, including the well-known treatise *Über Riemanns Theory der algebraischen Funktionen und ihrer Integrale* (On Riemann's theory of algebraic functions and their integrals) (1882) [58].

France, in the few years preceding the publication of Riemann's first memoir, saw the rise of a remarkable school of analysis whose major representative was Cauchy. Among the immediate followers of Cauchy, one has to mention Liouville, Puiseux, Hermite, Briot, and Bouquet, and then came another generation of analysts, including Jordan, Halphen, Goursat, Appell, Tannery, Lacour, Molk, Picard, Darboux, Simart, Fatou, and there are others. All these mathematicians had a great admiration for Riemann and had no doubt about the importance of his ideas, even if they did not fully make use of them in their works. Riemann's collected papers, translated into French, appeared in 1889, with a preface by Hermite [98], who starts with the following[1]:

> The work of Bernhard Riemann is the most beautiful and greatest one in analysis in our epoch. It has been consecrated by a unanimous admiration and will leave an imperishable mark in Science. [...] Never before that, in any mathematical publication, the gift of invention appeared with more power, never had anybody asked for such beautiful conquests in the most difficult questions in analysis.[2]

One notion which was particularly painful to accept by the French analysts is that of Riemann surface. Most of the treatises on the theory of functions of a complex

[1]In the present chapter, the translations from the French are mine.

[2]L'œuvre de Bernhard Riemann est la plus belle et la plus grande de l'Analyse à notre époque: elle a été consacrée par une admiration unanime, elle laissera dans la Science une trace impérissable. [...] Jamais, dans aucune publication mathématique, le don de l'invention n'était apparu avec plus de puissance, jamais on n'avait demandé autant de belles conquêtes dans les plus difficiles questions de l'analyse.

variable that were used in teaching in the French universities or at the École Polytechnique, in the few decades that followed Riemann's death, were based exclusively on the methods of Cauchy, missing the essential relevance of Riemann surfaces. As a general rule, Riemann's ideas were absorbed very slowly, and it was only around the turn of the twentieth century that the French treatises included the theory of Riemann surfaces in their full strength.

In the present chapter, we review this fascinating page of the history of complex analysis. This will also give us the occasion of surveying briefly the lives and works of several prominent mathematicians from this exceptional period, and of discussing the relations between the French and the German mathematical schools.

The plan of the rest of this chapter is the following.

In Sect. 2, we comment on the notion of Riemann surface and on Riemann's existence theorem and how these concepts were received when Riemann introduced them.

In Sect. 3, we review the way Riemann's ideas on this subject are presented in the famous French treatises on analysis, including those of Briot-Bouquet, Briot, Hermite, Jordan, Appell-Goursat, Goursat, Picard, Picard-Simart, Appel-Goursat-Fatou, Halphen, Tannery-Molk and Appell-Lacour. Elliptic functions constitute the central theme of several of these treatises. At the same time, we give some biographical information on the authors of these treatises, highlighting relations among them. The overall picture is that of a coherent group, forming a "school," which was probably the first French school of mathematics. Several doctoral dissertations were written under the same advisor, and the dissertation committees often consisted of the same persons: Darboux, Hermite, Bouquet, with some small variations.

In Sect. 4, we review the content of the doctoral dissertation of Georges Simart, which is entirely dedicated to a presentation of Riemann's work on Riemann surfaces and Abelian functions. To complete the picture, we have included a section, Sect. 5, in which we review a few French doctoral dissertations and other works of the period considered which contributed to the diffusion of other major ideas of Riemann: the zeta function, minimal surfaces and integration.

In Sect. 6, we take the opportunity of the topic discussed in this chapter to say a few words on the relationship between the French and the German schools of mathematics, in particular in the few years that followed the 1870 devastating French-German war.

The concluding section, Sect. 7, contains some additional notes on the relationship between the French and the German schools in the period considered.

2 Riemann Surfaces

In his doctoral dissertation [92], Riemann introduced Riemann surfaces as ramified coverings of the complex plane or of the Riemann sphere. He further developed his ideas on this topic in his paper on Abelian functions [94]. This work was motivated in particular by problems posed by multi-valued functions $w(z)$ of a complex variable z defined by algebraic equations of the form

$$f(w, z) = 0, \tag{1}$$

where f is a two-variable polynomial in w and z.

Cauchy, long before Riemann, dealt with such functions by performing what he called "cuts" in the complex plane, in order to obtain surfaces (the complement of the cuts) on which the various determinations of the multi-valued functions are defined. Instead, Riemann assigned to a multi-valued function a surface which is a ramified covering of the plane and which becomes a domain of definition of the function such that this function, defined on this new domain, becomes single-valued (or "uniform"). Riemann's theory also applies to transcendental functions. He also considered ramified coverings of surfaces that are not the plane.

Together with introducing Riemann surfaces associated with algebraic functions, Riemann considered the inverse problem: Given a Riemann surface obtained geometrically by gluing a certain finite number of pieces of the complex plane along some curves (which are equivalent to the "cuts" in the sense of Cauchy), can we find an algebraic relation such as (1) with which this Riemann surface is associated? This can also be formulated as the problem of finding on an arbitrary Riemann surface a meromorphic function with prescribed position and nature of its singularities (poles and branch points). The idea, contained in Riemann's 1851 dissertation [92], is natural, since a polynomial is described by its roots, and a rational function by its zeros and poles. Riemann showed that the general question has a positive answer, and in his solution to the problem, he proved that a meromorphic function is determined by its singularities. This result is one form of what is usually called the *Riemann existence theorem*, a theorem that had a tremendous impact on complex geometry. For instance, it was the main motivation for what became known as the Riemann–Roch theorem. In his paper on Abelian functions [94], Riemann proved one part of that theorem, namely, that given m points on a closed Riemann surface of genus p, the dimension of the complex vector space of meromorphic functions on this surface having at most poles of first order at the m points is $\geq m - p + 1$. In his paper [101] (1865), Gustav Roch, a student of Riemann, transformed this inequality into an equality, which became known as the Riemann–Roch theorem. Riemann's result relies on his existence theorem, the description of a meromorphic function by its singularities allowing a dimension count. The proof that Riemann gave of his inequality relies on the Dirichlet principle and it was considered non-rigorous. This initiated works by several mathematicians, some of them with the aim of finding alternative proofs Riemann's results that are based on this principle, and others with the goal of giving a solid foundation to the Dirichlet principle. Thus, an important activity was generated as an indirect consequence of Riemann's existence theorem.

The discussion around Riemann's existence theorem is spread in several sections of Riemann's doctoral dissertation [92] and his paper on Abelian function, [94], in particular in Section III of the preliminary part of the latter, entitled *Determination of a function of a complex variable magnitude by the conditions it fulfills relatively to the boundary and to the discontinuities*. Later in the same paper, an *existence result* is given in the case of functions defined by integrals of algebraic functions.

Riemann's use of the Dirichlet principle was harshly criticized by Weierstrass [109], and these critiques spread a doubt not only on the validity of Riemann's proof of his existence theorem but also of other theorems. It is important to emphasize this fact, because it explains in part why Riemann's results on Riemann surfaces were not used by his immediate followers. Klein writes in his *Development of mathematics in the 19th century* ([59] p. 247 of the English translation):

> With this attack by Weierstrass on Dirichlet's principle, the evidence to which Dirichlet, and after him, Riemann, had appealed, became fragile: Riemann's existence theorems[3] were left in the air.
>
> It is interesting to observe the positions mathematicians took with respect to Riemann's existence theorem and Weierstrass's critique.
>
> The majority of mathematicians turned away from Riemann; they had no confidence in the existence theorems, which Weierstrass's critique had robbed of their mathematical supports. They sought to salvage their investigations of algebraic functions and their integrals by again proceeding from a given equation $F(\zeta, z) = 0$ [...] Riemann's central existence theorem for algebraic functions on a given Riemann surface fell from its place, leaving only a vacuum.

It is also interesting to note Riemann's attitude toward Weierstrass's critique as recorded by Klein in the same book ([59] pp. 247–48 of the English translation):

> Riemann had a quite different opinion. He fully recognized the justice and correctness of Weierstrass's critique; but he said, as Weierstrass once told me "that he appealed to the Dirichlet principle only as a convenient tool that was right at hand, and that his existence theorems are still correct."

Concerning the notion of Riemann surface, Klein writes, in the same work ([59] p. 245 of the English translation):

> The most important point is that, according to Riemann's considerations, to any given Riemann surface there corresponds one (and only one) class (a "field") of algebraic functions (with their Abelian integrals). For Riemann a "class" of algebraic functions means the totality of functions $R(\zeta, z)$ that can be rationally expressed in terms of ζ and z; the term "field" was introduced later by Dedekind. This is a theorem that could not have been obtained in another way. At this point Riemann's theory remained, for the time being, ahead of all the others which started from the equation $F(\zeta, z) = 0$.

Riemann not only considered Riemann surfaces as associated with individual multi-valued functions or with meromorphic function in general, but he also considered them as objects in themselves, on which function theory can be developed

---

[3] The plural will be explained later, when we shall talk about Picard's work.

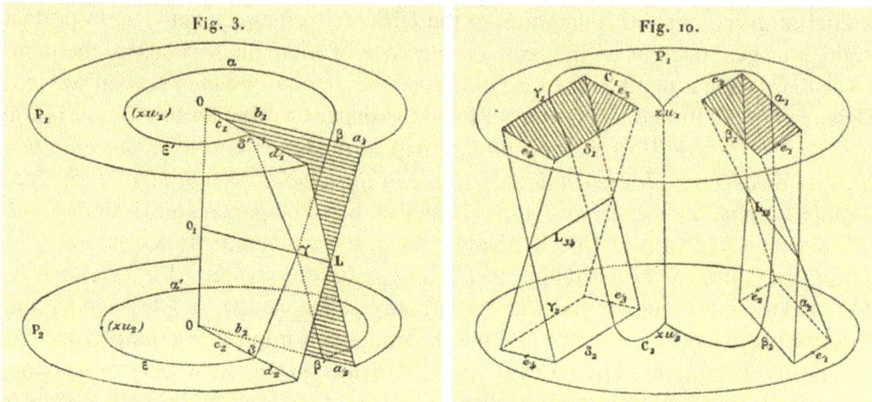

Fig. 1 A drawing of a Riemann surface, from the treatise *Théorie des fonctions algébriques* (1895) by Goursat and Appell

in the same way as the classical theory of functions is developed on the complex plane. Riemann's existence theorem for meromorphic functions with specified singularities on a Riemann surfaces is also an important factor in this setting of abstract Riemann surfaces. Riemann conceived the idea of an abstract Riemann surface, but his immediate followers did not. During several decades after Riemann, mathematicians (analysts and geometers) perceived Riemann surfaces as objects embedded in three-space, with self-intersections, instead of thinking of them abstractly. They tried to build branched covers by gluing together pieces of the complex plane cut along some families of curves, to obtain surfaces with self-intersections embedded in three-space. In his 1913 book *Idee der Riemannschen Fläche* (The concept of a Riemann surface), [110] (p. 16 of the English translation), Weyl writes about these spatial representations:

> In essence, three-dimensional space has nothing to do with analytic forms, and one appeals to it not on logical-mathematical grounds, but because it is closely associated with our sense-perception. To satisfy our desire for pictures and analogies in this fashion by forcing inessential representations on objects instead of taking them as they are could be called an anthropomorphism contrary to scientific principles.

Hilbert, in his 1903 paper [50], considers surfaces that are not embedded in a Euclidean space.[4]

The example of a Riemann surface in Fig. 1 is extracted from the treatise *Théorie des fonctions algébriques* (Theory of algebraic functions) by Paul Appell and Edouard Goursat (1895) in which the authors explain Riemann's ideas and on which we shall comment later in the present chapter. The authors explain that in the picture, the "sheets traverse each other," but that the reader should imagine that these "sheets are infinitely close to each other." We shall survey the treatise by Appell and Goursat in Sect. 3 below.

---

[4]I thank K. Ohshika for this reference.

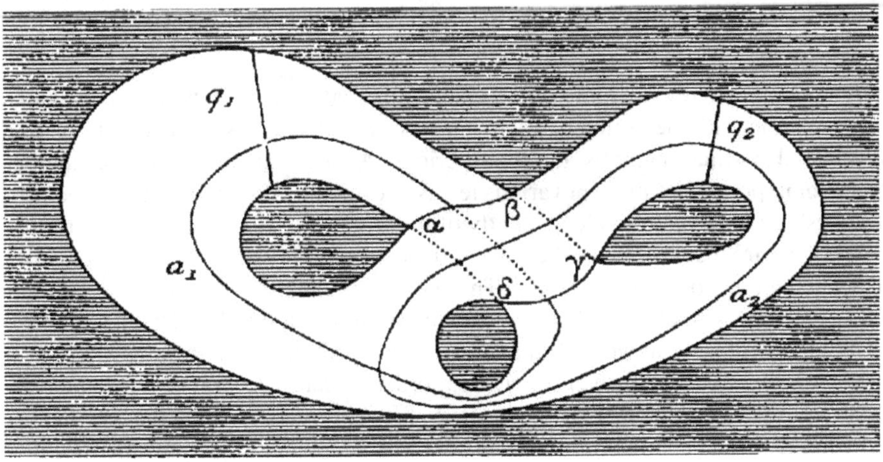

Fig. 2 A drawing from Riemann's paper on Abelian functions

In 1909, Hadamard, in his survey on topology entitled *Notions élémentaires sur la géométrie de situation* (Elementary notions of geometry of situation),[5] talking about Riemann surfaces, still considers lines along which the leaves cross each other (cf. [39] p. 204).

It was difficult to conceive these surfaces without the intersections of the sheets in 3-dimensional space. One had to wait several years before these surfaces were freed from their three-dimensional prison. Weyl, writes in his 1913 book ([110] p. 16 of the English translation): "The concept of 'two-dimensional manifold' or 'surface' will not be associated with points in three-dimensional space; rather it will be a much more general abstract idea." Figure 2 represents a more abstract drawing in the tradition of Riemann. It is extracted from the French version of Riemann's works [98].

Klein considers that around the year 1881, at least some of Riemann's important ideas were already understood in France. He writes in his *Development of mathematics in the 19th century* [59] p. 258:

> Working on the subject of automorphic functions, from 1881 on, I came into close touch with Poincaré; this was also the time when Riemann's modes of thoughts were transplanted to France and there found firm ground.

In the next section, we review the way Riemann surfaces are treated in some of the major French treatises on complex analysis that were published in the few years that followed Riemann's work.

[5]"Geometry of situation" was one of the various names given to topology, before the word "topology" became universally accepted.

3 The Nineteenth-Century French Treatises on Analysis

In this section, we review some of the nineteenth-century French treatises on analysis, in relation with the notion of Riemann surface and some associated notions like elliptic and Abelian integrals and their periods. As we shall see, there was a great variety of important treatises of various levels of difficulty, covering a large spectrum of topics. Let us note that independently of the work of Riemann, it is interesting to review these treatises, because these were the textbooks in which the French mathematicians of that epoch were trained. These mathematicians constituted a consistent and very strong school of analysis whose imprint is still felt today. The next table is a list of the treatises that we shall mention, in an approximate chronological order. It is difficult to make a precise chronological order, because several of these treatises consist of several volumes, with a time lapse of several years between the first and the last volume. In the commentary that follows this table, the order takes into account the connections between the ideas rather than the chronology.

Author	**Title**	**Year (1st ed.)**
Ch.-A. Briot and J.-C. Bouquet	Théorie des fonctions doublement périodiques et, en particulier, des fonctions elliptiques	1859
Ch. Hermite	Cours d'analyse de l'École Polytechnique	1873
Ch.-A. Briot	Théorie des fonctions Abéliennes	1879
C. Jordan	Cours d'analyse de l'École Polytechnique	1882–1897
Ch. Hermite	Cours à la faculté des sciences de Paris	1882
G.-H. Halphen	Traité des fonctions elliptiques et de leurs applications	1886–1891
É. Picard	Traité d'analyse	1891–1896
J. Tannery and J. Molk	Éléments de la théorie des fonctions elliptiques	1893–1902
P. Appell and É. Goursat	Théorie des fonctions algébriques et de leurs intégrales	1895
É. Picard and G. Simart	Théorie des fonctions algébriques de deux variables indépendantes	1897–1906
P. Appell and É. Lacour	Principes de la théorie des fonctions elliptiques et applications	1897
É. Goursat	Cours d'analyse mathématique	1902–1905
P. Appell, É. Goursat and P. Fatou	Étude des fonctions analytiques sur une surface de Riemann	1929

Briot and Bouquet

We start with the treatise *Théorie des fonctions doublement périodiques et, en particulier, des fonctions elliptiques* (Theory of doubly periodic functions, and in particular, elliptic functions) [17] by Briot and Bouquet. This treatise, whose first edition appeared in 1859, became one of the major references on the theory of functions of a complex variable in France during the second half of the nineteenth century. As the name of the treatise indicates, the stress is on elliptic functions and their generalizations to doubly periodic functions. We recall that elliptic functions have (at most) two independent periods; they are essentially functions defined on the torus. We start by recalling a few facts about these functions. For a glimpse into the history of elliptic integrals, which are at the origin of the general theory of elliptic functions, the reader is referred to Chap. 1 of the present volume [75].

Before Riemann, elliptic functions had occupied the greatest mathematicians: Euler, Gauss, Dirichlet, Legendre and others. In France, the first mathematician who made a thorough study of these functions is Legendre, who wrote treatises comprising several volumes on the subject, cf. [63, 64]. The subject became fashionable in France only after his death. It is interesting in this respect to quote a letter from Legendre to Jacobi, dated February 9, 1828, in which Legendre complains that in France, mathematicians, at his time, were not enough interested in elliptic functions. Responding to a letter in which Jacobi makes for him a summary of Abel's article *Recherches sur les fonctions elliptiques* (Researches on elliptic functions) [1] published in 1827, Legendre writes ([54], t. 1, p. 407):

> I was already aware of the beautiful work of Mr. Abel inserted in Crelle's Journal. But I was very pleased by the analysis you have given me in your own language, which is closer to mine. For me, it is a big satisfaction to see two young geometers, like you and him, cultivating with success a branch of analysis which has been for such a long time my favorite subject of study, and which is not as much welcome in my own country as it deserves to be.[6]

By the time of Briot and Bouquet published their treatise, that is, thirty years after this letter was written, the study of elliptic functions was already a very active field in France. Cauchy has already introduced line integrals in the field of functions of a complex variable, and elliptic integrals constituted a new class of functions with interesting properties. The known functions of a complex variable, before this class, were limited to polynomials, exponentials, logarithms, trigonometric functions, and some other special functions introduced by Euler. Several questions concerning these functions, motivated by the work of Legendre, Abel and Jacobi, constituted the basis of several research topics. Furthermore, elliptic functions were known to have numerous applications in geometry, number theory, mechanics and physics.

---

[6] J'avais déjà connaissance du beau travail de M. Abel inséré dans le *Journal de Crelle*. Mais vous m'avez fait beaucoup de plaisir de m'en donner une analyse dans votre langage qui est plus rapproché du mien. C'est une grande satisfaction pour moi de voir deux jeunes géomètres, comme vous et lui, cultiver avec succès une branche d'analyse qui a fait si longtemps l'objet de mes études favorites et qui n'a point été accueillie dans mon propre pays comme elle le méritait.

A few words about Briot and Bouquet may be useful, before talking about their treatise. Although they were great analysts and remarkable teachers, their names are rather unknown today.

In 1842, Charles-Auguste Briot (1817–1882) submitted at the Faculté des Sciences de Paris, a dissertation on mechanics whose title was *Sur le mouvement d'un corps solide autour d'un point fixe* (On the motion of a solid body around a fixed point) [15]. The aim of this dissertation was to provide complete proofs of results on mechanics that were stated by Poinsot in his memoir *Théorie nouvelle de la rotation des corps* (A new theory for the rotation of bodies) [91]. Briot then taught at the Sorbonne and at the École Normale Supérieure, but also, for several years, in two *lycées*[7] in Paris: Bourbon and Saint-Louis. These were among the famous lycées preparing for the highly competitive entrance examination of the École Polytechnique and the École Normale Supérieure. Having good teachers in such lycées was a tradition in France, and some of these teachers were excellent mathematicians.[8] Briot, like Riemann, Cauchy and many mathematicians of his generation, was highly interested in physics, in particular, heat, light and electricity, three topics which were particularly dear to Riemann. Briot's research in these fields was based on his theories of aether, and in his research on these topics he was strongly influenced by Louis Pasteur. He wrote a large number of textbooks for students, encompassing analysis, algebra,

[7]The lycées where Briot (and several other mathematicians we encounter in the present chapter) taught are high-schools whose curricula included an additional year of study after the high-school diploma (*baccalauréat*). During that year, called *Mathématiques spéciales*, the *élèves* (pupils) are prepared for the entrance examinations (*concours d'entrée*) to some highly competitive schools which, in the period we are interested in, were essentially the École Polytechnique and the École Normale Supérieure. In principle, only gifted and hard-working élèves were admitted in such classes.

Only a small percentage of the élèves were accepted into these schools (2–5%) at the first trial. The others usually returned to the lycée and spent another year in the class of Mathématiques spéciales where they deepened their knowledge and their training. The chances of entering one of the two schools after this second year were about 25%. Some of the élèves, after a second failure, repeated a third time the class of Mathématiques spéciales, and the chances of success, for those who tried the *concours d'entrée* after a third year, were about 50%. (These figures are extracted from the article [85] by Pierpont in which the author compares the French and the American mathematical education systems by the end of the nineteenth century.)

These classes still exist today in France, they are called *Classes préparatoires aux Grandes Écoles*, and include two years, *Mathématiques supérieures* and *mathématiques spéciales*. They prepare to the entrance examinations of a large number of schools.

[8]The list includes Briot, Bouquet, Darboux, Bertrand, Hoüel, Valiron, Châtelet, Tannery, Boutroux, Lacour, Lucas, Lichnerowicz, and there are others. The following story is related by Picard, in his *Eulogy* of Jules Tannery [84]: "Bouquet used to relate that after he graduated from the École [Normale Supérieure], and while he was in charge of the class of "mathématiques spéciales" at Marseille's lycée, he received the visit of the father of one of his élèves, who wanted that his son be prevented from working in mathematics, because they lead to noting good. He asked for a professor who would give a course which is enough bad so that his son does not enter the École Polytechnique, after which one gains less money than in business. [Bouquet aimait à raconter que, chargé à sa sortie de l'École, du cours de mathématiques spéciales au Lycée de Marseille, il avait eu la visite du père d'un de ses élèves, qui voulait qu'on empêchât son fils de travailler les mathématiques qui ne mènent à rien de bon. Il demandait que le professeur fit un assez mauvais cours pour que son fils n'entrât pas à l'École Polytechnique au sortir de laquelle on gagne moins d'argent que dans le commerce.]

analytic geometry, mechanics and physics. Having textbooks written by outstanding and devoted teachers was traditional in France in that period.

Jean-Claude Bouquet (1819–1885) defended his doctoral dissertation in 1842, the same year as Briot. The subject was the calculus of variations, and the title was *Sur la variation des intégrales doubles* (On the variation of double integrals) [14]. Bouquet first taught at a lycée in Marseille and then became, at the age of 26, professor at the University of Lyon. Seven years later he moved to Paris where he became professor at Lycée Bonaparte, and then Lycée Louis-le-Grand. In 1868, he became the successor of Puiseux at the École Normale Supérieure, and in 1885 the successor of Serret at the Chair of differential and integral calculus of the Faculté des Sciences de Paris. Bouquet's successor at that chair was Émile Picard.

Briot and Bouquet published, separately and as co-authors, several important articles and treatises on the theory of functions of a complex variable and on elliptic and Abelian functions. It might be useful to recall that in the period considered, joint mathematical works were rare, and for this reason the long-term collaboration of Briot and Bouquet stands as a singular spot in the history of mathematics. In 1856, Briot and Bouquet published a joint paper entitled *Étude des fonctions d'une variable imaginaire* (Study of functions of an imaginary variable) [16] in which they present in a comprehensive way Cauchy's theory of functions of a complex variable. In the introduction to that memoir, they write:

> This first memoir contains the principles of Cauchy's theory of an imaginary variable. We shall adopt the definition given by Mr. Cauchy, and we shall explain it by examples. We then study the properties of the functions defined by series ordered according to the increasing integer powers of the variable. This will allow us to establish, in a clear and precise manner, the necessary and sufficient conditions for a function to be expanded as a convergent series according to the increasing integer powers of the variable. In this way, we shall get rid of the clouds that still obfuscate the beautiful theorem of Mr. Cauchy.[9]

This paper, together with two other papers by Briot and Bouquet, became the bulk of their famous treatise *Théorie des fonctions doublement périodiques et, en particulier, des fonctions elliptiques* which we consider now. In that treatise, Cauchy's work is at the forefront. This treatise became famous especially by its second edition (1875), which carried the simpler name *Théorie des fonctions elliptiques*, cf. [19]. In the preface, the authors start by pointing out the importance of transcendental functions, recalling that Legendre spent almost all his life in trying to understand them. They then mention the works of Abel and Jacobi, declaring that Abel, around the year 1826, was the first to consider elliptic functions from the right point of view and to realize that these functions are doubly periodic. According to their account, Jacobi's *Fundamenta nova theoriæ functionum ellipticarum* [52], published three years later,

---

[9]Ce premier mémoire contient les principes de la théorie des fonctions d'une variable imaginaire. Nous adoptons la définition donnée par M. Cauchy, et nous l'expliquons par des exemples. Nous étudions ensuite les propriétés des fonctions définies par des séries ordonnées suivant les puissances entières et croissantes de la variable. Ceci nous permet d'établir, d'une manière nette et précise, les conditions nécessaires et suffisantes pour qu'une fonction se développe en série convergente suivant les puissances entières et croissantes de la variable. Nous faisons disparaître ainsi les nuages qui obscurcissent encore le beau théorème de M. Cauchy.

contains nothing essential which Abel had not discovered before. They declare that the difference between the two mathematicians is that Abel tried to prove the main results on the theory of elliptic functions from their double periodicity prorperty, whereas Jacobi did the same using algebraic reasonings which have the disadvantage of hiding the reason behind the results and which do not lead to interesting developments. Briot and Bouquet then write ([19] p. xviii of the Preface):

> Despite the remarkable works of these two great geometers, the theory of elliptic functions was still in the dark, and very complicated. Neither the double periodicity was recognized clearly, not the function itself was defined rigorously. To shed light on this theory, one had to introduce a new mathematical idea, and it is to the famous Cauchy that we owe this important progress.[10]

In this treatise, single-valued functions are called *monotropic* (monotropes) and multi-valued ones are called *polytropic* (polytropes). This terminology is introduced in the first pages of the second edition of the treatise (p. 9 and 11 of the 1875 edition). It indicates clearly that the authors think of these functions in terms of paths. (The word "tropos" in Greek means path.) Riemann's work (or, at least, its existence) is known to the authors, but they prefer to rely on Cauchy, completed by Puiseux. They write in the preface of the 1875 edition:

> In Cauchy's theory, the excursion of the imaginary variable is represented by the motion of a point in the plane. To represent the functions which acquire several values for the same value of the variable, Riemann used to look at the plane as composed of several sheets which are superposed and joined by weldings, in such a way that the variable can pass from a sheet to another by passing a junction line ("ligne de raccordement"). The conception of many-sheeted surfaces presents some difficulties; in spite of the beautiful results that Riemann reached by this method, it appeared to us that it has no advantage regarding the object we have in mind. Cauchy's idea is very well fit to the presentation of multiple functions; it suffices to attach to the value of the variable the corresponding value of the function, and, when the variable describes a closed curve and the value of the function changes, to indicate this change by an index.[11]

The authors acknowledge in the preface that they were influenced by Liouville's course at the Collège de France on elliptic functions, based on the double periodicity

---

[10]Malgré les remarquables travaux de ces deux grands géomètres, la théorie des fonctions elliptiques restait fort obscure et très-compliquée; ni la double périodicité n'avait été reconnue d'une manière nette, ni la fonction elle-même définie d'une manière rigoureuse. Il fallait, pour éclairer cette théorie, l'introduction d'une idée nouvelle en mathématiques, et c'est à l'illustre Cauchy que l'on doit cet important progrès.

[11]Dans la théorie de Cauchy, la marche de la variable imaginaire est figurée par le mouvement d'un point sur un plan. Pour représenter les fonctions qui acquièrent plusieurs valeurs pour une même valeur de la variable, Riemann regardait le plan comme formé de plusieurs feuillets superposés et réunis par des soudures, de manière que la variable puisse passer d'un feuillet à un autre en traversant une ligne de raccordement. La conception des surfaces à feuillets multiples présente quelques difficultés; malgré les beaux résultats auxquels Riemann est arrivé par cette méthode, elle ne nous a paru présenter aucun avantage pour l'objet que nous avions en vue. L'idée de Cauchy se prête très bien à la représentation des fonctions multiples; il suffit de joindre à la valeur de la variable la valeur correspondante de la fonction, et, quand la variable a décrit une courbe fermée et que la valeur de la fonction a changé, d'indiquer ce changement par un indice.

of these functions. A set of notes by Liouville on lectures he gave in 1847 on doubly periodic functions were published 33 years later,[12] cf. [66]. It seems that Liouville considered that Briot and Bouquet stole his ideas, and he treated them as "unworthy robbers," see [78], p. 232.

Bottazzini reports in [55] (p. 244) that in 1861, Riemann lectured on complex function theory following Cauchy's point of view as contained in Briot and Bouquet's treatise. A German translation of this treatise was published in 1862 [18].

Briot

In 1879, Briot published a treatise entitled *Théorie des fonctions abéliennes* (Theory of Abelian functions) [20]. His goal in this new book is to explain Riemann's theory of Abelian functions. These are integrals of algebraic differentials on Riemann surfaces that generalize elliptic functions (which are defined on surfaces of genus one, that is, tori), and they played a major role in the development of complex analysis and of algebraic geometry. In the introduction to his treatise, Briot recalls that Riemann was the first to study these functions, and that he found beautiful theorems concerning them. He nevertheless declares that the methods of Riemann present enormous difficulties and he describes them as lacking of clearness and rigor. He announces that, in his treatise, he relies on the works of Clebsch and Gordan,[13] but leaving aside some of their geometric considerations. Sofia Kovalevskaya did not like Briot's treatise. In a letter to Mittag-Leffler, sent on January 8, 1881 quoted by the latter in his 1900 Paris ICM talk [67], she writes:

> Isn't it surprising how, at the time being, the theory of Abelian functions with all the particularities of its own method and which make it rightly one of the most beautiful branches

[12]The notes were taken by C. W. Borchardt, the editor of the Journal für die reine und angewandte Mathematik. In a footnote to the article, Borchardt writes about these notes: "When, in the first half of the year 1847 I stayed in Paris at the same time of my late friend Ferdinand Joachimstahl, Mr. Liouville accepted to give, at his home, for the two of us, a few lessons on his method for treating the theory of doubly periodic functions. I collected Mr. Liouville's lessons, and when, back in Berlin, I have completed writing them up, I sent him a copy of my manuscript which he authorized me to communicated to Jacobi and Lejeune-Dirichlet. [...] In communicating to the geometers a work done more than thirty years ago and without the intention of publishing it, I think nevertheless that I can assure that in general my redaction reproduces faithfully the lessons of Mr. Liouville. [Lorsque dans la première moitié de l'année 1847 j'ai fait un séjour à Paris en même temps que mon ami bien regretté Ferdinand Joachimstahl, M. Liouville a bien voulu nous faire chez lui à nous deux quelques leçons sur sa méthode de traiter la théorie des fonctions doublement périodiques. [...] En communiquant aux géomètres un travail fait il y a plus de trente ans et sans l'intention de le faire imprimer, je crois néanmoins pouvoir assurer qu'en général ma rédaction reproduit fidèlement les leçons de M. Liouville.

[13]The work of Clebsch and Gordan which was a major reference at that time is their treatise *Theorie der Abelschen Funktionen* (Theory of Abelian functions), 1866 [24]. One of the major results of Clebsch is a classification of algebraic curves using Riemann's theory of Abelian functions and based on his notion of birational transformation. Clebsch's ideas were further developed by Brill and Noether.

> of analysis, is still poorly studied and poorly understood everywhere else than in Germany? I was really outraged in reading, for instance, the *Traité des fonctions abéliennes* by Briot, which I had not seen before. How can one present such beautiful material in such a dry and with so little benefits for the students? I am almost not surprised any more that our Russian mathematicians, who know this theory only through Neumann's[14] book and that of Briot, profess such a profound indifference to the study of these functions.[15]

This book by Briot is the only treatise that he authored alone. The book won the Poncelet prize.

The works of Briot and Bouquet were influential on Poincaré who, in his *Analysis* of his own works (*Analyse des travaux scientifiques de Henri Poincaré faite par lui-même*), [88], declares that the starting point of his research on differential equations—which was his first topic of investigation—were the works of Cauchy, Fuchs, Briot, Bouquet and Kovalevskaya.

Appell and Goursat

We now consider the treatise *Théorie des fonctions algébriques et de leurs intégrales* (Theory of Abelian functions and their integrals) by Appell and Goursat, [4]. This treatise was published in 1895, that is, thirty-six years after the first edition of Briot and Bouquet's *Théorie des fonctions doublement périodiques et, en particulier, des fonctions elliptiques*. The treatise carries the subtitle *Étude des fonctions analytiques sur une surface de Riemann* (A study of analytic functions on a Riemann surface). A few biographical notes on the authors are in order; both of them are important representatives of the nineteenth century French school of analysis.

Paul Appell (1855–1930) was born in Strasbourg. He started studying mathematics at the University of this city, but had to flee from there, in order to remain French, after the annexion of Alsace by Germany, in 1870.[16] His brother, who stayed in occupied Alsace, was later convicted for "anti-German activities." Appell wrote his

---

[14]The book by Neumann which is referred to in this quote is certainly his treatise *Vorlesungen über Riemann's Theorie der Abel'schen Integrale* (Lectures on Riemann's theory of Abelian integrals), published in 1865, [69]. Unlike the French treatises, Neumann's book was written in the spirit of Riemann.

[15]N'est-il pas étonnant vraiment comme, à l'heure qu'il est, la théorie des fonctions abéliennes avec toutes les particularités de la méthode qui lui sont propres et qui en font justement une des plus belles branches de l'Analyse, est encore peu étudiée et peu comprise partout ailleurs qu'en Allemagne ? J'ai été vraiment indignée en lisant, par exemple, le *Traité des fonctions abéliennes* par Briot, qui jusqu'à présent ne m'était pas tombé sous les yeux. Peut-on exposer une aussi belle matière d'une manière aussi aride et aussi peu profitable pour l'étudiant ? Je ne m'étonne presque plus que nos mathématiciens russes, qui ne connaissent toute cette théorie que par le livre de Neumann et celui de Briot, professent une indifférence aussi profonde pour l'étude de ces fonctions.

[16]In a chronicle on Appell which appeared in *Le petit parisien* (18/02/1929) it is reported that when he came back to Strasbourg, after the Second World War, he whispered: "I thought I was becoming crazy when I saw the French flag fleeting on our old cathedral. On that day, my life was filled. I could well have died." [Je croyais devenir fou en voyant le drapeau tricolore flotter sur notre chère cathédrale, murmure-t-il. Ce jour-là, ma vie était comblée. J'aurais pu mourir.]

doctoral dissertation under Chasles, on projective geometry. The title of this dissertation is *Sur la propriété des cubiques gauches et le mouvement hélicoïdal d'un corps solide* (On the properties of skew cubics and on the helocoidal motion of a solid body) [2]. The thesis was published in the Annales de l'École Normale Supérieure, [3]. Besides being a mathematician, Appell was the rector of the *Académie de Paris* from 1920 to 1925, and he became secretary general of France at the League of Nations. He is also the founder of the Paris *Cité Universitaire Internationale*. He married a niece of Bertrand and Hermite, and his daughter became the wife of Emile Borel. Appell, like many other French mathematicians of his generation (see Chap. 7 of the present volume, [77]), was profoundly religious.[17] There is an interesting correspondence between Appell and Poincaré, see [86].

Édouard Goursat (1858–1936) had as teachers Briot, Bouquet and Darboux. Goursat started as a teaching assistant ("agrégé préparateur") at the École Normale Supérieure in 1879, and one year later he was appointed at the Faculté des Sciences de Paris, taking over the position of Picard who was appointed at Toulouse. In 1881 he submitted a doctoral dissertation bearing the title *Sur l'équation différentielle linéaire qui admet pour intégrale la série hypergéométrique* (On the linear differential equation that admits as integral the hypergeometric series), [32]. The thesis committee consisted of Bouquet, Darboux and Tannery. It was published in the Annales de l'École Normale Supérieure [33]. This dissertation, written under Darboux, is based on results of Jacobi and Riemann, and it uses Cauchy's theory. Among other things, Goursat simplifies a proof of a theorem given by Riemann in his memoir of the hypergeometric function [97] (Second part of Goursat's dissertation). After his dissertation, he took a position at the Faculté des Sciences de Toulouse, as the successor of Picard who returned to Paris. In 1885, he came back to the École Normale Supérieure, replacing Bouquet. In 1897, he took over again Picard's position at the Chair of Differential and Integral Calculus at the Faculté des Sciences de Paris. The name of Goursat is attached to a theorem in complex function theory, which is usually referred to as the *Cauchy-Goursat* theorem. It says that given a holomorphic function on a simply connected domain in the plane, the integral of this function over a loop contained in the interior of the domain is zero. The first step of the proof is a lemma, called the Goursat lemma, which is a particular case of the theorem in which the loop bounds a rectangle. The result is contained in the 1814 paper of Cauchy [23] but under some unnecessary strong hypotheses on the function. Goursat's proof is contained in a paper that appeared in Acta Mathematica entitled "Proof of Cauchy's theorem" [36].

Unlike the case of the treatise of Briot and Bouquet, Riemann's theory is well present in the treatise *Théorie des fonctions algébriques et de leurs intégrales* by Goursat and Appell. Hermite wrote the preface of that treatise. In this preface, he starts by giving an overall summary of the work of Puiseux on algebraic functions,

[17]In a biography of Hermite, written by his grand-daughter (the manuscript, kept in the Archives of the Académie des Sciences de Paris) quoted in [48] p. 79, we read that Hermite told Appell once, "Can you imagine, my dear Appell, that after our death, we shall at last contemplate, face to face, the number π and the number e?" [Songez-vous, mon cher Appell, qu'après la mort nous contemplerons enfin face à face le nombre π et le nombre e ?].

which, he says "opened the field of research which led to the great discoveries of our epoch." He declares that this work transformed the field of analysis by giving it new bases.[18] Hermite, in his introduction, also mentions the influence of Cauchy. After that, he passes to the work of Riemann, praising this work and announcing that the treatise is based on the latter's ideas. Hermite writes in this introduction:

> The works of Puiseux were followed, in 1857, by those of Riemann, received with a unanimous admiration, as the most considerable event in analysis of our times. The present treatise is dedicated to the exposition of the work of this great geometer, and to the researches and the discoveries to which it led.
>
> A remarkably original concept is at their foundation. These are the surfaces to which is attached the name of their discoverer. They are constituted of superposed planes, whose number is equal to the degree of an algebraic equation, connected among themselves by crossing lines, which we obtain by joining in a certain manner the critical points. The establishment of these lines is a first question of great importance, which later on was made much simpler and easier by a beautiful theorem of Mr. Lüroth. After that, we are offered the notion of connected surfaces, their order of connection, the theorems on the lowering, using cuts, which lead the surface to a simply connected one. From these profound and delicate considerations follows a geometric representation, which is an element of the greatest power for the study of the algebraic functions. It would be too long to recall all the discoveries that carry the seal of the greatest mathematical genius, to which it led Riemann. [...][19]

In their treatise, Goursat and Appell present Riemann's topological theory of surfaces and their dissection, his theory of the complex-analytic Riemann surfaces, and his theory of Abelian integrals. Cauchy's calculus of residues is used, as well as Puiseux' method of dealing with multiple branch points of algebraic functions. The treatise also contains an exposition of Riemann–Roch's theorem, of the Brill-Noether law of reciprocity, of Abel's theorem and of the theory of moduli of algebraic curves. Jacobi's inversion problem of Abelian integrals, and a problem of Briot and Bouquet on the uniformization of solutions algebraic differential equations are addressed. W. F. Osgood published an extensive review of Appell and Goursat's treatise in the Bulletin of the AMS, see [72].

---

[18]The reader may find details on the work of Puiseux, and its relations to the works of Cauchy, Hermite and others, in Chap. 7 of the present volume [77].

[19]Aux travaux de Puiseux succèdent, en 1857, ceux de Riemann accueillis par une admiration unanime, comme l'événement le plus considérable dans l'analyse de notre temps. C'est à l'exposition de l'œuvre du grand géomètre, des recherches et des découvertes auxquelles elle a donné lieu qu'est consacré cet ouvrage.

Une conception singulièrement originale leur sert de fondement, celle des surfaces auxquelles est attaché le nom de l'inventeur, formées de plans superposés, en nombre égal au degré d'une équation algébrique, et reliés par des lignes de passage, qu'on obtient en joignant d'une certaine manière les points critiques. L'établissement de ces lignes est une première question de grande importance, rendue depuis beaucoup plus simple et plus facile par un beau théorème de M. Lüroth. S'offre ensuite la notion des surfaces connexes, de leurs ordres de connexion, les théorèmes sur l'abaissement par des coupures qui ramènent la surface à être simplement connexe. De ces considérations profondes et délicates résulte une représentation géométrique, qui est un instrument de la plus grande puissance pour l'étude des fonctions algébriques. Il serait trop long de rappeler toutes les découvertes portant l'empreinte du plus grand génie mathématique, auxquelles elle conduit Riemann [...].

Goursat

Goursat is mostly known today for his *Cours d'analyse mathématique* (A course in mathematical analysis) [34], a treatise which became a reference for all French students in mathematics. The first edition of that book, in two volumes, was published in 1902 and 1905. A second edition, in three volumes, appeared between 1910 and 1915, a third edition in 1917–1923, a fourth edition in 1923–1927, a fifth edition in 1933–1942, and there were several later editions after Goursat's death in 1936. The treatise was translated into English, cf. [35]. The whole treatise is a systematic treatment of analysis, including integration and differential equations. The subtitles of the various volumes of Goursat's *Cours* give an idea of the content. They are (in the final three-volume version): Volume I: *Dérivées et différentielles. Intégrales définies. Développements en séries. Applications géométriques.* (Derivatives and differentials. Definite integrals. Series expansions. Geometrical applications). Volume II: *Théorie des fonctions analytiques. Equations différentielles. Equations aux dérivées partielles du premier ordre.* (Theory of analytic functions. Differential equations. First order partial differential equations). Volume III: *Intégrales infiniment voisines. Équations aux dérivées partielles du second ordre. Équations intégrales. Calcul des variations* (Infinitely close integrals. Second order partial differential equations. Integral equations. Calculus of variations).

In his treatise, Goursat, in presenting the theory of functions of a complex variable, relies on Cauchy's methods on the theory of complex integration and on the existence of solutions for ordinary and partial differential equations. Weierstrass's methods are also presented, in particular for what concerns singular points and series of analytic functions, and the calculus of variations. Riemann's theories are briefly addressed in Volume III, Chap. XXVII, in relation with the Laplace equation. The author discusses, besides the methods of Riemann, those of Neumann, Schwarz and others, in relation with conformal mappings.

Osgood wrote two reviews for the Bulletin of the AMS, [73, 74], on Goursat's first edition (two volumes) of his treatise. As a conclusion to his review of Volume I, Osgood writes the following:

> When the future historian inquires how the calculus appeared to the mathematicians of the close of the nineteenth century, he may safely take Professor Goursat's book as an exponent of that which is central in the calculus conceptions and methods of this age.

Goursat's treatise lost its prestige with the advent of Bourbaki, and it was replaced in the French university curricula by the more rigorous (in the modern standards) treatises of Dieudonné, Cartan, Schwartz, etc.

Picard

Emile Picard (1856–1941) was one of those mathematicians whose work, encompassing a period straddling the nineteenth and the twentieth centuries, exerted an important influence on mathematics by giving it a new direction. In 1877, he submitted a doctoral dissertation on the geometry of Steiner surfaces, written under the guidance of Darboux. The title of the dissertation is *Application de la théorie des complexes linéaires à l'étude des surfaces et des courbes gauches* (Application of the theory of linear complexes to surfaces and skew curves) [80]. Picard's thesis was also published in the Annales de l'École Normale, [81]. Picard had a long career during which he worked on ordinary and partial differential equations, algebraic geometry, algebra, mechanics, elasticity, heat, electricity, relativity, astronomy and on other subjects of mathematics and theoretical physics. But he was above all an analyst. His name is attached in particular to two theorems he obtained in 1879 which exerted a tremendous influence on analysis. One of these theorems says that a non-constant entire function takes every complex value an infinite number of times, possibly with one exception. Picard's proof of this result uses Hermite's theory of elliptic modular functions. It is short, elegant but indirect. Giving simpler proofs and generalizations of that theorem gave rise to a large number of works done by several generations of mathematicians, including Borel, Hadamard, Montel, Julia, Bloch, Carathéodory, Landau, Lindelöf, Milloux, Schottky, Valiron, Nevanlinna, Ahlfors and several others. These works resulted in a thorough investigation of the nature of holomorphic functions and they led to a whole field of mathematics called *value distribution theory*. When the young Picard (he was 23) published his two theorems, he attracted the attention of Hermite, and they soon became friends. Two years later (in 1881), Picard married Hermite's daughter. Between 1895 and 1937, Picard taught mechanics at an engineering school in Paris, the École Centrale des Arts et Manufactures. Picard was also a philosopher and a historian of science. In 1917, Picard lost his son (who was therefore Hermite's grand-son) at the war.

In 1891, Picard published the first volume of his *Traité d'analyse* (Treatise on analysis) [79], a treatise in three volumes (the second volume was published in 1893 and the third one in 1896). This treatise was acclaimed as one of the important writings of its epoch. In a 27-page review of the first two volumes published by T. Craig in the Bulletin of the AMS, the author writes:

> One of the ablest of American mathematicians said to the writer not long ago, 'we have waited fifty years for the book!'

Cauchy's theory and all the introductory material on functions of a complex variable are presented in Volume I of Picard's *Traité* (1891). Riemann's ideas play a central role in Volume II (1893). Picard writes in the introduction to that volume:

> This volume contains the lessons I gave at the Sorbonne during the last two years. It is primarily dedicated to harmonic and to analytic functions. Without leaving aside Cauchy's point of view on the theory of analytic functions, I mainly dwell on a thorough study of harmonic functions, i.e., of the Laplace equation; a large section of this volume is dedicated to that famous equation, on which depends all the theory of analytic functions. I also dwell at length on the principle of Dirichlet, which plays such a big role in the works of Riemann, and which is as much important for mathematical physics as for analysis.
>
> Among the particular functions I study, I note the algebraic functions and the Abelian integrals. A chapter deals with Riemann surfaces, whose study has been too much left over in France. It is possible, by a convenient geometric representation, to make intuitive the main

results of this theory. Once this clear view of the Riemann surface is obtained, all the applications are conducted with the same facility as the classical Cauchy theory relative to the ordinary plane. But it is important to judge according to its real value the beautiful conception of Riemann. It would be an incomplete view to regard it only as a simplified method of presenting the theory of algebraic functions. No matter how important is the simplification brought in this study by the consideration of surfaces with many leaves, it is not there that the interest of Riemann's ideas lies. The essential point of his theory is the *a priori* conception of the connected surface formed by a finite number of plane leaves, and in the fact that to such a surface conceived in full generality corresponds a class of algebraic curves. Thus, we did not want to mutilate the profound thought of Riemann, and we have dedicated a chapter to the capital and difficult question of the existence of analytic functions on an arbitrarily given Riemann surface. The problem itself is susceptible of generalization, if we take an arbitrary closed surface in space and if we consider the corresponding Beltrami equation.[20]

Riemann surfaces are introduced in Chap. XIII of Volume II. They are associated with algebraic equations of the form $f(u, z) = 0$ where f is a polynomial in the two variables u and z. Their construction uses the method of paths and the analysis of permutations of roots developed by Puiseux which we describe in Chap. 7 of the present volume [77]. On the resulting Riemann surface, we have a single-valued function u of z. Picard writes that "the algebraic function u is uniform: to each point on that surface is associated a single value of u, which is the value corresponding to the leaf on which we find the point that we consider." He proves that the surface obtained by this construction is connected, and he spends some time explaining how one obtains a simply-connected surface from an arbitrary Riemann surface by performing a certain number of cuts. Picard refers to Riemann's article on Abelian functions [94], to Simart's dissertation [103] which we consider below, and to papers

[20]Ce second volume contient les leçons que j'ai faites à la Sorbonne ces deux dernières années. Il est principalement consacré aux fonctions harmoniques et aux fonctions analytiques. Sans négliger le point de vue de Cauchy dans la théorie de ces dernières fonctions, je me suis surtout attaché à une étude approfondie des fonctions harmoniques, c'est-à-dire de l'équation de Laplace; une grande partie de ce volume est consacrée à cette équation célèbre, dont dépend toute la théorie des fonctions analytiques. Je me suis arrêté longuement sur le principe de Dirichlet, qui joue un si grand rôle dans les travaux de Riemann, et qui est aussi important pour la physique mathématique que pour l'analyse.

Parmi les fonctions particulières que j'étudie, je signalerai les fonctions algébriques et les intégrales abéliennes. Un chapitre traite des surfaces de Riemann, dont l'étude a été laissée un peu trop de côté en France; on peut, par une représentation géométrique convenable, rendre intuitifs les principaux résultats de cette théorie. Cette vue claire de la surface de Riemann une fois obtenue, toutes les applications se déroulent avec la même facilité que dans la théorie classique de Cauchy relative au plan simple. Mais il importe de juger à sa véritable valeur la belle conception de Riemann. Ce serait une vue incomplète que de la regarder seulement comme une méthode simplificative pour présenter la théorie des fonctions algébriques. Si importante que soit la simplification apportée dans cette étude par la considération de la surface à plusieurs feuillets, ce n'est pas là ce qui fait le grand intérêt des idées de Riemann. Le point essentiel de sa théorie est dans la conception a priori de la surface connexe formée d'un nombre limité de feuillets plans, et dans le fait qu'à une telle surface conçue dans toute sa généralité correspond une classe de courbes algébriques. Nous n'avons donc pas voulu mutiler la pensée profonde de Riemann, et nous avons consacré un chapitre à la question difficile et capitale de l'existence des fonctions analytiques sur une surface de Riemann arbitrairement donnée; le problème même est susceptible de se généraliser, si l'on prend une surface fermée arbitraire dans l'espace et que l'on considère l'équation de Beltrami qui lui correspond.

by Clebsch and Lüroth. Chapter XIV of Volume II of Picard's treatise concerns periods of Abelian integrals, another topic which was dear to Riemann. Chapter XVI contains several results on meromorphic functions on Riemann surfaces, including the Riemann–Roch theorem. These are the famous *Riemann existence theorems*.[21] The title of this chapter is: "General theorems relative to the existence of functions on Riemann surfaces." Picard summarizes first the work he did in the previous chapters ([79], Vol. II, beginning of Chap. XVI). To an algebraic equation $f(x, y) = 0$ as above, a Riemann surface is associated, and on that surface, functions and integrals are studied. The problem addressed now is the converse: one starts with a connected Riemann surface which, Picard says, is defined a priori and "in a purely geometrical manner," taking a certain number of leaves and joining them by a certain number of "intersection curves" (*lignes de croisement*). One wishes to associate with such an abstract surface a class of algebraic curves, and to show a priori the existence of the functions of the type considered before. After formulating this problem, Picard writes: "We thus enter in the profound thought of Riemann." He declares that the previous chapters diverged from Riemann's ideas, in that one started there from a curve, or from an algebraic relation, whereas now, "the starting point is the m-sheeted Riemann surface." He adds (p. 459):

> Unfortunately, Riemann's method, which was so simple for establishing general existence theorems, does not have the rigor which we require today in the theory of functions. It relies on the consideration of the minimum of certain integrals which are very similar to those we already studied in the Dirichlet problem, and the same objections were addressed to him. Another way had to be found, and Mr. Neumann and Mr. Schwarz reached it independently.[22]

Picard mentions the references [69] (pp. 388–471) and [102] (p. 303), and from there he reconstructs completely the proof. In Sects. 6–13 of this chapter, the author studies the existence of harmonic functions on Riemann surfaces. These functions are used in the proof of the existence theorem. We note incidentally that for several decades, all the proofs of Riemann's existence theorem were based, like the one of Riemann, on potential theory. Picard states the main result of that chapter as a "fundamental theorem" ([79] Tome II, Chap. XVI, §18):

> *To an arbitrary Riemann surface there corresponds a class of algebraic curves.*

Another "fundamental theorem" is stated in §28 of the same chapter:

> *To a surface in space having p holes, corresponds uniformly an algebraic curve of genus p.*

Without entering into the technical definition of the genus of an algebraic curve, let us simply say that this is a birational invariant and that the equality between a

---

[21] Picard indeed uses the plural for Riemann's existence theorems.

[22] Malheureusement, la méthode si simple de Riemann pour établir les théorèmes généraux d'existence ne présente pas la rigueur qu'on exige aujourd'hui dans la théorie des fonctions. Elle repose sur la considération du minimum de certaines intégrales tout à fait analogues à celles que nous avons déjà étudiées dans le problème de Dirichlet et on lui a adressé les mêmes objections. Il a donc fallu chercher dans une autre voie. M. Neumann et M. Schwarz y sont parvenus, chacun de son côté.

notion from birational geometry and a topological notion is one of the major ideas of Riemann. It is interesting to read Picard's footnote to the theorem:

> This theorem was stated by Mr. Klein in his work which we quoted several times on the *Theory of Riemann surfaces*. The method of proof of Mr. Klein is extremely interesting, even though it does not pretend to be rigorous from the analytical viewpoint. The author borrows the elements of his proof to a fictive electrical experience performed on the surface. Thus, the existence of potential functions together with their various singularities is, in some way, proved experimentally.[23]

Section 5 of Chap. XVI concerns moduli of algebraic curves. Picard starts by addressing a preliminary question raised by Riemann: Suppose we are given in the complex plane of the variable z, the $2(m+p-1)$ ramification points of a Riemann surface of genus p with m sheets. (The count was carried on in §19 of Chap. XIII of Picard's treatise.) The question is to find the number of such surfaces. Picard notes that this number is finite, and that Hurwitz found it for small values of m. The question then is to find the number of arbitrary parameters on which a Riemann surface of some fixed genus p "essentially" depends. This is the famous moduli problem raised by Riemann and solved in a satisfactory manner by Teichmüller in his seminal paper [108]. Picard describes two methods, which are both due to Riemann, for computing these moduli. One of them relies on the Riemann–Roch theorem, and the other one uses a conformal representation of a Riemann surface onto a polygon, using an integral of the first kind, and a count of the number of periods of such integrals. The result of each of these methods is Riemann's count of the number of moduli, that is, $3p-3$, for a closed surface of genus p.

Picard concludes this important chapter by explaining how these ideas are used in the conformal representation of multiply-connected surfaces.

Picard-Simart

We now consider Picard and Simart's *Théorie des fonctions algébriques de deux variables indépendantes* (Theory of algebraic functions of two independent variables) [83], a treatise in two volumes, published in 1897 and 1906 respectively. The level of difficulty is higher than most of the other French treatises of the same period on the same subject, and the topics treated are more specialized. The introduction in each volume is written by Picard. In the introduction to the first volume, Picard declares that since a long time he had the intention to resume his ancient research on algebraic functions of two variables and to present them in a didactical form. He writes that he realized that, for more clarity, it was necessary to take into account the classical work

[23] Ce théorème a été énoncé par M. Klein dans son ouvrage déjà bien des fois cité sur la *Théorie des surfaces de Riemann*. Le mode de démonstration de M. Klein est extrêmement intéressant, quoi qu'il ne prétende pas à être rigoureux au point de vue analytique. C'est à une expérience électrique fictive faite sur la surface que l'illustre auteur emprunte les éléments de ses démonstrations. L'existence des fonctions potentielles avec leurs singularités diverses se trouve ainsi démontrée en quelque sorte expérimentalement.

of Mr. Noether as well as several works done in Italy on the same subject. The book contains indeed sections on invariants of algebraic surfaces and integrals of total differentials, including a study of the invariants introduced by Clebsch and Noether, and an exposition of the works of Castelnuovo and Enriques. Picard declares that his co-author and himself by all means "do not have the pretentiousness of going deeply into all the questions that are addressed in this "very difficult theory," but that their unique goal is "to give the state of the art on a question that deserves the effort of several researches."[24]

In the first volume, the authors develop Riemann's ideas on integrals of Abelian differentials and on Riemann surfaces, from the topological viewpoint. The title of the first chapter is *On multiple integrals of functions of several variables*. The theories of multiple integrals and integrals of total differentials constitute a link between several questions addressed in this treatise. They are generalizations of the Abelian integrals that were studied by Riemann, and they lead Picard and Simart to study hypersurfaces in a five-dimensional space. This is why the authors are led, in Chap. 2, to questions of topology in an n-dimensional space. Indeed, the second chapter is dedicated to *geometry of situation* (topology). By the time Picard and Simart's treatise was written, Poincaré had already published his famous paper with this title, two years before, in the Journal de l'École Polytechnique [87]. Picard and Simart show in particular that the genus of a Riemann surface is determined by the number of linear independent integrals of the first kind on such a surface. At the beginning of this chapter, they write (p. 19):

> This theory was founded by Riemann, who gave the name. In his study of Abelian functions, the great geometer considers only two-dimensional spaces, but later on he generalized his researches to an arbitrary number of dimensions, as is shown by his notes published after his death in the volume containing his Complete Works. Independently of Riemann, Betti studied various orders of connectivity in n-dimensional spaces, and he published a fundamental memoir on this subject.[25] In his memoir on algebraic functions of two variables, Mr. Picard showed the usefulness of such considerations in the study of algebraic surfaces. Very recently, Mr. Poincaré[26] took up in a general manner this question of *Analysis situs*, and after completing it and making more precise the results obtained by Betti, he drew attention to the considerable differences that the theories present, the two-dimensional and the higher-dimensional ones.[27]

[24] Nous n'avons certes pas la prétention d'approfondir toutes les questions qui se posent dans cette théorie difficile; notre seul but est de donner une idée de l'état actuel de la science sur un sujet dont l'étude mérite de tenter l'effort de nombreux chercheurs.

[25] Annali di Mathematica, t. IV (1870–71).

[26] Journal de Mathématiques (1899).

[27] Cette théorie a été fondée par Riemann, qui lui a donné ce nom; dans ses études sur les fonctions abéliennes, le grand géomètre ne considère que les espaces à deux dimensions, mais il a ensuite généralisé ses recherches pour un nombre quelconque de dimensions, comme le montrent les notes publiées après sa mort dans le volume renfermant ses œuvres complètes. Indépendamment de Riemann, Betti avait de son côté étudié les divers ordres de connexion dans les espaces à n dimensions, et publié un mémoire fondamental sur ce sujet. Dans son mémoire sur les fonctions algébriques, M. Picard avait montré l'intérêt que présentent des considérations de ce genre dans l'étude des surfaces algébriques. Tout récemment, M. Poincaré a repris d'une manière générale cette question

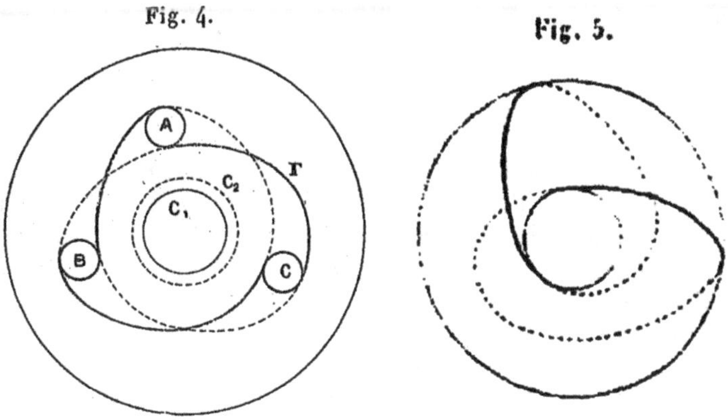

Fig. 3 Simple closed curves on surfaces, from the treatise by Picard and Simart [83]

On p. 22 of the same volume, the authors consider a Riemann surface as "a 2-dimensional manifold in 3-dimensional space," allowing the surfaces to traverse each other. The authors show that a surface defined by algebraic equations and inequalities is necessarily orientable. They introduce boundaries, Betti numbers, and their relations with multiple integrals. Homotopy classes of simple closed curves on orientable topological surfaces are drawn (cf. Fig. 3). The authors prove, at the end of Chap. 2, that for a general closed "multiplicity" (a word used by Riemann), the first and the last Betti numbers are equal, which is a special case of the result of Poincaré saying that two Betti numbers which are equidistant from the extreme ones are equal.

The 3rd chapter is dedicated to the extension of Cauchy's theorem to double integrals of functions of two variables, an extension due to Poincaré, and to residues of double integrals of rational functions. The 4th chapter concerns the reduction of singularities of an algebraic surface, and the study of its topological invariants. The authors prove in particular that any algebraic surface is birationally equivalent to a nonsingular surface embedded in the 5-dimensional space. Chapters 5 and 6 concern integrals of total differentials, and Chap. 7, double integrals.

In Volume II of the treatise, published nine years after the first one, the authors present the recent results, obtained by Picard, Castelnuovo, Enriques and others, on questions that were already addressed in the first volume and their extensions. In particular, the reduction theory for singularities of an algebraic surface is revisited, as well as the theory of double integrals of the second kind, in particular, their invariants and their periods.

(Footnote 27 continued)
dans l'*Analysis situs*, et, après avoir complété et précisé les résultats obtenus par Betti, a appelé l'attention sur les différences considérables que présentent ces théories, suivant qu'il s'agit d'un espace à deux dimensions ou d'un espace à un plus grand nombre de dimensions.

Appel-Goursat-Fatou

Riemann surfaces are also thoroughly studied in the first volume of the treatise *Théorie des fonctions algébriques et de leurs intégrales et des transcendantes qui s'y rattachent* (Theory of algebraic functions and their integrals, and their related transcendentals) [6] by Appell, Goursat and Fatou, which appeared in 1929. In reality, the treatise is a revised edition, by Fatou, of the treatise [4] by Appell and Goursat. Fatou was at the same time a mathematician and an astronomer. In 1906, he defended a thesis entitled *Séries trigonométriques et séries de Taylor* (Trigonometric series and Taylor series), [29, 30], whose subject is Lebesgue's integration theory, which in some sense is a refinement of Riemann's integration theory (see Sect. 5 below). It is in this thesis that we find the famous Fatou Lemma (also called the Fatou-Lebesgue Lemma) on the comparison between the integral of a lower limit of positive measurable functions and the lower limit of their integrals. The lemma is a key element in the proof of the Dominated Convergence Theorem. In the same year, Fatou started his work on the iteration of rational maps of the plane, a work that was revived in the last two decades of the twentieth century by Sullivan, Thurston and others. Fatou also worked on the dynamics of transcendental functions.

The title of the first volume of the treatise by Appell, Goursat and Fatou is *Étude des fonctions analytiques sur une surface de Riemann* (Study of analytic functions on a Riemann surface) [6]. In that treatise, Riemann surfaces are still represented, like in the 19th-century treatises, in an anthropomorphic fashion, (using Weyl's expression; see Sect. 2 of the present article). Figure 4 is extracted from that volume, and is already contained in the first edition by Appell and Goursat (Fig. 1 in Sect. 2 above). The authors declare, concerning the surface considered: "This surface is analogous to that represented in Fig. 10, with the difference that, in reality, the two leaves are infinitely close and the apertures are infinitely narrow.

Chapter III of this volume is entitled *Connexion des surfaces à deux feuillets. Périodicité des intégrales hyperelliptiques* (Connectivity of two-sheeted surfaces and periodicity of hyperelliptic integrals). The authors start by saying (p. 99):

> In what follows, we consider surfaces as leaves without thickness, in such a way that a point or a line drawn on that surface will be visible for a observer situated on one side or the other. These surfaces will be considered as *perfectly elastic and rip-stop*.[28]

Halphen

Among the other treatises that are related to Riemann surfaces, we mention Halphen's *Traité des fonctions elliptiques et de leurs applications* (A treatise on elliptic fonctions and their applications) in 3 volumes, published in 1886, 1888 and 1891 [41].

[28]Dans ce qui suit, nous considérons des surfaces comme des feuillets sans épaisseur, de sorte qu'un point ou une ligne tracée sur la surface seront visibles pour un observateur placé d'un côté ou de l'autre. Les surfaces seront en outre regardées comme *parfaitement élastiques et indéchirables*.

Cette surface est analogue à celle qui est représentée par la figure 10, avec cette différence que, dans la réalité, les deux feuillets sont infiniment rapprochés et les ouvertures infiniment

Fig. 10.

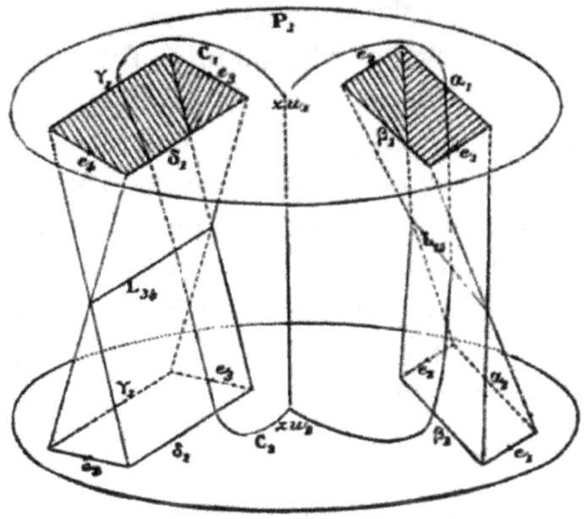

étroites. On a figuré la surface telle que la verrait un observateur debout sur le feuillet supérieur. Les droites joignant les

Fig. 4 A picture from the treatise by Appell, Goursat and Fatou [6]. One can read in the text above the figure: "This surface is analogous to the one presented in Fig. 10, with the difference that, in reality, the two sheets are infinitely close, etc.," and, below the figure: "We have represented the surface in the way an observer standing on the upper sheet would see it."

This treatise had a certain impact on students in algebra and analysis. Georges-Henri Halphen, was a graduate of the École Polytechnique,[29] and he started with a career in the army. He submitted a doctoral dissertation on 1878, titled *Sur les invariants différentiels* (On differential invariants) [40], in which he determined the invariants of planar or skew curves under projective transformations. His thesis committee consisted of Hermite, Bouquet and Darboux. Haplhen participated to the 1870 French-German war. In 1872, he was appointed *répétiteur*[30] at the École Polytechnique. He was a specialist, among other things, of differential invariants, elimination theory, and singularities of algebraic curves. Picard, in biography of Halphen [82], writes the following (p. x of the Introduction):

> Riemann, in his theory of Abelian functions, had introduced the major notion of genus of elliptic curves, and he classified them into different classes, two curves being in the same class whenever there is a uniform correspondence between them. The famous geometer, who liked the great horizons, passed quickly on more than one difficult point, in particular, for what concerns higher singularities. Halphen gave a general formula, which applies to all cases, for the determination of the genus of an algebraic curve. Then, passing to the study of curves belonging to the same class, he went deeper into a remarkable proposition of

[29]We remind the reader that the École Polytechnique is a military school.

[30]A kind of a teaching assistant.

> Mr. Nœther according to which one may find in every class curves that have only ordinary singularities [...][31]

The first part of Halphen's treatise concerns the general theory of elliptic functions. The second part makes this treatise special compared to the other treatises on the same subject: it concerns the applications of elliptic functions to various branches of mathematics and physics. The subtitle of that volume is *Applications à la mécanique, à la physique, à la géodésie, à la géométrie et au calcul intégral* (Applications to mechanics, physics, geodesy, geometry and integral calculus). It was known since the eighteenth century, that is, since the birth of the theory of elliptic functions, that these functions have many applications in physics. It suffices to recall in this respect that these functions are in some sense generalizations of the familiar trigonometric functions, and that they can be used to represent a large class of periodic phenomena. For instance, whereas the small oscillations of a pendulum are represented by the sine functions (which is the inverse function of the elliptic integral $\int_0^x \frac{dt}{\sqrt{1-t^2}}$), for large oscillations, one needs (inverses of) more general elliptic integrals. By the time of Riemann, elliptic integrals were used in problems of gravitation and electromagnetism. We recall in this respect that the famous treatise of Legendre, *Exercices de calcul intégral* (Exercises of integral calculus) [63] contains a substantial part on elliptic integrals and their applications to problems in geometry and mechanics. We also note that the subtitle of the first volume of Legendre's *Traité des fonctions elliptiques et des intégrales eulériennes* (Treatise on elliptic functions and Eulerian integrals) [64] is: *Contenant la théorie des fonctions elliptiques et son application à différents problèmes de géométrie et de mécanique* (Containing the theory of elliptic functions and its application to various problems of geometry and mechanics). One may also mention in this respect that expressions of the lengths of arcs of an ellipse (which are precisely given by elliptic integrals) are obviously useful in celestial mechanics, since Kepler's first law says that orbits of planets in the solar system are ellipses with the Sun at one of their two foci. His second law says that a segment joining a planet and the Sun sweeps out equal areas during equal intervals of time. We also recall that Gauss was also an astronomer, and his interest in elliptic functions was motivated by his work on the trajectories of planets. Finally, Abel's 1827 famous paper on elliptic functions that we already mentioned, starts by mentioning the "beautiful properties" of Abelian functions "and their applications." He writes ([1] p. 101):

> Since a long time, the logarithmic functions, and the exponential and circular functions were the only transcendental functions that attracted the attention of the geometers. It is only in

[31]Riemann, dans sa théorie des fonctions abéliennes, avait introduit la notion capitale du genre des courbes algébriques, et partagé celles-ci en différentes classes, deux courbes étant de la même classe quand elles se correspondent uniformément. L'illustre géomètre, qui aimait les grands horizons, avait peu insisté sur plus d'un point difficile, en particulier sur ce qui concerne les singularités élevées. Halphen donne une formule générale, applicable à tous les cas, pour la détermination du genre d'une courbe algébrique; puis, passant à l'étude des courbes d'une même classe, il approfondit une proposition remarquable donnée par M. Noether, d'après laquelle on peut trouver dans toute classe des courbes n'ayant que des singularités ordinaires [...].

recent times that some other functions started to be considered. Among them one has to distinguish the so-called elliptic functions, at the same time because of their beauty and of their use in the various branches of mathematics."[32]

The applications to geodesy mentioned by Halphen concern the geodesics on an ellipsoid of revolution whose ratio of major to minor axis is close to 1. Such a body is a representation of the shape of the Earth. It is also well known that Gauss was highly interested in geodesy. The applications of elliptic functions to geodesy were also considered by Jacobi in his paper [53]. In that paper, Jacobi solves a problem in geodesy which was addressed by Gauss. More details on elliptic functions are given in Chap. 1 of the present volume [76].

The third volume of Halphen's treatise contains fragments on elliptic functions which were collected after Halphen's death and published by Stieltjes.[33] The volume also contains Picard's biography of Halphen [82] which we already quoted. Picard declares there that Halphen was "one of the most eminent geometers in Europe."

Tannery and Molk

We now review the 4-volume treatise *Éléments de la théorie des fonctions elliptiques* (Elements of the theory of elliptic functions) [107] by Tannery and Molk. A few words on the authors are in order.

Jules Tannery (1848–1910) was a geometer, philosopher and writer. He edited the correspondence between Lagrange and d'Alembert.

In 1874, Tannery defended a doctoral dissertation whose title is *Propriétés des intégrales des équations différentielles linéaires à coefficients variables* (Properties of the integrals of linear differential equations with variable coefficients) [105] and [106]. The thesis committee consisted of Hermite, Briot and Bouquet. The dissertation starts with the following:

> The study of functions of an imaginary variable defined by an equation, a study which was substituted to the research, often unworkable, of the explicit form of these functions, profoundly renewed analysis in this century. It is well known that the glory of having shown this new way goes to Cauchy. The works of Mr. Puiseux on the solutions of algebraic equations, those of Messrs. Briot and Bouquet on doubly periodic functions and on differential equations, have largely proved the fertility of the idea of Cauchy in France. In Germany, the

---

[32]Depuis longtemps les fonctions logarithmiques, et les fonctions exponentielles et circulaires ont été les seules fonctions transcendantes qui ont attiré l'attention des géomètres. Ce n'est que dans les derniers temps qu'on a commencé à en considérer quelques autres. Parmi celles-ci il faut distinguer les fonctions, nommées elliptiques, tant pour leurs belles propriétés analytiques que pour leur application dans les diverses branches des mathématiques.

[33]Thomas Johannes Stieltjes (1856–1894) was Dutch but he decided to live in France. He acquired the French citizenship and in 1886 he became professor at the Faculté des Sciences de Toulouse. Stieltjes is known for several works on analysis and number theory, in particular on the so-called Stieltjes integral, elliptic functions, Dirichlet series, and is considered as the founder of the analytic theory of continued fractions. Stieltjes is also remembered for a failed attempt to prove the Riemann hypothesis, which he announced in his paper [104].

> beautiful discoveries of Riemann have accelerated the scientific movement which, since that time, did not slow down.
>
> Those who love science and who have too many reasons for distrusting their invention capacities, still have a useful role to play, that of clarifying the others' researches and disseminating them. This is what I tried to do in the present work.[34]

There is a beautiful biography of Tannery by Picard [84]. The latter, as the *secrétaire perpétuel* of the *Académie des Sciences* had to write several such biographies and reports, and many of them give us a lively image of the French mathematical life in France at his epoch. In his report on Tannery, describing his teachers—Puiseux, Bouquet and Hermite—at the École Normale, Picard writes, concerning the latter:

> What stroke Tannery above all in the teaching of Hermite is that he was able to give to mathematical abstractions color and life. He used to show how functions transform into one another, like a naturalist would do, in recounting the evolution of human beings.[35]

Jules Tannery was the thesis advisor of Hadamard. His brother, Paul Tannery, (1843-1904) was also a mathematician and (probably the most important French) historian of mathematics.

Jules Molk was Alsacian. He was born in 1857 in Strasbourg, where he studied at the Protestant Gymnasium founded by Jean Sturm in 1538. From 1874 to 1877 he studied at Zürich's Eidgenössische Technische Hochschule. His teachers there included Méquet, Geiser and Frobenius. After obtaining his diploma he went to Paris, where he followed courses by Hermite, Bouquet, Bonnet, Tisserand and Tannery. In 1882, he moved to Berlin, where he followed the courses of Weierstrass, Helmholtz, Kirchhoff and Kronecker. He obtained his doctorate in 1884 in Berlin under Kronecker. The title of his doctoral dissertation is: *Sur une notion qui comprend celle de la divisibilité et sur la théorie générale de l'élimination* (On a notion which included that of divisibility and on the general theory of elimination). The dissertation was published in Acta Mathematica, [68]. In the introduction, Molk writes that his goal is to unravel some points of Kronecker's memoir *Gründzüge einer arithmetischen Theorie der algebraischen Grössen* (Principles of an arithmetic theory of algebraic magnitudes) [51] published in 1882. He declares that this memoir seems to have been designed to give a new direction to algebra, and that his aim in his thesis is

[34]L'étude des fonctions d'une variable imaginaire définies par une équation, étude qui s'est substituée à la recherche, souvent impraticable, de la forme explicite de ces fonctions, a, dans notre siècle, profondément renouvelé l'analyse. C'est, comme on le sait, à Cauchy que revient la gloire d'avoir frayé cette voie nouvelle. Les travaux de M. Puiseux sur la recherche des racines des équations algébriques, ceux de MM. Briot et Bouquet sur les fonctions doublement périodiques et sur les équations différentielles ont, en France, amplement prouvé la fécondité de l'idée de Cauchy. En Allemagne, les belles découvertes de Riemann ont accéléré un mouvement scientifique qui, depuis lors, ne s'est pas ralenti.

Ceux qui aiment la science et qui ont trop de raisons pour se défier de leurs facultés d'invention, ont encore un rôle utile à jouer, celui d'élucider les recherches des autres et de les répandre: c'est ce que j'ai essayé de faire dans ce travail.

[35]Ce qui frappa surtout Tannery dans l'enseignement d'Hermite, c'est qu'il donnait aux abstractions mathématiques la couleur et la vie; il montrait les fonctions se transformant les unes dans les autres, comme l'eût fait un naturaliste retraçant l'évolution des êtres vivants.

to call the geometers to go thoroughly into Kronecker's difficult memoir. Molk died in Nancy in 1914. He was a specialist of elliptic functions, but he is mostly known for his collaboration with Klein to the edition of an encyclopedia of mathematics, which appeared in two versions, a German and a French one. The first volume of the German edition appeared in 1898 (Teubner, Leipzig) and the first volume of the French one in 1904 (Gauthier-Villars, Paris). The German name of the encyclopedia is *Encyklopädie der mathematischen Wissenschaften mit Einschluss ihre Anwendungen* (Encyclopedia of mathematical sciences including their applications). The French title is *Encyclopédie des sciences mathématiques pures et appliquées* (Encyclopedia of the pure and applied mathematical sciences). The French version comprises 22 volumes. More than a hundred mathematicians and physicists from Germany, France, Italy and England collaborated to the project. Their names include Abraham, Appell, Bauer, Borel, Boutroux, É. Cartan, Darwin, Ehrenfest, Enriques, Esclangon, Fano, Fréchet, Furtwängler, Goursat, Hadamard, Hilbert, Klein, Langevin, Montel, Painlevé, Pareto, Perrin, Runge, Schoenflies, Schwarzschild, Sommerfeld, Steinitz, Study, Vessiot, Zermelo, and there are others. The publication of the encyclopedia is a remarkable example, at the turn of the twentieth century, of a trans-border collaboration between mathematicians, especially French and German. The publication date also corresponds to the period where the International Congresses of Mathematicians started. The French edition is modeled on the German one, but it is not an exact translation of it. It contains several original articles, and several of the German articles, in the French version, are expanded. It is interesting to quote some excerpts from a letter from Molk to Poincaré, sent on December 12, 1901; cf. [90] pp. 188–189, in which he describes the project. This is also a testimony of the collaboration between mathematicians of the two countries.

> Our *Encyclopedia* will not be a translation of the German edition; it will be a *new edition* of that encyclopedia. We shall be free to insert new articles, to present the German articles according to our French habits, to add to them notes and complements. Each article will be published with the mark: exposed by (the French author) following (the German author), and the notes [or complemets] *added* by the French author will be, furthermore, mentioned in a special way, with the goal of *reserving our rights*, in the case where the French edition will be followed – which is most probable – by an English-American one, or a German one, or even other editions. [...] The Germans have very remarkable qualities in careful scholarship; we shall take advantage of those that they highlight in their German edition. Their exposition qualities may be less remarkable; we shall try to do our best in this regard. We shall may be succeed in helping them: this would be something! In any case, it would be dangerous to not to have in our country a research tool which is analogous to the one which is spreading more and more rapidly in their country [...] But there are also articles which *manifestly are missing* in the German edition. For instance, researches on the law of great numbers are hardly mentioned. Here, an additional article seems to be appropriate; the researches of Mr. Darboux, your own researches, those of Hadamard, should find their place in our edition. You will tell me if it is convenient for you to talk yourself about this subject, or if you find it appropriate to entrust this article to others.[36]

[36]Notre *Encyclopédie* ne sera pas une traduction de l'édition allemande; ce sera une *nouvelle édition* de cette encyclopédie. Nous serons libres d'intercaler de nouveaux articles, d'exposer, d'après nos habitudes françaises, les articles allemands, d'y ajouter des notes, des compléments. Chaque article sera publié avec la mention: exposé par (l'auteur français) d'après (l'auteur allemand), et les notes

Unfortunately, the French edition was interrupted during the First World War and the project was never resumed. We refer the reader who wishes to know more about this project to the article [31] by H. Gispert.

We now review the four volumes of the treatise *Éléments de la théorie des fonctions elliptiques* by Tannery and Molk [107]. They appeared in 1893, 1896, 1898 and 1902.

In the introduction, the authors explain why they "dared writing a book on elliptic functions, such a short time after the publication of Halphen's treatise." They say that they do not have any pretension of replacing or equating the work of the Master. But Halphen's work remained incomplete after his early death, and the missing part was long-awaited from the public. Tannery and Molk declare that the fragments edited by Stieltjes are difficult to be read by students and that their treatise is meant to compensate this fact. They write that their aim is that the student, after reading this treatise, becomes able to work on the applications—in particular those contained in the second volume of Halphen's treatise, and of reading without difficulty Schwarz's *Formeln und Lehrsäte zum Gebrauche der elliptischen Functionen* (Formulae and propositions for the use of elliptic functions)[37] which is based on the lessons and notes of Weierstrass, the fundamental memoirs of Abel and Jacobi, and the rest of the "rich and admirable literature on elliptic functions," in particular the researches of Kronecker and Hermite.

The first volume of the treatise by Tannery and Molk contains an exposition of infinite series and sums, with details on results of Weierstrass. The authors declare right at the beginning that they assume that Cauchy's theory of line integrals is known. The second volume is an exposition of ϑ functions and the general results on doubly periodic functions, deduced from the work of Hermite. The third volume is concerned with the problem of inversion of elliptic functions. One may recall here that the inverse functions of elliptic integrals are considered in some sense as a generalization of the familiar trigonometric functions. (The reader might recall that the the integral $\int_0^x \frac{dt}{\sqrt{1-t^2}}$ represents the inverse sine function.) The fourth

(Footnote 36 continued)
[ou compléments] *ajoutées* par l'auteur français seront, en outre, mentionnées d'une façon spéciale, afin de *réserver nos droits*, dans le cas où à l'édition française succéderait, ce qui est fort probable, une édition anglo-américaine, une nouvelle édition allemande, ou d'autres éditions encore. [...] Les Allemands ont des qualités d'érudition minutieuses très remarquables; nous profiterons de celles qu'ils ont mises en évidence dans leur édition allemande. Leurs qualités d'exposition sont peut-être moins remarquables; nous essayerons de faire mieux à cet égard. Nous parviendrons peut-être ainsi à leur rendre service; c'est quelque chose. Il serait en tous cas dangereux de ne pas avoir chez nous un instrument de recherche analogue à celui qui se répand de plus en plus rapidement chez eux. [...] Mais il y a aussi des articles qui *manquent manifestement* dans l'édition allemande. C'est à peine si l'on mentionne, par exemple, les recherches sur les lois des grands nombres. Là un article additionnel semblerait peut-être indiqué; les recherches de M. Darboux, les vôtres, celles d'Hadamard devraient trouver place dans notre édition. Vous me direz s'il vous convient d'en parler vous-même, ou si vous croyez bon de confier à d'autres cet article.

[37] Schwarz's treatise was also published in French, under the title *Formules et propositions pour l'emploi des fonctions elliptiques, d'après des leçons et des notes manuscrites de M. K. Weierstrass*, translated by Henri Padé, Gauthier-Villars, Paris, 1894. The translation was offered to Charles Hermite at the occasion of his seventieth birthday.

chapter of that volume is concerned with the applications. The authors declare in the introduction to Volume I (which serves as an introduction to the whole series) that the notation they use is that of Weierstrass. The fourth volume ends with a reprint of a long letter (9 pages), dated September 24, 1900, from Hermite to Tannery, preceded by a commentary (12 pages) by the authors on that letter. Hermite, in his letter, explains to the authors (at their demand) a result which he had published without proof in 1858, in two articles both entitled *Sur la résolution de l'équation du cinquième degré* [42, 43]. The authors refer to Hermite's result in their treatise, but they rely there on proofs by Weber and Dedekind, instead of the one of Hermite which was difficult to follow. They declare in their commentary that the reason for which they reproduce Hermite's proof is its beauty, and this explains the inclusion of that letter.

Jordan

We shall review Jordan's *Cours d'analyse de l'École Polytechnique* (Course in analysis of the École Polytechnique) [57] in three volumes, entitled respectively *Calcul différentiel* (Differential calculus), *Calcul intégral* (Integral calculus) and *Equations différentielles* (Differential equations). The first edition was published in 1882, 1883 and 1887 respectively. The courses given at the École Polytechnique had a large impact, because several French mathematicians were trained at that school. On the other hand, the *Cours* were intended to the students and had to comply with a specific official program, therefore they cannot be considered as a testimony of the research in mathematics that was conducted at that time. Still, the *Cours* by Jordan, like that by Hermite which we also consider below, contains enough interesting material related to the ideas of Riemann.

Jordan has been himself a student of the École Polytechnique (graduating in 1855). In 1860, he defended a doctoral dissertation entitled *Sur le nombre des valeurs des fonctions* (On the number of values of functions) [56]. The jury consisted of Duhamel, Serret and Puiseux. His second thesis[38] is entitled *Sur les périodes des fonctions inverses des intégrales des différentielles algébriques.* (On the periods of inverse functions of integrals of algebraic differentials). The subject was proposed to him by Puiseux. Jordan is mostly known for his results on topology and group theory, but he also worked on the theory of functions of a complex variable, and he was well aware of Riemann's work. Furthermore, he was among the first mathematicians to understand the impact of Galois' ideas, and he was also among the first who introduced group theory in the study of differential equations. Jordan was appointed examiner at the École Polytechnique in 1873, and then professor, at the chair of

[38]The French doctorate (until a reform which took place at the end of the 1980s) always involved a *second thesis*, on a subject which was proposed by the jury, about 3 months before the date of the thesis defense. The work done for that second thesis was not necessarily original, but it was an occasion for the student to familiarize himself with a subject which was not his main research subject.

analysis, in 1876. His last years were saddened by the loss of three of his sons in World War I.

Part of Jordan's *Cours d'analyse de l'École Polytechnique* is related to Riemann's theory. In fact, Jordan's treatise is concerned essentially with the (new) foundations of real analysis, but half of Volume II is on complex analysis. This volume is entitled *Calcul intégral* (Integral calculus). Chapter V (pp. 305–376) is on complex integration, Chap. VI (pp. 378–621) on elliptic functions, and Chap. VIII (pp. 619–693) on Abelian integrals.

Chapter V is an exposition of Cauchy's theory of integration, included in the new rigorous setting of analysis, with applications to algebraic functions. The theory is developed in the complex plane, and Riemann surfaces are not introduced. We refer the reader to Chap. 7 [77] of the present volume for a discussion of the relation between Cauchy's and Riemann's theories.

In Chap. VI, Jordan studies elliptic functions. He starts with the fact they have at most two (independent) periods. Group theory (in the language of "substitutions") is introduced in the study of linear transformations, and the language of determinants is used. Elliptic functions are considered, as in the modern point of view, as defined on the torus. Hermite's decomposition of elliptic functions into elementary functions is presented. This is an analogue of the decomposition theory of rational functions, and it is used in integration. Operations on elliptic functions (multiplication and division) are discussed in detail.

We now review Chap. VIII, on Abelian integrals. Jordan starts with a proposition which he attributes to Lüroth, concerning a canonical way of associating to an algebraic function a cut system of curves in the plane. He then introduces the connectivity of a Riemann surface in terms of such a canonical cut systems. The curves of such a system are called *retrosections*. The fact that a simple closed curves on a simply connected surface is homotopic to a point (Jordan says: "is equivalent to zero") is presented as a theorem. The definition of the genus of a surface is also given. The adjective *monodromic* ("one-path") for functions on a piece of a Riemann surface is introduced. A *synectic* function is monodromic with no critical point. A function is said to be *uniform* if it is synectic on the whole surface. Integrals of functions on Riemann surfaces are then introduced and studied. Using integrals, a function which is synectic on the whole Riemann surface is shown to be constant. A general expression is given for functions which are uniform on a Riemann surface and whose only critical points are poles. Abelian integrals are then studied, as integrals of the form $\int F dz$ where F is a rational function of two variables. Periods of these integrals are introduced, as integrals along certain paths. The number of times a rational function F takes a given value is independent of that value and is equal to the number poles of the function. From that, a proposition, called Abel's theorem, on the determination of Abelian integrals along some paths, is proved. Jordan gives then a theorem saying that an Abelian integral is determined up to a constant by some periods he calls the *first p cyclic periods*, and the location of its critical points together with some finite part of its expansion at each such point. Integrals of the first, second and third kind are introduced, and a strong form of Riemann's existence theorem, which Jordan calls the Riemann–Roch theorem, is obtained. ϑ functions and the inversion problem are introduced, and the solution of the inversion problem is presented. In particular, an expression of elementary integrals of the second and third type in terms of ϑ functions are given.

Appell and Lacour

In the treatise *Principes de la théorie des fonctions elliptiques et applications* (Principles of the theory of elliptic functions and applications) [5] (1897) by Appell and Lacour, the ideas of Riemann are hardly mentioned, but we include it in our series of commentaries because this treatise complements naturally those that we considered before.

Émile Lacour (1854–1913) was one of those good mathematicians who taught in the French *lycées*, namely, at the famous lycée Saint-Louis and at the fancy Parisian lycée Janson-de-Sailly. In 1895, he defended a thesis entitled *Sur des fonctions d'un point analytique à multiplicateurs exponentiels ou à périodes rationnelles* (On functions of an analytic point with exponential multipliers or with rational periods) [60]. The second thesis concerns the heat equation. The theory of Riemann surfaces of algebraic curves is used in this dissertation. The "analytic points" that are mentioned in the title are points on the Riemann surfaces of the functions considered. The "multipliers" are related to Riemann's theory of Abelian integrals, and they refer to the factors with which such an integral is multiplied when one traverses the cuts of a Riemann surface on which it is defined. In other words, they are periods. The functions considered (those that are referred to in the title) are generalizations of functions introduced by Appell which are analogues of the so-called doubly periodic functions of the third type. On of the simply connected surfaces obtained—in the tradition of Riemann—by cutting the Riemann surface along $2p$ arcs called "cuts", the multiplicative constants of the functions along the cuts are exponential, with an exponent being a linear function of p Abelian integrals of the first kind. The thesis contains results that make relations between, on the one hand, theorems of Abel on the zeros and singularities of algebraic functions and of Appell on the so-called "functions with multipliers," and on the other hand, results of Riemann on ϑ functions. We recall by the way that Riemann's solution of the inversion problem, given in his paper on Abelian functions, is based on the properties of the ϑ function in which the variables are replaced by the corresponding integrals of the first kind. The resulting functions become uniform when they are defined on their Riemann surfaces. In the last part of his dissertation, Lacour shows that the new functions he introduces are solutions of certain linear differential equations whose coefficients are rational functions.

In 1886, Lacour had Élie Cartan among his students, at the lycée Janson-de-Sailly. At the same time, he taught at the Faculté des Sciences de Paris. In 1901, he held the chair of differential and integral calculus at the University of Nancy, and he later moved to the University of Rennes. After Lacour left Nancy, he was replaced there by his former student Élie Cartan.

Appell and Lacour conceived their treatise as an elementary introduction to the subject, and as a preparation for the more advanced treatises (they refer to them as the "great treatises") of Briot-Bouquet, Halphen and Tannery-Molk. The treatise of Appell and Lacour also includes simple applications to geometry, mechanics and mathematical physics. The authors consider the theory of elliptic functions as a "higher-order trigonometry," in reference to the generalizations of the complex sine and cosine functions.

Hermite

To end this sequence of treatises, we say a few words on a treatise of Hermite, who was already mentioned several times in this chapter. This is his *Cours d'analyse de l'École Polytechnique*. We first mention a few biographical facts on Hermite, extracted from the Preface to Volume I of his collected works [49], written by Picard.

Charles Hermite (1822–1901) studied at the famous lycées Henri IV and Louis-le-Grand. His teacher at Louis-le-Grand was Richard, who, fifteen years before, had the young Galois as *élève*. Hermite, while he was still at Louis-le-Grand, used to go to the nearby library, the famous Bibliothèque Sainte-Geneviève, to read Lagrange's *Traité de la résolution des équations numériques*. He bought with his savings, in French translation, Gauss's *Recherches arithmétiques*. Later on, Hermite used to say that it was mainly in these two works that he learned algebra. In 1842, at the age of 20, Hermite entered the École Polytechnique, and the same year he published two papers in the new journal *Nouvelles annales de mathématiques*. One of these papers is on the impossibility of solving the fifth degree equation. A few months later, in January 1843, Hermite wrote to Jacobi, presenting his work on Abelian functions in which he extends results of Abel on the division of the argument of elliptic functions.The next year he sent another letter to Jacobi, on transformations on elliptic functions which included results on ϑ functions. Jacobi was so pleased by the letters of the young Hermite that he inserted them in his Collected Works. Later on, Hermite became mostly interested in number theory, and elliptic and Abelian functions continued to occupy his mind for the rest of his life. Jacobi's *Fundamenta nova* were always on his worktable. According to Picard, Hermite used to say that he will be until his last day a disciple of Gauss, Jacobi and Dirichlet.

Hermite taught at the École Polytechnique and he wrote, like many other professors at that school, a *Cours d'analyse de l'École Polytechnique* (1873) [44]. He also taught at the University of Paris, and lecture notes from his teaching, for the year 1882–1883, exist [45]. A large part of his course at the university is on elliptic integrals. The topics include the rectification of the parabola, ellipse and hyperbola, results of Fagnano, Graves and Chasles on arcs of ellipses whose difference is rectifiable (see Chap. 1 in the present volume for the work done on the rectifiability of these curves), and hyperelliptic integrals. Several results of Chebyshev are also presented together with Cauchy's theory on the dependence of a path integral on the homotopy class of the path. Riemann's method for the construction of holomorphic functions is also discussed, together with Green's theorem. Hermite also included in his course Riemann surfaces associated to multi-valued functions, periods of elliptic functions, doubly periodic functions, the transformation theory of elliptic functions, the ϑ function and other functions introduced by Jacobi.

4 Simart's Dissertation

Georges Simart (1846–1921) studied at the École Polytechnique. After that, he became a mathematician but he also worked as an officer in the Navy.[39] On the cover page of his doctoral dissertation, he is described as *Capitaine de vaisseau.*[40] On the one of his book with Picard, he is described as *Capitaine de frégate*[41] *et répétiteur*[42] *à l'École Polytechnique*. His dissertation is entitled *Commentaire sur deux mémoires de Riemann relatifs à la théorie générale des fonctions et au principe de Dirichlet* (A commentary on two memoirs of Riemann relative to the general theory of functions and to the principle of Dirichlet). It was defended on May 1, 1882, with a jury consisting of Hermite (acting as the president), Darboux and Bouquet. Simart had personal relations with Picard. In the introduction to Volume I of his *Traité d'analyse* [79], Picard writes that the volume was proof-read by Simart, "a dedicated friend and an invaluable collaborator" (un ami dévoué et un précieux collaborateur). We already mentioned the treatise that Picard and Simart wrote together, the *Théorie des fonctions algébriques de deux variables indépendantes* (Sect. 3). In the introduction to that work, Picard indicates that he wrote that book "with his friend, Georges Simart, who had helped him a lot in his *Traité d'analyse*."

Simart's thesis is a commentary on the two memoirs of Riemann on functions of a complex variable, namely, his doctoral dissertation [92] and his memoir on Abelian functions [94].

The first sentences of the thesis give us some hints on the status of Riemann's work among the French mathematicians at that epoch:

> We know the magnificent results obtained by Riemann in his two memoirs on the general theory of functions and on the theory of Abelian functions; but the methods he used, may be too briefly presented, are poorly known in France. On the other hand, reading these memoirs is particularly difficult and requires a heavy amount of work. Furthermore, the methods used by the famous geometer, and in particular his use of the Dirichlet principle, gave rise to several criticisms, whether in Germany or in France.[43]

---

[39] We remind the reader that the École Polytechnique is primarily a military school.

[40] A Captain in the Navy.

[41] A Frigate Captain. The progress is unusual because the rank of Capitaine de frégate is lower than that of Capitaine de vaisseau.

[42] See Footnote 30. From 1900 to 1906, Simart worked as a *répétiteur* at the École Polytechnique.

[43] On connaît les magnifiques résultats auxquels Riemann est parvenu dans ses deux mémoires relatifs à la théorie générale des fonctions et à la théorie des fonctions abéliennes; mais les méthodes qu'il a employées, peut-être trop succinctement exposées, sont peu connues en France. La lecture de ces mémoires est d'ailleurs singulièrement difficile et demande un travail approfondi. De plus,

The author then declares that his exposition is based on the works published in Germany by Königsberger, Neumann, Klein, Dedekind, Weber, Prym, Fuchs and a few others.[44] He declares that "reading these memoirs requires a knowledge of the so-called Riemann surfaces, whose use became classical in some German universities." He writes, at the end of the introduction, that at the moment he was achieving his work, he learnt about the existence of a booklet by Klein[45] in which the latter develops Riemann's ideas. Simart declares that Klein explains in that booklet that it is not necessary that Riemann surfaces be coverings of the plane ("des surfaces à plusieurs feuillets étendues sur le plan"), but that complex functions may be studied on arbitrary curved surfaces, in the same way as we do it on the plane. Simart also uses the work of Puiseux. We refer the reader to the description of the work of Puiseux given in Chap. 7 of the present volume, [77].

At the beginning of the dissertation, Simart shows how a Riemann surface is associated with an irreducible algebraic equation $F(s, z) = 0$ defining implicitly an algebraic function s of z. This surface is obtained using the distribution of the critical points and the poles, and it depends on the combinatorics of the (multi-)values of the function $s(z)$ at these points. This is considered as "the Riemann surface of the function s." This is the new domain on which the function s becomes uniform (that is, no more multi-valued). The construction of the surface is described on pp. 5–7 of the thesis. To the critical points (points z for which the given equation has multiple roots s) are associated products of cyclic transformations (permutations) obtained by winding around these values, in the tradition of Cauchy and Puiseux (see the review in [77]). The Riemann surface is obtained by gluing pieces of the complex plane using this combinatorial data. The pieces constitute the various "sheets" of the Riemann surface, which becomes a branched covering of the sphere. Each critical point gives rise to a certain number of ramification points of the covering, their number depending on the number of cyclic systems associated with the critical point. A ramification point of order μ corresponds to a cyclic permutation of $\mu + 1$ roots of the algebraic equation. Examples of gluing patterns for the various sheets are represented in Fig. 5. In this figure, the surface to the left (called Fig. 1 in the original drawing) represents a critical point of order 3, having a unique cycle. It corresponds to a unique ramification point of order 2. The surface in the middle (called Fig. 2) represents a critical point of order 4 having two cycles. It corresponds to two ramification points of order 1 each. The surface to the right (called Fig. 3) represents a critical point of order 4 having three cycles. It corresponds to three ramification points, one of order 1, and two others of order 0. The Riemann surface associated with the algebraic equation satisfies the following properties:

(Footnote 43 continued)
les procédés employés par l'illustre géomètre, en particulier l'application qu'il a faite du principe de Dirichlet, ont donné lieu à de nombreuses critiques tant en Allemagne qu'en France.

[44] Klein, in his *Development of mathematics in the 19th century* [59], gives a concise report on the contribution of these authors to the diffusion of Riemann's work.

[45] This should be Klein's *Über Riemanns Theorie der algebraischen Funktionen und ihrer Integrale* [58].

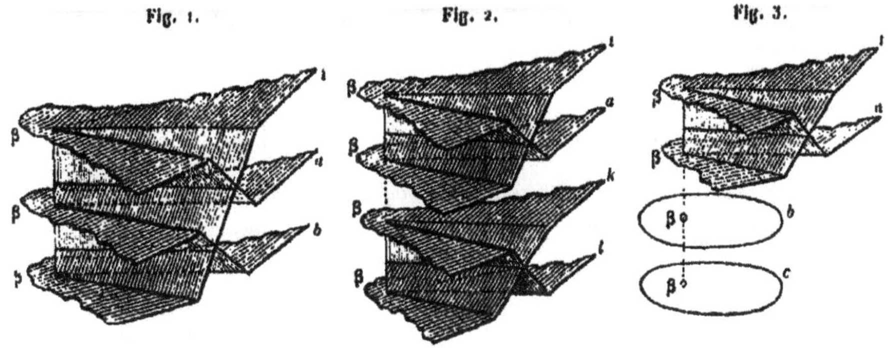

Fig. 5 Picture from Simart's thesis [103]

(1) Any rational function of s and z, when it is defined on the Riemann surface, is also a uniform function of z.
(2) The various integrals of the function s on this surface differ by a constant.

The rest of Part I of the dissertation is also purely topological. Simart recalls Riemann's definition and classification of surfaces according to their connectivity, and how an $n+1$-connected surface may be transformed into an n-connected one by performing cuts. He declares that this theory was outlined by Riemann, but that the details were worked out by Königsberger. Simart then proves that a connected $(n+1)$-connected surface is transformed by an arbitrary cut into an n-connected surface.

Part II of the dissertation concerns the study of the Laplace equation. We recall that Riemann, at the beginning of his doctoral dissertation, showed that if a function $w = u + iv$ of a complex variable $z = x + iy$ has the property that its derivative is independent of direction, then its real and imaginary parts satisfy the Laplace equation. This is one of the major tools that Riemann uses in the rest of his work. Using a system of coordinates that Riemann introduced in his dissertation and his memoir on Abelian functions, Simart proves an extension of Green's theorem to a region contained in an arbitrary Riemann surface bounded by an arbitrary finite number of curves. Riemann's use of the Dirichlet principle relies on that theorem. Simart gives the precise hypotheses on the functions which are concerned by Green's theorem, taking into account points of discontinuity and the points at infinity. The points of discontinuity of a function u are arranged, following Riemann's classification in §10 of his dissertation, into two species, according to whether the surface integral

$$\int\int ((\frac{\partial u}{\partial x})^2 + (\frac{\partial u}{\partial y})^2) dT$$

is finite or not on a piece of surface containing this point.

Simart proves the following theorem, which he attributes to Riemann (§10 of Riemann's dissertation):

Let u be a function defined on a simply connected Riemann surface with boundary satisfying the differential equation

$$\frac{\partial^2 u}{\partial x^2} + \frac{\partial^2 u}{\partial y^2} = 0$$

and suppose that the function satisfies furthermore the following conditions:

(1) *The set of points where this differential equation is not satisfied has dimension* ≤ 1.
(2) *The number of points where* u, $\frac{\partial u}{\partial x}$, $\frac{\partial u}{\partial y}$ *are discontinuous is finite.*
(3) *At these discontinuity points, the magnitudes* $\rho \frac{\partial u}{\partial x}$, $\rho \frac{\partial u}{\partial y}$ *are infinitely small compared with* ρ, *where* ρ *is the distance to the singular point.*
(4) *There are no isolated discontinuities of u which correspond to an instantaneous change in value.*

Then u as well as its partial derivatives are necessarily finite and continuous.

Simart then proves (§11 of Riemann's dissertation) the uniqueness of a function u satisfying the Laplace equation on the interior of a domain, with a given value on the boundary.

Part III of Simart's dissertation concerns the Dirichlet principle (§16–18 of Riemann's disssertation), in connection with Riemann's determination of the functions discussed in Part II. We recall that Riemann uses this principle in his proof of the so-called Riemann mapping theorem, stated as follows (§21 of Riemann's dissertation and p. 78 of Simart's dissertation):

Given a simply connected Riemann surface T with boundary, there exists a function $\zeta(z)$ *defined on this surface such that the image by* ζ *of T is the unit disc.*

Part IV concerns Abelian integrals, as an approach to the Riemann existence problem: "To determine a function knowing its ramification points, its discontinuity points and the way in which it is discontinuous." The analytic forms of the so-called integrals of the first kind are given as well as the Riemann–Roch theorem.

More precisely, Simart addresses in this part the following two problems, for which he gives a complete solution:

Problem 1 (p. 80) Given an irreducible algebraic equation $F(s, z) = 0$ defining a multi-valued algebraic function s of z, find the associated Riemann surface, that is:

(1) determine the critical points of the function s, the number of ramification points that are above these critical points, and the order of each of these ramification points;
(2) transform this surface, using Riemann's "cuts," into a simply connected surface, evaluate the number of cuts, and then determine the connection of the surface.

Problem 2 (p. 97) Let T be the closed surface associated with the function $s(z)$ defined in Problem 1, and assume it is $2p + 1$-connected. Let T' be the simply connected surface obtained from s using $2p$ cuts. Find a function $w(z)$ which is uniform on T, continuous on T' except at certain points and along certain lines, and satisfying the following:

(1) Along each cut, the difference of the function from one side of the cut to the other is a constant; the real parts of these constants are given in advance.
(2) The function is discontinuous at a certain finite number of points, and at such points it has a finite expression of the form

$$A \log r + Br^{-1} + Cr^{-2} + \cdots$$

where the constants $A, B, \ldots$ are given and r is an arbitrary function of z which at the given point is infinitely small of the first order.
(3) With the set of points in (2), the surface is no more closed, and one has to draw new cuts joining these points to the boundary of the surface T'. The difference of the function w along both sides of each of these new cuts is constant for each such cut and equal to $2\pi A$.

In the solution of Problem 1, the Puiseux expansions and the techniques of the Puiseux-Newton polygon are thoroughly used.

The second problem is one of the main problems that were addressed by Riemann in his memoir on Abelian functions. The proof that Simart gives uses, in the tradition of Riemann, the Dirichlet principle.

To each critical point corresponds a certain number of ramification points which are determined by the system of circular points formed around that point. A ramification point of order μ is a point around which $\mu + 1$ roots are permuted. A ramification point of order 1 is a point around which 2 roots are permuted, and it is called a simple ramification point. There is a relation between the *order* and the *degree* of a critical point, and the orders of the corresponding ramification points above it. These considerations are in the tradition of the work of Puiseux; cf. the exposition in Chap. 7 of the present volume [77].

Simart's dissertation is one of the important French writings that contributed to the understanding of Riemann's ideas by the French mathematicians.

5 Other French Dissertations and Other Works of Riemann

In this section, we review briefly a few other works done in France in which the authors explain some major ideas of Riemann, including his work on the zeta function, on minimal surfaces, and on integration.

The Zeta Function

Eugène Cahen, defended in 1895, at the Faculté des Sciences de Paris, a doctoral dissertation titled *Sur la fonction $\zeta(s)$ de Riemann et sur des fonctions analogues* (On Riemann's $\zeta(s)$ function and on analogous functions) [21]. The dissertation is dedicated to a generalization of Riemann's zeta function to functions of the form $\sum \frac{\alpha_n}{n^s}$, in particular for sequences α_n which are periodic, and to the development of a theory of Dirichlet series. The dissertation was criticized as being faulty, but it contains the kind of mistakes which were a ferment for further research. For instance, Cahen gives, with an incomplete proof, an asymptotic value of the sum of the logarithms of prime numbers which are smaller than x. In his paper [38], Hadamard writes:

> In his memoir which was previously quoted, Mr. Cahen presents a proof of the theorem stated by Halphen: *The sum of the logarithms of the prime numbers which are at most x is asymptotic to x*. However, his reasoning depends on Stieltjes' proposition concerning the realness of the roots of $\zeta(\frac{1}{2}+ti)=0$. We shall see that by modifying slightly the author's analysis, we can establish the same result in all rigor.[46]

The mistakes in Cahen's dissertation are analyzed in E. Landau's review [61]. Landau corrected some of them. Cahen's dissertation was published in the Annales de l'École Normale, [22].

It is interesting to recall that in 1891, the Paris *Académie des Sciences* announced a prize for a competition whose subject was: "The determination of the number of prime numbers smaller than a given quantity." When the competition was announced, it was thought that the prize would be attributed to Stieltjes, who had claimed a proof of the Riemann hypothesis, but his proof turned out to be wrong. The prize went in 1892 to Hadamard, for completing Riemann's proof of the prime number theorem. Here is how Hadamard relates his discoveries, in his report on his own works [37]:

> The last ring in the chain of deductions which started in my thesis and continued in my crowned memoir led to the clarification of the most important properties of Riemann's $\zeta(s)$ function.
>
> By considering this function, Riemann determines the frequency asymptotic law of prime numbers. But his reasoning assumes: 1) that the function $\zeta(s)$ has finitely many zeros; 2) that the successive moduli of these zeros grow roughly like $n \log n$; 3) that, in the expression of the auxiliary function $\xi(t)$ in prime factors, no exponential factor is introduced.
>
> Since these propositions remained without proof, Riemann's results remained completely hypothetical, and it was not possible to find others in the same trend. As a matter of fact, no effort has been attempted in this respect since Riemann's memoir, with the exception of: (1) Halphen's note which I mentioned earlier, which was, after all, a research project for the case where Riemann's postulates would be established; (2) a note by Stieltjes in which this geometer announced a proof of the realness of the roots of $\zeta(t)$, a proof which was never produced since.
>
> Nevertheless the propositions whose statements I recalled before are only a trivial application of general theorems contained in my memoir.
>
> Once these propositions are established, the analytic theory of prime numbers was able, after a break which lasted thirty years, to take a new boom; since that time, it continued to grow rapidly.

---

[46] Dans son mémoire précédemment cité, M. Cahen présente une démonstration du théorème énoncé par Halphen: *La somme des logarithmes des nombres premiers inférieurs à x est asymptotique à x*. Toutefois son raisonnement dépend de la proposition de Stieltjes sur la réalité des racines de $\zeta(\frac{1}{2}+ti)=0$. Nous allons voir qu'en modifiant légèrement l'analyse de l'auteur on peut établir le même résultat en toute rigueur.

This is how the knowledge of the genus[47] of $\zeta(s)$ allowed, first, Mr. von Mangoldt to establish in all rigor the final result of Riemann's memoir. Before that, Mr. Cahen had made a first step towards the solution of the problem addressed by Halphen; but he was not able to attain completely his goal: indeed, it was necessary, in order to achieve in an irrefutable way Halphen's reasoning, to prove once again that the ζ function has no zero on the line $R(s) = 1$.

I was able to overcome this difficulty in 1896, while Mr. de la Vallée-Poussin reached independently the same result. But the proof which I gave is much quicker and Mr. de la Vallée-Poussin adopted it in his later publications. It uses only the simple properties of $\zeta(s)$.

At the same time, I extended the reasoning to Dirichlet series and, consequently, I determined the distribution law for prime numbers in an arbitrary arithmetic progression, then I showed that this reasoning may be used as such for quadratic forms with negative determinant. Since then, the same general theorems on entire functions allowed Mr. de la Vallée-Poussin to complete this cycle of proofs by treating the case of forms with positive $b^2 - ac$.[48]

---

[47]Hadamard was studying, at the same period, a notion of genus for entire functions. In particular, he gave a formula for the growth of the moduli of the roots of such functions in terms of their power series expansion.

[48]Le dernier anneau de la chaîne de déductions commencée dans ma Thèse et continuée dans mon Mémoire couronné aboutit à l'éclaircissement des propriétés les plus importantes de la fonction $\zeta(s)$ de Riemann.

Par la considération de cette fonction, Riemann détermine la loi asymptotique de fréquence des nombres premiers. Mais son raisonnement suppose: (1) que la fonction $\zeta(s)$ a des zéros en nombre infini; (2) que les modules successifs de ces zéros croissent à peu près comme $n \log n$; (3) que, dans l'expression de la fonction auxiliaire $\xi(t)$ en facteurs primaires, aucun facteur exponentiel ne s'introduit.

Ces propositions étant restées sans démonstration, les résultats de Riemann restaient complètement hypothétiques, et il n'en pouvait être recherché d'autres dans cette voie. De fait, aucun essai n'avait été tenté dans cet ordre d'idées depuis le Mémoire de Riemann, à l'exception: (1) de la Note précédemment citée d'Halphen, qui était, en somme, un projet de recherches pour le cas où les postulats de Riemann seraient établis; (2) d'une Note de Stieltjes, où ce géomètre annonçait une démonstration de la réalité des racines de $\xi(t)$, démonstration qui n'a jamais été produite depuis.

Or les propositions dont j'ai rappelé tout à l'heure l'énoncé ne sont qu'une application évidente des théorèmes généraux contenus dans mon Mémoire.

Une fois ces propositions établies, la théorie analytique des nombres premiers put, après un arrêt de trente ans, prendre un nouvel essor; elle n'a cessé, depuis ce moment, de faire de rapides progrès.

C'est ainsi que la connaissance du genre de $\zeta(s)$ a permis, tout d'abord, à M. von Mangoldt d'établir en toute rigueur le résultat final du Mémoire de Riemann. Auparavant, M. Cahen avait fait un premier pas vers la solution du problème posé par Halphen; mais il n'avait pu arriver complètement au but: il fallait, en effet, pour achever de construire d'une façon inattaquable le raisonnement d'Halphen, prouver encore que la fonction ζ n'avait pas de zéro sur la droite $R(s) = 1$.

J'ai pu vaincre cette dernière difficulté en 1896, pendant que M. de la Vallée-Poussin parvenait de son côté au même résultat. La démonstration que j'ai donnée est d'ailleurs de beaucoup la plus rapide et M. de la Vallée-Poussin l'a adoptée dans ses publications ultérieures. Elle n'utilise que les propriétés les plus simples de $\zeta(s)$.

En même temps j'étendais le raisonnement aux séries de Dirichlet et, par conséquent, déterminais la loi de distribution des nombres premiers dans une progression arithmétique quelconque, puis je montrais que ce raisonnement s'appliquait de lui-même aux formes quadratiques à déterminant négatif. Les mêmes théorèmes généraux sur les fonctions entières ont permis, depuis, à M. de la Vallée-Poussin d'achever ce cycle de démonstrations en traitant le cas des formes à $b^2 - ac$ positif.

Minimal Surfaces

Regarding Riemann's work on minimal surfaces (see [95, 96] cf. also Chap. 5 of the present volume [111]), we mention the thesis defended at the Faculté des Sciences de Paris on May 27, 1880, by Boleslas-Alexandre Niewenglowski [71]. The title is *Exposition de la méthode de Riemann pour la détermination des surfaces minima de contour donné* (Exposition of Riemann's method for the determination of minimal surfaces with a given contour). The thesis committee consisted of Hermite, Bonnet and Tannery. The author declares there that Riemann, in his work on minimal surfaces, was inspired by Bonnet. He writes, in his introduction:

> I would like to clarify, if I can, a remarkable memoir of Riemann, relative to minimal surfaces. The famous author had briefly indicated most of the results he obtained; I hope that I established them in a satisfactory way.
>
> Riemann makes use of imaginary variables which we immediately reduce to the variables that were used before him by Mr. O. Bonnet, in several important memoirs on the general theory of surfaces. Indeed, the logarithm of the variable μ, chosen by Riemann, is equal to $y + x\sqrt{-1}$ and, therefore, the logarithm of the conjugate variable μ' is equal to $y - x\sqrt{-1}$, where x et y are the independent variables adopted by Mr. O. Bonnet. I think that I am not exaggerating at all in claiming that the scholarly research of Mr. O. Bonnet inspired that of Riemann.[49]

In §6 of his dissertation, Niewenglowski recalls the partial differential equation that Riemann obtains to show that a surface is minimal (that is, has zero mean curvature), and he shows that this equation is contained in Bonnet's memoir [7]. We note by the way that Bonnet wrote several other articles on minimal surfaces; cf. e.g. [8–12]. In the first section of the second part of his dissertation, titled *Applications*, Niewenglowski considers the special case of minimal surfaces that contain two non-planar surfaces. He notes that the only such surface that Riemann indicates in his article is a surface that was known since a long time (a surface Niewenglowski calls "hélicoïde gauche à plan directeur.") Niewenglowski notes that Serret showed that there are other surfaces that satisfy this requirement and he describes them. Other examples of minimal surfaces given by Riemann are described from a new point of view. Niewenglowski's dissertation was published in the Annales de l'École Normale Supérieure, [70].

---

[49] Je me propose d'élucider, s'il m'est possible, un mémoire remarquable de Riemann, relatif aux surfaces minima. L'illustre auteur a brièvement indiqué la plupart des résultats qu'il a obtenus; j'espère les avoir établis d'une manière satisfaisante.

Riemann se sert de variables imaginaires que l'on ramène immédiatement aux variables employées avant lui par M. O. Bonnet, dans plusieurs mémoires importants sur la théorie générale des surfaces. En effet, le logarithme népérien de la variable μ, choisie par Riemann, est égal à $y + x\sqrt{-1}$ et le logarithme de la variable conjuguée μ' est égal, par suite, à $y - x\sqrt{-1}$, x et y étant les variables indépendantes adoptées par M. O. Bonnet. Je pense ne rien exagérer en affirmant que les recherches savantes de M. O. Bonnet ont inspiré celles de Riemann.

The Riemann Integral

Finally, we talk about the fate of the Riemann integral in the French treatises on analysis of the period considered. It seems that it is only in the second edition of Jordan's *Cours d'analyse*, published in 1893, that this topic was considered for the first time. We note by the way that this second edition contains Jordan's theorem saying that a simple closed curve in the plane separates the plane into two regions.

Riemann introduced his theory of integration in his habilitation memoir on trigonometric series, *Über die Darstellbarkeit einer Function durch eine trigonometrische Reihe* (On the representability of a function by a trigonometric series) [93]. The text was written in 1853 but was published only after Riemann's death. Darboux, in a letter to Hoüel, who had just translated Riemann's memoir into French, dated March 30, 1873 and quoted in [28], writes the following:

> It is very kind of you to have finished the Riemann. There is a pearl which everybody will discover there, I hope. This is the definition of the definite integral. It is from here that I extracted a large quantity of functions which do not have a derivative.[50]

Darboux and Hoüel were the two editors of the *Bulletin des sciences mathématiques et astronomiques*, and we mention incidentally that Hoüel translated into French, and published, other memoirs of Riemann, including his two Habilitation works, *Über die Darstellbarkeit einer Function durch eine trigonometrische Reihe* (On the representability of a function by a trigonometric series) [93] and *Über die Hypothesen, welche der Geometrie zu Grunde liegen* (On the hypotheses that lie at the bases of geometry) [99].

Two years after he wrote that letter, Darboux published a memoir on discontinuous functions [25] in which he uses Riemann's ideas. His memoir starts as follows:

> Until the appearance Riemann's memoir on trigonometric series, no doubts were raised on the existence of a derivative for continuous functions. Excellent and famous geometers, among whom one must count Ampère, had tried to provide rigorous proofs for the existence of a derivative. These attempts were without doubt far from being satisfying. But I repeat it: no doubt was even formulated on the existence of a derivative for continuous functions.
>
> The publication of Riemann's memoir concluded the question in the opposite way. At the occasion of trigonometric series, the famous geometer presents his ideas on the principle of infinitesimal calculus: he generalizes, with one of these views that belong only to first order minds, the notion of definite integral; he shows that it applies to discontinuous functions on any interval, and he states the necessary and sufficient conditions under which a function, continuous or discontinuous, can be integrated. As we shall see, the sole fact that there exist discontinuous functions that can be integrated suffices to prove that there are discontinuous functions that have no derivative, and this consequence of Riemann's works was soon admitted by the German geometers.

[50] Vous êtes bien aimable d'avoir fini le Riemann. Il y a une perle que tout le monde y découvrira, je l'espère. C'est la définition de l'intégrale définie. C'est de là que j'ai tiré une foule de fonctions qui n'ont pas de dérivées.

> [...] In the work that will be read, I resume, providing all the necessary developments, the definitions of Riemann's definite integral after Riemann, and I show how this definition must lead to infinitely many continuous functions which have no derivative.[51]

Picard, in his *Notice* on Darboux, reports that the latter declared later on that his memoir "was poorly received by several among those who usually are interested by his works. They had dissuaded him to further cultivate this unproductive field of functions which do not have a derivative."[52]

Lebesgue, in a letter to Borel dated April 28, 1909, writes ([65] p. 189):

> I appreciate the last works of Riemann (I think he died young) as much as his dissertation on functions of a complex variable, whose importance, it seems to me, was exaggerated.[53]

One may mention here that the main idea that Lebesgue wanted to convey in that letter is that, from his point of view, the work of a mature mathematician is generally more important than the work he did when he was young. It is also true that Lebesgue found in Riemann's memoir on trigonometric series [93], which was written three years after his doctoral dissertation [92] (that is, he was more mature, in Lebesgue's wording), the bases of his integration theory, the work for which the name of Lebesgue is mostly remembered.

Lebesgue is the founder of measure theory, and he was inspired by Riemann's integration theory. In the introduction to his famous *Leçons sur l'intégration et la recherche des fonctions primitives* (Lessons on integration and on the search for primitive functions) [62], Lebesgue writes:

> [...] It is for the resolution of these problems, and not by love of complications, that I introduced in this book a definition of the integral which is more general than that of Riemann and which includes the latter as a special case.

---

[51] Jusqu'à l'apparition du mémoire de Riemann sur les séries trigonométriques aucun doute ne s'était élevé sur l'existence de la dérivée des fonctions continues. D'excellents, d'illustres géomètres, au nombre desquels il faut compter Ampère, avaient essayé de donner des démonstrations rigoureuses de l'existence de la dérivée. Ces tentatives étaient loin sans doute d'être satisfaisantes; mais je le répète, aucun doute n'avait été formulé sur l'existence même d'une dérivée pour les fonctions continues.

La publication du mémoire de Riemann a décidé la question en sens contraire. À l'occasion des séries trigonométriques, l'illustre géomètre expose ses idées sur le principe du Calcul Infinitésimal: il généralise, par une de ces vues qui n'appartient qu'aux esprits de premier ordre, la notion d'intégrale définie; il montre qu'elle est applicable à des fonctions discontinues dans tout intervalle, et il énonce les conditions nécessaires et suffisantes pour qu'une fonction, continue ou discontinue, soit susceptible d'intégration. Ce seul fait, qu'il existe des fonctions discontinues susceptibles d'intégration, suffit à prouver, comme on le verra, qu'il y a des fonctions continues n'ayant pas de dérivée, et cette conséquence des travaux de Riemann n'a pas tardé à être admise par les géomètres allemands.

[...] Dans le travail qu'on va lire, je reprends, en donnant tous les développements nécessaires, les définitions de l'intégrale définie d'après Riemann, et je montre comment cette définition doit conduire à une infinité de fonctions continues n'ayant pas de dérivée.

[52] Ce Mémoire avait été froidement accueilli par plusieurs de ceux qui habituellement s'intéressaient à ses travaux. Ils l'avaient dissuadé de labourer plus longtemps le champ stérile des fonctions qui n'ont pas de dérivée.

[53] J'apprécie autant les derniers travaux de Riemann (mort jeune je crois) que sa dissertation sur les fonctions de variable complexe dont l'importance m'a semblé parfois exagérée.

> I think that those who will read me carefully, even if they regret that things are not simpler, will grant me that this definition is necessary and natural. I dare say that in a certain sense it is simpler than that of Riemann, as much easy to grasp, and that only some previously acquired mental habits can make it appear more complicated. It is simpler because it highlights the most important properties of the integral, whereas Riemann's definition only highlights a computational mechanism. For this reason, it is almost always as much easy, and even easier, using the general definition of the integral, to prove a property for all the functions to which this definition applies, that is, the *summable* functions, than to prove it for all the integrable functions, relying on Riemann's definition. Even if one is only interested in the results relative to simple functions, it is therefore useful to be familiar with the notion of summable function because it suggests fast methods of proof.[54]

Chapter II of Lebesgue's treatise is entirely dedicated to Riemann's theory.

6 On the Relations Between the French and German Mathematicians

The impact of Riemann's work on the French mathematical school naturally leads to the question of the relation between the French and German schools of mathematics. We already addressed this issue, in particular in Sect. 3 above. The question has several sides, ranging from the attitude towards the so-called German tendency to abstraction, to the political aspect, taking into account the ravaging war that broke out 20 years after Riemann defended his dissertation. We recall that in 1870, a devastating war erupted between France and Germany, which resulted in the German annexion of the French provinces of Alsace and Moselle. This war clearly affected the relations between the two countries, but the French kept the great admiration they had for Riemann, Weierstrass and the German school of function theory. One must add that despite this admiration, some of Riemann's methods remained foreign to the French geometers. Darboux, in a letter to Hoüel, dated March 5, 1870, complains of the fact that the French mathematicians were still relying on the old methods. He writes ([26] p. 109):

[54][...] C'est pour la résolution de ces problèmes, et non par amour des complications, que j'ai introduit dans ce livre une définition de l'intégrale plus générale que celle de Riemann et comprenant celle-ci comme cas particulier.

Ceux qui me liront avec soin, tout en regrettant peut-être que les choses ne soient pas plus simples, m'accorderont, je le pense, que cette définition est nécessaire et naturelle. J'ose dire qu'elle est, en un certain sens, plus simple que celle de Riemann, aussi facile à saisir que celle-ci et que, seules, des habitudes d'esprit antérieurement acquises peuvent faire paraître plus compliquée. Elle est plus simple parce qu'elle met en évidence les propriétés les plus importantes de l'intégrale, tandis que la définition de Riemann ne met en évidence qu'un procédé de calcul. C'est pour cela qu'il est presque toujours aussi facile, parfois même plus facile, à l'aide de la définition générale de l'intégrale, de démontrer une propriété pour toutes les fonctions auxquelles s'applique cette définition, c'est-à-dire pour toutes les fonctions *sommables*, que de la démontrer pour toutes les fonctions intégrables, en s'appuyant sur la définition de Riemann. Même si l'on ne s'intéresse qu'aux résultats relatifs aux fonctions simples, il est donc utile de connaître la notion de fonction sommable parce qu'elle suggère des procédés rapides de démonstration.

> All our geometers, although very distinguished, seem to belong to another age. They are eminent scientists, belonging to a science which is twenty or thirty years old which they improve and develop with a lot of success, but all the modern branches remain inaccessible to them.[55]

One may naturally address the question of quoting the German mathematical literature by the French, and vice-versa, independently of the question of the difficulty of Riemann's ideas. Darboux, in another letter to Hoüel, complains about the fact that the Germans never quote Cauchy. In a letter written around the year 1870 (the letter does not carry a date), he writes (see [26] p. 89, Letter No. 3):

> People in France start studying extensively complex variables. It is odd that this theory, born in France with the work of Cauchy, received its most beautiful developments abroad, but, I don't know if you will be of the same opinion as me, I find that the Germans are not fair for what regards Cauchy. They take advantage of his work but never quote him.[56]

In another letter to Hoüel, talking again about the Germans ([26] p. 96, Letter No. 7, again with no date), Darboux writes:

> Their behavior concerning Cauchy is unworthy. All the copies of Cauchy['s writings] leave for Germany. Gauthier-Villars quite rightly said this to me. Nevertheless his work is never quoted.[57]

How was the situation in France? It is sometimes claimed that Poincaré was not keen on quoting the Germans. In a letter to Hermite (August 20, 1881), Mittag-Leffler ([46] p. 251, also quoted in Dugac [28], pp. 156–157), writes:

> Weiserstrass's work is prior to that of Merss. Briot and Bouquet, but Mr. Poincaré, who should have known this from the memoir of Mme Kowalewski—if ever he did not know about the work *Analytische Facultäten*—never said a word about it. Monsieur de Ramsey told me that he heard from Mr. Molk—the French student following Weierstrass's course in Berlin—that Mr. Poincaré hates the Germans, which I find very natural, and that he made it a principle to never quote any German author, which I find very bad if it were true.[58]

It is possible that Poincaré's passing over the German literature is simply due to his general ignorance about others' writings. Dieudonné, writes, in his article on Poincaré in the Dictionary of Scientific Biography ([27] Vol. 11, pp. 51–61):

---

[55]Tous nos géomètres, quoique tous fort distingués, semblent appartenir à un autre âge. Ce sont des savants éminents restés à la science d'il y a vingt ou trente ans qu'ils perfectionnent, développent avec beaucoup de succès, mais toutes les branches modernes sont pour eux très accessoires.

[56][...] on commence à s'occuper beaucoup en France des variables complexes. Il est singulier que cette théorie née en France par le travail de Cauchy ait reçu les plus beaux développements à l'étranger, mais je ne sais si vous serez de mon avis, je trouve que les Allemands ne sont pas justes envers Cauchy. Ils profitent de ses travaux mais ne le citent presque jamais.

[57]Leur conduite vis à vis de Cauchy est indigne. Tous les exemplaires de Cauchy partent pour l'Allemagne. Gauthier-Villars me l'a bien dit et cependant il n'est jamais cité.

[58]Le travail de Weierstrass est antérieur à celui de Messieurs Briot et Bouquet, mais M. Poincaré qui devait savoir ça par le mémoire de Madame Kowalewski—s'il n'a pas connu le travail *Analytische Facultäten*—n'en dit pas un mot. Monsieur de Ramsey m'a raconté qu'il a entendu par M. Molk—l'étudiant français qui suit le cours de M. Weierstrass à Berlin—que M. Poincaré déteste les Allemands, ce que je trouve fort naturel, et qu'il a pour principe de ne jamais citer un auteur allemand ce qui serait fort mal si c'était vrai.

> Poincaré's ignorance of the mathematical literature, when he started his researches, is almost unbelievable. He hardly knew anything on the subject beyond Hermite's work on the modular functions; he certainly had never read Riemann, and by his own account had not even heard of the Dirichlet principle.

This may also be due to Poincaré's lack of time, although the contrary may also be supported, that is, Poincaré had so much energy that it is unlikely that he could not find time to read others' writings, especially on topics on which he was working. The explanation may come from the fact that Poincaré belongs to this small category of a mathematician who reconstructs his background by himself, without reading others' works.

As we already mentioned, despite the war, the French mathematicians had an immense admiration for German mathematics, even though they considered it too abstract. Let us quote a few passages on this subject from the correspondence between Hermite and Mittag-Leffler. Hermite writes in a letter dated October 6, 1884, [47]:

> Abstraction, which is a charm for the Germans, is bothering us; it draws a kind of veil on the consequences which stays hidden to us in part, until we have taken, to attain it, a path which is more adapted to us.[59]

In other letters, Hermite expresses his highest esteem for the German mathematicians. For example, on January 14, 1892, he writes [48]:

> History of science keeps for ever the memory of the relations between Legendre and Jacobi; something good and affectionate emerges from the correspondence between these great geometers, which exerted its influence on their heirs.[60] No division ever emerged among mathematicians of these two countries. It is in entertaining friendly relations that they followed the same path in their works, and Appell's *mémoire couronné*[61] is a shining example, by its exceptional merit, by the new light it sheds on Riemann, of the ultimate alliance of the genius of the two nations, for the advancement of science."[62]

In another letter to Mittag-Leffler, dated July 10, 1893, Hermite writes [48]:

> I wrote to the French ambassador a letter which Appell read, at my request, with great care, and to which he gave his complete assessment. I was expressing, in a natural way, the

---

[59]L'abstraction, qui est un charme pour les Allemands, nous gêne et jette sur les conséquences comme un voile qui nous dérobe une partie jusqu'à ce que nous ayons fait pour y parvenir un chemin plus à notre convenance.

[60]The correspondence is reproduced in Jacobi's *Collected Works*, [54] t. I, pp. 385–461, and in Crelle's Journal, 80 (1875), pp. 205–279.

[61]This is Paul Appell's memoir *Sur les intégrales de fonctions à multiplicateurs et leur application au développement des fonctions Abéliennes en séries trigonométriques* (mémoire couronné par S. M. le roi Oscar II, le 21 janvier 1889).

[62]L'histoire de la science garde à jamais le souvenir des relations de Legendre et de Jacobi; quelque chose de bon et d'affectueux se dégage de la correspondance entre ces grands géomètres, qui a exercé son influence sur leurs successeurs. Aucune division ne s'est jamais montrée entre les mathématiciens des deux pays; c'est en entretenant des relations d'amitié qu'ils ont suivi la même voie dans leurs travaux, et le mémoire couronné d'Appell est un témoignage éclatant, par son mérite hors ligne, par le lustre nouveau qu'il jette sur Riemann, de l'intime alliance des génies des deux nations, pour la marche en avant de la science.

> sympathy and the admiration that all of us vow to the geometers that are the pride and the glory of German science.[63]

We quote, as the last example (there are many others) a letter from Hermite to Poincaré, dated November 27, 1880. We already mentioned that Poincaré was not keen on reading other's mathematical papers. Hermite writes ([89] pp. 169-170):

> [...] Allow me to urge you most of all to familiarize yourself with the works of Mr. Kronecker who infinitely surpassed me in this kind of research and to whom we owe the most remarkable and the most productive discoveries. The notions of class and of genus in the theory of quadratic forms were entirely linked to analysis by the eminent geometer [...] Some of the beautiful results discovered by Mr. Kronecker, and published in the *Monatsbericht*, were translated into French, at my request, and they appeared, around 1859 or 1860 in the *Annales de l'École Normale Supérieure*. But you must read in the same issue of the *Monatsbericht* of the Academy of Sciences of Berlin, and without omitting anythings of them, everything written by the hand of the great geometer.[64]

It is well known that Klein, at several places of his published talks, classifies mathematicians into logicians, formalists, and intuitives, and he claims that this has to do with the fact they are of Latin, Hebraic or German descent. Jules Tannery, whom we mentioned several times in this chapter, says that "Klein modestly related the gift of *envisioning*, which was so generously allocated to him, to the Teutonic race, whose natural power for intuition is supposed to be a pre-eminent attribute."[65] (quoted by Picard in [84] p. xxviii). This is an indication of the admiration that the French had for Klein. There are many other examples. Thus, to the question of whether French and German mathematicians ignored each other because of that war, the answer is clearly no.

7 In a Way of Conclusion

In this chapter, we tried to convey the idea that it took a certain amount of time for the notion of Riemann surface to be understood and used by French mathematicians. We also wanted to give a broad picture of the French mathematical community,

---

[63] [...] J'ai écrit à l'ambassadeur de France une lettre qu'Appell a lue avec grande attention à ma demande, et à laquelle il a donné son plus complet assentiment. J'exprimais naturellement les sentiments que nous éprouvons tous de sympathie et d'admiration pour les géomètres qui sont à l'honneur et la gloire de la science allemande.

[64] [...] Permettez-moi de vous engager à prendre surtout connaissance des travaux de Mr. Kronecker qui m'a infiniment dépassé dans ce genre de recherches et à qui l'on doit les découvertes les plus remarquables et les plus fécondes. Les notions de classes et de genres dans la théorie des formes quadratiques ont été entièrement rattachées à l'analyse par l'éminent géomètre [...] Quelques uns des beaux résultats découverts par Mr. Kronecker, et publiés dans les *Monatsbericht*, ont été à ma demande traduits en français et ont paru, vers 1859 ou 1860, dans les *Annales de l'École Normale Supérieure*. Mais il faut lire dans ce même recueil des *Monatsbericht* de l'Académie des Sciences de Berlin, et sans en rien omettre, tout ce qui est sorti de la plume du grand géomètre.

[65] Le don de *voir*, qui lui a été départi si généreusement, M. Klein le rapporte modestement à la race teutonique, dont la puissance naturelle d'intuition serait un attribut prééminent.

especially the branch on analysis, in the few decades following Riemann's work, and of the relations between the French mathematicians and their German colleagues. Let us quote again Hermite, from his preface to the French edition of Riemann's works [98], published in 1898. This is an interesting passage in which he summarizes the passage from Cauchy's ideas to Riemann's notion of Riemann surface.

> The notion of integration along a curve has been presented, in its simplest and easiest form, with numerous and important applications which showed their scope, since 1825, in a memoir by Cauchy entitled *Sur les intégrales définies prises entre des limites imaginaires* (On the definite integrals taken between imaginary limits). But it stays a property of the famous author. One had to wait for twenty-five years, until the works of Puiseux, Briot and Bouquet, so that it soars up and shines in Analysis. The profound notion of Riemann surface, whose access is very difficult, was soon introduced and it dominated Science, so as to remain there for ever.[66]

It is important to recall that in Germany, although Riemann's ideas were investigated since the beginning by several pre-eminent mathematicians, these ideas remained, to many, very cryptic. We may add that in Germany, Riemann's ideas were not always unanimously praised, and they were even subject to criticism. Bottazzini, in his ICM 2002 communication [13], reports on some notes written by Casorati during a visit he made to Berlin in 1864, at the time when Riemann was staying, for health reasons, in Italy (Pisa). Casorati writes ([13] p. 919) that "Riemann's things are creating difficulties in Berlin [...]" Bottazini quotes Casorati:

> Weierstrass claimed that "he understood Riemann, because he already possessed the results of his [Riemann's] research." As for Riemann surfaces, they were nothing other than "geometric fantasies." According to Weierstrass, "Riemann's disciples are making the mistake of attributing everything to their master, while many [discoveries] had already been made by and are due to Cauchy, etc.; Riemann did nothing more than to dress them in his manner for his convenience."

The mathematician and historian of science Leo Könisbsberger, who taught at the University of Heidelberg, recalls in his autobiography, *Mein Leben* (My life) published in 1919, that at the time he was a student in Berlin, the mathematics taught by Weierstrass was considered as the only mathematics that was rigorous. He writes: "All of us, the younger generation, had the impression that the ideas and methods of Riemann were not part of the rigorous mathematics of Euler, Lagrange, Gauss, Jacobi and Dirichlet" (p. 59). In his last course at the University of Berlin (1866), Weierstrass also declared that the theory of Riemann surfaces was a "pure fantasy." (From the manuscript course in the Humbolt-University in Berlin, quoted in [90], p. 131.) Regarding the same theory, Klein writes in his *Development of mathematics in the 19th century* (1926) ([59] p. 241):

---

[66]La notion de l'intégration le long d'une courbe avait été exposée, sous la forme la plus simple et la plus facile, avec de nombreuses et importantes applications qui en montraient la portée, dès 1825, dans un Mémoire de Cauchy ayant pour titre *Sur les intégrales définies prises entre des limites imaginaires*; mais elle reste dans les mains de l'illustre Auteur; il faut attendre vingt-cinq ans, jusqu'aux travaux de Puiseux, de Briot, de Bouquet, pour qu'elle prenne son essor et rayonne dans l'Analyse. La notion profonde des surfaces de Riemann, qui est d'un accès difficile, s'introduit sans retard et domine bientôt la Science pour y rester à jamais.

> Even today, the beginning student of Riemann surfaces faces great difficulties: The "winding points," around which the various "sheets" hang together, are essential; the curves proceeding from these points along which the sheets intersect, are not—they can be arbitrarily shifted, as long as their ends remain fixed, and in any case, they occur only because we involuntarily make the construction in three-dimensional space.

Riemann visited Paris in April 1860, on the invitation of French mathematicians. In a letter to his sister Ida, he describes a social atmosphere that was not in accord with his restrained character. He writes[67]:

> In general I am satisfied with the results of my trip, even if my expectations which I had earlier attached to the journey must remain unfulfilled, necessitated by the shortness of time. In this regard it would have been of little value if I had remained one or two weeks longer in Paris. And so I preferred to return to Göttingen at the right time.
>
> I can not complain at all about a lack of friendliness on the part of the Parisian scholars. The first social occasion, in which I took part, was a tea at Herr Serret's, who had become a member of the institute a few weeks before. Such a tea or "Réunion" contrasts sharply with our socials. It begins at 9:00 pm, really gets going at 10:00 and goes till 1 o'clock in the morning. During this time guests continually come and go; many come right from the theatre, which in Paris seldom closes before 12:30. They consist of nothing but teal ice cream and a variety of sweet-meats (?), namely, glazed fruits and other sweets of that sort. It cannot be denied that this unrestrained manner has perverted many.
>
> The social gathering at Serret's consisted of 30 to 40 ladies and gentlemen, among whom were also several Germans or rather speakers of German. I conversed chiefly with them.

Bottazzini declares in [55] p. 244 that during that stay in Paris, Riemann met, among others, Hermite, Puiseux, Briot and Bouquet.

The German mathematicians had in general a great consideration for the French. We quote a passage from a letter from Weierstrass to Kovalevskaya, sent on June 14, 1882, after the latter informed him that she met Hermite (the letter is reproduced in Mittag-Leffeler's ICM lecture [67]): "You should now also enter into a relationship with other mathematicians: the young ones, Appell, Picard, Poincaré will be extremely interesting for you."

Acknowledgements The author would like to thank Vincent Alberge, Marie Pascale Hautefeuille and Ken'ichi Ohshika for reading a preliminary version of this chapter.

References

1. N.H. Abel, Recherches sur les fonctions elliptiques. Journal für die reine und angewandte Mathematik **2**, 101–181 (1827)
2. P. Appell, Sur les propriétés des cubiques gauches et le mouvement hélicoïdal d'un corps solide. Thèse présentée à la Faculté des Sciences de Paris (Gauthier-Villars, Paris, 1876)
3. P. Appell, Sur les propriétés des cubiques gauches et le mouvement hélicoïdal d'un corps solide. Annales scientifiques de l'École Normale Supérieure, Série 2(5), 245–274 (1876)
4. P. Appell, É. Goursat, Théorie des fonctions algébriques et de leurs intégrales: Étude des fonctions analytiques sur une surface de Riemann (Gauthier-Villars, Paris, 1895)

[67]The translation is by Gallagher and M. Weissbach, cf. [100].

5. P. Appell, E. Lacour, *Principes de la théorie des fonctions elliptiques et applications* (Gauthier-Villars, Paris, 1897)
6. P. Appell, É. Goursat, P. Fatou, *Étude des fonctions analytiques sur une surface de Riemann* (Gauthier-Villars, 1929)
7. O. Bonnet, Mémoire sur la théorie générale des surfaces. Journal de l'école Polytechnique **19**, 1–133 (1848)
8. O. Bonnet, Sur la surface réglée dont les rayons de courbure principaux sont égaux et dirigés en sens contraires. Journal de l'école Polytechnique. Paris. **19**, 134–136 (1848)
9. O. Bonnet, Observations sur les surfaces minima. Comptes Rendus des Séances de l'Académie des Sciences. Paris, vol. 41 (1855), pp. 1057–1058
10. O. Bonnet, Sur la détermination des fonctions arbitraires qui entrent dans l'équation générale des surfaces à aire minimum. Comptes Rendus des Séances de l'Académie des Sciences. Paris, vol. 40 (1855), pp. 1107–1110
11. O. Bonnet, Nouvelles remarques sur les surfaces à aire minima. Comptes Rendus des Séances de l'Académie des Sciences. Paris, vol. 42 (1856), pp. 532–535
12. O. Bonnet, Sur la surface réglée minima. Bulletin des Sciences Mathématiques **9**, 14–15 (1885)
13. U. Bottazini, *Algebraic truths versus Geometric fantasies, Weierstrass' response to Riemann* ed. By T.T. Li, et al. *Proceedings ICM*, Beijing, China, August 20–28, 2002, Vol. III, Beijing, Higher Education Press, pp. 923–924
14. J.-C. Bouquet, Sur la variation des intégrales doubles. Thèse d'analyse présentée à la Faculté des sciences de Paris (Imprimerie Bachelier, Paris, 1843)
15. Ch.-A. Briot, Sur le mouvement d'un corps solide autour d'un point fixe. Présentée à la Faculté des Sciences de Paris (Imprimerie Bachelier, Paris, 1842)
16. Ch.-A. Briot, J.-C. Bouquet, Étude des fonctions d'une variable imaginaire (Premier mémoire). Jour. Éc. Polytechnique **21**, 85–121 (1856)
17. Ch.-A. Briot, J.-C. Bouquet, *Théorie des fonctions doublement périodiques et, en particulier, des fonctions elliptiques*, 1st edn. (Gauthier-Villars, Paris, 1859)
18. Ch.-A. Briot, J.-C. Bouquet, Theorie der doppelt-periodischen Functionen und insbesondere der elliptischen Transcendenten. Dargestellt von H. Fischer, Helle, (1862). Translation of [20]
19. Ch.-A. Briot, J.-C. Bouquet, *Théorie des fonctions elliptiques*, 2nd edn. of [19] (1875)
20. Ch.-A. Briot, *Théorie des fonctions abéliennes* (Gauthier-Villars, Paris, 1879)
21. E. Cahen, Sur la fonction $\zeta(s)$ de Riemann et sur des fonctions analogues, Thèse de doctorat présentée à la Faculté des Sciences de Paris (Gauthier-Villars, 1895)
22. E. Cahen, Sur la fonction $\zeta(s)$ de Riemann et sur des fonctions analogues, Annales scientifiques de l'École Normale Supérieure, 3e série **11**, pp. 75–164 (1894)
23. A. Cauchy, Mémoire sur les intégrales définies (1814), Mémoires présentés par divers savants à l'Académie royale des sciences de l'Institut de France 1(1827) pp. 611–799. in Œuvres, série 1, tome I, pp. 319–506
24. A. Clebsch, P.A. Gordan, *Theorie der Abelschen Funktionen* (Teubner, 1866)
25. G. Darboux, Mémoire sur les fonctions discontinues, Annales Sc. de l'École Normale Supérieure, 2e série, vol. 4 (1875), pp. 57–112
26. G. Darboux, J. Hoüel, La correspondance de G. Darboux avec J. Hoüel. Chronique d'un rédacteur (déc. 1869-nov., ed. By H. Gispert, Cahiers du séminaire d'histoire des mathématiques **8**(1987), pp. 67–202 (1871)
27. J. Dieudonné, *Article "Poincaré," Dictionary of Scientific Biography*, vol. XI (Charles Scribner's Sons, New York, 1975), pp. 51–61
28. P. Dugac, Éléments d'analyse de Karl Weierstrass. Arch. Hist. Exat Sci. **10**, 41–167 (1973)
29. P. Fatou, Séries trigonométriques et séries de Taylor (Université de Paris, Thèse de doctorat, 1906)
30. P. Fatou, Séries trigonométriques et séries de Taylor. Acta Math. **30**, 335–400 (1906)
31. H. Gispert, Les débuts de l'histoire des mathématiques sur les scènes internationales et le cas de l'entreprise encyclopédique de Félix Klein et Jules Molk. Hist. Math. **26**, 344–360 (1999)

32. É. Goursat, Sur l'équation différentielle linéaire qui admet pour intégrale la série hypergéométrique de Gauss, Annales Scientifiques de l'École Normale Supérieure, 2e série, 10, pp. 107–136 (1881)
33. É. Goursat, Sur l'équation différentielle linéaire, qui admet pour intégrale la série hypergéométrique. Annales scientifiques de l'École Normale Supérieure, Série 2(10), 3–142 (1881)
34. É. Goursat, Cours d'analyse mathématique, Paris, Gauthier-Villars, 1ère édition, 2 volumes, 1902–1905, 2e édition, 3 volumes, 1910–1913
35. É. Goursat, Translation of [41]: a course in mathematical analysis. New York, Ginn & Co., 3 volumes, 1904–1917. Reprint, New York, Dover, 1959. (Translated By E.R. Hedrick, O. Dunkel)
36. É. Goursat, Démonstration du théorème de Cauchy. Acta Math. **4**, 197–200 (1884)
37. J. Hadamard, *Notice sur les travaux scientifiques de M. Jacques Hadamard* (Gauthier-Villars, Paris, 1901)
38. J. Hadamard, Sur la distribution des zéros de la fonction $\zeta(s)$ et ses conséquences arithmétiques. Bulletin de la Société Mathématique de France **24**, 199–220 (1896)
39. J. Hadamard, Notions élémentaires sur la géométrie de situation, Ann. de math., 4e série, (1909) t. 9, pp. 193–235
40. H.-G. Halphen, *Sur les invariants différentiels. Thèse présentée à la Faculté des Sciences de Paris* (Gauthier-Villars, Paris, 1878)
41. G.-H. Halphen, *Traité des fonctions elliptiques et de leurs applications*, 3 volumes (Gauthier-Villars, Paris, 1886, 1888, 1891)
42. Ch. Hermite, Sur la résolution de l'équation du cinquième degré, Comptes Rendus des Séances de l'Académie des Sciences. Paris, vol. 46 A (1858), pp. 508–515
43. Ch. Hermite, Sur la résolution de l'équation du cinquième degré, Comptes Rendus des Séances de l'Académie des Sciences. Paris, vol. 46 A (1858), pp. 715–722
44. Ch. Hermite, *Cours d'analyse de l'École Polytechnique* (Gauthier-Villars, Paris, 1873)
45. Ch. Hermite, Cours de M. Hermite à la Faculté des Sciences de Paris, Rédigé en 1882 par M. Andoyer, 1st ed. 1882, 2nd ed. 1883, 3d edn. 1887, 4th ed. 1891. Hermann, Paris, 4th edn. (1891)
46. C. Hermite, Lettres de Charles Hermite à Gösta Mittag-Leffler (1874–1883). Cahiers du séminaire d'histoire des mathématiques, tome **5**, 49–285 (1984)
47. C. Hermite, Lettres de Charles Hermite à Gösta Mittag-Leffler (1884-1891). Cahiers du séminaire d'histoire des mathématiques, tome 6, 79–217 (1985)
48. C. Hermite, Lettres de Charles Hermite à Gösta Mittag-Leffler (1892-1900) Cahiers du séminaire d'histoire des mathématiques, tome 10, 1–82 (1989)
49. C. Hermite, *Œuvres de Charles Hermite, publiées sous les auspices de l'Académie des sciences par Émile Picard*, 4 volumes (Gauthier-Villars, Paris, 1905–1917)
50. D. Hilbert, Über die grundlagen der geometrie. Geom. Math. Annalen **56**, 381–422 (1903)
51. L. Kronecker, Grundzüge einer arithmetischen Theorie der algebraischen Grössen: festschrift zu Herrn Eduard Kummers fünzigjährigem Doctor-Jubiläum, 10., *Druck und verlag von G* (Reimer, Berlin, 1881). (Septembe, 1882)
52. C.G.J. Jacobi, Fundamenta nova theoriæ functionum ellipticarum, vol. 2, Regiomonti. Sumtibus fratrum Bornträger, 1829. in Collected works, tome 1, pp. 49–239
53. C.G.J. Jacobi, Solution nouvelle d'un problème fondamental de géodésie. Journal für die reine und angewandte Mathematik **53**, 335–341. Gesammelte Werke, t. II, 417–424 (1857)
54. C.G.J. Jacobi, Gesammelte Werke, Herausgegeben auf Veranlassung der königlich preussischen Akademie der Wissenschaften ed. By C.W. Borchardt, vol. 8 (New edition, Cambridge Library Collection, pp. 1881–1891)
55. H.N. Jahnke (ed.), in *A History of Analysis*, American Mathematical Society and London Mathematical Society History of Mathematics series, vol. 24 (2003)
56. C. Jordan, *Sur le nombre des valeurs des fonctions*. Thèse présentée à la Faculté des Sciences de Paris (Mallet-Bachelier, Paris, 1860)
57. C. Jordan, Cours d'analyse de l'École Polytechnique, vol. 3, Gauthier-Villars, Paris, 1st edn. 1882 (vol. I), 1883 (vol. II), 1887 (vol. III)

58. F. Klein, *Über Riemanns Theorie der algebraischen Funktionen und ihrer Integrale*, Teubner, Leipzig, 1882. English translation: *On Riemann's theory of algebraic functions and their integrals; a supplement to the usual treatises*. (Translated from the German by Frances Hardcastle, Macmillan and Bowes, Cambridge, 1893. Reprint: New York, Dover Publications, 1963)
59. F. Klein, Vorlesungen über die Entwicklung der Mathematik im vol. 19.2, Julius (Springer, Berlin, 1926). Development of mathematics in the 19th century, Math. Sci. Press, Brookline, Mass.,1979. (English translation By M. Ackermann)
60. É. Lacour, Sur des fonctions d'un point analytique à multiplicateurs exponentiels ou à périodes rationnelles. Thèse pour le doctorat de la Faculté des Sciences de Paris (Gauthier-Villars, Paris, 1895)
61. E. Landau, Handbuch der Lehre von der Verteilung der Primzahlen, Leipzig B.G. Teubner, 1909. (Reprint, Chelsea 1953)
62. H. Lebesgue, *Leçons sur l'intégration et la recherche des fonctions primitives* (Gauthier-Villars, Paris, 1904)
63. A.-M. Legendre, Exercices de calcul intégral sur divers ordres de transcendantes et sur les quadratures, 3 volumes, Paris, Courcier, 1811–1816
64. A.-M. Legendre, Traité des fonctions elliptiques et des intégrales eulériennes, avec des tables pour en faciliter le calcul numérique, 3 volumes, Paris, Huzard-Courcier, 1825–1828
65. H. Lebesgue, Lettres d'Henri Lebesgue à Émile Borel. Cahiers du séminaire d'histoire des mathématiques **12**, 1–506 (1991)
66. J. Liouville, Leçons sur les fonctions doublement périodiques, faites en 1847, Première partie. Théorie générale. Journal für die reine und angewandte Mathematik **88**, 277–310 (1880)
67. M.G. Mittag-Leffler, Une page de la vie de Weierstrass, in *Proceedings of the Second ICM, Paris, 1900* (Gauthiers-Villars, Paris, 1902), pp. 131–153
68. J. Molk, Sur une notion qui comprend celle de la divisibilité et sur la théorie générale de l'élimination. Acta Math. **6**, 1–165 (1885)
69. C. Neumann, Vorlesungen über Riemann's Theorie der Abel'schen Integrale, Leipzig, Teubner, 1865. 2nd revised edn. (1884)
70. A.B. Niewenglowski, Exposition de la méthode de Riemann pour la détermination des surfaces minima de contour donné, Annales scientifiques de l'École Normale Supérieure, 2e série **9**, pp. 227–300 (1880)
71. A.B. Niewenglowski, Exposition de la méthode de Riemann pour la détermination des surfaces minima de contour donné. Thesis submitted to the Faculté des Sciences (Gauthier-Villars, Paris, 1880)
72. W.F. Osgood, Review of Théorie des fonctions algébriques et de leurs intégrales by P. Appell, É. Goursat. Bull. Am. Math. Soc. 317–327 (1896)
73. W.F. Osgood, Review of Cours d'analyse mathématique by Édouard Goursat, Tome I, Gauthier-Villars, Paris. Bull. Am. Math. Soc. **9**(1902–1903), 547–555 (1902)
74. W.F. Osgood, Review of Cours d'analyse mathématique by Édouard Goursat, Tome II, Gauthier-Villars, Paris, 1905. Bull. Am. Math. Soc. **15**(3), 120–126 (1908)
75. A. Papadopoulos, Looking backward: euler and Riemann, this volume, Chap. 1, pp. 1–81
76. A. Papadopoulos, Physics in Riemann's mathematical papers, this volume, Chap. 6, pp. 151–199
77. A. Papadopoulos, Cauchy and Puiseux: Two precursors of Riemann, this volume, Chap. 7, pp. 209–232
78. J. Pfeiffer, Joseph Liouville (1809–1882): ses contributions à la théorie des fonctions d'une variable complexe. Revue d'histoire des sciences, t. **36**(3–4), 209–248 (1983)
79. É. Picard, *Traité d'analyse*, 3 volumes (Gauthier-Villars, Paris, 1891, 1893, 1896)
80. É. Picard, Application de la théorie des complexes linéaires à l'étude des surfaces et des courbes gauches. Thèse présentée à la Faculté des sciences de Paris (Gauthier-Villars, Paris, 1877)
81. É. Picard, Application de la théorie des complexes linéaires à l'étude des surfaces et des courbes gauches. Annales scientifiques de l'École Normale Supérieure, Série **2**(6), 329–366 (1877)

82. É. Picard, Notice sur G.-H. Halphen, in: G.-H. Halphen, Traité des fonctions elliptiques et de leurs applications, [48], vol. III
83. É. Picard, G. Simart, Théorie des fonctions algébriques de deux variables indépendantes, 2 volumes, Paris, Gauthier-Villars, 1897–1906 (Chelsea Publishing Co., Bronx, N.Y, Reprinted in one volume, 1971)
84. É. Picard, La vie et l'œuvre de Jules Tannery, Académie des Sciences, lecture faite dans la séance publique annuelle du14 décembre (1925)
85. J. Pierpont, Mathematical instruction in France. Bull. Amer. Math. Soc. **6**(6), 225–249 (1900)
86. H. Poincaré, Correspondance avec Paul Appell. Cahier du séminaire d'histoire des mathématiques **7**, 61–80 (1986)
87. H. Poincaré, Analysis situs, J. École Polytechnique (2) 1, 1895, pp. 1–121. Œuvres, Vol. VI, pp. 193–288
88. H. Poincaré, Analyse des travaux scientifiques de Henri Poincaré faite par lui-même, Acta Math. **38**, 36-135 (1921). (Œuvres, vol. I)
89. H. Poincaré, La correspondance avec des mathématiciens de A à H, ed. By P. Dugac, Cahiers du séminaire d'histoire des mathématiques, tome 10 (1986), pp. 59–219
90. H. Poincaré, La correspondance avec des mathématiciens de J à Z, Cahiers du séminaire d'histoire des mathématiques, ed. By P. Dugac, tome 10 (1989), pp. 83–229
91. L. Poinsot, Théorie nouvelle de la rotation des corps (Première et deuxième Parties). Journal de mathématiques pures et appliquées, 1ère série, **16**, 9–129 (1851)
92. B. Riemann, Grundlagen für eine allgemeine Theorie der Functionen einer veränderlichen complexen Grösse, (Göttingen, 1851), [109] pp. 3–48
93. B. Riemann, Über die Darstellbarkeit einer Function durch eine trigonometrische Reihe (On the representability of a function by a trigonometric series). Aus dem dreizehnten Bande der Abhandlungen der Königlichen Gesellschaft der Wissenschaften zu Göttingen (1867)
94. B. Riemann, Theorie der Abel'schen Functionen. Journal für die reine und angewandte Mathematik, 115–155 (1857). Reprinted in his *Gesammelte mathematische Werke* [109], pp. 88–144
95. B. Riemann, Über die Fläche vom kleinsten Inhalt bei gegebener Begrenzung, Abh. Königl. d. Wiss. Göttingen, Mathem. Cl., **13**, 3–52 (1867)
96. B. Riemann, Ein Beitrag zu den Untersuchungen über die Bewegung eines flüssigen gleichartigen Ellipsoides. Abhandlungen der Königlichen Gesellschaft der Wissenschaften zu Göttingen **9**, 3–36 (1861)
97. B. Riemann, Beiträge zur Theorie der durch die Gauss'sche Reihe $F(\alpha, \beta, \gamma, x)$ darstellbaren Functionen. Aus dem siebenten Band der Abhandlungen der Königlichen Gesellschaft der Wissenschaften zu Göttingen (1857), pp. 3–32
98. B. Riemann, Collected works translated into French: oeuvres mathématiques de Riemann, traduites par L. Laugel, avec une préface de C. Hermite, Paris, (Gauthier-Villars 1898)
99. B. Riemann, Über die Hypothesen, welche der Geometrie zu Grunde liegen, published By R. Dedekind, after Riemann's death, in Abhandlungen der Königlichen Gesellschaft der Wissenschaften zu Göttingen, vol. 13, 1867. French translation in [110], pp. 280–299
100. The Riemann letters at the Prussian cultural archive, mimeographed notes (1981). (translated By R. Gallagher, M. Weissbach)
101. G. Roch, Über die Anzahl der willkürlichen Constanten in algebraischen Functionen. J. Reine Angew. Math. **64**, 372–376 (1865)
102. H.A. Schwarz, Auszug aus einem Briefe an Herrn F. Klein, Complete Works, vol. II
103. G. Simart, *Commentaire sur deux mémoires de Riemann relatifs à la théorie générale des fonctions et au principe de Dirichlet*. Thèse de doctorat, Faculté des sciences, Paris (Gauthier-Villars, 1882)
104. T.J. Stieltjes, Sur une fonction uniforme. Comptes Rendus des Séances de l'Académie des Sciences. Paris, **101**, 153–154 (1885)
105. J. Tannery, Propriétés des intégrales des équations différentielles linéaires à coefficients variables. Thèse présentée à la Faculté ses sciences de Paris (Gauthier-Villars, 1874)
106. J. Tannery : Propriétés des intégrales des équations différentielles linéaires à coefficients variables. Annales Scientifiques de l'École Normale Supérieure, 2e série, **4**, 113–182 (1875)

107. J. Tannery, J. Molk, Éléments de la théorie des fonctions elliptiques, Paris (Gathier-Villars, 1893, 1896, 1898 and 1902)
108. O. Teichmüller, Extremale quasikonforme Abbildungen und quadratische Differentiale. *Abh. Preuss. Akad. Wiss., Math.-Naturw. Kl.* 22 (1940), 1–197. (English translation By G. Théret), Extremal quasiconformal mappings and quadratic differentials. In *Handbook of Teichmüller theory* (ed. A. Papadopoulos), vol. V (EMS Publishing House, Zürich 2015), pp. 321–483
109. K. Weierstrass, Über das sogenannte Dirichlet'sche Princip, Gelesen in der Königl. Akademie der Wissenschaften am 14. Juli 1870, Ausgewählte Kapitel aus der Funktionenlehre, Teubner-Archiv zur Mathematik vol. 9, 1988, pp. 184–189; published in Weierstrass' *Mathematische Werke*, 2, Mayer and Müller, Berlin (1895), pp. 49–54
110. H. Weyl, Die Idee, der Riemannschen Fläche. B.G. Teubner, Leipzig, (English translation By G.R. MacLane), The concept of a Riemann surface, Addison-Wesley, 1913). 1955
111. S. Yamada, Riemann on minimal surfaces. This volume, Chapter 5, pp. 135–150

Part II
Philosophy

The Origin of the Notion of Manifold: From Riemann's Habilitationsvortrag Onward

Ken'ichi Ohshika

1 Introduction

It goes without saying that the notion of manifold is a very important foundation of modern geometry, or even of modern mathematics in general. The invention of this notion is usually attributed to Riemann. In fact, the term "Mannigfaltigkeit", of which the word "manifold" is an English translation, appeared for the first time in the world of mathematics in Riemann's famous Habilitationsvortrag. There are other English translations such as "multiplicity" or "variety" in the literature. Prior to Riemann, the word "Mannigfaltigkeit" was already used in a non-mathematical context. There is even a poem by Schiller entitled "Mannigfaltigkeit." Still, if we read the text of Riemann today, we find that his description of manifolds is rather literary, and there is no clear, rigorous definition. On the other hand, his text is very suggestive and has fertile content. This is one of the reasons why his text inspired many mathematicians, and also philosophers. Even today, we can learn something new reading his text although it is not so easy to reach its depth.

New notions in mathematics quite often take much time for their maturation and consolidation. The notion of manifold is one such example, and it took nearly half a century from the first invention by Riemann to get to the modern definition we know today. The purpose of this article is to give an overview of this slow process where the notion of manifold was born, clarified and developed, taking a look at both mathematical and philosophical aspects. We shall start with the philosophical background of the time when Riemann's Habilitationsvortrag was written, focusing on Kant's metaphysics. Then we shall turn to Riemann's paper, showing how he described manifolds there and quoting some passages from his text directly. We

Partially supported by JSPS Grant-in-Aid 22244005.

K. Ohshika (✉)
Department of Mathematics, Graduate School of Science, Osaka University,
Toyonaka, Osaka, Japan
e-mail: ohshika@math.sci.osaka-u.ac.jp

L. Ji et al. (eds.), *From Riemann to Differential Geometry and Relativity*,
DOI 10.1007/978-3-319-60039-0_9

shall also see how Riemann's worldview diverges from the Kantian. It was Poincaré who put for the first time Riemann's idea into clear mathematical terms. We shall look at Poincaré's celebrated paper "Analysis Situs", where he gave two definitions of manifold, and also at his philosophical position, conventionalism. Finally, we shall see the process in which the notion of manifold became mathematically rigorous, and reached the notion we understand today. There are many mathematicians involved in this process, but we just pick up prominent ones among them, looking at the work of Hilbert, Weyl, Kneser, Veblen-Whitehead and Whitney.

The author would like to express his gratitude to Athanase Papadopoulos, who invited him to write this article, and drew his attention to Grothendieck's mention of "multiplicité modulaire" and to the Kantian use of the word "Mannigfaltigkeit."

2 Kantian Worldview

At the time when Riemann gave birth to the notion of manifold in the world, or before, what should be called a Kantian worldview was prevailing. This view was widely held by scientists and mathematicians. To understand the philosophical aspects of the papers of Riemann and Poincaré, it is very helpful first to take a glance at Immanuel Kant's work.

It is well known that Kant tried to give philosophical foundations to mathematics and the natural science. His position was quite different from naïve empiricism or even a "sceptical empiricism" à la David Hume, and also from continental rationalism. If we allow ourselves to use modern jargon, his work should be regarded as a trial to "aufheben" both.

He supposed that there are a priori things in human thinking which make natural science possible. In the Kritik der reinen Vernunft [7] (we use the abbreviation KrV from now on), he included space, which should be understood to be a three-dimensional Euclidean space or as subset of it, or the human recognition of space, into the category of 'synthetic a priori.' (Here the qualifier "synthetic" is opposed to "analytic", the latter referring to a tautological kind of propositions.) In spite of his non-empiricist view, in contrast to conventionalism which was to become preponderant afterwards, Kant regards space as neither a human invention nor a framework which we created, but something given to us from the beginning and preceding all other form of recognition and knowledge.

Let us take a look at how Kant describes "space" in the first chapter of KrV. For English translations of all quotations from KrV, we use the Cambridge version [8].

> Der Raum ist kein empirischer Begriff, der von äußeren Erfahrungen abgezogen worden.
>
> Space is not an empirical concept that has been drawn from outer experiences.

This is an outright negation of the empiricists' view that our notion of space and its perception derive from our experiences and observations. Moreover, he regarded space as a foundation of all our perceptions.

> Der Raum ist eine notwendige Vorstellung a priori, die allen äußeren Anschauungen zum Grunde liegt.

> Space is a necessary representation, a priori, which is the ground of all outer intuitions.

> Der Raum ist kein diskursiver oder, wie man sagt, allgemeiner Begriff von Verhältnissen der Dinge überhaupt sondern eine reine Anschauung. Denn erstlich kann man sich nur einen einigen Raum vorstellen, und wenn man von vielen Räumen redet, so versteht man darunter nur Teile eines und desselben alleinigen Raumes.

> Space is not a discursive or, as is said, general concept of relations of things in general, but a pure intuition. For, first, one can only represent a single space, and if one speaks of many spaces, one understands by that only parts of one and the same unique space.

The preceding claim is remarkable for us, mathematicians, although the present author is not sure whether Kant allowed this "unique space" to be disconnected.

> Der Raum wird als eine unendliche gegebene Größe vorgestellt.

> Space is represented as a given infinite magnitude.

The meaning of infinity for Kant is rather ambiguous if we consider it from a mathematical viewpoint, but it may be most natural to interpret this notion as the unboundedness of space with respect to some metric. Kant gave these claims with some justification, which looks like a kind of reductio ad absurdum.

In addition to these basic theses on space by Kant, it is worthwhile to look at what he claimed to be antinomies regarding the nature of space. He posed the following two theses which contradict each other.

1. Die Welt hat einen Anfang in der Zeit, und ist dem Raum nach auch in Grenzen eingeschlossen.
   The world has a beginning in time, and in space it is also enclosed in boundaries.
2. Die Welt hat keinen Anfang, und keine Grenzen im Raume, sondern ist, sowohl in Ansehung der Zeit, als des Raumes, unendlich.
   The world has no beginning and no bounds in space, but is infinite with regard to both time and space.

For today's mathematicians, what Kant means by words like "infinite" or "bounds" may seem rather cryptic. For instance, we may wonder in the same way as for the previous quotation if he meant by 'infinite' space non-compactness of a topological space or unboundedness of a metric space. Since the notion of topological space did not exist in the time of Kant and no metrics were known except for the Euclidean one, it is more natural to interpret this notion of infinity as unboundedness with respect to the Euclidean metric. Still, we are allowed to imagine that if Kant had known the existence of closed 3-manifolds, then he would have posed his antinomies in a quite different way.

It should be mentioned here that Kant also used the term "Mannigfaltigkeit" in KrV. For him, however, this word is an abstract noun which means a condition of things being "manifold" rather than an object, and hence may be translated as "manifoldness." In contrast, the word "Mannigfaltige," which is a noun derived from the adjective "mannigfaltig," appears in KrV in a sense a bit closer to the later use of "Mannigfaltigkeit" by Riemann etc., although its meaning is not mathematical. We cite here a couple of passages from KrV. The first is taken from a section where he talks about "time," and there he poses an infinite line as an analogy of time. The use of the word "Mannigfaltige" in this passage is comparable to Riemann's description of one-dimensional "Mannigfaltigkeit."

> Und, eben weil diese innere Anschauung keine Gestalt gibt, suchen wir auch diesen Mangel durch Analogien zu ersetzen, und stellen die Zeitfolge durch eine ins Unendliche fortgehende Linie vor, in welcher das Mannigfaltige eine Reihe ausmacht, die nur von einer Dimension ist, und schließen aus den Eigenschaften dieser Linie auf alle Eigenschaften der Zeit, außer dem einigen, daß die Teile der ersteren zugleich, die der letzteren aber jederzeit nacheinander sind.

> And just because this inner intuition yields no shape we also attempt to remedy this lack through analogies, and represent the temporal sequence through a line progressing to infinity, in which the manifold constitutes a series that is of only one dimension, and infer from the properties of this line to all the properties of time, with the sole difference that the parts of the former are simultaneous but those of the latter always exist successively.

The second passage is taken from a section where Kant explains what the transcendental logic means. Here the meaning of the word "Mannigfaltige" is more abstract.

> Dagegen hat die transzendentale Logik ein Mannigfaltiges der Sinnlichkeit a priori vor sich liegen, ... Raum und Zeit enthalten nun ein Mannigfaltiges der reinen Anschauung a priori, ...

> Transcendental logic, on the contrary, has a manifold of sensibility that lies before it a priori, ... Now space and time contain a manifold of pure a priori institution, ...

In the paragraph next to the one containing this passage, Kant also uses the word "Mannigfaltigkeit." In the Cambridge English translation, this word is translated as "manifoldness" as below.

> Ich verstehe aber unter Synthesis in der allgemeinsten Bedeutung die Handlung, verschiedene Vorstellungen zueinander hinzuzutun, und ihre Mannigfaltigkeit in einer Erkenntnis zu begreifen.

> By synthesis in the most general sense, I understand the action of putting different representations together with each other and comprehending their manifoldness in one cognition.

To sum up, as can be seen from these examples, Kant's use of the word "Mannigfaltige" or "Mannigfaltigkeit" is close to our daily use of words like "variety" and "diversity", and it does not indicate some concrete mathematical object as in Riemann's paper. Still, we can imagine that this notion motivated Riemann's choice of the word "Mannigfaltigkeit", presumably via the work of Herbart, who was an indirect successor of Kant in Königsberg.

3 Riemann's Habilitationsvortrag

Riemann's Habilitationsvortrag, which was published only posthumously in 1867, is the content of his habilitation lecture given in 1854. It was required for Riemann to submit three different topics for his habilitation, from which the faculty would choose one. Riemann proposed one work on trigonometric series, another one on a system of quadratic equations, and a third one entitled 'Über Hypothesen, welche der Geometrie zu Grunde liegen', and Gauss chose the last one.

3.1 *Philosophical Aspects*

Reading Riemann's Habilitationsvortrag [18] today, it is clear that this work has a strong philosophical connotation. Although Riemann did not mention the name of Kant there, he talked about the philosophy of Herbart. In the introduction of this paper, Riemann writes the following. (There are several English translations of Riemann's text. We shall mainly use the one by Spivak [20], but also refer to the one by Clifford [19] occasionally.)

> Diese Dunkelheit wurde auch von Euklid bis auf Legendre, um den berühmtesten neueren Bearbeiter der Geometrie zu nennen, weder von den Mathematikern, noch von den Philosophen, welche sich damit beschäftigten, gehoben. Es hatte dies seinen Grund wohl darin, dass der allgemeine Begriff mehrfach ausgedehnter Grössen, unter welchem die Raumgrössen enthalten sind, ganz unbearbeitet blieb.

> From Euclid to Lagrange, the most famous of the modern reformers of geometry, this darkness has been dispelled neither by the mathematicians nor by the philosophers who have concerned themselves with it. This is undoubtedly because the general concept of a multiply extended quantities, which includes spatial quantities, remains unexplored.

It is quite natural to imagine that Kant was included among the philosophers whom Riemann was talking about. With these words, Riemann began to elucidate the concept of space underlying geometry. The "multiply extended quantities" mentioned here would turn out to be "manifolds (Mannigfaltigkeiten)", whose study was supposed to be the subject of this Habilitationsvortrag.

There are phrases in Riemann's text which should be regarded as "anti-Kantian." The following is one example.

> ..., dass die Sätze der Geometrie sich nicht aus allgemeinen Grössenbegriffen ableiten lassen, sondern dass diejenigen Eigenschaften, durch welche sich der Raum von anderen denkbaren dreifach ausgedehnten Grössen unterscheidet, nur aus der Erfahrung entnommen werden können.
> ..., it is a necessary consequence that the theorems of geometry cannot be deduced from general notions of quantity, but that those properties that distinguish space from other conceivable triply extended quantities are only to be deduced from experience.

Riemann claimed that there are more possibilities for multiply extended quantities than we can usually imagine and that the space as we intuitively grasp it is just one of them. Consequently, this choice is not a priori, but based on our experience. This position should be contrasted with the Kantian 'a priorism', and has rather a flavour of empiricism. We shall see later that this position is also significantly different from that of Poincaré, whose doctrine is explicitly anti-empiricist.

3.2 *Mannigfaltigkeit*

Now we proceed to look at the main part of the Habilitationsvortrag. Riemann introduced the notion of "Mannigfaltigkeit" (manifold or multiplicity) as a kind of set which "Bestimmungsweisen" (instances or specialisations) form and he said that this Mannigfaltigkeit can be either discrete or continuous. Here is the quotation of the part where the word "Mannigfaltigkeit" appears for the first time in the text.

> Grössenbegriffe sind nur da möglich, wo sich ein allgemeiner Begriff vorfindet, der verschiedene Bestimmungsweisen zulässt. Je nachdem unter diesen Bestimmungsweisen von einer zu einer andern ein stetiger Uebergang stattfindet oder nicht, bilden sie eine stetige oder discrete Mannigfaltigkeit; ...

> Notions of quantity are possible only when there already exists a general concept that admits particular instances. These instances form either a continuous or a discrete manifold, depending on whether or not a continuous transition of instances can be found between any two of them; ...

Therefore, in modern terminology, this notion of manifold should be interpreted as a set parametrised by n-tuples of real numbers. There is no formal definition in this part, such as the one using charts contained in modern textbooks. We can interpret what Riemann had in mind in several ways. For instance, it is possible to imagine that he allowed singularities to exist in a Mannigfaltigkeit. Riemann talked about the set of colours as an example of a Mannigfaltigkeit, which was said to have three dimensions. Of course he also mentioned a "Riemann" surface as an example in mathematics. It would not be so anachronic to expect that he also regarded the moduli space of Riemann surfaces as an example. In fact, Grothendieck referred to the "mulitiplicités modulaires" in [3], apparently in homage to Riemann. Unfortunately we cannot find any hint of Riemann's thinking along this line in his papers.

To justify the possibility of thinking of n-fold extended Mannigfaltigkeit, Riemann gave a detailed explanation on how to construct the entity of dimension n as follows. He started from a one-dimensional manifold:

> Geht man bei einem Begriffe, dessen Bestimmungsweisen eine stetige Mannigfaltigkeit bilden, von einer Bestimmungsweise auf eine bestimmte Art zu einer andern über, so bilden die durchlaufenen Bestimmungsweisen eine einfach ausgedehnte Mannigfaltigkeit, deren wesentliches Kennzeichen ist, dass in ihr von einem Punkte nur nach zwei Seiten, vorwärts order rückwärts, ein stetiger Fortgang möglich ist.

> In a concept whose instances form a continuous manifold, if one passes from one instance to another in a well-determined way, the instances through which one has passed form a simply extended manifold, whose essential characteristic is that from any point in it a continuous movement is possible in only tow directions, forwards and backwards.

Then he observed that it is possible to increase the dimension one by one, and to construct an n-dimensional Mannigfaltigkeit inductively. Thus he wrote:

> Wenn man, anstatt den Begriff als bestimmbar, seinen Gegenstand als veränderlich betrachtet, so kann diese Construction bezeichnet werden als eine Zusammensetzung einer Veränderlichkeit von $n + 1$ Dimensionen aus einer Veränderlichkeit von n Dimensionen und aus einer Veränderlichkeit von Einer Dimension.
>
> If one considers the process as one in which the objects vary, instead of regarding the concept as fixed, then this construction can be characterised as a synthesis of a variability of $n + 1$ dimensions from a variability of n dimensions and a variability of one dimension.

Here again the description is rather intuitive, and there is no formal definition of dimension. Still, we can see that Riemann understood dimension as a number of parameters which can vary independently. It should be noted that n-dimensional Euclidean space itself had already been introduced by Grassmann [2] in 1844, as a vector space, using a basis. What is new in Riemann's argument lies in the fact that he considered dimension as something which can be applied to much more general spaces.

Riemann's definition of metrics on manifolds is more explicit than that of manifolds themselves. He considered how the lengths of curves can be measured, and he used the symbol ds to denote the length element in manifolds, as we still do today. He also defined (rather intuitively) the sectional curvature by considering a surface spanned by geodesics in the manifold, as follows.

> Um die Krümmungsmass einer n-fach ausgedehnten Mannigfaltigkeit in einem gegebenen Punkte und einer gegebenen durch ihn gelegten Flächenrichtung eine greifbare Bedeutung zu geben, muss man davon ausgehen, dass eine von einem Punkte ausgehende kürzeste Linie völlig bestimmt ist, wenn ihre Ansfangsrichtung gegeben ist. Hienach wird man eine bestimmte Fläche erhalten, wenn man sämmtliche von dem gegebenen Punkte ausgehenden und in dem gegebenene Flächenelement liegenden Anfangsrichtungen zu kürzesten Linien verlängert, und diese Fläche hat in dem gegebenen Punkte ein bestimmtes Krümmungsmass, welches zugleich das Krümmungsmass der n-fach ausgedehnten Mannigfaltigkeit in dem gegebenen Punkte und der gegebenen Flächenrichtung ist.
>
> To give a tangible meaning to the curvature of an n-fold extended manifold, at a given point, and in a given surface direction, we first mention that a shortest line emanating from a point is completely determined if its initial direction is given. Consequently, we obtain a certain surface if we prolong all the initial directions from the given point which lie in the given surface element, into shortest lines; and this surface has a definite curvature at the given point, which is equal to the curvature of the n-fold extended manifold at the given point, in the given surface direction.

4 Poincaré's Analysis Situs

In contrast to Riemann's paper, the definition of a manifold by Poincaré is clear-cut. In his celebrated paper "Analysis situs" which was published in the Journal d'Ecole Polytechnique in 1895 [13], Poincaré gave explicitly two different definitions of manifolds. This paper begins with an audacious declaration which reads (the translation is by the present author):

La Géométrie à n dimensions a un objet réel; personne n'en doute aujourd'hui. Les êtres de l'hyperespace sont susceptibles de définitions précises comme ceux de l'espace ordinaire, et si nous ne pouvons nous les représenter, nous pouvons les concevoir et les étudier. Si donc, par exemple, la Mécanique à plus de trois dimensions doit être condamnée comme dépourvue de tout objet, il n'en est pas de même de l'Hypergéométrie.

The geometry of n-dimensions has a real object, nobody doubts this today. The things in hyperspace are susceptible of precise definitions in the same way as those in the ordinary space, and even if we cannot represent them, we can conceive and study them. So if, for instance, Mechanics in more than three dimensions should be condemned as lacking any object, it is not the case for hypergeometry.

4.1 Poincaré's Definitions of Manifold

In the first two sections after the introduction, Poincaré gives his definitions of manifolds. The first definition is as follows. We consider the n-dimensional Euclidean space with coordinates $x_1, \ldots, x_n$. Suppose that we are given $p + q$ functions $F_1, \ldots, F_p; \varphi_1, \ldots, \varphi_q$ with $p \leq n$, which are assumed to be continuous and have continuous derivatives. Assume moreover that the rank of the Jacobian matrix

$$\begin{pmatrix} \frac{\partial F_1}{\partial x_1} & \frac{\partial F_1}{\partial x_2} & \cdots & \frac{\partial F_1}{\partial x_n} \\ \frac{\partial F_2}{\partial x_1} & \frac{\partial F_2}{\partial x_2} & \cdots & \frac{\partial F_2}{\partial x_n} \\ \vdots & & \ddots & \\ \frac{\partial F_p}{\partial x_1} & \frac{\partial F_p}{\partial x_2} & \cdots & \frac{\partial F_p}{\partial x_n} \end{pmatrix}$$

is p at every point in the domain where the F_i are defined. We then consider a system of equations and inequalities as follows,

$$\begin{cases} F_1(x_1, \ldots, x_n) = 0 \\ \qquad \vdots \\ F_p(x_1, \ldots, x_n) = 0 \\ \varphi_1(x_1, \ldots, x_n) > 0 \\ \qquad \vdots \\ \varphi_q(x_1, \ldots, x_n) > 0. \end{cases}$$

Poincaré defines a manifold of dimension $n - p$ as the set of points satisfying the above system of equations and inequalities. By the implicit function theorem, such a set constitutes a differentiable manifold in the modern sense. In fact, Poincaré himself showed that a manifold according to this definition is automatically a manifold according to the second definition which we shall describe below, using an argument involving a kind of implicit function theorem. Notice that the first definition does not use charts and local coordinates.

Now we turn to the second definition. Again, we work in the n-dimensional Euclidean space with coordinates $x_1, \ldots, x_n$. We consider n functions θ_i of m variables $y_1, \ldots, y_m$ such as the following,

$$\begin{cases} x_1 = \theta_1(y_1, \ldots, y_m) \\ x_2 = \theta_2(y_1, \ldots, y_m) \\ \ldots\ldots\ldots\ldots\ldots\ldots \\ x_n = \theta_n(y_1, \ldots, y_m). \end{cases}$$

These functions θ_n are assumed first to be continuous, but Poincaré claims that they can be assumed to be analytic, by approximating the original functions arbitrarily closely by analytic ones. Furthermore, the Jacobian matrix of these functions is assumed to have rank m at every point. We cut out from this an open region denoted by V by adding inequalities $\psi_j(y_1, \ldots, y_m) > 0$. For convenience, let us call such a V a local manifold.

Poincaré next considers another system of functions $x_i = \theta_i'(y_1, \ldots, y_m)$ for $i = 1, \ldots, n$ which defines another local manifold V', supposing that $V \cap V'$ is non-empty. Then he claims that we can extend V to a larger manifold $V \cup V'$ by analytic continuation. He also observes that in the same way, we can continue the process by gluing local manifolds $V_1, \ldots, V_n$.

This definition is closer to the modern one which usually starts with charts, but it is not clear whether Poincaré assumed that two local manifolds which are not adjacent to each other are disjoint. Without such an assumption, this is a definition of an immersed manifold rather than an embedded manifold. Anyway, it should be noted that in both definitions, a manifold always lies in a Euclidean space. Manifolds without this environment were to appear later, in the 20th century.

4.2 *Poincaré's Conventionalism*

It is well known that Poincaré's position regarding the philosophical aspects of geometry is radical. His position is often labelled as "conventionalism," which opposes him to both empiricists and neo-Kantians. Generally, conventionalism refers to a philosophical doctrine which regards (both judicial and natural) laws and morals as mere conventions. We can think of Poincaré's views as conventionalism with regard to geometry. This can be seen typically in his text on the foundation of geometry [16]. (Its English translation by T. J. McCormack, which we use here together with the original, appeared in [14] a long time before the original French version.)

> La géométrie n'est pas une science expérimentale; l'expérience n'est pour nous que l'occasion de réfléchir sur les idées géométriques qui préexistent en nous.
>
> Geometry is not an experimental science; experience forms merely the occasion for our reflecting upon the geometric ideas which pre-exist in us.

This shows that Poincaré's position is far from the empiricists' view. On the other hand, the following passage shows that his thinking is also quite different from Kantians (or neo-Kantians). It is about how we choose one geometry, say Euclidean geometry, among other possible geometries.

> Notre choix ne nous est donc pas imposé par l'expérience. Il est simplement guidé par l'expérience. Mais il reste libre: nous choisissons cette géométrie-ci plutôt que celle-là, non parce qu'elle est plus vraie, mais parce qu'elle est plus commode.
>
> Our choice is therefore not imposed by experience. It is simply guided by experience. But it remains free: we choose this geometry rather than that geometry, not because it is more true, but because it is more convenient.

Therefore, for him, geometry is something like a tool, and our preference of one geometry over another does not depend on its validity, but on its usefulness. His view is quite close to our modern thinking, and we have little difficulty in understanding it. But we can imagine that for people at the turn of the 20th century, this view should have been outrageous, as we explain now.

To see the impact and the novelty of his view, it is worthwhile to take a look at the debate between him and B. Russell. (See Nabonnand's paper [11] for more details and for the significance of this debate in wider contexts.) Russell, who was in his mid 20s at that time, published a book entitled "An essay on foundation of geometry" in 1897 [21], which was based on his doctoral dissertation. In this book, although Russell criticised Kantian a-priority of geometry, he tried to separate what is given a priori and what should be tested empirically in geometry. For Russell, a-priority means what precedes every experience and what makes experiences possible. To make a clear distinction between what is a priori and what depends on experience, he gave axioms for projective geometry, which is a common ground for both Euclidean and non-Euclidean geometries, and those of "metric geometry." Russell considered that axioms of projective geometry (except for the one concerning the dimension of space) and some of metric geometry are a priori, but there are others in metric geometry, for instance those which distinguish Euclidean geometry from non-Euclidean, which can be verified or falsified by experiences.

Poincaré criticised this approach of Russell in two ways, in his review of the book [15]. The first criticism is purely mathematical: he pointed out the insufficiency of Russell's axioms on projective geometry. The second criticism is more philosophical: he denied the a-priority of these axioms. Poincaré examined the axioms proposed by Russell one by one, and gave a harsh judgement declaring that for most of the cases, Russell failed to show that they were indispensable for experience. He said that although Russell's statements with unclear terms made it difficult to see through, once it was cleared, then the illusion he gave would also disappear.

Poincaré also criticised Russell's claim that some of his axioms (for metric geometry) are empirical. In particular Poincaré argued, in opposing Russell's claim that Euclidean geometry can be empirically tested, that none of our experiences verifies Euclidean geometry and falsifies hyperbolic geometry.

As can be seen in this polemic, Poincaré's view on the foundations of geometry was quite clearly posed, and we can say that it was quite ahead of time, even compared with that of a much younger philosopher like Russell.

5 Definitions Using Local Charts According to Hilbert, Weyl, Kneser and Veblen-Whitehead

As we saw in the previous section, although Poincaré gave two clear definitions of manifold, they are both different from the definition which we are familiar with. In this section, we shall see when and how the modern-day definition of topological manifold appeared for the first time.

Historically, the definition of a two-dimensional manifold precedes that for general dimensions. We first look at Hilbert's paper entitled "Ueber die Grundlagen der Geometrie" published in 1902 [5], where he tried to give axioms for a surface which he called "Ebene" i.e. a plane rather than a surface. Here are his axioms. A surface is defined to be a point set with bijections onto domains in the Cartesian plane with the following conditions. (The itemisation and the translation are by the present author.)

1. Zu jedem Punkte A unserer Ebene giebt es Jordan'sche Gebiete, in welchen der Bildpunkt von A liegt und deren sämmtliche Punkte ebenfalls Punkte unserer Ebene darstellen. Diese Jordan'schen Gebiete heissen Umgebungen des Punktes A.
2. Jedes in einer Umgebung von A enthaltene Jordan'sche Gebiet, welches den Punkt A einschliesst, ist wiederum eine Umgebung von A.
3. Ist B irgend ein Punkt in einer Umgebung von A, so ist diese Umgebung auch zugleich eine Umgebung von B.
4. Wenn A und B irgend zwei Punkte unserer Ebene sind, so giebt es stets eine Umgebung, die zugleich eine Umgebung von A und eine Umgebung von B ist.

1. For every point A on our surface, there is a Jordan domain in which the point corresponding to A lies and all of whose points represent points in our surface. These Jordan domains are called neighbourhoods of A.
2. Every Jordan domain contained in a neighbourhood of A which contains A is in turn a neighbourhood of A.
3. If B is any point contained in a neighbourhood U of A, then U is also a neighbourhood of B at the same time.
4. If A and B are any two points on our surface, then there is always a neighbourhood of A which is at the same time a neighbourhood of B.

We should keep in mind that this paper of Hilbert precedes the work of Hausdorff on abstract topological spaces, which appeared in his book "Grundzüge der Mengenlehre" [4]. Therefore, Hilbert could not start from a topological space, but from just a point set. Consequently, his axioms include those for neighbourhood systems in a topological space. Setting this part aside, these axioms are equivalent to the modern definition of topological manifold using charts. Thus we can say that two-dimensional topological manifolds as we know them today were formally defined for the first time by Hilbert in 1902.

There is one more remark which we should make: the last axiom implies that the surface is a Hausdorff space. This should be contrasted with the following definition by Weyl. (This subtle difference was also pointed out by Scholtz [22].)

Now, we turn to Weyl's book on Riemann surface [27]. A formal definition of a topological surface (a topological 2-dimensional manifold) which is close to the modern one is given in §4 of this book. It is somehow more common to attribute the first formal definition of a manifold using charts to this book rather than to Hilbert's paper. In Weyl's definition, the setting is quite similar to Hilbert's. We are given a point set $\mathfrak{F}$, and for each point $\mathfrak{p}$ of $\mathfrak{F}$, there is a system of subsets $\mathfrak{U}$ containing $\mathfrak{p}$ which are called neighbourhoods of $\mathfrak{p}$. For every neighbourhood $\mathfrak{U}_0$ of a point $\mathfrak{p}_0$ of $\mathfrak{F}$, there is a bijection from $\mathfrak{U}_0$ to an Euclidean open disc K_0 taking $\mathfrak{p}_0$ to the centre of K_0, and the following conditions hold.

1. ist $\mathfrak{p}$ irgend ein Punkt von $\mathfrak{U}_0$ und $\mathfrak{U}$ eine nur aus Punkten von $\mathfrak{U}_0$ bestehende Umgebung von $\mathfrak{p}$ auf $\mathfrak{F}$, so enthält das (durch jene Abbildung in K_0 entworfene) Bild von $\mathfrak{U}$ den Bildpunkt von $\mathfrak{p}$ im Innern; d. h. es läßt sich um den Bildpunkt p von $\mathfrak{p}$ eine Kreisfläche k beschreiben, sodaß jeder Punkt von k Bild eines Punktes von $\mathfrak{U}$ ist;
2. ist K das Innere irgend eines ganz in K_0 gelegenen Kreises mit dem Mittelpunkt p, so gibt es stets eine Umgebung $\mathfrak{U}$ von $\mathfrak{p}$ auf $\mathfrak{F}$, deren Bild ganz in K liegt.

1. If $\mathfrak{p}$ is any point of $\mathfrak{U}_0$ and $\mathfrak{U}$ is a neighbourhood of $\mathfrak{p}$ consisting only of points of $\mathfrak{U}_0$, then the image of $\mathfrak{U}$ (under the map from $\mathfrak{U}_0$ to K_0) contains that of $\mathfrak{p}$ as an interior point, i.e., there is an open disc k around the point p which is the image of $\mathfrak{p}$, such that every point of k is the image of a point of $\mathfrak{U}$.
2. If K is any circle with centre p contained entirely in K_0, then there is always a neighbourhood $\mathfrak{U}$ of $\mathfrak{p}$ on $\mathfrak{F}$ whose image entirely lies in K.

We see that if we define a topology on $\mathfrak{F}$ using the given neighbourhood systems, then these two conditions guarantee that the map from $\mathfrak{U}_0$ to K_0 is a homeomorphism. Thus, this definition is equivalent to a modern definition of topological two-manifold. There is one subtle point: the definition does not contain the axiom of separability, which means that the surface defined as such may not be a Hausdorff space. In a later revised version of the same book and its English translation, this part was substantially revised. The definition has been divided into two parts: the first part is the definition of a Hausdorff space using neighbourhood systems whereas the second part is almost the same as the two conditions above. In particular the subtlety concerning the separability disappeared.

There is one more difference with Hilbert's definition. Weyl assumed the image of the map from a neighbourhood to be an open disc in the plane, not a general Jordan domain. In fact Schoenflies proved that any Jordan domain is homeomorphic to an open disc in 1906; this was after Hilbert's paper was published but before Weyl wrote his book.

As we have seen above, the definitions by Hilbert and Weyl only deal with two-dimensional manifolds. The same kind of definition for higher dimensional manifolds was given for the first time by Kneser [9] in 1926. After giving axioms of neighbourhood systems for a Hausdorff space, Kneser added the following two conditions as axioms for an n-dimensional manifold.

(a) Zu jedem Punkt gibt es eine Umgebung, die sich topologisch auf die offene Vollkugel des n-dimensionalen Zaahlenraumes: $x_1^2 + \cdots + x_n^2 < 1$ abbilden läßt.
(b) Unter den Umgebungen des den topologischen Raum definierenden oder eines äquivalenten Systems befinden sich nur abzählbar unendlich viele verschiedene Mengen.

(a) For every point there exists a neighbourhood which is mapped homeomorphically to an open ball $x_1^2 + \cdots + x_n^2 < 1$ in the n-dimensional Euclidean space.
(b) Among neighbourhoods defining the topological space or those of an equivalent system, there are only countably many different sets.

Thus, Kneser defined a topological manifold of general dimension with an axiom of countability, which is precisely what we do today. This definition appears at the beginning of Kneser's paper. The main topic of the paper is rather the triangulability of manifolds and the uniqueness of triangulations (up to subdivision), i.e., the Hauptvermutung. Of course he did not prove this result, but he posed the conjecture in clear terms for topological manifolds and gave a precise definition of combinatorial manifolds, which made the meaning of the conjecture clear. The Hauptvermutung itself for simplicial complexes had been given earlier by Steinitz [23] and Tietze [24], and a definition of a manifold using simplices was first given by Brouwer [1]. The originality of Kneser was to consider this problem for topological manifolds, and to try to unify the approach using neighbourhood systems with the simplicial approach pioneered by Brouwer.

Up to this point, we only talked about the definition of a *topological* manifold. Although Weyl's book contains the notion of differentiability or analyticity of functions on the surface, hence also differentiable structures and complex structures, we did not touch upon this part. In the rest of this section, we shall see how the definition of a differentiable manifold as we understand it today appeared.

Veblen and Whitehead published a paper entitled "A set of axioms for differential geometry" in 1931 [25], whose aim was to define a space axiomatically where differential geometry can be developed without the aid of global coordinates as in Euclidean space. (An expanded and detailed version of this theory can be found in their book [26].) There they gave a set of axioms of C^u (class u) manifolds. This was done by defining a set of coordinate systems, which they call allowable coordinate systems, such that the transition between any two of them with an overlapping domain is C^r. They did not assume the manifold to be a topological space at the beginning, but they gave a topology using the domains of coordinate systems, which are assumed to satisfy the Hausdorff separability.

From this definition of a differentiable manifold, it became possible for the first time to define differentiation, etc. on manifolds without assuming they lie in Euclidean spaces. From the work of Whitney [28], it follows that every differentiable manifold can be regarded as lying in a Euclidean space. In fact Whitney showed that every n-dimensional C^r-manifold can be embedded into a $2n + 1$-dimensional Euclidean space by a C^r-map. This showed the equivalence between Poincaré's definition of manifolds and the modern one using charts. He also defined notions like submanifolds, C^r-maps between manifolds and so on. Thus we can say that at this stage the foundations of the notion of manifold were really consolidated.

6 Conclusion: Philosophical Significance

Now, we return to the philosophical viewpoint, with which we started our exposition. It is obvious that thanks to the development of manifold theory, our worldview (both in a cosmological sense and in a more practical sense) has been widened. If we take into account the possibility that our universe is compact, then Kantian antinomies must be seriously reconsidered, even in the framework of classical physics. The possibility of a non-simply connected universe poses further problems of epistemology. In fact, we do not have any way to distinguish a non-simply connected universe and its (non-trivial) coverings, only by observations.

It is quite well known that Einstein used Riemannian geometry for his theory of general relativity. Modern physics makes use of much more sophisticated manifold theory, typically seen in string/super-string theory. Such a development might lead to the necessity to redefine the "universe" and distinguish the world where physical theory should work from the three-dimensional manifold in which we believe that we live in our day-to-day recognition. On the other hand, what is striking is that there still remains a long way to go to the complete solution of a naïve question asking what is the topological structure of our day-to-day universe. By virtue of the resolution of Thurston's geometrisation conjecture by Perelman, once we know the (average) sectional curvature of the universe, we can make our list of possible topologies of the universe shorter. Still, to the author's knowledge, it is not known even whether the universe is compact or not, and although there is some data bounding the sectional curvature of our universe (see e.g. [12]), it is a far cry from determining the geometric structure of the universe. On the other hand, the development of manifold theory started by Riemann gave a strong impact and an inspiration to contemporary philosophy, which is often dubbed "post-modern". Some aspects of this kind of influence on philosophy are illustrated in Jedrzejewski [6] and Plotnitsky [17] in this volume.

The epistemological revolution with regard to the universe started by Riemann's Habilitationsvortrag is not completely achieved yet. It is still going on.

References

1. L.E.J. Brouwer, Über Abbildung von Mannigfaltigkeiten. Math. Ann. **69**, 97–115 (1911)
2. H. Grassmann, *Die Lineale Ausdehnungslehre, ein neuer Zweig der Mathematik* (Otto Wigand, Leipzig, 1844)
3. A. Grothendieck, Esquisse d'un programme, London Math. Soc. Lecture Note Ser., 242, Geometric Galois actions, 1, 5–48, (Cambridge University Press, Cambridge, 1997)
4. F. Hausdorff, *Grundzüge der Mengenlehre* (Veit & Comp, Leipzig, 1914)

5. D. Hilbert, Ueber die Grundlagen der Geometrie. Math. Ann. **56**, 381–422 (1902)
6. F. Jedrzejewski, Deleuze et la géométrie riemainnienne: une topologie des multiplicités. In this volume, pp. 311–327
7. I. Kant, Critik der reinen Vernunft, Johann Friedrich Hartknoch, Riga 1781, 2 Aufl 1787
8. I. Kant, *Critique of Pure Reason*, translated by P. Guyer, A.W. Wood (Cambridge University Press, 1998)
9. H. Kneser, Die Topologie der Mannigfaltigkeiten. Jahresbericht d. Deutschen Mathem. Vereinigung **34**, 1–14 (1926)
10. D. Laugwitz, *Bernhard Riemann 1826–1866: Wendepunkte in der Auffassung der Mathematik* (Birkhäuser, Basel, 1996)
11. P. Nabonnand, La polémique entre Poincaré et Russell au sujet du statut des axiomes de la géométrie. Revue d'histoire des mathématiques **6**, 219–269 (2000)
12. Planck Collaboration, Planck 2015 results. XIII. Cosmological parameters. arXiv:1502.01589
13. H. Poincaré, Analysis Situs, Journal de l'École Polytechnique, 1–121 (1895)
14. H. Poincaré, On the foundations of geometry (transl. by T.J. McCormack). The Monist **9**, 1–43 (1899)
15. H. Poincaré, Des fondements de la géométrie, à propos d'un livre de M. Russell, Revue de méthaphysique et de morale **3**, 630–646 (1895)
16. H. Poincaré, Les fondements de la géométrie. Bulletin des sciences mqthé-matiques **26**, 249–272 (1902)
17. A. Plotnitsky, Comprehending the connection of things: Bernhard Riemann and the architecture of mathematical concepts. In this volume, pp. 329–362
18. B. Riemann, Ueber die Hypothesen, welche der Geometrie zu Grunde liegen. Abhandlungen der Königlichen Gesellschaft der Wissenschaften zu Göttingen **13**, 133–150 (1867)
19. B. Riemann, translated by W.K. Clifford, Nature, 8, 14–17, 36–37 (1873)
20. B. Riemann, translated by Spivak in *A Comprehensive Introduction to Differential Geometry*, vol. 2 (Publish or Perish, 1999)
21. B. Russell, *An Essay on the Foundations of Geometry* (Cambridge University Press, Cambridge, 1897)
22. E. Scholz, *The concept of manifold, 1850–1950, in History of Topology* ed. by I.M. James (Elsevier, 1999), pp. 25–64
23. E. Steinitz, Beisträge zur Analysis situt. Sitzungsberichte Berliner Mathematisches Gesellschaft **7**, 28–49 (1908)
24. H. Tietze, Über die topologischen Invarianten mehrdimensionaler Mannigfaltigkeiten. Monat. Math. **19**, 1–118 (1908)
25. O. Veblen, J.H.C. Whitehead, A set of axioms for differential geometry. Proc. Natl. Acad. Sci. **10**, 551–561 (1931)
26. O. Veblen, J.H.C. Whitehead, *The Foundations of Differential Geometry* (Cambridge University Press, Cambridge, 1932)
27. H. Weyl, *Die Idee der Riemannsches Fläsche, Teubner, Leipzig, 1913, and 3* (Aufl, Teubner Stuttgart, 1955)
28. H. Whitney, Differentiable manifolds. Ann. Math. **37**, 645–680 (1936)

Deleuze et la Géométrie Riemannienne: Une Topologie des Multiplicités

Franck Jedrzejewski

1 Introduction

Bernhard Riemann est sans doute le mathématicien qui a eu la plus grande influence sur l'œuvre de Deleuze et Guattari. Non pas que ces philosophes aient commenté les textes de Riemann, mais ils les ont intégrés à leur pratique et à leur écriture à tel point que leur philosophie apparaît comme un modelage topologique de concepts, dans lequel le lieu des points, des voisinages, le support des singularités, l'ordre des plis, la dialectique du discret et du continu, du local et du global sont autant de points d'ancrage essentiels pour le cheminement de la pensée. Deleuze façonne ses concepts comme des objets mathématiques pris dans le maelstrom d'espaces dont la topologie échappe à notre connaissance. Dans le texte deleuzien, la référence à des notions de mathématiques est toujours présente à tel point que certains ont vu dans le concept de plan d'immanence une variété riemannienne. Cette notion de variété est esquissée dans le texte d'habilitation de Riemann qui commence par le constat que la géométrie assume à la fois la définition de l'espace et l'exposé axiomatique des principes de construction de cet espace sans toutefois exposer leur articulation[1] [15].

> Depuis Euclide jusqu'à Legendre, pour ne citer que le plus illustre des réformateurs modernes de la géométrie, nous dit Riemann, personne, parmi les mathématiciens, ni parmi des philosophes, n'est parvenu à éclaircir ce mystère. La raison en est que le concept général des grandeurs de dimensions multiples, comprenant comme cas particulier les grandeurs étendues, n'a jamais été l'objet d'aucune étude. [17]

C'est donc l'espace qui est au centre de la réflexion riemannienne, sa nature philosophique et ses implications structurales, mais aussi à travers l'espace, le con-

[1] Je ne parlerai pas ici des relations à la philosophie kantienne, ni de la notion de variété telle qu'elle apparaît dans l' *Analysis situs* de Poincaré. Elles sont étudiées de manière détaillée et exhaustive dans le chapitre écrit par Ken'ichi Ohshika.

F. Jedrzejewski (✉)
Commissariat à l'Énergie Atomique, Saclay, INSTN, 91191 Gif-sur-Yvette Cedex, France
e-mail: franckjed@gmail.com

L. Ji et al. (eds.), *From Riemann to Differential Geometry and Relativity*,
DOI 10.1007/978-3-319-60039-0_10

cept d'une grandeur de dimensions multiples qui donnera naissance à la notion de variété. Parmi ses sources, Riemann cite de "brèves indications dans le second *Mémoire sur les résidus biquadratiques*" de Gauss, mais déplore qu'il n'y ait pas eu de travaux antérieurs. Toutefois, et contrairement à Gauss, Riemann ne plonge pas les surfaces qu'il étudie dans l'espace tridimensionnel euclidien pour décider de leurs propriétés de dépendance, mais se place directement sur la surface et fait l'économie d'un plongement dans un espace euclidien, ce qui lui permet d'envisager le passage au local indépendamment de tout caractère euclidien. Toutes les notions de distances, de géodésiques prennent alors un sens intrinsèque et ne dépendent que de la variété sur laquelle elles ont été définies.

> Dans cette branche générale de la théorie des grandeurs étendues, où l'on ne suppose rien de plus que ce qui est déjà renfermé dans le concept de ces grandeurs, il nous suffira, pour notre objet actuel, de porter notre étude sur deux points, relatifs: le premier, à la génération du concept d'une variété de plusieurs dimensions; le second, au moyen de ramener les déterminations de lieu dans une variété donnée à des déterminations de quantité, et c'est ce dernier point qui doit faire clairement ressortir le caractère essentiel d'une étude de n dimensions. ([17], 283).

Pour que les concepts de grandeur soient possibles, il faut selon Riemann qu'il "existe un concept général qui permette différents modes de détermination", et qu'il soit possible de passer d'un de ces modes de détermination à un autre de manière continue. ([17], 282) À partir de ce constat, Riemann construit, par intégration successive des dimensions entières, le concept d'une variété de dimension quelconque. Il trouve que la détermination du lieu se ramène à n déterminations de grandeur et pose la question des rapports métriques dans une variété. Il se place sous deux hypothèses. La première est que les "lignes" sont indépendantes de leur position, et que toute ligne est mesurable par toute autre ligne. La seconde est que l'élément infinitésimal linéaire ds est "exprimable par la racine carrée d'une expression différentielle du second degré, c'est-à-dire que l'espace est une grandeur plane dans ses parties infinitésimales" ([17], 294).

2 Variété et multiplicité

Cette notion de variété qui naît dans le texte de Riemann, Deleuze l'appelle *multiplicité* traduisant l'allemand *mannigfaltigkeit* (pluralité, variété, que l'on traduit en anglais par *manifold*). Il le dit explicitement dans *Différence et répétition*:

> Les Idées sont des multiplicités, chaque Idée est une multiplicité, une *variété* [Je souligne]. Dans cet emploi riemannien du mot "multiplicité" (repris par Husserl, repris aussi par Bergson), il faut attacher la plus grande importance à la forme substantive: la multiplicité ne doit pas désigner une combinaison de multiple et d'un, mais au contraire une organisation propre au multiple en tant que tel, qui n'a nullement besoin de l'unité pour former un système. ([6], 236)

Plus loin, il ajoute:

> Une Idée est une multiplicité définie et continue, à *n* dimensions. La couleur, ou plutôt l'Idée de couleur est une multiplicité à trois dimensions. Par dimensions, il faut entendre les variables ou coordonnées dont dépend le phénomène; par continuité, il faut entendre l'ensemble des rapports entre les changements de ces variables, par exemple une forme quadratique des différentielles des coordonnées; par définition, il faut entendre les éléments réciproquement déterminés par ces rapports, qui ne peuvent pas changer sans que la multiplicité ne change d'ordre et de métrique. ([6], 236–237)

On pourra rapprocher ce passage du texte de Riemann: "Au contraire, les occasions qui peuvent faire naître les concepts dont les modes de détermination forment une variété continue sont si rares dans la vie ordinaire, que les lieux des objets sensibles et les couleurs sont à peu près les seuls concepts simples dont les modes de détermination forment une variété de plusieurs dimensions" ([17], 282). Parmi les exemples de variété cités par Deleuze, deux exemples, à savoir l'espace des sons et celui des couleurs sont cités par Hermann Weyl dans son livre *Espace, Temps, Matière* [20]. Il donne en outre la dimension de chaque variété: la variété des ellipses à isométrie près dont les dimensions sont les deux axes de l'ellipse, la variété des configurations physiques d'un gaz parfait dont les deux dimensions sont la température et la pression, la variété des sons purs avec pour dimension la fréquence et l'intensité et la variété des couleurs de dimension trois.

Mais c'est aussi en référence à Bergson que Deleuze parle de "multiplicité". Il reconnaît que le mot a été employé dès le chapitre 2 de l'*Essai sur les données immédiates* ([8], 604). Bergson y distingue deux multiplicités: "celle des objets matériels, qui forme un nombre immédiatement, et celle des faits de conscience, qui ne saurait prendre l'aspect d'un nombre sans l'intermédiaire de quelque représentation symbolique où intervient nécessairement l'espace" ([1], 65).

Si Bergson aborde la multiplicité dans son rapport au nombre, Riemann, quant à lui, classe les variétés en deux catégories. Il affirme: "Suivant qu'il est, ou non, possible de passer de l'un de ces modes de détermination à un autre, d'une manière continue, ils forment une variété continue ou une variété discrète" ([17], 282). Implicitement, c'est aussi une relation au nombre à laquelle renvoie le partage du continu et du discret, deux sortes de nombres réel ou entier, pris dans un continuum ou égrenés, solitaires. Comme le note Deleuze, "Bergson dégageait donc deux espèces bien différentes de multiplicité, l'une qualitative et de fusion, continue; l'autre, numérique et homogène, discrète" ([8], 604). La question qui se pose ici très certainement dans l'esprit de Deleuze est de savoir si la notion de variété topologique peut se passer du concept de nombre et représenter des parties de l'espace sensible indépendamment de tout nombre. Dans la même direction, Deleuze se demande si cette notion de variété topologique a besoin de structures logiques. Lorsqu'il se réfère au texte de Riemann, il sort la multiplicité de la logique et écrit sans équivoque:

> Ce fut un événement décisif lorsque le mathématicien Riemann arracha le multiple à son état de prédicat, pour en faire un substantif "multiplicité". C'était la fin de la dialectique, au profit d'une typologie et d'une topologie des multiplicités. Chaque multiplicité se définissait par *n* déterminations, mais tantôt les déterminations étaient indépendantes de la situation, tantôt en dépendaient. ([8], 602-603).

Pour comparer des sous-multiplicités entre elles, c'est-à-dire dans le vocabulaire de Riemann, pour comparer des parties de variétés—Ce que Riemann appelle des quanta—Il faut avoir un moyen de les mesurer. Si ce moyen n'existe pas, Riemann reconnaît que ces parties ne sont pas indépendamment de leur position, ni "exprimables au moyen d'une unité, mais sont comme des régions dans une variété" ([17], 283). On croise ici le point nodal où se coupent les interprétations de Deleuze, Bergson et Riemann et qui, pour Deleuze, se fondent sur la distinction entre espace lisse et espace strié. L'espace musical est un espace lisse structuré par des paramètres musicaux de différente nature. Deleuze écrit:

> Par exemple, on peut comparer la grandeur de la ligne verticale entre deux points et la grandeur de la ligne horizontale entre deux autres; on voit ici que la multiplicité est métrique, en même temps qu'elle se laisse strier, et que ces déterminations sont des grandeurs. En revanche, on ne peut comparer la différence entre deux sons de hauteur égale et d'intensité distincte avec deux sons d'intensité égale et de hauteur distincte; on ne peut dans ce cas comparer deux déterminations. ([8], 603).

La notion d'espace lisse apparaît à plusieurs endroits dans *Mille plateaux*. Contrairement aux mathématiciens qui emploient l'adjectif "lisse" pour désigner l'indéfiniment différentiable, Deleuze lui réserve le non-métrique. Est lisse ce qui ne peut se réduire au métrique. Dans un espace lisse, les règles glissent les unes sur les autres et il est impossible de comparer les lignes entre elles. Le désert, la steppe, la glace ou la mer, le champ de blé de Cézanne sont des espaces lisses, tout comme les objets fractals. "Ce sont des ensembles dont le nombre de dimensions est fractionnaire ou non entier, ou bien entier, mais avec variation continue de direction" ([8], 607). L'espace lisse n'a pas de structure interne ou d'excroissance qui viendrait le strier. "L'espace lisse se définit dès lors en ce qu'il n'a pas de dimension supplémentaire à ce qui le parcourt ou s'inscrit en lui: c'est en ce sens une multiplicité plate" ([8], 609). Ce que Deleuze appelle un agencement est "précisément cette croissance de dimensions dans une multiplicité qui change nécessairement de nature à mesure qu'elle augmente ses connexions" ([8], 15). Plus encore, l'espace lisse est l'objet d'une zone d'indiscernabilité où le hasard peut s'engouffrer: "un tel espace lisse, amorphe, se constitue par accumulation de voisinages, et chaque accumulation définit une zone d'indiscernabilité propre au "devenir" (plus qu'une ligne et moins qu'une surface, moins qu'un volume et plus qu'une surface)" ([8], 609).

Le geste deleuzien consistant à strier l'espace lisse est à mettre en regard de celui du mathématicien qui définit d'abord une variété topologique, où l'espace est pensé par les notions d'ouverts, de fermés et de voisinages, sans aucune considération de nombre (un espace lisse), puis munit (c'est-à-dire strie l'espace lisse) cette variété d'une structure différentielle pour en faire une variété différentiable, et par ajout d'un appareillage de mesures (justement une métrique) lui donne la possibilité de mesurer les longueurs entre les objets et la transforme en variété riemannienne. En ce sens, la variété topologique est un espace lisse, strié ensuite par des structures métriques et différentielles.

Dans le *Traité de nomadologie*, Deleuze reprend cette idée d'un espace strié par la matière et la gravitation. Il écrit:

> L'espace homogène n'est nullement un espace lisse, c'est au contraire la forme de l'espace strié. L'espace des piliers. Il est strié par la chute des corps, les verticales de pesanteur, la distribution de la matière en tranches parallèles, l'écoulement lamellaire ou laminaire de ce qui est flux. Ce sont ces verticales parallèles qui ont formé une dimension indépendante, capable de se communiquer partout, de formaliser toutes les autres dimensions, de strier tout l'espace dans toutes ses directions, et par là de le rendre homogène. ([8], 458)

Dans l'espace physique ordinaire, nous savons que la présence de matière déforme le trajet des rayons lumineux[2][16]. Les lignes tracées par la lumière se courbent naturellement au voisinage des grosses masses dont la présence modifie la structure de l'espace-temps. Celui-ci n'est plus euclidien et la non-nullité de la courbure de l'espace mesure sa non-platitude. Cette courbure s'impose par la gravité, le jeu des forces qui strie, selon Deleuze, l'espace ordinaire, homogène et isotrope. Au voisinage de chaque point, en se rapprochant indéfiniment du point, l'espace se confond localement avec un espace euclidien, ressemblant de plus en plus à son espace tangent en ce point. C'est pourquoi la courbure mesure le défaut des variétés riemanniennes d'être des espaces euclidiens. "L'espace euclidien dépend du célèbre postulat des parallèles, mais les parallèles sont d'abord gravifiques, et correspondent aux forces que la pesanteur exerce sur tous les éléments d'un corps supposé remplir cet espace" ([8], 458). Dans le cas où la courbure est constante, les mathématiciens ont établi que si la courbure est identiquement nulle, alors la variété de Riemann est localement isométrique à un espace euclidien. Si cette courbure est positive, la variété riemannienne est localement isométrique à une sphère et à un espace hyperbolique lorsqu'elle est négative. La courbure détermine essentiellement la nature géométrique de l'espace de Riemann. Dans le cas favorable où l'espace est simplement connexe, c'est-à-dire est un espace dans lequel tout lacet peut être déformé continûment en un point, sans trou ni poignée, alors la métrique est unique à un difféomorphisme près et l'isométrie est globale.

Pour ramener les déterminations de lieu aux déterminations de quantité, Riemann pose les conditions de la mesure: pour pouvoir mesurer, c'est-à-dire comparer les lignes entre elles, il faut "avoir un moyen de transporter la grandeur qui sert d'étalon pour les autres." Cette notion de transport est fondamentale en géométrie riemannienne. Elle a permis le développement de l'idée de parallélisme et de ce qu'on appelle le "transport parallèle". Lorsqu'un vecteur est transporté d'un point à un autre sur une variété selon deux chemins différents, les deux vecteurs obtenus au point terminal ne sont pas identiques et forment entre eux un angle qui caractérise la courbure de la variété. Cette notion a donné naissance à celle de connexion qui a été développée dans les années 1920 par Elie Cartan et Hermann Weyl. Non seulement la variété riemannienne est pourvue d'un ensemble de données géométriques comme les métriques, mais ces métriques $g(x)$ elles-mêmes induisent l'existence d'une mesure canonique, qui est grosso modo l'intégrale des mesures de Lebesgue sur les espaces tangents en un point x. Tout ceci n'est possible que grâce à une topologie qui est précisément donnée par les éléments géométriques de la variété riemannienne. Mais

[2] Il y aurait beaucoup à dire sur la conception de l'espace telle qu'elle est envisagée par les physiciens, mais aussi à travers la notion de topos introduite par Grothendieck. On pourra se référer à ce sujet, dans ce livre, au chapitre d'Arkady Plotnitsky.

les liens entre la topologie et la courbure, ou plus précisément entre la topologie et tout autre invariant que l'on peut construire, entre la courbure et les nombres de Betti, révèlent une situation bien plus profonde qu'il n'y paraît et l'étude des liens est souvent fort compliquée. Le plus souvent, il s'agit de contrôler la topologie par la construction de fonctions de ces invariants, autrement dit de passer du local au global et d'induire certaines propriétés à partir des propriétés d'invariants. La connaissance du signe de la courbure (donc de la nature géométrique de la variété riemannienne considérée) peut-elle, par exemple, nous renseigner sur le signe de cette variété ? Ces questions ont été étudiées par Heinz Hopf dans les années 1930 [11], et plus tard par d'autres. La plupart des résultats obtenus en géométrie riemannienne supposent que la variété est compacte ou qu'il existe des bornes sur certains paramètres, comme sur la courbure. Mais on peut se demander ce qui se passe lorsque la courbure tend vers l'infini. Les généralisations et les recherches actuelles, qui portent sur les variétés kählériennes, les variétés d'Alexandrov, les orbi-variétés, et bien d'autres objets encore, fournissent des résultats nouveaux sur les variétés riemanniennes et des applications importantes à la physique mathématique comme celle des variétés de Calabi-Yau et de la symétrie miroir.

3 Espaces, mesures et multiplicités

Sous cet appareillage conceptuel naissent des interrogations ontologiques en relation avec les mathématiques. Une des interrogations concerne les propriétés intrinsèques de la variété et son rapport à l'espace ambiant. Peut-on rapporter les propriétés induites par l'espace ambiant à une variété riemannienne à des propriétés purement intrinsèques à la variété ? Cette réduction de l'espace ambiant à l'intrinsèque où les propriétés de l'espace tendent à s'inscrire dans les structures de la variété par un procédé de décalque, d'osmose ou d'induction conduit à restaurer, comme le remarque A. Lautman, la vision de la monade leibnizienne ([13], 152).

La notion de connexion a été inventée par Ricci et Levi-Civita au début du XXe siècle [5]. En 1918, Hermann Weyl généralisa la notion de déplacement parallèle dans une variété riemannienne de Levi-Civita à celle d'une connexion affine, qui sera appelée plus tard par Cartan "connexion linéaire sans torsion". Comme Weyl, Élie Cartan a généralisé le transport parallèle à une classe de connexions infinitésimales [3]. Il a mis au point une méthode de recollement de voisinages infinitésimaux par la connexion donnant naissance à des structures qui ont été appelées, plus tard, les "géométries de Cartan".

Dans l'article de 1918 et dans son livre *Espace - Temps - Matière*, Hermann Weyl [21] conçoit l'espace comme une forme homogène, et se distingue de la tradition algébrique de Félix Klein et du programme d'Erlangen qui pensaient l'espace à travers son groupe de symétrie, c'est-à-dire comme un groupe de transformations agissant sur un ensemble de points et préservant ses structures. Weyl considère que le temps est la "forme du flux de la conscience" et l'espace "la forme de la réalité corporelle." Pour Weyl, l'homogénéité de l'espace est une propriété du monde extérieur. Si une chose physique est déplacée dans l'espace, aucune de ses propriétés essentielles n'est modifiée. Mais pour Weyl, l'homogénéité de l'espace n'est pas seulement relative à ses points, mais plus singulièrement, est relative à ses différents systèmes de coordonnées locales. De ce point de vue, l'espace et les relations spatiales deviennent des structures intuitives liées à la place du sujet en situation. C'est ainsi que l'espace peut devenir le cadre de la mesure physique que Weyl appelle le "principe de relativité du mouvement " (*Prinzip der Relativität der Bewegung*). Il pose que "les relations métriques ne sont pas des propriétés de l'espace en soi, mais de l'espace dans sa relation à son contenu matériel. " Dans une variété riemannienne, Weyl montre qu'il devient possible de choisir une échelle indépendamment de chaque point de l'espace, ce qui correspond à ce que nous appelons aujourd'hui une jauge, notion qui sera largement utilisée dans les développements de physique mathématique. Changer de cordonnées, recoller les systèmes de coordonnées, définir une mesure, une jauge, utiliser des mesures physiques sont des questions qui paraissent anodines dans l'espace ordinaire, mais qui ne le sont pas dans les variétés. Comme on le voit, à une conception de l'espace homogène s'oppose une conception de l'espace hétérogène, induite par la métrique de Riemann-Einstein et de l'espace physique, dans lequel la matière strie l'espace lisse du mathématicien. Si Einstein cherche une théorie qui réponde à l'exigence d'une métrique entièrement déterminée par la distribution de matière, Weyl, quant à lui, cherche à résoudre le conflit entre un espace naturel homogène et un espace hétérogène dépendant d'une métrique. L'idée que la mise en œuvre d'une mesure sur un espace produirait une déformation de la nature de cet espace est avancée sans être réellement comprise.

En opposant le concept qualitatif de distance à celui quantitatif de grandeur, Deleuze s'éloigne des thèses de Riemann, qui voyait dans la distance une sorte de grandeur. Selon Deleuze, les distances se laissent diviser, "mais, contrairement aux grandeurs, elles ne se divisent pas sans changer de nature à chaque fois" ([9], 603). L'allure d'un cheval que l'on divise en galop, trot et pas est une distance qui change de nature selon le type de division, sans qu'un de ces types entre dans la composition de l'autre. Pourtant, le passage d'un type (de *moment* dit Deleuze) à un autre se fait de façon continue: il n'y a pas de rupture dans la vitesse du cheval. C'est pourquoi Deleuze distingue les multiplicités de distance "inséparables d'un processus de variation continue" et les multiplicités de grandeur (ou multiplicités métriques) qui contrairement à celles de distance, "répartissent des fixes et des variables." Cette distinction est à rapprocher de celle de Riemann qui distingue les variétés continues et les variétés discrètes "suivant qu'il est possible de passer d'un mode de détermination à un autre." Mais aussi à rapprocher de celles de Bergson qui distinguait, lui aussi, deux types de multiplicités. La durée de Bergson est justement ce que Deleuze appelle une multiplicité de distance, qualitative, "qui ne se divise pas sans changer de nature à chaque division", tandis que l'étendue homogène est une multiplicité métrique, quantitative. Le physicien, lui, distingue des grandeurs intensives, indépendantes de la quantité comme la température, et des grandeurs extensives, proportionnelles à la quantité de la chose mesurée, comme la masse ou le volume. Seules les grandeurs extensives sont additives. Si l'on place une deuxième masse sur une première, le poids résultant sera la somme des deux masses. Alors que si l'on considère deux objets allant à la même vitesse, l'ensemble formé par les deux objets aura la même vitesse, et non une vitesse double. On dit que la vitesse est une grandeur intensive. Deleuze parle d'intensité, plutôt que de grandeur intensive, avec parfois le même sens que le physicien. Une intensité comme la température ou la pression n'est pas "composée de grandeurs additionnables et déplaçables." En appliquant les notions physiques de grandeurs intensives et extensives au champ philosophique, Deleuze les croise avec la qualité et la quantité. Dans *Différence et répétition*, il explique que l'intensité a trois caractères ([6], 299) et que, dans l'intensité, il appelle *différence* "ce qui est réellement impliquant, enveloppant" et *distance* "ce qui est réellement impliqué ou enveloppé." L'intensité est donc à la fois impliquante et impliquée, enveloppante et enveloppée. "C'est pourquoi l'intensité n'est ni divisible comme la quantité extensive, ni indivisible comme la qualité" ([6], 305–306). Ainsi se croisent les notions de distance, de grandeur et d'intensité, de qualité et de quantité, que Deleuze résume dans ce passage et plonge comme le dit joliment Anne Sauvagnargues "la critique kantienne transcendantale dans le bain dissolvant d'un empirisme renouvelé" [18]:

> L'espace en tant qu'intuition pure, *spatium*, est quantité intensive; et l'intensité comme principe transcendantal, n'est pas simplement l'anticipation de la perception, mais la source d'une quadruple genèse, celle des *extensio* comme schèmes, celle de l'*étendue* comme grandeur extensive, celle de la *qualitas* comme matière occupant l'étendue, celle du *quale* comme désignation d'objet. ([6], 298).

Il y a donc des multiplicités qualitatives, de distances, non métriques, faites de grandeurs intensives, correspondant à des espaces lisses et des multiplicités quantitatives, métriques, faites de grandeurs additives extensives, correspondant à des espaces striés. Il se pourrait que ce qui les distingue soit justement que le nombre appartienne uniquement aux seules multiplicités métriques. Mais Deleuze répond qu'il n'en est rien, que le nombre est bien "le corrélat de la métrique", mais qu'il pénètre aussi l'espace lisse dans des opérations locales, et qu'il peut aussi changer de nature: il distingue deux aspects du nombre: un nombre nombrant et un nombre nombré.

> Le nombre se distribue lui-même dans l'espace lisse, il ne se divise plus sans changer de nature à chaque fois, sans changer d'unité, dont chacune représente une distance et une grandeur. C'est le nombre articulé, nomade, directionnel, ordinal, le nombre nombrant qui renvoie à l'espace lisse, comme le nombre nombré renvoyait à l'espace strié. Si bien que, de toute multiplicité, on doit dire: elle est déjà nombre, elle est encore unité. Mais ce n'est ni le même nombre dans les deux cas, ni la même unité, ni la même manière dont l'unité se divise. ([8], 605).

Aristote distinguait déjà le *nombre nombrant* qui demeure extérieur à ce qu'il nombre, sans égards à l'espèce dénombrée. C'est une simple référence numérique abstraite qui indique un décompte d'unités. Mais si l'on cherche à savoir ce que sont ces unités, si on pense au nombre comme à une mesure, à un moyen de dénombrer des unités par rapport à une unité référente, le nombre dépend de la chose que l'on compte, il devient mesure de décompte, nombre de quelque chose, c'est-à-dire *nombre nombré.* Le nombre nombré renvoie à l'espace strié, car il est associé à la métrique qui autorise la mesure et la comparaison des longueurs entre elles, la mesure de la courbure de l'espace. Le nombre nombrant, abstrait, sans mesure de référence autre que le compte-pour-un renvoie à l'espace lisse, non métrique. Dans l'exemple du Stagirite "dix chevaux et dix chiens", "dix" est un nombre nombrant et "dix chevaux" un nombre nombré, différent de cet autre nombre nombré que sont "dix chiens." Chez Deleuze, il y a plus. Le nombre nombré de l'espace lisse rend compte des opérations locales et du divers des objets. Une expression comme "dix chevaux, trois vaches et deux picotins d'avoine" est un exemple simple de nombre nombré défini sur la variété de la ferme sans que l'on puisse précisément le caractériser par une suite de nombres réels, si ce n'est par (10, 3, 2), qui n'est pas suffisamment précis puisqu'on ne puisse attribuer au picotin une valeur précise. Le nombre nombré devient difficilement divisible lorsqu'on considère des déterminations comme la douleur. Il semble que chaque douleur soit propre à chaque individu, et que les divisions proposées, périphérique ou centrale, cyclique, persistante, aiguë, permanente ou intermittente, ne soient que des parties aux intersections instables et mouvantes. D'où l'idée d'un *nombre nomade* qui affecte les multiplicités lisses.

4 Typologies des multiplicités

De cette division en nombre nombré et en nombre nombrant, des quantités intensives et des quantités extensives, Deleuze déduit qu'il n'existe que deux types de multiplicités, aux ramifications plurielles: "les multiplicités implicites et les explicites, celles dont la métrique varie avec la division et celles qui portent le principe invariable de leur métrique" ([6], 307).

Cette idée qu'il n'existe que deux types de détermination se trouve déjà chez Husserl (*Philosophie de l'arithmétique*) et chez Bergson (*Essai sur les données immédiates de la conscience*). Remarquant que les sons d'une cloche que l'on entend ne se comptent pas, mais que l'on retient des impressions successives, qualitatives, Bergson distingue la multiplicité d'un nombre qui représente une collection d'unités et celle plus "confuse de sentiments et de sensations", comme celle des sons de cloche, qui bien que formant une unité n'en est pas moins une succession d'instants plus ou moins distincts. "D'où résulte enfin qu'il y a deux espèces de multiplicité: celle des objets matériels, qui forment un nombre immédiatement, et celle des faits de conscience, qui ne saurait prendre l'aspect d'un nombre sans l'intermédiaire de quelque représentation symbolique, où intervient nécessairement l'espace" ([1], 65).

Il y a donc deux multiplicités: une multiplicité quantitative et homogène (du nombre en acte dans le vocabulaire d'Aristote) et une multiplicité qualitative et hétérogène (qui serait en quelque sorte, une multiplicité du nombre en puissance) caractéristique de la durée, deux conceptions, qui ne peuvent se défaire de la médiation de l'espace et naissent de leur rapport différent entre l'un et le multiple. Mais, chez Deleuze, la différence entre ces multiplicités a bien d'autres caractéristiques.

> Il nous est souvent arrivé de rencontrer toutes sortes de différences entre deux types de multiplicités: métriques, et non métriques; extensives, et qualitatives; centrées, et acentrées; arborescentes, et rhizomatiques; numéraires, et plates; dimensionnelles, et directionnelles; de masse, et de meute; de grandeur, et de distance; de coupure, et de fréquence; striées, et lisses. Non seulement, ce qui peuple un espace lisse, c'est une multiplicité qui change de nature en se divisant – ainsi les tribus dans le désert: distances qui se modifient sans cesse, meutes qui ne cessent pas de se métamorphoser – mais l'espace lisse lui-même, désert, steppe, mer ou glace, est une multiplicité de ce type, non métrique, acentrée, directionnelle, etc. ([8], 604).

Deleuze envisage deux aspects des multiplicités lisses ou non métriques. Le premier aspect renvoie à la question: "Comment une détermination peut être en situation de faire partie d'une autre, sans qu'on puisse assigner de grandeur exacte, ni d'unité commune, ni d'indifférence à la situation. C'est le caractère enveloppant ou enveloppé de l'espace lisse" ([8], 605). Et le second aspect se pose "quand la situation même de deux déterminations exclut leur comparaison" ([8], 606). Ces deux aspects déterminent, dit-il, le *nomos* de l'espace lisse.

> Nous définissons donc un double caractère positif de l'espace lisse en général: d'une part, lorsque les déterminations qui font partie l'une de l'autre renvoient à des distances enveloppées ou à des différences ordonnées, indépendamment de la grandeur; d'autre part, lorsque surgissent des déterminations qui ne peuvent pas faire partie l'une de l'autre, et qui se connectent par des processus de fréquence ou d'accumulation, indépendamment de la métrique. ([8], 606)

Deleuze caractérise le lisse et le strié, non seulement selon les critères d'intensivité, mais aussi par recollement comme le fait Riemann. Il cite d'ailleurs Albert Lautman, qui lui-même cite Cartan:

> Les espaces de Riemann sont au contraire dépourvus de toute espèce d'homogénéité. Chacun d'eux est caractérisé par la forme de l'expression qui définit le carré de la distance de deux points infiniment voisins. Cette expression est ce qu'on appelle une forme quadratique qui généralise la formule euclidienne de la distance de deux points: $ds^2 = du_1^2 + du_2^2$. Le ds^2 riemannien à deux dimensions est de la forme suivante
>
> $$ds^2 = g_{11}du_1^2 + g_{12}du_1du_2 + g_{21}du_2du_1 + g_{22}du_2^2$$
>
> Dans une variété à n dimensions, on a la formule générale:
>
> $$ds^2 = \sum_{i,j}^{n} g_{ij}du_idu_j$$
>
> Les g_{ij} sont des coefficients absolument quelconques et qui varient de point en point. Il en résulte, comme dit M. Cartan, que "deux observateurs voisins peuvent repérer dans un espace de Riemann les points qui sont dans leur voisinage immédiat, mais ils ne peuvent pas sans convention nouvelle se repérer l'un par rapport à l'autre" [2] Chaque voisinage est donc comme un petit bout d'espace euclidien, mais le raccordement d'un voisinage au voisinage suivant n'est pas défini et peut se faire d'une infinité de manières; l'espace de Riemann le plus général se présente ainsi comme une collection amorphe de morceaux juxtaposés sans être rattachés les uns aux autres. [13]

Cette idée de recollements de domaines d'un espace pour en créer un autre plus vaste est souvent prise par Deleuze comme la caractérisation des espaces riemanniens. Deleuze soutient qu'il est difficile d'utiliser des déterminations scientifiques hors de leur contexte, qui se résument souvent à une "métaphore arbitraire" et dangereuse, mais conçoit une autre voie d'utilisation des concepts scientifiques. Il l'explique dans *Cinéma 2*:

> Mais peut-être, ces dangers sont conjurés si l'on se contente d'extraire des opérateurs scientifiques tel ou tel caractère conceptualisable qui renvoie lui-même à des domaines non scientifiques, et converge avec la science sans faire application ni métaphore. C'est en ce sens qu'on peut parler d'espaces riemanniens chez Bresson, dans le néo-réalisme, dans la nouvelle vague, dans l'école de New-York, d'espaces quantiques chez Robbe-Grillet, d'espaces probabilitaires et topologiques chez Resnais, d'espaces cristallisés chez Herzog et Tarkovski. Nous disons par exemple qu'il y a espace riemannien lorsque le raccordement des parties n'est pas prédéterminé, mais peut se faire de multiples façons: c'est un espace déconnecté, purement optique, sonore ou même tactile (à la manière de Bresson). ([9], 169)

Ce qui caractérise donc un espace riemannien est bien qu'il y ait raccordement, mais que de surcroit ce raccordement soit non prédéterminé. Deleuze ne dit pas que ce raccordement est aléatoire, mais simplement joue de son caractère non prédéterminé. Le mathématicien comprend ce raccordement de manière inverse à sa construction. C'est un atlas, un ensemble de cartes qui ont été recollées, mais rien ne dit comment ces cartes ont été confectionnées. Le découpage aurait pu être un peu plus à l'Ouest ou un plus au Nord. Rien ne prédétermine la façon dont l'atlas a été fait. D'ailleurs, le mathématicien en accord avec les principes de la topologie ne s'en soucie guère. Pour le mathématicien, que la Chine, les États-Unis, la Russie ou l'Afrique soient au centre de l'atlas n'a pas l'importance que lui accorde le géographe ou le philosophe, car le caractère riemannien du recollement qui caractérise ces espaces agit de la même manière quel que soit le découpage. Pour prendre un exemple cinématographique, la caractéristique des films de Bresson et de la Nouvelle vague est bien de morceler à l'extrême le récit sans que les recollements qui s'opèrent au fil du temps ne soient prédéterminés. C'est en cela qu'ils sont riemanniens. Chez Bresson, dit Deleuze, le dialogue "est traité comme s'il était rapporté par quelqu'un d'autre: d'où la célèbre voix bressonienne, la voie du "modèle", par opposition à la voix de l'acteur de théâtre" ([9], 315). Cette distance qui s'institue entre le dialogue et l'autre est le garant de la non-prédétermination du recollement.

Dans les recollements entre les parties d'espaces, il y a très certainement chez Deleuze une complexité du collage qui s'inspire des théories mathématiques. Pour Deleuze, le principe de connexion et d'hétérogénéité des espaces riemanniens peut induire des collages entre structures rhizomiques. Dans l'ordre des discours ciné matographiques, les espaces narratifs et non narratifs déterminent des chaînes sémantiques qui elles-mêmes déterminent des séries de gestes, de phonèmes, et de divers éléments qui se recollent comme les parties d'un espace riemannien.

Le rhizome devient chez Deleuze le symbole de la complexité topologique des espaces philosophiques et de leurs recollements. Des espaces non euclidiens avec des structures rhizomiques jalonnent tout l'espace de réflexion de Deleuze et Guattari. Le rhizome est une structure complexe qui balaye à la fois les structures en arbre et en table, et dans laquelle toutes les connexions entre parties existent.

> N'importe quel point d'un rhizome peut être connecté avec n'importe quel autre, et doit l'être. C'est très différent de l'arbre ou de la racine qui fixent un point, un ordre. L'arbre linguistique à la manière de Chomsky commence encore à un point S et procède par dichotomie. Dans un rhizome au contraire, chaque trait ne renvoie pas nécessairement à un trait linguistique: des chaînons sémiotiques de toute nature y sont connectés à des modes d'encodage très divers, chaînons biologiques, politiques, économiques, etc., mettant en jeu non seulement des régimes de signes différents, mais aussi des statuts d'états de choses. ([8], 13).

Pour Deleuze, le rhizome consiste en des multiplicités qui s'agencent pour connecter et produire du nouveau. Du coup, le rhizome devient une méthode (ou une anti-méthode) pour penser l'ontologie des multiplicités, c'est-à-dire les modalités d'être de ces objets ou de ces variétés. Les mathématiques offrent de nombreux exemples d'un objet qui peut se dire de plusieurs manières, sans que l'on perçoive de prime abord la similitude de ces façons de l'approcher. C'est le cas de la sphère d'homologie de Poincaré qui se définit par un procédé de recollement d'une figure géométrique, mais aussi de bien d'autres manières dont nous n'en citerons que trois (R.C. Kirby et M.G. Scharlemann donnent huit manières de la construire dans [12]). Un objet et trois modalités d'être. Dans la variété de Poincaré, les faces du dodécaèdre sont des pentagones réguliers que l'on recolle deux à deux par leurs faces opposées. La variété ainsi obtenue est un espace topologique quotient qui a été pris par certains astrophysiciens comme modèle global d'univers [14]. Lorsqu'on sort par une des faces du dodécaèdre, on rentre par la face opposée, mais en ayant subi une rotation de $\pi/5$ due au recollement des pentagones. Cet espace est la représentation de la sphère de Poincaré qui est l'une des sphères d'homologie en dimension 3 (tous ses nombres de Betti sont nuls sauf b_0 et b_3 et n'a pas de coefficient de torsion). Lorsque tous les pentagones ont été recollés, le recollement s'arrête de lui-même puisque toutes les faces ont été traitées. La sphère de Poincaré se construit aussi par le quotient du groupe $SO(3)$ par le groupe alterné A_5 ou par chirurgie de Dehn le long du nœud de trèfle. Comme on le voit sur ces exemples mathématiques, un objet peut se dire d'une multitude de points de vue.

Deleuze généralise ces espaces à la sphère philosophique et les prolonge à travers les plissements du monde ou le feuilletage induit par les plis d'un espace ou d'une région. Dans son cours sur Leibniz, il dit que le monde est plié, que le pli a une certaine inflexion ou courbure et que cette courbure, comme dans le cas des foyers d'une ellipse, détermine un ou plusieurs points de vue. "La courbure des choses exige le point de vue." Et le point de vue devient du coup, la "condition de surgissement ou de manifestation d'une vérité dans les choses." Il s'ensuit que le pli pensé comme un réseau de lieux singuliers, de points de vue, induit en retour, une vérité monadique des choses. Mais ce que cherche Deleuze est précisément d'échapper à ce dualisme de l'un et du multiple, qu'il dénonce chez Bergson, de la monade où se nouent les turbulences du monde et les représentations de l'être et de l'âme, mais aussi échapper à toute axiomatique. Il le dit clairement dans le livre sur *Foucault*:

> C'est Riemann qui a formé la notion de "multiplicité", et de genres de multiplicités, en rapport avec la physique et les mathématiques. L'importance philosophique de cette notion apparaît ensuite chez Husserl dans *Logique formelle et Logique transcendantale*, et chez Bergson dans l'*Essai* (quand Bergson s'efforce de définir la durée comme un genre de

> multiplicité qui s'oppose aux multiplicités spatiales, un peu comme Riemann distinguait les multiplicités discrètes et continues). Mais dans ces deux directions, la notion avorta, soit parce que la distinction des genres venait la cacher en restaurant un simple dualisme, soit parce qu'elle tendait vers le statut d'un système axiomatique. L'essentiel de la notion, c'est pourtant la constitution d'un substantif tel que "multiple" cesse d'être un prédicat opposable à l'Un, ou attribuable à un sujet repéré comme un. ([10], 22-23)

Mais qu'on ne s'y trompe pas: la notion philosophique avorta dans ses deux acceptions, l'une bergsonienne et l'autre husserlienne et logiciste. Deleuze ne condamne pas la multiplicité riemannienne. Au contraire, il revient sur cet aspect que nous avons déjà vu, que Riemann sort la multiplicité de la logique en faisant du "multiple" un substantif. Il ajoute:

> La multiplicité reste tout à fait indifférente aux problèmes traditionnels du multiple et de l'un, et surtout au problème d'un sujet qui la conditionnerait, la penserait, la dériverait d'une origine, etc. Il n'y a ni un ni multiple, ce qui serait, de toute manière, renvoyer à une conscience qui se reprendrait dans l'un et se développerait dans l'autre. Il y a seulement des multiplicités rares, avec des points singuliers, des places vides pour ceux qui viennent un moment y fonctionner comme sujets, des régularités cumulables, répétables et qui se conservent en soi. La multiplicité n'est ni axiomatique ni typologique, mais topologique. ([10], 23)

Soulignons ce caractère topologique, sous lequel se place la philosophie de Deleuze et fera la fortune des variétés riemanniennes. Car il s'agit de la pâte des variétés et de leur forme, que la philosophie entend travailler, et pas seulement leur horizon mathématique. "Il y a tant de multiplicités. Non seulement le grand dualisme des multiplicités discursives, et non discursives ; mais, parmi les discursives, toutes les familles ou formations d'énoncés, dont la liste est ouverte et varie à chaque époque" ([10], 27). Ces multiplicités sont au cœur de cette science que Deleuze appelle la "science nomade". Elle se distingue de la "science royale" par la manière qu'elle a de participer à l'organisation du champ social. Toute science participe de ce champ, mais de manière différente. La science royale est liée à ce modèle hylémorphique qui a été étudié par Gilbert Simondon [19]. Deleuze y voit "une forme organisatrice pour la matière, et une matière préparée pour la forme" ([8], 457) où en suivant la dichotomie hjelmslévienne de l'expression et du contenu, croisée à celle de matière et forme, "toute la matière est mise du côté du contenu, tandis que toute la forme passe dans l'expression" ([8], 457). La science nomade est, quant à elle, plus sensible "à la connexion du contenu et de l'expression pour eux-mêmes, chacun de ces deux termes ayant forme et matière" ([8], 457).

> Ainsi, du point de vue de cette science qui se présente aussi bien comme art et comme technique, la division du travail existe pleinement, mais n'emprunte pas la dualité forme-matière (même avec des correspondances biunivoques). Elle suit plutôt les connexions entre des singularités de matière et des traits d'expression, et s'établit au niveau de ces connexions, naturelles ou forcées. C'est une autre organisation du travail, et du champ social à travers le travail. ([8], 457)

C'est donc deux modèles différents que Deleuze oppose, qu'il appelle l'un *compars* et l'autre *dispars*. Le *compars* correspond à l'espace homogène strié, tandis que le *dispars* est associé à l'espace hétérogène, lisse.

Le compars est le modèle légal ou légaliste emprunté par la science royale. La recherche des lois consiste à dégager des constantes, même si ces constantes sont seulement des rapports entre variables (équations). Une forme invariable des variables, une matière variable de l'invariant, c'est ce qui fonde le schéma hylémorphique. Mais le dispars comme élément de la science nomade renvoie à matériau-forces plutôt qu'à matière-forme. Il ne s'agit plus exactement d'extraire des constantes à partir de variables, mais de mettre les variables elles-mêmes en état de variation continue. S'il y a encore des équations, ce sont des adéquations, des inéquations, des équations différentielles irréductibles à la forme algébrique, et inséparables pour leur compte d'une intuition sensible de la variation. Elles saisissent ou déterminent des singularités de la matière au lieu de constituer une forme générale. ([8] 457–458).

5 Conclusion

On n'insistera sans doute jamais assez sur l'importance dans la philosophie de Deleuze de ces singularités qui structurent les multiplicités, et qui, comme chez Gilles Châtelet [4], forment autant de points de passage entre l'actuel et le virtuel. Ces singularités, par un processus d'auto-unification, forment des séries, et souvent par groupement, des séries de séries, qui bifurquent sous l'effet de comportements aléatoires, et par leur nature topologique, expliquent cette synthèse de l'hétérogène, que l'on trouve dans cette version philosophique des variétés riemanniennes de la science nomade. La pensée de Deleuze se constitue donc autour de notions de topologie, de compréhension des multiplicités, qu'il puise tant dans les mathématiques, que chez d'autres philosophes, Husserl et Bergson, mais aussi Simondon et Foucault. Il reconnaît que l'*Archéologie du savoir* "représente le pas le plus décisif dans une théorie-pratique des multiplicités" ([10], 23). Chez Deleuze, tout le savoir et son cortège de points singuliers sont aussi des multiplicités:

Bref, une science se localise dans un domaine de savoir qu'elle n'absorbe pas, dans une formation qui est, par elle-même, objet de savoir et non pas de science. Le savoir n'est pas science, ni même connaissance, il a pour objet les multiplicités précédemment définies, ou plutôt la multiplicité précise qu'il décrit lui-même, avec ses points singuliers, ses places et ses fonctions. ([10], 28)

Tout le savoir, y compris le savoir de Deleuze, se fractionne en multiplicités. Ce savoir est lui-même une multiplicité, c'est-à-dire dans le vocabulaire très connoté de termes physico-mathématiques employés, une variété indépendante de l'espace dans lequel il est plongé. Il sera donc étudié pour lui-même et les singularités seront autant de points de jonction de cette théorie rationalisée du sujet connaissant que l'on trouve déjà chez Simondon. Deleuze va plus loin encore: il emprunte à la physique la notion de champ qu'il trouve chez Monge ([8], 459) et qui épouse les variétés riemanniennes.

L'espace lisse est justement celui du plus petit écart: aussi n'a-t-il d'homogénéité qu'entre points infiniment voisins, et le raccordement des voisinages se fait indépendamment de toute voie déterminée. C'est un espace de contact, de petites actions de contact, tactile ou manuel, plutôt que visuel comme était l'espace strié d'Euclide. L'espace lisse est un champ sans conduits ni canaux. Un champ, un espace lisse hétérogène épouse un type particulier de multiplicités: les multiplicités non métriques, acentrées, rhizomatiques, qui occupent l'espace "sans compter", et qu'on ne peut "explorer qu'en cheminant sur elles" . Elles ne répondent pas à la condition visuelle d'être observées d'un point de l'espace extérieur à elles: ainsi le système des sons, ou même des couleurs, par opposition à l'espace euclidien. ([8], 459-460)

Deleuze déploie de nouveau le cortège des multiplicités et de leurs singularités dans l'espace hétérogène lisse, où elles forment un réseau rizhomatique caractéristique du dispar, qui ne peut s'observer que de l'intérieur. Le monde s'organise autour de singularités nomades, et les individus se constituent au voisinage de ces singularités. Seule, dit Deleuze, une théorie des singularités peut dépasser "la synthèse de la personne et l'analyse de l'individu telles qu'elles sont (ou se font) dans la conscience" ([7], 125). Le champ transcendantal, préindividuel et impersonnel explique ainsi la genèse des individus et des sujets. Pour Deleuze, les singularités sont les vrais événements transcendantaux. Elles sont plongées dans un potentiel, qui "ne comporte ni Moi, ni Je, mais qui les produit en les actualisant" ([7], 125). Ce sont des séries hétérogènes organisées en système métastable dont la surface est le lieu du sens.

Tels sont sommairement esquissés, quelques éléments de mathématiques ou d'inspiration mathématique que l'on trouve dans la philosophie de Deleuze. Le concept de multiplicité y est présenté comme un concept qui n'est "ni axiomatique, ni typologique, mais topologique". Il décrit une longue courbe, dont l'origine remonte à la notion de variété riemannienne, métamorphosée au contact de Bergson et de Husserl, puis au voisinage de Simondon et de Foucault, qui toujours croise plusieurs niveaux, actualise des unités possibles dans le divers des singularités.

6 Extended English Abstract

Bernhard Riemann is arguably the mathematician who had the greatest influence upon the work of Deleuze and Guattari. I would like, in this essay, to explore some reasons for this influence and in particular for the genesis of the Deleuzian concept of "*multiplicity*", its links with smooth and striated spaces, and the Riemannian concept of "*manifold*" (*mannigfaltigkeit*). In the book about *Foucault*, Deleuze says: "It was Riemann in the field of physics and mathematics who dreamed up the notion of *multiplicity* and different kinds of multiplicities. The philosophical importance of this notion then appeared in Husserl's *Formal and Transcendental Logic*, and in Bergson's *Essay on the Immediate Given of Awareness* (where he tries to define duration as a type of multiplicity to be contrasted with spatial multiplicities, rather as Riemann had distinguished between discrete and continuous multiplicities). But the notion died out in these two areas, either because it became obscured by a newly restored simple dualism arising from a distinction made between genres, or because it tended to assume the status of an axiomatic system."[3] In *Différence et répétition*, Deleuze argues that ideas are multiplicities, and each idea is a multiplicity, a manifold in the sense of Riemann. Following Riemann and Weyl, he gives two examples: the space of sounds and the space of colours. He recognizes that modes of determination are important for the definition of a manifold, and wonders whether the concept of number is also useful to define a manifold. In the same way, he asks if logical structures are essential for manifolds. We know that Deleuze did not like logic very much. He writes: "It was a decisive event when the mathematician Riemann uprooted the multiple from its predicate state." Given the philosophical developments of the twentieth century, much of which is dedicated to logic and language, we understand why Deleuze prefers the mathematical concept of manifold to the concept of Grothendick topos used by Alain Badiou. Deleuze chooses topology, because Riemann "marked the end of dialectics and the beginning of a typology and topology of multiplicities." Each multiplicity is defined by n determinations. Some multiplicities are metric because the magnitude between two points from a line with the magnitude of two other points from another line can be compared. In this case, the underlying space is said to be striated. Some others multiplicies are not metric, because they can only be measured by indirect means. Two sounds of equal pitch and different intensity cannot be compared with two sounds of equal intensity and different pitch. In this case, the underlying space is said to be smooth. The smoothness is not the infinitely differentiability of mathematicians. For Deleuze, smooth means non-metric. The desert, steppe, ice, sea, Cézanne's wheat field and fractal objects are smooth spaces. At this point, in relation with smoothness, we try to comment this extract, involving the Aristotelician notions of *numbering number* and *numbered number*: "The number distributes itself in smooth space; it does not divide without changing nature each time, without changing units, each of which represents a distance and not a magnitude. The ordinal, directional, nomadic, articulated number, the numbering number, pertains to smooth space, just as the numbered number pertains to striated space. So we may say of every multiplicity that it is already a number, and still a unit. But the number and the unit, and even the way in which the unit divides, are different in each case."[4] Summarizing all kinds of multiplicities, the concepts of striated and smooth spaces in connection with the thought of Riemann is questioned. "We have on numerous occasions—says Deleuze—encountered all kinds of differences between two types of multiplicities: metric and nonmetric; extensive and qualitative; centered and acentered; arborescent and rhizomatic; numerical and flat; dimensional and directional; of masses and of packs; of magnitude and of distance; of breaks and of frequency; striated and smooth. Not only is that which peoples a smooth space a multiplicity that changes in nature when it divides—such as tribes in the desert: constantly modified distances, packs that are always undergoing metamorphosis—but smooth space itself, desert, steppe, sea, or ice, is a multiplicity of this type,

[3]G. Deleuze, *Foucault*, Translated by Seán Hand, University of Minnesota Press, 1988, 13.

[4]G. Deleuze, F. Guattari, *A Thousand Plateaus*, Translation and foreword by Brian Massumi, University of Minnesota Press, 1987, 484.

non-metric, acentered, directional, etc."[5] It is also instructive to follow Deleuze's distinction between the royal science and the nomad science, two models of science, inspired from Plato's *Timaeus*, called by Deleuze the *Compars* and the *Dispars*. The Compars corresponds to the homogeneous striated space, and the Dispars is linked to the heterogeneous smooth space. "The compars is the legal or legalist model employed by royal science. The search for laws consists in extracting constants, even if those constants are only relations between variables (equations). An invariable form for variables, a variable matter of the invariant: such is the foundation of the hylomorphic schema. But for the dispars as an element of nomad science the relevant distinction is material-forces rather than matter-form. Here, it is not exactly a question of extracting constants from variables but of placing the variables themselves in a state of continuous variation. If there are still equations, they are adequations, inequations, differential equations irreducible to the algebraic form and inseparable from a sensible intuition of variation. They seize or determine singularities in the matter, instead of constituting a general form."[6] Compars and Dispars are linked to nomos and logos, but also to singularities. For Deleuze, the world is organized around nomad singularities, and individuals are in the vicinity of these singularities. The problem of multiplicities and their singularities has a lot of philosophical consequences. Deleuze moves the philosophic thought from logic to topology. This is a breakthrough and its most innovative aspect. "Multiplicity remains completely indifferent to the traditional problems of the multiple and the one, and above all to the problem of a subject who would think through this multiplicity, give it conditions, account for its origins, and so on. There is neither one nor multiple, which would at all events entail having recourse to a consciousness that would be regulated by the one and developed by the other. There are only rare multiplicities composed of particular elements, empty places for those who temporarily function as subjects, and cumulable, repeatable and self-preserving regularities. Multiplicity is neither axiomatic nor typological, but topological."[7]

References

1. H. Bergson, *Essai sur les données immédiates de la conscience*, F. Alcan (1889)
2. É. Cartan, *Les récentes généralisations de la notion d'espace*, Bulletin de la société mathématique **48**, 294–320 (1924)
3. É. Cartan, Espaces à connexion affine, projective et conforme. Acta Math. **48**, 1–42 (1926)
4. G. Châtelet, *Les Enjeux du mobile* (Seuil, Paris, 1993)
5. T. Levi-Civita, G. Ricci, Méthodes de calcul différentiel absolu et leurs applications. Math. Ann. B **54**, 125–201 (1900)
6. G. Deleuze, *Différence et répétition* (Presses Universitaires de France, 1968)
7. G. Deleuze, *Logique du sens* (Editions de Minuit, 1969)
8. G. Deleuze, *Mille Plateaux* (Editions de Minuit, 1980)

[5] *A Thousand Plateaus, Op. Cit.,* 484.

[6] *A Thousand Plateaus, Op. Cit.* 369.

[7] *Foucault, Op. Cit.*, 14.

9. G. Deleuze, *Cinéma 2. L'image-temps* (Editions de Minuit, 1983)
10. G. Deleuze, *Foucault,* (Editions de Minuit, 1986)
11. H. Hopf, Differential geometrie und topologische gestalt. Jahresbericht der Deutschen Mathematiker Vereinigung **41**, 209–229 (1932)
12. R.C. Kirby, M.G. Scharlemann, Eigth faces of the Poincaré homology 3-sphere, in *Geometric Topology* (Academic Press, 1979), pp. 113–146
13. A. Lautman, *Essai sur les notions de structure et d'existence en mathématiques*, repris dans A. Lautman, *Les mathématiques, les idées et le réel physique*, Vrin, p. 136
14. J.-P. Luminet, J. Weeks, A. Riazuelo, R. Lehoucq, J.-P. Uzan, Dodecahedral space topology as an explanation for weak wide-angle temperature correlations in the cosmic microwave background. Nature **425** (6958), pp. 593–595 (2003)
15. K. Ohshika, The origin of the notion of manifold: from Riemann's Habilitationsvortrag onward, in *From Riemann to Differential Geometry and Relativity*, ed. by L. Ji, A. Papadopoulos, S. Yamada, p. 295 (Springer, 2016)
16. A. Plotnitsky, Comprehending the connection of things: Bernhard Riemann and the architecture of mathematical concepts, in *From Riemann to Differential Geometry and Relativity*, ed. by L. Ji, A. Papadopoulos, S. Yamada, pp. 329–362 (Springer, 2017)
17. B. Riemann, *Über die Hypothesen, welche der Geometrie zu Grunde liegen*. Habilitationsschrift, 1854, publiée en 1867. Traduction française *Sur les hypothèses qui servent de fondement à la géométrie*, traduction de L. Laugel, in Œuvres Mathématiques, Gauthier Villars, 1908, réédition Jacques Gabay, 1990
18. A. Sauvagnargues, *Deleuze L'empirisme transcendantal* (Presses Universitaires de France, 2009)
19. G. Simondon, *L'individu et sa genèse physico-biologique* (Presses Universitaires de France, 1964), pp. 42–56
20. H. Weyl, *Temps, Espace, Matière. Leçons sur la théorie de la relativité générale*. Traduit de la quatrième édition allemande *Raum — Zeit – Materie* par, G. Juvet, R. Leroy. Paris Blanchard, 1922. Réimprimé (Blanchard, 1958)
21. H. Weyl. Reine Infinitesimalgeometrie. Mathematische Zeitschrift **2**, 384–411 (1918) (R.W, Herman. Gesammelte Abhandlungen, in vol. 4, ed. by K. Chandrasekaran (Springer II, Berlin, 1968), pp. 1–28

Comprehending the Connection of Things: Bernhard Riemann and the Architecture of Mathematical Concepts

Arkady Plotnitsky

Abstract This chapter is an essay on the conceptual nature of Riemann's thinking and its impact, as conceptual thinking, on mathematics, physics, and philosophy. In order to fully appreciate the revolutionary nature of this thinking and of Riemann's practice of mathematics, one must, this chapter argues, rethink the nature of mathematical or scientific concepts in Riemann and beyond. The chapter will attempt to do so with the help of Deleuze and Guattari's concept of philosophical concept. The chapter will argue that a fundamentally analogous concept of concept is also applicable in mathematics and science, specifically and most pertinently to Riemann, in physics, and that this concept is exceptionally helpful and even necessary for understanding Riemann's thinking and practice, and creative mathematical and scientific thinking and practice in general.

1 Introduction

This chapter is an essay on the conceptual nature of Bernhard Riemann's thinking and its impact, as conceptual thinking, on mathematics, physics, and philosophy. In order to fully appreciate the revolutionary nature of this thinking and of Riemann's practice of mathematics, one must, I argue, rethink the nature and structure, architecture, of mathematical or scientific concepts in Riemann and beyond. I shall attempt to do so here with the help of Gilles Deleuze and Félix Guattari's concept of philosophical concept, as defined in *What Is Philosophy*? [8], the culminating work of Deleuze's philosophy, on which I shall comment presently. I argue that a fundamentally analogous concept of concept is also applicable in mathematics and science, specifically and most pertinently to Riemann, in physics, and that this concept is exceptionally helpful and even necessary for understanding Riemann's thinking and practice, and creative mathematical and scientific thinking and practice in general. While I shall address Riemann's work in physics, I shall, given my scope, be less concerned with

A. Plotnitsky (✉)
Theory and Cultural Studies Program, Purdue University, W. Lafayette, IN 47907, USA
e-mail: plotnits@purdue.edu

L. Ji et al. (eds.), *From Riemann to Differential Geometry and Relativity*,
DOI 10.1007/978-3-319-60039-0_11

physical concepts as such.[1] The concept of concept in question here is discussed in Sect. 2, which follows this Introduction. Section 3 considers Riemann's concept of manifold(ness) [*Mannigfaltigkeit*] and his understanding of space and geometry, grounded in this concept. Finally, Sect. 4 addresses the relationships between mathematics and physics in Riemann. Thus, Sect. 2 is more philosophical, Sect. 3 is more mathematical, and Sect. 4 more physical. When it comes to Riemann, however, philosophy, mathematics, and physics are never far from each other; and the workings of all three in Riemann's thought will be addressed throughout this article.

I would like to begin, by way of a prologue, with the 1907 article, "*La logique et lintuition en mathématique,*" by Émile Borel, who questioned the logicists' philosophy of mathematics, which theorized mathematics as an extension of logic and which, championed by, among others, Bertrand Russell, was in vogue at the time. Borel opens his argument by considering polynomial identities, such as $(x+1)(x-1) = x^2 + 1$ familiar to all of us since our school days. One can imagine such identities to be produced mechanically and entirely correctly, by a logical mechanism or machine. One can then imagine one of these identities, say, number 35,427, being $4P^3 = Q^2 + 27R^2$, satisfied when $P = x^2 - x + 1$, $Q = (2x-1)(x+1)(x-2)$, and $R = x(x-1)$. While the previous or the following one in this sequence may be entirely uninteresting mathematically, this one is interesting and even special, first, because it expresses a cube as a (weighted) sum of two squares, and, although Borel did not mention this fact, this identity is also important in the theory of elliptic functions [2, pp. 273–274] [12, pp. 402–403]. Logic, Borel argues, cannot capture the importance of this identity. Borel gave other examples of this failure of logic to capture the essence of mathematical thought, such as the fact that the formula expressing the invariance of the cross-ratio of four points on a line under a perspectivity is easy to find, but it took a Chasles to see in it the key to projective geometry. Borel also considered, with the same aim, the case of another polynomial identity (the icosahedral equation) that could only have been found to be valuable if discovered by a nonmechanical route, namely through Felix Klein's unification of Evariste Galois's theory with the theory of the symmetries of the icosahendron. This is, I think, one of those findings in which the essence of modern mathematics is manifested, not the least because it brings together concepts and calculation, very much in the spirit of Riemann, which often guided Klein's work. This finding and its generalizations have profound connections to Riemann's ideas concerning Riemann surfaces, in the so-called Belyi theorem and Grothendieck's *Dessin d'Enfants* theory. Borel's view, then, was that a truly fertile invention in mathematics consists of the discovery of a new point of view from which to classify and interpret the facts, followed by a search for the necessary proofs by plausible reasoning (later considered by George Pólya [30]), and only in the third and final stage does logic take over. As Jeremy Gray notes, "Borel's criticisms are not quite the staples they might seem, and not just because they had a specific resonance in the France of the time. They point quite clearly toward a problem that has not gone away in philosophers' treatment of mathematics: a tendency to reduce it to some essence that not only deprives it of

[1] I have considered physical concepts from this perspective in [27, pp. 2–11].

purpose but is false to mathematical practice. The logical enterprise, even if it had succeeded, would only have been an account of part of mathematics—its deductive skeleton" [11, pp. 202–203].

Nor, I would contend, would one do much better on that score by using the axiomatic approach to foundations of mathematics, sometimes juxtaposed to logicism, or by using formalism, which assumes that mathematics is not a body of propositions representing an abstract sector of reality but is much more akin to a game, and as such, allows one to capture more of creative mathematical thinking [34]. Henri Poincaré, it is worth noting, was as skeptical as Borel, and on similar grounds, concerning the capacity of these approaches to capture creative mathematical thinking [11, 29, pp. 261–265, 390–391]. However, as I argue here, things are different when it comes to the conceptual aspects of mathematics, which may well be primarily, even if not solely, responsible for truly fertile inventions in mathematics, if we understand properly what mathematical concepts are and how they work. Riemann's thinking and practice provide a particularly helpful guidance for this understanding, especially, I would like to propose, if one sees Riemann's understanding and practice of concepts through the optics of Deleuze and Guattari's concept of a philosophical concept in *What Is Philosophy*? [8]. Gilles Deleuze (1925–1995), the main driving force behind this concept of concept, was one of the most creative, as well as the most controversial, French philosophers of the second half of the twentieth century, both in his own work and in his collaboration with Félix Guattari (1930–1992). Deleuze's thinking, including that concerning the nature of concepts, was influenced by Riemann, especially by Riemann's concept of manifold and the resulting rethinking of spatiality, which Deleuze saw as heralding a revolutionary change in philosophy, mathematics, and physics alike, and, I would add, in the relationships among them [7, p. 483].[2]

Deleuze and Guattari's concept of philosophical concept can and here will be transferred, partially and against their own grain, into our understanding of mathematical or scientific, such as physical, concepts. Riemann's contribution to physics in this regard and in general was, if more limited, nearly as revolutionary as his contribution to mathematics. In this chapter, I shall primarily focus on Riemann's mathematics, to which I shall mainly refer at the moment as well. Most of my argument concerning concepts, will equally apply to Riemann's physics, which I shall consider in Sect. 4, or, by definition, to Riemann's philosophical thinking. Riemann's mathematical thinking was, I argue, fundamentally defined by the invention of new concepts in Deleuze and Guattari's sense, now applied to mathematical rather than philosophical concepts. This transfer of their concept of concept to mathematics is partial because mathematics cannot be defined only by the invention of mathematical concepts, given the roles of other aspects of mathematical thinking, such as logical and axiomatic reasoning, which Deleuze and Guattari see as most defining in mathematics. Accordingly, it becomes a matter of the relative precedence of these components at different junctures, with, however, the invention of concepts having

[2]On this influence, see F. Jedrzejewski's contribution to this volume [14] and an earlier article by the present author [26].

the main role in creative mathematical or scientific thought. But then, one could argue that philosophy cannot be limited to the invention of new concepts either, and that there, too, it is a matter of the relative precedence of different aspects of philosophical thinking at different junctures, with the invention of concepts having the primary role in creative philosophical thought. This transfer is against Deleuze and Guattari's own grain, because, following Georg W. F. Hegel and other post-Kantians, Deleuze and Guattari associate their concept of concept primarily and even uniquely with philosophical thinking. The concept [in this sense], they say, belongs to philosophy and only to philosophy [8, pp. 11–12, 33–34].

At the same time, as noted above, Riemann's ideas and specifically his concept of manifoldness had a major impact on their philosophy, in part in juxtaposition to Hegel, in a different context, that of dialectical thinking, overturned, they argued, by thinking in terms of manifold-like concepts in philosophy [7, p. 483]. This is not inconsistent. Hegel is close to Riemann when it comes to their respective concepts of concept, although Hegel, too, associates his concept of concept [*Begriff*] primarily and even uniquely with philosophical thought. My argument here, however, is only against Deleuze and Guattari's, or Hegel's, own grain, not against their thinking. Neither Deleuze and Guattari nor Hegel (who made an even stronger claim to this effect) are ultimately able to deny mathematical and scientific concepts most of the essential features of the architecture that they associate with philosophical concepts, and are, more expressly, Deleuze and Guattari, compelled to reinstate this architecture to mathematical and scientific concepts [8, pp. 217–218]. Riemann, by contrast, fuses philosophy, mathematics, and physics in his thinking and in the architecture of his concepts, and thus also offers us a better way of understanding the nature and workings of concepts in all three domains, and in the relationships among them.

The conceptual character of Riemann's mathematics has been noted and even emphasized in commentaries on Riemann, especially in contrast to mathematics based in calculations (e.g. [9, 16]). These commentaries have not, however, paid sufficient attention to the architecture of Riemann's concepts, or of fundamental mathematical concepts in general. They have not asked the question "What is a concept?" They either take the concept of concept for granted or adopt a conventional view of concepts as generalizations from particulars. As a result, they miss the architectural complexity of Riemann's concept of concepts and, along with it, the deeper nature of Riemann's conceptual thinking, which was not defined most essentially by its juxtaposition to calculation, but rather by his concept of concept.[3] It is not, it is worth noting, that Riemann did not do calculations. But he gave priority to concepts (in his sense), even in doing calculations, grounded in and made more effective by his use of concepts. This is manifested, for example, in his work on functions of a complex variable, where the invention of new concepts, such as that of Riemann surface, created new possibilities for calculations, the potential of which has been explored throughout the subsequent history and is still far from exhausted. It may indeed be inexhaustible.

[3] K. Ohshika's chapter in this volume [20] is a notable exception as a reflection on the architecture of Riemann's concepts.

Although arising from Riemann's work in general, my argument here is based most essentially on Riemann's Habilitation lecture, "On the Hypotheses That Lie at the Foundations of Geometry" [*Ueber die Hypothesen, welche der Geometrie zu Grunde liegen*] [31], given in 1854 but first published by Richard Dedekind only in 1868, two years after Riemann's death.[4] The lecture offered a radical rethinking of space and geometry, as against the preceding history of both, from before Euclid to the discovery of non-Euclidean geometry around 1830. This rethinking was based on the concept of manifold or manifoldness [*Mannigfaltigkeit*], a major mathematical innovation. Establishing the possibility of non-Euclidean geometry was a great mathematical discovery, with profound implications for mathematics, physics, and philosophy, and indeed culture. However, as Riemann argued in his lecture, the pre-Riemannian conceptions of non-Euclidean geometry did not sufficiently depart from previous thinking concerning space and geometry, with a possible exception of certain ideas of Karl Friedrich Gauss, Riemann's teacher and precursor. The role of the discovery of the non-Euclidean geometry of Nikolai Lobachevsky and János Bolyai in Riemann's thinking is not clear, and the consensus appears to be that it was not important. Indeed, part of my argument here concerns some of the reasons why it might not have been. Lobachevsky's and Bolyai's work did not figure in Riemann's lecture and, unlike Gauss, neither was mentioned there. The question of the fifth (parallel) postulate of Euclid, central to the history that led to the discovery of non-Euclidean geometry, played no role in Riemann's argument either. Riemann pursued a different way of thinking, in part following Gauss's ideas concerning the curvature of two-dimensional surfaces. Riemann's thinking was also, and correlatively, problematic rather than axiomatic-theorematic (as was that of Lobachevsky and Bolyai), the juxtaposition that I explain below, merely noting for the moment that the axioms of geometry were neither Riemann's starting point nor figured significantly, if at all, in his investigation. Riemann's approach led him beyond a single alternative to Euclidean geometry to an uncontainable multiplicity of geometries and, in principle, to an even greater multiplicity of possible spaces, because some among them would not admit geometry. The latter circumstance became important for the development of topology, which, unlike geometry, deals with the architecture of a given space and associating numerical and algebraic entities with this architecture, rather than, as geometry, with the measurement of distances. Riemann's ideas, beginning with his concept of Riemann surfaces, which are two-dimensional manifolds with a particular type of topological structure, played a major role in the emergence of topology as a mathematical discipline.

The implications for physics, extending those of the discovery of a single non-Euclidean geometry, were dramatic as well, and as indicated above and as will be discussed in detail in Sect. 4, Riemann's contribution to physics in the lecture was nearly as revolutionary as his contribution to mathematics. For the moment, before the discovery of non-Euclidean geometry, one and only one geometry would be available for a geometrical description of physical space (which is, again, how the term space

[4]On Riemann's subsequent developments of his geometrical ideas and their application to physics and beyond, see Athanase Papadopoulos's contribution to this volume [22].

was understood at the time) if one assumes that space or, in Riemann's more rigorous phrasing, "the reality underlying space" could be described geometrically [31, p. 33]. This had been a grounding assumption of modern physics, from Galileo and Newton on, or of modern philosophy, from Descartes on. Kant's epistemology of phenomena (referring to appearances or representations constructed by our minds) vs. noumena or things-in-themselves (referring to how things, material or mental, exist independently of how we perceive or represent them) qualified this assumption [15]. While important, this qualification did not change the essential import of this assumption for physics. Kant's epistemology does not affect measurements that allow us to ascertain the observable properties of space, specifically whether the observable space is Euclidean (flat) or non-Euclidean (curved), or is three-dimensional (the only space we can phenomenally imagine) or not. Kant's epistemology even helps. Possible deviations from the flatness of physical space (or in any event, from what we refer to as physical space) could be established indirectly by using measuring instruments, while our phenomenal experience of space could still be Euclidean. Riemann reflected on this situation in his lecture, leaving the question of the ultimate nature of physical space or of the reality underlying space to the future physics. Kant, by contrast, ultimately assumed that physical space (or, again, whatever can be possibly referred to as physical space) is Euclidean and three-dimensional, or again, that it was unlikely to be anything else. He allowed that such alternatives are logically possible, but saw them as improbable, in part on theological grounds [23, pp. 3–4, 7]. These grounds have not deterred Riemann, who was originally trained in theology, which might, however, have affected his mathematical and philosophical thinking otherwise, possibly even his concept of manifold.[5] In Riemann's view, that space was a three-dimensional Euclidean manifold was a hypothesis, reasonably well confirmed by the measurements performed at the time, but a hypothesis nevertheless, the truth of which cannot be ascertained by reason alone, as Kant ultimately thought possible, although his position has further complexities, which cannot be addressed here.

In any event, the situation changed with the discovery of non-Euclidean geometry. That actual physical space may not be Euclidean (the three-dimensional nature of space was not contested) made this discovery a major event, even though no measurements made at the time showed any deviation from Euclidean geometry. The situation changed even more radically with Riemann, because his argument implied that an infinite number of possible geometries or, in principle, even topologies (some of which or rather the corresponding mathematical entities, which we now call spaces, would not possess geometry) could be associated with physical space or, again, the reality underlying space, that is, the reality underlying what appears to us as physical

[5]Although the term "*Mannigfaltigkeit*" was not uncommon in German philosophical literature, including in Leibniz and Kant, it is worth noting that the German word for the Trinity is "*Dreifaltigkeit*," thus, etymologically, suggesting a kind of "three-folded-ness," which could not have been missed by Riemann, or, for that matter, Leibniz and Kant. See [20] on the use of the term "*Mannigfaltigkeit*" in Kant vs. Riemann.

space.[6] Any such association is a hypothesis (this is what the hypotheses of Riemann's title refers to) and as such is subject to testing, verification, qualification, refinement, and so forth, which can rule out some among possible geometries or require different geometries at different scales, as indeed happens in modern physics. Thus, with the help of Einstein's general relativity (his non-Newtonian theory of gravity), we know reasonably well certain local physical geometries, say, the one, curved, in the vicinity of the solar system, and even more global geometries, say, that (on average flat) in the Milky Way. Or, to speak more rigorously and, arguably, closer to Riemann's way of thinking, the corresponding argumentation works well in physics and astronomy as things stand now. It is, however, much more difficult to be sure concerning the ultimate geometry of the Universe, although the current data seems to suggest that it is, on the average, flat, as far as we can observe it. Locally, space could be curved by gravity, in accordance with general relativity. This disregards possible quantum aspects of the reality underlying space, for example, as permeated by quantum fields, which may make this reality discrete, the possibility Riemann entertained in his lecture. This discreteness remains conjectural and it is not inherent in quantum field theory as currently constituted (in most interpretations), but it is envisioned in some versions of it, or its extrapolations beyond its current scope, for example, to the Planck scale. Still other versions or interpretations of quantum field theory or even quantum mechanics (the currently standard nonrelativistic, low energy, form of quantum theory) suggests that the reality of the ultimate constitution of nature, including possibly the reality underlying space are beyond representation or even conception, discrete or continuous, the possibility on which I shall further comment below. It is also possible (there are physical considerations that suggest this as well) that there are other Universes with different geometries and topologies. Riemann did not envision this possibility, which in part arises from quantum considerations concerning the Big-Bang origin of the Universe that we observe. In fact, he rejected, perhaps too hastily, any consideration of the Universe on such a scale, "in the infinitely large," as an idle speculation [31, p. 23]. However, his discovery of an infinite number of possible geometries is in accord with this idea, the genealogy of which goes back to Leibniz's concept of (com)possible worlds, although for Leibniz there is only one, the best one, in which we live and which is monadologically reflected in our thought. On the other hand, Riemann argued that "the reality underlying space" "in the infinitely small" is an important question: this reality may have a dimension higher than three (possibly even be infinite-dimensional), be discrete rather than continuous, and so forth. This question, he argued, could only be answered by physics, because this reality is defined by matter and forces acting upon it, rather than on the basis of purely philosophical considerations or "traditional prejudices" [31, p. 33]. This question, which was given new dimensions by quantum theory, is still with us,

[6]It is true that Riemann never considered or even mentioned this possibility, arguably, first expressly investigated by Poincaré, and it is not my aim to make a historical claim to the contrary. My point instead is that this possibility and, as will be explained below, the concept of topological space may be seen as conceptual implications of his argument. It is conceivable, especially given his concept of a Riemann surface, that Riemann entertained this type of idea, just as he (admittedly, expressly) entertained the idea that the reality underlying space may be discrete.

testifying to the remarkable reach and lasting significance of Riemann's thought for mathematics, physics, and philosophy.

2 Philosophy: Planes of Thought and the Architecture of Concepts

In *What Is Philosophy*? Deleuze and Guattari view thought [*la pensée*] as a confrontation between the brain and chaos. On the surface, this view is hardly surprising: much of our thinking (in the sense of mental states and processes) may be understood as this type of confrontation. Deleuze and Guattari, however, have in mind a special form of this confrontation, defined by their conception of thought as different from merely thinking and manifested especially in philosophy, art, and mathematics and science. While unremittingly at war with chaos, thought is also working together with chaos, rather than only protecting us against chaos, as do certain other forms of thinking, in particular, opinion. Deleuze and Guattari see chaos (which they also understand in a particular way, explained below) not only as an enemy but also as a friend of thought, its greatest friend and its best ally in a yet greater struggle, that against opinion, an enemy only, "like a sort of umbrella that protects us from chaos." As they say:

> [The] struggle against chaos does not take place without an affinity with the enemy, because another struggle develops and takes on more importance—the struggle against opinion, which claims to protect us from chaos... [T]he struggle with chaos is only the instrument in a more profound struggle against opinion, for the misfortune of people comes from opinion... But art, science, and philosophy require more: they cast planes over chaos. These three disciplines are not like religions that invoke dynasties of gods, or the epiphany of a single god, in order to paint the firmament on the umbrella, like the figures of an Urdoxa from which opinions stem. Philosophy, science, and art want us to tear open the firmament and plunge into chaos. And what would thinking be if it did not confront chaos? [8, pp. 203, 206, 202].

Chaos itself is "defined [by them] not so much by its disorder as by the infinite speed with which every form taking shape in it vanishes. It is a void that is not a nothingness but a virtual, containing all possible particles and drawing out all possible forms, which spring up only to disappear immediately, without consistency or reference, without consequence. Chaos is an infinite speed of birth and disappearance" [8, p. 118]. This is an unusual conception of chaos. Indeed, it does not appear to have been previously used in philosophy. It originates in quantum field theory and the concept of virtual particle formation there, as is suggested by the terms "particle" and "virtual," although "virtual" is also Deleuze's own philosophical concept [24]. This conception gives a particular form to thought's interaction with chaos. Thought extracts more stable forms of order from speedily disappearing forms of order inhabiting chaos, analogously to the way our measuring technology in high-energy physics extracts "real particles," as they are called, from the "foam" of continuously transforming "virtual particles": electrons into positrons or electron-positron pairs, either to photons, and for forth, in the case of quantum electrodynamics. The picture becomes still more complex (involving neutrinos, electroweak bosons, quarks, Higgs bosons, and so forth) in higher-energy quantum regimes, governed by other forms of quantum field theory [27, pp. 226–238].

Given the essentially creative nature of thought, thus defined, it is not surprising that philosophy, art, and science are among the primary means, and for Deleuze and Guattari are even the primary means, for thinking to become thought [8, p. 208]. Philosophy engages with chaos by creating concepts and planes of immanence; art by creating affects and planes of composition; and mathematics and science by creating functions and propositions, and planes of reference or coordination in science. These conceptions are intricate, and their fuller meaning will become apparent in the course of the discussion to follow. It suffices to say for the moment that the corresponding planes of immanence, composition, and reference, are defined by the movement of thought in each field, while concepts in philosophy, compositions in art, and functions (or other mathematical entities) and propositions in mathematics and science emerge from and are created by this movement.

The specificity of the workings of thought in each field makes them different from each other; and part of the project of *What Is Philosophy*? is to explore this specificity and this difference, in order to answer or (it might not be possible to ever answer it) to pose the title question of the book more sharply. However, the affinities and relationships among art, science, and philosophy are just as significant, and reflect a more complex landscape of thought, in which these fields and the interactions among them are positioned. Here, I shall address conceptual thought in philosophy and mathematics and science. I argue, again, against the grain of Deleuze and Guattari's argument, that creative thought in mathematics and science, and Riemann's thought in particular, are defined as much by the invention of new mathematical and scientific (in Riemann's case, specifically physical) concepts as is creative thought in philosophy by the invention of new philosophical concepts.[7] This argumentation does not negate that of Deleuze and Guattari. First of all, planes of reference, and mathematical entities, such as functions, or logical propositions, are unavoidable in and crucial to mathematics and science. Secondly, as noted from the outset, Deleuze and Guattari are ultimately unable to unconditionally maintain this distinction either. In particular, they are compelled to address the interferences among philosophy, art, and science, interferences essential even for the workings of any single field itself [8, pp. 216–218]. The present argument, which moves beyond only such interferences (found in Riemann's thinking as well), makes this distinction even less definitive and by doing so becomes even more open to the interactions between these fields (again, leaving art aside for the moment).

I shall now explain Deleuze and Guattari's *concept* of a philosophical concept. A concept is not only a generalization from particulars (which is commonly assumed to define concepts) or merely "a [single] general or abstract idea," although a concept may contain such generalizations and abstract ideas [8, pp.11–12, 24]. (Abstract

[7]Mathematics, science, and philosophy also involve the creation of compositions, found in artistic thought, and the latter may, conversely, involve planes of immanence and the creation of concepts, or planes of reference. For one thing, concepts thus defined are composed. More pertinently here, Riemann's concept of manifold is compositional because it defines a manifold as composed of local spaces [26].

ideas invoked here are not the same as abstract mathematical formations, which are, in the present view, concepts in Deleuze and Guattari's sense.) A concept is a multi-component entity, defined by the specific *organization* of its components, which may be general or particular, and some of these components are concepts in turn: "there are no simple concepts. Every concept has components and is defined by them. It is a multiplicity. There is no concept with only one component" [8, p. 16]. Each concept is a multi-component conglomerate of concepts (in their conventional senses), figures, metaphors, and so forth, which are conjoint in a heterogeneous, but interactive, architecture, and this multiplicity that does not amount to a unity, even if it is the unity of the multiple [8, pp.12–13]. It is the *relational organization* of a concept's components that defines it. The role of the multiple in the architecture of concepts is thus crucial. Some unification could take place within the architecture of a concept, but, again, without necessarily fully encompassing the multiplicity, at least a potential multiplicity, of this architecture. It is rare for a concept to have only one component, and ultimately impossible to do so. A single-component concept is only a product of a provisional cut-off of its multi-component organization. In practice, there are always cut-offs in delineating a concept, which results from assuming some of the components of this concept to be primitive entities whose structure is not specified. These components could, however, be specified by alternative delineations, leading to a new overall concept, containing a new set of primitive (unspecified) components. The history of a concept, and every concept has a history, is a history of such successive specifications and changes in previous specifications [8, p.17].

Consider the concept of "bird," beginning with its use in daily life. On the one hand, it may be seen as a single generalization. On the other hand, what makes this concept that of "bird" is the implied presence of components or sub-concepts, such as "wings," "feathers," and "beak," and the relationships among them. The concept acquires further features and components, and thus becomes a different concept, in zoology or biology (as reflected, for example, in the evolutionary relationships between birds and theropod dinosaurs). A philosophical concept of a bird is yet something else. According to Deleuze and Guattari: "a [philosophical] concept of a bird is found not in its genus or species but in the composition of its postures, colors, and songs, something indiscernible that is not so much synesthetic as syneidetic" (a product of the synthesis of the eidos, form, of each concept it absorbs) [8, p. 20].

Each concept is also defined as a problem (as multifaceted as the concept is), a definition that has a mathematical genealogy. A problem is not something that, like a theorem (in the direct sense of the term), is derived from assumed axioms by means of strict logical rules, but is something that is posed, created, along with a concept. A mathematical theorem could also be a problem, when it arises, as in Riemann, from mathematical concepts, rather than from axioms. A problem in this sense, while it must be solved, does not disappear in its solutions: it is "determined at the same time as it is solved" and is "at once both transcendent and immanent in relation to its solution," insofar as it leads to ever-new problems and concepts [6, p. 163]. This persistence helps to make a problem and the concept associated to it "always new," to live on [8, p. 5]. The invention and exploration of new, "always new," concepts, has, Deleuze and Guattari argue, defined the practice of philosophical thought from the pre-Socratics on.

I contend that the same type of argument could be made for the concepts invented in creative mathematics and science. Each mathematical concept (1) emerges from the cooperative confrontation between mathematical thought and chaos; (2) is multi-component; (3) is related to or is a problem; and (4) has a history. Thus, consider the concept of space, historically suspended between mathematics and physics (provisionally putting its philosophical aspects aside), with its constitutive concept-components, point, line, plane, distance, and so forth, each of which, just as the overall concept of space, has a long history of modifications, transformations, redefinitions, and so forth. To mark some of its junctures, by symbolically placing Riemann at the center of this history, this history extends from Euclid (who does not define space, but defines the components just listed) to René Descartes (a coordinate space) to Riemann (a space defined as a manifold) to Felix Hausdorff (topological space) to Alexandre Grothendieck (topos).

It is sometimes difficult to perceive this multi-component architecture of concepts in mathematics and science, because this complexity could be circumvented in their technical practice, in this respect in contrast to philosophy. A more conventional understanding of concepts (such as a generalization from particulars), joined with mathematical and scientific formulas and propositions, tends to suffice. This may be one of the reasons why Deleuze and Guattari (almost) deny that concepts in their sense are found in mathematics and science. They even declare (I think, quite misleadingly) that "it is pointless to say that there are concepts in science [including mathematics]" and adding "even when science is concerned with the same 'objects' [as philosophy] it is not from the view point of the concept; it is not by creating concepts" [8, p. 33]. This includes Riemann's thinking concerning manifolds and spatiality, even as they, at the same time, invoke "a Riemannian concept of space peculiar to philosophy," possibly also in Riemann's philosophical thought, but not, as I argue here, his mathematical or physical thought [8, p. 61]. For them, mathematical and scientific thought is limited to planes of reference, linked to the invention of functions (or other mathematical entities, for example, again, in Riemann) and propositions, and lacks planes of immanence, which make philosophical concepts possible [8, pp. 33–34, 132, 161]. Although their view is more ambivalent and complex than this brief summary and these unequivocal statements by them suggest, they do not extend their concept of concept or their conception (it is not quite a concept in their sense) of the plane of immanence to mathematics and science. By contrast, I argue that planes of immanence and the creation of concepts in this type of sense play central roles not only in creative philosophical thought but also in creative mathematical and scientific thought. In fact, very little of what they say about the *architecture* of philosophical concepts does not apply to mathematical and scientific concepts. Mathematics or science, certainly that of Riemann, is concerned with its objects (shared with philosophy or not) *from the viewpoint of concepts, by creating concepts.*

I am not disputing that mathematical and scientific thought also works with planes of reference, and, via planes of reference, with functions, propositions, and so forth.

Planes of reference give rise to these formations, which define the disciplinary nature of mathematics and science, and essentially shape mathematical and scientific thinking and practice—essentially, but, I would argue, not completely or even most centrally, at least in creative mathematics and science. One might say that creative or, to adopt Thomas Kuhn's language, revolutionary, thought in philosophy and mathematics or science alike is defined by planes of immanence and creation of concepts in Deleuze and Guattari's sense, which always carry individual signatures underneath them [8, p. 50]. Just as there are Plato's ideas, Descartes's cogito, and Leibniz's monads, there are Gauss's curvature, Riemann's manifolds, Dedekind's ideals, and Grothendieck's topoi in mathematics, or Einstein's spaces curved by gravity in general relativity and Heisenberg's matrix variables or Dirac's spinors in quantum physics. This is true even though the functioning of mathematical and scientific concepts does require planes of reference, functions (or other formal mathematical entities), propositions, and so forth.

The difference between philosophy and mathematics or science may instead be, to stay with Kuhn's idiom, in the nature of *normal*, rather than *revolutionary*, practice in each domain. In philosophy, the normal practice consists primarily in understanding, interpreting, and commenting on concepts, while in mathematics and science, the normal practice consists primarily in creating, by means of planes of references, frames of reference, functions and other mathematical or scientific formations, propositions, and so forth. It is true that for Deleuze and Guattari creative, revolutionary philosophical practice is the only true philosophy. However, leaving aside an arguably too restrictive character of this view of philosophy, this is not in conflict with the view of creative mathematics or science advocated here.

Deleuze and Guattari do allow that creative mathematical and scientific thought, such as that of Riemann (one of their primary examples), could have philosophical or artistic, compositional, aspects. But they appear to associate these aspects with *philosophical* or, in Deleuze's language, *inexact* (but philosophically rigorous) thought within mathematical or scientific thought. This philosophical thought is either operative alongside mathematical and scientific thought or enters by way of interference (in the positive sense of interfering wave fronts rather than in the more negative sense of inhibition) between mathematics and philosophy [8, pp. 217–218]. Both types of association are pertinent and important, certainly in Riemann's case. My argument is different, however. I argue that creative *technical*, *exact* mathematical and scientific thought is defined by planes of immanence and multi-component mathematical or scientific concepts, the architecture of which is analogous to that of philosophical concepts in Deleuze and Guattari's sense. That is, even apart from their philosophical strata, *mathematical* and *scientific* planes of immanence and the nature of mathematical or scientific concepts are *analogous* to philosophical thought. It is not only a matter of mathematical and scientific thought becoming philosophical at certain junctures, but a matter of the mathematical and philosophical thought creating parallel homomorphic (partially corresponding to each other), although not isomorphic (fully corresponding to each other), architectures of mathematical and philosophical concepts. They are not isomorphic because of technical, exact, aspects of mathematical and scientific concepts, demanded by the disciplinary nature of mathematics and science, aspects, generally, not found in philosophical concepts.

To bring this point home, I need to say more about Deleuze and Guattari's conceptions of planes of immanence and reference. The plane of immanence, as the plane of the movement of thought, is not "a concept that is or can be thought," but is "the image of thought, the image that thought gives itself of what it means to think" [8, p. 37]. As they say: "Concepts are like multiple waves, rising and falling, but the plane of immanence is the single wave that roles them up and unrolls them" [8, p. 36]. The present argument aims to extend, rather than to juxtapose, the plane of immanence (and the relationship between it and concepts) to mathematical and scientific thought, and to join this plane with the plane of reference. For Deleuze and Guattari, mathematics or science "relinquishes the infinite, infinite speed [of thought], in order to a gain *a reference able to actualize the virtual* [of chaos]. [It] gives reference to the virtual, a reference that actualizes the virtual through functions [or other mathematical objects]" [8, p. 118; translation modified]. Thought's enactment of this process constitutes a plane of reference. Planes of reference do play a major, indeed irreducible, role in mathematical and scientific thinking, especially in the disciplinary functioning of mathematics and science. This, however, is not inconsistent with the view that planes of immanence and the creation of concepts are found in mathematics and science. Mathematical and scientific thought combines both planes (sometimes, as does philosophy, also adding planes of composition) and creates its concepts from this fusion. The processes of thought defined the plane of immanence and specifically the creation of concepts (in Deleuze and Guattari's sense) are equally found in mathematics and science, and, again, define creative thinking there most essentially, analogously to the way it happens in philosophy. Such mathematical planes of immanence and the concepts they give rise to may coexist and interact with philosophical planes of immanence and concepts, but they are not reducible to philosophical planes and concepts. This is because of the equally irreducible interaction between these mathematical planes and concepts with technical, exact aspects of mathematical and scientific thinking.

Thus, in his rethinking of spatiality and geometry, Riemann, not only laid out a philosophical plane of immanence that gives rise to philosophical concepts and architectures, as Deleuze and Guattari rightly argue [7, pp. 483–486] but also introduced *a new mathematical plane of immanence*, which gives rise to multicomponent mathematical concepts, alongside *a mathematical plane of reference*. Riemann's thought shaped the plane of immanence of modern mathematical thought arguably more than that of any other mathematician (although Newton, Gauss, and Galois before Riemann, and Poincaré and Hilbert after him offer some competition). This plane extends well beyond geometry. Riemann made major transformative contributions, especially of a conceptual nature, to many areas of modern mathematics: geometry, topology, analysis, algebra, and number theory–not the least by bringing these fields to bear on each other, and his contributions to physics or philosophy were part of this interactive thinking and practice.

The interactive heterogeneity of Riemann's practice is a crucial aspect of Riemann's mathematics, and it extends beyond mathematics and shapes its mathematical operation from this exterior. Thus, the plane of immanence of Riemann's thought also has a more strictly philosophical dimension and thus creates more strictly philosophical concepts, emphasized by Deleuze and Guattari, rather than only mathematical concepts analogous to philosophical concepts by virtue of their multi-component architecture, stressed here. Riemann's philosophical thought was uncommonly and even nearly uniquely significant for his mathematical or physical thought, although one could think of a few competing cases, such as Weyl, who might have been inspired by Riemann in this respect as well, as he was by many other aspects of Riemann's thought. So were the physical dimensions of Riemann's thought, and this, too, is shared by Weyl, or, earlier Gauss, although this is more common. This role of physics in their work is also essentially connected to the role of philosophy there, which is, again, uncommon, if one speaks of such essential connections, making one's philosophical thinking a constitutive part of one's mathematical thinking, rather than of general philosophical reflections concerning mathematics and science, on the part of mathematicians and scientists.[8] The creation of new mathematics in Riemann was enabled by all three dimensions—mathematical, physical, and philosophical—of Riemann's thought and concepts. Riemann lays out a new plane of immanence of mathematical thought by changing both how to think about geometry or space and how to pursue thinking differently mathematically, via bringing together different mathematical fields and combining them with physical and philosophical concepts. A similar claim could also be made about his thinking concerning physics.

Riemann's Habilitation lecture is a magnificent unfolding of this plane (a geometrical metaphor of "plane" is fitting here). It is a major contribution not only to mathematics but also to physics and philosophy, especially to the philosophy of mathematics and physics, but far from exclusively so, as, for example, Deleuze and Guattari's use of Riemann's concepts shows. It is difficult to overestimate the significance and impact of Riemann's thinking concerning spatiality and geometry in mathematics and physics, in shaping the planes of immanence of thought of both. In mathematics the list of even major areas of impact is nearly inexhaustible, and I shall only mention a few of, arguably, the most important ones. First of all, Riemann's rethinking of geometry in terms of manifoldness crucially expanded the idea of the multiplicity of spaces and geometries themselves. Riemann's view of geometry as, in Deleuze and Guattari's language, topology and typology of manifolds led from the late-nineteenth century on to the extraordinary (a still ongoing) progress of geometry, beginning with the work of Sophus Lie and Felix Klein, and a bit later Élie Cartan [7, p. 483]. This work also connected differential geometry of manifolds and the theory of groups, specifically Lie groups. These connections eventually proved to have a major significance for quantum theory, especially in the theory of elementary

[8]Poincaré's extensive (much more extensive than Riemann's) philosophical works (e.g. [28, 29]), while influenced, as were his mathematical works in geometry, by Riemann's Habilitation lecture, may be seen along these lines. I realize that this claim may be challenged, and make it with caution. I would, nevertheless, argue that Riemann's philosophical thinking plays a greater constitutive role in his mathematical thinking than Poincaré's philosophical thinking in his mathematical thinking. The situation is of course different when it comes to physics, which is a major part of Riemann's and Poincaré's mathematical thinking alike, and both made major contributions to physics.

particles, which are classified by using Lie groups as symmetry groups, although in connection with the infinite-dimensional spaces. The concept of manifold was also crucial for the development of topology. Initially, it was Riemann's earlier work on Riemann surfaces that had a greater impact. Eventually, topology came to be defined by understanding, which, as I shall explain, follows Riemann, of topological spaces as composed of local neighborhoods of points and (open) subspaces of a given space. It is true that Riemann did not have a concept of topological space. My main concern, however, is the development, transformation of Riemann's concepts (such as manifold) leading to new concepts, such as topological space, which makes Riemann's concepts alive, makes them "live on" in new concepts.

3 Mathematics: Space, Geometry, and the Concept of Manifold

The significance of Riemann's lecture for mathematics, physics, and philosophy, and its impact in all three fields were immense. This impact was delayed until its publication, in 1868, two years after Riemann's death, and fourteen years after it was presented in 1854, although some of Riemann's key ideas contained there and in Riemann's related works became known and had their impact earlier. One can only surmise (a tantalizing surmise!) the consequences for the history of mathematics and physics if the lecture was published more immediately. Riemann opens as follows:

> As is well known, geometry presupposes the concept of space, as well as assuming the basic principles for construction in space. It gives only nominal definitions of these things, while their essential specification appears in the form of axioms. The relationship between these presuppositions is left in the dark; we do not see whether, or to what extent, any connection between them is necessary, or a priori whether any connection between them is even possible.
>
> From Euclid to Lagrange this darkness has been dispelled neither by the mathematicians nor the philosophers who have concerned themselves with it. The reason [ground] [*Grund*] for this is undoubtedly because the general concept of multiply extended magnitudes [*Grösse*], which includes spatial magnitudes, remains completely unexplored. I have therefore first set myself the task of constructing the concept of a multiply extended magnitude from general notions of magnitude. It will be shown that a multiply extended magnitude is susceptible of various metric relations, so that space constitutes only a special case of a triply extended magnitude. From this, however, it is a necessary consequence that the theorems of geometry cannot be deduced from general notions of magnitude, but that those properties that distinguish space from other conceivable triply extended magnitudes can only be deduced from experience. Thus arises the problem of seeking out the simplest data from which the metric relations of space can be determined, a problem that by its very nature is not completely determined, for there may be several systems of simple data that suffice to determine the metric relations of space; for the present purposes, the most important system is that laid down as a foundation of geometry by Euclid. These data are—like all data—not logically necessary, but only of empirical certainty, they are hypotheses [*Hypothesen*]; one can therefore investigate their likelihood, which is certainly very great within the bounds of observation, and afterwards decide on the legitimacy of extending them beyond the bounds of observation, both in the direction of the immeasurably large [*Unmessbargrosse*] and in the direction of the immeasurably small [*Unmessbarkleinen*]. [31, p. 23; translation modified]

These introductory reflections are already profound and far-reaching, and Riemann develops the ideas suggested here quite a bit further in the lecture. First of all, Riemann does not appear to be interested in the axiomatic approach to geometry and is even suspicious of axioms of geometry. In contrast to most previous works on non-Euclidean geometry, the parallel postulate is not the starting point of his investigation. The reason is clearly that the concepts of space and of measuring distances in space are not adequately defined. This is what leaves "the relationships between [axioms] in the dark. We do not see whether, or to what extent, any connection between them is necessary, or a priori whether any connection between them is even possible." This questioning of the axioms of geometry was unusual at Riemann's time, and it was much deeper than customary doubts concerning the parallel postulate. The axiomatic approach to non-Euclidean geometry is, thus, abandoned and even implicitly questioned by Riemann, even though it was this approach that led to the discovery of non-Euclidean geometry by Lobachevsky and Bolyai. This may have been one of the reasons why only Gauss, rather than either Lobachevsky (highly regarded by Gauss) or Bolyai, was expressly invoked in the lecture.[9] It is true that Gauss was ultimately unable to establish a possible existence (in the sense of logical consistency) of non-Euclidean geometry, because, unlike Lobachevsky and Bolyai, he did not do this for the three-dimensional case. In spirit, however, Gauss's work on the geometry of two-dimensional surfaces and his concept of curvature as intrinsic to a given surface was much closer to Riemann. For Riemann, Gauss was pursuing a trajectory of thought better suited for the foundations of geometry and more fruitful for its conceptual development. In his famous extraordinary theorem (*theorema egrerium*, as he called it), Gauss proved that the curvature of a surface, which he defined as well, was intrinsic to the surface. More precisely, the theorem states that the Gaussian curvature of a surface does not change if one only bends the surface but does not stretch it. This means that curvature can be fully determined by measuring angles and distances on the surface itself, without considering the way it is embedded in the ambient (three-dimensional) Euclidean space or, in Riemann's terms, manifold, which makes the Gaussian curvature an intrinsic invariant of a given surface.

One nearly has here a two-dimensional conceptual architecture that suggests and even approaches that of Riemann, although major additional thinking is necessary to give this architecture the type of generality Riemann is able to do. Gauss's theorem suggests that one could see the surface as an independent curved manifold and then to generalize this concept to higher dimensions via "the concept of a multiply extended magnitude," mentioned in this passage and the concept of manifold, which is what Riemann did. These concepts also helped him to generalize to higher dimensions Gauss's concept of curvature. To do so required yet another new concept, another

[9]There were other, more extrinsic, reasons, beginning with the fact that Gauss was Riemann's mentor and the chair of his Habilitation committee. Indeed, Gauss selected this topic among three proposed by Riemann (following the rules). The philosopher R. H. Lotze, a fervent opponent of non-Euclidean geometry, was a member of the philosophy faculty, to which Riemann's Habilitation was presented. Later on, Lotze criticized Riemann's approach anyway, as part of his general critique of non-Euclidean geometry (see [16, pp. 222–226] and [33, pp. 97–112]).

great invention of Riemann, the tensor of curvature, and a new form of differential calculus, tensor calculus on manifolds, a generalization of differential calculus. This calculus was fully developed later on, and it played a major role in Einstein's general relativity. The independence of the curvature or of geometry of a given manifold of any dimension from its embedding also allows one to determine, at least in principle, intrinsically whether this space is flat or curved, which is crucial for determining the nature of the physical space we inhabit. For example, as indicated earlier, space is curved in the immediate vicinity of the solar system, because of the gravity of the Sun, planets, and other material entities in the solar system (or locally around other stars), but appears to be on the overage flat on the scale of the observable Universe.[10] In any event, by virtue of inventing his concepts and by using them, Riemann provided a general rigorous grounding to all geometry, Euclidean or non-Euclidean, rather than merely establishing, as Lobachevsky and Bolyai did, the logical possibility of the non-Euclidean geometry of (constant) negative curvature in three dimensions. The absence of a concept, such as that of manifold, leaves us in "the dark," because "we do not see whether, or to what extent, any connection between [axioms] is necessary, or a priori whether any connection between them is even possible."

Riemann's thinking is conceptual-problematic rather than axiomatic-theorematic: he grounds mathematics, as well as physics, in hypotheses and concepts, concepts-problems, such as the concept of manifold and its subconcepts (distance, curvature, tensor, and so forth). The problem of space and geometry is now posed in terms of finding concepts that are necessary to define space and to give it geometry. A general concept of manifold was assumed to be applicable to any possible space, while specific manifolds, flat or curved, define different spaces or subspaces. In this respect, as foundational thinking, that of Riemann is different from that conforming to Hilbert's concept of "foundations" [*Grundlagen*] in the sense of axiomatic foundations, an idea that Hilbert also tried to apply to physics. One of the problems of his famous 1900 list (Problem 6) was in fact that of the development of (mathematized) axiomatic foundations, system(s) of axioms, for physics, on the model of his own *Foundations of Geometry* [13].[11] That such a system is possible is inevitably a hypothesis. It became clear subsequently that the existence of such a system is a hypothesis even in mathematics, and in view of Gödel's incompleteness theorems in 1931, ultimately a wrong hypothesis, insofar as such a system cannot be proven to be free of contradiction, once it is large enough to include arithmetic, as geometry is.

---

[10] As I qualified earlier, at least this is a workable and widely accepted view, widely but not universally. It has never been established definitively or, in any event, agreed upon whether such a determination is ever rigorously possible, as opposed to having a practically effective and possibly, within its proper limits, the best available theory or, as Poincaré would have it, "convention," without making a real claim concerning "the reality underlying space" [31, p. 33]. Einstein had his doubts too, although he was ultimately inclined to accept the possibility of such a determination, at least in principle, as, it appears, was Riemann, but not Poincaré, with whose position Einstein, nevertheless, had to contend and which he tried to accommodate within his own (e.g. [11, pp. 324–328]).

[11] For the development of Hilbert's ideas, as reflected in different editions of the *Grundlagen*, see [4]. In the first version of the book, Hilbert was closer to Riemann, and he later returned to a more Riemannian view of geometry in the wake of general relativity to which he made important contributions.

While Hilbert's foundational thinking aimed to bring physics closer to mathematics, even to make it mathematics, by giving physics an axiomatic form, that of Riemann brings mathematics closer to physics by grounding it in hypotheses and concepts, rather than in axioms.[12]

Riemann's approach and his concepts arise from the plane of immanence, at once, mathematical, physical, and philosophical, a plane defined by thinking in terms of multi-component concepts, such as that of multiply extended magnitude and manifoldness, rather than in terms of axioms and propositions, an approach that has defined most thinking concerning geometry before and even after Riemann. Riemann's thought is defined by a plane of immanence by virtue of giving rise to multicomponent concepts, rather than only creating frames of reference, functions, or propositions, from a plane of reference, although this is necessary as well. In Riemann, both planes are joined, as they must be in the creation of mathematical or scientific concepts. Thus, functions define both local neighborhoods (as infinitesimally Euclidean) and how local spaces are connected or pass into each other, and metrical relations and curvature, although curvature is ultimately defined by tensors, which are more complex entities.

The parallel postulate is, again, never mentioned in the lecture, although Euclidean geometry is invoked there as the geometry of "flat space," merely a particular and very special case of geometry, where metrical relations take an especially simple form, defined by the Pythagorean theorem. In Gray's words, "[In Riemann] geometry no longer starts with Euclidean geometry" [11, p. 52]. It is the metrical relation characterizing a given space that defines a possible geometry. In other words, the character of this relation is a hypothesis on which a geometry could be based, a hypothesis to be tested physically in order to establish whether such a space corresponds to the actual physical space. Non-Euclidean geometry (Riemann, again, does not use the term) is introduced as such a possible case of geometry, that of a curved, rather than flat, space of either negative or positive curvature, defined by Riemann by the corresponding type of quadratic form determining the metric. Riemann, thus, not only introduces a more general concept of geometry, but also gives a more rigorous conceptual grounding to non-Euclidean geometry of either negative or positive curvature. While the non-Euclidean geometry of negative curvature (hyperbolic geometry) was discovered before Riemann, Riemann's lecture introduced the three-dimensional non-Euclidean geometry of positive curvature (elliptical geometry), keeping in mind that space for Riemann meant the three-dimensional physical space. (The two-dimensional spherical geometry was considered well before Riemann.) Although eclipsed by Riemann's overall achievement in the lecture, this

[12]One might challenge this argument on historical grounds because it would have been difficult, if not impossible, to present a concept such as that of manifold in axiomatic form at the time of Riemann's lecture. That may be true. My point, however, is that Riemann's alternative, conceptual-problematic rather than axiomatic-theorematic, thinking, could still be contrasted to that of Hilbert and lead to a different type of mathematical thinking. It is difficult to say how Riemann would have approached the foundations of geometry if he had means of axiomatizing his concepts. On the other hand, it is possible to argue, as I do here, *on historical grounds*, that Riemann, unlike his predecessors, Lobachevsky and Bolyai, was not pursuing an axiomatic approach to geometry.

was a major mathematical discovery with important cosmological implications, for example, in its anticipation of the idea, later considered by Einstein, that the universe may be unbounded and yet finite.[13]

Riemann defines the concept of space, again, understood as physical space (as against, the concept of manifold, which is mathematical), as a three-dimensional instance of the concept of continuous manifoldness, in accordance with the hypotheses that he assumed as likely given the experimental data then available. (There is still no definitive data to refute this view, unless perhaps in the very small, say, at the Planck scale.) While a manifold, as defined by Riemann, may be either discrete or continuous, the concept of continuous manifoldness has a richer and more complex architecture, and most of Riemann's lecture is devoted to it. Technically, continuous manifolds considered by Riemann were differentiable manifolds, which means that one can define differential calculus on them. Indeed, they are metrical manifolds, now called Riemannian, which allow for the concept of distance between any two points and thus for geometry.[14] I shall, however, speak of continuous manifolds, following Riemann and his juxtaposition between continuous and discrete manifoldness. In modern use, the term manifold more customarily refers to continuous (but not necessarily differentiable) manifolds, although one also refers to discrete manifolds, which have topological dimension zero, as zero-manifolds. As defined by Riemann, discrete and continuous manifolds do not appear to have that much in common, and in effect form two different concepts.[15]

Riemann's concept of continuous manifoldness was a new concept of geometrical multiplicity. It is a multiplicity of local subspaces, most specifically those, "neighborhoods," associated with each point, out of which a given space is com-

---

[13]It would be similar to the three-dimensional sphere. As I explained, the currently dominant view or hypothesis (which appears to be confirmed by cosmological measurements) is that the universe is on average flat and is expanding.

[14]As most of his contemporaries, Riemann did not distinguish continuous and differentiable manifolds. It became eventually clear, however, that not all continuous (also called topological) manifolds are differentiable. There are topological manifolds with no differentiable structure, and some with multiple non-diffeomorphic differentiable structures. Thus, there is a continuum of non-diffeomorphic differentiable structures of $\mathbf{R}^4$.

[15]These two concepts could, especially in modern understanding, be subsumed under the same concept. This is because all zero-dimensional manifolds, which are discrete manifolds in Riemann's terms, are continuous (topological) manifolds. In fact they are also differentiable manifolds, because transition functions for them are constant functions, which are continuous and even differentiable. In modern terminology, the distinction between continuous and discrete manifolds in Riemann's lecture would be interpreted as that of zero-dimensional manifolds and positive dimensional manifolds. I am grateful to Ken'ichi Ohshika for helping to clarify this point. It is not inconceivable that Riemann's thought along similar lines, which would explain his choice of the term manifold for both discrete and continuous manifolds, although the term had a more general use at the time. (Georg Cantor, possibly influenced by Riemann's lecture, initially referred to sets as *Mannigfaltigkeiten* but eventually switched to *Mengen*.) It is, however, difficult to be certain on the basis of his lecture or his other writings. I would argue that the difference between these two types of manifolds is still crucial, both in general and for Riemann, especially for his analysis of physical space and geometry. Riemann stressed the significance of the relationships between continuity and discontinuity for mathematics, physics, and philosophy (e.g. [32, pp. 515–524]; [9, pp. 77–80]).

posed. A continuous (differentiable) manifold is understood by Riemann on the model of two-dimensional surfaces, which, as explained earlier, were defined by Gauss in terms of their intrinsic geometry. Riemann defines first the concept of "n-dimensional magnitude," which allows one to determine a position in a manifold by n numerical determinations, in the same way a position is determined by coordinates in the Euclidean space of n dimensions. Riemann is rigorous to extend (scale down) the concept of manifold to one-dimensional manifolds, curves, which, however, also helps him to built up the concept n-dimensional manifold by analogy. He starts with the concept of "a simply extended [one-dimensional] manifold, whose essential characteristic is that from any point in it a continuous movement is possible in only two directions, forwards and backwards." Then, he defines a two-dimensional or "a doubly-extended manifold" by saying that "if one now imagines that this [one-dimensional manifold] passes to another, completely different one, and once again in a well-determined way, that is, so that every point passes to a well-determined point of the other, then the instances for, similarly, a double extended manifold" [31, p. 25]. In other words, one continuously "fills" a surface with curves. Then, one similarly defines a triply extended manifold by imagining a similar continuous passing of a doubly extended manifold to another, thus continuously filling a three-dimensional object with two dimensional-ones, and so forth. "This construction," Riemann says, "can be characterized as a synthesis of a variability of $n + 1$ dimensions from a variability of n dimensions and a variability of one dimension" [31, p. 25]. This construction may have been one of the reasons for his use of the term: a manifold is literally a continuous fold(er) of manifolds of lower dimensions. Conversely, one can unfold a variability of n dimensions, which allows one to determine a position in a manifold by n numerical determinations, generalizing the way a position is determined by coordinates in the Euclidean space of n dimensions.

The most defining feature of the concept of manifold (under the assumption than one can measure the length of line-segments, straight or curved) is that it is conceived as infinitesimally Euclidean. This makes a continuous manifold into a conglomerate of local, continuously connected, small open neighborhoods around each point. The concept of neighborhood, again, assumed to be infinitesimally flat and Euclidean, is a component-concept of the concept of manifoldness. This concept of manifold as composed out of local neighborhoods is extendable to a still more general concept of topological space, in which case local neighborhood need no longer be Euclidean and can be defined with a great degree of generality. It is true that Riemann did not have a concept of topological space, in contrast to his concept of a Riemann surface, which had a more direct and immediate impact on the development of topology as an independent mathematical discipline. I would argue, however, that the conceptual architecture defining topological spaces is a generalization of that of Riemann's conceptual architecture of manifoldness as a "space" composed of neighborhoods, or generalizing it even further to other "spaces," a conception that, as will be seen presently, extends to Grothendieck's topos theory.[16] Admittedly, this architecture is presented here in the spirit of modern axiomatic thinking rather than in the spirit of Riemann's conceptual-problematic thinking, but it does, I think, inherit Riemann's conceptual architecture defining his concept of manifold. The theory of Riemann surfaces, too, came to be recast in terms of manifolds. Weyl, who, as his title *The Concept of a Riemann Surface* stated, considered a Riemann surface to be a concept, was the first to expressly define Riemann surfaces as manifolds [20, 35]. Riemann, however, undoubtedly realized that they were manifolds, and they were part of the genealogy of the concept of manifold.

In the case of Riemannian manifolds, while each neighborhood is infinitesimally flat, Euclidean, the manifold as a whole is, in general, not, except in the special case of flat, Euclidean manifolds. A manifold may be negatively or positively curved, and, which is another major innovation of Riemann, this curvature can also be variable. Riemann defined the metric form as a quadratic differential form, by the only formula in his lecture (discounting the coordinate expression for the line element), and assumed that the transition from one local coordinate system to another was differentiable. Thus, he, again, de facto, considered differentiable manifolds with positive definite metrics, Riemannian manifolds. In modern terms, such a manifold is defined by using a differentiable section of positive-definite quadratic forms on the tangent-bundle. While, however, modern technical language can bring out deeper mathematical aspects of Riemann's concepts, it can also displace how Riemann thought, mathematically, physically, and, especially, philosophically, a displacement sometimes found in twentieth-century English translations of Riemann's works, including his Habilitation lecture. One is, accordingly, always in complex negotiations between Riemann's and contemporary technical language, even though and because Riemann is so often ahead of his time, so much our contemporary.

Another important and equally future-oriented conceptual aspect of Riemann's approach is that it allows one to define a geometrical or, more generally, topological space (in modern terminology) not as a multiple, say, a set of points, but as a space that could be covered by maps (Euclidean in the case of manifolds) and in its relation to other spaces. As just explained, in part following Riemann's way of thinking, topology describes a given space not only in terms of its points, continuously connected to each other, but also and most essentially in terms of its open neighborhoods around each point. These neighborhoods are subspaces of this space, the idea that, again, underlies Riemann's concept of manifold, in this case, however, giving each neighborhood a Euclidean geometry. The approach, again, enabled Riemann to define manifolds of any dimension, even infinite-dimensional ones, in terms of its inner properties rather than in relation to the ambient Euclidean space, where a manifold could be placed, against the flat Euclidean background. It is true that, if one appeals, as is usual even in considering Riemann, to open sets, this concept of

[16]It is also worth recalling in this connection that Grothendieck's initial primary areas of mathematical research concerned topological vector spaces, which suggests yet another genealogical line in the history of the (broadly) Riemannian problematics in question here.

space retains the concept of set (of points) as a primitive concept.[17] Riemann's way of thinking concerning manifolds, however, also suggests a possibility of thinking of and even defining a space in terms of its relations to other spaces, which allows one to use this structure as more primordial by replacing covering a space by *open sets (of points)* with covering it by *open spaces*. A topological space, defined above in set-theoretical terms, becomes a collection of open spaces as sub-spaces with certain (algebraic) rules for the relationships between them.

This way of defining space by its relation to other spaces (as opposed to their constitution as sets of points) leads all the way to Grothendieck's topos theory, inspired by Riemann's ideas of manifolds and of the so-called covering spaces, originating in Riemann's theory of Riemann surfaces. Although it extends far beyond the question of spatiality, including mathematical logic, topos theory is arguably the farthest and most abstract extension of the concept of spatiality available, if one can rigorously speak of spatiality in this case, given an essentially algebraic nature of the concept. It does, however, give important new dimensions to our understanding of spatiality, when we deal in mathematics and elsewhere with objects or concepts that are considered in spatial terms.

It would not be possible here to present topos theory in its proper abstractness and rigor, sometimes prohibitive even for those not trained in the field of algebraic geometry or mathematical logic, where the concept is used as well (e.g. [18]). The essential philosophical ideas involved may, however, be sketched, as an example of both a rich mathematical concept in its own terms and of Riemann's influence on modern mathematics.[18] First, very informally, consider the following way of endowing a space with a structure, generalizing the definition of topological space. One begins with an arbitrarily chosen space, X, potentially any given space, which may initially be left unspecified in terms of its properties and structure. What would be specified are the relationship between spaces applicable to X, such as mapping or covering one or a portion of one, by another. One calls this structure the arrow structure $Y \to X$ (X is the space under consideration), where the arrow designates the relationship(s) in question. One can also generalize the notion of neighborhood or of an open subspace of (the topology of) a topological space in this way, by defining it as a relation between a given point and space (a generalized neighborhood or open subspace) associated with it. This procedure enables one to specify a given space not

[17]Thus, Ferreirós's discussion of Riemann in [11, pp. 39–80] appears to me to displace Riemann's thinking into the axiomatic and set-theoretical register, dominant in the wake of Cantor, a displacement arguably due to Ferreirós's insufficient attention to the nature of Riemann's mathematical concepts, to Riemann's concept of mathematical concept. In fairness, Ferreirós does relate Riemanns view of axioms to his concept of "hypothesis" and distinguishes it from the understanding of axioms developed in the twentieth-century philosophy of mathematics and mathematical logic. It does not appear to me, however, that Riemann thinks either in terms of axioms or, especially, in terms of sets (of points), as Ferreirós contends, although it could be and subsequently has been translated into these terms (e.g. [22]). See also Note 12 above.

[18]It would be instructive on both counts, to consider, as part of this genealogy, Dedekind's and Noether's work in algebra, reflecting the impact of Riemann's work on modern algebra, and even apart from his work on the distribution of primes and his famous hypothesis concerning the ζ-function. See [19] on Noether's work in this connection.

in terms of its intrinsic structure (e.g., a set of points with relations among them) but sociological[ly], throughout its relationships with other spaces of the same category, say, that of Riemannian spaces as manifolds [17], p. 7]. Some among such spaces may play a special role in defining the initial space, *X*, and algebraic structures (such as homotopy and cohomology, as Riemann realized in the case of covering spaces over Riemann surfaces. Indeed, the concept of covering space was one of the main inspirations for Grothendieck's concept of topos. The so-called *étale* topos (of a scheme), one of the main motivations for the concept of topos, is directly linked to the concept of covering space, as the term *étale* suggests.

To make this scheme more rigorous and to explain (albeit still quite informally) the concept of topos, I need to explain in my own words category theory. It was introduced in as part of the cohomology theory in algebraic topology in 1940 and later extensively used by Grothendieck in his approach to algebraic geometry, leading to the concept of topos. Category theory considers multiplicities (which need not be sets) of mathematical objects conforming to a given concept, such as the category of Riemannian manifolds, and the arrows or morphisms, the mappings between these objects that preserve this structure. Studying morphisms allows one to learn about the individual objects involved, often to learn more than we would by considering them only or primarily individually. In a certain sense, by appealing to the conceptual determination of each manifold, Riemann already thinks categorically. Thus, one does not have to, and Riemann does not, start with a Euclidean space, whether seen in terms of sets of points or otherwise. Instead the latter is just one specifiable object of a large categorical multiplicity, here that of the category of Riemannian manifolds, an object marked by a particularly simple way we can measure the distance between any two points. Categories themselves may be viewed as such objects, and in this case one speaks of "functors" rather than "morphisms." Topology relates topological or geometrical objects, such as manifolds, to algebraic ones, especially, as in the case of homotopy and cohomology theories, groups, a concept, it is true, not used by Riemann, as against Poincaré, who made it central to his geometrical and topological thinking, which established his uniquely significant role in the rise of algebraic topology. Thus, in contrast to geometry (which relates its spaces to algebraic aspects of measurement), topology, almost by its nature, deals with functors between categories of topological objects, such as manifolds, and categories of algebraic objects, such as groups.

Now, a topos in Grothendieck's sense is a category of spaces and arrows over a given space, used especially for the purpose of allowing one to define richer algebraic structures associated with this space, as explained above. There are certain additional conditions such categories must satisfy, but this is not essential at the moment. To give a simple example, for any topological space *S*, the category of sheaves on *S* is a topos. The concept of topos is, however, very general, and extends far beyond spatial or space-like mathematical objects (thus, the category of finite sets is a topos); indeed it replaces the latter with a more algebraic structure of categorical and topos-theoretical relationships between objects. On the other hand, it derives from the properties of and (arrow-like) categorical relationships between properly topological objects, such as Riemann surfaces or manifolds. The conditions, mentioned above, that categories that form topoi must satisfy have to do with these connections. The concept of topos is especially suited to deal in the way we do with standard manifolds with objects, such as certain (discrete) algebraic varieties, which are solutions of polynomial equations and are space-like, that cannot be meaningfully defined otherwise sufficiently analogously to continuous spaces, specifically in order to define nontrivial cohomology or homotopy groups for them. This had been an outstanding problem of algebraic geometry, arising from the so-called Weil conjectures for algebraic varieties over finite fields, which was solved with the help of topos theory, specifically the concept of *étale* topos, mentioned above (e.g. [1]). What both types of objects now share are analogously defined topoi associated with them and, as a result, analogously defined algebraic structures associated with them, equally enabling the necessary functoriality in both cases.

Topos theory allows for such esoteric constructions as non-trivial or non-punctual single-point "spaces" or, conversely, spaces (topoi) without points (first constructed by Pierre Deligne), sometimes slyly referred to by mathematicians as "pointless topology." Philosophically, this notion is far from pointless, especially if considered within the overall topos-theoretical framework. In particular, it amplifies a Riemannian idea that "space," especially is defined by its relation to other spaces, as a more primary object than a "point" or, again, a "set of points." Space becomes a Leibnizean, "monadological" concept, insofar as points in such a space (when it has points) may themselves be seen as a kind of monads, thus also giving a non-trivial structure to single-point spaces. These monads are certain elemental but structured entities, spaces, rather than structure-less entities (classical points), or at least as entities defined by (spatial) structures associated to and defining them [3]. Naturally, my appeal to monads here is qualified and metaphorical. Leibniz's monads are elemental souls, the atoms of soul-ness, as it were. But one might say that the space thus associated to a given point is the soul of this point, which defines its nature or structure, not unlike an infinite-dimensional Hilbert space associated with an elementary particle, such as an electron, in quantum mechanics and enabling us to predict its behavior. In other words, not all points are alike insofar as the mathematical (and possibly philosophical) nature of a given point may depend on the nature or structure of the space or topos to which it belongs or with which it is associated in the way just described. This approach also gives a much richer architecture to spaces with multiple points, such as Riemann's manifold (in which this architecture is inherent), and one might see (with caution) such spaces as analogous to Leibniz's universe composed by monads. It also allows for different (mathematical) universes associated with a given space, possibly a single-point one, in which case a monad and a universe would coincide. Grothendieck's topoi are such possible universes, possible worlds, or even com-possible worlds in Leibniz's sense, without assuming, like Leibniz (in dealing with the physical world), the existence of only one of them, the best possible one.

The outcome of Riemann's investigation into the foundations of geometry was, thus, a new mathematics of great generality, power, and potential, which involved not only new geometry, but also new topology and analysis (the tensor calculus on manifolds). Although it was Riemann's theory of continuous or, again, differentiable manifolds that had the greatest impact, the concept of discrete manifolds was important for Riemann's argument, and it is important for the modern understanding of both spatiality and geometry in mathematics, physics, and philosophy. While a discrete manifold has topological dimension zero, it may still be seen as multiply extended, if defined as forming a very fine lattice with very small intervals between points, which can be "filled," as it were, to form a continuous space of the corresponding topological dimensions. It is also possible to introduce metrical relations for discrete manifolds. This concept is important in the context of the relationships between physical, dynamical forces in nature and the nature of space or, to return to Riemann's terms, the physical "reality underlying space," although Riemann is cryptic on such metrical relations. Crucially, however, he does allow for the possibility that the physical "reality underlying space" might be "a discrete manifold" [31, p. 33]. This possibility has been entertained even before Riemann and has been even more often considered since, especially more recently. Also, mathematically, finite geometries were beginning to be developed, usually in more axiomatic ways, around Riemann's time as well, later on also under the impact of his geometrical thinking.[19] As indicated earlier, Riemann saw the relationships between continuity and discontinuity as foundationally central to mathematics, physics, and philosophy (e.g., [32, pp. 515–524]; [10, pp. 77–80]), a view confirmed by the subsequent developments in the foundations of mathematics, from Dedekind and Cantor on, and quantum physics. The latter uses continuous (technically, differential) mathematics to predict, in probabilistic terms, irreducibly discrete phenomena, that is, phenomena that are not, and that possibly cannot be, assumed to be connected to each other by a continuous physical process [27, pp. 232].

4 Physics: "The Reality Underlying Space"

As noted from the outset, Riemann's contribution to physics in his lecture was as important as his contribution to mathematics there. Riemann's terms space and geometry refer to the three-dimensional space and its geometry, in accordance with the use of these terms at the time, although this was soon to change. We now speak not only of manifolds of any dimensions, as Riemann does, but also of their geometry and refer to them as "spaces," or of discrete spaces and geometries, without necessarily assuming any connections between these objects and physical (or phenomenal) spatiality. While Riemann allows that "the reality underlying space" may prove to be discrete at a very small scale, this is not the same as extending, as was done subsequently, the terms "space" and "geometry" to finite entities defined by certain geometrical-like properties. On the other hand, closer to Riemann's view of what "the reality underlying space" could in principle be, some physical theories, dealing with such connections, suggest that the ultimate reality underlying space might be discrete or, as in superstring and brane theories, that physical space has a dimension

[19] For the discussion of some of these developments, see the chapter by V. Pambuccian, H. Struve, and R. Struve [21] and other chapters in the part of this volume that addresses later developments of Riemann's work.

higher than three (most commonly nine). While there is no experimental evidence thus far to support either claim, there are legitimate theoretical considerations in their favor.[20] Quantum theory also uses spaces of infinite dimensions, Hilbert spaces over complex numbers, although, as indicated above, for the purposes of predicting quantum events without representing physical space or physical processes in space and time. This is, admittedly, not the type of the relationships between mathematics and physics that Riemann appears to have entertained, but it may still be seen as in the spirit of Riemann's thinking concerning these relationships.[21] All this was to come later, however.

In Riemann's view, mathematically, one can define a general concept of manifold and the concept of metric relations in this manifold. These relations define a flat or curved nature of a given manifold, unless a manifold is discrete, in which case the metric relations, which could be defined for them, are no longer related to flatness or curvature of the manifold. This concept can then be used, suitably specified, to represent physical space and geometry there. Riemann considers this situation in more detail in the last chapter of the lecture, entitled "Applications to Space." As we have seen, however, he makes clear from the outset of the lecture that any such use can only be based on hypotheses that we form concerning space, and which we can then test. These hypotheses ground our thinking concerning space and geometry, although they may also reciprocally arise from this thinking from our previous hypotheses that we have tested or what we assumed, axiomatically, as self-evident. Hence, the fact "that a multiply extended magnitude is susceptible of various metric relations, so that space constitutes only a special case of a triply extended magnitude" implies that "those properties that distinguish space from other conceivable triply extended magnitudes [manifolds] can only be deduced from experience" [31, p. 23]. By "experience" Riemann means an experimental determination of the nature of physical space, rather than our phenomenal experience, although the latter may and even must play a role in this determination.

In other words, Riemann argues as follows, by both, in a very modern or even, *avant la lettre*, "modernist" way, separating mathematics from physics, making it independent, and then reconnecting them, an approach adopted by Einstein, expressly following Riemann, in creating general relativity [11, pp. 325–327]. There is math-

[20]The higher-dimensional spaces of superstring theory have been extensively discussed in literature and can be safely bypassed here, pertinent as their geometrical and topological features are. I would like, however, to mention a recent investigation, along quantum-informational lines, of the possibility that the reality underling space is discrete at the Planck scale, with a radical implication that the Lorentz invariance and hence special relativity is broken at the Planck scale as well [5] [27, pp. 259–262]. The article is also innovative mathematically in its use of geometric group theory, which emerged from Gromov's realization, Riemannian in spirit, that mathematical objects, such as groups, defined in algebraic terms, can be considered as geometric objects and studied with geometric techniques. This argument is still hypothetical, however, as, again, are all arguments thus far to the effect that the reality underlying space is discrete. If one accept what may be called the strong Copenhagen view, following Bohr, this "reality" may be beyond conception altogether and, hence, be neither continuous nor discontinuous [27, pp. 11–22]. I return to this possibility below.

[21]I have discussed the connections between Riemann's Habilitation lecture and quantum theory in [25].

ematics, which he introduced in the lecture, suitable for our description of physical space, via our phenomenal experience. This suitability allows one to have a geometry based on this mathematics. However, as grounded in the concept of manifold, this mathematics, the conceptual architecture of this mathematics, is sufficiently general both to be developed independently in mathematics itself quite apart from physics (it is has done subsequently as well) and to account for various possible forms of physical spatiality. This mathematics then needs to be adjusted in accordance with the hypotheses that we make concerning physical space, some of which may acquire the status of experimental evidence, possibly long-standing but not guaranteed to be permanent. Thus, such hypotheses may concern whether physical space is flat or curved, or whether this curvature is positive or negative, or (a question, again, never posed before Riemann) whether this curvature is constant or variable, or whether the ultimate (small-scale) "reality underlying space" is continuous or discrete, "beyond the bounds of observation." We can then "investigate the likelihood [of such hypotheses], which [in the case of Euclidean geometry] is certainly very great within the bounds of observation [in Riemann's time], and afterwards decide on the legitimacy of extending them beyond the bounds of observation, both in the direction of the immeasurably large and in the direction of the immeasurably small."

Riemann's line of reasoning here both follows and goes beyond Kant, in part by adopting Johann Herbart's argument, which questioned Kant's view (e.g. [11], pp. 77–99). Riemann follows Kant insofar as he sees our observations, always defined by our phenomenal experiences, as the basis of possible hypotheses concerning nature, the truth of which is possible but is not assured. He goes beyond Kant insofar as he views these hypotheses as only having an established validity within the bounds of observation with one or another degree of certainty. These hypotheses may not be applicable at all if we extend our investigation of the nature of space arbitrarily far in either direction, that of the infinitely large and that of the infinitely small. Kant, by contrast, believes in the absolute validity of (the hypothesis of) Euclidean geometry or Newton's physics. On the other hand, as already noted, unlike Riemann or Herbart, Kant does not believe that space, or time, is an empirical concept, whose validity, either as a general concept or in any of its instantiation, is established by experience.[22] Kant sees it as an a priori given concept that we use to frame our experience, a claim persistently challenged from the time it was made, including by Herbart and Riemann. Nature may have a different form of spatiality from what our phenomenal concept of space tells us, or nature may not have spatial aspects to it at all. By the end of the lecture, in considering the question of space in the infinitely small, Riemann comes closer to a more Kantian (although not quite Kant's own) view that "the reality underlying space" may be different from our phenomenal intuition of spatiality or may not be spatial in our phenomenal sense. For example, this reality may be discrete or be beyond the reach of any concept, discrete or continuous, available to us. In this latter view, this reality, while still real, would be beyond any representation and thus

[22]How our phenomenal experience of space emerges is separate question, psychological, physiological, or now neurological. Remarkably, Riemannian geometry is used in recent neurological research, as in the work of Jean Petitot (neurogeometry).

beyond realism [27, pp. 11–22]. It is, as indicated earlier, doubtful that Riemann entertained, anymore than did Kant [27, pp. 17–21], so radical a hypothesis, which emerged only in the wake of quantum mechanics. This hypothesis may, nevertheless, be seen as an implication of Riemann's closing reflections in the lecture and possibly Kant's epistemology [25].[23]

Thus far, Riemann only spoke of *physical space* rather than of *physics* in the sense of material forces, bodies, and motion. In closing, however, he brings physics into consideration. He argues that it is physics that defines the nature of space in the immeasurably small. Thus, while space may be assumed—this was a plausible hypothesis at the time and still is now–to be a three-dimensional manifold, what kind of manifold it is will be defined by physics. According to Weyl: "Riemann rejects the opinion that had prevailed up to his own time, namely, that the metrical structure of space is fixed and is inherently independent of the physical phenomena for which it serves as a background, and that the real [physical] content takes possession of it as a residential flat" [36, p. 98]. This was a revolutionary move on Riemann's part, later furthered by Einstein, who rigorously connected Riemannian geometry to the physics of gravity. For Riemann and Einstein, on this point following Leibniz (who, it is true, did not appear to have contemplated non-Euclidean geometry), matter defines the character of space, say, as flat or curved, while for Newton, space preexists matter, as an absolute space, a *flat* residential flat. Earlier arguments for non-Euclidean geometry had only changed the Newtonian view of space insofar as they imply that space might not be flat, which, however, still leaves open whether or not the Euclidean or non-Euclidean nature of space is defined by matter. Weyl adds: "*[Riemann] asserts, on the contrary, that space is itself nothing more than a three-dimensional manifold devoid of all form; it acquires a definite form only through the advent of the material content filling it and determining its metric relations*" [36, p. 98; Weyl's italics]. This is not quite what Riemann says. Weyl's statement may suggest (although Weyl does not appear to intend this) that space, as "a manifold, devoid of all form," preexists a given form, which form is then determined by matter, "*through the advent of the material content filling it and determining its metric relations*." For Riemann, as for Leibniz and Einstein, and ultimately for Weyl, matter preexists space, or, more accurately, it reciprocally co-exists with space and defines its character as a manifold. In addition, to return to Riemann's more precise language, "the reality underlying space" may be different from our phenomenal or, depending on scale, even physical experience of space, in particular, as flat, Euclidean space [31, p. 33]. It may reveal itself to be curved and have a varying curvature (which is to say, to be assumed to conform to the corresponding hypothesis), as in general relativity, which proved Riemann's insights especially prescient, and his concept of manifold especially capacious. Physics may also find that this reality requires manifolds of different types (including possibly, discrete) on different scales. One may, accordingly, modify Weyl's statement by saying that a general form of space may be assumed, hypothesized, to be, say, a three-dimensional continuous manifold

[23]Cf. [20], on the epistemological differences between Kant and Riemann.

or a three-dimensional discrete lattice, while its specific form (local or global) is determined by matter and forces acting upon it.

In approaching the subject, Riemann first states that "the questions about the immeasurably large [*Unmessbargrosse*] are idle questions for the explanation of nature [*die Naturerklärung*]," an assessment, for which Riemann offers no further justification and which one might question now, at least insofar as very large scales as concerned [31, p. 32]. From the present-day perspective, the question of the character of space on a very large cosmic scale is far from idle, although the idea, the hypothesis, of the infinite cosmic space poses conceptual difficulties, and it is possible that Riemann sensed some of them in making his assessment.[24] Be it as it may, the subject could be put aside, given that these are Riemann's reflections concerning "the questions about the immeasurably small [*Unmessbarkleine*]" that are most important for the present argument. These questions, Riemann argues, are "not idle ones:"

> Upon the exactness with which we pursue phenomena into the infinitely small [*Unendlichkleine*] does our knowledge of their causal connections essentially depends. The progress of recent centuries in understanding the mechanisms of Nature depends almost entirely on the exactness of construction which has become possible through the invention of the analysis of the infinite and through the simple principles discovered by Archimedes, Galileo, and Newton, which modern physics makes the use of. By contrast, in the natural sciences where the simple principles for such constructions are still lacking, to discover causal connections one follows phenomenon into the spatially small, just so far as the microscope permits. Questions about the metric relations of space in the immeasurably small are thus not idle ones [31, p. 32].

Riemann, thus, sees the mathematical representation of space, or time, or physical processes in space and time, offered by classical physics, as defined by the kinematical and dynamical principles established by the figures he mentions here. Riemann also sees physics as based, mathematically, on the principles of differential calculus, which is an analysis of the infinitely small. This is not the same as the immeasurably small [*Unmessbarkleine*], but it provides the proper mathematical representation of the physical concepts just mentioned, which explains Riemann's shift in this paragraph from "the immeasurably small" [*Unmessbarkleine*] to "the infinitely small" [*Unendlichkleine*]. More generally, as Weyl noted, "The principle of gaining knowledge of the external world from the behavior of its infinitesimal parts is the mainspring of the theory of knowledge in infinitesimal physics as in Riemann's geometry, and, indeed, the mainspring of all the eminent work of Riemann" [36, p. 92]. As the mathematics of the infinitely small, differential calculus also allows one to relate classical physics to causality, indeed is correlative to causality (which is one of the physical principles in question, defined by the fact that the state of a given system at a given moment of time determines its state at any other moment of time).[25] Riemann was, again, aware, as was Weyl, that the reality underlying space, in the immeasurable small, may be discrete and hence, at that scale, no longer subject to a continuous analysis. Hence, again, there is the difference between the immeasurably

[24]On some of these difficulties, see [12, pp. 31–42].

[25]Riemann offered important reflections on causality, which he linked to continuity [32, p. 522].

small [*Unmessbarkleine*], also in its direct sense of that which cannot be measured, and the infinitely small [*Unendlichkleine*], which is an important point, especially, as became apparent later, in the context of quantum theory. As noted above, however, quantum theory, while dealing with discrete phenomena, does not generally assume or imply that "the reality underlying space" is discrete, and if anything, suggests that the ultimate reality of nature may be beyond any possible representation of even conception (discrete or continuous) [27, pp. 11–22].[26] By the same token, the theory is no longer causal, but is irreducibly probabilistic even in dealing with elemental individual quantum processes (always assumed to be causal in classical physics or relativity). This fact is reflected in Heisenberg's uncertainty relations, which prevent us from ever simultaneously determining both the position and the momentum of a quantum object, which is necessary in order to maintain causality. In any event, it is clear that "questions about the metric relations of space in the immeasurably small are not idle ones." They connect the hypotheses that lie at the foundations of geometry to those that lie at the foundations of physics. Riemann says next:

> If one assumes that bodies exist independently of position, then the curvature is everywhere constant, and it then follows from astronomical measurements that it cannot be different from zero; or at any rate its reciprocal must be an area in comparison with which the range of our telescopes can be neglected. But if such an independence of bodies from position does not exist, then one cannot draw conclusions from metric relations in the infinitely small from those in the large; at every point the curvature can have arbitrary values in three directions, provided only that the total curvature of every measurable portion of space is not perceptibly different from zero [31, p. 32].

It took Einstein's general relativity to give, for the first time, a rigorous physical content to these insights by bringing together the physics of gravitation and Riemannian geometry. The curvature of a manifold representing the physical reality underlying space not only may not be zero, but may also not be constant, which is, again, a powerful new mathematical concept and, as possible physics is concerned, a tremendous physical insight of Riemann. It is generally not constant in a gravitational field, and establishing this fact in rigorous terms is an equally tremendous contribution of Einstein. We do know now that the hypothesis of Euclidean geometry or even non-Euclidean geometry of constant curvature, do not apply to the ultimate nature of space, or again, the physical reality underlying space, except perhaps on average on a very large scale, as current observations suggest. Nor, in part correlatively, do the hypotheses of classical physics, specifically those that ground Newton's law of gravity, apply at any scale, except as an approximation, workable within very large limits as this approximation is. Newton's law of gravity is incorrect even within its proper scope, as was first exemplified by the aberrant precession of the perihelion of

[26] We cannot conceive of entities that are simultaneously continuous and discontinuous, the difficulty handled in quantum mechanics by means of Bohr's concept of complementarity. Complementarity reflects the fact that continuous and discontinuous *quantum phenomena* (defined by what is observed in measuring instruments) are always mutually exclusive, while *quantum objects* themselves, responsible for these phenomena through their impacts on measuring instruments, are, again, assumed to be beyond any representation or even conception, continuous or discontinuous. For a full treatment, see [27, pp. 107–172].

Mercury. The principles of calculus, used in tensor calculus, still apply in the infinitely small in general relativity as a way of providing the mathematics, the mathematical model, of space as defined by gravity. Riemann then adds:

> Still more complicated relations can occur if the line element cannot be represented, as was presupposed, as the square root of a differential expression of the second degree. Now it seems that the empirical notions on which the metrical determinations of space are based, the concept of a solid body and that of a ray of light, lose their validity in the infinitely small; it is therefore quite definitely conceivable that the metric relations of space in the infinitely small do not conform to the hypotheses of geometry; and in fact one ought to assume this as soon as it permits a simpler way of explaining phenomena [31, p. 32].

Riemann, thus, envisions not only that space in the infinitely small, or, as it would, again, be more accurate to say, the immeasurably small may not conform to the hypothesis of Euclidean geometry, but also that it may not conform even to Riemannian geometry, used by Einstein in general relativity. In the latter case, the concept of metric relations still applies, although they are non-Euclidean and allow for a variable curvature. Relativity merely modifies, albeit radically, Euclidean concepts in view of the relativistic contraction of bodies and of the curving of light in the vicinity of a heavy body, such as the Sun. Bringing together gravity and quantum theory (a still outstanding problem) may change this. Riemann is about to suggest that "the reality underlying space" may be "a discrete manifold:"

> The question of the validity of the hypotheses of geometry in the infinitely [the immeasurably?] small is bound up with the question of the basis for the metric relations of space. In connection with this question, which may indeed still be ranked as part of the study of space, the above remark is applicable, that in a discrete manifold the principle of metric relations is already contained in the concept of the manifold, but in a continuous one it must come from something else. Therefore, either the reality underlying space must form a discrete manifold, or the basis for the metric relations must be sought outside it, in binding forces that act upon it.
>
> An answer to these questions can be found only by starting from the conception of phenomena which has hitherto been approved by experience, and for which Newton laid the foundation, and gradually modifying it under the compulsion of facts that cannot be explained by it. Investigations like the one just made here, which begin from general concepts, can *only* serve to insure that this work is not hindered by unduly restricted concepts and that progress in comprehending the connection of things is not obstructed by traditional prejudices. This leads us away into the domain of another science, the realm of physics, into which the nature of the present occasion does not allow us to enter. [31, p. 33]

Riemann, again, differs from Kant, insofar as Riemann assumes that this reality must at least be established by physical experiments, even if not perceived phenomenally, rather than is given a priori, indeed by definition because we do not phenomenally perceive space as discrete. On the other hand, as explained earlier, Kant, who is often misunderstood on this point, does not assume that the physical reality underlying space is given a priori or is phenomenal otherwise. With this qualification in mind, Riemann's formulation becomes close to Kant, except perhaps that Riemann believes that our hypotheses concerning the character of "the reality underlying space" could be tested so as to bring us closer to knowing this reality. I qualify by "perhaps," because Kant might have even agreed on this point as well. In addition, as also explained earlier, just as Riemann, Kant would likely have been hesitant to call this reality space, if it is different from the continuous three-dimensional space of our phenomenal experiences. The main point here is that one needs physics, as an experimental-mathematical science of nature, to establish the facts that would enable us to test and, to begin with, to form hypotheses concerning the reality underlying space, and have geometry of this space, possibly a higher-dimensional or discrete geometry.

Einstein's relativity justified Riemann's view that we must proceed "by starting from the conception of phenomena which has hitherto been approved by experience, and for which Newton laid the foundation, and gradually modifying this conception under the compulsion of facts that cannot be explained by it." The facts at stake in relativity can no longer be explained by the conception of physical phenomena provided by Newton's physics; and, as is clear from this elaboration, Riemann saw this conception as likely to be insufficient. This is not surprising given his investigations into electromagnetism and the contemporary development of this field and of physics in general, even though Maxwell's electromagnetic theory was not yet in place at the time of the lecture [22]. Riemann's subsequent work on electromagnetism suggests intriguing affinities with that of Maxwell and then that of Einstein (e.g. [16, pp. 257–271]). The subject would require a separate treatment. It may, however, be fitting to note that, extending its role in general relativity, Riemannian geometry also served as the basis for several early projects of establishing a unification of gravity and electromagnetism, the first form of the unified field theory, pursued, in particular, by Einstein, Hilbert, and Weyl. While they set into motion the program that still dominates fundamental physics, these attempts, all essentially along the lines of classical-like field theory (on the model of Maxwell's electromagnetic theory), were unsuccessful. This was in part because such a theory appears unlikely to be developed without taking into account quantum aspects of electromagnetism or, by now, of other strong and weak forces, covered by quantum field theory, within the so-called standard model of all known forces of nature, except for gravity, with which the standard model is incompatible. Both Einstein and Weyl made attempts, again, unsuccessful, to incorporate Dirac's 1928 relativistic theory of the electron into their unified-field-theoretical schemes, still governed, however, by a classical-like field theoretical thinking, which was in a manifested conflict with the principles behind Dirac's theory [27, pp. 207-226].[27] It is, accordingly, not surprising, at least in retrospect, that these attempts did not succeed. Unsuccessful as they have been, they, nevertheless, showed the fruitfulness of Riemann's thinking in geometry for foundational thinking in physics, and Riemann's foundational thinking in geometry was, again, also a foundational thinking in physics.

[27] Dirac's famous equation also introduced spinors into physics. Although the name itself was coined (by Paul Ehrenfest) in 1929, following Dirac's theory, the concept, still enigmatic and uneasily suspended between geometry and algebra, existed in mathematics previously and was extensively studied by Cartan, for example. It belongs to the post-Riemannian evolution of geometrical thinking, also as extending beyond geometry, for example, to algebra, although spinors are important for geometry as well.

The search for such a unified theory is still ongoing, now in attempting to unify all forces of nature, although, thus far, even within the standard model, we only have the electroweak unification, which is, besides, quite different in nature from the way such a unification was envisioned previously. As indicated above, the incompatibility between the standard model, as quantum theory, and general relativity is one of the great outstanding problems of the present-day fundamental physics, perhaps the greatest one. Superstring and brane theories, which have been around for quite a while now, are still generally seen as the best candidate, although the skepticism that has always shadowed them has become, for both mathematical and physical reasons, even more pronounced more recently. But then, that currently available alternatives, such as loop quantum gravity, will succeed appears no more likely. Both programs, that of superstring and brane and that of loop quantum gravity, have Riemannian genealogies, loop quantum gravity more immediately, via Einstein's general relativity, and superstring and brane theories, which originated in quantum field theory, via a more complex history of development, by virtue of using the so-called Calabi-Yau manifolds.[28] Which among these or other currently available programs, such as those along the lines of quantum information theory, are likely to succeed in approaching the reality underlying, to borrow Weyl's famous title, "space, time, and matter" is impossible to predict, perhaps none of them, given, thus far, the immense physical and mathematical difficulties they pose. These difficulties, however, also open new possibilities, both inside these programs, which might succeed after all, and for new, possibly as yet unimaginable, alternatives. Whatever the future holds, just as does mathematics, fundamental physics continues to return us to Riemann and to show that the manifold of connection[s] of things, mathematical, physical, and philosophical, that Riemann's thought brought into existence is inexhaustible.

5 Conclusion

I return, in closing, to Riemann's assessment of his own project in his lecture: "Investigations like the one just made here, which begin with general concepts, can *only* serve to insure that this work [of developing new physics] is not hindered by unduly restricted concepts and that progress in comprehending the connection of things is not obstructed by traditional prejudices" (emphasis added). "Only" is hardly necessary here. We do need physics to test and indeed to form our hypotheses concerning space. But we also need—physics proves that we do!—mathematical and philosophical plane of thought and new, richer concepts to counteract "unduly restricted concepts" and to be able "in comprehending the connection of things," which is the aim of thought in its cooperative confrontation with chaos, and not to be obstructed by traditional prejudices or rigid, dogmatic opinions, the danger of which is, Deleuze and Guattari warn us, as Riemann does here, a constant threat to thought. Riemann created such planes and such concepts. This does not mean that we need to stop with Riemann, who never stopped. The subsequent history of mathematics and physics

[28]On these connections, see [22].

has proven that we need to go beyond Riemann. Otherwise, his concepts cannot continue to live on, to remain "always new," to be concepts of the future.

Acknowledgements I am grateful to Franck Jedrzejewski, Ken'ichi Ohshika, and Athanase Papadopoulos for reading the original draft of the article and helpful suggestions for improving it.

References

1. M.A. Artin, A. Grothendieck, J.-L. Verdier, *Séminaire de Géométrie Algébrique du Bois Marie - 1963–64 - Théorie des topos et cohomologie étale des schémas - (SGA 4)*, vol. 1 (Springer, Berlin, New York, 1972)
2. E. Borel, La logique et lintuition en mathématique. Revue de Métaphysique et de Morale **15**(3), 273–283 (1907)
3. P. Cartier, A Mad Days Work: From Grothendieck to Connes and Kontsevitch. The Evolution of Concepts of Space and Symmetry. B. Am. Math. Soc. **38**(4), 389–408 (2001)
4. L. Corry, Axiomatics, empiricism, and anschauung in hilberts conception of geometry: between arithmetic and general relativity, in *The Architecture of Modern Mathematics: Essays in History and Philosophy*, ed. by J. Ferreirós, J.J. Gray (Oxford University Press, Oxford, 2006), pp. 133–156
5. G.M. D'Ariano, P. Perinotti, Derivation of the Dirac equation from principles of information processing. Phys. Rev. A **90**, 062106 (2014)
6. G. Deleuze, *Difference and Repetition* (Columbia University Press, New York, 1994). Translated by P. Patton
7. G. Deleuze, F. Guattari, *A Thousand Plateaus* (University of Minnesota Press, Minneapolis, MN, 1987). Translated by B. Massumi
8. G. Deleuze, F. Guattari, *What Is Philosophy?* (Columbia University Press, New York, 1994). Translated by H. Tomlinson and G. Burchell
9. J. Ferreirós, Riemann's habilitationsvortrag at the crossroads of mathematics, physics, and philosophy, in *The Architecture of Modern Mathematics: Essays in History and Philosophy*, ed. by J. Ferreirós, J.J. Gray (Oxford, Oxford University Press, 2006), pp. 67–96
10. J. Ferreirós, *Labyrinth of Thought: A History of Set Theory and its Role in Modern Mathematics* (Birkhäuser, Boston, MA, 2010)
11. J. Gray, *Plato's Ghost: The Modernist Transformation of Mathematics* (Princeton University Press, Princeton, 2008)
12. B. Greene, *The Hidden Reality: Parallel Universes and the Deep Laws of the Cosmos* (Vintage, New York, 2011)
13. D. Hilbert, *Foundations of Geometry*, (10th ed.) (Translated by L. Unger. Open Court, La Salle, IL, 1979)
14. F. Jedrzejewski, Deleuze et la géométrie Riemannienne: Une topologie des multiplicitités, this volume, pp. 311–327
15. I. Kant, *Critique of Pure Reason* (Cambridge University Press, Cambridge, 1997). Translated by P. Guyer and A. W. Wood
16. D. Laugwitz, *Bernhard Riemann: Turnings Points in the Conception of Mathematics* (Birkhäuser, Boston, MA, 1999). Translated by A. Shenitzer
17. Y. I. Manin, Georg Cantor and His Heritage, arXiv.math.AG/0209244 v1. Accessed 19 Sept 2002
18. C. McLarty, *Elementary Categories* (Elementary Toposes, Clarendon, Oxford, 1992)
19. C. McLarty, Emmy noether's set-theoretical topology: From Dedekind to the rise of functors, in *The Architecture of Modern Mathematics: Essays in History and Philosophy*, ed. by J. Ferreirós, J.J. Gray (Oxford University Press, Oxford, 2006), pp. 211–236

20. K. Ohshika, The origin of the notion of manifold: Riemann's Habilitations vortrag onward, this volume, pp. 295–308
21. V. Pambuccian, H. Struve, R. Struve, Metric Geometries in Axiomatic Perspective, this volume, pp. 413–451
22. A. Papadopoulos, Physics in Riemann's Mathematical Papers, this volume, pp. 151–199
23. P. Pesic, Introduction, in *Beyond Geometry: Classic Papers from Riemann to Einstein*, ed. by P. Pesic (Dover, Mineola, NY, 2007), pp. 1–22
24. A. Plotnitsky, Chaosmologies: Quantum field theory, chaos, and thought in deleuze and guattaris what is philosophy? Paragraph **29**(2), 40–56 (2006)
25. A. Plotnitsky, On foundational thinking in fundamental physics: From Riemann to Heisenberg. in G. M. D'Ariano et al. (ed). Foundations of Probability in Physics 6. Melville, NY: American Institute of Physics, Melville, NY, 2012, 282–304
26. A. Plotnitsky, The spacetimes of nympheas: Matter and multiplicity in Einstein, Monet, and Deleuze and Guattari, in *Revisiting Normativity with Deleuze*, ed. by R. Braidotti, P. Pisters (Bloomsbury, London, 2012), pp. 37–50
27. A. Plotnitsky, *The Principles of Quantum Theory, From Planck's Quantum to the Higgs Boson: The Nature of Quantum Theory and the Spirit of Copenhagen* (Springer, New York, 2016)
28. H. Poincaré, *La Science et l'hypothèse* (Flammarion, Paris, 1902)
29. H. Poincaré, *Science et methodé* (Flammarion, Paris, 2008)
30. G. Polya, *Mathematics and Plausible Reasoning, Volume 1: Induction and Analogy in Mathematics* (Princeton, Princeton University Press, 1958, rpt. 1990)
31. B. Riemann, B. On the Hypotheses That Lie at the Foundations of Geometry (1854). In P. Pesic (ed.) *Beyond Geometry: Classic Papers from Riemann to Einstein* (Dover, Mineola, NY, Dover, 2007), pp. 23–40
32. B. Riemann, *Gesammelte mathematische Werke und wissenschaftlicher* Nachlass. ed. by R. Narasimhan (Springer, New York, 1991)
33. B. Russell, *An Essay on the Foundations of Geometry*. (Cambridge University Press, 1997). (rpt. 2012)
34. A. Weir, Formalism in the Philosophy of Mathematics. The Stanford Encyclopedia of Philosophy (Spring 2015 Edition), E. N. Zalta (ed.), http://plato.stanford.edu/archives/spr2015/entries/formalism-mathematics/
35. H. Weyl, *The Concept of a Riemann Surface*. Translated G. L. MacLane. Dover, Mineola, NY, 1913. (rpt. 2013)
36. H. Weyl, *Space-Time-Matter* (Dover, Mineola, NY, 1952). Translated H. L. Brose

Part III
Some Recent Developments

The Riemann Mapping Theorem and Its Discrete Counterparts

Feng Luo

Abstract We introduce some of the recent work on discrete versions of the Riemann mapping theorem and the uniformization theorem.

Keywords Conformal maps · Riemann mapping · Uniformization theorem · Circle packing · Discrete conformality · Polyhedral surfaces

2000 Mathematics Subject Classification: 52C26 · 58E30 · 53C44

1 Introduction

The Riemann mapping theorem was formulated by B. Riemann in 1851. It states that given any two simply connected open sets U_1, U_2 in the complex plane $\mathbb{C}$ with $U_i \neq \mathbb{C}$, there exists an analytic bijection (i.e., *conformal*) map $f : U_1 \to U_2$. In particular, if one takes U_2 or U_1 to be the open unit disk, then the map f is called a *Riemann mapping*. The Riemann mapping theorem is one of the most important results in complex analysis. It relates geometry (e.g. open sets) to analysis (e.g. complex analytic functions).

The uniformization theorem of Poincaré and Koebe generalizes the Riemann mapping theorem to Riemann surfaces. By definition, a Riemann surface is a connected orientable surface Σ with a special collection of charts (analytic charts) covering Σ so that the transitions functions are complex analytic maps. The essential feature of Riemann surfaces is that one can measure angles between curves on them. Riemann surfaces are ubiquitous in mathematics. For instance connected open sets in $\mathbb{C}$, smooth orientable surfaces with Riemannian metrics, smooth algebraic curves and polyhedral surfaces are naturally Riemann surfaces. In 1907, Poincaré and Koebe independently proved the uniformization theorem which states that any simply con-

Work partially supported by the NSF DMS 1222663, DMS 1405106 and DMS 1207832.

F. Luo (✉)
Department of Mathematics, Rutgers University, Piscataway, NJ 08854, USA
e-mail: fluo@math.rutgers.edu

L. Ji et al. (eds.), *From Riemann to Differential Geometry and Relativity*,
DOI 10.1007/978-3-319-60039-0_12

nected Riemann surface is conformally diffeomorphic to the complex plane $\mathbb{C}$, the open unit disk $\mathbb{D}$, or the Riemann sphere $\mathbb{S}^2$. The conformal diffeomorphism is called a *uniformization map*. This result is a pillar in mathematics and has a wide range of applications within and outside mathematics.

Computing the Riemann mapping or the uniformization mapping is not easy. For instance, the boundary of a tetrahedron is naturally a Riemann surface. Here the analytic charts consist of unions of two open triangle faces together with their common open edges and the orientation preserving isometric embedding, and charts at vertices are of the form $(U, z^{2\pi/\alpha})$ where U is a small neighborhood of a vertex of cone angle α. Using the uniformization theorem, one concludes that it is conformal to the Riemann sphere $\mathbb{S}^2$ with four marked points $\{0, 1, \infty, z\}$ corresponding to the four vertices. However, there is no algorithm to compute the conformal invariant z directly from the 6 edge lengths of the tetrahedron. There are powerful algorithms computing the Riemann mapping for simply connected domains. For instance the Schwarz–Christoffel algorithm developed by Trefethen and Driscoll [33] and the circle packing algorithm developed by Thurston and Stephenson [31] are powerful tools. However, computing the uniformization map for a simply connected surface with a non-flat Riemannian metric has been difficult. Our recent work [11, 18, 19] produces an algorithm to compute the uniformization maps, and shows that the uniformization maps are computable.

Over the years, there have been many research activities on establishing various discrete versions of the uniformization theorem and the Riemann mapping theorem. The purpose of this chapter is to introduce some of these works and their proofs. We will also discuss several open problems in the discrete setting.

The following two topics will be discussed in this chapter. These are: (1) the Koebe–Andreev–Thurston's circle packing version of the Riemann mapping theorem and (2) our recent work with Gu, Sun, Wu and Guo ([11, 12, 18, 19]) on a discrete uniformization theorem for polyhedral surfaces.

We remark that this is not a survey of works on discrete Riemann mapping theorems and we have left many important works untouched.

The chapter is organized as follows. Section 2 discusses circle packings and Sect. 3 covers a discrete uniformization theorem for polyhedral surfaces.

We thank A. Papadopoulos for comments and suggestions on improving the writing of the paper.

2 Koebe–Andreev–Thurston's Circle Packing Theorem

We will discuss a simple form of the circle packing theorem in this section. For more details on circle packing, one may consult the nice book by Stephenson [31].

A *circle packing* on the Riemann sphere or the plane is a collection of closed round disks $D_1, \ldots, D_k$ with disjoint interiors. Its *nerve* is a finite graph on the 2-sphere $\mathbb{S}^2 = \mathbb{C} \cup \{\infty\}$ or the plane $\mathbb{C}$ with one vertex for each disk D_i and an edge between two vertices if the corresponding disks are tangent (Fig. 1).

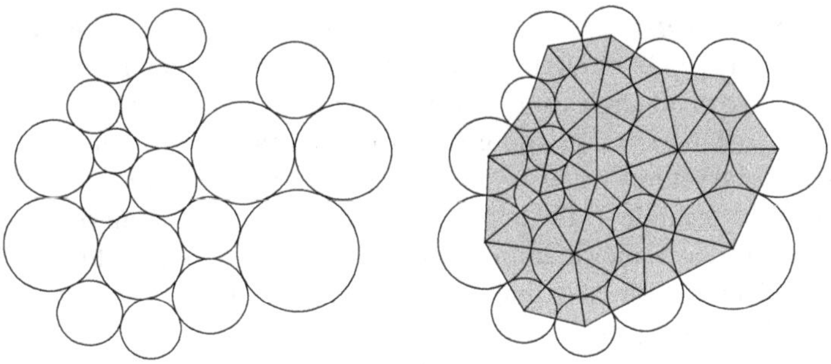

Fig. 1 A circle packing and its nerve. The picture is produced by K. Stephenson

Theorem 2.1 (Koebe–Andreev–Thurston) *Suppose $\mathcal{T}$ is a simplicial triangulation of the 2-sphere $\mathbb{S}^2$. There exists a circle packing $D_1, \ldots, D_n$, unique up to Möbius transformations, on the Riemann sphere $\mathbb{S}^2$ such that its nerve is isomorphic to the 1-skeleton $\mathcal{T}^{(1)}$ of $\mathcal{T}$.*

The theorems proved by Andreev and Thurston are more general allowing circles to intersect at angles at most $\pi/2$. For more details, see [3, 25, 35] or others.

There are many proofs of Theorem 2.1. See [3, 21, 31, 35] and others. Below we give a proof using ideas from [7, 21].

Following Marden-Rodin [21], we first reduce the circle packing on $\mathbb{S}^2$ to a circle packing on the plane $\mathbb{C}$. Removing a triangle face τ_0 from the triangulation $\mathcal{T}$, one produces a simplicial triangulation $\mathcal{T}_1$ of the (topological) triangle $T = \mathbb{S}^2 - int(\tau_0)$. To prove Theorem 2.1, it suffices to produce a circle packing on the plane whose nerve is the 1-skeleton $\mathcal{T}_1^{(1)}$. Indeed, if $D_1, \ldots, D_n$ is a circle packing on the plane whose nerve is the 1-skeleton $\mathcal{T}_1^{(1)}$, then $D_1, \ldots, D_n$ is a circle packing on $\mathbb{S}^2$ whose nerve is $\mathcal{T}^{(1)}$. Conversely suppose $D_1, \ldots, D_n$ is a circle packing on $\mathbb{S}^2$ whose nerve is $\mathcal{T}^{(1)}$ such that D_1, D_2, D_3 correspond to the three vertices of τ_0. Applying a Möbius transformation to $\{D_1, \ldots, D_n\}$ so that infinity is in the triangle region in $\mathbb{S}^2 - \cup_i D_i$ bounded by the circles $\partial D_1, \partial D_2, \partial D_3$, then the circle packing $\{D_1, \ldots, D_n\}$ on the plane $\mathbb{C}$ has nerve $\mathcal{T}_1^{(1)}$.

It is known that given three pairwise tangent closed disks D_1, D_2 and D_3 in the plane, there exists a Möbius transformation sending D_1, D_2 and D_3 to three disks of radii 1. Therefore, Theorem 2.1 is equivalent to producing a circle packing on $\mathbb{C}$ whose nerve is isomorphic to the 1-skeleton of a triangulation $\mathcal{T}_1$ of a triangle $T = \Delta v_1 v_2 v_3$ so that the three circles corresponding to three vertices v_i are of radii 1.

Thurston's approach to Theorem 2.1 uses polyhedral metrics on surfaces. Let V and E be the sets of all vertices and edges in $\mathcal{T}_1$ so that $v_1, v_2, v_3 \in V$ are the boundary vertices (i.e., vertices of τ_0). To produce a circle packing, Thurston assigns each vertex v a positive number $r(v)$, called the radius. The radius assignment is a function $r : V \to \mathbb{R}_{>0}$. For each radius assignment r, construct a polyhedral metric d on the triangulated triangle $(T, \mathcal{T}_1)$ by making each triangle in $\mathcal{T}_1$ a Euclidean triangle of edge lengths $l(vv') = r(v) + r(v')$ where $v, v' \in V$ and $vv' \in E$. The *discrete curvature* of d is the function $K_d : V \to (-\infty, 2\pi)$ sending each vertex $v \in V - \{v_1, v_2, v_3\}$ to 2π minus the sum of all angles at v and sending v_i $(i = 1, 2, 3)$ to π minus the sum of all angles at v_i. It is well known that the Gauss-Bonnet theorem holds, i.e., $\sum_{v\in V} K_d(v) = 2\pi$. The goal is to find a radius assignment $r \in \mathbb{R}^V_{>0}$ so that its discrete curvatures at all $v \in V - \{v_1, v_2, v_3\}$ are zero, i.e., (T, d) is a flat surface. Since the triangle T is simply connected, the developing map for the flat structure produces an isometric immersion $\Phi : (T, d) \to \mathbb{C}$ where the plane has the standard Euclidean metric. The map Φ sends the boundary ∂T to a triangle in $\mathbb{C}$. In particular, $\Phi|_{\partial T}$ is injective. This implies that $\Phi : (T, d) \to \mathbb{C}$ is an isometric embedding. Let the images of V under Φ be $\{v'_1, v'_2, ..., v'_m\}$ on the plane $\mathbb{C}$. Then by the construction, the circle packing $\{B(v'_1, r(v_1)), ..., B(v'_m, r(v_m))\}$ has nerve isomorphic to $\mathcal{T}_1^{(1)}$ where $B(c, r)$ is the ball of radius r centered at c.

The above discussion shows that Theorem 2.1 is a consequence of the following:

Proposition 2.2 *Suppose $\mathcal{T}_1$ is a triangulation of a triangle $T = \Delta v_1 v_2 v_3$ such that there are only three vertices v_1, v_2, v_3 of $\mathcal{T}_1$ in the boundary ∂T. Then there exists a unique radius assignment $r : V \to \mathbb{R}_{>0}$ with $r(v_i) = 1$ for $i = 1, 2, 3$ such that the associated circle packing metric on T has zero discrete curvatures at all $v \in V - \{v_1, v_2, v_3\}$.*

2.1 A Variational Principle Associated to Circle Packing

The following variational principle was first established by Colin de Verdière in [7].

Proposition 2.3 *(Colin de Verdière) Let $\Delta A_1 A_2 A_3$ be a Euclidean triangle such that the length of edge $A_i A_j$ is $e^{x_i} + e^{x_j}$ and the angle at A_i is $\theta_i = \theta_i(x_1, x_2, x_3)$. Let $x = (x_1, x_2, x_3)$. Then*

(a) $\sum_{i=1}^3 \theta_i(x)dx_i$ is a closed 1-form such that $\frac{\partial \theta_i}{\partial x_j} > 0$ for $i \neq j$;

(b) the function $f(x) = \int_0^x \sum_{i=1}^3 \theta_i(x)dx_i$ is a well defined concave function in $x \in \mathbb{R}^3$ such that $\frac{\partial f}{\partial x_i} = \theta_i$ and f is strictly concave when restricted to the plane $P_c = \{x \in \mathbb{R}^3 | x_1 + x_2 + x_3 = c\}$ for any $c \in \mathbb{R}$;

(c) if $a_1, a_2, a_3 > 0$ such that $a_1 + a_2 + a_3 = \pi$, then $g(x) = \int_0^x \sum_{i=1}^3 (\theta_i(x) - a_i)dx_i$ satisfies that $\lim_{\max_{i,j} |x_i - x_j| \to \infty} g(x) = -\infty$ and $g(x + (t, t, t)) = g(x)$ for all $t \in \mathbb{R}$.

In [6, 17], this variational principle is generalized to the case of three circles intersecting at angles and more general polyhedral surfaces.

Proof Recall that the cosine law for triangles states that $\cos(\theta_i) = \frac{y_j^2 + y_k^2 - y_i^2}{2 y_j y_k}$ where y_k is the length of $A_i A_j$ and $\{i, j, k, \} = \{1, 2, 3\}$. Let the area of the triangle $\Delta A_1 A_2 A_3$ be A. Taking derivatives of the cosine law, we obtain (see Lemma A-1 in the appendix of [6])

$$\frac{\partial \theta_i}{\partial y_i} = \frac{y_i}{2A} > 0, \tag{2.1}$$

$$\frac{\partial \theta_i}{\partial y_k} = -\frac{\partial \theta_i}{\partial y_i} \cos(\theta_j). \tag{2.2}$$

Now to see part (a), it suffices to show that $\frac{\partial \theta_i}{\partial x_j} = \frac{\partial \theta_j}{\partial x_i} > 0$. By definition $y_i = e^{x_i} + e^{x_k}$. Therefore,

$$\frac{\partial \theta_i}{\partial x_j} = \frac{\partial \theta_i}{\partial y_i}\frac{\partial y_i}{\partial x_j} + \frac{\partial \theta_i}{\partial y_k}\frac{\partial y_k}{\partial x_j} = \frac{\partial \theta_i}{\partial y_i} e^{x_j} - \frac{\partial \theta_i}{\partial y_i}\cos(\theta_j)e^{x_j} = \frac{(1-\cos(\theta_j))y_i e^{x_j}}{2A}.$$

Let R be the radius of the inscribed circle in the triangle. Then $e^{x_j} = R\cot(\theta_j/2)$. Using the relation $1 - \cos(\theta_j) = 2\sin^2(\theta_j/2)$, we see that

$$\frac{(1-\cos(\theta_j))y_i e^{x_j}}{2A} = \frac{R\sin(\theta_j)y_i}{2A} = \frac{R}{y_k} > 0$$

and that $\frac{\partial \theta_i}{\partial x_j}$ is symmetric in i, j.

To see part (b), since $\sum_{i=1}^3 \theta_i dx_i$ is closed in $\mathbb{R}^3$, the integral $\int_0^x \sum_{i=1}^3 \theta_i dx_i$ is independent of the choice of paths and therefore $f(x)$ is well defined. Furthermore, $\frac{\partial f}{\partial x_i} = \theta_i$ follows from the definition of f. The Hessian of f is the 3×3 matrix $[h_{rs}]$ $(h_{rs} = \partial\theta_r/\partial x_s)$ which satisfies the condition that $h_{ij} = h_{ji} > 0$ and $h_{1i} + h_{2i} + h_{3i} = \frac{\partial(\theta_1+\theta_2+\theta_3)}{\partial x_i} = \frac{\partial \pi}{\partial x_i} = 0$. It follows that the matrix $-[h_{rs}]$ is a diagonally dominated matrix whose kernel consists of vectors $\lambda[1, 1, 1]^t$. Hence $[h_{rs}]$ is negative semi-definite. This implies that the function $f(x)$ is concave in $\mathbb{R}^3$ and is strictly concave when restricted to the affine plane P_c.

To see part (c), given a_1, a_2, a_3, there exists a Euclidean triangle $\Delta B_1 B_2 B_3$ whose inner angles are a_1, a_2, a_3. Let C be the inscribed circle to $\Delta B_1 B_2 B_3$ and e^{u_i} be the distance from B_i to $C \cap B_i B_j$. Then by construction, the length of $B_i B_j$ is $e^{u_i} + e^{u_j}$. This shows that the point (u_1, u_2, u_3) is a critical point of the function $g(x)$ on $\mathbb{R}^3$ since $\frac{\partial g}{\partial x_i}(u) = \theta_i - a_i = 0$. Since $g(x)$ is strictly concave with a critical point in the plane P_c where $c = u_1 + u_2 + u_3$, it follows that $\lim_{x \in P_c, x \to \infty} g(x) = -\infty$. On the other hand, for any $b \in \mathbb{R}$, by definition and $\theta_i(x + (b, b, b)) = \theta_i(x)$, we have $g(x + (b, b, b)) = g(x)$. To see this,

$$g(x + (b, b, b)) - g(x) = \int_x^{x+(b,b,b)} \sum_{i=1}^3 (\theta_i - a_i) dx_i$$

$$= \int_0^1 \sum_{i=1}^3 (\theta_i(x + t(b, b, b)) - a_i) b dt = b \sum_{i=1}^3 (\theta_i(x) - a_i) = 0.$$

For each vector $v \in \mathbb{R}^3$, let $\Pi(v) = v + (t, t, t) \in P_c$ be the orthogonal projection to P_c. Then a sequence of vectors $x(n) = (x_1(n), x_2(n), x_3(n)) \in \mathbb{R}^3$ satisfies $\max_{i,j} |x_i(n) - x_j(n)| \to \infty$ if and only if $\pi(x(n)) \to \infty$. Thus $g(x(n)) = g(\pi(x(n))) \to -\infty$ when $\max_{i,j} |x_i(n) - x_j(n)| \to \infty$. □

2.2 A Proof of Koebe–Andreev–Thurston's Theorem

We now prove Proposition 2.2 using Colin de Verdière's variational principle (see [7]).

To set up an appropriate variational framework, one needs the concept of an *angle structure* on a triangulated surface introduced in [7]. Suppose $(S, \mathcal{T})$ is a triangulated surface. An angle structure on $(S, \mathcal{T})$ assigns each vertex v in each triangle $\tau \in \mathcal{T}$ a positive number $a(v, \tau) \in \mathbb{R}_{>0}$, called the angle, such that (a) the sum of the three angles in each triangle is π and (b) the sum of all angles at each interior vertex v is 2π. Using linear programming, Colin de Verdière ([7]) proved that each simplicial triangulation of the triangle admits an angle structure. Another way to see it is to note that each geometric triangulation of a flat surface has a natural angle structure, i.e., $a(v, \tau)$ is the angle of the Euclidean triangle τ at v.

Lemma 2.4 *If $\mathcal{T}_1$ is an abstract simplicial triangulation of a triangle T with three vertices in ∂T, then there exists a geometric triangulation $\mathcal{T}'$ of an equilateral Euclidean Δ such that $\mathcal{T}'$ is isomorphic to $\mathcal{T}$.*

This lemma follows easily from Steinitz's theorem ([37]) that any 3-connected graph on $\mathbb{S}^2$ can be realized as the 1-skeleton of a compact convex polytope in $\mathbb{R}^3$. Indeed, by Steinitz's theorem, there exists a compact convex polytope P whose boundary with an open 2-cell Q removed is isomorphic $\mathcal{T}_1$. Project $\partial P - Q$ onto a plane from a point outside P and close to Q. The result is a geometric triangulation $\mathcal{T}''$ of a triangle such that $\mathcal{T}''$ is isomorphic to $\mathcal{T}_1$. Finally sending the triangle to the equilateral triangle Δ by an affine map produces the required $\mathcal{T}'$.

Label triangles in $\mathcal{T}'$ by $\Delta_1, \Delta_2, ..., \Delta_m$, let the vertices of Δ_i be v_{i1}, v_{i2}, v_{i3} and the inner angle of Δ_i at v_{ij} be a_{ij}, i.e., $\{a_{ij}\}$ is an angle structure on $\mathcal{T}'$. For each $x \in \mathbb{R}^V$, define

$$W(x) = \sum_{i=1}^{m} g_{\Delta_i}(x)$$

where $g_{\Delta_i}(x) = \int_0^{(x(v_{i1}),x(v_{i2}),x(v_{i3}))} \sum_{j=1}^{3} (\theta_{ij} - a_{ij}) dx(v_{ij})$ is the Colin de Verdière's function in Proposition 2.3 associated to the triangle Δ_i with radius assignment $e^{x(v_{i1})}$, $e^{x(v_{i2})}$, and $e^{x(v_{i3})}$ such that the angle in Δ_i at v_{ij} is θ_{ij}.

By definition and Proposition 2.3, the function $W(x)$ is concave in $\mathbb{R}^V$ since it is a sum of concave functions. Also, $W(x + t(1, 1, 1, ..., 1)) = W(x)$ due to Proposition 2.3(c). Furthermore, since each g_{Δ_i} is bounded from above, $W(x)$ is bounded from above. We claim that W is a proper function when restricted to

$P = \{x \in \mathbb{R}^V \mid \sum_{v \in V} x(v) = 0\}$, i.e., $\lim_{x \in P, x \to \infty} W(x) = -\infty$. Indeed, if $x \in P$ such that $x \to \infty$, then $\max_{i,j,j'} |x(v_{ij}) - x(v_{ij'})|$ converges to ∞. Therefore, by Proposition 2.3(c), we see that $W(x) \to -\infty$. This shows that $W|_P$ has a critical point $u \in P$. Since $W(x + (t, t, t, ..., t)) = W(x)$, this shows the point u is a critical point of W.

For this critical point u, suppose v_i, $i > 3$, is an interior vertex and $x_i = x(v_i)$. Then by Proposition 2.3(a), $\frac{\partial W}{\partial x_i}(u) = \sum_j (\theta_{n_i,j} - a_{n_i,j}) = -K(v_i)$ where $\theta_{n_i,j}$ and $a_{n_i,j}$ are the angles at the vertex v_i and $\sum_j a_{n_i,j} = 2\pi$. This shows that the circle packing metric associated to u is flat. At vertices v_i with $i = 1, 2, 3$, the same calculation shows $K(v_i) = 2\pi/3$ due to the choices of a_{ij} (i.e., Δ is an equilateral triangle). This implies $u_1 = u_2 = u_3$.

To prove uniqueness, if $\tilde{u} \in \mathbb{R}^V_{>0}$ comes from the radii of a circle packing whose nerve is isomorphic to $\mathcal{T}_1$ such that the associated polyhedral surface is an equilateral triangle, then the above calculation shows that $\tilde{u}$ is a critical point of W. Since W is concave, all critical points of W are maximum points. Therefore, it suffices to prove that the restriction of the function W to P is strictly concave. Indeed, otherwise there exist two distinct points $x, y \in P$ such that the function $h(t) = W(tx + (1-t)y)$ is linear in $t \in [0, 1]$. This implies that for each triangle Δ_i, $g_{\Delta_i}(tx + (1-t)y)$ is linear in t. By Proposition 2.3, this implies there is a vector $u_i(1, 1, 1) \in \mathbb{R}^3$, one for each triangle Δ_i, such that

$$(x(v_{i1}), x(v_{i2}), x(v_{i3})) = (y(v_{i1}), y(v_{i2}), y(v_{i3})) + u_i(1, 1, 1). \tag{2.3}$$

We claim that $u_i = u_j$ for all i, j. Indeed, consider two triangles Δ_i and Δ_j sharing a vertex v. Then (2.3) at v shows $u_i = u_j$. Since any two triangles Δ_i and Δ_j can be linked by a sequence of triangles $\Delta_{n_1} = \Delta_i, \Delta_{n_2}, ..., \Delta_{n_k} = \Delta_j$ such that Δ_{n_r} and $\Delta_{n_{r+1}}$ share a common vertex, we see that $u_i = u_j$. It follows that the two vectors x, y differ by a vector of the form $t(1, 1, ..., 1) \in \mathbb{R}^V$. On the other hand, both $x, y \in P$, therefore $t = 0$, i.e., $x = y$ which contradicts the choice of x, y.

2.3 *Thurston's Conjecture on Circle Packing And Rodin-Sullivan's Work*

The relationship between the Koebe–Andreev–Thurston's theorem and the Riemann mapping theorem was explored by W. Thurston in early 1980s. The basic idea is that since conformal maps send infinitesimal circles (circles in the tangent space) to circles, a circle packing should be a good approximation to conformal maps.

Here is Thurston's conjecture which was proved by Rodin-Sullivan in [29].

Given a bounded simply connected domain Ω in the complex plane $\mathbb{C}$ and a point $p \in \Omega$, for each large integer n, let P_n be a maximum (hexagonal) circle packing by disks of radii $1/n$ inside Ω and p_n be a circle in P_n within distance $1/n$ to p. Here maximum means that one cannot add another $1/n$ radius disk in Ω to P_n such that

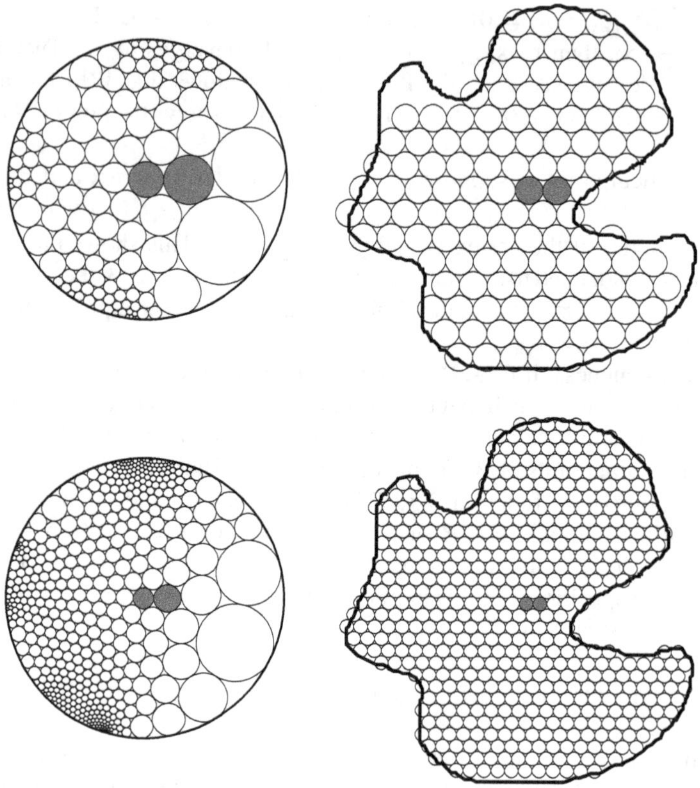

Fig. 2 Thurston's conjecture, Rodin-Sullivan's theorem, on convergence of circle packing to the Riemann mapping. The picture is produced by K. Stephenson

its nerve is the 1-skeleton of a topological triangulation $\mathcal{T}_n$ of a disk. Let p_n' be the circle in P_n adjacent to p_n from the right. Modify $\mathcal{T}_n$ to be a triangulation $\mathcal{T}_n^*$ of the 2-sphere $\mathbb{S}^2$ by adding one vertex v_∞ and edges from v_∞ to all boundary vertices of $\mathcal{T}_n$. Now by Koebe–Andreev–Thurston's theorem, there exists a circle packing Q_n of the Riemann sphere such that (a) its nerve is isomorphic to the 1-skeleton of $\mathcal{T}_n^*$; (b) the disk corresponding to v_∞ is the complement of the unit disk $\mathbb{D}$; (c) the disk corresponding to p_n is centered at 0, (d) the disk in Q_n corresponding to p_n' is centered in the positive x-axis (Fig. 2).

Let f_n be the piecewise linear map constructed as follows. It sends the center of a circle in P_n to the center of the corresponding circle in Q_n and f_n is linear on each triangles. Thurston's conjecture, proved by Rodin-Sullivan, is that as $n \to \infty$, f_n converges to the Riemann mapping $f : \Omega \to D$ uniformly on compact subsets of Ω.

Rodin-Sullivan's proof of convergence is beautiful and elegant. The readers are strongly recommended to read the original paper [29]. There are two steps involved in the proof. In the first step, they showed that there exists a constant $K > 0$ so that all approximation functions f_n are K-quasi-conformal. This is a consequence of Rodin-Sullivan's ring lemma which states that in a hexagonal circle packing, the ratio of the radii of any two adjacent circles is at most 1000. One can deduce the ring lemma by inspection. Now uniform K-quasiconformality follows since inner angles in a Euclidean triangle of edge lengths $r_1 + r_2, r_2 + r_3, r_3 + r_1$ with $\frac{r_i}{r_j} \leq 1000$ cannot be too small. Since f_n are uniformly K-quasi-conformal, it has a convergent subsequence. Let f be the limit of the subsequence. The claim is that f is the Riemann mapping. To establish conformality of f, Rodin-Sullivan proved that the hexagonal circle packing in the plane is unique. To be more precise, if $\{D_i\}$ is a locally finite collection of disks in $\mathbb{C}$ with disjoint interiors such that each D_i is tangent to exactly six other disks D_j's and $\mathbb{C} - \cup D_i$ is a disjoint union of open triangles whose boundary are in $\cup \partial D_i$, then all D_i have the same size.

Rigidity of hexagonal circle packing is the first rigidity theorem proved for infinite circle packing. This work has inspired and initiated many research activities. For instance Schramm [30] proved that any locally finite infinite circle packing of $\mathbb{C}$ is rigid. See also the works of He [13], He-Schramm [14] and many others.

3 A Discrete Uniformization Theorem

One form of the uniformization theorem states that each Riemann surface admits a complete Riemannian metric of constant curvature -1, 0, or 1 within its conformal class. Furthermore, the metric is unique unless the Riemann surface is conformal to the complex plane $\mathbb{C}$, the punctured plane $\mathbb{C} - \{0\}$, the sphere $\mathbb{S}^2$, or tori $\mathbb{C}/(\mathbb{Z} + \tau\mathbb{Z})$ for some $\tau \notin \mathbb{R}$. In this section, we introduce our recent work on discrete conformal geometry for compact polyhedral surfaces and discuss a discrete version of uniformization theorem for compact polyhedral surfaces.

Polyhedral surfaces are ubiquitous due to digitization (e.g. 3D scan). Classifying them according to some discrete conformality should be useful in organizing polyhedral surfaces. Circle packing can be considered as a discrete conformality if one allows the changing of radii. However not all polyhedral surfaces can be canonically packed by circles. A discrete conformality for all polyhedral surfaces was introduced in [11, 12]. The main features of the discrete conformality are the following. First, the discrete conformality is algorithmic; second the corresponding discrete uniformization theorem holds for compact surfaces; third there exists a finite dimensional (convex) variational principle to find the discrete uniformization metric; and fourth discrete conformality is closely related to the convex ideal hyperbolic polyhedra in the 3-dimensional hyperbolic space $\mathbb{H}^3$. Similar to Thurston's conjecture on the convergence of circle packing metrics, we have recently proved a convergence result [18] which shows that the discrete conformality converges to smooth conformality when the triangulations are suitably chosen. Several conjectures about a discrete uniformization for non-compact polyhedral surfaces will be discussed at the end of this section.

3.1 Discrete Conformality of Polyhedral Surfaces

A closed surface S together with a non-empty finite subset of points $V \subset S$ will be called a *marked surface*. A triangulation $\mathcal{T}$ of a marked surface (S, V) is a topological triangulation of S such that the vertex set of $\mathcal{T}$ is V. We use $E = E(\mathcal{T}), V = V(\mathcal{T})$ to denote the sets of all edges and vertices in $\mathcal{T}$ respectively. A (Euclidean) *polyhedral metric* on (S, V), to be called a *PL metric* on (S, V) for simplicity, is a flat cone metric on (S, V) with cone points contained in V. We call the triple (S, V, d) a polyhedral surface. All PL metrics are obtained by isometric gluing of Euclidean triangles along pairs of edges. For instance boundaries of convex polytopes are PL metrics on $(\mathbb{S}^2, V)$. The *discrete curvature* of a PL metric d is the function $K_d : V \to (-\infty, 2\pi)$ sending a vertex v to 2π minus the cone angle at v. For a closed surface S, it is well known that the Gauss-Bonnet theorem $\sum_{v\in V} K_d(v) = 2\pi\chi(S)$ holds. If $\mathcal{T}$ is a triangulation of (S, V) with a PL metric d for which all edges in $\mathcal{T}$ are geodesic, we say $\mathcal{T}$ is *geometric* in d and d is a PL metric on $(S, V, \mathcal{T})$. In this case, we can represent the PL metric d by the length function $l_d : E(\mathcal{T}) \to \mathbb{R}_{>0}$ sending an edge to its length. Thus the polyhedral surface (S, V, d) can be represented by $(S, \mathcal{T}, l)$. This is a way of coding a polyhedral surface by a finite-dimensional vector $l_d \in \mathbb{R}^E$.

In general, a polyhedral surface (S, V, d) admits infinitely many different geometric triangulations. However, each polyhedral surface (S, V, d) has a natural *Delaunay triangulation* $\mathcal{T}_d$ which is a geometric triangulation with vertices V such that for each edge, the sum of two angles facing e is at most π. Delaunay triangulations are the most commonly used triangulations in scientific computing. It can be constructed from the Voronoi decomposition $\{R(v)|v \in V\}$ of (S, V, d) as follows. Here a Voronoi 2-cell $R(v)$ for $v \in V$ is defined to be $\{x \in S|d(x, v) \le d(x, v'), \forall v' \in V\}$. The Delaunay tessellation of (S, V, d) is the dual cell decomposition of $\{R(v)|v \in V\}$ whose vertices are V and each 1-dimensional connected component of $R(v) \cap R(v')$ corresponds to a (geodesic) edge from v to v'. A Delaunay triangulation is a subdivision of the Delaunay tessellation into triangles without introducing extra vertices. Any two Delaunay triangulations of (S, V, d) are related by a sequence of Delaunay triangulations such that adjacent ones differ by a diagonal switch along an edge. See for instance [4].

Suppose d is a PL metric on a triangulated surface $(S, \mathcal{T})$ whose edge length function is $l_d : E(\mathcal{T}) \to \mathbb{R}_{>0}$. For a positive function $u : V(\mathcal{T}) \to \mathbb{R}_{>0}$, the vertex scaling of l_d by u is the new function $u * l_d : E(\mathcal{T}) \to \mathbb{R}_{>0}$ such that $u * l_d(vv') = u(v)u(v')l_d(vv')$ where vv' is an edge with end points v, v'. If d, d' are two PL metrics on $(S, \mathcal{T})$, then they differ by a *vertex scaling* if $l_d = u * l_{d'}$ for some $u : V \to \mathbb{R}_{>0}$. The notation of vertex scaling change of PL metrics was introduced in [28] and in [16].

The definition of discrete conformality involves Delaunay triangulations and vertex scaling.

Definition 3.1 ([11]) Two PL metrics d and d' on a marked closed surface (S, V) are *discrete conformal* if there is a sequence of PL metrics $d_1 = d, d_2, ..., d_n = d'$ and a sequence of triangulations $\mathcal{T}_1, \mathcal{T}_2, ..., \mathcal{T}_n$ of (S, V) such that

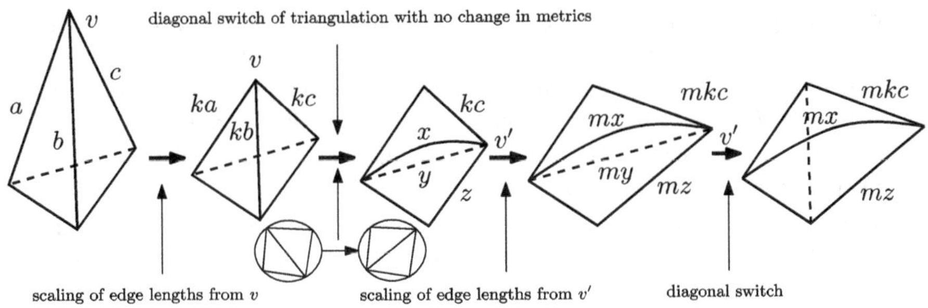

Fig. 3 Discrete conformal change of PL metrics from an arbitrary tetrahedron to one with constant curvature π. All triangulations involved are Delaunay

(a) each $\mathcal{T}_i$ is Delaunay in d_i,

(b) if $\mathcal{T}_i \neq \mathcal{T}_{i+1}$, then there is an isometry h_i from (S, V, d_i) to (S, V, d_{i+1}) such that h_i is homotopic to the identity map on (S, V), and

(c) if $\mathcal{T}_i = \mathcal{T}_{i+1}$, there is a function $u_i : V \to \mathbb{R}_{>0}$ such that for each edge $e = vv'$ in $\mathcal{T}_i$, the lengths $l_{d_i}(vv')$ and $l_{d_{i+1}}(vv')$ of e in d_i and d_{i+1} are related by

$$l_{d_{i+1}}(vv') = u_i(v)u_i(v')l_{d_i}(vv'), \tag{3.1}$$

i.e., $l_{d_{i+1}} = u_i * l_{d_i}$.

The original motivation in [16] for introducing vertex scaling $u * l_d$ as an approximation to conformal change is the following. Since a PL polyhedral metric l_d on $(S, \mathcal{T})$ is a discretization of a Riemannian metric g and a function $u : V(\mathcal{T}) \to \mathbb{R}_{>0}$ is a discretization of a positive function λ on S, the conformal Riemannian metric λg should be approximated by the PL metric defined by $u * l_d$. The deeper reason for $u * l_d$ to be a discrete conformal change is due to the following observation in Riemannian geometry ([18]). Given a Riemannian metric g on a compact connected manifold M and $\lambda : M \to \mathbb{R}_{>0}$, there exists a constant $C = C(M, g, \lambda)$ such that for any $p, q \in M$, we have

$$|d_{\lambda^4 g}(p, q) - \lambda(p)\lambda(q)d_g(p, q)| \leq C d_g(p, q)^3$$

where $d_g(p, q)$ is the distance between p, q in the metric g.

The relationship between discrete conformal geometry and hyperbolic geometry is the following [5, 11]. Given a Delaunay triangulated polyhedral surface $(S, \mathcal{T}, d)$ with $V = V(\mathcal{T})$, one can naturally associate to d a cusped hyperbolic metric d^* on $S - V$. Here is the construction. Take a Euclidean triangle τ in $(\mathcal{T}, d)$ considered as the Euclidean convex hull of vertices $v_1, v_2, v_3 \in \mathbb{C}$. Let τ^* be the hyperbolic convex hull $C_{\mathbb{H}}(v_1, v_2, v_3)$ of v_1, v_2, v_3 in the hyperbolic 3-space $\mathbb{H}^3$. Here we consider $\mathbb{C}$ to be in the sphere at infinity of the upper half-space model $\mathbb{C} \times \mathbb{R}_{>0}$ of $\mathbb{H}^3$. Now if σ and τ are two Euclidean triangles in $\mathcal{T}$ glued by a Euclidean isometry f along an edge, since each isometry f of $\mathbb{C}$ extends naturally to an isometry f^* of $\mathbb{H}^3$, we glue

τ^* and σ^* along the corresponding edge by the isometry f^*. In this way, we obtain a complete finite area hyperbolic metric d^* on $S - V$. It follows from the construction that d^* is independent of the choices of the Delaunay triangulations $\mathcal{T}$ used in the construction. It is proved in [11, Theorem 43] that two PL metrics d_1 and d_2 on a closed marked surface (S, V) are discrete conformal in the sense of Definition 3.1 if and only if their associated hyperbolic metrics d_1^* and d_2^* are isometric by an isometry homotopic to the identity (respecting V). Conversely, if S is a closed surface and $\hat{d}$ is a complete finite area hyperbolic metric on $S - V$, then there exists a polyhedral metric d on (S, V) such that $d^* = \hat{d}$. Thus for closed surfaces, there exists a bijection between the space of all discrete conformal classes of polyhedral metrics on (S, V) and the Teichmüller space of cusped metrics on $S - V$.

By this construction, if $\mathcal{T}$ is a Delaunay triangulation of the plane $(\mathbb{C}, d_{st})$ with $V = V(\mathcal{T})$ and d_{st} being the standard flat metric on $\mathbb{C}$, then the associated hyperbolic metric d_{st}^* is the boundary of the convex hull $C_{\mathbb{H}}(V)$ of V in $\mathbb{H}^3$. To see this, we note that codimension-1 faces of $C_{\mathbb{H}}(V)$ correspond to the circum-disks of triangles $\tau \in \mathcal{T}$ due to the Delaunay condition. This shows the relationship between discrete conformal geometry and convex hull construction in the hyperbolic 3-space $\mathbb{H}^3$ and the essential role of Delaunay condition in discrete conformality.

The main theorems proved in [11, 12] are:

Theorem 3.2 ([11]) *Given any PL metric d on a closed marked surface (S, V) and any $K^* : V \to (-\infty, 2\pi)$ such that $\sum_{v \in V} K^*(v) = 2\pi\chi(S)$, there exists a PL metric d^* on (S, V), unique up to scaling and isometries homotopic to the identity map on (S, V), such that*

(a) d^* is discrete conformal to d, and

(b) the discrete curvature K_{d^*} is equal to K^*.

Furthermore, the PL metric d^* can be found by a finite-dimensional variational principle.

For the constant function $K^* = 2\pi\chi(S)/|V|$ in Theorem 3.2, we obtain a constant curvature PL metric d^*, unique up to scaling and isometries homotopic to the identity, discrete conformal to d. We call d^* *the discrete uniformization metric* associated to d. The existence and uniqueness of d^* is a discrete version of the uniformization theorem for closed surfaces.

Theorem 3.2 for the torus $S = \mathbb{S}^1 \times \mathbb{S}^1$ with $K^* = 0$ is equivalent to a theorem of Fillastre [9]. Theorem 3.2 shows that every polyhedral torus $(\mathbb{S}^1 \times \mathbb{S}^1, V, d)$ is discrete conformal to a flat torus $(\mathbb{S}^1 \times \mathbb{S}^1, V, d_{flat})$. Translating it into the language of hyperbolic metrics, we can replace d by any cusped hyperbolic metric $\hat{d}$ on the punctured torus $\mathbb{S}^1 \times \mathbb{S}^1 - V$. The hyperbolic metric associated to $(\mathbb{S}^1 \times \mathbb{S}^1, V, d_{flat})$ is constructed as follows. Take a lattice $L = \mathbb{Z} + \tau\mathbb{Z}$ in $\mathbb{C}$ and consider the boundary $\partial C_{\mathbb{H}}(V^*)$ of the convex hull of V^* in $\mathbb{H}^3$ where V^* is a discrete set in $\mathbb{C}$ invariant under the action of L. Then by the discussion above, d_{flat}^* is isometric to the cusped hyperbolic metric $\partial C_{\mathbb{H}}(V^*)/L$. Furthermore, the lattice L is unique up to complex linear transformations. This is the result proved in [9]. To be more precise, Fillastre proved the following version of convex embedding theorem. For any cusped hyperbolic metric $\hat{d}$ on $\mathbb{S}^1 \times \mathbb{S}^1 - V$, there exist a lattice $L \subset \mathbb{C}$ and a finite set V' in the conformal infinite of the hyperbolic manifold $\mathbb{H}^3/L$ such that $\hat{d}$ is isometric to the boundary of the convex hull of V' in $\mathbb{H}^3/L$.

This shows a close connection between discrete conformal geometry and the convex isometric embedding program of Weyl, Alexandrov, Nirenberg, Pogorelov and others.

Theorem 3.3 ([12]) *Given two PL metrics on a closed marked surface (S, V) such that the lengths of edges are algebraic numbers, there exists an algorithm to decide if they are discrete conformal.*

Theorem 3.3 is proved in our joint work with Ren Guo in [12]. The counterpart of Theorem 3.2 for hyperbolic polyhedral surfaces is also proved in [12].

An important question is whether discrete conformality defined above approximates smooth conformality. To this end, let us recall discrete conformal maps between polyhedral surfaces [5, 18]. Given a closed polyhedral surface (S, V, d). Let d^* be the hyperbolic metric on $S - V$ associated d constructed using ideal hyperbolic triangles associated to Euclidean triangles. The vertical projection of the ideal hyperbolic triangle $\tau^* = C_{\mathbb{H}}(v_1, v_2, v_3)$ to the Euclidean triangle $\tau = C_{\mathbb{E}}(v_1, v_2, v_3)$ produces a piecewise projective homeomorphism Φ_d from $(S - V, d^*)$ to $(S - V, d|_{S-V})$. If d_1, d_2 are two discrete conformal PL metrics on (S, V), then the *discrete conformal map* from (S, V, d_1) to (S, V, d_2) is defined to be (the extension to S) of the composition $\Phi_{d_2} \circ \Psi \circ \Phi_{d_1}^{-1}$ where $\Psi : (S - V, d_1^*) \to (S - V, d_2^*)$ is the isometry homotopic to the identity. Discrete conformal maps are piecewise projective.

Our recent work with Sun and Wu [18] shows that discrete conformality does converge to the smooth conformality. Given a simply connected marked polygonal domain with a PL metric (D, V, d) and three boundary vertices $p, q, r \in V$, let the metric double of (D, V, d) along the boundary be the marked polyhedral 2-sphere $(\mathbb{S}^2, V', d')$. Using Theorem 3.2, one produces a new polyhedral surface $(\mathbb{S}^2, V', d^*)$ such that (1) $(\mathbb{S}^2, V', d^*)$ is discrete conformal to $(\mathbb{S}^2, V', d')$, (2) the discrete curvatures of d^* at p, q, r are $4\pi/3$, (3) the discrete curvatures of d^* at all other vertices are zero and (4) its area is $\sqrt{3}/2$. Thus $(\mathbb{S}^2, V', d^*)$ is the metric double of an equilateral triangle ΔABC of edge length 1. Here A, B, C correspond to p, q, r. Let $F : (\mathbb{S}^2, V', d') \to (\mathbb{S}^2, V', d^*)$ be the associated discrete conformal map sending $\{p, q, r\}$ to $\{A, B, C\}$ respectively. Due to the uniqueness part of Theorem 3.2, we see that $f = F|_D : D \to \Delta ABC$ sending p, q, r to the vertices A, B, C respectively. We call f the *discrete uniformization map* associated to $(D, \mathcal{T}, l, \{p, q, r\})$

Theorem 3.4 ([18]) *Suppose Ω is a Jordan domain in the complex plane with three distinct points p, q, r in the boundary of Ω. There exists a sequence of simply connected polygonal domains $(\Omega_n, \mathcal{T}_n, \{p_n, q_n, r_n\})$ with triangulations $\mathcal{T}_n$ by equilateral triangles of edge lengths converging to 0 where p_n, q_n, r_n are three boundary vertices such that the following hold*

(i) $\Omega_n \subset \Omega_{n+1}$ and $\Omega = \cup_{n=1}^{\infty} \Omega_n$,

(ii) the discrete uniformization maps f_n associated to $(\Omega_n, \mathcal{T}_n, \{p_n, q_n, r_n\})$ converge uniformly on compact sets to the Riemann mapping $f : (\Omega, \{p, q, r\}) \to (\Delta ABC, \{A, B, C\})$.

3.2 Vertex Scaling and Its Associated Variational Principle

A key property established in [16] for the vertex scaling is the following variational principle (See Lemma 3.5(a) below).

Lemma 3.5 *Suppose $\Delta v_1 v_2 v_3$ is a Euclidean triangle of edge lengths l_1, l_2, l_3 such that v_i is opposite to the edge of length l_i. Let $l_1 e^{x_2+x_3}, l_2 e^{x_1+x_3}, l_3 e^{x_1+x_2}$ be the edge lengths of a vertex scaled Euclidean triangle whose inner angle at v_i is $\theta_i = \theta_i(x_1, x_2, x_3)$.*

(a)([16]) There exists a locally concave function $F(x_1, x_2, x_3)$ such that

$$\frac{\partial F}{\partial x_i} = \theta_i \tag{3.2}$$

and the kernel of the positive semidefinite symmetric matrix $[\frac{\partial \theta_i}{\partial x_j}]_{3\times 3}$ consists of column vectors $(a, a, a)^t$ and

(b) If $e^{x_1} \to \infty$ and e^{x_2} is bounded, then e^{x_3} is bounded and $\theta_1(x) \to 0$.

Proof To see part (a), it suffices to show that the 3×3 matrix $[\frac{\partial \theta_r}{\partial x_s}]_{3\times 3}$ is symmetric and negative semi-definite. Let the area of the triangle $\Delta v_1 v_2 v_3$ be A and $y_i = l_i e^{x_j + x_k}$ be the length of the edge $v_j v_k$ where $\{i, j, k\} = \{1, 2, 3\}$. Note that $\frac{\partial y_i}{\partial x_j} = y_i$ and $y_j = y_k \cos(\theta_i) + y_i \cos(\theta_k)$. By (2.1) and (2.2), we have

$$\frac{\partial \theta_i}{\partial x_j} = \frac{\partial \theta_i}{\partial y_i}\frac{\partial y_i}{\partial x_j} + \frac{\partial \theta_i}{\partial y_k}\frac{\partial y_k}{\partial x_j} = \frac{y_i(y_i - y_k\cos(\theta_j))}{2A} = \frac{y_i y_j \cos(\theta_k)}{2A} = \cot(\theta_k)$$

and

$$\frac{\partial \theta_i}{\partial x_i} = \frac{\partial \theta_i}{\partial y_j}\frac{\partial y_j}{\partial x_i} + \frac{\partial \theta_i}{\partial y_k}\frac{\partial y_k}{\partial x_i} = -\frac{\partial \theta_i}{\partial y_i}(y_j\cos(\theta_k) + y_k\cos(\theta_j)) = -\frac{y_i^2}{2A} - \frac{\sin\theta_i}{\sin(\theta_j)\sin(\theta_k)}.$$

This shows that the matrix $[\frac{\partial \theta_r}{\partial x_s}]_{3\times 3}$ is symmetric and can be written as $-DGD^t$ where $G = [g_{rs}]_{3\times 3}$ is the Gram matrix of the triangle and D is a diagonal matrix. Here $g_{ii} = -1$ and $g_{ij} = -\cos(\theta_k)$. Let n_i be the unit outward normal vector to the triangle at edge $v_j v_k$ and (u, v) be the inner product in $\mathbb{R}^3$. Then the Gram matrix G is the same as $[(n_r, n_s)]_{3\times 3}$ which is well known to be positive semi-definite with kernel (t, t, t). Thus part (a) follows.

To see part (b), note that the triangle of edge lengths $e^{x_2+x_3} l_1, e^{x_1+x_3} l_2, e^{x_1+x_2} l_3$ is similar to the Euclidean triangle of edge lengths $l_1 e^{-x_1}, l_2 e^{-x_2}, l_3 e^{-x_3}$. In particular, we have the triangle inequality that $l_2 e^{-x_2} < l_1 e^{-x_1} + l_3 e^{-x_3}$. This implies that x_3 must be bounded. Since $l_1 e^{-x_1} \to 0$ and $l_2 e^{-x_2}, l_3 e^{-x_3}$ are bounded away from 0, it follows $\theta_1 \to 0$. □

The identity in Lemma 3.5(a) can be considered as a 2-dimensional analogue of the Schlaefli formula. This variational principle has been generalized in the work of [5]. In particular, an explicit formula for the function F was found in [5] using Lobachevsky function $-\int_0^x \ln(|2\sin(t)|)dt$.

3.3 Basic Idea of the Proof of Theorem 3.2

There are two steps involved in the proof. The first step is to understand discrete conformality using hyperbolic metrics. The goal is to show that given any PL metric d on (S, V), the space $DC([d])$ of all PL metrics on (S, V) discrete conformal to d is C^1-diffeomorphic to the Euclidean space $\mathbb{R}^V$. The second step is to show that the discrete curvature map $K : DC([d]) \to \{x \in (-\infty, 2\pi)^V | \sum_{v\in V} K(v) = 2\pi\chi(S)\}$ is a bijection up to scalings. This is achieved by showing that the discrete curvature map K is the gradient of a convex function using Lemma 3.5(a) and the work of [1].

The first step is achieved by using Penner's theorey of decorated Teichmüller space. Let us first recall that two PL metrics on (S, V) are *Teichmüller equivalent* if they are isometric by an isometry homotopic to the identity in (S, V). For instance the condition (b) in Definition 3.1 says that (S, V, d_i) is Teichmüller equivalent to (S, V, d_{i+1}). The PL Teichmüller space $T_{pl} = T_{pl}(S, V)$ is the space of all Teichmüller equivalence classes of PL metrics on (S,V). The space $T_{pl}(S, V)$ is known to be a real analytic manifold diffeomorphic to a Euclidean space by the work of Troyanov [34]. The discrete conformality is an equivalence relation on $T_{pl}(S, V)$. The discrete curvature $K : T_{pl}(S, V) \to (-\infty, 2\pi)^V$ is a real analytic map. There exists a natural action of the set of positive real numbers $\mathbb{R}_{>0}$ on $T_{pl}(S, V)$ by scaling. The discrete curvature is well defined on the quotient space $K : T_{pl}(S, V)/\mathbb{R}_{>0} \to \{x \in (-\infty, 2\pi)^V | \sum_{v\in V} x(v) = 2\pi\chi(S)\}$.

Given a metric $[d] \in T_{pl}(S, V)$, let $DC([d]) = \{[d'] \in T_{pl} | \ d'$ is discrete conformal to $d\}$ be the discrete conformal class associated to $[d]$. Theorem 3.2 is equivalent to the statement that the restriction of the discrete curvature map K to $DC([d])/\mathbb{R}_{>0}$ is a bijection from $DC([d])/\mathbb{R}_{>0}$ onto $\{x \in (-\infty, 2\pi)^V | \sum_v x(v) = 2\pi\chi(S)\ \}$. We prove that K is a C^1 diffeomorphism.

Let $T(S - V)$ be the Teichmüller space of complete hyperbolic metrics of finite area on $S - V$ and $T_D = T(S - V) \times \mathbb{R}^V_{>0}$ be Penner's decorated Teichmüller space [23]. Recall that a decorated hyperbolic metric on $S - V$ is a complete finite area hyperbolic metric together with a horoball at each cusp. By measuring the lengths of the boundaries of the horoballs, one can write a decorated hyperbolic metric as a pair (d, u) where $u \in \mathbb{R}^V_{>0}$. This shows that the space of all decorated hyperbolic metrics modulo the natural equivalence relation is $T(S - V) \times \mathbb{R}^V_{>0}$. Decorated hyperbolic metrics on an ideal triangulated surface $(S - V, \mathcal{T})$ can be constructed by isometrically gluing decorated ideal hyperbolic triangles along edges. Here a decorated ideal hyperbolic triangle is an ideal triangle with a horoball at each vertex. Since all ideal hyperbolic triangles are isometric, a decorated ideal triangle is determined up to isometries preserving decoration by the three lengths of horocycles inside it. Another way to parameterize a decorated ideal triangle is to use the edge lengths. If e is an edge of a decorated ideal triangle, then the length $l(e)$ of e is the distance between the two horoballs B_1, B_2 at its end points if $B_1 \cap B_2 = \emptyset$, and is the negative of the length of the interval $e \cap (B_1 \cap B_2)$ if $B_1 \cap B_2 \neq \emptyset$. Penner defines the λ-length of an edge e is defined to be $e^{l(e)/2}$. Given any three positive real numbers, there exists a unique decorated ideal triangle whose λ-lengths are the given numbers. In particular, given any Euclidean triangle σ of edge lengths l_1, l_2, l_3, one can associate a decorated ideal triangle σ^* of λ-length l_1, l_2, l_3 to σ. Given a PL metric d represented as $(S, \mathcal{T}, l)$ (i.e., $\mathcal{T}$ is geometric in d), one assigns a decorated hyperbolic metric $\Phi_{\mathcal{T}}(d)$ on $S - V$ as follows. Each Euclidean triangle $\sigma \in \mathcal{T}$ is replaced by its decorated ideal triangle counterpart σ^*. These decorated ideal triangles are glued along edges by isometries preserving decorations. The resulting decorated hyperbolic metric is $\Phi_{\mathcal{T}}(d)$. See [5]. We prove the following theorem.

Theorem 3.6 ([11]) *For any closed marked surface (S, V) such that $(S, V) \neq (\mathbb{S}^2, \{p\})$ or $(\mathbb{S}^2, \{p, q\})$, there exists a C^1 smooth diffeomorphism Φ from the PL Teichmüller space $T_{pl}(S, V)$ to the decorated Teichmüller space $T(S - V) \times \mathbb{R}^V_{>0}$ such that two PL metrics d and d' are discrete conformal if and only if the projections of $\Phi(d)$ and $\Phi(d')$ to the Teichmüller space $T(S - V)$ are the same.*

The map Φ is constructed in a piecewise smooth manner on the natural cell decompositions of T_{pl} and T_D. For each triangulation $\mathcal{T}$ of (S, V), define $D_{pl}(\mathcal{T})$ (and $D(\mathcal{T})$) to be the set of all PL metrics (and decorated hyperbolic metrics) $[d]$ in T_{pl} (and T_D) such that $\mathcal{T}$ is isotopic to a Delaunay triangulation in d. The important works of Rivin [26] and Penner [23] show that $D_{pl}(\mathcal{T})$ and $D(\mathcal{T})$ are cells and $T_{pl} = \cup_{\mathcal{T}} D_{pl}(\mathcal{T})$ and $T_D = \cup_{\mathcal{T}} D(\mathcal{T})$ are cell decompositions of the Teichmüller spaces invariant under the action of the mapping class group. The definition of Φ goes as follows. For each triangulation $\mathcal{T}$, define $\Phi_{\mathcal{T}} : D_{pl}(\mathcal{T}) \to T_D(S, V)$ by sending a PL metric $(S, \mathcal{T}, l)$ to $\Phi_{\mathcal{T}}(S, \mathcal{T}, l)$. By definition the two decorated metrics $\Phi_{\mathcal{T}}(S, \mathcal{T}, l)$ and $\Phi_{\mathcal{T}}(S, \mathcal{T}, w * l)$ have the same underlying hyperbolic metrics and differ only in decorations.

It is a straightforward calculation to see that Euclidean Delaunay condition is mapped to hyperbolic Delaunay condition, i.e., $\Phi_{\mathcal{T}}(D_{pl}(\mathcal{T})) \subset D(\mathcal{T})$. Penner observed that hyperbolic Delaunay condition implies the triangle inequality for (Euclidean) edge lengths, i.e., $\Phi_{\mathcal{T}}(D_{pl}(\mathcal{T})) = D(\mathcal{T})$. Furthermore, Penner's result that the Ptolemy identity holds for λ-lengths of decorated ideal quadrilaterals implies that for different triangulations $\mathcal{T}$ and $\mathcal{T}'$ of (S, V),

$$\Phi_{\mathcal{T}}|_{D_{pl}(\mathcal{T}) \cap D_{pl}(\mathcal{T}')} = \Phi_{\mathcal{T}'}|_{D_{pl}(\mathcal{T}) \cap D_{pl}(\mathcal{T}')}.$$

Thus these maps $\Phi_{\mathcal{T}}$ can be glued together to produce a homeomorphism $\Phi = \cup_{\mathcal{T}} \Phi_{\mathcal{T}} : T_{pl} \to T_D$. Note that the complete finite area hyperbolic metric $d_{\mathbb{H}}$ on $S - V$ associated to a PL metric d on (S, V) is $P \circ \Phi([d])$ where $P : T(S - V) \times \mathbb{R}^V_{>0} \to T(S - V)$ is the projection.

We prove that Φ is a C^1 diffeomorphism by using the following lemma on quadrilaterals.

Lemma 3.7 *Let Q be a convex Euclidean quadrilateral whose four edge lengths are x, y, z, w labelled cyclically and the length of a diagonal be u. Let $A(x, y, z, w, u)$ be the length of second diagonal and $B(x, y, z, w, u) = \frac{xz+yw}{u}$. If a point (x, y, z, w, u) satisfies $A(x, y, z, w, u) = B(x, y, z, w, u)$, i.e., Q is inscribed in a circle, then $DA(x, y, z, w, u) = DB(x, y, z, w, u)$ where DA is the derivative of A.*

In the second step, we examine the restriction $K|$ of the discrete curvature map to the space of discrete conformal classes $DC([d])$. By Theorem 3.6, $DC([d])$ is naturally a Euclidean space. Using Lemma 3.5(a), we show that the discrete curvature map on $DC([d])/\mathbb{R}_{>0}$ is the gradient of a strictly convex function. Thus, $K| : DC([d])/\mathbb{R}_{>0} \to Y := \{x \in (-\infty, 2\pi)^V | \sum_v x(v) = 2\pi\chi(S)\}$ is injective. On the other hand, by using Lemma 3.5(b) and a result of Akiyoshi [1] we show that the image $K(DC([d]))$ is closed in Y. Since both $DC([d])/\mathbb{R}_{>0}$ and Y are connected manifolds of the same dimension, we conclude that $K|$ is a homeomorphism and thus prove Theorem 3.2.

3.4 Basic Ideas of the Proof of Theorem 3.3

Suppose d, d' are two PL metrics on a marked closed surface (S, V) such that d, d' are given by the edge length functions $l_d : E(\mathcal{T}) \to \mathbb{A}$ and $l_{d'} : E(\mathcal{T}') \to \mathbb{A}$ where $\mathbb{A}$ is the set of all real algebraic numbers. Our goal is to use these two vectors l_d and $l_{d'}$ to decide whether d, d' are discrete conformal or not.

Using a well-known algorithm from computational geometry, we may assume that both $\mathcal{T}$ and $\mathcal{T}'$ are Delaunay in d and d' respectively. Now consider the associated decorated hyperbolic metrics $y = \Phi_{\mathcal{T}}(d)$ and $y' = \Phi_{\mathcal{T}'}(d')$ in Penner's decorated Teichmüller space. By Theorem 3.6, it suffices to check if y, y' have the same underlying hyperbolic metric. To this end, we use a theorem of Thurston and Mosher [22] that there is an algorithm which produces a sequence of triangulations $\mathcal{T}_1 = \mathcal{T}, \mathcal{T}_2, ..., \mathcal{T}_n = \mathcal{T}'$ of (S, V) such that for each i, $\mathcal{T}_i$ and $\mathcal{T}_{i+1}$ differ by a diagonal switch. Combining with Penner's Ptolemy identity, we find algorithmically the λ-length coordinates z $(= y)$ and z' of the decorated metrics y, y' in the same triangulation $\mathcal{T}$. For a triangulation $\mathcal{T}$, it is known by Penner's work that z, z' represent the same underlying hyperbolic metric if and only if their associated shear coordinates in the triangulation $\mathcal{T}$ are the same. Here the shear coordinate of $z : E(\mathcal{T}) \to \mathbb{R}_{>0}$ is the function $\phi(z) : E(\mathcal{T}) \to \mathbb{R}$ given by $\phi(z)(e) = \frac{ab}{cd}$ where a, b, c, d are the values of z at the four edges, ordered cyclically, of the quadrilateral in $\mathcal{T}$ formed by the two triangles adjacent to e. Therefore, we can check algorithmically if z and z' have the same underlying hyperbolic structure.

3.5 Basic Idea of the Proof of Convergence Theorem 3.4

The proof of Theorem 3.4 follows the basic strategy appeared in Rodin-Sullivan's work [29]. Namely, we prove that the approximating discrete conformal maps f_n are K-quasi-conformal for some K independent of n and a rigidity result about the hexagonal triangulations of the plane. Finally since Delaunay triangulations may change due to flip operations, we choose the approximation triangulations nicely to ensure that no flips occur.

The K-quasi-conformality is relatively easy to establish and is based on a ratio lemma first appeared in [36] and a non-degeneration lemma. The conditions which ensure no flips are technical and will not be addressed here. We will discuss the rigidity result in more details.

The rigidity theorem that we proved is the following,

Theorem 3.8 ([18]) *Suppose $(\mathbb{C}, \mathcal{T}, l)$ is a geometric Delaunay triangulation of an open set in the complex plane $\mathbb{C}$ such that (i) each vertex is adjacent to exactly six triangles and (ii) there exists a function $w : V(\mathcal{T}) \to \mathbb{R}_{>0}$ satisfying $l(vv') = w(v)w(v')$ for all edges vv'. Then the triangulation is the regular hexagonal triangulation, i.e., w is a constant function.*

This should be compared with Rodin-Sullivan's rigidity theorem for circle packing metric which can be stated as,

Theorem 3.9 (Rodin-Sullivan [29]) *Suppose $(\mathbb{C}, \mathcal{T}, l)$ is a geometric triangulation of an open set in the complex plane $\mathbb{C}$ such that (i) each vertex is adjacent to exactly six triangles and (ii) there exists a function $w : V \to \mathbb{R}_{>0}$ satisfying $l(vv') = w(v) + w(v')$ for all edges vv'. Then the triangulation is the regular hexagonal triangulation, i.e., w is a constant function.*

Recall that a PL metric on a triangulated surface $(S, \mathcal{T})$ can be represented by the edge length function $l : E(\mathcal{T}) \to \mathbb{R}_{>0}$ so that the triangle inequality $l(e_i) + l(e_j) >$

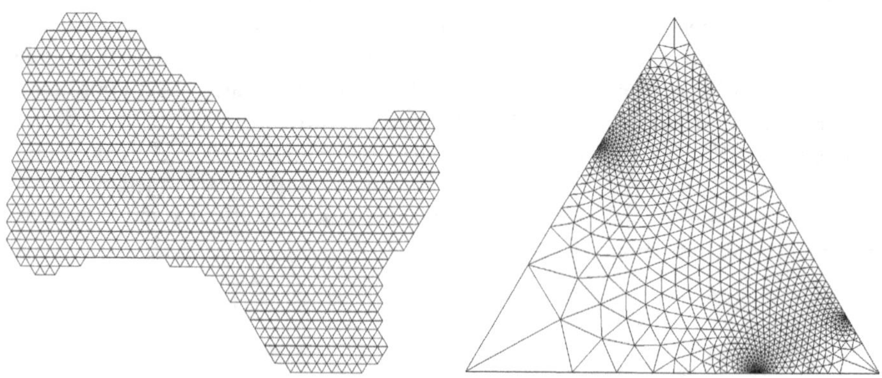

Fig. 4 Convergence of discrete conformality and approximation of the Riemann mapping

$l(e_k)$ holds for three edges e_i, e_j, e_k of a triangle. A *generalized PL metric* on $(S, \mathcal{T})$ is map $l : E(\mathcal{T}) \to \mathbb{R}_{>0}$ so that $l(e_i) + l(e_j) \geq l(e_k)$ holds for three edges e_i, e_j, e_k of a triangle. Since the edge lengths $l(e) > 0$ in a generalized PL metric, the inner angles, discrete curvatures and Delaunauy conditions are still defined for generalized PL metrics. A generalized PL metric is called *flat* if its curvature is zero at each vertex (Fig. 4).

The idea of the proof of Theorem 3.8 is as follows. Suppose otherwise that w is not a constant, we will derived a contradiction by using a maximum principle and a spiral lemma.

Let $V = \mathbb{Z} + e^{2\pi i/3}\mathbb{Z}$ be the set of vertices of the standard hexagonal triangulation $\mathcal{T}_{st}$ with $l_{st} : V \to \{1\}$ being the edge length function. Consider those $u : V \to \mathbb{R}_{>0}$ so that $u * l_{st}$ are generalized PL metrics, i.e., $u * l(v_1v_2) + u * l(v_2v_3) \geq u * l(v_3v_1)$ for vertices $\{v_1, v_2, v_3\}$ of a triangle. The maximum principle says if $u : V \to \mathbb{R}_{>0}$ is a function so that $u * l_{st}$ is a flat generalized PL metric and has a maximum point, then u is a constant. This is essentially a consequence of Lemma 3.5(a). The ratio lemma says if $u * l_{st}$ is flat, then $\frac{u(v)}{u(v')} \leq 6$ for each edge $vv' \in \mathcal{T}$. The spiral lemma says for any non-constant linear function $\ln(u) : V \to \mathbb{R}$ so that $u * l_{st}$ is a generalized PL metric, then the metric $e^u * l_{st}$ is flat and furthermore, if $u * l_{st}$ contains a triangle of positive area, then the developing map for the $u * l_{st}$ metric sends two triangles to two triangles in $\mathbb{C}$ with overlapping interiors. Using these lemmas, one proves Theorem 3.8 as follows. We may assume without loss of generality that $\lambda = \sup\{\frac{w(v)}{w(v\pm 1)} | v \in V\} > 1$. By the ratio lemma, we know $\lambda < \infty$. Choose a sequence of vertices $v_n \in V$ so that, say, $\frac{w(v_n)}{w(v_n+1)} \to \lambda$. Now using the symmetry of the lattice $\mathbb{Z} + e^{2\pi i/3}\mathbb{Z}$, we produce a new sequence of function $w_n : V \to \mathbb{R}_{>0}$ obtained by shifting $v_n \in V$ to 0 and re-scaling so that $\{w_n\}$ contains a convergent subsequence converging to $w_\infty : V \to \mathbb{R}_{>0}$. The generalized PL metric $w_\infty * l_{st}$ is still flat since flatness is a closed condition. By the maximum principle applied to the generalized flat PL metric $w'_\infty * l_{st}$ where $\alpha'(v) = \alpha(v)/\alpha(v+1) : V \to \mathbb{R}_{>0}$, we see that $w_\infty(v) = \lambda w_\infty(v+1)$ for all $v \in V$. By the same argument applied to $\delta = sup\{\frac{w(v)}{w(v \pm e^{2\pi i/3})} | v \in V\}$ and taking subsequence of the subsequence, we can improve the result to a limit function $w_\infty : V \to \mathbb{R}_{>0}$ so that $w_\infty(v) = \lambda w_\infty(v+1)$ and $w_\infty(v) = \delta w_\infty(v + e^{2\pi i/3})$ for all v. Therefore, $\ln(w_\infty) : V \to \mathbb{R}$ is a non-constant linear function. We show that there exists a triangle in $w_\infty * l_{st}$ of positive area. By the spiral lemma, the developing map for the flat generalized PL metric $w_\infty * l_{st}$ sends two triangles to two triangles with overlapping interiors. On the other hand, by the construction, $w_\infty * l_{st}$ is the limit of $w_n * l_{st}$ which is a geodesic triangulation of $\mathbb{C}$ obtained from $(\mathbb{C}, \mathcal{T}, w * l_{st})$ by shifting base points and re-scaling. In particular, the developing map D_∞ of $w_\infty * l_{st}$ is the limit of injective maps where the convergence is uniform on compact sets. This shows that D_∞ cannot send two triangles to two triangles with overlapping interiors. The contradiction shows Theorem 3.8 holds.

Our proof of Theorem 3.8 also gives a new proof of Rodin-Sullivan's Theorem 3.9 since the similar maximum principle, the ratio lemma and the spiral lemma hold in the circle packing case. The spiral lemma in the circle packing case was first discovered by Peter Doyle and the phenomena is called the Doyle spiral.

The rigidity theorem proved in [18] also holds for any lattice in $\mathbb{C}$ instead of the regular hexagonal lattice.

3.6 Discrete Uniformization for Non-compact Simply Connected Polyhedral Surfaces

An essential step in Poincaré's and Koebe's proofs of the uniformization theorem is to establish that every simply connected non-compact Riemann surface is conformal to the plane $\mathbb{C}$ or the unit disk $\mathbb{D}$. The corresponding statement for discrete uniformization is that every non-compact simply connected polyhedral surface (S, V, d) is discrete conformal to $(\mathbb{C}, V', d_{st})$ or $(\mathbb{D}, V', d_{st})$ for some discrete set $V' \subset \mathbb{C}$ or $V' \subset \mathbb{D}$ and the set V' is unique up to Möbius transformations. Here d_{st} is the standard flat Euclidean metric. Given a closed set $X \subset \mathbb{S}^2$, the convex hull of X in the hyperbolic 3-space $\mathbb{H}^3$ is denoted by $C_{\mathbb{H}}(X)$. Using geometry, discrete uniformization is equivalent to the statement that a hyperbolic metric d^* on $S - V$ (with cusp ends at each $v \in V$) is isometric to the boundary of the convex hull $\partial C_{\mathbb{H}}(V')$ or $\partial C_{\mathbb{H}}(V' \cup (\mathbb{S}^2 - \mathbb{D}))$. Furthermore, the set V' is unique up to Möbius transformations.

Recall that a closed set X in the Riemann sphere $\mathbb{S}^2$ is of *circle type* if each connected component of X is either a point or a closed round disk. For instance if V' is a discrete subset of $\mathbb{D}$, then $(\mathbb{S}^2 - \mathbb{D}) \cup V'$ is a circle type closed set. The generalization of the above discrete uniformization conjecture is the following. For any complete hyperbolic surface (Σ, g) of genus zero, there exists a circle type closed set Y, unique up to Möbius transformations, such that (Σ, g) is isometric to the boundary of the convex hull of $C_{\mathbb{H}}(Y)$ in $\mathbb{H}^3$. This can be rephrased using a theorem of Alexandrov [2] that any genus zero hyperbolic surface (Σ, g) is isometric to $\partial C_{\mathbb{H}}(X)$ for some closed set $X \subset \mathbb{S}^2$. Therefore, we have,

Conjecture 1 ([18]) *Given any closed set $X \subset \mathbb{S}^2$ with $\mathbb{S}^2 - X$ connected, there exists a circle type closed set Y such that the boundaries of $C_{\mathbb{H}}(X)$ and $C_{\mathbb{H}}(Y)$ are isometric.*

In particular, Conjecture 1 for X to be $V \cup \{\infty\}$ or $(\mathbb{S}^2 - \mathbb{D}) \cup V$ where V is discrete in $\mathbb{C}$ or $\mathbb{D}$ is the existence part of the discrete uniformization for non-compact simply connected polyhedral surfaces. In [19] we proved that

Theorem 3.10 *Conjecture 1 holds if the given closed set X has countably many connected components. In particular, the existence part of the discrete uniformization theorem holds.*

Conjecture 1 is a geometric form of the Koebe conjecture that any genus zero Riemann surface S is conformal to $\mathbb{S}^2 - Y$ for a circle type closed set Y.

Conjecture 2 ([18]) *Suppose X and Y are two circle type closed sets in* $\mathbb{S}^2$ *such that the boundary of* $C_{\mathbb{H}}(X)$ *is isometric to the boundary of* $C_{\mathbb{H}}(Y)$*. Then X and Y differ by a Möbius transformation.*

Here are some evidences supporting Conjectures 1 and 2. If the given set X is finite, Rivin [27] proved that both Conjectures 1 and 2 hold. If X is a disjoint union of a finite number of closed round disks, then Schlenker [30] proved that both Conjectures 1 and 2 hold. See also [20] for the case of a union of a closed round disk with a finite set of points. Theorem 3.8 is a very special case of Conjecture 2.

References

1. H. Akiyoshi, Finiteness of polyhedral decompositions of cusped hyperbolic manifolds obtained by the Epstein-Penner's method. Proc. Amer. Math. Soc. **129**(8), 2431–2439 (2001). electronic
2. A. D. Alexandrov, in *A. D. Alexandrov Selected Works. Part II. Intrinsic Geometry of Convex Surfaces*, ed. by S.S. Kutateladze. Translated from the Russian by S. Vakhrameyev (Chapman & Hall/CRC, Boca Raton, 2006)
3. E.M. Andreev, Convex polyhedra of finite volume in Lobacevskii space. Mat. Sb. (N.S.) **83**(125), 256–260 (1970)
4. A. Bobenko, B. Springborn, A discrete Laplace-Beltrami operator for simplicial surfaces. Discrete Comput. Geom. **38**(4), 740–756 (2007)
5. A. Bobenko, U. Pinkall, B. Springborn, Discrete conformal maps and ideal hyperbolic polyhedra. Geom. Topol. **19**(4), 2155–2215 (2015)
6. B. Chow, F. Luo, Combinatorial Ricci flows on surfaces. J. Differ. Geom. **63**(1), 97–129 (2003)
7. Y. Colin de Verdière, Un principe variationnel pour les empilements de cercles. Invent. Math. **104**(3), 655–669 (1991)
8. D.B.A. Epstein, A. Marden, *Convex Hulls in Hyperbolic Space, A Theorem of Sullivan, and Measured Pleated Surfaces*. Analytical and Geometric Aspects of Hyperbolic Space (Coventry/Durham 1984), London Math. Soc, Lecture Note Ser. 111 (Cambridge Univ. Press, Cambridge, 1987), pp. 113–253
9. F. Fillastre, Polyhedral hyperbolic metrics on surfaces. Geom. Dedicata. **134**, 177–196 (2008)
10. R. Guo, F. Luo, Rigidity of polyhedral surfaces. II. Geom. Topol. **13**(3), 1265–1312 (2009)
11. D. Gu, F. Luo, J. Sun, T. Wu, *A discrete uniformization theorem for polyhedral surfaces*. J. Diff. Geometry (2017)
12. D. Gu, R. Guo, F. Luo, J. Sun, and T. Wu, *A discrete uniformization theorem for polyhedral surfaces II*, J. Diff. Geometry (2017)
13. Z. He, *Rigidity of infinite disk patterns*. Ann. of Math. (2) **149**(1), 1–33 (1999)
14. Z. He, O. Schramm, Rigidity of circle domains whose boundary has s-finite linear measure. Invent. Math. **115**(2), 297–310 (1994)
15. P. Koebe, Kontaktprobleme der konformen Abbildung. Ber. Verh. Schs. Akad. Leipzig **88**, 141–164 (1936)
16. F. Luo, Combinatorial Yamabe flow on surfaces. Commun. Contemp. Math. **6**(5), 765–780 (2004)
17. F. Luo, Rigidity of polyhedral surfaces I. J. Differ. Geom. **96**(2), 241–302 (2014)
18. F. Luo, J. Sun, and T. Wu, *Convergence of discrete conformal geometry on polyhedral surfaces*, (2016). preprint
19. F. Luo and T. Wu, *Discrete conformal geometry and convex geometry in hyperbolic 3-space*, (2017). preprint
20. F. Luo, S. Tillmann, and T. Wu, *Rigidity of convex surfaces of finite topological type in the hyperbolic 3-space*. In preparation

21. A. Marden, B. Rodin, in *On Thurston's Formulation and Proof of Andreev's theorem*. Computational methods and function theory (Valparalo, 1989), Lecture Notes in Math. vol. 1435 (Springer, Berlin, 1990), pp. 103–115
22. L. Mosher, Tiling the projective foliation space of a punctured surface. Trans. Amer. Math. Soc. **306**(1), 1–70 (1988)
23. R. Penner, The decorated Teichmüller space of punctured surfaces. Comm. Math. Phys. **113**(2), 299–339 (1987)
24. A. V. Pogorelov, *Extrinsic Geometry of Convex Surfaces*. Translated from the Russian by Israel Program for Scientific Translations. Translations of Mathematical Monographs, Vol. 35. (American Mathematical Society, Providence, R.I., 1973). pp. vi+669
25. R. Roeder, J. Hubbard, W. Dunbar, Andreev's theorem on hyperbolic polyhedra. Ann. Inst. Fourier (Grenoble) **57**(3), 825–882 (2007)
26. I. Rivin, Euclidean structures on simplicial surfaces and hyperbolic volume. Ann. Math. (2) **139**(3), 553–580 (1994)
27. I. Rivin, *Intrinsic Geometry of Convex Ideal Polyhedra in Hyperbolic 3-Space*. Analysis, algebra, and computers in mathematical research, Lecture Notes in Pure and Appl. Math, vol. 156 (Dekker, New York, 1994), pp. 275–291
28. M. Roček, R.M. Williams, The quantization of Regge calculus. Z. Phys. C **21**(4), 371–381 (1984)
29. B. Rodin, D. Sullivan, The convergence of circle packings to the Riemann mapping. J. Differ. Geom. **26**(2), 349–360 (1987)
30. O. Schramm, Rigidity of infinite (circle) packings. J. Amer. Math. Soc. **4**(1), 127–149 (1991)
31. K. Stephenson, in *Introduction to Circle Packing*. The theory of discrete analytic functions (Cambridge University Press, Cambridge, 2005), pp. xii+356
32. J. Sun, T. Wu, and X. Zhu, *Discrete uniformization for spherical polyhedral on the 2-sphere*, (2016) preprint
33. L. Trefethen and T. Driscoll, *Schwarz-Christoffel Mapping in the Computer Era*. Proceedings of the International Congress of Mathematicians, Vol. III. Doc. Math. 1998, Extra Vol. III, (Berlin, 1998) pp. 533–542
34. M. Troyanov, Les surfaces euclidiennes à singularitïés coniques. Enseign. Math. (2) **32**(1-2), 79–94 (1986)
35. W. Thurston, *Geometry and topology of 3-manifolds*, http://library.msri.org/books/gt3m/
36. T. Wu, D. Gu, J. Sun, Rigidity of infinite hexagonal triangulation of the plane. Trans. Amer. Math. Soc. **367**(9), 6539–6555 (2015)
37. G. Ziegler, *Lectures on Polytopes, Graduate Texts in Mathematics* (Springer, New York, 1995)

The Riemann–Roch Theorem

Norbert A'Campo, Vincent Alberge and Elena Frenkel

Abstract We sketch here a proof of the Riemann–Roch theorem.

1 Introduction

The aim of this chapter is to provide the general idea of the proof of a modern version of the Riemann–Roch theorem in the case of closed Riemann surfaces of genus at least 2. This version, as the original one, combines the concepts of topology and analysis. We shall recall and apply notions of holomorphic line bundle, sheaf cohomology and divisor that are used in the statement and in the proof of this modern version.

Let us recall briefly the story of the Riemann–Roch theorem. It is Riemann who first established in [8] an inequality in order to prove the existence of non-constant meromorphic functions on closed Riemann surfaces. This inequality is now called the *Riemann inequality* and says that the dimension of the vector space of meromorphic functions on a closed Riemann surface of genus g with at most d simple poles is at least $d - g + 1$. The reader may deduce that if d is large enough, that is, $d > g$, then the existence of non-constant meromorphic functions follows. Moreover, it maybe important to add that Riemann obtained such an inequality by using the so-called *Riemann theorem* (see Theorem 17 below) which asserts that the genus can be seen as the dimension of the vector space of holomorphic 1-forms on such a Riemann surface.

N. A'Campo (✉)
Mathematisches Institut, Universität Basel, Spiegelgasse 1, 4051 Basel, Switzerland
e-mail: norbert.acampo@unibas.ch

V. Alberge
Mathematics Department, Fordham University, 441 East Fordham Road,
Bronx, NY 10458, USA
e-mail: valberge@fordham.edu

E. Frenkel
Université de Strasbourg et CNRS, IRMA, 7 rue René Descartes,
67084 Strasbourg Cedex, France
e-mail: frenkel@math.unistra.fr

L. Ji et al. (eds.), *From Riemann to Differential Geometry and Relativity*,
DOI 10.1007/978-3-319-60039-0_13

After Riemann proved his inequality, Gustav Roch, one of his students, extended this result in his doctoral thesis to an equality by giving a correction term which is the dimension of a certain subspace of holomorphic forms (cf. [9]). Furthermore, according to J. Gray in [5], the term *Riemann–Roch theorem* was first used by Brill and Noether in [4]. We also refer the reader to Chap. 2 of [10] for more descriptions and details.

First, let us give here two equivalent versions of the Riemann–Roch theorem. The notions used in the following statements will become clearer later on in this chapter.

Theorem 1 *Let X be a closed Riemann surface of genus $g \geq 2$ and $L \xrightarrow{\pi} X$ a holomorphic line bundle over X. Then*

$$\dim H^0\left(X, \underline{L}\right) - \dim H^1\left(X, \underline{L}\right) = \deg\,(L) - g + 1. \tag{1}$$

Theorem 1 highlights an equality between an analytic quantity (the left-hand side of (1)) and a topological quantity. This statement uses the formalism of sheaf theory, introduced by Leray, which was developed much later than the Riemann–Roch theorem. Before giving a (more) classical statement of this theorem, we mention that the formalism of sheaves permits to generalize this theorem to any finite dimensional compact complex manifold. This generalisation is known as the Hirzebruch–Riemann–Roch theorem (cf. [7]).[1] Furthermore, as we shall see, the analytic quantity is the index of an elliptic differential operator and then Theorem 1 (as well as the Hirzebruch–Riemann–Roch theorem) is a particular case of the so-called *Atiyah–Singer index* theorem. Now, let us state an equivalent statement.

Theorem 2 *Let X be a closed Riemann surface of genus $g \geq 2$ and D a divisor on X. Then*

$$l\,(D) - l\,(K_X - D) = \deg\,(D) - g + 1. \tag{2}$$

Theorem 2 is a more classical statement that uses the notion of divisors on a Riemann surface. It is "classical" in the sense that in most books on the subject, the Riemann–Roch theorem appears in that form. The notion of divisor originates in algebraic geometry.

This chapter is organized as follows. The main part is devoted to ideas that form the background of Theorem 1. In particular, we shall start by recalling the notion of line bundle and, based on this notion, we shall introduce sheaf cohomology. Then, we shall restate Theorem 1 (cf. Theorem 19 below) and give a sketch of its proof. After that we shall recall the notion of divisor on a Riemann surface and show the equivalence between the two theorems stated above. Finally, we shall close this chapter by explaining the importance of the Riemann–Roch theorem in Teichmüller's work. This was one of the earliest uses of this theorem in geometry.

[1] In 1957 Grothendieck gave in Princeton another generalization of Riemann–Roch's theorem that includes the result of Hirzebruch (cf. [3]). This generalisation is known as the *Grothendieck–Hirzebruch–Riemann–Roch* theorem.

We assume throughout this chapter that X is a closed Riemann surface of genus $g \geq 2$, when no other specifications are made.

2 Line Bundles

2.1 General Setting

Recall that X is a *Riemann surface* if it is a Hausdorff topological space and if there exist an open covering $\{U_i\}_i$ and a system of charts $\{\phi_i : U_i \to \phi_i(U_i) \subset \mathbb{C}\}_i$ such that for every non-empty intersection $U_i \cap U_j$ the transition functions $\phi_{ij} = \phi_j \circ \phi_i^{-1} : \phi_i(U_i \cap U_j) \to \phi_j(U_i \cap U_j)$ are biholomorphic.

Given an onto mapping $f : Y \to X$ between two spaces Y and X, then for any subset $U \subset X$, we set $Y_{|_U} = f^{-1}(U)$ and we call for any $x \in X$, $Y_{|_x} = f^{-1}(x)$ the *fiber of f at x*. We use this notation in the following definition.

Definition 3 A *holomorphic line bundle* over X is a pair (L, π), where L is a complex manifold and $\pi : L \to X$ a holomorphic surjective mapping such that the following two properties are satisfied:

(1) (Local triviality) There exist an open covering $\{V_i\}_{i\in I}$ of X and a system of biholomorphisms $\{h_i : L_{|_{V_i}} \to V_i \times \mathbb{C}\}_i$ such that the following diagram is commutative:

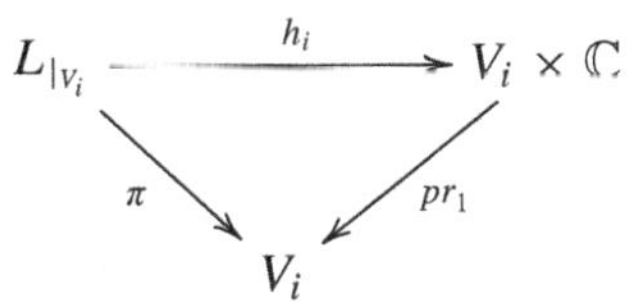

where $pr_1 : V_i \times \mathbb{C} \to V_i$ is the canonical projection onto the first coordinate.

(2) For each $p \in X$, the fiber $L_{|_p}$ has the structure of a 1-dimensional complex vector space such that the restriction of h_i to $L_{|_p}$ is a vector space isomorphism.

We denote a line bundle by $L \xrightarrow{\pi} X$, and we say that the space L is the *total space*, X the *base* and $\{V_i, h_i\}_i$ a *system of local trivializations*. Moreover, it can be easily seen that such a system implies that L is a complex surface.

Important examples of holomorphic line bundles are the trivial line bundle $X \times \mathbb{C} \xrightarrow{pr_1} X$, the holomorphic tangent bundle TX and the holomorphic cotangent bundle T^*X. In order to keep our notation close to conventions in complex geometry, let us call the cotangent bundle the *canonical line bundle*[2] over X and denote it by K_X.

[2]This label is more meaningful for the case of complex manifolds of higher dimensions. In this case, the canonical bundle is the *determinant* bundle of T^*X.

Let $\{V_i, h_i\}_i$ be a system of local trivialization of $L \xrightarrow{\pi} X$. The definition above implies that the coordinate changes are of the following form:

$$\begin{aligned} h_{ij} = h_j \circ h_i^{-1} : (V_i \cap V_j) \times \mathbb{C} &\to (V_i \cap V_j) \times \mathbb{C} \\ (p, u) &\mapsto (p, c_{ij}(p) \cdot u), \end{aligned} \tag{3}$$

where $V_i \cap V_j \ni p \mapsto c_{ij}(p) \in \mathbb{C}^*$ is holomorphic. We call $\{c_{ij}\}_{ij}$ a *system of transition functions* of $L \xrightarrow{\pi} X$ associated with the covering $\{V_i\}_i$. Moreover, we observe that such a system $\{c_{ij}\}_{ij}$ satisfies the so-called *cocycle relation*

$$c_{ij} \cdot c_{jk} = c_{ik} \tag{4}$$

whenever $V_i \cap V_j \cap V_k \neq \emptyset$. Any system of functions satisfying this relation is called a *cocycle*.

It is interesting to note that from a cocycle, we can construct a line bundle over X. Indeed, if a covering $\{V_i\}_i$ of X is given and if a system of holomorphic non-vanishing functions $\{c_{ij} : V_i \cap V_j \to \mathbb{C}^*\}_{i,j}$ satisfying Relation (4) is given, then

$$L = \left(\coprod_i (V_i \times \mathbb{C})\right) / \sim_c \tag{5}$$

defines a holomorphic line bundle over X, where for $p \in V_i$ and $q \in V_j$

$$(p, u) \sim_c (q, v) \iff p = q \text{ and } v = c_{ij}(p) \cdot u. \tag{6}$$

Let us consider some examples of line bundles using this construction. For a fixed $p \in X$, we define a line bundle Z_p in the following way. We take an open neighbourhood of $p \in X$, denoted by U_0, sufficiently small such that it is entirely contained in an open chart of X. We then consider the open covering of X formed by U_0 and $U_1 = X - \{p\}$. We set $c_{01} : U_0 \cap U_1 \ni q \mapsto z(q) \in \mathbb{C}^*$, where z is the coordinate of U_0 centered at p. Hence, Z_p is the holomorphic line bundle obtained by the cocycle $\{c_{01}\}$. Analogously, we define P_p as the line bundle determined by the same covering and the cocycle d_{01}, given by $q \mapsto \frac{1}{z(q)}$.

Another example of line bundle arises from taking the *tensor product* $L_1 \otimes L_2$ of two line bundles $L_1 \xrightarrow{\pi_1} X$ and $L_2 \xrightarrow{\pi_2} X$. Taking a sufficiently small covering, we can assume that $\{U_i\}_i$ is a covering of X which induces trivialization systems for these two line bundles. We then denote the transition functions of $L_1 \xrightarrow{\pi_1} X$ and $L_2 \xrightarrow{\pi_1} X$ that are defined on $U_i \cap U_j$ by $\{a_{ij}\}_{ij}$ and $\{b_{ij}\}_{ij}$, respectively. By (5), we define $L_1 \otimes L_2$ as the holomorphic line bundle determined by the cocycle $\{a_{ij} b_{ij}\}_{ij}$.

Let $L \xrightarrow{\pi} X$ be a line bundle whose transition functions are given by $\{c_{ij}\}_{ij}$. Then we define $L^{\otimes -1}$ as the line bundle determined by $\left\{\frac{1}{c_{ij}}\right\}_{ij}$. For example, we can then see for any $p \in X$, the line bundle P_p as $Z_p^{\otimes -1}$.

2.2 *Isomorphism and Section*

Let us turn to the definition of morphism between two line bundles over the same Riemann surface.

Definition 4 A *line bundle morphism* between two line bundles $L_1 \xrightarrow{\pi_1} X$ and $L_2 \xrightarrow{\pi_2} X$ is a mapping $f : L_1 \to L_2$ such that:

(1) f is fiber-preserving, i.e. $\pi_2 \circ f = \pi_1$,
(2) f is $\mathbb{C}$-linear on each fiber.

Moreover, f is called a *line bundle isomorphism* if f is an invertible line bundle morphism.

More concretely, if $\{a_{ij}\}_{ij}$ and $\{b_{ij}\}_{ij}$ are transition functions for $L_1 \xrightarrow{\pi_1} X$ and $L_2 \xrightarrow{\pi_1} X$, respectively, then an isomorphism f determines a set $\{f_i\}_i$ of non-vanishing holomorphic functions satisfying the following relation:

$$a_{ij}^{-1} \cdot f_i \cdot b_{ij} = f_j \iff \frac{f_i}{f_j} = \frac{a_{ij}}{b_{ij}}. \tag{7}$$

Conversely, if for transition functions $\{a_{ij}\}_{ij}$ and $\{b_{ij}\}_{ij}$, there exists a collection of non-vanishing holomorphic functions $\{f_i\}_i$ which satisfies Relation (7), then the two line bundles determined by $\{a_{ij}\}_{ij}$ and $\{b_{ij}\}_{ij}$ are isomorphic. In particular, this proves that $L \otimes L^{\otimes -1}$ is isomorphic to the trivial line bundle $X \times \mathbb{C}$.

We can equip the set of holomorphic line bundles up to isomorphism with a group structure by taking the tensor product as composition law and the equivalence class of the trivial line bundle as the identity element. This group is called the *Picard group* of X and is denoted by Pic (X). Using notions introduced in Sect. 3, we shall see that Pic (X) is isomorphic to $H^1\left(X, \mathcal{O}_X^*\right)$, the first cohomology group with values in the sheaf of holomorphic functions without zeros.

Let us now introduce the notion of section of a line bundle.

Definition 5 We say that s is a *section* of the line bundle $L \xrightarrow{\pi} X$, if $s : X \to L$ satisfies $\pi \circ s = \mathrm{id}_X$.

Since locally a line bundle looks like the trivial line bundle, the image of $p \in X$ by a section s can be thought as a pair $(p, s(p))$.

Moreover, as we did for isomorphisms between line bundles, we can describe a section in terms of a system of transition functions. Indeed, if $\left\{c_{ij}\right\}_{ij}$ is a system of transition functions for $L \xrightarrow{\pi} X$, then a section s can be seen as a collection of functions $\{s_i\}_i$ which satisfy the following relation:

$$s_j = c_{ij} s_i. \tag{8}$$

Furthermore, a section s is said to be *holomorphic* if the functions $\{s_i\}_i$ are holomorphic. We denote the set of such sections by $\Gamma_{\text{hol}}(X, L)$. In an analogous way, we can define a *meromorphic section* or a *$\mathcal{C}^\infty$-section*.

Now let us make an important remark.

Remark 6 By Relations (7) and (8), a line bundle is isomorphic to the trivial bundle if and only if it admits a holomorphic section that never vanishes.

Let us explain the meaning of the set $\Gamma_{\text{hol}}\left(X, L \otimes Z_p\right)$ for $p \in X$. Taking if needed sufficiently small trivializing open sets $\{V_i\}_i$, we can assume that V_0 is the only set of the covering that contains p and is entirely contained in an open chart. We denote by z a local coordinate on V_0 centered at p. Therefore, a holomorphic section σ of $L \otimes Z_p$ is given by a collection of holomorphic functions $\{\sigma_i\}_i$ such that for any non-empty intersection $V_i \cap V_j$ we have

$$\sigma_j = c_{ij}\sigma_i,$$

and

$$\sigma_i = c_{0i} \cdot z \cdot \sigma_0.$$

Setting $\widetilde{\sigma} = \{\widetilde{\sigma}_i\}_i$ where $\widetilde{\sigma}_0 = z \cdot \sigma_0$ and $\widetilde{\sigma}_i = \sigma_i$ for any $i \neq 0$, we get an element of $\Gamma_{\text{hol}}(X, L)$, which vanishes at p with order at least 1. Conversely, in the same way we obtain that a holomorphic section of L with zero at p defines an element of $\Gamma_{\text{hol}}\left(X, L \otimes Z_p\right)$. Hence, $\Gamma_{\text{hol}}\left(X, L \otimes Z_p\right)$ is in one-to-one correspondence with the set of holomorphic sections that vanish at p. Due to this fact, the notation Z_p for "zero at p" makes sense.

Arguing in the same manner, we show that $\Gamma_{\text{hol}}\left(X, L \otimes P_p\right)$ is in one-to-one correspondence with the set of meromorphic sections of L that have a pole at p of order at most 1.

We can introduce now the notion of degree of a line bundle, the last notion of this section.

2.3 *Degree*

Throughout this chapter, we shall meet three definitions of the degree of a line bundle $L \xrightarrow{\pi} X$. The first is the analytical one that uses a meromorphic section of L. The second one is a purely topological definition that uses the $\mathcal{C}^\infty$-structure of L. In particular, the presence of the second definition justifies the fact that on the right-hand side of the Riemann–Roch theorem (Relation (1)) we see a topological quantity. The third one arises from cohomology theory. We shall give it in Sect. 3 (cf. Relation (19)).

Let L be a line bundle and let σ a meromorphic section of L. The existence of such a section will be justified in Sect. 6. Since the section σ is meromorphic and X is supposed to be closed, it has a finite number of poles and zeros with order (or multiplicity). It then leads to the first definition of degree.

Definition 7 Given a meromorphic section σ of a line bundle $L \xrightarrow{\pi} X$, the *degree* of L, denoted by deg (L) is

$$\deg(L) = \#Z(\sigma) - \#P(\sigma), \tag{9}$$

where $Z(\sigma)$ and $P(\sigma)$ are the sets of zeros and poles of σ counted with multiplicity, respectively.

We first need to check that such a definition does not depend on the meromorphic section σ. This is clear because if δ is another meromorphic section of L, then using Relation (8) we deduce that $\frac{\sigma}{\delta}$ is a meromorphic function on X, which by compactness of X has the same number of poles and zeros.

This definition of degree implies directly that for $p \in X$

$$\deg\left(P_p\right) = -\deg\left(Z_p\right) = +1. \tag{10}$$

Indeed, it suffices to consider a meromorphic section σ of Z_p given by $\{\sigma_i\}_{i=0,1}$, where $\sigma_1 \equiv 1$ and $\sigma_0(q) = \frac{1}{z(q)}$ (where z is a coordinate of U_0 centered at p). Due to the transition function for Z_p, it is indeed a meromorphic section of Z_p. Furthermore, this section has one simple pole at p, which implies that the degree of Z_p is -1. Using the same strategy for P_p, we can set $\tau_1 \equiv 1$ on U_1 and $\tau_0(q) = z(q)$ on U_0 and observe that $\{\tau_i\}_{i=0,1}$ is a holomorphic section of P_p with a zero at p of order 1. This implies that the degree of P_p is 1.

As we mentioned before, there exists another definition of the degree of a line bundle, which involves only the differentiable structure of that line bundle. We know that if $L \xrightarrow{\pi} X$ is a holomorphic line bundle, then it can be interpreted as a $\mathcal{C}^\infty$-vector bundle of rank 2 over X, where X is considered as an orientable smooth manifold of dimension 2. The orientation is given by the complex structure of X.

Let us choose a $\mathcal{C}^\infty$-section $s : X \to L$. Such a section always exists because of the existence of partitions of unity for paracompact manifolds. We say that the image of s intersects transversally the zero section s^0 at p, if $s(p) = (p, 0)$ and $ds_p : T_pX \to \mathbb{R}^2$ is one-to-one. Furthermore, since an orientation on L is prescribed by the one of X, we assign $+1$ (resp. -1) at p if ds_p preserves (resp. reverses) the orientation.

We are now able to give the topological definition of the degree.

Definition-Theorem 8 Let $L \xrightarrow{\pi} X$, and $s : X \to L$ be a $\mathcal{C}^\infty$-section. We call *degree* of $L \xrightarrow{\pi} X$, the integer

$$\deg\ (L) = \sum_{y \in \mathrm{Im}(s) \pitchfork \mathrm{Im}\left(s^0\right)} \pm 1.$$

We admit that this definition does not depend on the choice of the transverse section.

An immediate application of this definition is

$$\deg(T_X) = \chi(X) = 2 - 2g, \tag{11}$$

which is also known as the *Poincaré-Hopf index formula.*

Moreover, let us give the idea for the passage from the meromorphic definition of the degree to the topological one. We will restrict our arguments to the case of the line bundle Z_p. For this purpose, we deal with the meromorphic section $\sigma = \{\sigma_i\}_{i=0,1}$, which was used earlier, in order to prove that the degree of Z_p is -1. Let us recall that $\sigma_1 \equiv 1$, and $\sigma_0(q) = \frac{1}{z(q)}$, where z is a local coordinate centered at p. We define a $\mathcal{C}^\infty$-section $\widetilde{\sigma}$ as follows. We modify σ_0 in order to obtain something which is well-defined at p. Since p is a pole, $|\sigma_0(q)|$ diverges to $+\infty$ as $q \to p$ and then (up to rescaling) we can assume that U_0 contains the unit disc. We set $\widetilde{\sigma}_0(q) = \overline{z(q)}$ whenever $|z(q)| \le 1$ and $\widetilde{\sigma}_0(q) = \sigma_0(q)$, otherwise. We set $\widetilde{\sigma}_1$ such that on $U_0 \cap U_1$, $\widetilde{\sigma}_1 = z \cdot \widetilde{\sigma}_0$ and on $X \backslash (U_0 \cap U_1)$, $\widetilde{\sigma}_1 \equiv 1$. It is then elementary to check that $\widetilde{\sigma} = \{\widetilde{\sigma}_i\}_{i=0,1}$ is a $\mathcal{C}^\infty$-section, which intersects transversally the zero section at p and reverses the orientation.

3 Sheaf Cohomology

This section is just a reminder about sheaf cohomology and is essentially based on Sect. 3 of Chap. 0 from [6].

3.1 General Definition

We give here the general definition of a sheaf along with a few examples.

Definition 9 A *sheaf* $\mathcal{F}$ of abelian groups on X is a mapping which associates to every non-empty open set U of X an abelian group $\mathcal{F}(U)$ such that

(1) for every pair of open sets $V \subset U$ in X there exists a group homomorphism $r_V^U : \mathcal{F}(U) \to \mathcal{F}(V)$, called the *restriction morphism*, satisfying

$$\begin{cases} r_U^U &= \mathrm{id}_U, \\ r_W^U &= r_W^V \circ r_V^U \text{ for all open sets } W \subset V \subset U \text{ in } X, \end{cases}$$

(2) for every collection of non-empty open sets $\{U_i\}_i$ of X, setting $U = \bigcup_i U_i$ we have the following:

(a) given a collection $(s_i)_i$ of $s_i \in \mathcal{F}(U_i)$, satisfying the *compatibility relations*, namely, that if for any pair (s_i, s_j) corresponding to open sets with non-empty intersection $U_i \cap U_j \neq \emptyset$ we have

$$r^{U_i}_{U_i \cap U_j}(s_i) = r^{U_j}_{U_i \cap U_j}(s_j),$$

then there exists $s \in \mathcal{F}(U)$ such that $r^U_{U_i}(s) = s_i$ for all i;

(b) if $s, t \in \mathcal{F}(U)$ are such that $r^U_{U_i}(s) = r^U_{U_i}(t)$ for all i, then $s = t$.

We call an element of $\mathcal{F}(U)$ a *section of* $\mathcal{F}$ over U.

Following the same pattern, we can define the notion of sheaves of vector spaces, modules, rings etc. In our exposition, for simplicity, we shall use the term "sheaf" for sheaf of abelian groups, as well as for sheaf of vector spaces or modules.

An important class of examples are sheaves of functions, i.e. we consider $\mathcal{F}(U)$ as a space of functions on U and the restriction morphisms r^U_V as the usual restrictions of functions to the subset V of U. The second condition in the definition of a sheaf says that given a family of locally defined functions $\{f_i\}_i$, as long as the natural conditions for existence of a global function f on U are satisfied, the global function f exists and is unique. Note that the definition of a sheaf is "neutral" in the sense that it does not require the existence of such a global function in $\mathcal{F}(U)$.

Let us introduce sheaves that we will use in our setting. We denote by $\underline{\mathbb{Z}}$ the *sheaf of locally constant functions* on X taking values in $\mathbb{Z}$, $\mathcal{E}_X$ (resp. $\mathcal{E}^*_X$) the *sheaf of smooth functions* on X with values in $\mathbb{C}$ (resp. $\mathbb{C}^*$) and analogously, $\mathcal{O}_X$ (resp. $\mathcal{O}^*_X$) the *sheaf of (non-vanishing) holomorphic functions* on X. More precisely, the sheaf $\underline{\mathbb{Z}}$ is defined by setting $\underline{\mathbb{Z}}(U) = \mathbb{Z}$, the sheaf $\mathcal{O}_X$ is defined by setting $\mathcal{O}_X(U) = \mathcal{O}(U)$, where $\mathcal{O}(U)$ denotes the set of holomorphic functions on U with values in $\mathbb{C}$ and so on.

Let us close this subsection by introducing two more sheaves. For a fixed $p \in X$, we define $\mathcal{S}_p$, the so-called *skyscraper sheaf* at p, by setting for any open set U of X

$$\mathcal{S}_p(U) = \begin{cases} \mathbb{C} & \text{if } p \in U, \\ \{0\} & \text{otherwise.} \end{cases}$$

The last but important example arises from line bundles. Let $L \xrightarrow{\pi} X$ be a line bundle. We define the *sheaf of holomorphic sections with values in* L and denote it by $\underline{L}$, by setting

$$\underline{L}(U) = \{s : U \to L \mid \pi \circ s = \mathrm{id}_U \text{ and } s \text{ holomorphic}\}.$$

3.2 Sheaf Cohomology

Let $\mathcal{F}$ be a sheaf on X. Fix $\mathcal{U} = \{U_i\}_{i \in I}$, an open covering of X. For $k \in \mathbb{N}$, we define the group $\mathcal{C}^k(\mathcal{U}, \mathcal{F})$ of *k-cochains* as the set of $s = \left\{s_{i_0 \cdots i_k}\right\}_{i_0 \cdots i_k \in I^{k+1}}$ such that

$$s_{i_0 \cdots i_k} \in \mathcal{F}\left(U_{i_0} \cap \cdots U_{i_k}\right)$$

and

$$\forall \sigma \in \mathfrak{S}_{k+1},\ s_{i_{\sigma(0)} \cdots i_{\sigma(k)}} = \epsilon(\sigma)\, s_{i_0 \cdots i_k},$$

where $\mathfrak{S}_n$ is the group of all permutations on the first n integers and $\epsilon(\cdot)$ is the signature. We define the *coboundary* operator as the group morphism

$$\check{d} : \mathcal{C}^k(\mathcal{U}, \mathcal{F}) \to \mathcal{C}^{k+1}(\mathcal{U}, \mathcal{F})$$

such that for any $s \in \mathcal{C}^k(\mathcal{U}, \mathcal{F})$

$$\left(\check{d}s\right)_{i_0 \cdots i_{k+1}} = \sum_{j=0}^{k+1} (-1)^j\, r_{U_{i_0} \cap \cdots \cap U_{i_{k+1}}}^{U_{i_0} \cap \cdots \cap \widehat{U_{i_j}} \cap \cdots \cap U_{i_{k+1}}} \left(s_{i_0 \cdots \widehat{i_j} \cdots i_{k+1}}\right).$$

For $k \in \mathbb{N}$, we introduce the group of *k-cocycles*

$$Z^k(\mathcal{U}, \mathcal{F}) = \left\{s \in \mathcal{C}^k(\mathcal{U}, \mathcal{F}) \mid \check{d}s = 0\right\},$$

and the group of *k-coboundaries*

$$B^k(\mathcal{U}, \mathcal{F}) = \begin{cases} \{0\} & \text{if } k = 0, \\ \left\{s \in \mathcal{C}^k(\mathcal{U}, \mathcal{F}) \mid \exists \tilde{s} \in \mathcal{C}^{k-1}(\mathcal{U}, \mathcal{F}) \text{ with } \check{d}\tilde{s} = s\right\} & \text{if } k \geq 1. \end{cases}$$

We can check that for all $k \in \mathbb{N}$, $B^k(\mathcal{U}, \mathcal{F}) \subset Z^k(\mathcal{U}, \mathcal{F})$. We now have all the elements for the following definition.

Definition 10 Let $k \in \mathbb{N}$. We call the k-th cohomology group of $\mathcal{F}$ relative to $\mathcal{U}$, the quotient

$$\check{H}^k(\mathcal{U}, \mathcal{F}) = Z^k(\mathcal{U}, \mathcal{F}) \,/\, B^k(\mathcal{U}, \mathcal{F}).$$

This definition depends on the choice of the covering. For this reason we define an order relation $\leq$ on the set $\mathcal{R}$ of open coverings of X. Given two elements $\mathcal{U} = \{U_i\}_i$ and $\mathcal{V} = \left\{V_j\right\}_j$ of $\mathcal{R}$, we say that $\mathcal{V} \leq \mathcal{U}$ if $\mathcal{V}$ is finer than $\mathcal{U}$, meaning that for any i, there exists j so that $U_i \subset V_j$. This relation allows us to define cohomology groups which do not depend on the choice of the covering. This definition is as follows.

Definition 11 Let $k \in \mathbb{N}$. We call the k-th *cohomology group* of $\mathcal{F}$ on X, the projective limit

$$H^k(X, \mathcal{F}) = \varinjlim_{\mathcal{U} \in \mathcal{R}} \check{H}^k(\mathcal{U}, \mathcal{F}).$$

In practice, in order to compute a cohomology group of a sheaf, it suffices to consider a sufficiently fine covering. This is justified by the following theorem that we admit.

Theorem 12 (Leray) *Let* $\mathcal{U} = \{U_i\}_{i \in I} \in \mathcal{R}$ *such that*

$$\forall k > 0 \text{ and } \forall i_0 \cdots i_q, \quad H^k\left(U_{i_0} \cap \cdots \cap U_{i_q}\right) = \{0\}.$$

Then

$$\forall k \in \mathbb{N}, \quad H^k(X, \mathcal{F}) = H^k(\mathcal{U}, \mathcal{F}). \tag{12}$$

3.3 Examples of Cohomology Groups

The easiest cohomology group to compute is the zero cohomology group. Indeed, if we consider the sheaf $\underline{L}$ associated with the line bundle $L \xrightarrow{\pi} X$, then

$$H^0\left(X, \underline{L}\right) = \Gamma_{\text{hol}}(X, L). \tag{13}$$

The cohomology groups of skyscraper sheaves $\mathcal{S}_p$ for $p \in X$ are also easy to understand. Indeed, by taking a covering of X determined by two open sets such that one of them does not contain p, we show that

$$H^0\left(X, \mathcal{S}_p\right) = \mathbb{C} \text{ and } \forall k \geq 1, \; H^k\left(X, \mathcal{S}_p\right) = \{0\}. \tag{14}$$

We can also show, using the existence of partitions of unity that for any $k > 0$,

$$H^k(X, \mathcal{E}_X) = \{0\}. \tag{15}$$

Still, again by the existence of partitions of unity, Eq. (15) remains true if one replaces $\mathcal{E}_X$ by an arbitrary sheaf of $\mathcal{E}_X$-modules.

Moreover, for any holomorphic line bundle $L \xrightarrow{\pi} X$, we define for $p \in \mathbb{N}$, the $\mathbb{C}$-vector space of smooth $(0, p)$-forms with values in L, denoted by $\Omega_X^{0,p}(L)$. We note that if $p \geq 2$, then such a vector space is reduced to 0. Moreover, such a vector space can be seen as a sheaf of $\mathcal{E}_X$-modules over X, denoted by $\underline{\Omega_X^{0,p}(L)}$ and then whenever $q \geq 1$, we have

$$H^q\left(X, \underline{\Omega_X^{0,p}(X, L)}\right) = \{0\}. \tag{16}$$

4 Further Preparations

4.1 Properties of Cohomology Groups

We can justify now that Pic (X) is canonically isomorphic to $H^1\left(X, \mathcal{O}_X^*\right)$. Indeed, Relation (5) shows that a cocycle, i.e. an element of $H^1\left(X, \mathcal{O}_X^*\right)$, determines a line bundle; and Relation (7) shows that two isomorphic line bundles define two identical cocycles up to a coboundary. In other terms, two holomorphic line bundles that are isomorphic define the same element in $H^1\left(X, \mathcal{O}_X^*\right)$. By an analogous argument we can show that $H^1\left(X, \mathcal{E}_X^*\right)$ represents the isomorphism classes (in the $\mathcal{C}^\infty$-sense) of line bundles of class $\mathcal{C}^\infty$.

Recall that a sequence of groups and group homomorphisms

$$G_0 \xrightarrow{g_0} G_1 \xrightarrow{g_1} \ldots \xrightarrow{g_{n-1}} G_n$$

is called *exact*, if Im $g_{k-1} = \ker g_k$ holds for all $k = 1 \cdots n$. A *short exact sequence* is a sequence of the form $0 \to A \xrightarrow{f} B \xrightarrow{g} C \to 0$. In this case, f is injective and g is surjective. The *long exact sequence* is a sequence with infinitely many groups and group homomorphisms. Similarly to sequences of groups, we define sequences of sheaves.

Let us recall, without proof, an important result on group cohomology.

Property 13 (Snake Lemma) *Let $\mathcal{F}$, $\mathcal{G}$ and $\mathcal{H}$ be three sheaves such that*

$$0 \longrightarrow \mathcal{F} \longrightarrow \mathcal{G} \longrightarrow \mathcal{H} \longrightarrow 0$$

is an exact sequence of sheaves. Then we have a long exact cohomology sequence, that is,

$$\begin{array}{l} \cdots \longrightarrow H^k(X,\mathcal{F}) \longrightarrow H^k(X,\mathcal{G}) \longrightarrow H^k(X,\mathcal{H}) \\ \longrightarrow H^{k+1}(X,\mathcal{F}) \longrightarrow H^{k+1}(X,\mathcal{G}) \longrightarrow H^{k+1}(X,\mathcal{H}) \longrightarrow \cdots. \end{array}$$

4.2 Chern Class and Degree of a Line Bundle

One of the immediate applications of Property 13 is the definition of the Chern class of a holomorphic line bundle. Indeed, we have the following short exact sequence

$$O \longrightarrow \underline{\mathbb{Z}} \longrightarrow \mathcal{O}_X \xrightarrow{\exp(2i\pi\cdot)} \mathcal{O}_X^* \longrightarrow 0 \tag{17}$$

that induces the following long exact sequence

$$0 \dashrightarrow H^1(X, \mathcal{O}_X) \longrightarrow H^1\left(X, \mathcal{O}_X^*\right) \xrightarrow{c} H^2\left(X, \underline{\mathbb{Z}}\right) \longrightarrow \cdots . \tag{18}$$

Thus, for any holomorphic line bundle $L \xrightarrow{\pi} X$, we can consider its class in Pic (X), denoted by $[L]$, which can be seen as an element of $H^1\left(X, \mathcal{O}_X^*\right)$. The *Chern class* of L is then

$$c_1(L) = c([L]) \in H^2\left(X, \underline{\mathbb{Z}}\right). \tag{19}$$

It is known that $H^2\left(X, \underline{\mathbb{Z}}\right)$ is identical to $\mathbb{Z}$. Recalling that $H^1(X, \mathcal{E}_X)$ represents the $\mathcal{C}^\infty$-isomorphism classes of complex line bundles over X, we define in an analogous way the Chern class of a complex line bundle over X. This is justified by the following short exact sequence

$$0 \longrightarrow \underline{\mathbb{Z}} \longrightarrow \mathcal{E}_X \xrightarrow{\exp(2i\pi\cdot)} \mathcal{E}_X^* \longrightarrow 0. \tag{20}$$

The Chern class is a topological invariant for line bundles in the sense that two complex (not necessarily holomorphic) line bundles have the same Chern class if and only if they are isomorphic in the $\mathcal{C}^\infty$ sense. Indeed, since all holomorphic functions are $\mathcal{C}^\infty$, using (15), (17) and (20), we obtain the following diagram

$$\begin{array}{ccccccc}
H^1(X, \mathcal{O}_X) & \longrightarrow & H^1\left(X, \mathcal{O}_X^*\right) & \longrightarrow & \mathbb{Z} & \longrightarrow & H^2(X, \mathcal{O}_X) \\
 & & \big\downarrow & \circlearrowleft & \big\| & & \\
0 & \longrightarrow & H^1\left(X, \mathcal{E}_X^*\right) & \longrightarrow & \mathbb{Z} & \longrightarrow & 0.
\end{array} \tag{21}$$

Finally, we admit here that the Chern class corresponds to the degree defined in Sect. 2.

4.3 Cauchy–Riemann Operator and Dolbeault Cohomology

We recall that for any holomorphic line bundle $L \xrightarrow{\pi} X$, we can define the *Cauchy–Riemann* operator, denoted by $\overline{\partial_L}$ and the operator ∂_L. With such notation, the associated exterior operator d_L is equal to $\partial_L + \overline{\partial_L}$. Some properties of the Cauchy–Riemann operator will be recalled in the following section, but one of the main properties is that this operator induces a short exact sequence of sheaves, that is,

$$0 \longrightarrow \underline{L} \longrightarrow \underline{\Omega_X^{0,0}(L)} \xrightarrow{\overline{\partial_L}} \underline{\Omega_X^{0,1}(L)} \xrightarrow{\overline{\partial_L}} 0. \tag{22}$$

The exactness of the sequence is given by the following lemma:

Lemma 14 ($\bar{\partial}$-Poincaré Lemma) *Let U be an open subset of $\mathbb{C}$ and $f \in \mathcal{C}^\infty(U)$. Then for all $p \in U$, there exist an open subset $V_p \subset U$ containing p and $g \in \mathcal{C}^\infty(V_p)$ such that*

$$\forall z \in V_p, \quad f(z) = \frac{\partial g}{\partial \bar{z}}(z).$$

Even if this theorem is well known to the reader, we shall give here a sufficiently original proof.

Proof For simplicity, we assume that $U = \mathbb{C}$. We will show this lemma only in the case of $p = 0$. Let R be the square centered at 0, whose sides are of lengths 1 and parallel to the axes. By a classical construction, we can find a $\mathcal{C}^\infty$ function ρ on $\mathbb{C}$ such that

$$\forall |z| < r_1, \ \rho(z) = 1$$

and

$$\forall |z| > r_2 > r_1, \ \rho(z) = 0$$

with $r_1 < r_2 < 1$. Identifying the parallel sides of R, we can consider $\rho \cdot f$ as a $\mathcal{C}^\infty$ function on the torus $\mathbb{S}^1 \times \mathbb{S}^1$. By general results on Fourier series we can say that

$$\forall z = x + iy \in R, \ (\rho \cdot f)(z) = \sum_{(n,m)} a_{n,m} e^{2i\pi nx} e^{2i\pi my}.$$

We then define for any $z = x + iy \in R$,

$$g(z) = a_{0,0}\bar{z} + \sum_{(n,m) \neq (0,0)} \frac{a_{n,m}}{i\pi n - \pi m} e^{2i\pi nx} e^{2i\pi my}.$$

Recalling that $\frac{\partial}{\partial \bar{z}} = \frac{1}{2}\left(\frac{\partial}{\partial x} + i\frac{\partial}{\partial y}\right)$, we have

$$\forall |z| < r_1, \ \frac{\partial g}{\partial \bar{z}}(z) = f(z)$$

and then the lemma. □

The zero cohomology group being the set of global sections, applying Property 13 to Sequence (22), we obtain

Theorem 15 (Dolbeault Theorem) *Let $L \xrightarrow{\pi} X$ be a holomorphic line bundle over X. Then*

$$H^1\left(X, \underline{L}\right) \simeq H^{(0,1)}\left(X, L\right),$$

where $H^{(0,1)}(X, L) = \Omega_X^{0,1}(X) \Big/ \overline{\partial_L}\left(\Omega_X^{0,0}(L)\right)$ is called the $(0,1)$-Dolbeault cohomology group of the line bundle $L \xrightarrow{\pi} X$.

Furthermore, we note that there exists a natural "pairing" between $\Omega_X^{0,1}(L)$ and $\Gamma_{\text{hol}}\left(X, K_X \otimes L^{\otimes -1}\right)$ defined in the following manner:

$$\begin{aligned} \Omega_X^{0,1}(L) \times \Gamma_{\text{hol}}\left(X, K_X \otimes L^{\otimes -1}\right) &\to \mathbb{C} \\ (s_1, s_2) &\mapsto \int_X s_1 \wedge s_2. \end{aligned} \tag{23}$$

Indeed, we can see an element of $\Omega_X^{0,1}(L)$ as the product $\sigma \times \omega$, where σ is a $\mathcal{C}^\infty$-section of the line bundle $L \xrightarrow{\pi} X$ and ω is a $(0,1)$-form on X. In the same manner, an element of $\Gamma_{\text{hol}}\left(X, K_X \otimes L^{\otimes -1}\right)$ can be seen as the product $\delta \times \nu$, where $\delta \in \Gamma_{\text{hol}}\left(L^{\otimes -1}\right)$ and ν is a holomorphic 1-form on X. We then have by Relation (8) and the definition of $L^{\otimes -1}$ that $\sigma \times \delta \in \mathcal{C}^\infty(X)$. Therefore, if $s_1 \in \Omega_X^{0,1}(L)$ and $s_2 \in \Gamma_{\text{hol}}\left(X, K_X \otimes L^{\otimes -1}\right)$, then locally $s_1 \wedge s_2$ can be written as $s_1(z, \overline{z}) \times s_2(z)\, d\overline{z} \wedge dz$. The map given by (23) is then well defined. Moreover, we note that if $s_1 = \overline{\partial_L}\sigma$ where $\sigma \in \Omega_X^{0,0}(L)$, then applying the Stokes theorem we have for any $s \in \Gamma_{\text{hol}}(X, L)$

$$\int_X \overline{\partial_L}\sigma \wedge s = 0.$$

Hence, the map passes to the quotient, and by Theorem 15 we obtain the well-defined map

$$H^1\left(X, \underline{L}\right) \times H^0\left(X, \underline{K_X \otimes L^{\otimes -1}}\right) \to \mathbb{C}. \tag{24}$$

We shall recall below that the operator $\overline{\partial_L}$ is Fredholm. This implies that the considered cohomology groups are of finite dimension. However, we can prove that the map (24) is non-degenerate in order to obtain

Theorem 16 (Serre Duality)

$$H^1\left(X, \underline{L}\right) \simeq H^0\left(X, \underline{K_X \otimes L^{\otimes -1}}\right).$$

Another result that we shall use for the proof of Theorem 1 is the following.

Theorem 17 (Riemann Theorem)

$$H^0\left(X, \underline{K_X}\right) \simeq \mathbb{C}^g.$$

This theorem is justified in the following way. Knowing the de Rham cohomology, we know that $H^1_{\mathrm{dR}}(X) \simeq \mathbb{R}^{2g}$. Moreover, we can show that this group is canonically isomorphic (in the real sense) to the vector space of harmonic 1-forms on X.[3] Thus, the vector space of harmonic 1-forms is isomorphic to the vector space of holomorphic 1-forms on X. The latter vector space corresponds by (13) to $H^0(X, \underline{K_X})$, and therefore Theorem 17 follows.

4.4 Index of the Cauchy–Riemann Operator

We have introduced for a line bundle $L \xrightarrow{\pi} X$, the differential operator $\overline{\partial_L}$. Theorem 15 can be restated in the form of the following exact sequence

$$\begin{array}{ccccccc} 0 \longrightarrow \Gamma_{\mathrm{hol}}(X, L) & \hookrightarrow & \Omega^{0,0}(X, L) & \xrightarrow{\overline{\partial_L}} & \Omega^{0,1}(X, L) & & \\ & & & & \downarrow & & \\ & & & & H^1(X, \underline{L}) & \longrightarrow & 0. \end{array} \tag{25}$$

In other terms, we have $\ker \overline{\partial_L} = \Gamma_{\mathrm{hol}}(L) = H^0(X, \underline{L})$ and $\operatorname{coker} \overline{\partial_L} = H^1(X, \underline{L})$. Moreover, it is known that this operator is elliptic and thus is *Fredholm*.

We recall that given two vector bundles E and F over X, a differential operator $D : \mathcal{C}^\infty(X, E) \to \mathcal{C}^\infty(X, F)$, where $\mathcal{C}^\infty(X, E)$ (resp. $\mathcal{C}^\infty(X, F)$) denotes the set of $\mathcal{C}^\infty$-sections with values in E (resp. F), is *Fredholm* if

- $\ker D$ and $\operatorname{coker} D$ are of finite dimension,
- $\mathrm{Im}(D)$ is closed.

To such an operator D, we can associate an integer called the *index* of D which is defined as follows:

$$\mathrm{Ind}\ (D) = \ker D - \operatorname{coker}\ D.$$

Moreover, and we assume it here, adding an operator of sufficiently small norm to a Fredholm operator does not change the index.

The Cauchy–Riemann operator $\overline{\partial_L}$ being a Fredholm operator, its kernel and cokernel are of finite dimension and thus by (25), we have

$$\mathrm{Ind}\ \overline{\partial_L} = \dim H^0(X, \underline{L}) - \dim H^1(X, \underline{L}). \tag{26}$$

Another important result is the following.

---

[3]In the literature, this result, also true in higher dimensions, is called the *Hodge theorem*.

Property 18 *Let $L_0 \overset{\pi_0}{\to} X$ and $L_1 \overset{\pi_1}{\to} X$ be two line bundles such that*

$$\deg(L_0) = \deg(L_1).$$

Then

$$\operatorname{Ind} \overline{\partial_{L_0}} = \operatorname{Ind} \overline{\partial_{L_1}}.$$

Indeed, if $L_0 \overset{\pi_0}{\to} X$ and $L_1 \overset{\pi_1}{\to} X$ have the same degree, then they belong to the same connected component of Pic (X). We can thus find a family depending on a parameter $\{L_t\}_{0\leq t\leq 1}$ of holomorphic line bundles of the same degree. We can thus consider for all $t \in [0, 1]$, the operator $\overline{\partial_{L_t}}$. We can show that for t sufficiently close to 0, the norm of the operator $\overline{\partial_{L_t}}$ is very close to the one of $\overline{\partial_{L_0}}$ and thus since the operators are Fredholm, the corresponding indices do not change. Choosing as needed a sufficiently fine subdivision of $[0, 1]$, we deduce Property 18.

5 The Riemann–Roch Theorem

Now we are ready to state and prove Theorem 1.

Theorem 19 *Let X be a closed Riemann surface of genus $g \geq 2$ and $L \overset{\pi}{\to} X$ a holomorphic line bundle over X. Then*

$$\operatorname{Ind} \overline{\partial_L} = \deg(L) - g + 1. \tag{27}$$

Proof We will argue by induction on the degree of the line bundle. If the holomorphic line bundle is of degree 0, then by Property 18, the index of the corresponding Cauchy–Riemann operator is the same as the one associated with the trivial bundle $X \times \mathbb{C}$. We can thus assume that L is the trivial bundle. Since X is closed and holomorphic sections with values in the trivial bundle are exactly holomorphic functions on X, they are constants. Hence, we obtain $H^0\left(X, \underline{X \times \mathbb{C}}\right) = \mathbb{C}$. Moreover, by Serre duality (see Theorem 16), $H^1\left(X, \underline{X \times \mathbb{C}}\right) \simeq H^0\left(X, \underline{K_X}\right)$ and thus by Theorem 17 we deduce that

$$\dim_{\mathbb{C}} H^0\left(X, \underline{X \times \mathbb{C}}\right) - \dim_{\mathbb{C}} H^1\left(X, \underline{X \times \mathbb{C}}\right) = 1 - g.$$

The degree of the trivial bundle being zero, the theorem is verified.

We fix from now on some $p \in X$.

Let us assume that the theorem is true for all line bundles of degree $n \geq 0$. Let $L \overset{\pi}{\to} X$ be a holomorphic line bundle of degree $n + 1$. Due to notions that were introduced in the previous sections, we have a short exact sequence of sheaves

$$0 \longrightarrow \underline{L \otimes Z_p} \longrightarrow \underline{L} \overset{v}{\longrightarrow} \mathcal{S}_p \longrightarrow 0, \tag{28}$$

where υ is the map which assigns for any open set U and any holomorphic section σ, the complex number $\sigma(p)$ if $p \in U$, and 0 otherwise. Sequence (28) induces a long exact cohomology sequence and then using Relations (14) we have

$$\text{Ind } \overline{\partial_L} = \text{Ind } \overline{\partial_{L \otimes Z_p}} + 1. \tag{29}$$

Moreover, by the additivity of the degree (or of the Chern class) we have by (10)

$$\deg\left(L \otimes Z_p\right) = \deg(L) - 1. \tag{30}$$

Applying the induction hypothesis to the line bundle $L \otimes Z_p$, we have

$$\text{Ind } \overline{\partial_{L \otimes Z_p}} = \deg\left(L \otimes Z_p\right) - g + 1$$

and then by (29) and (30) we deduce that

$$\text{Ind } \overline{\partial_L} = \deg(L) - g + 1,$$

and then the theorem.

Let us assume finally that the theorem holds for all line bundles of degree $n \leq 0$. We will argue as before, except for the fact that we shall consider for $L \xrightarrow{\pi} X$ a line bundle of degree $n - 1$, and the following short exact sequence:

$$0 \longrightarrow \underline{L} \longrightarrow \underline{L \otimes P_p} \xrightarrow{Res} \mathcal{S}_p \longrightarrow 0, \tag{31}$$

where *Res* is the mapping that assigns to every locally holomorphic section of $L \otimes P_p$ (and then to every meromorphic section with at most one simple pole at p), the residue at p. This sequence allows us to increase the index by 1 and then, again by the additivity of the degree, we conclude the proof of Theorem 19. □

We can justify now that the set $\mathcal{M}(X)$ of meromorphic functions is non-trivial. Indeed, we have:

Proposition 20 *Any Riemann surface of genus $g \geq 2$ admits an infinite number of non-constant meromorphic functions.*

In particular, any holomorphic line bundle over a Riemann surface of genus $g \geq 2$ admits an infinite number of meromorphic sections.

Proof For some arbitrary $p \in X$ and for some holomorphic line bundle $L \xrightarrow{\pi} X$, we have, taking if necessary $k \in \mathbb{N}$ big enough, that

$$\deg\left(L \otimes P_p^{\otimes k}\right) - g + 1 \geq 2$$

and then by the Riemann–Roch theorem

$$\dim H^0\left(X, \underline{L \otimes P_p^{\otimes k}}\right) \geq 2.$$

By (13), we deduce that there exist two linearly independent meromorphic sections σ_1 and σ_2 of L, which proves the second statement of the proposition. Furthermore, σ_1/σ_2 defines a meromorphic (non-constant) function on X. □

6 Divisors and the Riemann–Roch Theorem

In this section we shall restate Theorem 1 in a formalism more familiar to algebraic geometers. For this purpose, we shall recall the notion of a divisor on X and how the notions of holomorphic line bundle and divisor on X are related to each other.

We recall that a *divisor* D on X is a formal sum

$$D = \sum_{p \in X} n_p \cdot (p),$$

where $(n_p)_p \subset \mathbb{Z}$ are zero except for a finite number of points.

The set of divisors on X, denoted by Div (X), is naturally equipped with a commutative group structure.

To a divisor $D = \sum_{p \in X} n_p \cdot (p)$ we define its *degree* deg (D), as

$$\deg(D) = \sum_{p \in X} n_p.$$

We shall say that a divisor D is *effective*, if its coefficients n_p are non-negative. We shall denote this by $D \geq 0$.

The first examples of divisors on X are given by meromorphic functions on X. Indeed, if f is such a function, then

$$(f) = \sum_{p \in X} \mathrm{ord}_p(f) \cdot (p)$$

defines a divisor.

We call a *principal divisor*, a divisor that arises from a meromorphic function and we denote by Div $_P(X)$ the subgroup of such divisors. Let us recall again that by compactness of X, any principal divisor has degree 0.

There exists a "natural" correspondence between divisors on X and line bundles over X. Such a correspondence is given by the following construction. Let $D = \sum_i n_i p_i$ be a divisor on X. We set

$$L_D = \bigotimes_i P_{p_i}^{n_i} = \bigotimes_{n_p>0} P_p^{\otimes n_p} \bigotimes \bigotimes_{n_p<0} Z_p^{\otimes -n_p}, \tag{32}$$

which is a line bundle over X whose degree is exactly the same as D.

This correspondence descends to a group homomorphism between $\mathrm{Div}\,(X)\,/\mathrm{Div}_{\,P}\,(X)$ and the group of holomorphic line bundles up to isomorphism, that is, Pic (X). Actually, we shall see that thanks to the two following properties this homomorphism is an isomorphism.

Proposition 21 *L_D is isomorphic to a trivial line bundle if and only if D is a principal divisor.*

Proof Assume that L_D is isomorphic to a trivial line bundle via g. Using Relation (32) and a similar construction as what we did in order to prove Relation (10), we can construct a meromorphic section σ_D of L_D with $(\sigma_D) = D$. Therefore, $g \circ \sigma_D$ can be seen as a meromorphic function on X with zeros and poles prescribed by D, and then $D = (g \circ \sigma_D)$ is a principal divisor.

Conversely, let $D = (f)$ be a principal divisor associated with the meromorphic function f. In addition, as previously, L_D admits a meromorphic section σ_D such that $(\sigma_D) = D = (f)$, i.e. σ_D and f have the same number of zeros and poles. Then, f/σ_D defines an isomorphism between L_D and $X \times \mathbb{C}$. □

Proposition 22 *Any line bundle induces a divisor.*

Proof Let $L \xrightarrow{\pi} X$ be a line bundle. By Proposition 20, there exists a meromorphic section σ with values in this line bundle. Similarly as in the case of functions, we can assign to σ a divisor (σ) and then another line bundle, denoted by $L_{(\sigma)}$. Using the same method as previously, we show that σ induces an isomorphism between $L_{(\sigma)}$ and L. Indeed, let us assume for simplicity that σ admits a simple zero at p and a pole of order 2 at q. Taking sufficiently small open sets, we can assume that the covering $\{U_i\}_i$ induces a system of trivializations of L with transition functions $\{c_{ij}\}$ such that U_1 et U_2 are two disjoint open sets containing p and q respectively. We thus have a family of meromorphic functions $\{\sigma_i\}_i$ such that whenever $U_i \cap U_j \neq \emptyset$,

$$\sigma_j = c_{ij}\sigma_i.$$

By hypothesis on the section σ, we have $\sigma_1(z) = z\widetilde{\sigma}_1(z)$ and $\sigma_2(z) = \frac{1}{z^2}\widetilde{\sigma}_2(z)$. Let us set for all $i > 2$, $\widetilde{\sigma}_i = \sigma_{|U_i}$. We then have, by Relation (7), that the family $\{\widetilde{\sigma}_i\}$ defines an isomorphism between $L_{(\sigma)} = P_p \otimes Z_q^{\otimes 2}$ and L. □

Let us show now that Theorems 1 and 2 are equivalent. In order to do this, let us fix a divisor D on X. We define the $\mathbb{C}$-vector space

$$\mathcal{L}(D) = \{f \in \mathcal{M}(X) \mid (f) + D \geq 0\} \cup \{0\} \tag{33}$$

and we set

$$l(D) = \dim \mathcal{L}(D) . \tag{34}$$

We have a natural isomorphism between $\mathcal{L}(D)$ and $H^0\left(X, \underline{L_D}\right)$. Indeed, even if the argument that follows has already been used, it may be important to recall it. For simplicity, we suppose that $D = (p)$. We then have two open sets U_0 and U_1 that define the line bundle $L_D = P_p$. Let us consider a meromorphic function f from the set $\mathcal{L}(D)$. Therefore, f admits at most one simple pole at p, from where on U_1, $f(z) = \frac{1}{z}\widetilde{f}(z)$, with $\widetilde{f}$ holomorphic. The set of $\{g_i\}_{i=1,2}$, where $g_1 = \widetilde{f}$ and $g_2 = f_{|U_2}$ defines a holomorphic section of L_D and justifies the isomorphism. Furthermore, let us take a meromorphic 1-form ω_0, that is an element of K_X and set for any divisor D

$$l(K_X - D) = l((\omega_0) - D) .$$

This is well defined since the quotient of two meromorphic 1-forms is a meromorphic function. We then deduce that

$$l(K_X - D) = \dim H^0\left(X, \underline{L_{K_X - D}}\right) = \dim H^0\left(X, \underline{L_{K_X} \otimes L_D^{\otimes -1}}\right)$$

which by Theorem 16 proves the second version of the Riemann–Roch theorem, namely Theorem 2.

7 The Use of the Riemann–Roch Theorem in Teichmüller's Work

Let us close this chapter by writing a few words about Teichmüller's work. For this purpose we assume that X is a closed Riemann surface of genus $g \geq 0$, with n distinguished inner points, b boundary components, and k distinguished boundary points. For such a Riemann surface, Teichmüller introduced in [11] the set of *topologically determined principal regions* which is now called *Teichmüller space* and which is a non-singular covering of the so-called *moduli space*. He proved that it is a differentiable manifold and justified[4] (using extremal quasiconformal mappings) that such a space is homeomorphic to a space of *quadratic differentials*.

[4] We say "justified" because there is a gap in his justification in [11] which is filled in for closed Riemann surfaces in his other paper [13].

By using the same formalism as in Sect. 6 for the Riemann–Roch theorem, namely Theorem 2, Teichmüller first deduced in Chap. 13 (named the Rieman–Roch theorem) the (real) dimension of the Teichmüller space of a closed Riemann surface of genus $g \geq 2$ is exactly $6g - 6$, confirming an assertion[5] made by Riemann about the number of parameters which entirely caracterizes the moduli space of such a surface. Then, by a "doubling" process he deduced the dimension of the Teichmüller space of X at the last part of Chap. 21. The dimension is deduced by the following sentence (see p. 409 of the English translation):

> The difference between the maximal number of real-linearly independent quadratic differentials that have at most first-order poles at the distinguished points and the maximal number of real-linearly independent everywhere finite inverse differentials that vanish at the distinguished points is always equal to
>
> $$-6 + 6g + 2n + 3b + k.$$

Indeed, the real vector space of such quadratic differentials is homeomorphic to the Teichmüller space and the dimension of the vector space of such inverse differentials can be seen as the "number of parameters for the continuous group of conformal self-mappings" of X. This consititutes one of the first major applications of the Riemann–Roch theorem.

We finally say that Teichmüller in [12] gave a generalization of the Riemann–Roch theorem using Lie theory. See [2] for the corresponding commentary.

Acknowledgements The authors would like to express their sincere gratitude to Athanase Papadopoulos for giving them the opportunity for writing this chapter and especially for his patience, kindness, and constructive remarks. This work was partially supported by the French ANR grant FINSLER (Géométrie de Finsler et applications, ANR-12-BS01-0009). The last author acknowledges support from U.S. National Science Foundation grants DMS 1107452, 1107263, 1107367 "RNMS: GEometric structures And Representation varieties" (the GEAR Network).

References

1. N. A'Campo, L. Ji, A. Papadopoulos, On the early history of moduli and Teichmüller spaces, in *Lipman Bers, A life in Mathematics*, eds. by L. Keen, I. Kra, R.E. Rodriguéz (Amer. Math. Soc., Providence, 2015), pp. 175–261
2. N. A'Campo, A. Papadopoulos, A commentary on Teichmüller's paper Über Extremalprobleme der konformen Geometrie, in *Handbook of Teichmüller theory*, vol. VI, ed. by A. Papadopoulos (EMS Publishing House, Zürich 2016), pp. 595–600
3. A. Borel, J.-P. Serre, Le théorème de Riemann-Roch (d'après Grothendieck). Bull. Math. Soc. Fr. **86**, 97–136 (1958)
4. A. Brill, M. Noether, Über die algebraischen Functionen und ihre Anwendung in der Geometrie. Math. Ann. **7**, 269–316 (1874)
5. J.J. Gray, The Riemann–Roch theorem and geometry, 1854–1914. *Doc. Math., J. DMV* **3**, 811–822 (1998)

[5]This assertion is known as the moduli problem and for more historical details we refer the reader to [1].

6. P. Griffiths, J. Harris, *Principles of Algebraic Geometry*. Pure and Applied Mathematics (A Wiley-Interscience Publication, New York etc., 1978)
7. F. Hirzebruch, Arithmetic genera and the theorem of Riemann-Roch for algebraic varieties. Proc. Natl. Acad. Sci. USA **40**, 110–114 (1944)
8. B. Riemann, Theorie der Abel'schen functionen. J. Reine Angew. Math. **54**, 115–155 (1857)
9. G. Roch, Über die Anzahl der willkürlichen constanten in algebraischen functionen. J. Reine Angew. Math. **64**, 372–376 (1865)
10. H.P. de Saint-Gervais, *Uniformisation des Surfaces de Riemann* (Retour sur un théorème centenaire. ENS Éditions, Lyon, 2010)
11. O. Teichmüller, Extremale quasikonforme Abbildungen und quadratische Differentiale. Abh. Preuss. Akad. Wiss., Math.-Naturw. Kl. **22**, 1–197 (1939). English translation by G. Théret, Extremal quasiconformal maps and quadratic differentials, in *Handbook of Teichmüller theory*, ed. by A. Papadopoulos, vol. V (EMS Publishing House, Zürich, 2015), pp. 321–484
12. O. Teichmüller, Über Extremalprobleme der konformen Geometrie. Deutsche Math. **6**, 50–77 (1941). English translation by M. Karbe, On extremal problems of conformal geometry, in *Handbook of Teichmüller theory*, A. Papadopoulos, ed., vol. VI (EMS Publishing House, Zürich, 2016), pp. 567–594
13. O. Teichmüller, Bestimmung der extremalen quasikonformen Abbildungen bei geschlossenen orientierten Riemannschen Flächen. Abh. Preuss. Akad. Wiss., Math.-Naturw. Kl. **4**, 1–42 (1943). English translation by A. A'Campo, Determination of extremal quasiconformal mappings of closed oriented Riemann surfaces, in *Handbook of Teichmüller theory*, A. Papadopoulos, ed., vol. V (EMS Publishing House, Zürich, 2016), pp. 533–567

Metric Geometries in an Axiomatic Perspective

Victor Pambuccian, Horst Struve and Rolf Struve

Abstract In his 1854 *Habilitationsvortrag* Riemann presented a new concept of space endowed with a metric of great generality, which, through specification of the metric, gave rise to the spaces of constant curvature. In a different vein, yet with a similar aim, J. Hjelmslev, A. Schmidt, and F. Bachmann, introduced axiomatically a very general notion of plane geometry, which provides the foundation for the elementary versions of the geometries of spaces of constant curvature. We present a survey of these *absolute* geometric structures and their first-order axiomatizations, as well as of higher-dimensional variants thereof. In the 2-dimensional case, these structures were called *metric planes* by F. Bachmann, and they can be seen as the common substratum for the classical plane geometries: Euclidean, hyperbolic, and elliptic. They are endowed with a very general notion of orthogonality or reflection that can be specialized into that of the classical geometries by means of additional axioms. By looking at all the possible ways in which orthogonality can be introduced in terms of polarities, defined on (the intervals of a chain of subspaces of) projective spaces, one obtains a further generalization: the Cayley-Klein geometries. We present a survey of projective spaces endowed with an orthogonality and the associated Cayley-Klein geometries.

V. Pambuccian (✉)
School of Mathematical and Natural Sciences, Arizona State University — West Campus, P. O. Box 37100, Phoenix, AZ 85069-7100, USA
e-mail: pamb@asu.edu

H. Struve
Institut für Mathematikdidaktik, Universität zu Köln, Gronewaldstraße 2, 50931 Cologne, Germany
e-mail: h.struve@uni-koeln.de

R. Struve
SIGNAL IDUNA Gruppe, Joseph-Scherer-Straße 3, 44139 Dortmund, Germany
e-mail: rolf.struve@arcor.de

L. Ji et al. (eds.), *From Riemann to Differential Geometry and Relativity*,
DOI 10.1007/978-3-319-60039-0_14

1 Introduction

With his *Habilitationsvortrag* of 1854, Riemann opened up a great number of vistas. Its first paragraph indicates Riemann's disagreement with the conventional, Euclidean approach to the foundations of geometry:

> Bekanntlich setzt die Geometrie sowohl den Begriff des Raumes, als die ersten Grundbegriffe für die Constructionen im Raume als etwas Gegebenes voraus. Sie giebt von ihnen nur Nominaldefinitionen, während die wesentlichen Bestimmungen in Form von Axiomen auftreten. Das Verhältniss dieser Voraussetzungen bleibt dabei in Dunkeln; man sieht weder ein, ob und in wie weit ihre Verbindung nothwendig, noch a priori, ob sie möglich ist.[1]

Here we find, for the first time (as noticed by Ferreirós [21, p. 69]), a statement of the requirement that the *consistency* of an axiom system be proved ("ob ihre Verbindung [die Verbindung der Voraussetzungen] möglich ist" ("whether their [the assumptions'] association is possible"). Expressed in a modern language unavailable in his time, Riemann would ask, given axioms $\alpha_1, \ldots, \alpha_n$, for a proof that their *Verbindung* ("association") $\alpha_1 \wedge \ldots \wedge \alpha_n$ is *satisfiable*. There is a requirement of the *independence* of the axioms implicit in the phrase "ob [...] ihre Verbindung nothwendig ist" ("whether [...] their association is necessary"), as well as one asking for the *structure of the models of independence* of an axiom, implicit in the request to find out "in wie weit ihre Verbindung nothwendig [ist]" ["to what extent their association is necessary"].

It will become apparent during this survey that Riemann's critique no longer applies to the modern axiomatization of geometry, that the "assumptions" have been weakened, producing a very general notion of "geometry," and that the "relations between" these "assumptions" are no longer "obscure."

The great breakthroughs the *Habilitationsvortrag* contains, pertaining to the concept of a differentiable manifold and to that of Riemannian geometry, have been analyzed in detail in [11–14, 21, 54, 82–84], and will not be referred to in the sequel.

There is no doubt in our minds regarding the visionary value of the *Habilitationsvortrag* or of its importance for the foundations of geometry. Our goal is to indicate how the *aims* Riemann had in mind when he provided a solid foundation for geometry, leading to spaces of constant curvature, can be reached from a different point of view, using advances in abstract algebra, logic, and our understanding of the structure of the Universe, none of which were present in any usable form in 1854.

Riemann's overall *aim* can be read from his critique of past attempts, and from his own proposal. What he dislikes in the old axiomatic approach is the fact that one lonely geometry appears at the end of a list of axioms, making it very hard, if not

[1]"As is well known, geometry assumes as given both the notion of space and the fundamental notions for constructions in space. If offers merely nominal definitions for these notions, whereas the essential determinations appear in the form of axioms. In the process, the relation between these assumptions remains obscure; we neither realize whether and to what extent their association is necessary, nor a priori, whether it is possible." (all translations are by V. Pambuccian).

impossible, to judge the contribution of the individual axioms. Nor is there a fundamental concept of space available, which allows, by adding new axioms, (*hypotheses* as Riemann would say, to emphasize their arbitrary nature, as opposed to *axioms*, which were thought to be *true*), to obtain a wide variety of geometries. It wasn't quite true that this did not exist at all in 1854, but it is apparent that Riemann had no knowledge of it, since he cites only Euclid and Legendre as precursors in the foundations of geometry. In §15 of his *Appendix* of 1832, J. Bolyai had defined *absolute geometry*, a geometry worthy of being considered a *scientiam spatii*, that did allow the addition of further *hypotheses* to reach more specific geometries. Riemann's aim, of starting with an n-dimensional manifold, and then adding a metric, is certainly much more bold, and offers a far more general notion of space. However, that approach, although very general from one point of view, is extraordinarily restrictive from another point of view. It stipulates that space be continuous, and be modeled by the real numbers. Riemann is very well aware that this is a *hyothesis*, i.e., that it is not a self-evident truth. He writes:

> Die Frage über die Gültigkeit der Voraussetzungen der Geometrie im Unendlichkleinen hängt zusammen mit der Frage nach dem innern Grunde der Massverhältnisse des Raumes. Bei dieser Frage, welche wohl noch zur Lehre vom Raume gerechnet werden darf, kommt die obige Bemerkung zur Anwendung, dass bei einer discreten Mannigfaltigkeit das Princip der Massverhältnisse schon in dem Begriffe dieser Mannigfaltigkeit enthalten ist, bei einer stetigen aber anders woher hinzukommen muss. Es muss also entweder das dem Raume zu Grunde liegende Wirkliche eine discrete Mannigfaltigkeit bilden, oder der Grund der Massverhältnisse ausserhalb, in darauf wirkenden bindenen Kräften, gesucht werden.[2]

This concern with the discrete in the context of "metric relations" is highly unusual for the prevailing zeitgeist. Riemann very likely expressed it due to the acknowledged influence the philosopher Johann Friedrich Herbart (1776–1841) had on his own philosophical outlook (the only other acknowledged influence being that of Gauß).[3] Among the five ideas from Herbart's works that "gave rise to many of Riemann's epoch-making speculations," ([77, p. 63]) Bertrand Russell lists Herbart's "general preference for the discrete above the continuous." ([77, pp. 62–63]).

Riemann was interested in the connection between geometry and physics, in particular the spatial structure of the Universe. The recent realization that space very likely exhibits a granular structure (see [76] for a contemporary point of view, and [35, p. 705] for the same realization a century ago, in Hilbert's words "ein homogenes Kontinuum, das die fortgesetzte Teilbarkeit zuliesse und somit das Unendlich-Kleine realiseren würde, [wird] in der Wirklichkeit nirgends angetroffen. Die unendliche

---

[2]The question of the validity of the hypotheses of geometry in the infinitely small is connected with the question of the intrinsic reasons for the metric relations of space. It is in this last question, which may still be regarded as belonging to the doctrine of space, that the remark made above finds its application, viz. that in the case of a discrete manifold, the principle of its metric relations are already contained in the very notion of this manifold, whereas in the case of a continuous manifold, this principle must come from somewhere else. Thus either the underlying reality of space must form a discrete manifold, or else we must seek the reason for its metric relations outside it, in binding forces acting upon it.

[3]See [111, 2.2.10] for more on the influence of Herbart.

Teilbarkeit eines Kontinuums ist nur eine in Gedanken vorhandene Operation, nur eine Idee, die durch unsere Beobachtungen der Natur und die Erfahrungen der Physik und Chemie widerlegt wird."[4]) makes a search for a foundation of geometry that would not be completely dependent on the real numbers desirable from this point of view as well.

Most mathematicians, in Riemann's time and at present, work inside well-established theories and objects, inside a social consensus of what is acceptable and what not. A very small minority, among whose ranks one could, in his own time, consider J. Bolyai and N. I. Lobachevsky to belong, develop a firm belief in the validity of a well-reasoned vision that does not yet have a place in the world of socially accepted mathematical practice. An idealist strain is at work here, one for which what is born of exact thought is primordial, one for which "in the beginning was the word." This approach looks at mathematics as the art of producing new insights from a few basic principles, that have been singled out as "hypotheses," central to the envisioned realm of discourse. The idealist approach originates with the ancient Greeks, and it is no wonder that some of its early practitioners in modern times, in particular Russell and Hilbert, were under their spell.

As the author of the *Habilitationsvortrag*, Riemann is an accomplished idealist. He emphasizes the hypothetical character of the assumptions, finds it necessary to justify at length the choice of continuous rather than discrete scales, and—much like Lobachevsky and Bolyai before him—has no problem believing in the truth of his "Riemannian manifolds" more than 100 years before they were shown to exist globally in the sense concrete mathematical practice would deem convincing, namely as submanifolds of a Euclidean space with the induced metric. That was the way 2-dimensional manifolds had been first considered by Gauß and the representation of a part of the hyperbolic plane in that manner by Beltrami in 1868 was the turning point eventually leading to the social acceptance of hyperbolic geometry.

Few abstractly existing entities, without a concrete model, had been put forward before (such as complex numbers), and certainly none of that level of complexity.

Approaches that were not accessible in 1854 were those involving *groups*, whose abstract concept had only appeared that same year, in Cayley's papers. Nor was there any awareness of the complexities involved in providing a foundation ex nihilo for the real numbers. With our current knowledge, we know that the real numbers, if they are to receive a foundation from the ground up, in the idealist manner, require the language of set theory and its axioms. As Skolem had emphasized since 1923, and as has become commonly accepted (see [18]) for the past 70 years or so, in the words of J. Ferreirós,"if we are interested in producing an axiomatic system, we can *only* use first-order logic." [20, p. 472]) And within first-order logic, the real numbers

[4]A homogeneous continuum, that would allow indefinite divisibility and would thus achieve the infinitely small, cannot be encountered anywhere in nature. The infinite divisibility of the continuum is an operation existing only in thought, only an idea, which is refuted by our observations of nature and by the experience drawn from physics and chemistry.

cannot be axiomatized on the basis of addition, multiplication, some constants, and the order relation. They do require all of set theory, which means the foundation for all of mathematics.

Riemann had no reason to doubt the fundamental nature of the real numbers, as their only competition in the world of *Grössenbegriffe*, which Riemann mentions when referring to discrete or continuous *Bestimmungsweisen*, were the natural numbers. In that foundationally innocent time before 1854, the notion of a field was non-existent, nor was there any doubt that Archimedeanity is a fundamental attribute of any *Grössenbegriff*.

We should mention from the start of our alternative approach—one that bears no direct relation to Riemann's *Habilitationsvortrag*—that no elementary account of the foundations of geometry (i.e., no first-order axiomatization) can ever hope to provide a foundation for the bewildering variety of Riemannian manifolds, i.e., to have all of them among its models. What we *can* do inside our elementary approach is to do justice to Riemann's desire of starting with a very general notion of space that allows, through a step by step process (often referred to as a *Stufenaufbau* in German), for the foundation of the essentially geometric scaffolding of spaces of constant curvature (the differential geometric story of which has been told elsewhere, see [114]).

Despite the apparent differences in these two approaches, there are historical connections inexorably leading from Riemann to Hilbert, and then to Hjelmslev, to Hilbert's student Arnold Schmidt, and finally to Bachmann. In the fourth appendix to his *Grundlagen der Geometrie*, Hilbert starts with "the studies of Riemann and Helmholtz on the foundations of geometry",[5] which have led ("veranlaßten") Lie to approach the problem by using the concept of a group. He then proposes his own version of what we call today the Riemann-Helmholtz-Lie space problem in both topological and group-theoretical terms. On the last page of that appendix, Hilbert points out that the difference between the approach mixing topology with group theory and that of the "main part" ("Hauptteil") of the book lies in the place occupied by the axiom of continuity in the scaffolding of the axiom system. In the fourth appendix *Über die Grundlagen der Geometrie*, continuity is assumed from the start, so comes first, whereas in the axiom system of the *Grundlagen der Geometrie* it comes last, to allow for a continuity-independent development of elementary geometry. Bachmann's approach, which will be followed closely in this survey, can be seen as combining the group-theoretical aspect present in Hilbert's fourth appendix, and originating in the Riemann-Helmholtz-Lie space problem, with the continuity-independent approach found in the elementary foundation of geometry—along ancient Greek lines going back to Aristotle's *Posterior Analytics*—in the "main part" of the *Grundlagen der Geometrie*. It was precisely the Aristotelian approach that had been completely forgotten, and was nowhere present in the mathematics of the first half of the 19th century. Up until the modern axiomatization of arithmetic and geometry, the zeitgeist was one of the belief in the unity of all mathematics, in the spirit of Plato (see [46]). There was no sense to be made out of Aristotle's

[5]Die Untersuchungen von *Riemann* und *Helmholtz* über die Grundlagen der Geometrie.

Posterior Analytics, A 7, 75a38-b20, the first half of which reads (in the translation of Theophilos Kouremenos):

> It follows that it is impossible to prove something by passing to it from another kind, e. g. to prove a geometrical truth with arithmetic. For there are three elements in demonstration: what is proved, the conclusion (which is an attribute belonging to a kind in itself); the axioms (which are premises of the proof); third, the underlying kind whose attributes and properties that hold of it in itself are revealed by the demonstration. The axioms, which are premises of demonstration, may be identical in two or more sciences: in the case of two different kinds such as arithmetic and geometry, however, you cannot fit arithmetical demonstration to the attributes of magnitudes, unless the magnitudes in question are numbers; how this is possible in certain cases I will explain later. Arithmetical proof always has its own kind, and so do the proofs in the other sciences. Thus, if a proof is to cross from one science to another, the kind must be the same absolutely or to some extent. Otherwise crossing is evidently impossible since the extreme and the middle terms must come from the same kind; for, if they do not hold in themselves, they hold incidentally.

At the start of the modern axiomatic approach we have Pasch, who in 1882 provided a modern axiomatic foundation for ordered geometry (see [67] for details on their possible axiomatizations). These can be considered, in a certain sense, the elementary version of differentiable manifolds, given that there is only a topology present (the one induced by the order relation), but no metric, i.e., no notion of orthogonality or congruence. In dimensions ≥ 3, these spaces are, however, much more rigidly structured than manifolds, given that they have to be Desarguesian, i.e., that they can be embedded in projective spaces over ordered skew fields. In trying to mimic Riemann's approach, one could start with ordered spaces and then add a notion of orthogonality or congruence to obtain elementary versions of spaces of constant curvature. This would roughly correspond to the approach present in Hilbert's *Grundlagen der Geometrie* of 1899. There the axioms are divided into groups. The first group consists of incidence axioms, the second group of order axioms, the third group of congruence axioms. Taken together, the three groups axiomatize an elementary (i.e., first-order) version of J. Bolyai's *absolute geometry*.

We will follow instead a different *Stufenaufbau*, that starts with a bare bones orthogonality structure, in which there is neither order nor the various forms of *free mobility* that Riemann asks of his geometry, nor the possibility of embedding the structures in Euclidean spaces over the real numbers. This originates, to a certain extent, in Hilbert's work (see also [68]), for he states, in the conclusion of his *Grundlagen der Geometrie*, that he was led throughout by the fundamental principle (*Grundsatz*):

> eine jede sich darbietende Frage in der Weise zu erörtern, daß wir zugleich prüften, ob ihre Beantwortung auf einem verschiedenen Wege mit gewissen eingeschränkten Hilfsmitteln möglich ist.[6]

[6]To treat any question that might arise in a manner which also allowed us to check whether its answer is possible by a different route with certain restricted means.

This purely metric[7] treatment of geometry started with the 2-dimensional case, and is due, in large measure, to J. Hjelmslev. It was he who, in [36], had the deep insight that line-reflections have certain properties that are independent of any assumption regarding parallels, and thus *absolute*. Line-reflections—and with them the crucial *three-reflection theorem*, stating that the composition of three reflections in lines which have a common perpendicular or a common point must be a line-reflection—had been the subject of earlier studies, such as [31, 33, 89, 113]. However, in these works, line-reflections were treated inside the particular geometry at hand (Euclidean, hyperbolic, or elliptic), and not independently of it, as they were by Hjelmslev, who carried on this line of research in [37]. Many more geometers—whose contributions are chronicled in [7, 38]—have helped build up geometry in terms of line-reflections. Their work helped remove order or free mobility assumptions. What is left after the removal work was done consists of the three-reflections theorem, beside very basic axioms stating that there are at least two points, that there is exactly one line incident with two distinct points, that perpendicular lines intersect, and that through every point there is a perpendicular to any line, which is unique if the point and the line are incident. The final touch in carving this austere axiom system came from Bachmann [4], who showed that two axioms from the axiom system of Hilbert's student Schmidt [80] are superfluous.

Later, several of Bachmann's students and other geometers extended the reflection-geometric axiomatization to higher-dimensional and to dimension-free geometries. It is these geometries that we consider to be the elementary (first-order) counterpart of Riemannian manifolds. They share the following characteristics: (i) they are both defined as abstract structures, which can be shown—with great effort—to be embeddable in some Euclidean space (in the case of Riemannian manifolds) or in some projective-metric space (in the case of reflection geometries); (ii) they both allow the definition of a notion of orthogonality (in the case of Riemannian manifolds on the tangent space of each point), defined by a bilinear symmetric map (which is given a priori in the Riemannian case, while it is discovered through the hard work of a representation theorem in the case of reflection geometries). Since the symmetric bilinear map is left unspecified, except for the restriction that the radical (orthogonal complement) of the quadratic space it determines be ≤ 1, in the case of reflection geometries, the notion of space thus created is one of wide generality.

Much like in the case of Riemannian manifolds, some of which were known in the 2-dimensional case as surfaces in 3-dimensional Euclidean space, variants of the reflection-geometrically defined geometries had been studied earlier as inhabitants of projective-metric spaces. Understanding a metric geometry inside a projective space originated in the discovery of Cayley [15] and Klein [42] that projective geometry allows the introduction of metric concepts. By distinguishing an absolute figure (the *absolute*) in a real projective manifold, they were able to introduce a projective measure (*Maßbestimmung*). Metrical properties became properties of the relation of

[7]Throughout this paper *metric* will always refer to a structure with an orthogonality relation or in which one such relation can be defined. It is in no way related to metrics defined as distances with real values.

a figure to the absolute and the projective *Maßbestimmung* "blazed a convenient road through [the] jungle undergrowth of Lobachevsky's computations."[8]

In the Euclidean case the absolute is a degenerate imaginary conic, consisting of a couple of complex points, called "the circular points at infinity" (see [92, II §8] or [42]). If the two complex circular points are replaced by a real non-degenerate conic then the associated geometry is the geometry of Bolyai and Lobachevsky, commonly referred to, following Klein, as *hyperbolic*. If the absolute non-degenerate conic is imaginary, then the associated geometry is, again following Kleinian terminology, *elliptic*. Elliptic geometry is the spherical geometry of Riemann if antipodal points are identified, so that any two points have a unique joining line. The incidence structure of an elliptic plane, i.e., an elliptic plane in which one "forgets" the metric structure, is a projective plane. Elliptic lines are unbounded[9] but of finite length—a distinction which Riemann emphasized in section III.2 of his *Habilitationsvortrag* with the words:

> Bei der Ausdehnung der Raumconstructionen in's Unmessbargrosse ist Unbegrenztheit und Unendlichkeit zu scheiden; jene gehört zu den Ausdehnungsverhältnissen, diese zu den Massverhältnissen.[10]

Klein made a systematic analysis to determine all projective measures of a projective space and described the associated Euclidean and non-Euclidean geometries which are nowadays commonly referred to as *Cayley-Klein geometries*. He himself was initially reluctant to refer to them as *geometries*, for although they have "from a logical point of view equal rights beside Euclidean geometry",[11] "they are in part not usable for measurements in the outside world",[12] so he preferred to refer to them as *Maßbestimmungen*. With the advent of relativity theory, he changed his mind and pointed out that all the geometries underlying the newly proposed models of "space", be they Minkowski space or de Sitter space, were among the "geometries" for which he had reserved the more modest term "projective measure".

Cayley and Klein showed that both Euclidean and hyperbolic geometry are subordinate to projective geometry, and that the only difference—from a projective point of view—is in the choice of the absolute. Moreover, all Cayley-Klein geometries are independent entities in their own right, in the sense that they do not need to be considered as geometries embedded in projective geometry. The fact that, even if the

---

[8] "bahnt eine bequeme Straße durch ...[das] Urwaldgestrüpp der Lobatschefskijschen Rechnungen" [42, p. 277].

[9] In the sense that there are no boundaries to a line, that one can travel along one without ever reaching anything remotely resembling an end, or, in Euclid's own formulation, in Postulate 2 of Book I of the *Elements*, it is always possible "To produce a finite straight line continuously in a straight line.".

[10] When space-constructions are extended toward the unmeasurably large, one must distinguish between unboundedness and infinitude; the former belongs to the realm of extension, the latter to the that of measure.

[11] "stellen sich vom logischen Standpunkte aus gleichberechtigt neben die euklidische Geometrie" [42, p. 164].

[12] "da sie zum Teil nicht für Messungen in der Außenwelt verwendbar sind." [42, p. 164].

Cayley-Klein geometries are abstractly defined, they end up being embeddable in a projective space with a projective metric, is a remarkable result, referred to as the *Begründung* (*grounding*) of a geometry. It is worth mentioning that such a *Begründung* cannot proceed by constructing a model in Euclidean space, such as a sphere model of elliptic geometry, for such constructions presuppose Euclidean geometry rather than happen inside the neutrality of the projective setting.

This Cayley-Klein approach was not without its critics from an epistemological point of view. Its fundamental problem, as pointed out by Russell in his *Foundations of Geometry* [77, p. 31], was one of circularity (see also [26, Chaps. 1–3]):

> But what are projective coordinates, and how are they introduced? This question was not touched upon in Cayley's Memoir, and it seemed, therefore, as if a logical error were involved in using coordinates to define distance. For coordinates, in all previous systems, had been deduced from distance; to use any existing coordinate system in defining distance was, accordingly, to incur a vicious circle.[13]

This criticism asks one to justify in a purely geometrical manner the introduction of coordinates in geometry to validate the Cayley-Klein approach. This problem had been left unanswered for a very long time after Descartes showed that geometry can be practised inside a coordinate structure without providing reasons why synthetically given geometry can be coordinatized. In other words, the problems was to indicate how numbers or magnitudes show up in a realm like that of synthetic geometry, in which they do not belong to the vocabulary of its axiom system. This question was first answered by Schur [89, 90], and made widely known for Euclidean geometry by Hilbert in the *Grundlagen der Geometrie* with his arithmetic of line segments (*Streckenrechnung*). It is also addressed in [42, Kap. V] for the projective case to ensure that the construction of non-Euclidean geometries does not depend upon the specifically Euclidean coordinatization process. For absolute geometry, where the task is significantly more complex, it was Hjelmslev and Bachmann who provided the coordinatization of an abstractly presented geometry by means of their *calculus of reflections*. This method turns out to be applicable to all Cayley-Klein geometries.

Our aim is to survey results of what can be considered the modern axiomatic foundation of geometry. This will bring to light the little known fact that this is a field of research with its own challenging problems, rather than one of largely historical interest.

Given the axiomatic nature of our undertaking, one needs a language in which to write the axioms, and a logic by means of which to deduce consequences from those axioms. Based on the work of Skolem, Hilbert and Ackermann, Gödel, and Tarski, a consensus had been reached by the end of the first half of the 20th century that, as Skolem had emphasized since 1923, "if we are interested in producing an axiomatic system, we can only use first-order logic" (cp. [20, p. 472]).

Given that symbolic logic is not within the comfort zone of a majority of present-day mathematicians, each axiom that is phrased in formal logic is followed by a plain

---

[13]Russell's question is rhetorical in nature. He answers it on the next page, pointing out that the work of von Staudt, with its introduction of coordinates in a metric-free manner, removes all doubts regarding the independence of projective coordinates from distances.

English description of what it says. This allows the reader to skip the formal part of an axiom without losing the thread of the story.

We will proceed by first presenting the theory of metric planes in its group-theoretical axiomatization in Sect. 2.1, followed by a more traditional, synthetic geometric, axiomatization in Sect. 2.2, and a partial algebraic characterization of the models in Sect. 2.3. Next come the introduction of order and free mobility in Sects. 2.4 and 2.5, turning metric planes into Hilbert planes (planes satisfying the axioms for absolute geometry presented by Hilbert in [34]), as well as Pejas's algebraic characterization of Hilbert planes. We next pause to reflect in Sect. 2.6 on the methodological advantages of this approach, and mention the generalizations of metric planes proposed in the literature in Sect. 2.7. Sections 3 and 4 are devoted to n-dimensional and dimension-free generalizations of metric planes. Given that the orthogonality relation of metric planes or of higher-dimensional metric spaces is induced by a polarity defined on a subspace of a projective plane or space, we turn in Sect. 5 to the study of all possible orthogonality relations that are induced by polarities. The 2-dimensional case is treated in Sect. 5.1, the finite-dimensional case in Sect. 5.2. While the 2-dimensional case has a venerable history, going back to Cayley [15] and Klein [42], the higher-dimensional case has been systematically dealt with only recently in [106, 108], and offers a better understanding of the manner in which Cayley-Klein geometries, which are dealt with in Sect. 6, come into existence. The 2-dimensional case of Cayley-Klein geometries is treated in greater detail in Sect. 6.1, with a novel reflection-geometric axiomatization presented in Sect. 6.3. Remarks concerning finite plane Cayley-Klein geometries and on the connection between Cayley-Klein spaces and differential geometry can be found in Sects. 6.2 and 6.4. We append a reasonably comprehensive list of references.

2 Metric Planes

2.1 The Group-Theoretical Approach

We now present *metric planes* as they appear in [7]. There, however, they are presented as structures living inside groups generated by a set of involutions. This is not a first-order axiomatization (a fact Bachmann knew all too well, having written his thesis and done research in formal logic), but rather a convenient language in which the theory should be presented to a wider audience. That the theory could be phrased in first-order logic he no doubt knew. We choose to present the theory of metric planes in formal logic just to show that it can be done, that it is an *elementary* theory, far removed from the concept of set.[14] Our language will be a one-sorted one, with variables to be interpreted as "rigid motions," containing a unary predicate symbol G, with $G(x)$ to be interpreted as "x is a line-reflection," a constant symbol 1,

[14] The axiom system inside group theory can be found, with $n = 2$, in Sect. 3.

to be interpreted as "the identity," and a binary operation $\circ$, with $\circ(a, b)$, which we shall write as $a \circ b$, to be interpreted as "the composition of a with b."

To improve the readability of the axioms, we introduce the following abbreviations:

$$\begin{aligned} a^2 &= a \circ a, \\ \iota(g) &:\Leftrightarrow g \neq 1 \wedge g^2 = 1, \\ a \mid b &:\Leftrightarrow G(a) \wedge G(b) \wedge \iota(a \circ b), \\ J(abc) &:\Leftrightarrow \iota((a \circ b) \circ c), \\ pq \mid a &:\Leftrightarrow p \mid q \wedge G(a) \wedge J(pqa). \end{aligned}$$

Thus $\iota(g)$ stands for "g is an involutory element;" $a \mid b$ for "a and b are orthogonal lines;" $J(abc)$ stands for "a, b, and c lie in a pencil;" $pq \mid a$ stands for "the line a and the orthogonal lines p and q lie in a pencil." The axioms are (we omit universal quantifiers whenever the axioms are universal sentences):

M 1 $(a \circ b) \circ c = a \circ (b \circ c)$
M 2 $(\forall a)(\exists b)\, b \circ a = 1$
M 3 $1 \circ a = a$
M 4 $G(a) \rightarrow \iota(a)$
M 5 $G(a) \wedge G(b) \rightarrow G(a \circ (b \circ a))$
M 6 $(\forall abcd)(\exists g)\, a \mid b \wedge c \mid d \rightarrow G(g) \wedge J(abg) \wedge J(cdg)$
M 7 $ab \mid g \wedge cd \mid g \wedge ab \mid h \wedge cd \mid h \rightarrow (g = h \vee a \circ b = c \circ d)$
M 8 $\bigwedge_{i=1}^{3} pq \mid a_i \rightarrow G(a_1 \circ (a_2 \circ a_3))$
M 9 $\bigwedge_{i=1}^{3} g \mid a_i \rightarrow G(a_1 \circ (a_2 \circ a_3))$
M 10 $(\exists ghj)\, g \mid h \wedge G(j) \wedge \neg j \mid g \wedge \neg j \mid h \wedge \neg J(jgh)$
M 11 $(\forall x)(\exists ghj)\, G(g) \wedge G(h) \wedge G(j) \wedge (x = g \circ h \vee x = g \circ (h \circ j))$

Since $a \circ b$ with $a \mid b$ represents a point-reflection, we may think of an unordered pair (a, b) with $a \mid b$ as a *point*, an element a with $G(a)$ as a *line*, two lines a and b for which $a \mid b$ as a pair of *perpendicular* lines, and say that a point (p, q) is *incident* with the line a if $pq \mid a$. With these figures of speech in mind, the above axioms make the following statements: **M1**, **M2**, and **M3** are the group axioms for the operation $\circ$; **M4** states that line-reflections are involutions; **M5** states the invariance of the set of line-reflections, **M6** states that any two points can be joined by a line, which is unique according to **M7**; **M8** and **M9** state that the composition of three reflections in lines that have a common point or a common perpendicular is a line-reflection; **M10** states that there are three lines g, h, and j, such that g and h are perpendicular, j is perpendicular to neither g nor h, nor does j go through the intersection point of g and h; **M11** states that every motion is the composition of two or three line-reflections. It is this fact, that every element of the group generated by line-reflections can be written as the product of at most three line-reflections, that made the first-order axiomatization of the group of motions of a metric plane possible. Notice that, in the presence of **M11**, **M4**, **M1**, and **M3**, the statement regarding the existence of the inverse, **M2**, becomes superfluous. We have listed it nevertheless, given that **M1–M3** will be used in axiom systems that appear later. We will denote by $\mathcal{M}$ the axiom system {**M1–M11**} for metric planes.

From here on, there are two options, according to the answer the question "Is it possible for a product of an odd number of line-reflections to be the identity?" receives. If the answer is yes, which means—given that any product of an odd number of line-reflections can be reduced to a product of three line-reflections—that

El 1 $(\exists abc)\, G(a) \wedge G(b) \wedge G(c) \wedge a \circ (b \circ c) = 1$

then we have an axiom system for *elliptic planes* (the geometry first mentioned by Riemann in his *Habilitationsvortrag* as a geometry with positive constant curvature). **El1** states that the composition of three line-reflections can be the identity.

If the answer is no, meaning that ¬**El1** holds, then we have a non-elliptic metric plane. The presence of ¬**El1** ensures that the perpendicular from a point not on a line to that line is unique.

Within the theory of metric planes we can separate the *hypotheses* regarding the *nature* of the metric (Euclidean or non-Euclidean (hyperbolic, elliptic)) from those regarding *free mobility* (with subdivisions into the free mobility of points (every point-pair has a midpoint) and the free mobility of lines (every pair of intersecting lines has an angle bisector)), and from those regarding the order of the plane. These three requirements are almost completely distinct, in the sense that a metric plane may satisfy, within limits, a variety of combinations of them. The two cases in which one *hypothesis* leads to another are the case in which the metric is hyperbolic, in which the order comes for free, and the case of a Euclidean metric, in which the free mobility of points, i. e., the universal existence of midpoints, is ensured.

2.1.1 The Elliptic Case

There are simpler axiom systems for elliptic planes than $\mathcal{M} \cup \{\mathbf{El1}\}$. The first in-depth study of an axiomatization in terms of reflections for elliptic planes goes back to Baer [10]. After proving that one of Baer's axioms is superfluous and re-writing Baer's axiom system, Heimbeck [28] showed that {**M1–M3, El2, El3, El4**} is an axiom system for elliptic planes in a one-sorted language with one binary operation symbol $\circ$. The specifically elliptic axioms are:

El2 $(\forall g)(\exists i)(\forall x)\, g \neq 1 \rightarrow (\iota(i) \wedge (\iota(x) \rightarrow (\iota(x \circ g) \leftrightarrow \iota(x \circ i)))$
El3 $(\forall g)(\exists h)\, \iota(g) \rightarrow g \circ h \neq h \circ g$
El4 $(\exists g)\, g \neq 1$

El2 states that, for all elements $g \neq 1$ of the group, there is an involution i of that group, such that the set of all involutions x for which $x \circ g$ is an involution coincides with the set of all involutions x for which $x \circ i$ is an involution. **El3** states that no involution commutes with all elements of the group, and **El4** that the group is not trivial.

2.1.2 The Hyperbolic Case

Two lines a and b are called *non-connectable*, to be denoted by $\omega(a, b)$ if a and b neither intersect nor have a common perpendicular, i.e.,

$$\omega(a, b) \Leftrightarrow (\forall gh) \neg(gh \,|\, a \wedge gh \,|\, b) \wedge \neg(g \,|\, a \wedge g \,|\, b).$$

To obtain an axiom system for *hyperbolic planes* from metric planes, one just needs to add two axioms to {**M1–M11**}, namely (addition in the indices being modulo 3)

H 1 $(\exists ab)\, \omega(a, b)$
H 2 $(\forall a_1 a_2 a_3 mng) \left(\bigwedge_{i=1}^{3} mn \,|\, a_i \wedge \omega(a_i, g)\right) \rightarrow \left(\bigvee_{i=1}^{3} a_i = a_{i+3}\right)$

H1 states that there are two lines that are non-connectable. **H2** states that through a given point (m, n) there can be at most two lines a_i that are non-connectable with a given line g. The theory axiomatized by $\mathcal{M} \cup$ {**H1, H2**} was studied by Klingenberg [43], who showed that all of its models are isomorphic to Beltrami-Cayley-Klein unit disk models of hyperbolic geometry built over arbitrary ordered fields. To get to the elementary version of plane hyperbolic geometry, first axiomatized by Hilbert [33], one needs to add to $\mathcal{M} \cup$ {**H2**} an axiom stronger than **H1**, namely one that states that from a point (p_1, p_2) to a line g not through (p_1, p_2) there are two distinct lines non-connectable with g, i.e.,

H 3 $(\forall p_1 p_2 g)(\exists a_1 a_2)\, p_1 \,|\, p_2 \wedge \neg(p_1 p_2 \,|\, g) \rightarrow a_1 \neq a_2 \wedge \bigwedge_{i=1}^{2}(p_1 p_2 \,|\, a_i \wedge \omega(a_i, g))$

It forces the arbitrary ordered coordinate field of the models of $\mathcal{M} \cup$ {**H1, H2**} to be *Euclidean*, i.e., one in which all positive elements must have square roots. As shown in [43, 4.5], one can replace **H3** with the requirement that every point-pair has a midpoint

H 4 $(\forall a_1 a_2 b_1 b_2)(\exists c_1 c_2)\, a_1 \,|\, a_2 \wedge b_1 \,|\, b_2$
$\rightarrow c_1 \,|\, c_2 \wedge ((c_1 \circ c_2) \circ (a_1 \circ a_2)) \circ (c_1 \circ c_2) = b_1 \circ b_2$

to get another axiom system, $\mathcal{M} \cup$ {**H1, H2, H4**}, for Hilbert's plane elementary hyperbolic geometry. Another, simpler axiom system in terms of line-reflections and their composition can be found in [9, Satz 7].

2.1.3 The Euclidean Case

There are two particular behaviors that may be deemed as *Euclidean*. One is purely metric and can be expressed by either requiring the existence of a rectangle, i.e.,

E 1 $(\exists abcd)\, a \,|\, c \wedge b \,|\, c \wedge a \,|\, d \wedge b \,|\, d \wedge a \neq b \wedge c \neq d$

or by asking that a quadrilateral with three right angles is a rectangle,

E 2 $a \,|\, c \wedge b \,|\, c \wedge a \,|\, d \rightarrow b \,|\, d$

It turns out that, in the presence of $\mathcal{M}$, **E1** and **E2** are equivalent (see [7, p. 306]).

Alternatively, one may think of the behavior of parallels as being quintessentially *Euclidean* and ask that two distinct lines either intersect or have a common perpendicular, i.e.,

E3 $(\forall ab)(\exists mn)\, a \neq b \rightarrow (mn\,|\,a \wedge mn\,|\,b) \vee (m\,|\,a \wedge m\,|\,b)$

That **E1** (or **E2**) describe a phenomenon different from that **E3** postulates became apparent only after Dehn's [16] investigation, at Hilbert's suggestion, of the matter. Dehn found out that, even if both order and free mobility were present, **E1** and **E2** do *not* imply **E3**. On the either hand, neither is the reverse implication **E3**$\rightarrow$**E1** valid in the presence of $\mathcal{M}$ (see [7, p. 124]).

Put differently, a metric plane satisfies both **E1** and **E3** if and only if Playfair's form of the Euclidean parallel postulate—"There is exactly one line through P that does not intersect l, whenever P is a point not on the line l"—holds in it.

2.2 The Synthetic Approach

The axiom system $\mathcal{M}$ we have presented for metric planes appears to be one for its group of motions, not for the geometry itself. It turns out, however, that the information contained in the group of motions of a metric plane, in which we know which of the involutory elements are to be considered as line-reflections, contains enough information to enable the recovery of the underlying geometry. That underlying geometry can be axiomatized, as shown in [7, §2,3], in a more traditional, *synthetic*, manner, in which the individual variables are the usual *points* and *lines*, and the primitive notions are *incidence*, *line orthogonality*, and *reflections in lines*. Technically speaking, the axiom system is one inside a bi-sorted logic, given that there are two distinct kinds of variables, with points and lines to be denoted by upper-case, respectively lower-case letters of the Latin alphabet. Point-line incidence, a binary relation with point variables in the first place and line variables in the second, will be denoted by I, and we will write $P\,\mathrm{I}\,l$ instead of $\mathrm{I}\,(P, l)$. Line orthogonality, a binary relation among lines, will be denoted by $\perp$, and we will write $g \perp h$ instead of $\perp (g, h)$. Reflections in lines are binary operations—the first argument of which are line variables, whereas the second argument and its value are of the same sort (that is, both line variables or both point variables)—are denoted by σ.

An axiom system logically equivalent to $\mathcal{M}$ thus is (addition in the indices being modulo 3):

O 1 $(\exists AB)\, A \neq B$
O 2 $(\forall g)(\exists A_1A_2A_3)\, \bigwedge_{i=1}^{3} A_i \neq A_{i+1} \wedge \bigwedge_{i=1}^{3} A_i\,\mathrm{I}\,g$
O 3 $(\forall AB)(\exists^{=1} g)\, A \neq B \rightarrow A\,\mathrm{I}\,g \wedge B\,\mathrm{I}\,g$
O 4 $(\forall ab)\, a \perp b \rightarrow b \perp a$
O 5 $(\forall ab)(\exists P)\, a \perp b \rightarrow P\,\mathrm{I}\,a \wedge P\,\mathrm{I}\,b$
O 6 $(\forall Pg)(\exists h)\, P\,\mathrm{I}\,h \wedge h \perp g$

O 7 $P\,\mathrm{I}\,g \wedge P\,\mathrm{I}\,m \wedge g \perp m \wedge P\,\mathrm{I}\,n \wedge g \perp n \to m = n$
O 8 $\sigma(g, \sigma(g, h)) = h \wedge \sigma(g, \sigma(g, P)) = P$
O 9 $(P\,\mathrm{I}\,h \to \sigma(g, P)\,\mathrm{I}\,\sigma(g, h)) \wedge (m \perp n \to \sigma(g, m) \perp \sigma(g, n))$
O 10 $(\forall P g a_1 a_2 a_3)(\exists b)(\forall X x)\,[(\bigwedge_{i=1}^{3} P\,\mathrm{I}\,a_i) \vee (\bigwedge_{i=1}^{3} g \perp a_i)]$
$\to \sigma(a_1, \sigma(a_2, \sigma(a_3, x)) = \sigma(b, x) \wedge \sigma(a_1, \sigma(a_2, \sigma(a_3, X)) = \sigma(b, X)$

Here **O1** states that there are two distinct points; **O2** that every line has at least three points on it; **O3** that any two distinct points are incident with a unique line; **O4** that line-orthogonality is a symmetric relation; **O5** that orthogonal lines intersect; **O6** and **O7** that there is, through any given point P a perpendicular h to any given line g, which is unique if P is on g; **O8** states that, for each line g, the mapping $\alpha_g(\cdot)$, defined by $\alpha_g(\cdot) := \alpha(g, \cdot)$ is an involution (and thus a bijection) on the set of points and lines; **O9** states that, for any line g, σ_g preserves both incidence and orthogonality; **O10** is the three-reflection theorem, stating that the composition of reflections in three lines with a common point or a common perpendicular is a line reflection.

There is also, as shown in [64], an axiom system for metric planes that can be expressed, in terms of ∀∃-axioms (axioms in which all universal quantifiers precede all existential quantifiers) stated in a language with points and the single ternary relation of orthogonality—with $\perp(abc)$ to be read as abc is a right triangle with right angle at a—as primitive notions.

Another synthetic axiomatization, as well as one in terms of groups operating on sets (all in first-order logic) have been proposed for non-elliptic metric planes in [61] and in [52], and their logical equivalence to the group theoretic axiomatization $\mathcal{M} \cup \{\neg$**El1** was spelled out in [63].

2.3 *Algebraic Characterization*

Metric planes, being embeddable in projective planes satisfying the Pappus axiom, can be, to a certain degree, characterized algebraically. To do so we recall a few notions from analytic projective geometry.

By a *projective-metric* coordinate plane $\mathfrak{P}(K, \mathfrak{f})$ over a field K of characteristic $\neq 2$, with $\mathfrak{f}$ a symmetric bilinear form, which may be chosen to be defined by

$$\mathfrak{f}(\mathbf{x}, \mathbf{y}) = \lambda x_1 y_1 + \mu x_2 y_2 + \nu x_3 y_3, \tag{1}$$

with $\lambda\mu \neq 0$, for $\mathbf{x}, \mathbf{y} \in K^3$ (where $\mathbf{u}$ always denotes the triple (u_1, u_2, u_3), line or point, according to context), we understand a set of points and lines—the former to be denoted by (x, y, z) the latter by $[u, v, w]$ (determined up to multiplication by a non-zero scalar, not all coordinates being allowed to be 0)—endowed with a notion of incidence—point (x, y, z) being incident with line $[u, v, w]$ if and only if $xu + yv + zw = 0$—and an orthogonality of lines defined by $\mathfrak{f}$, under which lines $\mathbf{g}$ and $\mathbf{g}'$ are orthogonal if and only if $\mathfrak{f}(\mathbf{g}, \mathbf{g}') = 0$.

The reflection of a line $\mathbf{u} = [u_1, u_2, u_3]$ in a line $\mathbf{v} = [v_1, v_2, v_3]$ is the line

$$2\mathbf{v}\frac{\mathfrak{f}(\mathbf{v}, \mathbf{u})}{\mathfrak{f}(\mathbf{v}, \mathbf{v})} - \mathbf{u}.$$

Every model of a metric plane (i.e., of $\mathcal{M}$) can be represented as a *locally-complete* subplane (i.e., one containing with every point all the lines of the projective-metric plane that are incident with it) that contains the point $(0, 0, 1)$ of a projective-metric coordinate plane $\mathfrak{P}(K, \mathfrak{f})$, from which it inherits the collinearity and orthogonality relations.

The problem of conveniently describing algebraically the possible point-sets of metric planes inside projective-metric planes, also known as the *Umkehrproblem*, is hopeless in this generality.[15] For several classes of metric planes satisfying additional axioms, however, the *Umkehrproblem* was solved.

If the metric plane satisfies **E1** and **E3** (in which case it is called a *Euclidean plane*), then the point-set is precisely the affine plane over some field K of characteristic $\neq 2$ (i.e., the projective plane mentioned above, from which the line $[0, 0, 1]$ has been removed), and in (1) we have $\nu = 0$ and $\mathfrak{f}(\mathbf{x}, \mathbf{x}) \neq 0$ for $\mathbf{x} \neq \mathbf{0}$. The models can be described more conveniently in terms of a constant k, with $-k$ not a square in K, as having the point and line set of the affine plane over K, i.e., points are pairs (x, y) of elements from K, lines are triples $[u, v, w]$, point-line incidence is given by $ux + vy + w = 0$, whereas the orthogonality of the lines $[u, v, w]$ and $[u', v', w']$ is given by

$$kuu' + vv' = 0. \tag{2}$$

If a metric plane satisfies only **E1** (in which case it is called a *metric-Euclidean plane*), then it can be embedded in a Euclidean plane. There is a large literature providing alternative axiomatization of Euclidean planes [27, 57, 60, 81] and of metric-Euclidean planes [3, 7, 59], as well as a detailed description of their models.

In a metric plane which satisfies **El1**, we have $\lambda\mu\nu \neq 0$ and $\mathfrak{f}(\mathbf{x}, \mathbf{x}) = 0$ holds only for $\mathbf{x} = \mathbf{0}$ in (1).

In a metric plane which satisfies **H1**, we have that K is an ordered field, $\lambda\mu\nu \neq 0$, there is $\mathbf{x} \neq \mathbf{0}$ such that $\mathfrak{f}(\mathbf{x}, \mathbf{x}) = 0$ in (1). The points of the metric plane are all the points inside the *absolute* (which is the set of solutions of $\mathfrak{f}(\mathbf{x}, \mathbf{x}) = 0$).

2.3.1 Free Mobility

A metric plane is said to possess *free mobility* if any two intersecting lines g and h have an angle bisector w (i.e., if $(w \circ g) \circ w = h$ holds), and any two points (a_1, a_2) and (b_1, b_2) (recall that points are pairs of orthogonal lines) have a midpoint (c_1, c_2)

[15] In [7, p. 339] one finds the only known algebraic characterization and in [7, Satz 1 on p. 286] a geometric characterization of these point-sets. Both are far from the specificity obtained in the actual solution of the *Umkehrproblem* for restricted classes of metric planes.

(i.e., if $((c_1 \circ c_2) \circ (a_1 \circ a_2)) \circ (c_1 \circ c_2) = b_1 \circ b_2$ holds). The rather intricate algebraic structure of these metric planes has been described in [17].

2.4 Order

To introduce order in metric planes, we need an additional predicate, a ternary one, Z, among points, with $Z(ABC)$ standing for "B lies between A and C." To simplify the statement of the axioms, it is useful to have a name for the collinearity predicate, so we introduce the following abbreviation

$$L(ABC) \Leftrightarrow (\exists g)\, A \mathrm{I} g \wedge B \mathrm{I} g \wedge C \mathrm{I} g,$$

with $L(ABC)$ to be read as "A, B, and C are collinear points."

Z 1 *If A, B, and C are three different collinear points, then $Z(ABC)$ or $Z(BCA)$ or $Z(BCA)$.*
Z 2 *If $Z(ABC)$, then A, B, and C are collinear points.*
Z 3 *If $Z(ABC)$, then $Z(CBA)$.*
Z 4 *If $Z(ABC)$, then $Z(ACB)$ does not hold.*
Z 5 *If $Z(ACB)$ and $Z(ABD)$, then $Z(CBD)$.*
Z 6 *If $Z(CAB)$ and $Z(ABD)$, then $Z(CBD)$.*
Z 7 *If $C \neq D$, $Z(ABC)$, and $Z(ABD)$, then $Z(BCD)$ or $Z(BDC)$).*
Z 8 *For all $A \neq B$ there exists a point C such that $Z(ABC)$.*
Z 9 *If A, B, and C are three non-collinear points and D and E are two points such that $Z(ADC)$, E is such that it is neither collinear with A and C nor with D and B, then there exists a point F collinear with E and D, such that $Z(AFB)$ or $Z(BFC)$.*

Z1 ensures that any three points on any line are in some order; **Z2** that only collinear points are ordered, **Z3–Z7** are linear order axioms, **Z8** states that the order is unending. **Z9** is the Pasch axiom, stating that the line determined by D and E, which intersects the side AC of triangle ABC, must intersect one of the sides AB or BC as well. Ordered metric planes, i.e., the models of {**O1–O10**, **Z1–Z9**}, are well-understood in case the metric is Euclidean, that is, whenever the plane satisfies **E1**. There is an algebraic characterization of ordered metric planes with a non-Euclidean metric, due to Pejas [72], which is, however, not very helpful in establishing the validity of a given statement.

2.5 Order and Free Mobility

Metric planes endowed with both order and free mobility are, historically speaking, at the origin of the term *absolute*, coined by J. Bolyai. They are the models of the plane axioms of the groups I, II, and III (of incidence, order, and congruence) in the second and in all later editions of Hilbert's *Grundlagen der Geometrie*. One of the greatest achievements of the reflection-geometric foundation of geometry has been the algebraic characterization of the models of these planes, also called *Hilbert planes*. It happens to be a very useful characterization, in the sense that one can often accomplish much more and much easier with the algebraic description than with synthetic geometry.

Let K be again a field of characteristic $\neq 2$, and k an element of K, to be referred to as the *orthogonality constant* (or the *metric constant*). By the *affine-metric plane* $\mathfrak{A}(K,k)$ (cf. [32, p. 215]) we mean the projective plane $\mathfrak{P}(K)$ over the field K from which the line $[0,0,1]$, as well as all the points on it have been removed (and we write $\mathfrak{A}(K)$ for the structure with the remaining point-set, the corresponding line-set, with their incidence and orthogonality relations), for whose points of the form $(x,y,1)$ we shall write (x,y) (which is incident with a line $[u,v,w]$ if and only if $xu+yv+w=0$), together with a notion of orthogonality, the lines $[u,v,w]$ and $[u',v',w']$ being orthogonal if and only if

$$uu'+vv'+kww'=0. \tag{3}$$

If K is an ordered field, then one can order $\mathfrak{A}(K)$ in the usual way.

The algebraic characterization of the Hilbert planes consists in specifying a point-set E of an affine-metric plane $\mathfrak{A}(K,k)$, which is the universe of the Hilbert plane. The Hilbert plane will thus inherit the order relation Z from $\mathfrak{A}(K)$. We can also define a notion of congruence of two segments **ab** and **cd**, which will be given, in case $E\subset\mathfrak{A}(K,0)$, by the usual Euclidean formula

$$(a_1-b_1)^2+(a_2-b_2)^2=(c_1-d_1)^2+(c_2-d_2)^2$$

and, in case $E\subset\mathfrak{A}(K,k)$ with $k\neq 0$, by

$$\frac{F(\mathbf{a},\mathbf{b})^2}{Q(\mathbf{a})Q(\mathbf{b})}=\frac{F(\mathbf{c},\mathbf{d})^2}{Q(\mathbf{c})Q(\mathbf{d})}, \tag{4}$$

where

$$F(\mathbf{x},\mathbf{y})=k(x_1y_1+x_2y_2)+1,\;\; Q(\mathbf{x})=F(\mathbf{x},\mathbf{x}),\;\; \text{and } \mathbf{x}=(x_1,x_2), \mathbf{y}=(y_1,y_2).$$

Let now K be an ordered Pythagorean field, R the ring of *finite* elements, i.e., $R=\{x\in K:(\exists n\in\mathbf{N})\,|x|<n\}$ and P the ideal of *infinitely small* elements of K, i.e., $P=\{0\}\cup\{x\in K : x^{-1}\notin R\}$. All Hilbert planes are isomorphic to a plane of the following three types:

Type 1 $E = \{(a, b) : a, b \in M\} \subset \mathfrak{A}(K, 0)$, *where M is an R-module $\neq (0)$;*
Type 2 $E = \{(a, b) : a, b \in M\} \subset \mathfrak{A}(K, k)$ *with $k \neq 0$, where M is an R-module $\neq (0)$ included in $\{a \in K \mid ka^2 \in P\}$, that satisfies the condition*

$$a \in M \Rightarrow ka^2 + 1 \in K^2;$$

Type 3 $E = \{\mathbf{x} : Q(\mathbf{x}) > 0, Q(\mathbf{x}) \notin J\} \subset \mathfrak{A}(K, k)$ *with $k < 0$, where $J \subseteq P$ is a prime ideal of R that satisfies the condition*

$$ka^2 + 1 > 0, ka^2 + 1 \notin J \Rightarrow ka^2 + 1 \in K^2,$$

with K satisfying

$$\{a \in K : ka^2 \in R \setminus P\} \neq \emptyset.$$

The meaning of k in this context can be best described by mentioning that its sign is the same as that of $\alpha + \beta + \gamma - \pi$, where α, β, and γ are the measures of the three angles of a triangle.

Thus, a Hilbert plane is either (i) a part of an ordered Euclidean plane with free mobility (Type 1), thus a plane of Dehn-type, as these were first described in [16], or (ii) an infinitely small neighborhood of the origin in a plane with an arbitrary orthogonality constant, or (iii) a generalized Beltrami-Cayley-Klein model, consisting of the interior of a circle, with, possibly, an infinitely small collar around its circumference removed.

This characterization, due to Pejas [71], may be considered to be one of the most impressive early applications of the reflection-geometric approach. It may look strange that a first-order theory (the geometry of Hilbert planes) has models that require second-order notions (such as R and P which occur in the representation theorem) for their description. These models can, indeed, be expressed completely inside first-order logic, but we chose the original expression of Pejas for its intuitive character.

2.6 Methodological Reflections

One may, at this point, ask what is gained by this approach, other than Pejas's representation theorem, which answered an existing open question. Do metric planes, in themselves, have an interesting geometry, the way Riemannian manifolds have?

There are, indeed, a wide variety of universal statements σ, expressed solely in terms of incidence, orthogonality, and segment congruence, that are commonly encountered as theorems in Euclidean geometry, and which turn out to be either true in all metric planes or else logically equivalent to **E1** (i.e. we have either $\mathcal{M} \vdash \sigma$ or $\mathcal{M} \vdash \sigma \leftrightarrow$ **E1**). In the first category we have the theorem stating the concurrence of the altitudes of a triangle (a concurrence re-interpreted to mean that the composition of the reflections in the three altitudes is a line-reflection[16])— in fact, less is needed for that theorem, as shown in [6]—and the one stating that the medians of a triangle lie in a pencil. In the same category is the theorem stating that a triangle with two congruent medians is isosceles (see [69]). These theorems are significantly harder to prove inside the theory of metric planes than inside Euclidean geometry. However, their proofs reveal the *true* reasons for the validity of these theorems, which their better-known proofs in an affine setting obscure.

In the second category, that of statements equivalent to **E1**, we have, as shown in [66], the statement that, in a non-isosceles triangle ABC, with $AB \not\equiv AC$, with M, N and P the midpoints of AB, AC, and BC respectively, and R the midpoint of MN, the points A, R, and P are collinear. In the "not yet decided" category is a theorem that raised Hilbert's interest (see [68]), called *Three Chord Theorem*. It states that, if three circles pairwise intersect in two points, then the three lines joining those two points (to be referred in the sequel as "chords") lie in a pencil. It probably belongs to the first category, theorems true in $\mathcal{M}$. What is important, though, is the profoundly changed nature of the *questions* asked. Whereas previously the question was whether a statement σ holds in Euclidean geometry, or, more generally holds in $\mathcal{M} \cup \{\mathbf{E1}\}$, the question is now whether that statement is one of metric geometry or whether its validity is characterizing the Euclidean nature of the metric of the metric plane. There is no theorem stating that a purely metric universal statement σ, true in Euclidean geometry, must be in one of the two categories. It just so happens that this is the case for all known instances in which the question has been answered.

If we are presented with a universal statement σ, true in Euclidean geometry, but involving order, then there are more options. One of them is that the statement holds without additional assumptions regarding the nature of the metric. In that case, an additional question arises, namely whether the statement is true in all ordered metric planes, or whether it holds only inside all *standard* ordered metric planes, which are those in which the foot of the altitude to the hypotenuse of a right triangle always lies between the endpoints of the hypotenuse. While the number of universal statements involving both order and metric notions known to be true in Hilbert planes is large, there are very few known to hold in all ordered metric planes. A version of the Steiner-Lehmus theorem holds in all standard metric planes, as shown in [69]. It is very likely that all universal statements that hold in all Hilbert planes are true in all standard ordered metric planes, but no such theorem has been proved. Candidates for sentences that are likely true in all ordered metric planes are: (i) Urquhart's theorem, usually referred to as "the most 'elementary' theorem of Euclidean geometry" (see [70, 110]), when stated as a universal sentence; (ii) Gergonne's theorem, stating that the lines joining the vertices of a triangle with the points of tangency of the inscribed circle are concurrent; (iii) the Steiner-Lehmus theorem. A candidate for a universal statement which, given the theory of ordered metric planes, is equivalent to **E1**, is Morley's trisector theorem.

Another option is that σ holds in some Hilbert planes and not in others. For example, its validity may depend on the sign of the orthogonality constant k, as is the

---

[16]This kind of "concurrence" of three lines will be referred to as "the three lines lie in a pencil".

case of the Erdős-Mordell inequality, whose validity is equivalent to the statement that $k \leq 0$ (as shown in [65]). Or it may hold only in planes of Type 1 and 2, as is the case with the universal statement equivalent to Bachmann's *Lotschnittaxiom* ("A quadrangle with three right angles closes," see [5]), stated in [58]. Yet the change in perspective is the same as in the case of metric planes without order. Instead of asking for the truth of a certain statement, we are asking for its relative strength vis-à-vis the theory of ordered metric planes, for its *strength* as a *hypothesis*, in case it is not a theorem holding in all ordered metric planes.

2.7 Generalizations of Metric Planes

There are even more general notions than that of a metric plane, in which the above questions can be asked. The weakest is that of a *generalized metric plane*, whose properties are analyzed in [7, §2,4–5], and which can be considered as axiomatized by the axioms **O1–O9** (so no form of the three reflections axiom is assumed). Hardly any theorem of interest holds in them, although it is not easy to prove that a certain theorem does not hold in generalized metric planes, given that there is no useful description of their models. The theorem stating that the altitudes of a triangle lie in a pencil is known not to hold in them, as it is equivalent to the validity of the three reflections theorem for lines with a common perpendicular, as shown in [6]. Generalized metric planes that do satisfy the three reflections theorem for lines with a common perpendicular, referred to in [9] as *semi-absolute planes*, are the next stage in the hierarchy of generalizations of metric planes. It is not easy to determine which particular theorems that hold in metric planes already do so in semi-absolute planes.

Another generalization is that of the *Hjelmslev planes*, in which both the existence and the uniqueness of the line joining two points may be omitted. Their properties have been studied in [8] and they are reasonably well understood. Closely related are the plane Cayley-Klein geometries, that we will turn to in Sect. 6. A further generalization, to pre-Hjelmslev groups, can be found in [78] (see also [45]).

An independent level of generalization is that of the *S-planes*, introduced by Lingenberg, which are based on a certain relaxation of the three reflections axiom. Their properties were presented in monograph form in [48], and most theorems valid in metric planes, sometimes with slight modifications, hold in S-planes as well.

3 Higher-Dimensional Metric Spaces

The question regarding higher-dimensional analogues of metric planes was first raised for the 3-dimensional case. The first reflection-geometric axiom system, in the style of $\mathcal{M}$, was put forward in [1]. One in the style of the O-axioms in Sect. 2.2, in terms of points, planes, point-plane incidence, plane orthogonality, and reflections in planes, logically equivalent to that in [1], was put forward in [79].

Ahrens's [1] axiom system has been extended by Kinder [40] to one for n-dimensional metric spaces, for any $n \geq 2$. In analogy to the 2-dimensional case, their models can be embedded in projective-metric n-dimensional *spaces*, where the metric is given, again, by a symmetric bilinear form. We will no longer write its axioms in symbolic language, as it is by now plain how the English of the axiom systems expressed inside group theory with a distinguished set of generators can be translated into first-order logic, provided that every element of the group can be written as the composition of an a priori bounded number of generators.

The *fundamental assumption* of n-dimensional metric geometry thus on (G, S) is that G is a group (written multiplicatively) and that S is a set of involutory elements of S which generates G, and such that $bab \in S$, for all $a, b \in S$. The elements of S will be denoted by lowercase Latin letters and will be called *reflections in hyperplanes* (also referred to simply as *hyperplanes*). As before, we will write, for any two involutory elements of G, α and β, $\alpha \mid \beta$ whenever $\alpha\beta$ is involutory. We also write $\alpha_{11}, \ldots \alpha_{1n_1} \mid \alpha_{21}, \ldots, \alpha_{2n_2} \mid \ldots \mid \alpha_{m1}, \ldots \alpha_{mn_m}$ to mean that, for all $i < k$, we have $\alpha_{ij} \mid \alpha_{kl}$. An involutory product $a_1a_2 \ldots a_n$, with $a_1 \mid a_2 \mid \ldots \mid a_n$ will be referred to as a *point reflection* (or simply as a *point*), and will be denoted by uppercase Latin letters. In addition to the fundamental assumption, Kinder postulates the following:

K 1 *Given $a_1, \ldots a_{n-1}, A$, there is an a such that $a \mid a_1, \ldots, a_{n-1}, A$.*

K 2 *Given $a_1, \ldots a_{n-2}, A, B$, with $a_1 \mid \ldots \mid a_{n-2} \mid A, B$ there is an a such that $a \mid a_1, \ldots, a_{n-2}, A, B$.*

K 3 If $a_1 \mid \ldots \mid a_{n-2} \mid a, b \mid A, B$, *then $a = b$ or $A = B$.*

K 4 *Given $a_1, \ldots a_{n-2}, A, a, b, c$, with $a_1 \mid \ldots \mid a_{n-2}, A \mid a, b, c$ and $a_{n-2} \neq A$, there is a d with $ab = dc$.*

K 5 *Given $a_1, \ldots a_{n-1}, a, b, c$, with $a_1 \mid \ldots \mid a_{n-1} \mid a, b, c$, there is a d with $ab = dc$.*

K 6 *There are n hyperplane reflections $a_1, \ldots a_n$ with $a_1 \mid \ldots \mid a_n$.*

K 7 *Given $a_1, \ldots a_n$, with $a_1 \mid \ldots \mid a_n$, there is an a with $a \mid a_1, \ldots, a_{n-1}$, as well as $a \neq a_n$ and $a \nmid a_n$.*

For $n = 2$ this axiom system is equivalent to $\mathcal{M}$, and for $n = 3$ to the axiom system of Ahrens.

As in the 2-dimensional case, one can add additional axioms to specify the nature of the metric (i.e., the nature of the symmetric bilinear form). Thus two hyperplanes a and b will be called *non-connectable* if there is neither a point A with $A \mid a, b$, nor a line Γ with $\Gamma \mid a, b$. Here a *line* is a product $a_1 \ldots a_{n-1}$ of $n - 1$ many hyperplanes, with $a_1 \mid \ldots \mid a_{n-1}$.

Among the additional axioms we have

P$_n$ (Existence of a polar simplex) *There are $a_1, \ldots a_{n+1}$, with $a_1 \mid \ldots \mid a_{n+1}$.*

E$_n$ (Existence of a rectangle) *There are $a_1, \ldots a_{n-2}, a, b, c, d$, with $a_1 \mid \ldots \mid a_{n-2} \mid a, b \mid c, d$.*

H$_n$ (The hyperbolic metric axiom) *There are non-connectable hyperplanes.*

C$_n$ (The completeness axiom) *If $a_1 \mid \ldots \mid a_{n-2} \mid a, b_1, b_2, b_3, P$, as well as $b_1, b_2, b_3 \mid P$, and, for $i = 1, 2, 3$, the hyperplanes a and b_i are non-connectable, then one of $b_1 = b_2, b_2 = b_3, b_3 = b_1$ must hold.*

In the presence of the fundamental assumption, of **K1–K7** (the models of which will be referred to as *n-dimensional metric spaces*) and of $\mathbf{C}_n$: (i) adding $\mathbf{E}_n$ we get n-dimensional Euclidean geometry[17]; (2) adding $\mathbf{H}_n$ we get n-dimensional hyperbolic geometry; (3) adding $\mathbf{P}_n$ we get n-dimensional elliptic geometry (which has received an alternative axiomatization in [41]). These names are meant to express the fact that these are the n-dimensional generalizations of the 2-dimensional cases presented in Sects. 2.1.1–2.1.3 (the hyperbolic case being the n-dimensional generalization of Klingenberg's generalized hyperbolic geometry, in which, just like in the 2-dimensional case, the coordinate fields needs only be ordered). More on these geometries and those obtained in the absence of $\mathbf{C}_n$, as well as models of n-dimensional Euclidean geometries, are found in [39].

As in the 2-dimensional case, these metric spaces can be seen as subspaces of projective-metric spaces, but, just like in the 2-dimensional case (if not more so), the question of describing algebraically the possible point-sets of metric spaces within the projective-metric space is hopeless.

In the important special case in which we add free mobility axioms, the order axioms **Z1–Z8**, as well as Peano's form of the Pasch axiom (which asks that a line l that intersects the extension of side AB of a triangle ABC in D, with $Z(ABD)$, and side BC in E, must also intersect side AC in a point F) to the axiom system for metric spaces, the models are, as shown by Klopsch [44], similar to the models in Pejas's [71] characterization of models of Hilbert planes. A more in-depth analysis of the *Umkehrproblem* for metric spaces can be found in [30].

The question we raised in the 2-dimensional case, regarding the revolutionary nature of this approach, the complete change of perspective, is best illustrated with two examples.

The first looks at the following theorem of 3-dimensional Euclidean geometry: "The points of tangency of a skew quadrilateral, whose sides are tangent to a sphere, are co-planar." This statement is, as can be easily seen, one of the 3-dimensional metric space axiomatized by Ahrens and Scherf (and the $n = 3$ case of Kinder's axiom presented above). It is likely that it holds in all 3-dimensional metric spaces.

The second example looks at a problem requiring order besides metric notions for its statement. The problem of the thirteen spheres in Euclidean three-space, going back, as a conjecture, to Newton (and a disagreeing Gregory), states that the largest number of non-overlapping unit spheres that can be arranged such that they each touch another given unit sphere is 12. This is also called the *kissing number* in dimension 3. It was proved in [91] (see also [49]). There are two statements the problem makes: (i) that there are 12 non-overlapping unit spheres that can be arranged such that they each touch another given unit sphere, and (ii) that no 13 non-overlapping unit spheres can be arranged such that they each touch another given unit sphere. A similar question, known as the *kissing number problem*, can be asked in any finite dimension, and the precise values are known only for $n = 4, 8, 24$. This problem can, in any dimension, be stated inside the theory of ordered metric spaces, raising the question: "In which 3-dimensional metric spaces is the kissing number 12?" Similarly for

[17] A different axiomatization for the geometry obtained by adding $\mathbf{E}_3$ has been provided in [74].

higher dimensions. It is very likely that (ii) does not hold in the hyperbolic case. One can see this by reasoning along the following lines: in three-dimensional hyperbolic space over the real numbers, (ii) is certainly false, as can be seen from the Pizzetti-Toponogov triangle comparison theorem, which states that if O is the center of the original unit sphere $\mathfrak{U}$ and A and B two points of tangency of outside spheres with $\mathfrak{U}$, and A' and B' the reflections of O in A and B respectively, then the distance between A' and B' is greater in hyperbolic space than in Euclidean space, and the difference can be made very large by choosing a large "unit." One expects this kind of behavior to be present in the much more austere world of ordered metric spaces satisfying $\mathbf{H}_3$. So, the question would become: "what is part (ii) of the thirteen sphere problem equivalent to?" Is it $\neg\mathbf{H}_3$? Does part (i) hold in all ordered metric spaces? This is by no means trivial, as the "sphere" in our 3-dimensional metric spaces may have far fewer points on its "surface" than in the real Euclidean case.

A generalization of n-dimensional metric spaces along the lines of Lingenberg's generalization of metric planes was carried out for $n = 3$ in [55] and for all $n \geq 2$ in [56].

There are generalizations of metric spaces, in which, just like in the 2-dimensional case, one asks only for basic orthogonality axioms and for the existence of reflections, but no three-reflections theorem. They can be obtained in the 3-dimensional case by dropping the three-reflections axiom in Scherf's axiom system. In the dimension-free case, to which we turn, they were considered in [95].

4 The Dimension-Free Case

What if we do not want to specify the dimension of the space, but just know that it is at least 2?

This question was first raised and answered by Smith [93, 94], in the synthetic tradition—with point, lines, planes, incidence, line-orthogonality, reflections in points and in lines as primitive notions—by extending the work of Lenz [47] on incidence and orthogonality. Later, Smith [97], provided another synthetic axiom system for the non-elliptic case in terms of points, orthocomplemented hyperplanes, incidence and orthogonality as primitive notions.

The reflection-geometric approach was provided by Ewald's [19] axiom system for the groups of motions of such spaces, in terms of point-reflections and line-reflections. He showed that those geometries can be embedded in projective-metric spaces. Alternative embeddings were provided in [22, 23]. Ewald's axiom system was simplified by Heimbeck [29], and it is that axiom system that we present here.

The *fundamental assumption* is this time that $\mathfrak{G}$ is a group with invariant complexes $\mathfrak{P}$ and $\mathfrak{L}$ of involutions, which together generate $\mathfrak{G}$.

Here, by "invariant" we mean that, for all $g \in \mathfrak{G}$, $p \in \mathfrak{P}$, $l \in \mathfrak{L}$, we have $g^{-1}pg \in \mathfrak{P}$ and $g^{-1}lg \in \mathfrak{L}$. The elements of $\mathfrak{P}$ are called "points" (or "point-reflections"), those of $\mathfrak{L}$ "lines" (or "line-reflections"), the former to be denoted by upper-case Latin letters, the latter by lower-case Latin letters. The sign $|$ has the same meaning as before, and we say that line g *connects* the distinct points P and Q if and only if $P, Q \mid g$ and $X \mid P, Q \Rightarrow g^{-1}Xg = X$. We say that P is *incident* with g (and write $P \,\mathrm{I}\, g$) if and only if g connects P with a point $Q \neq P$. We say that line g is "orthogonal" to line h (and write $g \perp h$) if and only if $g \mid h$ and there is a point P incident with both g and h. We say that the lines g, h, and k *lie in a pencil* if and only if ghk is a line and there is a point P incident with each of g, h, k, and ghk. We denote by $\langle P, g\rangle$ the set of points X for which $X = P$ or else g connects P and X. We denote by $\langle Pg\rangle$ the set of points X for which $X \mid Pg$. The axioms are:

E-H 1 *Any two distinct points P and Q have a unique line (P, Q) connecting them.*
E-H 2 *If P, Q, R, and S are four different points, and if (P, Q), (P, R), (P, S) lie in a pencil, then so do (R, Q), (R, P), (R, S).*
E-H 3 *For all Q with $Q \notin \langle P, g\rangle$ there is a point $R \in \langle P, g\rangle$, with $(Q, R) \perp g$.*
E-H 4 *If Q and R belong to $\langle P, g\rangle$, then PQR is a point.*
E-H 5 *If $P \,\mathrm{I}\, g$, then $\langle P, g\rangle \cap \langle Pg\rangle = \{P\}$.*
E-H 6 *There are three different lines. There are three different points incident with every line.*

One gets an axiom system for elliptic geometry by stipulating that

Ell *There are different points P and Q with $PQ = QP$.*

To get an axiom system equivalent to that of Ewald one needs an additional axiom,

E-H 7 *If $P \,\mathrm{I}\, g$, $P' \,\mathrm{I}\, g'$, $\langle Pg\rangle = \langle P'g'\rangle$, then $Pg = P'g'$.*

If these geometries satisfy an additional, quite technical axiom, stated in [25], whose intuitive meaning is very simple, namely that all the points should not lie in a finite-dimensional subspace of the entire space, then $\mathfrak{G}$ is isomorphic to a subgroup of a projective-metric space. In the absence of that axiom, the same can be said only about a factor group of $\mathfrak{G}$.

In the dimension-free elliptic case, a mixed synthetic and reflection-theoretic axiom system can be found in [96], and another reflection-theoretic one in [24].

Axiom systems for the dimension-free Euclidean case can be found in [85, 86].

A broad generalization of the concept of (dimension-free) metric geometry has been proposed by E. M. Schröder in [87, 88].

5 Projective-Metric Geometry

5.1 Projective-Metric Planes

A projective plane is a triple $(\mathcal{P}, \mathcal{L}, \mathrm{I})$, consisting of a set $\mathcal{P}$ of points, a set $\mathcal{L}$ of lines, and a (symmetric) incidence relation I, with the property that any two distinct points are incident with a unique line and any two distinct lines are incident with a unique point. The only existence assumption it must satisfy is that it contains a quadrangle and a quadrilateral.

Plane projective geometry enjoys the property referred to as the *principle of duality*: Every definition remains valid and every theorem remains true if we consistently interchange the words "point" and "line" (the incidence relation being symmetric, it is self-dual, and thus need no change).

Central problems of the foundations of geometry, such as the introduction of numbers and the role of three-dimensional space for plane geometry, find conclusive answers in the projective setting. A projective plane can be coordinatized by a skew field (resp. a commutative field) of characteristic $\neq 2$ if and only if the configuration theorem of Desargues (resp. Pappus) and the Fano axiom hold. A projective plane is embeddable in a projective space (of dimension ≥ 3) if and only if the theorem of Desargues holds.

Introduction of a metric.

In a projective plane $(\mathcal{P}, \mathcal{L}, \mathrm{I})$ a metric can be introduced by an orthogonality relation on the set of lines (which we denote by $a, b, \ldots$) and on the set of points (which we denote by $A, B, \ldots$). Let $\perp$ be a binary relation on $\mathcal{L}$ with $a \perp b$ to be read as "a and b are orthogonal lines" and let $\top$ be a relation on $\mathcal{P}$ with $A \top B$ to be read as "A and B are orthogonal (or polar) points."

A point A is a *pole* of a line a if every line through A is orthogonal to a. Dually, a line b is a polar of a point B if every point on b is polar to B.

Following Struve and Struve [104], we call $(\mathcal{P}, \mathcal{L}, \mathrm{I}, \perp, \top)$ a *projective-metric plane* if the following axioms and the dual ones (which we do not explicitly state) hold:

PM1. *Every line a has a pole A.*
PM2. *Every triangle has altitudes which intersect in a common point.*
PM3. *A point A is the pole of a line a if and only if a is the polar of A.*
PM4. *There are lines a, b with $a \not\perp b$ and points A, B with $A \not\top B$.*

To get the dual axioms, just interchange the words point and line and the relations $\perp$ and $\top$. Notice that the axioms **PM3** and **PM4** are self-dual.

Given that the axiom system is self-dual (i.e., it contains the dual of each of its axioms), the principle of duality can be extended to projective-metric geometry: every definition remains valid, and every theorem remains true, if we consistently interchange the words "point" and "line" and the relations $\perp$ and $\top$.

There are seven types of projective-metric planes. They can be classified based on the properties of the following sets: (i) the set $\mathcal{L}_r$ of radical lines (which are orthogonal to every line), (ii) the set $\mathcal{L}_i$ of isotropic lines (which are orthogonal to themselves), (iii) the set $\mathcal{P}_r$ of radical points (which are polar to every point), and (iv) the set $\mathcal{P}_i$ of isotropic points (which are polar to themselves):

(1) planes with an *elliptic* metric: $|\mathcal{L}_r| = 0$ and $\mathcal{L}_i = \mathcal{L}_r$;
(2) planes with a *hyperbolic* metric: $|\mathcal{L}_r| = 0$ and $\mathcal{L}_i \neq \mathcal{L}_r$;
(3) planes with an *Euclidean* metric: $|\mathcal{L}_r| = 1$ and $\mathcal{L}_i = \mathcal{L}_r$;
(4) planes with a *Minkowskian* metric: $|\mathcal{L}_r| = 1$ and $\mathcal{L}_i \neq \mathcal{L}_r$;
(5) planes with a *co-Euclidean* metric: $|\mathcal{L}_r| \geq 2$ and $\mathcal{P}_r| \leq 1$ and $\mathcal{P}_i = \mathcal{P}_r$;

(6) planes with a *co-Minkowskian* metric: $|\mathcal{L}_r| \geq 2$ and $\mathcal{P}_r| \leq 1$ and $\mathcal{P}_i \neq \mathcal{P}_r$;
(7) planes with a *Galilean* metric: $|\mathcal{L}_r| \geq 2$ and $\mathcal{P}_r| \geq 2$.

Algebraic models.

Every projective-metric plane can be represented as a *projective-metric coordinate plane* $\mathfrak{P}(K, \mathfrak{f})$ over a field K of characteristic $\neq 2$ and a (non-trivial) symmetric bilinear form f. If V is a three-dimensional vector space over K and f a non-null symmetric bilinear form on V, then elements x and y of V are called orthogonal if $f(x, y) = 0$. If T is a subspace of V, then $T^{\perp} = \{x \in V : f(x, y) = 0 \text{ for all } y \in T\}$ is a subspace of V; subspaces T_1 and T_2 are called *orthogonal*, which we denote by $T_1 \vdash T_2$, if $T_1 \cap T_2^{\perp} \neq \{o\}$ and $T_1^{\perp} \cap T_2 \neq \{o\}$, where o stands for be the null vector.

$(\mathcal{P}, \mathcal{L}, \mathrm{I}, \perp, \top)$ is a projective-metric coordinate plane if

- $\mathcal{P}$ is the set of all i-dimensional subspaces of V with $i \in \{1, 2\}$;
- $\mathcal{L}$ is the set of all j-dimensional subspaces of V with $j \in \{1, 2\}$ and $j \neq i$;
- I is the set-theoretic inclusion restricted to $(\mathcal{P} \times \mathcal{L}) \cup (\mathcal{L} \times \mathcal{P})$;
- $\perp$ is the relation $\vdash$ restricted to $\mathcal{L} \times \mathcal{L}$;
- $\top$ is the relation $\vdash$ restricted to $\mathcal{P} \times \mathcal{P}$.

The seven types of projective-metric planes correspond to different dimensions of the radical and of the Witt index of the vector space V.

5.2 *Projective-Metric Spaces of Arbitrary Finite Dimension*

There are various ways to axiomatize projective geometry of higher dimensions. Veblen's classical axiomatization [112] is based on the terms of "point" and "line" and a binary relation of incidence. Inside that setting, higher dimensional subspaces are defined as sets of points.

We will follow Menger [50, 51], who noticed that projective geometry can be considered as a theory about joins and meets of linear subspaces (*Geometrie des Verbindens und Schneidens*). He axiomatized projective spaces in a first-order language with one sort of individual variables, to be referred to as "subspaces" or "flats" and denoted by lowercase Greek letters, two binary operations $\vee$ and $\wedge$, called "join" and "meet", and two constant symbols $\mathsf{0}$ and $\mathsf{1}$, which are called "element zero" and "element one".

The axioms are simple postulates about the joining and intersecting of geometric subspaces. They state that the operations $\vee$ and $\wedge$ are commutative and associative with neutral elements 0 and 1, and that the absorption laws hold. The models of this axiom system, $\mathcal{L} = (\mathsf{L}, \vee, \wedge, 0, 1)$, are lattices with $\mathsf{0}$ and $\mathsf{1}$.

To characterize projective spaces of dimension $n \geq 3$, one needs to add to the above-mentioned axioms the requirements that the lattice $\mathcal{L}$ be complemented and irreducible and that the maximal length of a chain of $\mathcal{L}$ be $n + 1$. In an algebraic language this can be summarized by the statement that $\mathcal{L}$ is an irreducible projective lattice.

The dual of each axiom holds, so the principle of duality holds in projective-metric geometry of arbitrary finite dimension.

The notion of a "point", which is the basic concept of analytic geometry and particularly of Riemannian geometry, is not even mentioned in the axiom system. The complete elimination of the notion of point from geometry was one of the ideas of von Neumann's *continuous geometries*.

In complete harmony with Euclid's first words in the *Elements*, "A point is that which has no part", the elements α of $\mathcal{L}$ for which $\varepsilon \leq \alpha \rightarrow \varepsilon = 0$ are called "points" of a projective space (i.e., the points are precisely the atoms of the projective lattice).

Since points and lines are no longer distinguished from subspaces of other dimensions, the sentences of projective geometry are statements about finite sets of elements of the basic class of subspaces, without any need for either a multi-sorted language or of set-theoretical definitions of subspaces. Unlike the first modern axiomatizations of geometry, by Pasch, Peano, Pieri, and Hilbert, which were expressed in languages which contained only relation symbols, but no operation symbol, the above axiomatization, with two operation symbols and no relation symbol, is much closer in spirit to those of arithmetic or of algebraic theories.

Algebraic models.

Every projective space of dimension $n \geq 3$ can be represented as the lattice of subspaces of a finite dimensional vector space over a skew field (division ring) with the set-theoretic inclusion $\subseteq$ as $\leq$-relation of the associated partially ordered set of subspaces.

Introduction of a metric.

Much like in the 2-dimensional case, in a projective space of dimension $n \geq 3$ a metric can be introduced by an orthogonality relation. The metric is called *singular* if there are radical subspaces and *ordinary* otherwise.

In the ordinary case, the orthogonality relation is a binary relation, which is defined on the set of hyperplanes (subspaces of dimension $n - 1$) and on the set of points, and which satisfies mutatis mutandis the axioms for projective-metric planes noted in Sect. 5.1. The orthogonality relation can be described algebraically by a non-degenerate symmetric bilinear form which is a hyperbolic polarity if there are self-polar points, an elliptic polarity otherwise.

The classical example of a projective space with a singular metric is the projective closure of a Euclidean space. The orthogonality relation of Euclidean subspaces induces on the hyperplane ε at infinity an elliptic metric (in the sense of Sect. 5.1).

In the general case, the hyperplane at infinity may as well be endowed with a hyperbolic metric (as in Minkowskian geometry) or with a Euclidean metric (as in a Galilean geometry)—to mention only two alternatives—and the subspace at infinity need not be a hyperplane but may be a subspace of arbitrary dimension.

This general situation is captured in the following definition which is formulated in an algebraical setting (an axiomatic definition can be given along the lines of Sect. 5.1). A metric in a projective space is given by a flag $0 < \varepsilon_1 < \ldots < \varepsilon_r < 1$ of subspaces and a (hyperbolic or elliptic) polarity on each of the associated intervals $[0, \varepsilon_1], ..., [\varepsilon_r, 1]$.

$(\mathcal{L}, ((\varepsilon_1, \pi_1), ..., (\varepsilon_r, \pi_r), (1, \pi_{r+1})))$ *with* $r \geq 0$ *is a projective space with Cayley-Klein metric (or Cayley-Klein spacefor short) of dimension* $n \geq 0$ *if the following assumptions hold*:

(1) $\mathcal{L}$ is a projective lattice of finite dimension n.
(2) $\varepsilon_1, ..., \varepsilon_r$ are subspaces of $\mathcal{L}$ with $0 < \varepsilon_1 < \ldots < \varepsilon_r < 1$.
(3) π_k is a hyperbolic or elliptic polarity on the interval $[\varepsilon_{k-1}, \varepsilon_k]$ with $1 \leq k \leq r+1$ and $\varepsilon_0 = 0$ and $\varepsilon_{r+1} = 1$.

For notational convenience we denote a Cayley-Klein space by $\mathcal{CK}(\varepsilon_0, ..., \varepsilon_{r+1})$, if the underlying polarities π_k are of no special concern. If $\mathcal{CK}(\varepsilon_0, ..., \varepsilon_{r+1})$ is a Cayley-Klein space, then the Cayley-Klein space $\mathcal{CK}(\varepsilon_i, ..., \varepsilon_k)$ (for $0 \leq i < k \leq r+1$) is called *ordinary* if $k = i+1$ and *singular* otherwise.

That the *principle of duality* can be extended from projective geometry to projective-metric geometry (Cayley-Klein spaces) can be seen by noticing that the dual of an interval $[\varepsilon_{k-1}, \varepsilon_k]$ of a projective lattice $\mathcal{L}$ is an interval of the dual projective lattice $\mathcal{L}^*$, and that the dual of a polarity (on an interval of $\mathcal{L}$) is a polarity (on an interval of $\mathcal{L}^*$).

Of special interest are Cayley-Klein spaces which are *self-dual*, i.e., isomorphic to their dual structures. $\mathcal{CK}(\varepsilon_0, ..., \varepsilon_{r+1})$ is self-dual if and only if $\mathcal{CK}(\varepsilon_k, \varepsilon_{k+1})$ and $\mathcal{CK}(\varepsilon_{r-k}, \varepsilon_{r+1-k})$ are isomorphic (for $0 \leq k \leq r$).

Every ordinary Cayley-Klein space is self-dual. Further examples are the projective closure of a Galilean plane over a field of characteristic $\neq 2$ and the projective closure of the Desargues configuration which can be embedded in the projective plane with an elliptic metric over the field of order 5.

As mentioned in Sect. 5.1, there are seven Cayley-Klein spaces of dimension 2. There are eighteen Cayley-Klein spaces of dimension 3. For a detailed classification see Struve and Struve [106].

Metric concepts like the *pole-polar-theory* of quadratic spaces can be extended to Cayley-Klein spaces. A subspace β is a *polar* of a subspace α if the projections of α and β into the intervals $[\varepsilon_k, \varepsilon_{k+1}]$[18] map α and β onto polar elements of the ordinary Cayley-Klein spaces $\mathcal{CK}(\varepsilon_k, \varepsilon_{k+1})$.

If β is a polar of α, then α is a polar of β. In an n-dimensional Cayley-Klein space, the sum of the dimensions of a subspace and of its polar is equal to $n-1$, a formula which is well known for projective spaces with an elliptic or a hyperbolic metric. Every subspace of a Cayley-Klein space has at least one polar. A subspace α with a unique polar is called *regular*. This is equivalent to the existence of an integer k with $\alpha \wedge \varepsilon_k = 0$ and $\alpha \vee \varepsilon_{k+1} = 1$.

Subspaces α and β are *orthogonal* if there are subspaces α^* and β^* which are polar to α respectively β and satisfy $\alpha \leq \beta^*$ and $\beta \leq \alpha^*$.

Let β be a polar of α with $\alpha \wedge \beta = 0$ (i.e., let α and β be complements). The harmonic homology $\sigma_{\alpha\beta}$ with α and β as center and axis leaves the subspaces ε_k

[18] i.e., the elements $(\alpha \vee \varepsilon_k) \wedge \varepsilon_{k+1}$ and $(\beta \vee \varepsilon_k) \wedge \varepsilon_{k+1}$ (if $\wedge$ and $\vee$ denote the lattice operations).

invariant and induces an automorphism on $\mathcal{CK}(\varepsilon_k, \varepsilon_{k+1})$. Hence $\sigma_{\alpha\beta}$ is an involution of the group of automorphism of the Cayley-Klein space which we call a *(projective) reflection* in α respectively β. The group which is generated by all reflections $\sigma_{\alpha\beta}$ is called the *group of motions*.

This highlights the special role of reflections in projective geometry: They can be used to single out motions within the group of all projective collineations.

Bachmann (see [7, §20,11]) carried this idea a step further. He showed that projective-metric geometry can be formulated in the group of motions of a projective-metric space (a quadratic space). Geometric relations like incidence and orthogonality correspond to group-theoretical relations between elements of the group of motions (e.g., projective subspaces are orthogonal respectively incident if the product of the associated reflections is involutory). This correspondence allows not only the proof of geometric theorems by group-theoretical calculations but also group-theoretical characterizations of orthogonal groups (see Bachmann [7, §20,8 and §20,11]).

The full group $\mathcal{G}$ of projective automorphisms of a Cayley-Klein space has been analyzed in Struve and Struve [107]. In the ordinary case, $\mathcal{G}$ can be represented as the orthogonal group of the associated quadratic space.

In the singular case, an element φ of $\mathcal{G}$ is called a *dilatation*[19] if φ is the identity on the ordinary Cayley-Klein spaces $\mathcal{CK}(\varepsilon_k, \varepsilon_{k+1})$ with $0 \leq k \leq r$. The group of dilatations is a normal subgroup of $\mathcal{G}$.

Every element of $\mathcal{G}$ is up to a dilatation uniquely determined by its operation on the ordinary Cayley-Klein spaces $\mathcal{CK}(\varepsilon_k, \varepsilon_{k+1})$, and conversely every automorphism of a Cayley-Klein space $\mathcal{CK}(\varepsilon_k, \varepsilon_{k+1})$ can be extended to an element of $\mathcal{G}$ (in a trivial way). Hence the group $\mathcal{G}$ is the semi-direct product of the (normal) group of dilatations and the subgroup of $\mathcal{G}$ which is generated by the (extensions) of the automorphisms of the Cayley-Klein spaces $\mathcal{CK}(\varepsilon_k, \varepsilon_{k+1})$.

This representation theorem of $\mathcal{G}$ generalizes theorems which are well-known in metric affine geometry (i.e., in Euclidean, Minkowskian, and Galilean geometry).

6 Cayley-Klein Geometries

In the approach of Cayley and Klein non-Euclidean geometries are introduced as geometries living inside of a projective space which is endowed with a projective metric.

Following this approach we consider in this section real projective spaces, which are endowed with a Cayley-Klein metric, and single out substructures which define Cayley-Klein geometries.

[19]This concept of a dilatation generalizes the notion of a dilatation which is given in incidence geometry (as a transformation which preserves direction) and in similarity geometry (as a transformation which preserves circles resp. the angular measure).

Following Klein, these substructures are called *Eigentlichkeitsbereiche*[20] and the associated points, lines, planes etc. "proper subspaces". Geometric relations such as incidence and orthogonality are inherited from the associated Cayley-Klein space.

There are different ways to distinguish substructures of a projective-metric space. Klein's famous model of hyperbolic geometry, for example, is defined as a substructure of the real projective plane $\mathbf{P}^2$, where a projective metric is given by a hyperbolic polarity π. Points of the hyperbolic plane $\mathbf{H}^2$ are the points of $\mathbf{P}^2$ which are interior to the "absolute conic" of self-conjugate points of π. Lines of $\mathbf{H}^2$ are the lines of $\mathbf{P}^2$ which are incident with at least one interior point. The projective reflections in points and in lines of $\mathbf{H}^2$ generate the group of motions of $\mathbf{H}^2$ (which is in fact isomorphic to the full group of automorphisms of $\mathbf{P}^2$).

The set of points of $\mathbf{P}^2$ which are *exterior* to the "absolute conic" of the polarity π (any two of these points have according to Klein a real positive distance) are the set of points of the co-hyperbolic geometry (see Sect. 6.1).

This shows that to a given Cayley-Klein space there may exist several substructures which are Cayley-Klein geometries. Necessary conditions for a substructure to be an n-dimensional Cayley-Klein geometry are:

(1) Subspaces with the same dimension are "of the same kind".
(2) The substructure contains with a subspace all subspaces which have the same dimension and which are of the same kind.
(3) There is a flag which contains subspaces of dimension $0, 1, 2, \ldots, n$.

A classification of subspaces of a Cayley-Klein space into elements "of the same kind" can be done in various ways. For example, in a projective space with a hyperbolic polarity—as in Klein's model of a hyperbolic plane—the set of points is the union of the set of isotropic points (which form a conic κ) and the set of points which are internal resp. external with respect to κ. A point which is not incident with a tangent to κ is an internal point. A point which is not an internal point and not on κ is an external point. The geometric classification into internal and external points corresponds on the algebraic side to the distinction between points with signature 0 or 1 (signature[21] of a point with respect to the bilinear form which describes κ).

Similarly, the set of lines is the union of the set of isotropic lines (tangents to κ) and the sets of lines which are incident with two resp. none of the points of κ (secants and non-secants), a classification which corresponds on the algebraic side to the distinction between lines with signature 1 or 2.

Condition (2) ensures that a Cayley-Klein geometry is maximal with respect to the property which defines the classification of subspaces of equal dimension. So, in Klein's model of a hyperbolic plane, every interior point of the absolute conic κ is a point of the model.

---

[20] cp. Klein [42], Bachmann [7], Klopsch [44], Hessenberg and Diller [32] and Karzel and Kroll [38].

[21] According to Sylvester's law of inertia all maximal positive definite subspaces of a (real) quadratic space V, i.e., of a vector space endowed with a quadratic form, have the same dimension, which is called the *signature* of the quadratic space (the term "signature" is used in the literature in different ways; we follow Snapper and Troyer [99]). The signature of a subspace U of V is the signature of U with respect of the restriction of the quadratic form of V to U, see [108].

According to condition (3), there is a chain of subspaces which contains elements of every dimension of the projective space. This ensures that the dimension of the Cayley-Klein geometry is n.

These considerations lead to the following model-theoretic definition of (real) ordinary Cayley-Klein geometries:

If $0 < \alpha_1 < ... < \alpha_n < 1$ is a maximal flag of subspaces of an ordinary Cayley-Klein space, then the set of subspaces β which have the same dimension and signature as one of the elements α_k is the set of subspaces of a Cayley-Klein geometry.

The general (not necessarily ordinary) case can be reduced to the ordinary one since a Cayley-Klein space $\mathcal{CK}(\varepsilon_0, ..., \varepsilon_{r+1})$ is build up from the ordinary Cayley-Klein spaces $\mathcal{CK}(\varepsilon_k, \varepsilon_{k+1})$. This leads to the following general model-theoretic definition of (real) *Cayley-Klein geometries* :

If $0 < \alpha_1 < ... < \alpha_n < 1$ is a maximal flag of subspaces of a Cayley-Klein space $\mathcal{CK}(\varepsilon_0, ..., \varepsilon_{r+1})$ which contains $\varepsilon_0, ..., \varepsilon_{r+1}$ as subspaces, then the set of subspaces β which have a polar with the same dimension and signature as one of the elements α_k is the set of subspaces of a Cayley-Klein geometry.

The dual structure of a Cayley-Klein geometry is a Cayley-Klein geometry, i.e., the principle of duality can be extended to metric geometry. So, for example, the dual geometry of n-dimensional hyperbolic geometry is co-hyperbolic geometry. Elliptic geometry is self-dual.

The number of real Cayley-Klein geometries of dimension n is 3^n (with $n \geq 1$). For a more detailed discussion of plane Cayley-Klein geometries we refer to Sect. 4. The number of real ordinary Cayley-Klein geometries of dimension n is 2^n.

Cayley-Klein geometries have properties which are well known from Euclidean, hyperbolic and elliptic geometry: Every subspace α of a Cayley-Klein geometry is regular. There exists one and only one projective reflection σ_α in α. The set of reflections σ_α generates the *group of motions* and the calculus of reflections allows the axiomatization and the coordinatization of a Cayley-Klein geometry.

Remark This definition of a Cayley-Klein geometry is based on the algebraic notion of the signature of a subspace. This corresponds, as we indicated above, to geometric properties which are more complex (like a classification in interior and exterior points or in secants and non-secants) and which may depend on properties of the underlying field of coordinates. The algebraic notion of signature of a subspace allows, on the other hand, a simple definition, which only assumes that the field of coordinates allows the introduction of a half-order (i.e., of a homomorphism from the multiplicative group of K into the cyclic group $(\{1, -1\}, \cdot)$ of order two; see [108]). This is satisfied in particular by all fields which are orderable or of finite order. The concept of a Cayley-Klein geometry is hence not restricted to the real or complex case.

6.1 Plane Cayley-Klein Geometries

Let $\mathbf{P}_3(\mathbb{R})$ be the 3-dimensional projective space over the field of real numbers and $\mathcal{Q}$ a non-degenerate quadric of $\mathbf{P}_3(\mathbb{R})$, i.e., a quadric with the property that there exists a plane section which is a non-degenerate conic.

As is well known, up to projective equivalence, there are three quadrics of this kind, namely, the *sphere*, which has no generators (there are no lines lying entirely in the quadric), the *cone*, where every point (with the exception of the vertex) is incident with exactly one generator, and the ruled surface, where every point is incident with exactly two generators. The vertex of a cone is incident with all generators and is called a *singular point*.

We call a line g a *secant* (or secant line) of $\mathcal{Q}$ if g is incident with exactly two points of $\mathcal{Q}$. The line g is called a *tangent* (or a tangent line) of $\mathcal{Q}$ if g and $\mathcal{Q}$ have one and only one non-singular point of intersection.

A plane ε is called a *secant plane* of $\mathcal{Q}$ if the points of intersection of ε and $\mathcal{Q}$ are the points of a non-degenerate conic. The secant planes through a point A of the projective space can be divided into three classes, depending on whether they contain one, two or no tangent line to $\mathcal{Q}$.

In every secant plane ε of $\mathcal{Q}$ there is a *projective reflection*, i.e., an involutory projective collineation leaving $\mathcal{Q}$ invariant, and ε and the pole of ε (with respect to $\mathcal{Q}$) pointwise fixed. In every secant line g of $\mathcal{Q}$ there is a projective reflection, i.e., an involutory projective collineation leaving $\mathcal{Q}$ invariant, and g and the polar of g (with respect to $\mathcal{Q}$) pointwise fixed.

With these concepts in mind we now give a model-theoretic characterization of all nine plane Cayley-Klein geometries.

$(\mathcal{P}, \mathcal{L}, G)$ is called a *plane Cayley-Klein geometry* if there is a point A and a non-degenerate quadric $\mathcal{Q}$ of $\mathbf{P}_3(\mathbb{R})$ and a number $n \in \{0, 1, 2\}$ such that

- $\mathcal{P}$ is the set of secant lines through A.
- $\mathcal{L}$ is the set of secant planes through A which contain n tangents to $\mathcal{Q}$.
- G is the group of projective collineations generated by reflections in the elements of $\mathcal{P}$ and $\mathcal{L}$ (restricted to $\mathcal{P} \cup \mathcal{L}$).

The elements of $\mathcal{P}$ are the *points* and the elements of $\mathcal{L}$ are the *lines* of the plane Cayley-Klein geometry. The incidence relation between points and lines is inherited from the projective space.

The elements of G are called *motions*. In each point A and in each line g of the plane Cayley-Klein geometry there exists a unique reflection, which is the restriction of the associated projective reflection in A (resp. g) to $\mathcal{P} \cup \mathcal{L}$.

Metric concepts can be defined in the following way: Two pairs (B, C) and (D, E) of points (which can be thought of as *segments*) are called *congruent* if there is a motion α with $B^\alpha = D$ and $C^\alpha = E$. Dually, two pairs (b, c) and (d, e) of lines (which can be thought of as *angles*) are called congruent if there is a motion α with $b^\alpha = d$ and $c^\alpha = e$.

The *type* of a plane Cayley-Klein geometry is a pair of natural numbers (m, n) with $m, n \in \{0, 1, 2\}$ where m denotes the number of generators through a non-singular point of the quadric $\mathcal{Q}$ and n the number of lines of the elements of $\mathcal{L}$ which are incident with A and tangent to $\mathcal{Q}$.

The value of m is 0, 1 or 2 depending on whether $\mathcal{Q}$ is a sphere, a cone, or a ruled surface. If A is a point of $\mathcal{Q}$ then $n = 1$. If A is an interior point of $\mathcal{Q}$ then $n = 0$ and if A is an exterior point then $n = 2$.

According to Struve and Struve [102] there are nine real plane Cayley-Klein geometries which are presented (name and type) in the following table.

elliptic (0, 0)	*Euclidean* (0, 1)	*hyperbolic* (0, 2)
co − Euclidean (1, 0)	*Galilean* (1, 1)	*co − Minkowskian* (1, 2)
co − hyperbolic (2, 0)	*Minkowskian* (2, 1)	*doubly hyperbolic* (2, 2)

The points and lines of a plane Cayley-Klein geometry $(\mathcal{P}, \mathcal{L}, G)$ are lines and planes through a point A of a projective space $\mathbf{P}$. Hence $(\mathcal{P}, \mathcal{L}, G)$ can be extended to a projective ideal plane: ideal points are the lines through A, ideal lines are the planes through A, and the incidence relation is inherited from $\mathbf{P}$. The motions of a plane Cayley-Klein geometry (which are induced by collineations of $\mathbf{P}$, which have A as a fixed point) can be extended to collineations of the projective ideal plane.

To represent the points and lines of $(\mathcal{P}, \mathcal{L}, G)$ by points and lines of $\mathbf{P}$ we consider the intersection of the elements of $\mathcal{P}$ and $\mathcal{L}$ with a secant plane of $\mathcal{Q}$ which is not incident with A. In this way one gets *Klein models* of the Cayley-Klein geometries.

1. The Klein model of an *elliptic* plane is a projective plane.
2. The Klein model of a *Euclidean* plane is an affine plane.
3. The Klein model of a *hyperbolic* plane contains the interior points of a non-degenerate conic κ and the lines which are incident with at least one interior point of κ.
4. The Klein model of a *co-Euclidean* plane is obtained from a projective plane by the removal of a point A and of all lines which are incident with A.
5. The Klein model of a *Galilean* plane is obtained from an affine plane by the removal of a pencil of parallel lines.
6. The Klein model of a *co-Minkowskian* plane contains exactly all points of an affine plane which lie between two parallel lines a and b as well as all lines which are not parallel to a or b.
7. The Klein model of a *co-hyperbolic* plane contains exactly the exterior points of a non-degenerate conic κ and the lines which have no common point with κ.
8. The Klein model of a *Minkowskian* plane is obtained from an affine plane by the removal of two pencils of parallel lines.
9. The Klein model of a *doubly hyperbolic* plane contains exactly the exterior points of a non-degenerate conic κ and the lines which are incident with at least one interior point of κ.

As mentioned above, in each point A and in each line g of a plane Cayley-Klein geometry there exists a unique reflection, which is the restriction of the associated projective reflection in A (resp. g) to $\mathcal{P} \cup \mathcal{L}$. This shows that metric geometry in the sense of Cayley and Klein can be formulated in the group of motions. For an axiomatization and coordinatization of plane Cayley-Klein geometries over fields of characteristic $\neq 2$ we refer to Sect. 6.3.

The geometry of plane sections of a quadric $\mathcal{Q}$ is the circle geometry of Möbius, Laguerre or Minkowski depending on whether $\mathcal{Q}$ is a sphere, a cone, or a ruled surface. The points of the circle geometries are the non-singular points of $\mathcal{Q}$ and the circles are the plane sections of $\mathcal{Q}$. The points and lines of a plane Cayley-Klein geometry can be represented as point-pairs and circles of the above-mentioned circle geometries. The group of motions of a Cayley-Klein geometry is isomorphic to a group of circle transformations. In this way one gets *Poincaré models* of the Cayley-Klein geometries.

6.2 Finite Cayley-Klein Geometries

The model-theoretic characterization of plane Cayley-Klein geometry, given in Sect. 6.1, allows the transfer of Riemann's idea of an elliptic plane to the realm of finite geometries.

In the 3-dimensional projective space over the finite field $\mathbf{GF(n)}$ of order n there exist three non-degenerate quadrics $\mathcal{Q}$ (i.e., quadrics with the property that there exists a plane section which is a non-degenerate conic): the sphere without generators, the cone with one generator through every point distinct from the vertex, and the ruled surface with two generators through every point of the quadric.

Let A be an arbitrary point of the projective space. The set of secant lines through A and the set of secant planes through A with n tangents to $\mathcal{Q}$ for a number $n \in \{0, 1, 2\}$ are the set $\mathcal{P}$ of points and the set $\mathcal{L}$ of lines of a plane Cayley-Klein geometry (if both sets are non-empty). The group of projective collineations generated by reflections in the elements of $\mathcal{P}$ and $\mathcal{L}$ (restricted to $\mathcal{P} \cup \mathcal{L}$) is the group of motions of the Cayley-Klein geometry.

As in the real case, there are nine plane Cayley-Klein geometries over any finite field of characteristic $\neq 2$. Among these finite geometries there are well-known configurations: The configurations of Desargues, Pappus, and Petersen (with their groups of automorphisms) can be represented by the elliptic plane over $\mathbf{GF(5)}$, the Galilean plane over $\mathbf{GF(3)}$, and the hyperbolic plane over $\mathbf{GF(5)}$. This is in stark contrast to the theory of metric planes, presented in Sect. 2.1, for which there are finite models *only* in the case of the Euclidean metric, i.e., only if **E1** holds (see [7, §6,12]).

Every finite plane Cayley-Klein geometry can be represented as a Klein model and as a Poincaré model. For the number of points and lines of a finite Cayley-Klein geometry and a uniform representation of the groups of motions we refer to Struve and Struve [103].

6.3 Cayley-Klein Geometries and Reflection Geometry

According to the table (in Sect. 6.1) there are nine types of plane Cayley-Klein geometries. Elliptic, Euclidean, and hyperbolic planes are metric planes in the sense of Bachmann, which were characterized in Sect. 2.1.

For a characterization of all types of plane Cayley-Klein geometries, several aspects of Bachmann's notion of a metric plane have to be broadened. The most important aspect is the principle of duality: the dual of a Cayley-Klein geometry is also a Cayley-Klein geometry.

Thus the set S of line reflections will no longer play a distinguished role in the group of motions G. S no longer needs to be a set of generators of G, and the set P of point reflections can no longer be defined as the set of involutions of $S^2 = \{ab : a, b \in S\}$.

This corresponds to new geometric phenomena which are unknown in the setting of classical plane absolute geometry. In a Cayley-Klein geometry there may be motions which are involutions without being point or line reflections. A rotation which is not the identity (the product of the reflections in lines a and b with a unique point of intersection) may have several fixed points, and the product of the reflections in three lines a, b, c which are the sides of a non-degenerate triangle may be a line reflection.

On the other hand, well known axioms of classical plane absolute geometry, such as the uniqueness of a joining line, the existence of a perpendicular (in a self-dual form), and the three reflections theorems (in a dual form), continue to hold.

We generalize the axiom system for metric planes based on the following principles:

- The axiom system is satisfied by the metric planes of Sect. 2.1.
- The axiom system is satisfied by all types of plane Cayley-Klein geometries (for reasons of simplicity with the exception of the doubly hyperbolic case).
- The axiom system allows a formulation in a first-order language.
- The axioms are statements about points and lines with a direct geometric interpretation and without any non-geometric assumptions about the type or structure of the underlying group G (such as $Z(G) = 1$).
- The axiom system contains with each axiom the dual one.

(G, S, P) is called a *Cayley-Klein group*[22] if the following Basic Assumption and axioms hold (see [109])[23]:

---

[22]or more precisely a *non-doubly hyperbolic Cayley-Klein group*.

[23]We recall from Sect. 3: Elements $a, b, c, \ldots$ of S are called *lines* and elements $A, B, C, \ldots$ of P *points*. The "stroke relation" $\alpha \mid \beta$ is an abbreviation for the statement that α, β and $\alpha\beta$ are involutory elements (i.e., group elements of order 2). The statement $\alpha, \beta \mid \delta$ is an abbreviation of $\alpha \mid \delta$ and $\beta \mid \delta$. A point A and a line b are *incident* if $A \mid b$. Lines $a, b \in S$ are *orthogonal* if $a \mid b$. A *quadrangle* is a set of four points A, B, C, D and four lines a, b, c, d with $a \mid A, B$ and $b \mid B, C$ and $c \mid C, D$ and $d \mid D, A$.

Basic Assumption *Let G be a group and S and P invariant subsets of involutions of G such that*

N 1 *If $a \mid b$ then $ab \in P$.*
N 2 *If $A \mid B$ then $AB \in S$.*
N 3 *For every pair (A, b) there exists (a, B) with $a \mid A$ and $B \mid b$ and $Aa = bB$ and if $A \neq b$ then (a, B) is unique.*
N 4 *If $A, B \mid c, d$ then $A = B$ or $c = d$.*
N 5 *If $A, B, C \mid d$ then $ABC \in P$.*
N 6 *If $a, b, c \mid D$ then $abc \in S$.*
N 7 *If $A \mid a$ and $B \mid b$ and $C \mid c$ and $Aa = Bb = Cc$ then $ABC \in P$ and $abc \in S$.*
N 8 *There exists a quadrangle.*

According to axiom **N1**, orthogonal lines a, b intersect in the point ab. **N2** is the dual axiom which states that polar points A, B are incident with the line AB. Axiom **N3** states that, if A is a point and b a line, then there exists a line a through A and a point B on b with $Aa = bB$ (a "perpendicular" from A to b with foot B) and that (a, B) is unique if A is not the pole of b. According to **N4**, two different points have at most one joining line and two different lines have at most one common point. **N5** states that, if A, B, C are collinear points, then ABC is a point (the *fourth reflection point*). **N6** is the dual axiom, which states that, if a, b, c are copunctual lines, then abc is a line, the *fourth reflection line*. **N7** is a self-dual axiom which is a special generalization of the theorem of three reflections. According to **N8**, there exists at least a quadrilateral (the assumption of the existence of a triangle—cp. axiom **M10** in Sect. 2.1—does not hold in every Cayley-Klein geometry, e. g., in the Minkowskian plane over **GF**(**3**)).

The metric planes of Sect. 2 are exactly those plane Cayley-Klein geometries which satisfy the axiom of the existence of a joining line (as Euclidean, hyperbolic and elliptic planes). The Galilean, co-Minkowskian, and co-Euclidean planes satisfy the *dual parallel axiom*.

> If A is not incident with b, then there is a unique point on b which has no joining line with A.

By dualization one gets the following two statements: The elliptic, co-Euclidean and co-hyperbolic planes are dual metric planes, i.e., plane Cayley-Klein geometries with the property that any two lines have a point of intersection. The Euclidean, Galilean and Minkowskian planes satisfy the *parallel postulate*

> If A is not incident with b, then there is a unique line through A which has no point of intersection with b.

A plane Cayley-Klein geometry which satisfies the parallel axiom is *singular*, i.e., the set of translations forms a group (or equivalently, in any quadrilateral with three right angles the fourth angle is a right one).

The hyperbolic and co-Minkowskian planes satisfy the *hyperbolic parallel axiom* which states that through a given point A there are at most two lines a and b that have neither a common point nor a common line with a given line g (cp. axiom H2 in

Sect. 2.1.2). The co-hyperbolic and Minkowskian planes satisfy the *dual hyperbolic parallel axiom*.

For axiomatizations of Cayley-Klein geometries in terms of reflections we refer for Minkowskian planes to Wolff [115], for Galilean planes to Struve [100], for co-Minkowskian and co-Euclidean planes to Struve and Struve [101, 105] (cp. Bachmann [8]), and for metric planes to the references in Sect. 2.1.

Methodological reflections.

In the axiomatic approach to geometry, a *Begründung* has the important function of providing a convenient means of ensuring the consistency of that geometry's axiom system. The latter can be reduced by means of an embedding in a projective-metric space (with respect to both the incidence and the metric structure) to the consistency of the algebraic structure that coordinatizes that projective-metric space. Those algebraic structures are fields with some additional properties. Since the axiom systems of those fields can be extended to that of the theory of real-closed fields, which we know to be consistent (see [62, p. 68]), any fragment thereof must be consistent as well.

Begründungen in this sense were provided first by Hilbert in his *Grundlagen der Geometrie* [34] and then in *Neue Begründung der Bolyai-Lobatschefskyschen Geometrie* [33], then by Hjelmslev in *Neue Begründung der ebenen Geometrie* [36], by Podehl and Reidemeister in *Eine Begründung der elliptischen Geometrie* [73], by Bachmann in *Eine Begründung der absoluten Geometrie in der Ebene* [2] and by many other geometers who worked in the foundations of geometry.

It is worth emphasizing that a geometry's *Begründung* (i.e., its embedding in a projective-metric space) not only ensures its "existence" from a logical point of view in Hilbert's sense, but also its authenticity from a projective-geometric point of view championed by Klein, as a geometry in its own right, in no way inferior or subservient to the Euclidean one.

6.4 Cayley-Klein Spaces and Differential Geometry

Cayley-Klein manifolds.

There are many natural connections between Riemannian manifolds and Cayley-Klein spaces. The tangent space of an n-dimensional Riemannian manifold is an n-dimensional vector space endowed with a (positive definite) quadratic form, which corresponds—from a geometric point of view—to an $(n-1)$-dimensional Cayley-Klein space with an elliptic metric. The elements of the Cayley-Klein space can be represented by the set of Euclidean subspaces through the point of contact of the tangent space (if the manifold is embedded in a Euclidean space). If the quadratic form of a manifold is not positive definite, then the metric of the associated Cayley-Klein space is hyperbolic and the manifold is called *pseudo-Riemannian*.

The second connection we want to point to is that a Riemannian manifold which is embedded in a Euclidean space is also embeddable in the projective closure of that Euclidean space. This means that the concept of a Riemannian manifold can be generalized by considering manifolds which are embedded in arbitrary Cayley-Klein spaces (and whose tangent spaces can be Cayley-Klein spaces of any type). Such manifolds are called by Rosenfeld [75] *quasi-Riemannian* or *quasipseudo-Riemannian* (see also [116]). Perhaps a more appropriate name would be *Cayley-Klein manifold*. The groups of motions of Cayley-Klein spaces are examples for Cayley-Klein manifolds.

References

1. J. Ahrens, Begründung der absoluten Geometrie des Raumes aus dem Spiegelungsbegriff. Math. Z. **71**, 154–185 (1959)
2. F. Bachmann, Eine Begründung der absoluten Geometrie in der Ebene. Math. Ann. **113**, 424–451 (1936)
3. F. Bachmann, Geometrien mit euklidischer Metrik, in denen es zu jeder Geraden durch einen nicht auf ihr liegenden Punkt mehrere Nichtschneidende gibt. I, II, III. *Math. Z.* **51** (1949), 752–768, 769–779, *Math. Nachr.* **1**, 258–276 (1948)
4. F. Bachmann, Zur Begründung der Geometrie aus dem Spiegelungsbegriff. Math. Ann. **123**, 341–344 (1951)
5. F. Bachmann, Zur Parallelenfrage. Abh. Math. Semin. Univ. Hamburg **27**, 173–192 (1964)
6. F. Bachmann, Der Höhensatz in der Geometrie involutorischer Gruppenelemente. Can. J. Math. **19**, 895–903 (1967)
7. F. Bachmann, *Aufbau der Geometrie aus dem Spiegelungsbegriff.* Zweite ergänzte Auflage (Springer-Verlag, Berlin, 1973)
8. F. Bachmann, *Ebene Spiegelungsgeometrie* Eine Vorlesung über Hjelmslev-Gruppen (BI-Wissenschaftsverlag, Mannheim, 1989)
9. O. Bachmann, Zur spiegelungsgeometrischen Begründung von Geometrien. Geom. Dedicata **5**, 497–516 (1976)
10. R. Baer, The group of motions of a two dimensional elliptic geometry. Compositio Math. **9**, 241–288 (1951)
11. L. Boi, L'espace: concept abstrait et/ou physique; la géométrie entre formalisation mathématique et étude de la nature. L. Boi, D. Flament, J.-M. Salanskis (eds.), *1830–1930: A century of geometry (Paris, 1989)*, 65–90, Springer, Berlin, 1992
12. L. Boi, Die Beziehungen zwischen Raum, Kontinuum und Materie im Denken Riemanns; die Äthervorstellung und die Einheit der Physik. Das Entstehen einer neuen Naturphilosophie. *Philos. Natur.* **31** (1994), 171–216
13. L. Boi, Le concept de variété et la nouvelle géométrie de l'espace dans la pensée de Bernhard Riemann: l'émergence d'une nouvelle vision des mathématiques et de ses rapports avec les sciences fondamentales. Arch. Internat. Hist. Sci. **45**(134), 82–128 (1995)
14. L. Boi, *Le problème mathématique de l'espace* (Springer-Verlag, Berlin, 1995)
15. A. Cayley, A sixth memoir upon quantics. Philos. Trans. R. Soc. Lond. **149**, 61–90 (1859) (also *Collected Mathematical Papers*, Vol. 2, pp. 561–592, Cambridge University Press, Cambridge 1889)
16. M. Dehn, Die Legendre'schen Sätze über die Winkelsumme im Dreieck. Math. Ann. **53**, 404–439 (1900)
17. J. Diller, Eine algebraische Beschreibung der metrischen Ebenen mit ineinander beweglichen Geraden. Abh. Math. Semin. Univ. Hamburg **34**, 184–202 (1970)
18. M. Eklund, On how logic became first-order. Nordic J. Philos. Logic **1**, 147–167 (1996)
19. G. Ewald, Spiegelungsgeometrische Kennzeichnung euklidischer und nichteuklidischer Räume beliebiger Dimension. Abh. Math. Sem. Univ. Hamburg **41**, 224–251 (1974)

20. J. Ferreirós, The road to modern logic–an interpretation. Bull. Symb. Logic **7**, 441–484 (2001)
21. J. Ferreirós, Riemann's Habilitationsvortrag at the crossroads of mathematics, physics, and philosophy, in *The Architecture of Modern Mathematics*, ed. by J. Ferreirós, J.J. Gray (Oxford University Press, Oxford, 2006), pp. 67–96
22. H.R. Friedlein, Generalizacion de snail maps a geometrias metricas de cualquier dimension, e immersion de estas geometrias. Gac. Mat. (Madrid) **29**(1–2), 10–21 (1977)
23. H.R. Friedlein, E.S. Saez, Immersion de una geometria metrica en una geometria proyectiva. Gac. Mat. (Madrid) **29**(3–4), 53–60 (1977)
24. H.-R. Friedlein, Elliptische Bewegungsgruppen und orthogonale Gruppen vom Index 0. Abh. Math. Sem. Univ. Hamburg **49**, 140–154 (1979)
25. H.-R. Friedlein, Über das Zentrum metrischer Bewegungsgruppen nach Ewald. J. Geom. **22**, 153–157 (1984)
26. S. Gandon, *Russell's Unknown Logicism*, A study in the history and philosophy of mathematics (Palgrave Macmillan, Basingstoke, 2012)
27. M. Grochowska, Euclidean two dimensional equidistance theory. Demonstratio Math. **17**, 593–607 (1984)
28. G. Heimbeck, Zur gruppentheoretischen Kennzeichnung elliptischer Bewegungsgruppen. Arch. Math. (Basel) **28**, 374–377 (1977)
29. G. Heimbeck, Zum Aufbau der absoluten Geometrie nach Ewald. Abh. Math. Sem. Univ. Hamburg **50**, 70–88 (1980)
30. G. Hellwig, *Modelle der absoluten Geometrie*. Dissertation, Universität Kiel, Kiel, 1991
31. G. Hessenberg, Neue Begründung der Sphärik. S.-Ber. Berlin. Math. Ges. **4**, 69–77 (1905)
32. G. Hessenberg, J. Diller, *Grundlagen der Geometrie*, 2nd edn. (Walter de Gruyter, Berlin, 1967)
33. D. Hilbert, Neue Begründung der Bolyai-Lobatschefskyschen Geometrie. Math. Ann. **57**, 137–150 (1903)
34. D. Hilbert, *Grundlagen der Geometrie* (Leipzig, Teubner,1899) (translated by L. Unger, Open Court, La Salle, Ill., under the title: *Foundations of Geometry*, 1971)
35. D. Hilbert, *David Hilbert's lectures on the foundations of arithmetic and logic, 1917–1933*, eds. by W. Ewald, W. Sieg (Springer, Berlin, 2013) (Collaboration with U. Majer, D. Schlimm)
36. J. Hjelmslev, Neue Begründung der ebenen Geometrie. Math. Ann. **64**, 449–474 (1907)
37. J. Hjelmslev, Einleitung in die allgemeine Kongruenzlehre. I, II., Mat.-Fys. Medd. K. Dan. Vidensk. Selsk. 8, Nr. **11**, 1–36; 10 (1929). Nr. **1**, 1–28 (1929)
38. H. Karzel, H.-J. Kroll, *Geschichte der Geometrie seit Hilbert* (Wissenschaftliche Buchgesellschaft, Darmstadt, 1988)
39. G. Hübner, Klassifikation n-dimensionaler absoluter Geometrien. Abh. Math. Semin. Univ. Hamburg **33**, 165–182 (1969)
40. H. Kinder, *Begründung der n-dimensionalen absoluten Geometrie aus dem Spiegelungsbegriff*. Dissertation, Christian-Albrechts-Universität zu Kiel, Kiel, 1965
41. H. Kinder, Elliptische Geometrie endlicher Dimension. Arch. Math. (Basel) **21**, 515–527 (1970)
42. F. Klein, *Vorlesungen über nicht-euklidische Geometrie* (Springer, Berlin, 1928)
43. W. Klingenberg, Eine Begründung der hyperbolischen Geometrie. Math. Ann. **127**, 340–356 (1954)
44. P. Klopsch, Algebraische Kennzeichnung angeordneter Bachmann-Räume. Geom. Dedicata **18**, 249–259 (1985)
45. F. Knüppel, Reflection groups—On pre-Hjelmslev groups and related topics, in *Generators and Relations in Groups and Geometries*, eds. by A. Barlotti, E.W. Ellers, P. Plaumann, K. Strambach (Kluwer, Dordrecht, 1991), pp. 125–164
46. T. Kouremenos, *The Unity of Mathematics in Plato's Republic* (Franz Steiner Verlag, Stuttgart, 2015)
47. H. Lenz, Inzidenzräume mit Orthogonalität. Math. Ann. **146**, 369–374 (1962)
48. R. Lingenberg, *Metric Planes and Metric Vector Spaces* (Wiley, New York, 1979)

49. H. Maehara, The problem of thirteen spheres–a proof for undergraduates. Eur. J. Combin. **28**, 1770–1778 (2007)
50. K. Menger, Bemerkungen zu Grundlagenfragen. Jahresber. Deutsch. Math.-Verein. **37**, 309–325 (1928)
51. K. Menger, New foundations of affine and projective geometry. Ann. Math. **37**, 456–482 (1936)
52. H. Müller, *Zur Begründung der ebenen absoluten Geometrie aus Bewegungsaxiomen*. Dissertation, Technische Hochschule München, München 1966
53. J. von Neumann, Continuos geometry. Proc. Natl. Acad. Sci. USA **22**, 92–100 (1936)
54. G. Nowak, Riemann's Habilitationsvortrag and the synthetic a priori status of geometry, in *The History of Modern Mathematics*, vol. I, ed. by D.E. Rowe, J. McCleary (Academic Press, Boston, MA, 1989), pp. 17–46
55. W. Nolte, Zur Begründung der absoluten Geometrie des Raumes. Math. Z. **94**, 32–60 (1966)
56. W. Nolte, Metrische Räume mit dreiseitverbindbaren Teilräumen. I, II. J. Geom. **4**, 53–90, 91–117 (1974)
57. V. Pambuccian, Ternary operations as primitive notions for constructive plane geometry IV. Math. Log. Q. **40**, 76–86 (1994)
58. V. Pambuccian, Zum Stufenaufbau des Parallelenaxioms. J. Geom. **51**, 79–88 (1994)
59. V. Pambuccian, Ternary operations as primitive notions for constructive plane geometry V. Math. Log. Q. **40**, 455–477 (1994)
60. V. Pambuccian, Fragments of Euclidean and hyperbolic geometry. Sci. Math. Jpn. **53**, 361–400 (2001)
61. V. Pambuccian, Constructive axiomatization of non-elliptic metric planes. Bull. Polish Acad. Sci. Math. **51**, 49–57 (2003)
62. V. Pambuccian, Review of R. Hartshorne, Geometry: Euclid and beyond. (2000) (Springer, New York) Am. Math. Mon. **110**, 66–70 (2003)
63. V. Pambuccian, Groups and plane geometry. Studia Logica **81**, 387–398 (2005)
64. V. Pambuccian, Orthogonality as single primitive notion for metric planes. With an appendix by H. and R. Struve. *Beiträge Algebra Geom.* **48**, 399–409 (2007)
65. V. Pambuccian, The Erdős-Mordell inequality is equivalent to non-positive curvature. J. Geom. **88**, 134–139 (2008)
66. V. Pambuccian, R. Struve, On M.T. Calapso's characterization of the metric of an absolute plane. J. Geom. **92**, 105–116 (2009)
67. V. Pambuccian, The axiomatics of ordered geometry I. Ordered incidence spaces. Expo. Math. **29**, 24–66 (2011)
68. V. Pambuccian, *Essay review of David Hilbert's lectures on the foundations of geometry, 1891–1902*, eds. by M. Hallett, U. Majer (Springer-Verlag, Berlin, 2004). Philos. Math. (III) **21**, 255–277 (2013)
69. V. Pambuccian, H. Struve, R. Struve, The Steiner-Lehmus theorem and "triangles with congruent medians are isosceles" hold in weak geometries. Beiträge Algebra Geom. **57**, 483–497 (2016)
70. D. Pedoe, The most "elementary" theorem of Euclidean geometry. Math. Mag. **49**, 40–42 (1976)
71. W. Pejas, Die Modelle des Hilbertschen Axiomensystems der absoluten Geometrie. Math. Ann. **143**, 212–235 (1961)
72. W. Pejas, Eine algebraische Beschreibung der angeordneten Ebenen mit nichteuklidischer Metrik. Math. Z. **83**, 434–457 (1964)
73. E. Podehl, K. Reidemeister, Eine Begründung der ebenen elliptischen Geometrie. Abh. Math. Sem. Univ. Hamburg **10**, 231–255 (1934)
74. G. Richter, R. Schnabel, Euklidische Räume. J. Geom. **58**, 164–178 (1997)
75. B. Rosenfeld, *Geometry of Lie Groups* (Kluwer, Dordrecht, 1997)
76. C. Rovelli, *Quantum Gravity* (Cambridge University Press, Cambridge, 2004)
77. B. Russell, *Foundations of Geometry (with an foreword of M. Kline)* (Dover, New York, 1956) (first edition 1897)

78. E. Salow, Singuläre Hjelmslev-Gruppen. Geom. Dedicata **1**, 447–467 (1973)
79. H. Scherf, *Begründing der hyperbolischen Geometrie des Raumes*. Dissertation, Christian-Albrechts-Universität zu Kiel, Kiel, 1961
80. A. Schmidt, Die Dualität von Inzidenz und Senkrechtstehen in der absoluten Geometrie. Math. Ann. **118**, 609–625 (1943)
81. R. Schnabel, *Euklidische Geometrie* (Universität Kiel, Kiel, Habilitationsschrift, 1978)
82. E. Scholz, *Geschichte des Mannigfaltigkeitsbegriffs von Riemann bis Poincaré* (Mass, Birkhäuser, Boston, 1980)
83. E. Scholz, Riemanns frühe Notizen zum Mannigfaltigkeitsbegriff und zu den Grundlagen der Geometrie. Arch. Hist. Exact Sci. **27**, 213–232 (1982)
84. E. Scholz, Riemann's vision of a new approach to geometry, in *1830–1930: A century of geometry (Paris, 1989)*, eds. by L. Boi, D. Flament, J.-M. Salanskis (Springer, Berlin, 1992), pp. 22–34
85. E.M. Schröder, Eine Kennzeichnung der regulären euklidischen Geometrien. Abh. Math. Sem. Univ. Hamburg **49**, 95–117 (1979)
86. E.M. Schröder, Zur Kennzeichnung Fanoscher affin-metrischer Geometrien. J. Geom. **16**, 56–62 (1981)
87. E.M. Schröder, On foundations of metric geometries. In: A. Barlotti, M. Marchi, G. Tallini (eds), Atti del convegno "Geometria combinatoria e di incidenza: fondamenti e applicazioni" (La Mendola, 1982), *Rend. Sem. Mat. Brescia* **7**, 583–601 (1984)
88. E.M. Schröder, Aufbau metrischer Geometrie aus der Hexagrammbedingung. Atti Sem. Mat. Fis. Univ. Modena **33**, 183–217 (1984)
89. F. Schur, Über den Fundamentalsatz der projectiven Geometrie. Math. Ann. **51**, 401–409 (1899)
90. F. Schur, Über die Grundlagen der Geometrie. Math. Ann. **55**, 265–292 (1902)
91. K. Schütte, B.L. van der Waerden, Das Problem der dreizehn Kugeln. Math. Ann. **125**, 325–334 (1953)
92. J.G. Semple, G.T. Kneebone, *Algebraic Projective Geometry* (Oxford University Press, Oxford, 1998) (Reprint of the 1952 original)
93. J.T. Smith, Orthogonal geometries I. II. Geom. Dedicata **1**(221–235), 334–339 (1973)
94. J.T. Smith, Metric geometries of arbitrary dimension. Geom. Dedicata **2**, 349–370 (1973)
95. J.T. Smith, Generalized metric geometries of arbitrary dimension. Geom. Dedicata **2**, 485–497 (1974)
96. J.T. Smith, Group theoretic characterization of elliptic geometries of arbitrary dimension. Math. Nachr. **67**, 265–272 (1975)
97. J.T. Smith, Point-hyperplane axioms for orthogonal geometries. Geom. Dedicata **17**, 279–286 (1985)
98. J.T. Smith, Group theoretic characterization of metric geometries of arbitrary dimension. Geom. Dedicata **17**, 287–296 (1985)
99. E. Snapper, R.J. Troyer, *Metric Affine Geometry* (Academic Press, New York, 1971)
100. H. Struve, Ein spiegelungsgeometrischer Aufbau der Galileischen Geometrie. Beiträge Algebra Geom. **17**, 197–211 (1984)
101. H. Struve, R. Struve, Ein spiegelungsgeometrischer Aufbau der cominkowskischen Geometrie. Abh. Math. Sem. Univ. Hamburg **54**, 111–118 (1984)
102. H. Struve, R. Struve, Eine synthetische Charakterisierung der Cayley-Kleinschen Geometrien. Z. Math. Logik Grundlag. Math. **31**, 569–573 (1985)
103. H. Struve, R. Struve, Endliche Cayley-Kleinsche Geometrien. Arch. Math. (Basel) **48**, 178–184 (1987)
104. H. Struve, R. Struve, Zum Begriff der projektiv-metrischen Ebene. Z. Math. Logik Grundlag. Math. **34**, 79–88 (1988)
105. H. Struve, R. Struve, Coeuklidische Hjelmslevgruppen. J. Geom. **34**, 181–186 (1989)
106. H. Struve, R. Struve, Projective spaces with Cayley-Klein metrics. J. Geom. **81**, 155–167 (2004)

107. H. Struve, R. Struve, Lattice theory and metric geometry. Algebra Universalis **58**, 461–477 (2008)
108. H. Struve, R. Struve, Non-euclidean geometries: the Cayley-Klein approach. J. Geom. **98**, 151–170 (2010)
109. R. Struve, An axiomatic foundation of Cayley-Klein geometries. J. Geom. **107**, 225–248 (2016)
110. G. Szekeres, Kinematic geometry: an axiomatic system for Minkowski space-time. J. Aust. Math. Soc. **8**, 134–160 (1968)
111. R. Torretti, *Philosophy of Geometry from Riemann to Poincaré* (D. Reidel, Dordrecht, 1976)
112. O. Veblen, J.W. Young, *Projective Geometry*, vols. 1 and 2 (Ginn and Co, Boston, 1910 and 1918)
113. H. Wiener, *Sechs Abhandlungen über das Rechnen mit Spiegelungen, nebst Anwendungen auf die Geometrie der Bewegungen und auf die projective Geometrie* (Breitkopf & Härtel, Leipzig, 1893)
114. J.A. Wolf, *Spaces of Constant Curvature* (McGraw-Hill, New York, 1967)
115. H. Wolff, Minkowskische und absolute Geometrie. Math. Ann. **171**, 144–193 (1967)
116. I.M. Yaglom, B.A. Rosenfeld, E.U. Yasinskaya, Projective metrics (Russian). *Uspehi Mat. Nauk* **19** No. 3 (119), 51–113 (1964) (also in *Russ. Math. Surveys* **19**, No. 5, 49–107 (1964))

Generalized Riemann Sums

Toshikazu Sunada

Abstract The primary aim of this chapter is, commemorating the 150th anniversary of Riemann's death, to explain how the idea of *Riemann sum* is linked to other branches of mathematics. The materials I treat are more or less classical and elementary, thus available to the "common mathematician in the streets". However one may still see here interesting inter-connection and cohesiveness in mathematics.

Keywords Constant density · Coprime pairs · Primitive pythagorean triples · Quasicrystal · Rational points on the unit circle

1 Introduction

On Gauss's recommendation, Bernhard Riemann presented the paper *Über die Darstellbarkeit einer Function durch eine trigonometrische Reihe* to the Council of Göttingen University as his Habilitationsschrift at the first stage in December

---

The author is partially supported by JSPS Grant-in-Aid for Scientific Research (No. 15H02055).

T. Sunada (✉)
School of Interdisciplinary Mathematical Sciences, Meiji University,
Nakano 4-21-1, Nakano-ku, Tokyo 164-8525, Japan
e-mail: sunada@meiji.ac.jp

T. Sunada
Advanced Institute for Materials Research, Tohoku University,
2-1-1 Katahira, Aoba-ku, Sendai 980-8577, Japan

L. Ji et al. (eds.), *From Riemann to Differential Geometry and Relativity*,
DOI 10.1007/978-3-319-60039-0_15

of 1853.[1] As the title clearly suggests, the aim of his essay was to lay the foundation for the theory of trigonometric series (Fourier series in today's term).[2]

The record of previous work by other mathematicians, to which Riemann devoted three sections of the essay, tells us that the Fourier series had been used to *represent* general solutions of the wave equation and the heat equation without any convincing proof of convergence. For instance, Fourier claimed, in his study of the heat equation (1807, 1822), that if we put

$$a_n = \frac{1}{\pi}\int_{-\pi}^{\pi} f(x)\sin nx\,dx, \quad b_n = \frac{1}{\pi}\int_{-\pi}^{\pi} f(x)\cos nx\,dx, \tag{1}$$

then

$$f(x) = \frac{1}{2}b_0 + (a_1\sin x + b_1\cos x) + (a_2\sin 2x + b_2\cos 2x) + \dots \tag{2}$$

without any restrictions on the function $f(x)$. But this is not true in general as is well known. What is worse (though, needless to say, the significance of his paper as a historical document cannot be denied) is his claim that the integral of an "arbitrary" function is meaningful as the area under/above the associated graph.

L. Dirichlet, a predecessor of Riemann, was the first who gave a solid proof for convergence in a special case. Actually he proved that the right-hand side of (2) converges to $\frac{1}{2}\big(f(x+0)+f(x-0)\big)$ for a class of functions including piecewise monotone continuous functions (1829). Stimulated by Dirichlet's study, Riemann made considerable progress on the convergence problem. In the course of his discussion, he gave a precise notion of integrability of a function,[3] and then obtained a condition for an integrable function to be representable by a Fourier series. Furthermore he proved that the Fourier coefficients for any integrable function a_n, b_n converge to zero as $n \to \infty$. This theorem, which was generalized by Lebesgue later to a broader class of functions, is to be called the Riemann-Lebesgue theorem, and is of importance in Fourier analysis and asymptotic analysis.

What plays a significant role in Riemann's definition of integrals is the notion of *Riemann sum*, which, if we use his notation (Fig. 1), is expressed as

$$S = \delta_1 f(a+\epsilon_1\delta_1) + \delta_2 f(x_1+\epsilon_2\delta_2) + \delta_3 f(x_3+\epsilon_3\delta_3) + \cdots + \delta_n f(x_{n-1}+\epsilon_n\delta_n).$$

[1]Habilitationsschrift is a thesis for qualification to become a lecturer. The famous lecture *Über die Hypothesen welche der Geometrie zu Grunde liegen* delivered on 10 June 1854 was for the final stage of his Habilitationsschrift.

[2]The English translation is "On the representability of a function by a trigonometric series". His essay was published only after his death in the Abhandlungen der Königlichen Gesellschaft der Wissenschaften zu Göttingen (Proceedings of the Royal Philosophical Society at Göttingen), vol. 13, (1868), pages 87–132.

[3]See Sect. 4 in his essay, entitled "Über der Begriff eines bestimmten Integrals und den Umfang seiner Gültigkeit" (On the concept of a definite integral and the extent of its validity), pages 101–103.

Ueber den Begriff eines bestimmten Integrals und den Umfang seiner Gültigkeit.

4.

Die Unbestimmtheit, welche noch in einigen Fundamentalpunkten der Lehre von den bestimmten Integralen herrscht, nöthigt uns, Einiges voraufzuschicken über den Begriff eines bestimmten Integrals und den Umfang seiner Gültigkeit.

Also zuerst: Was hat man unter $\int_a^b f(x)\,dx$ zu verstehen?

Um dieses festzusetzen, nehmen wir zwischen a und b der Grösse nach auf einander folgend, eine Reihe von Werthen $x_1, x_2, \ldots, x_{n-1}$ an und bezeichnen der Kürze wegen $x_1 - a$ durch δ_1, $x_2 - x_1$ durch $\delta_2, \ldots$, $b - x_{n-1}$ durch δ_n und durch ε einen positiven ächten Bruch. Es wird alsdann der Werth der Summe

$$S = \delta_1 f(a + \varepsilon_1 \delta_1) + \delta_2 f(x_1 + \varepsilon_2 \delta_2) + \delta_3 f(x_2 + \varepsilon_3 \delta_3) + \cdots + \delta_n f(x_{n-1} + \varepsilon_n \delta_n)$$

von der Wahl der Intervalle δ und der Grössen ε abhängen. Hat sie nun die Eigenschaft, wie auch δ und ε gewählt werden mögen, sich einer festen Grenze A unendlich zu nähern, sobald sämmtliche δ unendlich klein werden, so heisst dieser Werth $\int_a^b f(x)\,dx$.

Fig. 1 Riemann's paper

Here $f(x)$ is a function on the closed interval $[a, b]$, $a = x_0 < x_1 < x_2 < \cdots < x_{n-1} < x_n = b$, and $\delta_i = x_i - x_{i-1}$ $(i = 1, 2, \ldots, n)$. If S converges to A as $\max_i \delta_i$ goes to 0 whatever ϵ_i with $0 < \epsilon_i < 1$ $(i = 1, \ldots, n)$ are chosen (thus $x_{k-1} + \epsilon_k \delta_k \in [x_{k-1}, x_k]$), then the value A is written as $\int_a^b f(x)dx$, and $f(x)$ is called *Riemann integrable*. For example, every continuous function is Riemann integrable as we learn in calculus.

Compared with Riemann's other supereminent works, his essay looks unglamorous. Indeed, from today's view, his formulation of integrability is no more than routine. But the harbinger must push forward through the total dark without any definite idea of the direction. All he needs is a torch of intelligence.

The primary aim of this chapter is *not* to present the subsequent development after Riemann's work on integrals such as the contribution by C. Jordan (1892),[4] G. Peano (1887), H. L. Lebesgue (1892), T. J. Stieltjes (1894), and K. Ito (1942),[5] but

[4]Jordan introduced a measure (Jordan measure) which fits in with the Riemann integral. A bounded set is Jordan measurable if and only if its indicator function is Riemann integrable.

[5]Ito's integral (or stochastic integral) is a sort of generaization of the Stieltjes integral. Stieltjes defined his integral $\int f(x)d\varphi(x)$ by means of a modified Riemann sum.

to explain how the idea of Riemann sum is linked to other branches of mathematics; for instance, some counting problems in elementary number theory and the theory of quasicrystals, the former having a long history and the latter being an active field still in a state of flux.

I am very grateful to Xueping Guang for drawing attention to Ref. [10] which handles some notions closely related to the ones in the present chapter.

2 Generalized Riemann Sums

The notion of Riemann sum is immediately generalized to functions of several variables as follows.

Let $\Delta = \{D_\alpha\}_{\alpha\in A}$ be a partition of $\mathbb{R}^d$ by a countable family of bounded domains D_α with piecewise smooth boundaries satisfying

(i) $\mathrm{mesh}(\Delta) := \sup_{\alpha\in A} d(D_\alpha) < \infty$, where $d(D_\alpha)$ is the diameter of D_α,

(ii) there are only finitely many α such that $K \cap D_\alpha \neq \emptyset$ for any compact set $K \subset \mathbb{R}^d$.

We select a point $\boldsymbol{\xi}_\alpha$ from each D_α, and put $\Gamma = \{\boldsymbol{\xi}_\alpha \mid \alpha \in A\}$. The Riemann sum $\sigma(f, \Delta, \Gamma)$ for a function f on $\mathbb{R}^d$ with compact support is defined by

$$\sigma(f, \Delta, \Gamma) = \sum_{\alpha\in A} f(\boldsymbol{\xi}_\alpha)\mathrm{vol}(D_\alpha),$$

where $\mathrm{vol}(D_\alpha)$ is the volume of D_α. Note that $f(\boldsymbol{\xi}_\alpha) = 0$ for all but finitely many α because of Property (ii).

If the limit

$$\lim_{\mathrm{mesh}(\Delta)\to 0} \sigma(f, \Delta, \Gamma)$$

exists, independently of the specific sequence of partitions and the choice of $\{\boldsymbol{\xi}_\alpha\}$, then f is said to be Riemann integrable, and this limit is called the (d-tuple) Riemann integral of f, which we denote by $\displaystyle\int_{\mathbb{R}^d} f(\mathbf{x})d\mathbf{x}$.

In particular, if we take the sequence of partitions given by $\Delta_\epsilon = \{\epsilon D_\alpha \mid \alpha \in A\}$ $(\epsilon > 0)$, then, for every Riemann integrable function f, we have

$$\lim_{\epsilon\to +0} \sum_{\alpha\in A} \epsilon^d f(\epsilon\boldsymbol{\xi}_\alpha)\mathrm{vol}(D_\alpha) = \int_{\mathbb{R}^d} f(\mathbf{x})d\mathbf{x}, \tag{3}$$

where we should note that $\mathrm{vol}(\epsilon D_\alpha) = \epsilon^d \mathrm{vol}(D_\alpha)$.

Now we look at Eq. 3 from a different angle. We think that $\omega(\boldsymbol{\xi}_\alpha) := \mathrm{vol}(D_\alpha)$ is a *weight* of the point $\boldsymbol{\xi}_\alpha$, and that Eq. 3 is telling how the *weighted discrete set* (Γ, ω) is distributed in $\mathbb{R}^d$; more specifically we may consider that Eq. 3 implies *uniformity*, in a weak sense, of (Γ, ω) in $\mathbb{R}^d$. This view motivates us to propose the following definition.

In general, a weighted discrete subset in $\mathbb{R}^d$ is a discrete set $\Gamma \subset \mathbb{R}^d$ with a function $\omega : \Gamma \to \mathbb{C}\backslash\{0\}$. Given a compactly supported function f on $\mathbb{R}^d$, define the (generalized) *Riemann sum associated with* (Γ, ω) by setting

$$\sigma(f, \Gamma, \omega) = \sum_{\mathbf{z}\in\Gamma} f(\mathbf{z})\omega(\mathbf{z}).$$

In addition, we say that (Γ, ω) has *constant density* $c(\Gamma, \omega) \neq 0$ (Marklof and Strömbergsson [9]) if

$$\lim_{\epsilon\to+0} \sigma(f^\epsilon, \Gamma, \omega)\Big(= \lim_{\epsilon\to+0} \sum_{z\in\Gamma} \epsilon^d f(\epsilon z)\omega(z)\Big) = c(\Gamma, \omega) \int_{\mathbb{R}^d} f(\mathbf{x})d\mathbf{x} \qquad (4)$$

holds for any bounded Riemann integrable function f on $\mathbb{R}^d$ with compact support, where $f^\epsilon(x) = \epsilon^d f(\epsilon x)$; thus the weighted discrete set associated with a partition $\{D_\alpha\}$ and $\{\boldsymbol{\xi}_\alpha\}$ has constant density 1. In the case $\omega \equiv 1$, we write $\sigma(f, \Gamma)$ for $\sigma(f, \Gamma, \omega)$, and $c(\Gamma)$ for $c(\Gamma, \omega)$ when $\Gamma = (\Gamma, \omega)$ has constant density.

In connection with the notion of constant density, it is perhaps worth recalling the definition of a *Delone set*, a qualitative concept of "uniformity". A discrete set Γ is called a Delone set if it satisfies the following two conditions (Delone [3]).

(1) There exists $R > 0$ such that every ball $B_R(x)$ (of radius R whose center is x) has a nonempty intersection with Γ, i.e., Γ is *relatively dense*;

(2) there exists $r > 0$ such that each ball $B_r(x)$ contains at most one element of Γ, i.e., Γ is *uniformly discrete*.

The following proposition states a relation between Delone sets and Riemann sums.

Proposition 2.1 *Let Γ be a Delone set. Then there exist positive constants c_1, c_2 such that*

$$c_1 \int_{\mathbb{R}^d} f(\mathbf{x})d\mathbf{x} \leq \underline{\lim}_{\epsilon\to+0}\, \sigma(f^\epsilon, \Gamma) \leq \overline{\lim}_{\epsilon\to+0}\, \sigma(f^\epsilon, \Gamma) \leq c_2 \int_{\mathbb{R}^d} f(\mathbf{x})d\mathbf{x}$$

for every nonnegative-valued function f.

Proof In view of the Delone property, one can find two partitions $\{D_\alpha\}$ and $\{D'_\beta\}$ consisting of rectangular parallelotopes satisfying

(i) Every D_α has the same size, and contains at least one element of Γ;
(ii) every D'_β has the same size, and contains at most one element of Γ.

Put $c_1 = \mathrm{vol}(D_\alpha)^{-1}$ and $c_2 = \mathrm{vol}(D'_\beta)^{-1}$. We take a subset Γ_1 of Γ such that every D_α contains just one element of Γ_1, and also take $\Gamma_2 \supset \Gamma$ such that every D'_β contains just one element of Γ_2. We then have $\sigma(f^\epsilon, \Gamma_1) \leq \sigma(f^\epsilon, \Gamma) \leq \sigma(f^\epsilon, \Gamma_2)$. Therefore using Eq. 3, we have

$$
\begin{aligned}
c_1 \int_{\mathbb{R}^d} f(\mathbf{x}) d\mathbf{x} &= \lim_{\epsilon \to +0} \sigma(f^\epsilon, \Gamma_1) \le \underline{\lim}_{\epsilon \to +0} \sigma(f^\epsilon, \Gamma) \\
&\le \overline{\lim}_{\epsilon \to +0} \sigma(f^\epsilon, \Gamma) \le \lim_{\epsilon \to +0} \sigma(f^\epsilon, \Gamma_2) = c_2 \int_{\mathbb{R}^d} f(\mathbf{x}) d\mathbf{x},
\end{aligned}
$$

where we should note that $\sigma_\epsilon(f, \Gamma_1)$ and $\sigma_\epsilon(f, \Gamma_2)$ are ordinary Riemann sums. □

One might ask "what is the significance of the notions of generalized Riemann sum and constant density?" Admittedly these notions are not so much profound (one can find more or less the same concepts in plural references). It may be, however, of great interest to focus our attention on the constant $c(\Gamma, \omega)$. In the subsequent sections, we shall give two "arithmetical" examples for which the constant $c(\Gamma)$ is explicitly computed.

3 Classical Example 1

Let $\mathbb{Z}^d_{\text{prim}}$ $(d \ge 2)$ be the set of *primitive lattice points* in the d-dimensional standard lattice $\mathbb{Z}^d$, i.e., the set of *lattice points visible from the origin* (note that $\mathbb{Z}^2_{\text{prim}}$ is the set of $(x, y) \in \mathbb{Z}^2$ such that $(|x|, |y|)$ is a coprime pair of positive integers, together with $(\pm 1, 0)$ and $(0, \pm 1)$).

Theorem 3.1 $\mathbb{Z}^d_{\text{prim}}$ *has constant density* $\zeta(d)^{-1}$; *that is,*

$$
\lim_{\epsilon \to +0} \sum_{\mathbf{z} \in \mathbb{Z}^d_{\text{prim}}} \epsilon^d f(\epsilon \mathbf{z}) = \zeta(d)^{-1} \int_{\mathbb{R}^d} f(\mathbf{x}) d\mathbf{x}.
$$

Here $\zeta(s) = \sum_{n=1}^{\infty} n^{-s}$ *is the zeta function.*

The proof, which is more or less known as a sort of folklore, will be indicated in Sect. 5.

Noting that $\zeta(2) = \pi^2/6$ and applying this theorem to the indicator function f for the square $\{(x, y) | \, 0 \le x, y \le 1\}$, we obtain the following well-known statement (Fig. 2).

Corollary 3.1 *The probability that two randomly chosen positive integers are coprime is* $6/\pi^2$. *More precisely*

$$
\lim_{N \to \infty} \frac{1}{N^2} \left| \{(a, b) \in \mathbb{N} \times \mathbb{N} | \, \gcd(a, b) = 1, \ a, b \le N\} \right| = \frac{6}{\pi^2},
$$

where $\gcd(a, b)$ *stands for the greatest common divisor of* a, b.

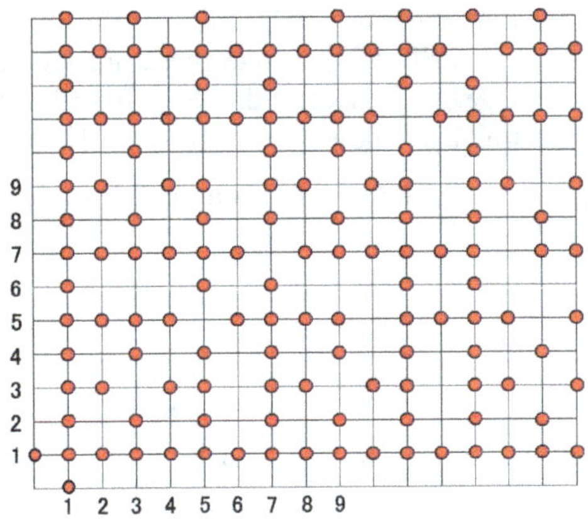

Fig. 2 Coprime pairs

Remark 3.1 (1) Gauss's *Mathematisches Tagebuch*[6] (Mathematical Diary), a record of the mathematical discoveries of C. F. Gauss from 1796 to 1814, contains 146 entries, most of which consist of brief and somewhat cryptical statements. Some of the statements which he never published were independently discovered and published by others often many years later.[7]

The entry relevant to Corollary 3.1 is the 31st dated 1796 September 6:

"Numero fractionum inaequalium quorum denomonatores certum limitem non superant ad numerum fractionum omnium quarum num[eratores] aut denom[inatores] sint diversi infra limitem in infinito ut $6 : \pi\pi$"

This vague statement about counting (irreducible) fractions was formulated in an appropriate way afterwards and proved rigorously by Dirichlet (1849) and Ernesto Cesàro (1881). As a matter of fact, because of its vagueness, there are several ways to interpret what Gauss was going to convey.[8]

(2) In connection with Theorem 3.1, it is perhaps worthwhile to make reference to the *Siegel mean value theorem* ([14]).

Let $g \in \mathrm{SL}_d(\mathbb{R})$. For a bounded Riemann integrable function f on $\mathbb{R}^d$ with compact support, we consider

$$\Phi(g) = \sum_{\mathbf{z}\in\mathbb{Z}^d\setminus\{\mathbf{0}\}} f(g\mathbf{z}), \quad \Psi(g) = \sum_{\mathbf{z}\in\mathbb{Z}^d_{\mathrm{prim}}} f(g\mathbf{z}).$$

[6]See vol. X in Gauss's *Werke*.

[7]The first entry, the most famous one, records the discovery of the construction of a heptadecagon by ruler and compass. The diary was kept by Gauss's bereaved until 1899. It was Stäckel who became aware of the existence of the diary.

[8]For instance, see Ostwald's Klassiker der exakten Wissenschaften; Nr. 256. The 14th entry dated 20 June, 1796 for which Dirichlet gave a proof is considered a companion of the 31st entry. The Yagloms [21] refer to the question on the probability of two random integers being coprime as "Chebyshev's problem".

Both functions Φ and Ψ are $\mathrm{SL}_d(\mathbb{Z})$-invariant with respect to the right action of $\mathrm{SL}_d(\mathbb{Z})$ on $\mathrm{SL}_d(\mathbb{R})$, so that these are identified with functions on the coset space $\mathrm{SL}_d(\mathbb{R})/\mathrm{SL}_d(\mathbb{Z})$. Recall that $\mathrm{SL}_d(\mathbb{R})/\mathrm{SL}_d(\mathbb{Z})$ has finite volume with respect to the measure dg induced from the Haar measure on $\mathrm{SL}_d(\mathbb{R})$. We assume $\int_{\mathrm{SL}_d(\mathbb{R})/\mathrm{SL}_d(\mathbb{Z})} 1\, dg = 1$. Then the Siegel theorem asserts

$$\int_{\mathrm{SL}_d(\mathbb{R})/\mathrm{SL}_d(\mathbb{Z})} \Big(\sum_{\mathbf{z}\in\mathbb{Z}^d\setminus\{\mathbf{0}\}} f(g\mathbf{z})\Big) dg = \int_{\mathbb{R}^d} f(\mathbf{x})d\mathbf{x},$$

$$\int_{\mathrm{SL}_d(\mathbb{R})/\mathrm{SL}_d(\mathbb{Z})} \Big(\sum_{\mathbf{z}\in\mathbb{Z}^d_{\mathrm{prim}}} f(g\mathbf{z})\Big) dg = \zeta(d)^{-1} \int_{\mathbb{R}^d} f(\mathbf{x})d\mathbf{x}.$$

□

4 Classical Example 2

A *Pythagorean triple*,[9] the name stemming from the Pythagorean theorem for right triangles, is a triple of positive integers (ℓ, m, n) satisfying the equation $\ell^2 + m^2 = n^2$. Since $(\ell/n)^2 + (m/n)^2 = 1$, a Pythagorean triple yields a *rational point* $(\ell/n, m/n)$ on the unit circle $S^1 = \{(x, y) |\, x^2 + y^2 = 1\}$. Conversely any rational point on S^1 is derived from a Pythagorean triple. Furthermore the well-known parameterization of S^1 given by $x = (1-t^2)/(1+t^2)$, $y = 2t/(1+t^2)$ tells us that the set of rational points $S^1(\mathbb{Q}) = S^1 \cap \mathbb{Q}^2$ is dense in S^1 (we shall see later how rational points are distributed from a quantitative viewpoint).

A Pythagorean triple (x, y, z) is called *primitive* if x, y, z are pairwise coprime. "Primitive" is so named because any Pythagorean triple is generated trivially from the primitive one, i.e., if (x, y, z) is Pythagorean, there are a positive integer ℓ and a primitive (x_0, y_0, z_0) such that $(x, y, z) = (\ell x_0, \ell y_0, \ell z_0)$.

The way to produce primitive Pythagorean triples (PPTs) is described as follows: If (x, y, z) is a PPT, then there exist positive integers m, n such that

(i) $m > n$,
(ii) m and n are coprime,
(iii) m and n have different parity,
(iv) $(x, y, z) = (m^2 - n^2, 2mn, m^2 + n^2)$ or $(x, y, z) = (2mn, m^2 - n^2, m^2 + n^2)$.

Conversely, if m and n satisfy (i), (ii), (iii), then $(m^2 - n^2, 2mn, m^2 + n^2)$ and $(2mn, m^2 - n^2, m^2 + n^2)$ are PPTs.

[9] Pythagorean triples have a long history since the Old Babylonian period in Mesopotamia nearly 4000 years ago. Indeed, one can read 15 Pythagorean triples in the ancient tablet, written about 1800 BCE, called Plimpton 322 (Weil [19]).

In the table below, due to M. Somos [15], of PPTs (x, y, z) enumerated in ascending order with respect to z, the triple (x_N, y_N, z_N) is the N-th PPT (we do not discriminate between (x, y, z) and (y, x, z)).

N	x_N	y_N	z_N
1	3	4	5
2	5	12	13
3	15	8	17
4	7	24	25
5	21	20	29
6	35	12	37
7	9	40	41
8	45	28	53
9	11	60	61
10	63	16	65

N	x_N	y_N	z_N
11	33	56	65
12	55	48	73
13	77	36	85
14	13	84	85
15	39	80	89
16	65	72	97
17	99	20	101
18	91	60	109
19	15	112	113
20	117	44	125

…

N	x_N	y_N	z_N
1491	4389	8300	9389
1492	411	9380	9389
1493	685	9372	9397
1494	959	9360	9409
1495	9405	388	9413
1496	5371	7740	9421
1497	9393	776	9425
1498	7503	5704	9425
1499	6063	7216	9425
1500	1233	9344	9425

What we have interest in is the *asymptotic behavior* of z_N as N goes to infinity. The numerical observation tells us that the sequence $\{z_N\}$ almost linearly increases as N increases. Indeed $z_{100}/100 = 6.29$, $z_{1000}/1000 = 6,277$, $z_{1500}/1500 = 6.28333\ldots$, which convinces us that $\lim_{N\to\infty} z_N/N$ exists (though the speed of convergence is very slow), and the limit is expected to be equal to $2\pi = 6.2831853\ldots$. This is actually true as shown by Lehmer [8] in 1900, though his proof is by no means easy.

We shall prove Lehmer's theorem by counting coprime pairs (m, n) satisfying the condition that $m - n$ is odd. A key of our proof is the following theorem.

Theorem 4.1 $\mathbb{Z}^{2,*}_{\text{prim}} = \{(m.n) \in \mathbb{Z}^2_{\text{prim}} |\ m - n$ *is odd*$\}$ $(= \{(m.n) \in \mathbb{Z}^2_{\text{prim}} |\ m - n \equiv 1$ *(mod* $2)\})$ *has constant density* $4/\pi^2$; *namely*

$$\lim_{\epsilon\to+0} \sum_{\mathbf{z}\in\mathbb{Z}^{2,*}_{\text{prim}}} \epsilon^2 f(\epsilon\mathbf{z}) = \frac{2}{3}\zeta(2)^{-1}\int_{\mathbb{R}^2} f(\mathbf{x})d\mathbf{x} = \frac{4}{\pi^2}\int_{\mathbb{R}^2} f(\mathbf{x})d\mathbf{x}. \tag{5}$$

We postpone the proof to Sect. 5, and apply this theorem to the indicator function f for the set $\{(x, y) \in \mathbb{R}^2 | x \geq y,\ x^2 + y^2 \leq 1\}$. Since

$$\sum_{\mathbf{z}\in\mathbb{Z}^{2,*}_{\text{prim}}} \epsilon^2 f(\epsilon\mathbf{z}) = \epsilon^2 \big|\{(m, n) \in \mathbb{N}^2 |\ \gcd(m, n) = 1,\ m > n,$$
$$m^2 + n^2 \leq \epsilon^{-2},\ m - n \text{ is odd}\}\big|,$$

we obtain

$$\lim_{N\to\infty} \frac{1}{N} \big|\{(m, n) \in \mathbb{N}^2 |\ \gcd(m, n) = 1,\ m > n,\ m^2 + n^2 \leq N, \tag{6}$$
$$m - n \text{ is odd}\}\big| = \frac{2}{3}\cdot\frac{6}{\pi^2}\cdot\frac{\pi}{8} = \frac{1}{2\pi}.$$

Note that $\left|\{(m,n)\in\mathbb{N}^2|\ \gcd(m,n)=1,\ m>n,\ m^2+n^2\le N,\ m-n \text{ is odd}\}\right|$ coincides with the number of PPT (x,y,z) with $z\le N$. This observation leads us to

Corollary 4.1 (Lehmer) $\displaystyle\lim_{N\to\infty}\frac{z_N}{N}=2\pi.$

Remark 4.1 Fermat's theorem on sums of two squares,[10] together with his *little theorem* and the formula $(a^2+b^2)(c^2+d^2)=(ac\pm bd)^2+(ad\mp bc)^2$, yields the following complete characterization of PPTs which is substantially equivalent to the result stated in the letter from Fermat to Mersenne dated 25 December 1640 (cf. Weil [19]).

An odd number z is written as m^2+n^2 by using two coprime positive integers m,n (thus automatically having different parity) if and only if every prime divisor of z is of the form $4k+1$. In other words, the set $\{z_N\}$ coincides with the set of odd numbers whose prime divisors are of the form $4k+1$. Moreover, if we denote by $\nu(z)$ the number of distinct prime divisors of z, then $z=z_N$ in the list is repeated $2^{\nu(z)-1}$ times. □

Theorem 4.1 can be used to establish

Corollary 4.2 *For a rational point $(p,q)\in S^1(\mathbb{Q})(=S^1\cap\mathbb{Q}^2)$, define the height $h(p,q)$ to be the minimal positive integer h such that $(hp,hq)\in\mathbb{Z}^2$. Then for any arc A in S^1, we have*

$$\left|\{(p,q)\in A\cap\mathbb{Q}^2|\ h(p,q)\le h\}\right|\sim\frac{2\cdot\text{length}(A)}{\pi^2}h\quad(h\to\infty),$$

and hence rational points are equidistributed on the unit circle in the sense that

$$\lim_{h\to\infty}\frac{\left|\{(p,q)\in A\cap\mathbb{Q}^2|\ h(p,q)\le h\}\right|}{\left|\{(p,q)\in S^1\cap\mathbb{Q}^2|\ h(p,q)\le h\}\right|}=\frac{\text{length}(A)}{2\pi}.$$

In his paper [4], W. Duke suggested that this corollary can be proved by using tools from the theory of L-functions combined with Weyl's famous criterion for *equidistribution on the circle* ([20]). Our proof below relies on a generalization of Eq. 6.

Given α,β with $0\le\alpha<\beta\le 1$, we put

$$P(N;\alpha,\beta)=\{(m,n)\in\mathbb{N}\times\mathbb{N}|\ \gcd(m,n)=1,\alpha\le n/m\le\beta,$$
$$m-n \text{ is odd},\ m^2+n^2\le N\}.$$

Namely we count coprime pairs (m,n) with odd $m-n$ in the circular sector

$$\{(x,y)\in\mathbb{R}^2|\ x,y>0,\ \alpha x\le y\le\beta x,\ x^2+y^2\le N\}.$$

[10] Every prime number $p=4k+1$ is in one and only one way a sum of two squares of positive integers.

Since the area of the region $\{(x, y) \in \mathbb{R}^2 | \, x, y > 0, \ \alpha x \leq y \leq \beta x, \ x^2 + y^2 \leq 1\}$ is $\frac{1}{2} \arctan \frac{\beta - \alpha}{1 + \alpha\beta}$, applying again Eq. 5 to the indicator function for this region, we obtain

$$\lim_{N \to \infty} \frac{1}{N} \big| P(N; \alpha, \beta) \big| = \frac{2}{\pi^2} \arctan \frac{\beta - \alpha}{1 + \alpha\beta}.$$

Now we sort points $(p, q) \neq (\pm 1, 0), (0, \pm 1)$ in $S^1(\mathbb{Q})$ by 4 quadrants containing (p, q), and also by parity of x when we write $|p| = x/z$, $|q| = y/z$ with a PPT (x, y, z). Here we should notice that $h(p, q) = z = m^2 + n^2$. Thus counting rational points with the height function $h(p, q)$ reduces to counting PPTs.

Put

$$\begin{aligned} S^1_{\mathbb{Q}}(\text{odd}) &= \big\{(p, q) \in S^1(\mathbb{Q}) | \, x \text{ is odd}\big\}, \\ S^1_{\mathbb{Q}}(\text{even}) &= \big\{(p, q) \in S^1(\mathbb{Q}) | \, x \text{ is even}\big\}. \end{aligned}$$

Then

$$S^1(\mathbb{Q}) = S^1_{\mathbb{Q}}(\text{odd}) \cup S^1_{\mathbb{Q}}(\text{even}) \cup \big\{(\pm 1, 0), (0, \pm 1)\big\} \quad \text{(disjoint)}.$$

Note that the correspondence $(p, q) \mapsto (q, p)$ interchanges $S^1_{\mathbb{Q}}(\text{odd})$ and $S^1_{\mathbb{Q}}(\text{even})$. Therefore, in order to complete the proof, it is enough to show that

$$\begin{aligned} &\Big|\big\{(p, q) \in S^1_{\mathbb{Q}}(\text{odd}) | \, \theta_1 \leq \theta(p, q) < \theta_2, \ h(p, q) \leq h\big\}\Big| \\ &\sim \frac{1}{\pi^2}(\theta_2 - \theta_1)h \quad (h \to \infty), \end{aligned}$$

where $(p, q) = \big(\cos\theta(p, q), \sin\theta(p, q)\big)$. Without loss of generality, one may assume $0 \leq \theta_1 < \theta_2 \leq \pi/2$. Since

$$\tan\theta(p, q) = \frac{q}{p} = \frac{2mn}{m^2 - n^2} = \frac{2\frac{n}{m}}{1 - \left(\frac{n}{m}\right)^2},$$

if we define $\Theta(m, n) \in [0, \pi/2)$ by $\tan\Theta(m, n) = n/m$, then $\tan\theta(p, q) = \tan 2\Theta(m, n)$, and hence $\theta(p, q) = 2\Theta(m, n)$. Therefore

$$\begin{aligned} &\Big|\big\{(p, q) \in S^1_{\mathbb{Q}}(\text{odd}) | \, \theta_1 \leq \theta(g) < \theta_2, \ h(p, q) \leq h\big\}\Big| \\ &= \big|P(h; \arctan\theta_1/2, \arctan\theta_2/2)\big| \\ &\sim \frac{1}{\pi^2}(\theta_2 - \theta_1)h, \end{aligned}$$

as required.

Remark 4.2 Interestingly, $S^1(\mathbb{Q})$ (and hence Pythagorean triples) has something to do with crystallography. Indeed $S^1(\mathbb{Q})$ with the natural group operation is an example of *coincidence symmetry groups* that show up in the theory of crystalline interfaces and grain boundaries[11] in polycrystalline materials (Ranganathan [11], Zeiner [22]). This theory is concerned with partial coincidence of lattice points in two identical crystal lattices. See [17] for the details, and also [16] for the mathematical theory of crystal structures. □

5 The Inclusion-Exclusion Principle

The proof that the discrete sets $\mathbb{Z}^d_{\text{prim}}$ and $\mathbb{Z}^{2,*}_{\text{prim}}$ have constant density relies on the identities derived from the so-called *Inclusion-Exclusion Principle* (IEP), which is a generalization of the obvious equality $|A \cup B| = |A| + |B| - |A \cap B|$ for two finite sets A, B. Despite its simplicity, the IEP is a powerful tool to approach general *counting problems* involving aggregation of things that are not mutually exclusive (Comtet [1]).

To state the IEP in full generality, we consider a family $\{A_h\}_{h=1}^{\infty}$ of subsets of X where X and A_h are not necessarily finite. Let f be a real-valued function with finite support defined on X. We assume that there exists N such that if $h > N$, then $A_h \cap \operatorname{supp} f = \emptyset$, i.e. $f(x) = 0$ for $x \in A_h$. In the following theorem, the symbol A^c means the complement of a subset A in X.

Theorem 5.1 (Inclusion-Exclusion Principle)

$$\sum_{x\in\bigcap_{h=1}^{\infty} A_h^c} f(x) = \sum_{k=0}^{\infty}(-1)^k \sum_{h_1<\cdots<h_k} \sum_{x\in A_{h_1}\cap\cdots\cap A_{h_k}} f(x) \tag{7}$$

$$\left(= \sum_{k=0}^{N}(-1)^k \sum_{h_1<\cdots<h_k} \sum_{x\in A_{h_1}\cap\cdots\cap A_{h_k}} f(x)\right),$$

where, for $k = 0$, the term $\sum_{h_1<\cdots<h_k} \sum_{x\in A_{h_1}\cap\cdots\cap A_{h_k}} f(x)$ *should be understood to be* $\sum_{x\in X} f(x)$.

For the proof, one may assume, without loss of generality, that X is finite, and it suffices to handle the case of a finite family $\{A_h\}_{h=1}^{N}$. The proof is accomplished by induction on N.

Making use of the IEP, we obtain the following theorem (this is actually an easy exercise of the IEP; see Vinogradov [18] for instance).

[11] Grain boundaries are interfaces where crystals of different orientations meet.

Theorem 5.2

$$\sum_{\mathbf{z}\in\mathbb{Z}^d_{\rm prim}} f(\mathbf{z}) = \sum_{k=1}^{\infty} \mu(k) \sum_{\mathbf{w}\in\mathbb{Z}^d\setminus\{\mathbf{0}\}} f(k\mathbf{w}),$$

where f is a function on $\mathbb{R}^d$ with compact support (thus both sides are finite sums), and $\mu(k)$ is the Möbius function:

$$\mu(k) = \begin{cases} 1 & (k=1) \\ (-1)^r & (k = p_{h_1}\cdots p_{h_r};\ h_1 < \cdots < h_r) \\ 0 & (otherwise), \end{cases}$$

where $p_1 < p_2 < \cdots$ are all primes enumerated into ascending order.

The proof goes as follows. Consider the case that

$$X = \mathbb{Z}^d\setminus\{\mathbf{0}\}, \quad A_h = \{(x_1,\ldots,x_d)\in X \mid p_h|x_1,\ldots,p_h|x_d\}.$$

Then $\bigcap_{h=1}^{\infty} A_h^c = \mathbb{Z}^d_{\rm prime}$. We also easily observe

$$A_{h_1}\cap\cdots\cap A_{h_k} = p_{h_1}\cdots p_{h_k}X.$$

Applying Eq. 7 to this case, we have

$$\begin{aligned}\sum_{\mathbf{z}\in\mathbb{Z}^d_{\rm prime}} f(\mathbf{z}) &= \sum_{k=0}^{\infty}(-1)^k \sum_{h_1<\cdots<h_k}\ \sum_{\mathbf{w}\in\mathbb{Z}^d\setminus\{\mathbf{0}\}} f(p_{h_1}\cdots p_{h_k}\mathbf{w}) \\ &= \sum_{k=1}^{\infty}\mu(k) \sum_{\mathbf{w}\in\mathbb{Z}^d\setminus\{\mathbf{0}\}} f(k\mathbf{w}).\end{aligned}$$

Proof of Theorem 3.1 Applying Theorem 5.2 to f^{ϵ}, we have

$$\sum_{\mathbf{z}\in\mathbb{Z}^d_{\rm prim}} \epsilon^d f(\epsilon\mathbf{z}) = \sum_{k=1}^{\infty}\mu(k) \sum_{\mathbf{w}\in\mathbb{Z}^d\setminus\{\mathbf{0}\}} \epsilon^d f(\epsilon k\mathbf{w}),$$

What we have to confirm is the exchangeability of the limit and summation:

$$\lim_{\epsilon\to+0}\sum_{k=1}^{\infty}\left(\mu(k)\sum_{\mathbf{w}\in\mathbb{Z}^d\setminus\{\mathbf{0}\}} \epsilon^d f(\epsilon k\mathbf{w})\right) = \sum_{k=1}^{\infty}\lim_{\epsilon\to+0}\left(\mu(k)\sum_{\mathbf{w}\in\mathbb{Z}^d\setminus\{\mathbf{0}\}} \epsilon^d f(\epsilon k\mathbf{w})\right).$$

If we take this for granted, then we easily get the claim since

$$\lim_{\epsilon\to+0}\sum_{\mathbf{w}\in\mathbb{Z}^2\setminus\{\mathbf{0}\}}\epsilon^d f(\epsilon k\mathbf{w}) = k^{-d}\lim_{\delta\to+0}\sum_{\mathbf{w}\in\mathbb{Z}^d\setminus\{\mathbf{0}\}}\delta^d f(\delta\mathbf{w}) = k^{-d}\int_{\mathbb{R}^d} f(\mathbf{x})d\mathbf{x},$$

and $\sum_{k=1}^{\infty}\mu(k)k^{-d} = \zeta(d)^{-1}$. As a matter of fact, the exchangeability does not follow from Weierstrass' M-test in a direct manner. One can check it by a careful argument. □

In the case of Theorem 4.1, we consider

$$(\mathbb{Z}^{\mathrm{odd}})^2_{\mathrm{prim}} = \{(m,n)\in\mathbb{Z}^{\mathrm{odd}}\times\mathbb{Z}^{\mathrm{odd}}|\ \gcd(m,n)=1\},$$

where $\mathbb{Z}^{\mathrm{odd}}$ is the set of odd integers. Then

$$\mathbb{Z}^{2,*}_{\mathrm{prim}} = \mathbb{Z}^2_{\mathrm{prim}}\setminus(\mathbb{Z}^{\mathrm{odd}})^2_{\mathrm{prim}}.$$

Therefore it suffices to show that $(\mathbb{Z}^{\mathrm{odd}})^2_{\mathrm{prim}}$ has constant density $2/\pi^2$. This is done by using the following theorem for which we need a slightly sophisticated use of the IEP.

Theorem 5.3

$$\sum_{\mathbf{z}\in(\mathbb{Z}^{\mathrm{odd}})^2_{\mathrm{prim}}} f(\mathbf{z}) = \sum_{k=1}^{\infty}\mu(k)\sum_{h=0}^{\infty}\sum_{\mathbf{w}\in(\mathbb{Z}^{\mathrm{odd}})^2} f(k2^h\mathbf{w}).$$

For the proof, we put

$$X = \coprod_{\ell=1}^{\infty}\ell(\mathbb{Z}^{\mathrm{odd}})^2_{\mathrm{prim}},\quad A_h = \{(x,y)\in X|\ p_h|x \text{ and } p_h|y\}.$$

Lemma 5.1 $A_{h_1}\cap\cdots\cap A_{h_k} = \coprod_{h=0}^{\infty} p_{h_1}\cdots p_{h_k}2^h(\mathbb{Z}^{\mathrm{odd}})^2.$

Proof It suffices to prove that $A_{h_1}\cap\cdots\cap A_{h_k} = p_{h_1}\cdots p_{h_k}X$ since any positive integer ℓ is expressed as $2^i\times$ odd. Clearly $A_{h_1}\cap\cdots\cap A_{h_k}\supset p_{h_1}\cdots p_{h_k}X$. Let $(x,y)\in A_{h_1}\cap\cdots\cap A_{h_k}$. Then one can find $(a,b)\in\mathbb{Z}^2$ such that $x = p_{h_1}\cdots p_{h_k}a$ and $y = p_{h_1}\cdots p_{h_k}b$. Moreover there exist $\ell\in\mathbb{N}$ and $(m,n)\in(\mathbb{Z}^{\mathrm{odd}})^2_{\mathrm{prim}}$ such that $x=\ell m$, $y=\ell n$, so $p_{h_1}\cdots p_{h_k}|\gcd(\ell m,\ell n)=\ell$. Therefore $(x,y)\in p_{h_1}\cdots p_{h_k}X$. □

Lemma 5.2 $\left(\bigcup_{h=1}^{\infty}A_h\right)^c = (\mathbb{Z}^{\mathrm{odd}})^2_{\mathrm{prim}}.$

Proof Obviously $\coprod_{\ell=2}^{\infty} \ell(\mathbb{Z}^{\text{odd}})^2_{\text{prim}} = \bigcup_{h=1}^{\infty} A_h$, from which the claim follows. □

Theorem 5.3 is a consequence of the above two lemmas.

Now using Theorem 5.3, we have

$$\begin{aligned}\sum_{\mathbf{z}\in(\mathbb{Z}^{\text{odd}})^2_{\text{prim}}} \epsilon^2 f(\epsilon\mathbf{z}) &= \sum_{k=0}^{\infty}(-1)^k \sum_{h_1<\cdots<h_k} \sum_{h=0}^{\infty} \sum_{\mathbf{w}\in(\mathbb{Z}^{\text{odd}})^2} \epsilon^2 f(\epsilon p_{h_1}\cdots p_{h_k} 2^h\mathbf{w})\\ &= \sum_{k=1}^{\infty}\mu(k)\sum_{h=0}^{\infty}\sum_{\mathbf{w}\in(\mathbb{Z}^{\text{odd}})^2} \epsilon^2 f(\epsilon k 2^h \mathbf{w}).\end{aligned}$$

We also have

$$\lim_{\epsilon\to+0} \sum_{\mathbf{z}\in(\mathbb{Z}^{\text{odd}})^2} \epsilon^2 f(\epsilon\mathbf{z}) = \frac{1}{4}\int_{\mathbb{R}^2} f(\mathbf{x})d\mathbf{x},$$

since the left-hand side is the ordinary Riemann sum associated with the partition by the squares with side length 2, and hence

$$\lim_{\epsilon\to+0} \sum_{\mathbf{w}\in(\mathbb{Z}^{\text{odd}})^2} \epsilon^2 f(\epsilon k 2^h\mathbf{w}) = \frac{1}{(k2^h)^2}\frac{1}{4}\int_{\mathbb{R}^2} f(\mathbf{x})d\mathbf{x}.$$

Thus

$$\begin{aligned}&\lim_{\epsilon\to+0} \sum_{\mathbf{z}\in(\mathbb{Z}^{\text{odd}})^2_{\text{prim}}} \epsilon^2 f(\epsilon\mathbf{z})\\ &= \zeta(2)^{-1}\sum_{h=0}^{\infty}\frac{1}{4^h}\int_{\mathbb{R}^2} f(\mathbf{x})d\mathbf{x} = \frac{6}{\pi^2}\cdot\frac{1}{3}\int_{\mathbb{R}^2} f(\mathbf{x})d\mathbf{x} = \frac{2}{\pi^2}\int_{\mathbb{R}^2} f(\mathbf{x})d\mathbf{x},\end{aligned}$$

as desired (this time, the exchangeability of the limit and summation is confirmed by Weierstrass' M-test). □

Remark 5.1 Historically IEP was, for the first time, employed by Nicholas Bernoulli (1687–1759) to solve a combinatorial problem related to permutations.[12] More specifically he counted the number of *derangements*, that is, permutations such that none of the elements appears in its original position.[13] His result is pleasingly phrased, in a similar fashion as in the case of coprime pairs, as "the probability that randomly chosen permutations are derangements is $1/e$" (e is the base of natural logarithms). □

[12] The probabilistic form of IEP is attributed to de Moivre (1718). Sometimes IEP is referred to as the formula of Da Silva, or Sylvester.

[13] This problem ("problème des rencontres") was proposed by Pierre Raymond de Montmort in 1708. He solved it in 1713 at about the same time as did N. Bernoulli.

6 Generalized Poisson Summation Formulas

Generalized Riemann sums appear in the theory of *quasicrystals*, a form of solid matter whose atoms are arranged like those of a crystal but assume patterns that do not exactly repeat themselves.

The interest in quasicrystals arose when in 1984 Schechtman et al. [12] discovered materials whose X-ray diffraction spectra had sharp spots indicative of long range order. Soon after the announcement of their discovery, material scientists and mathematicians began intensive studies of quasicrystals from both the empirical and theoretical sides.[14]

At the moment, there are several ways to mathematically define quasicrystals (see Lagarias [7] for instance). As a matter of fact, an official nomenclature has not yet been agreed upon. In many reference, however, the Delone property for the discrete set Γ representing the location of atoms is adopted as a minimum requirement for the characterization of quasicrystals. In addition to the Delone property, many authors assume that a *generalized Poisson summation formula* holds for Γ, which embodies the patterns of X-ray diffractions for a real quasicrystal.

Let us recall the classical Poisson summation formula. For a *lattice group* L, a subgroup of $\mathbb{R}^d$ generated by a basis of $\mathbb{R}^d$, we denote by L^* the dual lattice of L, i.e., $L^* = \{\eta \in \mathbb{R}^d |\ \langle \eta, \mathbf{z} \rangle \in \mathbb{Z} \text{ for every } \mathbf{z} \in L\}$, and also denote by D_L a fundamental domain for L. We then have

$$\sum_{\mathbf{z}\in L} f(\mathbf{z})e^{2\pi i\langle \mathbf{z},\eta\rangle} = \mathrm{vol}(D_L)^{-1} \sum_{\xi\in L^*} \hat{f}(\xi - \eta) \quad (i = \sqrt{-1}), \tag{8}$$

in particular,

$$\sum_{\mathbf{z}\in L} f(\mathbf{z}) = \mathrm{vol}(D_L)^{-1} \sum_{\xi\in L^*} \hat{f}(\xi), \tag{9}$$

which is what we usually call the Poisson summation formula. Here $\hat{f}$ is the *Fourier transform* of a rapidly decreasing smooth function f:

$$\hat{f}(\xi) = \int_{\mathbb{R}^d} f(\mathbf{x})e^{-2\pi i\langle \mathbf{x},\xi\rangle} d\mathbf{x}.$$

Note that the left-hand side of Eq. 8 is the Riemann sum $\sigma(f, L, \omega_\eta)$ for the weighted discrete set (L, ω_η), where $\omega_\eta(\mathbf{z}) = e^{2\pi i\langle \mathbf{z},\eta\rangle}$.

[14] As will be explained below, the theoretical discovery of quasicrystal structures was already made by R. Penrose in 1973. See Senechal and Taylor [13] for an account on the theory of quasicrystals at the early stage.

Having Eq. 9 in mind, we say that a generalized Poisson formula holds for Γ if there exist a countable subset $\Lambda \subset \mathbb{R}^d$ and a sequence $\{a(\boldsymbol{\xi})\}_{\boldsymbol{\xi}\in\Lambda}$ such that

$$\sum_{\mathbf{z}\in\Gamma} f(\mathbf{z}) = \sum_{\boldsymbol{\xi}\in\Lambda} a(\boldsymbol{\xi})\hat{f}(\boldsymbol{\xi}) \tag{10}$$

for every compactly supported smooth function f.

What we must be careful about here is that the set Λ is allowed to have accumulation points, so that one cannot claim that the right-hand side of Eq. 10 converges in the ordinary sense. Thus the definition above is rather formal. One of the possible justifications is to assume that there exist an increasing family of subsets $\{\Lambda_N\}_{N=1}^{\infty}$ and functions $a_N(\boldsymbol{\xi})$ defined on Λ_N such that

(i) $\displaystyle\bigcup_{N=1}^{\infty} \Lambda_N = \Lambda$,
(ii) $\displaystyle\sum_{\boldsymbol{\xi}\in\Lambda_N} a_N(\boldsymbol{\xi})\hat{f}(\boldsymbol{\xi})$ converges absolutely,
(iii) $\displaystyle\lim_{N\to\infty} a_N(\boldsymbol{\xi}) = a(\boldsymbol{\xi})$,
(iv) $\displaystyle\sum_{\mathbf{z}\in\Gamma} f(\mathbf{z}) = \lim_{N\to\infty} \sum_{\boldsymbol{\xi}\in\Lambda_N} a_N(\boldsymbol{\xi})\hat{f}(\boldsymbol{\xi})$.

We shall say that a discrete set Γ is a *quasicrystal of Poisson type* if a generalized Poisson formula holds for Γ.[15]

A typical class of quasicrystals of Poisson type is constructed by the *cut and project method*.[16] Let L be a lattice group in $\mathbb{R}^N = \mathbb{R}^d \times \mathbb{R}^{N-d}$ $(N > d)$, and let W be a compact domain (called a *window*) in $\mathbb{R}^{N-d}$. We denote by p_d and p_{N-d} the orthogonal projections of $\mathbb{R}^N$ onto $\mathbb{R}^d$ and $\mathbb{R}^{N-d}$, respectively. We assume that $p_{N-d}(L)$ is dense, and p_d is invertible on $p_d(L)$. Then the quasicrystal (called a *model set*) Γ associated with L and W is defined to be $p_d\big(L \cap (\mathbb{R}^d \times W)\big)$.

We put $\Lambda = p_d(L^*)$. It should be remarked that for each $\boldsymbol{\xi} \in \Lambda$, there exists a unique $\boldsymbol{\xi}' \in \mathbb{R}^{N-d}$ such that $(\boldsymbol{\xi}, \boldsymbol{\xi}') \in L^*$. Indeed, if $(\boldsymbol{\xi}, \boldsymbol{\xi}'') \in L^*$, then $(\mathbf{0}, \boldsymbol{\xi}' - \boldsymbol{\xi}'') \in L^*$, and hence $\mathbb{Z} \ni \langle(\mathbf{0}, \boldsymbol{\xi}' - \boldsymbol{\xi}''), \boldsymbol{\alpha}\rangle = \langle\boldsymbol{\xi}' - \boldsymbol{\xi}'', p_{N-d}(\boldsymbol{\alpha})\rangle$ for every $\boldsymbol{\alpha} \in L$. Since $p_{N-d}(L)$ is dense, we conclude that $\boldsymbol{\xi}' - \boldsymbol{\xi}'' = \mathbf{0}$.

Let us write down a generalized Poisson formula for Γ in a formal way. Let f be a compactly supported smooth function on $\mathbb{R}^d$, and let χ_W be the indicator function of the window $W \subset \mathbb{R}^{N-d}$. Define the compactly supported function F on $\mathbb{R}^N$ by setting $F(\mathbf{x}, \mathbf{x}') = f(\mathbf{x})\chi_W(\mathbf{x}')$ $(\mathbf{x} \in \mathbb{R}^d, \mathbf{x}' \in \mathbb{R}^{N-d})$. Applying the Poisson summation formula to F, we obtain

$$\sum_{\mathbf{z}\in\Gamma} f(\mathbf{z}) = \sum_{\boldsymbol{\alpha}\in L} F(\boldsymbol{\alpha}) = \mathrm{vol}(D_L)^{-1} \sum_{\boldsymbol{\beta}\in L^*} \widehat{F}(\boldsymbol{\beta}),$$

---

[15] Some people use the term “Poisson comb” in a bit different formulation.

[16] This method was invented by de Bruijn [2], and developed by many authors.

which is, of course, a "formal" identity because the right-hand side does not necessarily converge. Pretending that this is a genuine identity and noting

$$\widehat{F}(\boldsymbol{\beta}) = \hat{f}(\boldsymbol{\xi})\widehat{\chi_W}(\boldsymbol{\xi}') \quad (\boldsymbol{\beta} = (\boldsymbol{\xi}, \boldsymbol{\xi}') \in \mathbb{R}^d \times \mathbb{R}^{N-d}),$$

we get

$$\sum_{\mathbf{z}\in\Gamma} f(\mathbf{z}) = \sum_{\boldsymbol{\xi}\in\Lambda} a(\boldsymbol{\xi})\hat{f}(\boldsymbol{\xi}), \tag{11}$$

where, for $\boldsymbol{\xi} \in \Lambda$, we put

$$a(\boldsymbol{\xi}) = \mathrm{vol}(D_L)^{-1}\widehat{\chi_W}(\boldsymbol{\xi}') \quad ((\boldsymbol{\xi}, \boldsymbol{\xi}') \in L^*).$$

We may justify Eq. 11 as follows. Let $U_{1/N}(W)$ be the $1/N$-neighborhood of W, and take a smooth function g_N on $\mathbb{R}^{N-d}$ satisfying $0 \le g_N(\mathbf{x}') \le 1$ and

$$g_N(\mathbf{x}') = \begin{cases} 1 & (\mathbf{x}' \in W) \\ 0 & (\mathbf{x}' \in U_{1/N}(W)^c) \end{cases}.$$

Put $F_N(\mathbf{x}, \mathbf{x}') = f(\mathbf{x})g_N(\mathbf{x}')$. If we take $N \gg 1$, we have $(\mathrm{supp} f \times U_{1/N}(W)) \cap L = (\mathrm{supp}\, f \times W) \cap L$, so that, if $f(\mathbf{z})g_N(\mathbf{z}') \neq 0$ for $\boldsymbol{\alpha} = (\mathbf{z}, \mathbf{z}') \in L$, then $(\mathbf{z}, \mathbf{z}') \in (\mathrm{supp}\, f \times U_{1/N}(W)) \cap L = (\mathrm{supp}\, f \times W) \cap L$, and hence $\mathbf{z} \in \Gamma$ and $F_N(\boldsymbol{\alpha}) = f(\mathbf{z})$. We thus have

$$\begin{aligned} \sum_{\mathbf{z}\in\Gamma} f(\mathbf{z}) &= \sum_{\boldsymbol{\alpha}\in L} F_N(\boldsymbol{\alpha}) = \mathrm{vol}(D_L)^{-1} \sum_{\boldsymbol{\beta}\in L^*} \widehat{F_N}(\boldsymbol{\beta}) \\ &= \mathrm{vol}(D_L)^{-1} \sum_{(\boldsymbol{\xi},\boldsymbol{\xi}')\in L^*} \hat{f}(\boldsymbol{\xi})\widehat{g_N}(\boldsymbol{\xi}') \\ &= \sum_{\boldsymbol{\xi}\in\Lambda} a_N(\boldsymbol{\xi})\hat{f}(\boldsymbol{\xi}), \end{aligned}$$

where $a_N(\boldsymbol{\xi}) = \mathrm{vol}(D_L)^{-1}\widehat{g_N}(\boldsymbol{\xi}')$. Obviously $\lim_{N\to\infty} a_N(\boldsymbol{\xi}) = a(\boldsymbol{\xi})$.

A typical example of model sets is the set of nodes in a *Penrose tiling* discovered by R. Penrose in 1973/1974, which is a remarkable non-periodic tiling generated by an aperiodic set of prototiles (see de Bruijn [2] for the proof of the fact that a Penrose tiling is obtained by the cut and projection method) (Fig. 3).

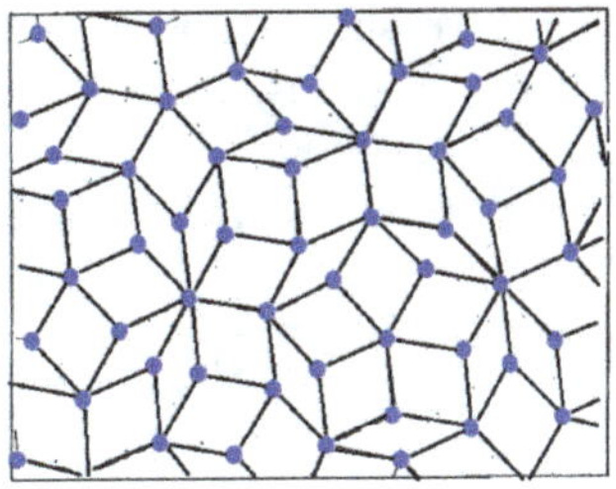

Fig. 3 A Penrose tiling

7 Is $\mathbb{Z}^d_{\mathrm{prim}}$ a Quasicrystal?

It is natural to ask whether $\mathbb{Z}^d_{\mathrm{prim}}$ is a quasicrystal. The answer is "No." However $\mathbb{Z}^d_{\mathrm{prim}}$ is *nearly* a quasicrystal of Poisson type.

To see this, take a look again at the identity

$$\sum_{\mathbf{z}\in\mathbb{Z}^d_{\mathrm{prim}}} f(\mathbf{z}) = \sum_{k=1}^{\infty} \mu(k) \sum_{\mathbf{w}\in\mathbb{Z}^d\setminus\{\mathbf{0}\}} f(k\mathbf{w}).$$

Suppose that $\operatorname{supp} f \subset B_N(\mathbf{0})$. Then applying the Poisson summation formula, we obtain

$$\begin{aligned}
\sum_{\mathbf{z}\in\mathbb{Z}^d_{\mathrm{prim}}} f(\mathbf{z}) &= \sum_{k=1}^{\infty} \mu(k) \sum_{\mathbf{w}\in\mathbb{Z}^d\setminus\{0\}} f(k\mathbf{w}) \\
&= \sum_{k=1}^{N} \mu(k)\Big[\sum_{\mathbf{w}\in\mathbb{Z}^d} f(k\mathbf{w}) - f(\mathbf{0})\Big] \\
&= \sum_{k=1}^{N} \mu(k)k^{-d} \sum_{\boldsymbol{\xi}\in k^{-1}\mathbb{Z}^d} \hat{f}(\boldsymbol{\xi}) - \Big(\sum_{k=1}^{N} \mu(k)\Big) f(\mathbf{0}).
\end{aligned}$$

Now for $\boldsymbol{\xi} \in \mathbb{Q}^d$, we write

$$\boldsymbol{\xi} = \Big(\frac{b_1}{a_1}, \ldots, \frac{b_d}{a_d}\Big), \quad \gcd(a_i, b_i) = 1, a_i > 0,$$

and put $n(\boldsymbol{\xi}) = \operatorname{lcm}(a_1, \ldots, a_d)$. Then $\boldsymbol{\xi} \in k^{-1}\mathbb{Z}^d \iff n(\boldsymbol{\xi})|k$, and hence

$$\sum_{\mathbf{z}\in\mathbb{Z}^d_{\mathrm{prim}}} f(\mathbf{z}) = \sum_{k=1}^{N} \mu(k)k^{-d} \sum_{\substack{\boldsymbol{\xi}\in\mathbb{Q}^d \\ n(\boldsymbol{\xi})|k}} \hat{f}(\boldsymbol{\xi}) - \Big(\sum_{k=1}^{N} \mu(k)\Big) f(\mathbf{0}),$$

where we should note that the first term in the right-hand side is an absolutely convergent series. To rewrite the right-hand side further, consider

$$\begin{aligned}\mathbb{Q}^d_N &= \{\boldsymbol{\xi} \in \mathbb{Q}^d \,|\, n(\boldsymbol{\xi}) \le N\},\\ A &= \{(k, \boldsymbol{\xi}) \,|\, k = 1, \dots, N,\ \boldsymbol{\xi} \in \mathbb{Q}^d,\ n(\boldsymbol{\xi})|k\},\\ B &= \{(\ell, \boldsymbol{\xi}) \,|\, 1 \le \ell \le N n(\boldsymbol{\xi})^{-1},\ \boldsymbol{\xi} \in \mathbb{Q}^d_N\}.\end{aligned}$$

Then the map $(k, \boldsymbol{\xi}) \mapsto (kn(\boldsymbol{\xi})^{-1}, \boldsymbol{\xi})$ is a bijection of A onto B. Therefore we get

$$\sum_{\mathbf{z}\in\mathbb{Z}^d_{\mathrm{prim}}} f(\mathbf{z}) = \sum_{\boldsymbol{\xi}\in\mathbb{Q}^d_N} \sum_{1\le\ell\le N/n(\boldsymbol{\xi})} \frac{\mu(\ell n(\boldsymbol{\xi}))}{(\ell n(\boldsymbol{\xi}))^d} \hat{f}(\boldsymbol{\xi}) - \Big(\sum_{k=1}^{N} \mu(k)\Big) f(\mathbf{0}).$$

Clearly

$$\mu(\ell n(\boldsymbol{\xi})) = \begin{cases} \mu(\ell)\mu(n(\boldsymbol{\xi})) & (\gcd(\ell, n(\boldsymbol{\xi})) = 1)\\ 0 & (\gcd(\ell, n(\boldsymbol{\xi})) > 1).\end{cases}$$

Therefore putting

$$\begin{aligned} a_N(\boldsymbol{\xi}) &= \frac{\mu(n(\boldsymbol{\xi}))}{n(\boldsymbol{\xi})^d} \sum_{\substack{1\le\ell\le N/n(\boldsymbol{\xi})\\ \gcd(\ell, n(\boldsymbol{\xi}))=1}} \frac{\mu(\ell)}{\ell^d},\\ \Lambda_N &= \{\boldsymbol{\xi} \in \mathbb{Q}^d_N \,|\, \mu(n(\boldsymbol{\xi})) \ne 0\},\end{aligned}$$

we get

$$\sum_{\mathbf{z}\in\mathbb{Z}^d_{\mathrm{prim}}} f(\mathbf{z}) = \sum_{\boldsymbol{\xi}\in\Lambda_N} a_N(\boldsymbol{\xi}) \hat{f}(\boldsymbol{\xi}) - \Big(\sum_{k=1}^{N} \mu(k)\Big) f(\mathbf{0}).$$

Furthermore, if we put

$$\begin{aligned}\Lambda &= \{\boldsymbol{\xi} \in \mathbb{Q}^d \,|\, \mu(n(\boldsymbol{\xi})) \ne 0\},\\ a(\boldsymbol{\xi}) &= \frac{\mu(n_{\boldsymbol{\xi}})}{n(\boldsymbol{\xi})^d} \zeta(d)^{-1} \prod_{p|n(\boldsymbol{\xi})} \big(1 - p^{-d}\big)^{-1} \quad (\boldsymbol{\xi} \in \Lambda),\end{aligned}$$

then

$$\Lambda = \bigcup_{N=1}^{\infty} \Lambda_N, \qquad \lim_{N\to\infty} a_N(\boldsymbol{\xi}) = \frac{\mu(n(\boldsymbol{\xi}))}{n(\boldsymbol{\xi})^d} \sum_{\substack{\ell=1\\ \gcd(\ell, n(\boldsymbol{\xi}))=1}}^{\infty} \frac{\mu(\ell)}{\ell^d} = a(\boldsymbol{\xi}).$$

This implies that if the "extra term" $\left(\sum_{k=1}^{N}\mu(k)\right)f(\mathbf{0})$ is ignored, then the set $\mathbb{Z}^d_{\text{prim}}$ looks like a quasicrystal of Poisson type. This is the reason why we say that $\mathbb{Z}^d_{\text{prim}}$ is *nearly* a quasicrystal of Poisson type.

Remark 7.1 (1) Applying Theorem 5.3, we obtain

$$\sum_{\mathbf{z}\in(\mathbb{Z}^{\text{odd}})^2_{\text{prim}}} f(\mathbf{z}) = \sum_{\boldsymbol{\xi}\in\mathbb{Q}^2_{2N}} \Big(\sum_{\substack{k2^h\le N \\ k\ge1,h\ge0 \\ n(\boldsymbol{\xi})|k2^{h+1}}} \frac{\mu(k)}{k^2}\frac{1}{2^{2h+2}} e^{\pi i k 2^{h+1}\langle\boldsymbol{\xi},\mathbf{1}\rangle}\Big)\hat{f}(\boldsymbol{\xi}),$$

where $\operatorname{supp} f \subset B_N(\mathbf{0})$ and $\mathbf{1} = (1, 1)$. This implies that $(\mathbb{Z}^{\text{odd}})^2_{\text{prim}}$ is a quasicrystal of Poisson type. The reason why no extra terms appear in this case is that $(\mathbb{Z}^{\text{odd}})^2 = 2\mathbb{Z}^2 + \mathbf{1}$ is a full lattice.

(2) In much the same manner as above, we get

$$\sum_{\mathbf{z}\in\mathbb{Z}^d_{\text{prim}}} f(\mathbf{z})e^{2\pi i\langle\mathbf{z},\boldsymbol{\eta}\rangle} = \sum_{\boldsymbol{\xi}\in\Lambda_N} a_N(\boldsymbol{\xi})\hat{f}(\boldsymbol{\xi}-\boldsymbol{\eta}) - \Big(\sum_{k=1}^{N}\mu(k)\Big)f(\mathbf{0}).$$

Using this identity, we can show

$$\lim_{\epsilon\to+0}\sigma(f^{\epsilon}, \mathbb{Z}^d_{\text{prim}}, \omega_{\boldsymbol{\eta}}) = a(\boldsymbol{\eta})\int_{\mathbb{R}^d} f(\mathbf{x})\,d\mathbf{x},$$

that is, $(\mathbb{Z}^d_{\text{prim}}, \omega_{\boldsymbol{\eta}})$ has constant density for $\boldsymbol{\eta}$ with $\mu(n(\boldsymbol{\eta})) \neq 0$.

(3) An interesting problem related to quasicrystals comes up in the study of *nontrivial zeros* of the Riemann zeta function (thus we come across another Riemann's work,. which were to change the direction of mathematical research in a most significant way).

We put

$$\Gamma^{\text{zero}} = \{\operatorname{Im} s \in \mathbb{R} \mid \zeta(s) = 0,\ 0 < \operatorname{Re} s < 1\}.$$

Under the Riemann Hypothesis (RH), one may say that Γ^{zero} is *nearly* a quasicrystal of Poisson type of 1-dimension (cf. Dyson [5]). Actually a version of Riemann's explicit formula looks like a generalized Poisson formula (see Iwaniec and Kowalski [6]):

$$\sum_{\rho} f\Big(\frac{\rho-1/2}{i}\Big) = f\Big(\frac{1}{2i}\Big) + f\Big(-\frac{1}{2i}\Big)$$
$$+\frac{1}{2\pi}\int_{-\infty}^{\infty} f(u)\mathrm{Re}\frac{\Gamma'}{\Gamma}\Big(\frac{1}{4}+\frac{iu}{2}\Big)du$$
$$-\frac{1}{2\pi}\hat{f}(0)\log\pi - \frac{1}{2\pi}\sum_{m=1}^{\infty}\sum_{p}\frac{\log p}{p^{m/2}}\Big(\hat{f}\Big(\frac{\log p^m}{2\pi}\Big) + \hat{f}\Big(-\frac{\log p^m}{2\pi}\Big)\Big).$$

where $\{\rho\}$ is the set of zeros of $\zeta(s)$ with $0 < \mathrm{Re}\rho < 1$, $\sum_{p}$ is the sum over all primes, and Γ'/Γ is the logarithmic derivative of the gamma function. Notice that, under the RH, the sum in the left-hand side is written as $\sum_{\mathbf{z}\in\Gamma^{\mathrm{zero}}} f(\mathbf{z})$.[17] What we should stress here is that the test function $f(s)$ is not arbitrary, and is supposed to be analytic in the strip $|\mathrm{Im}s| \leq 1/2+\epsilon$ for some $\epsilon > 0$, and to satisfy $|f(s)| \leq (1+|s|)^{-(1+\delta)}$ for some $\delta > 0$ when $|\mathrm{Re}\, s| \to \infty$. This restriction on f together with the extra terms in the formula above says that Γ^{zero} is not a *genuine* quasicrystal of Poisson type. Furthermore Γ^{zero} does not have the Delone property. □

References

1. L. Comtet, *Advanced Combinatorics: The Art of Finite and Infinite Expansions*, (Springer, New York, 1974)
2. N.G. de Bruijn, Algebraic theory of Penrose's nonperiodic tilings of the plane, I, II. Indagationes Mathematicae (Proceedings) **84**(39–52), 53–66 (1981)
3. B.N. Delone, Neue Darstellung der geometrischen Kristallographie. Z. Kristallographie **84**, 109–149 (1932)
4. W. Duke, Rational points on the sphere. Ramanujan J. **7**, 235–239 (2003)
5. F. Dyson, Birds and frogs. Not. AMS **56**, 212–223 (2009)
6. H. Iwaniec and E. Kowalski, *Analytic number theory*, Amer. Math. Soc. Colloq. Publ. 53 (American Mathematical Society, Providence, RI, 2004)
7. J.C. Lagarias, *Mathematical Quasicrystals and The Problem of Diffraction, Directions in Mathematical Quasicrystals*, ed by M. Baake, V. Robert (American Mathematical Soc, Moody, 2000), pp. 61–94
8. D.N. Lehmer, Asymptotic evaluation of certain totient sums. Amer. J. Math. **22**, 293–335 (1900)
9. J. Marklof, A. Strömbergsson, Visibility and directions in quasicrystals. Int. Math. Res. Not. doi:10.1093/imrn/rnu140. first published online September 2, 2014
10. P. Mattila, *Geometry of Sets and Measures in Euclidean Spaces: Fractals and Rectifiability*, (Cambridge University Press, 1999)
11. S. Ranganathan, On the geometry of coincidence-site lattices. Acta Cryst. **21**, 197–199 (1966)
12. D. Shechtman, I. Blech, D. Gratias, J.W. Cahn, Metallic phase with long-range orientational order and no translational symmetry. Phys. Rev. Lett. **53**, 1951–1953 (1984)
13. M. Senechal, J. Taylor, Quasicrystals: The view from les houches. Math. Intel. **12**, 54–64 (1990)

[17]The *simple zero conjecture* says that all zeros ρ are simple. In the case that we do not assume this conjecture, we think of Γ^{zero} as a weighted set with the weight $\omega(\rho) = \mathrm{ord}_{\rho}(\zeta)$.

14. C.L. Siegel, A mean value theorem in geometry of numbers. Annals of Math. **46**, 340–347 (1945)
15. M. Somos, http://grail.cba.csuohio.edu/~somos/rtritab.txt
16. T. Sunada, *Topological Crystallography—with a view towards Discrete Geometric Analysis—*, (Springer, New York, 2012)
17. T. Sunada, *Topics on mathematical crystallography*, to appear in "Groups, Graphs and Random Walks," ed. by T. Ceccherini-Silberstein, M. Salvatori, and Ecaterina Sava-Huss, London Mathematical Society Lecture Notes Series 436, pp. 475–519 (2017)
18. I.M. Vinogradov, *Elements of Number Theory, Mineola* (Dover Publications, NY, 2003)
19. A. Weil, *Number Theory*, An approach through history from Hammurapi to Legendre, (Birkhäuser, 1984)
20. H. Weyl, Über die Gleichverteilung von Zahlen mod. Eins. Math. Ann. **77**, 313–352 (1916)
21. A. M. Yaglom and I. M. Yaglom, *Challenging Mathematical Problems with Elementary Solutions*, Vol 1, (Dover, 1987)
22. P. Zeiner, Symmetries of coincidence site lattices of the cubic lattice. Z. Kristallgr. **220**, 915–925 (2005)

From Riemannian to Relativistic Diffusions

Jacques Franchi

Abstract We first introduce Euclidean and Riemannian Brownian motions. Then considering Minkowski space, we present the Dudley relativistic diffusion. Finally we construct a family of covariant relativistic diffusions on a generic Lorentz manifold, the quadratic variation of which can be locally determined by the curvature (which allows the interpretation of the diffusion effect on a particle by its interaction with the ambient space-time). Examples are considered, in some classical space-time models: Schwarzschild, Gödel and Robertson-Walker manifolds.

1 Introduction

One of the four celebrated brilliant articles Einstein published in 1905 was devoted to Brownian Motion. He was seeing it as a consequence of the kinetic theory of gases: infinitely many small shocks on a given tiny particle move it in a Brownian way. Together with Langevin, Einstein then made the relation with the heat transport.

The mathematical construction of Brownian Motion, especially in terms of its law on $\mathbb{R}^d$-valued continuous paths, was performed by N. Wiener (1925–30). Namely, since Wiener this is a rigorously defined continuous stochastic process which has independent and homogeneous Gaussian increments, also called the Wiener process. Its trajectories are nowhere differentiable, which corresponds to the infinitely many small shocks specified by Einstein.

This gave rise to a huge literature, by among so many others P. Lévy, K. Itô, G. A. Hunt, J. L. Doob, S. R. S. Varadhan, M. Yor, to quote only very few probabilists, about $\mathbb{R}^d$-valued Brownian Motion and its relations to martingales, potential theory, heat equation and kernel, etc. Defining stochastic (Itô) integrals and then solving stochastic differential equations led in particular to the larger notion of so-called "diffusion" (after the physical corresponding phenomenon), namely continuous Markov process:

J. Franchi (✉)
Institut de Recherche Mathématique Avancée, Université de Strasbourg et CNRS,
7 rue René Descartes, 67084 Strasbourg Cedex, France
e-mail: jacques.franchi@math.unistra.fr

L. Ji et al. (eds.), *From Riemann to Differential Geometry and Relativity*,
DOI 10.1007/978-3-319-60039-0_16

the evolution of such a process, from a given state, depends only on this precise state and on the underlying law, but not on the past.

Thus Brownian Motion appeared soon as a very important physical and mathematical object, related to several theories as well as having its own interest. It originated in Biology (and was named after the biologist Brown) and soon enough, after Bachelier, progressively took a big importance in mathematical finance (and then insurance) too.

A following step was the extension of Brownian Motion to Riemannian manifolds. This was performed around 1970 by D. Elworthy (with J. Eells) and P. Malliavin, using Itô's Calculus and the Cartan moving frame method. It is worth underlining that this construction yields the stochastic parallel transport as well, and also stresses the intimate relation between Brownian Motion and the Laplace-Beltrami operator Δ. An important feature is that the latter makes the Riemannian Brownian Motion into a geometrical object, which is covariant with respect to the isometries of the underlying manifold, as well as a physical one, in the sense of Einstein (and Langevin).

A considerable amount of work has been achieved since, and still goes on, to exploit this relationship between probability theory and Riemannian geometry. To give only some examples and references: functional inequalities—such as isoperimetric, concentration or log-Sobolev—curvature-dimension inequalities, the study of harmonic maps [10], estimates about the heat kernel and its gradient [3, 16, 17], geometry of paths [9], gradient flows, optimal mass transportation [25], new proofs of the Gauss-Bonnet and Atiyah-Singer index Theorems (by Patodi, Bismut). Time-evolving Riemannian metrics are also considered now [2], in connection with Ricci flows.

Thus the extension of Brownian motion from a Euclidean to a Riemannian object allows us to understand it as a geometrical object, and explains its repeated use as a geometrical tool. To a certain extent, at this stage it remains a physical object too, as Einstein had in mind already in 1905, since its description in terms of kinetics of gases remains valid in a Riemannian context as well.

In this spirit, it is fairly natural to ask what counterpart Brownian motion might have in the relativistic framework, thereby bringing together two brilliant contributions by Einstein in 1905. The answer is not obvious, since a priori the property of Brownian paths to have an unbounded mean velocity (without any instantaneous velocity in the strict sense, since Brownian paths are nowhere differentiable) looks contradictory with the relativistic constraint of never exceeding the velocity of light.

During a long time several attempts were made, without any success, in order to define a reasonable "relativistic Brownian motion". The first real progress in this direction arrived in 1965, when R. M. Dudley showed that a relativistic diffusion, i.e., a Lorentz-covariant Markov diffusion process, cannot exist on the base space, even in the Minkowski framework of special relativity. This is indeed the precise mathematical counterpart of the fact that in a relativistic setting a Brownian motion cannot be physical any longer, since it can run at arbitrary large mean velocities. On the contrary, Dudley [8] showed that a relativistic diffusion makes sense at the level of the tangent bundle of Minkowski space, and he then specified the asymptotic behaviour of his (Dudley) diffusion. Moreover he showed that his relativistic diffusion is unique (in law), hence canonical as well as in the Euclidean setting, under the natural (at least geometrically) constraint to be covariant under the action of the Lorentz group.

A similar construction in the generic framework of General Relativity, that is, on the unit tangent bundle of a generic Lorentzian manifold, thereby attempting to relate further two major contributions by Einstein, was then made by Y. Le Jan and the author ([13], 2007). The related relativistic diffusion can be seen as a random perturbation of the geodesic flow, as well as the stochastic geometric development of the Dudley diffusion over a fixed tangent space of the Lorentzian manifold. It still enjoys the covariance with isometries, but therefore cannot be seen as resulting from a kinetic theory of gases, contrary to the maybe more physical process of [7], which is not covariant.

As in the Riemannian non-flat case, other intrinsic diffusions exist in the Lorentzian non-flat case. They enjoy the same geometrical invariance in law as the basic one, and could maybe be seen as more physical, as their quadratic variation is locally determined by their velocity and the curvature of the space, and vanishes in flat or in Ricci-flat (empty) regions ([14], 2011).

An important difference between the Riemannian and the Lorentzian (i.e., relativistic) settings is that the former gives naturally rise to elliptic and often self-adjoint infinitesimal generators, whereas the latter produces only hypoelliptic and non-self-adjoint generators, the analysis of which is much harder. Moreover, in the Riemannian framework the fibre of the frame bundle is compact, whereas it is not any longer in the Lorentzian one.

Lorentzian geometry is also at the heart of [19, 23, 24], in this same volume. Namely, in this relativistic framework, Hermann and Humbert and Nicolas address equations of hyperbolic type ([24] deals mainly with the wave equation) and elliptic type ([19] deals with the Yamabe equation); whereas the present chapter is concerned by equations of parabolic type, since diffusion processes are strongly associated with heat equations. Of course, tensor fields on space-time and the same Einstein equation are central for these four chapters.

In order to understand what relativistic diffusions look like, the best is to study them in some basic examples of General Relativity models, which exactly solve the Einstein equations, beyond the Minkowski space. The maybe most known such models are the following ones: the Schwarzschild space-time, which is intended to describe the physical space surrounding an isolated black hole or very massive star; the Robertson-Walker manifolds, which are intended to model an expanding (or shrinking) universe resulting from a "Big-Bang", as ours; the Gödel universe, which is a striking model where global causality does not hold (rending theoretically possible to return into the past after a long trajectory). Note that [19, 23, 24] (in this same volume) also particularize at some extend to the same basic examples of Minkowski and Schwarzschild.

The use of relativistic diffusions to address geometrical questions about Lorentzian geometry or analysis is still at its very beginning [4, 6, 12], and seems to be much harder than in the elliptic (Riemannian) case.

This chapter is intended to be a survey, relying mainly on [11, 13–15], written on the kind request of the editors Lizhen Ji, Athanase Papadopoulos and Sumio Yamada, for the volume of Springer "From Riemann to differential geometry and relativity". It is addressed not only to probabilists, and hopefully could also interest geometers and mathematical physicists. The proofs are omitted here, but can be found in the above quoted references.

2 Euclidean Brownian Motion

Basically, this is a continuous $\mathbb{R}^d$-valued stochastic process which has independent and homogeneous Gaussian increments. A precise definition (for $d = 1$ first) is as follows.

Definition 2.1 A real Brownian motion (or Wiener process) is a real valued continuous process $(B_t)_{t\geq 0}$ such that for any $n \in \mathbb{N}^*$ and $0 = t_0 < \cdots < t_n$, the random variables $(B_{t_j} - B_{t_{j-1}})$ are independent, and the law of $(B_{t_j} - B_{t_{j-1}})$ is $\mathcal{N}(0, t_j - t_{j-1})$, i.e., centred Gaussian with variance $(t_j - t_{j-1})$.

A slightly different formulation of the second part of the definition is:
The increments of (B_t) are independent, and stationary: $(B_t - B_s) \overset{\text{law}}{\equiv} B_{t-s}$ for any $s \leq t \in \mathbb{R}_+^*$, and moreover the law of B_t is $\mathcal{N}(0, t)$.

The construction of (B_t) can be done either as a limit of symmetrical conveniently normalized random walks, or by means of a multi-scale series, for example the Fourier expansion $\big($in terms of independent standard $\mathcal{N}(0, 1)$ Gaussian variables $(\xi_k\,,\ k \in \mathbb{N})\big)$:

$$B_t \;=\; \xi_0\, t + \frac{\sqrt{2}}{\pi} \sum_{k\in\mathbb{N}^*} \xi_k\, \frac{\sin(\pi k t)}{k}\;, \quad \text{for any } 0 \leq t \leq 1\,.$$

The (probability) law of such a process is clearly unique, and is known as the *Wiener measure* on the space of real continuous functions indexed by $\mathbb{R}_+$ and vanishing at 0.

The following property is straightforward from the definition, since the law of a Gaussian process is prescribed by its mean and its covariance.

Proposition 2.2 *The standard real Brownian motion (B_t) is the unique real process which is Gaussian centred with covariance function $\mathbb{R}_+^2 \ni (s, t) \longmapsto \mathbb{E}(B_s B_t) = \min\{s, t\}$.*

The processes $t \mapsto B_{a+t} - B_a\,,\ t \mapsto c^{-1} B_{c^2 t}\,,\ t \mapsto t\, B_{1/t}\,,$ and $t \mapsto (B_T - B_{T-t})$ (for $0 \leq t \leq T$) satisfy the same conditions. We therefore deduce the following properties:

Corollary 2.3 *The standard real Brownian motion (B_t) satisfies*

(1) the Markov property: for all $a \in \mathbb{R}_+$, $(B_{a+t} - B_a)$ is also a standard Brownian motion, and is independent from the "past" σ-field $\mathcal{F}_a := \sigma\{B_s \,|\, 0 \le s \le a\}$;
(2) the self-similarity: for any $c > 0$, $(c^{-1} B_{c^2 t})$ is also a standard real Brownian motion ;
(3) $(-B_t)$ and $(t\, B_{1/t})$ are also standard real Brownian motions;
(4) for any fixed $T > 0$, $(B_T - B_{T-t})_{0\le t\le T}$ is also a standard real Brownian motion.

An $\mathbb{R}^d$-valued process $B_t := \big(B_t^1, \ldots, B_t^d\big)$ made of d independent standard Brownian motions (B_t^j) is called a *d-dimensional Brownian motion*. For $v \in \mathbb{R}^d$, $(v + B_t)$ is also called a d-dimensional Brownian motion, starting at v. The law of a d-dimensional Brownian motion is covariant with respect to Euclidean isometries of $\mathbb{R}^d$: if f is such an isometry then $f(v + B_t)$ is another d-dimensional Brownian motion, starting at $f(v)$.

The fundamental formula of Stochastic Calculus is due to K. Itô (around 1945).

Theorem 2.4 (Itô's Formula) *Let $B \equiv (B^1, \ldots, B^d)$ be a Brownian motion in $\mathbb{R}^d$, and F a C^2 function on $\mathbb{R}^d$. Then $F \circ B$ is a so-called semi-martingale, and precisely, we almost surely have: for all $t \in \mathbb{R}_+$,*

$$F(B_t) = F(B_0) + \sum_{j=1}^{d} \int_0^t \partial_j F(B_s)\, dB_s^j + \tfrac{1}{2} \int_0^t \Delta F(B_s)\, ds\ .$$

The half Laplacian $\frac{1}{2}\Delta$ appears naturally here, as a particular case of *infinitesimal generator*, namely that of the Brownian motion B. The stochastic so-called Itô integrals $\int_0^t \partial_j F(B_s)\, dB_s^j$ constitute the (local) *martingale part* of the above right hand side. They are pairwise orthogonal in L^2, and obey the following fundamental isometric Itô identity (physicists often understand "$(dB_s^j)^2 = ds$"):

$$\mathbb{E}\left[\left(\int_0^t \varphi_j(B_s)\, dB_s^j\right)^2\right] = \mathbb{E}\left[\int_0^t \varphi_j(B_s)^2\, ds\right].$$

Accordingly, the finite-variation process $\int_0^t \varphi_j(B_s)^2\, ds$ is called the *quadratic variation* of the *martingale* $\int_0^t \varphi_j(B_s)\, dB_s^j$ (which can be approached by stochastic Riemann-like sums of type $\sum_k Z_{s_{k-1}}(B_{s_k}^j - B_{s_{k-1}}^j)$, where each $Z_{s_{k-1}}$ is a functional of $\{B_s^j \,|\, 0 \le s \le s_{k-1}\}$).

Furthermore, the above class of *semi-martingales*, i.e., the sums of a (local) continuous (Brownian) martingale and of a *drift* term having finite variation, is closed under C^2 mappings, and a similar Itô formula holds, with a second order elliptic differential operator (generalizing $\frac{1}{2}\Delta$) as the infinitesimal generator. A given infinitesimal generator, together with a given starting point, specify the whole law of an associated diffusion. See for example [20] (for an exhaustive exposition) or [15] (for a shorter one).

3 Riemannian Brownian Motion

Let $\mathcal{M}$ be a d-dimensional oriented smooth Riemannian manifold, equipped with its Levi-Civita connection ∇. Denote by $O\mathcal{M}$ its direct orthonormal frame bundle, whose fibers are modelled on the special orthogonal group SO(d). Let $H_1, \ldots, H_d$ be the canonical horizontal vector fields on $O\mathcal{M}$, and π denote the canonical projection from $O\mathcal{M}$ onto $\mathcal{M}$. The Bochner horizontal Laplacian is $\mathcal{G} := \sum_{j=1}^{d} H_j^2$. The proofs relating to this section can be found for example in [9, 17, 22]. The following simple fact is crucial.

Lemma 3.1 *The Bochner horizontal Laplacian $\mathcal{G}$ acts on C^2 functions on $\mathcal{M}$, and induces the Beltrami Laplacian Δ: for any $F \in C^2(\mathcal{M})$, we have $\mathcal{G}(F \circ \pi) = (\Delta F) \circ \pi$ on $O\mathcal{M}$. Besides, in local coordinates (x^i, e_j^k), with $e_j = e_j^k \frac{\partial}{\partial x^k}$, denoting by Γ_{jk}^{ℓ} the Christoffel coefficients of the Levi-Civita connexion ∇, for $0 \le i, j \le d$ we have:*

$$\nabla_{\frac{\partial}{\partial x^i}} \frac{\partial}{\partial x^j} = \Gamma_{ij}^k(x) \frac{\partial}{\partial x^k} \quad \textit{and} \quad H_j = e_j^k \frac{\partial}{\partial x^k} - e_j^k e_i^m \Gamma_{km}^{\ell}(x) \frac{\partial}{\partial e_i^{\ell}} .$$

The construction of the Riemannian Brownian motion uses the Cartan moving frame method, by means of a stochastic development, to produce a stochastic flow on the frame bundle $O\mathcal{M}$, putting white noises dB_s^j on the horizontal (velocity) vectors. The resulting diffusion will project to a diffusion on the base manifold $\mathcal{M}$, due to Lemma 3.1.

To proceed, let us fix $\Phi_0 \in O\mathcal{M}$ and an $\mathbb{R}^d$-valued Brownian motion $B = (B_s^j)$. The following theorem defines the Riemannian Brownian motion (X_s), with possibly some positive explosion time.

Theorem 3.2 (see [9, 17, 22]) (i) *The $O\mathcal{M}$-valued Stratonovitch stochastic differential equation*

$$(*) \qquad \Phi_s = \Phi_0 + \int_0^s \sum_{j=1}^{d} H_j(\Phi_t) \circ dB_t^j$$

defines a Riemannian Brownian motion $(X_s) := \pi(\Phi_s)$ on $\mathcal{M}$ (starting from $\pi(\Phi_0)$): this is a continuous Markovian (i.e., diffusion) process whose infinitesimal generator is $\frac{1}{2}\Delta$.
(ii) *The stochastic parallel transport of a vector $V_0 \in T_{X_0}\mathcal{M}$ along the Brownian path (X_s) is given by $V_s = \Phi_s V_0 \in T_{X_s}\mathcal{M}$.*

Remark 3.3 (o) The Stratonovitch integral (using it, Itô's Formula takes on the usual chain rule form of classical calculus) is deduced from the Itô one by the following defining rule:

$$\int_0^t \varphi(B_s^1, \ldots, B_s^d) \circ dB_s^j := \int_0^t \varphi(B_s^1, \ldots, B_s^d)\, dB_s^j + \tfrac{1}{2} \int_0^t \partial_j \varphi(B_s^1, \ldots, B_s^d)\, ds\,.$$

(i) In local coordinates (x^i, e_j^k), $\Phi_s = \big(X_s\,;\, e_1(s), .., e_d(s)\big)$, Equation $(*)$ reads:

$$dX_s^i = e_j^i(s) \circ dB_s^j\ ;\quad de_j^k(s) = -\Gamma_{il}^k(X_s) e_j^i(s) e_m^l(s) \circ dB_s^m\,.$$

This means that the Riemannian Brownian motion (X_s) is the stochastic development of the $T_{X_0}\mathcal{M}$-valued Brownian motion $(\Phi_0 B_s)$.
(ii) The $O\mathcal{M}$-valued diffusion (Φ_s) admits the half Bochner horizontal Laplacian $\frac{1}{2}\mathcal{G}$ as its infinitesimal generator: for any $F \in C_b^2(O\mathcal{M})$, $F(\Phi_s) - \frac{1}{2}\int_0^s \mathcal{G}F(\Phi_t)\, dt$ is a martingale.
(iii) In the Itô form and in local coordinates, we have the following differential equation:

$$dX_s^i = (g^{-1/2})_j^i(X_s)\, dB_s^j - \tfrac{1}{2}\, g^{kl}(X_s)\, \Gamma_{kl}^i(X_s)\, ds\,.$$

(iv) The Riemannian Brownian motion (X_s) is covariant with respect to the isometries of $\mathcal{M}$: for any such isometry f, the process $(f \circ X_s)$ is another Riemannian Brownian motion, starting at $f(X_0)$.
(v) This construction offers the strong advantage of providing the *stochastic parallel transport* (Φ_s) along the Brownian curve (X_s) together with the Brownian motion itself.

4 The Relativistic Dudley Diffusion in Minkowski Space

Let us consider an integer $d \geq 2$, the Minkowski space $\mathbb{R}^{1,d} := \{\xi = (\xi^o, \vec{\xi}\,) \in \mathbb{R} \times \mathbb{R}^d\}$, endowed with its canonical basis $(e_0, \ldots, e_d)$ and the Minkowski pseudo-metric $\langle \xi, \xi \rangle := |\xi^o|^2 - \|\vec{\xi}\,\|^2$.

Let $G = \mathrm{PSO}(1, d)$ denote the Lorentz-Möbius group, i.e., the connected component of the identity in the pseudo-orthogonal group $\mathrm{O}(1, d)$ (of linear mappings preserving $\langle \cdot, \cdot \rangle$), and denote by $\mathbb{H}^d := \{p \in \mathbb{R}^{1,d} \mid p^o > 0 \text{ and } \langle p, p \rangle = 1\}$ the positive half of the unit pseudo-sphere.

The opposite of the Minkowski pseudo-metric induces a Riemannian metric on $\mathbb{H}^d$, namely the hyperbolic one, so that $\mathbb{H}^d$ is a model for the d-dimensional hyperbolic space. A convenient parametrization of $\mathbb{H}^d$ is $(\varrho, \theta) \in \mathbb{R}_+ \times \mathbb{S}^{d-1}$, given by $\varrho := \operatorname{argch}(p^o)$ and $\theta := \vec{p} \big/ \sqrt{|p^o|^2 - 1}$. In these polar coordinates the hyperbolic metric reads $d\varrho^2 + \mathrm{sh}^2\varrho\,|d\theta|^2$, and the hyperbolic Laplacian is $\Delta^{\mathbb{H}} := \dfrac{\partial^2}{\partial\varrho^2} + (d-1)\coth\varrho\,\dfrac{\partial}{\partial\varrho} + \mathrm{sh}^{-2}\varrho \times \Delta_\theta$, Δ_θ denoting the Laplacian of $\mathbb{S}^{d-1}$. The associated volume measure is $|\mathrm{sh}\,\varrho|^{d-1}d\varrho\,d\theta$.

The group G acts isometrically on $\mathbb{R}^{1,d}$ and on $\mathbb{H}^d$, and the Casimir operator $\mathcal{C}$ on G induces the hyperbolic Laplacian on $\mathbb{H}^d$.

Fix $\sigma > 0$, and denote by $\mathcal{L}_\sigma$ the σ-relativistic Laplacian, defined on $\mathbb{R}^{1,d} \times \mathbb{H}^d$ by

$$\mathcal{L}_\sigma f(\xi, p) := p^o \frac{\partial f}{\partial \xi^o}(\xi, p) + \sum_{j=1}^d p^j \frac{\partial f}{\partial \xi^j}(\xi, p) + \tfrac{\sigma^2}{2}\,\Delta^{\mathbb{H}}_{(p)} f(\xi, p),$$

that is to say,

$$\mathcal{L}_\sigma f := \langle p, \mathrm{grad}_{(\xi)} f\rangle + \tfrac{\sigma^2}{2}\,\Delta^{\mathbb{H}}_{(p)} f .$$

This is a hypoelliptic operator.

Given any $(\xi_0, p_0) \in \mathbb{R}^{1,d} \times \mathbb{H}^d$, there exists a unique (in law) diffusion process (ξ_s, p_s), $s \in \mathbb{R}_+$, such that for any compactly supported $f \in C^2(\mathbb{R}^{1,d} \times \mathbb{H}^d)$,

$$f(\xi_s, p_s) - f(\xi_0, p_0) - \int_0^s \mathcal{L}_\sigma f(\xi_t, p_t)\,dt \text{ is a martingale.}$$

Note that p_s is a hyperbolic Brownian motion, and that $\xi_s = \xi_0 + \displaystyle\int_0^s p_t\,dt$.

Remark 4.1 (1) The relativistic trajectories $(\xi_s \mid s \in \mathbb{R}_+)$ we get in Minkowski space are fully causal: since their spacetime velocities $\frac{d\xi_s}{ds} = p_s$ belong to $\mathbb{H}^d$, they are timelike, hence locally causal; moreover they satisfy $\frac{d\xi_s^o}{ds} = p_s^o > 0$, which ensures that $t(s) = \xi_s^o$ increases always strictly. Hence they are globally causal: in the terminology of [18], they satisfy the "causality condition": they cannot be closed.
(2) Note that ξ_s is parametrized by its arc length. Mechanically, ξ_s describes the trajectory of a relativistic particle of small mass indexed by its proper time, submitted to a white noise acceleration (in proper time). Its law is invariant under any Lorentz transformation.

If we denote by (e_j^*) the dual base of the canonical base $(e_0, e_1, .., e_d)$ (with respect to $\langle\ ,\ \rangle$), the matrices $E_j := e_0 \otimes e_j^* + e_j \otimes e_0^*$ belong to the Lie algebra $\mathrm{so}(1,d)$ of G, and generate the so-called *boost* transformations. Given d independent real Wiener processes w_s^j, $p_s = (p_s^o, \vec{p}_s)$ can be defined by $p_s := \Lambda_s e_0$, where the matrix $\Lambda_s \in G$ is defined by the following stochastic differential equation:

$$\Lambda_s = \Lambda_0 + \sigma \sum_{j=1}^d \int_0^s \Lambda_t\,E_j \circ dw_t^j .$$

This means that the relativistic diffusion process (ξ_s, p_s) is in fact the projection of some diffusion process having independent increments, namely a Brownian motion with drift, living in the Poincaré group. This group is the analogue in the present Lorentz-Minkowski setup of the classical group of rigid motions, and can be seen as the group of $(d+2, d+2)$ real matrices having the form $\begin{pmatrix} \Lambda & \xi \\ 0 & 1 \end{pmatrix}$, with $\Lambda \in G$, $\xi \in \mathbb{R}^{1,d}$ (written as a column), and $0 \in \mathbb{R}^{1+d}$ (written as a row). Its Lie algebra is the set of matrices $\begin{pmatrix} \beta & x \\ 0 & 0 \end{pmatrix}$, with $\beta \in \text{so}(1,d)$ and $x \in \mathbb{R}^{1,d}$. The Brownian motion with drift we consider on the Poincaré group solves the stochastic differential equation $d\begin{pmatrix} \Lambda_s & \xi_s \\ 0 & 1 \end{pmatrix} = \begin{pmatrix} \Lambda_s & \xi_s \\ 0 & 1 \end{pmatrix} \circ d\begin{pmatrix} \beta_s & e_0\, s \\ 0 & 0 \end{pmatrix}$, where $\left(\beta_s = \sigma \sum_{j=1}^{d} E_j\, w_s^j\right)$ is a Brownian motion on $\text{so}(1,d)$. This equation is equivalent to $d\Lambda_s = \Lambda_s \circ d\beta_s$ and $d\xi_s = \Lambda_s e_0\, ds$, so that (Λ_s) is a Brownian motion on G. On functions of $p = \Lambda e_0$, its infinitesimal generator $\sum_{j=1}^{d} (\mathcal{L}_{E_j})^2$ coincides with a Casimir operator, and induces the hyperbolic Laplacian, so that $(p_s = \Lambda_s e_0)$ is a Brownian motion on $\mathbb{H}^d$, as required.

Then it is well known that $\theta_s := \vec{p}_s \Big/ \sqrt{|p_s^o|^2 - 1}$ converges almost surely in $\mathbb{S}^{d-1}$ to some random limit θ_∞, and that p_s^o increases to infinity. We also set $\varrho_s := \text{argch}(p_s^o)$.

The Euclidean trajectory $Z(t)$ is defined by $\vec{\xi}_{s(t)}$, where $s(t)$ is determined by $\xi_{s(t)}^o = t$.
Let us note that the Euclidean velocity $dZ(t)/dt = \theta_{s(t)}\, \text{th}\, \varrho_{s(t)}$ has norm <1, 1 being here the velocity of light, beyond which the relativistic diffusion cannot actually go. Moreover we have the following.

Remark 4.2 The mean Euclidean velocity $Z(t)/t$ converges almost surely to $\theta_\infty \in \mathbb{S}^{d-1}$.

Proof We have $\lim\limits_{t \nearrow \infty} s(t) = +\infty$, so that $\text{th}\, \varrho_{s(t)} = \sqrt{1 - (p_{s(t)}^o)^{-2}}$ approaches 1.
Thus we get almost surely $\lim\limits_{t \to \infty} \dfrac{dZ(t)}{dt} = \theta_\infty$, and the result follows easily. ◇

The Poisson boundary of Minkowski's space has been determined, as follows.

Theorem 4.3 ([4]) (i) *As proper time s goes to infinity, the quantity* $\langle \xi_s, e_0 + \theta_\infty \rangle$ *converges almost surely to a random variable* ζ_∞.
(ii) *The limiting random variable* $(\theta_\infty, \zeta_\infty)$ *contains all the asymptotic information regarding the Dudley diffusion* $(\xi_s, \dot{\xi}_s \equiv p_s)$, *that is, generates its invariant* σ*-field and its tail* σ*-field as well. Equivalently, the bounded* $\mathcal{L}_\sigma$*-harmonic functions are precisely the functions which admit a Choquet representation* $(\xi, p) \mapsto \mathbb{E}_{(\xi,p)}\big[h(\theta_\infty, \zeta_\infty)\big]$, *for some bounded measurable function h.*

5 The Lorentzian Frame Bundle $G(\mathcal{M})$ over $(\mathcal{M}, g)$

We aim at presenting the extension of the relativistic diffusion, from the Minkowski space to a generic Lorentzian manifold, framework for any General Relativity model. This will be as well the Lorentzian counterpart of the Riemannian Brownian motion, as the Dudley diffusion was the relativistic counterpart of the Euclidean Brownian motion.

The leading idea is to proceed similarly as what was done to get from the Euclidian setting to the Riemannian one, that is to say, to rely again on the Cartan moving frame method (recall Sect. 3). To begin, in this section we introduce the necessary geometrical material and background, the frames and the canonical vector fields being somewhat more complicated in the Lorentzian (or pseudo-Riemannian) setting than in the Riemannian one.

Let $\mathcal{M}$ be a C^∞ time-oriented $(1+d)$-dimensional Lorentz manifold, with pseudo-metric g having signature $(+, -, \ldots, -)$, Levi-Civita connection ∇, and let $T^1\mathcal{M}$ denote the positive half of its pseudo-unit tangent bundle. Let $G(\mathcal{M})$ be the bundle of direct pseudo-orthonormal frames, with first element in $T^1\mathcal{M}$ and with fibers modelled on the Lorentz-Möbius group G. Let $\pi_1 : u \mapsto \big(\pi(u), e_0(u)\big)$ denote the canonical projection from $G(\mathcal{M})$ onto the unit tangent bundle $T^1\mathcal{M}$, which to each frame $\big(e_0(u), \ldots, e_d(u)\big)$ associates its first vector $e_0(u)$.

The action of $\mathrm{SO}(d)$ on $(e_1, \ldots, e_d)$ induces the identification $T^1\mathcal{M} \equiv G(\mathcal{M})/\mathrm{SO}(d)$.

Let $H_0, H_1, \ldots, H_d$ be the canonical horizontal vector fields on $G(\mathcal{M})$, and $V_{e_i\wedge e_j}$ (for $0 \le i < j \le d$) the canonical vertical vector fields on $G(\mathcal{M})$. In particular, we have $T\pi(H_k) = e_k$. To abbreviate the notation, we shall write V_j for $V_{e_0\wedge e_j}$, i.e., the vector field associated with the previous matrix $E_j \in \mathrm{so}(1, d)$.

The canonical vectors H_k, $V_{e_i\wedge e_j}$ span $TG(\mathcal{M})$, the horizontal (resp. vertical) sub-bundle of $TG(\mathcal{M})$ being spanned by the H_ks (resp. the $V_{e_i\wedge e_j}$s). Note that H_0 generates the geodesic flow, that $V_1, \ldots, V_d$ generate the *boosts*, and that the $V_{e_i\wedge e_j}$ $(1 \le i, < j \le d)$ generate rotations. We have:

$$[V_{e_i\wedge e_j}, H_k] = \langle e_i, e_k\rangle\, H_j - \langle e_j, e_k\rangle\, H_i\,, \qquad \text{for } 0 \le i, j, k \le d\,,$$

and

$$[H_i, H_j] = \sum_{0\le k<\ell\le d} \mathcal{R}_{ij}{}^{k\ell}\, V_{e_k\wedge e_\ell}\,,$$

where the $((\mathcal{R}_{ij}{}^{k\ell}))$ are the entries of the Riemann *curvature tensor*. The associated *curvature operator* satisfies: for any C^1 vector fields X, Y, Z, A,

$$\langle \mathcal{R}(X\wedge Y), A\wedge Z\rangle = \big\langle \big([\nabla_X, \nabla_Y] - \nabla_{[X,Y]}\big)\, Z\,,\ A\big\rangle_g\,.$$

The *Ricci tensor* and *Ricci operator* are defined, for $0 \le i, k \le d$, by:

$$R_i^k := \sum_{j=0}^{d} \mathcal{R}_{ij}{}^{kj}\,, \quad \text{and} \quad \mathrm{Ricci}_\xi\big(e_i(u)\big) := \sum_{k=0}^{d} R_i^k\, e_k(u)\,, \quad \text{for any } u \in \pi^{-1}(\xi)\,.$$

The *scalar curvature* is: $R := \sum_{k=0}^{d} R_k^k$.

The indices of the curvature tensor $((\mathcal{R}_{ij}{}^{k\ell}))$ and of the Ricci tensor $((R_i^k))$ are lowered or raised by means of the Minkowski tensor $((\eta_{ab} := \langle e_a, e_b\rangle))$ and its inverse $((\eta^{ab}))$. For example, we have: $R_{ij} = R_i^k\, \eta_{kj}$.

The *energy-momentum* tensor $((T_j^k))$ and operator T_ξ are defined as:

$$T_j^k := R_j^k - \tfrac{1}{2}\, R\, \delta_j^k \quad \text{and} \quad T_\xi := \mathrm{Ricci}_\xi - \tfrac{1}{2}\, R\,. \tag{1}$$

Note that $\sum_{j=0}^{d} T_j^j = -\frac{d-1}{2}\, R$. The *energy* at any line-element $(\xi, \dot\xi) \in T^1\mathcal{M}$ is

$$\mathcal{E}(\xi, \dot\xi) := \langle T_\xi(\dot\xi), \dot\xi\rangle_{g(\xi)} = T_{00}(\xi, \dot\xi)\,. \tag{2}$$

The *weak energy condition* (see [18]) stipulates that $\mathcal{E}(\xi, \dot\xi) \geq 0$ on the whole $T^1\mathcal{M}$. This is also the content of ([21], (94, 10)).

5.1 Expressions in Local Coordinates

Consider local coordinates (ξ^i, e_j^k) for $u = (\xi, e_0, \ldots, e_d) \in G(\mathcal{M})$, with $e_j = e_j^k\, \dfrac{\partial}{\partial \xi^k}$.

As in the Riemannian case, for $0 \leq i, j \leq d$ we have:

$$\nabla_{\frac{\partial}{\partial \xi^i}} \frac{\partial}{\partial \xi^j} = \Gamma_{ij}^k(\xi)\, \frac{\partial}{\partial \xi^k} \quad \text{and} \quad H_j = e_j^k\, \frac{\partial}{\partial \xi^k} - e_j^k\, e_i^m\, \Gamma_{km}^\ell(\xi)\, \frac{\partial}{\partial e_i^\ell}\,.$$

The Christoffel coefficients of ∇ are computed by: $\Gamma_{ij}^k = \frac{1}{2}\, g^{k\ell}\Big(\dfrac{\partial g_{\ell j}}{\partial \xi^i} + \dfrac{\partial g_{i\ell}}{\partial \xi^j} - \dfrac{\partial g_{ij}}{\partial \xi^\ell}\Big)$, or equivalently, by the fact that geodesics solve $\ddot\xi^k + \Gamma_{ij}^k\, \dot\xi^i\, \dot\xi^j = 0$.

Then $V_{e_i \wedge e_j} = e_i^k\, \dfrac{\partial}{\partial e_j^k} - e_j^k\, \dfrac{\partial}{\partial e_i^k}$ and $V_j = e_0^k\, \dfrac{\partial}{\partial e_j^k} + e_j^k\, \dfrac{\partial}{\partial e_0^k}$, for $1 \leq i, j \leq d$.

The curvature operator is expressed in a local chart as: for $0 \leq m, n, p, q \leq d$,

$$\widetilde{\mathcal{R}}_{mnpq} := \Big\langle \mathcal{R}\left(\tfrac{\partial}{\partial \xi^m} \wedge \tfrac{\partial}{\partial \xi^n}\right), \tfrac{\partial}{\partial \xi^p} \wedge \tfrac{\partial}{\partial \xi^q}\Big\rangle_g = g_{mr}\left(\Gamma_{ps}^r\, \Gamma_{nq}^s - \Gamma_{qs}^r\, \Gamma_{np}^s + \frac{\partial \Gamma_{nq}^r}{\partial \xi^p} - \frac{\partial \Gamma_{np}^r}{\partial \xi^q}\right). \tag{3}$$

Then, the Ricci operator can be computed similarly, as: for $0 \le m, p \le d$,

$$\tilde{R}_{mp} := \Big\langle \mathrm{Ricci}(\tfrac{\partial}{\partial \xi^m}), \tfrac{\partial}{\partial \xi^p} \Big\rangle_g = \widetilde{\mathcal{R}}_{mnpq}\, g^{nq} = \Gamma^n_{nq} \Gamma^q_{mp} - \Gamma^n_{pq}\, \Gamma^q_{mn} + \frac{\partial \Gamma^n_{mp}}{\partial \xi^n} - \frac{\partial \Gamma^n_{mn}}{\partial \xi^p} . \tag{4}$$

The scalar curvature and the energy-momentum operator can be computed by:

$$R = \tilde{R}_{ij}\, g^{ij} \quad \text{and} \quad \tilde{T}_{\ell m} = \tilde{R}_{\ell m} - \tfrac{1}{2}\, R\, g_{\ell m} \text{ (Einstein equations).} \tag{5}$$

To summarize, the Riemann curvature tensor $(\!(\mathcal{R}_{ij}{}^{k\ell})\!)$ is made of the coordinates of the curvature operator $\mathcal{R}$ in an orthonormal moving frame, and its indices are lowered or raised by means of the Minkowski tensor $(\!(\eta_{ab})\!)$, while the curvature tensor $(\!(\tilde{\mathcal{R}}_{mnpq})\!)$ is made of the coordinates of the curvature operator in a local chart, and its indexes are lowered or raised by means of the metric tensor $(\!(g_{ab})\!)$.
To go from one tensor to the other, note that by (3) we have
$\mathcal{R}(\frac{\partial}{\partial \xi^m} \wedge \frac{\partial}{\partial \xi^n}) = \frac{1}{2}\, \widetilde{\mathcal{R}}_{mn}{}^{ab}\, \frac{\partial}{\partial \xi^a} \wedge \frac{\partial}{\partial \xi^b}$, whence: $e_i^k\, e_j^\ell\, \widetilde{\mathcal{R}}_{k\ell}{}^{pq} = \mathcal{R}_{ij}{}^{mn}\, e_m^p\, e_n^q$, or equivalently:

$$\mathcal{R}_{ijab} = \widetilde{\mathcal{R}}_{k\ell rs}\, e_i^k\, e_j^\ell\, e_a^r\, e_b^s , \qquad \text{or also :} \quad \widetilde{\mathcal{R}}^{rspq} = \mathcal{R}^{abmn}\, e_a^r\, e_b^s\, e_m^p\, e_n^q .$$

5.2 Example of a Perfect Fluid

The energy-momentum tensor T (of (1), or equivalently $\tilde{T}$, recall (5)) is associated to a *perfect fluid* (see [18]) if it has the form:

$$\tilde{T}_{k\ell} = q\, U_k\, U_\ell - p\, g_{k\ell} , \tag{6}$$

for some C^1 field U in $T^1\mathcal{M}$ (which represents the velocity of the fluid), and some C^1 functions p, q on $\mathcal{M}$. By Einstein's equations (5), (6) is equivalent to:

$$\tilde{R}_{k\ell} = q\, U_k\, U_\ell + \tilde{p}\, g_{k\ell} , \quad \text{with} \quad \tilde{p} = (2p - q)/(d-1), \tag{7}$$

or also, by (4), to:

$$\langle \mathrm{Ricci}(V), V\rangle_\eta = q \times g(U,V)^2 + \tilde{p} \times g(V,V) , \quad \text{for any } V \in T\mathcal{M} . \tag{8}$$

The quantity $\langle U(\xi_s), \dot{\xi}_s\rangle$ is the hyperbolic cosine of the distance, on the unit hyperboloid at ξ_s identified with the hyperbolic space, between the space-time velocities of the fluid and of the path; it will be denoted by $\mathcal{A}_s$ or $\mathcal{A}(\xi_s, \dot{\xi}_s)$. Note that necessarily $\mathcal{A}_s \ge 1$. By Formulas (2) and (6), the energy equals:

$$\mathcal{E}(\xi, \dot{\xi}) = q(\xi)\, \mathcal{A}(\xi, \dot{\xi})^2 - p(\xi). \tag{9}$$

The energy of the fluid is simply: $\tilde{T}_{k\ell}\, U^k\, U^\ell = q - p$, and the scalar curvature equals $R = 2\,[(d+1)\,p - q]/(d-1)$. By (9), the weak energy condition reads here: $q \geq p^+$.

6 The Basic Relativistic Diffusion

The following (where $\mathcal{C}$ denotes the Casimir operator) is analogous to Lemma 3.1.

Lemma 6.1 *The operators* $H_0\,,\ \sum_{j=1}^d V_j^2\,,\ \mathcal{C}\,,\ H_0 + \frac{\sigma^2}{2}\sum_{j=1}^d V_j^2$ *do act on* C^2 *functions on the pseudo-unit tangent bundle* $T^1\mathcal{M}$, *inducing respectively: the vector field* $\mathcal{L}_0$ *generating the geodesic flow on* $T^1\mathcal{M}$, *the so-called vertical Laplacian* Δ_v *(i.e., the Laplacian on* $T^1_\xi\mathcal{M}$ *equipped with the hyperbolic metric induced by* $g(\xi)$*),* Δ_v *again, and the generator* $\mathcal{H}^1 := \mathcal{L}_0 + \frac{\sigma^2}{2}\,\Delta_v$. *More precisely, for any* $F \in C^2(T^1\mathcal{M})$, *on* $G(\mathcal{M})$ *we have:*

$$(\mathcal{L}_0 F)\circ\pi_1 \;=\; H_0(F\circ\pi_1)\,,\quad (\Delta_v F)\circ\pi_1 \;=\; \mathcal{C}(F\circ\pi_1)\,.$$

Besides, in local coordinates (x^i, e^k_j) *such that* $e_j = e^k_j\,\frac{\partial}{\partial x^k}$ *we have* $V_j = e^k_j\,\frac{\partial}{\partial e^k_0} + e^k_0\,\frac{\partial}{\partial e^k_j}$, *and denoting the inverse matrix of the pseudo-Riemannian metric of* $\mathcal{M}$ *by* (g^{kl}) *in these coordinates, we have:*

$$(\Delta_v F)\circ\pi_1 = \sum_{j=1}^d V_j^2(F\circ\pi_1) = \Big((e^k_0 e^l_0 - g^{kl})\,\frac{\partial^2}{\partial e^k_0\partial e^l_0} + d\,e^k_0\,\frac{\partial}{\partial e^k_0}\Big)F\circ\pi_1\,.$$

Now, according to Sect. 4, the relativistic motion we will consider lives on $T^1\mathcal{M}$ and admits as infinitesimal generator the operator $\mathcal{H}^1 = \mathcal{L}_0 + \frac{\sigma^2}{2}\,\Delta_v$ of Lemma 6.1 above. If $\mathcal{M}$ is the Minkowski flat space of special relativity, it coincides with the (Dudley) diffusion defined in Sect. 4 above.

To construct this general relativistic diffusion, we use a kind of stochastic development to produce a stochastic flow on the bundle $G(\mathcal{M})$, as for the Riemannian Brownian motion in Sect. 3. But we have now to project on $T^1\mathcal{M}$ and no longer on the base manifold $\mathcal{M}$ (this cannot work here), and to put the white noises on the acceleration, i.e., on the vertical vectors, and no longer on the velocity, i.e., on the horizontal vectors.

To proceed, let us fix $\Psi_0 \in G(\mathcal{M})$ and an $\mathbb{R}^d$-valued Brownian motion $w = (w^j_s)$. By Lemma 6.1, the stochastic flow (Ψ_s) defined by $(**)$ in the theorem below, possibly till some explosion time, commutes with the action of SO(d) on $G(\mathcal{M})$, thereby allowing to project it on $T^1\mathcal{M}$. This projection is precisely the relativistic diffusion we intended to define and construct. The vector field $\mathcal{L}_0$ denotes the generator of the geodesic flow, which operates on the position ξ-component, and

Δ_v denotes the vertical Laplacian (restriction to $T^1\mathcal{M}$ of the Casimir operator on $G(\mathcal{M})$), which operates on the velocity $\dot{\xi}$-component.

Theorem 6.2 ([13]) *(i) The $G(\mathcal{M})$-valued Stratonovitch stochastic differential equation*

$$(**)\qquad \Psi_s = \Psi_0 + \int_0^s H_0(\Psi_t)\,dt + \sigma \int_0^s \sum_{j=1}^d V_j(\Psi_t)\circ dw_t^j$$

defines a diffusion $(\xi_s, \dot{\xi}_s) := \pi_1(\Psi_s)$ on $T^1\mathcal{M}$, with generator $\mathcal{H}^1 = \mathcal{L}_0 + \frac{\sigma^2}{2}\,\Delta_v$.
(ii) If $\overleftarrow{\xi}(s) : T_{\xi_s}\mathcal{M} \to T_{\xi_0}\mathcal{M}$ denotes the inverse parallel transport along the C^1 curve $(\xi_{s'}\,|\,0 \le s' \le s)$, then $\zeta_s := \overleftarrow{\xi}(s)\,\dot{\xi}_s$ is a hyperbolic Brownian motion on $T_{\xi_0}\mathcal{M}$.
Therefore the path (ξ_s) is almost surely the development of a relativistic (Dudley) diffusion path in the Minkowski space $T_{\xi_0}\mathcal{M}$.

The infinitesimal generator of the $G(\mathcal{M})$-valued relativistic diffusion (Ψ_s) is $H_0 + \frac{\varrho^2}{2}\sum_{j=1}^d V_j^2$, which by Lemma 6.1 projects under π_1 to the infinitesimal generator of the relativistic diffusion $(\xi_s, \dot{\xi}_s)$; namely the relativistic operator expressed by:

$$\mathcal{H}^1 = \mathcal{L}^0 + \frac{\varrho^2}{2}\,\Delta^v = \dot{\xi}^k \frac{\partial}{\partial \xi^k} + \left(\frac{d\,\varrho^2}{2}\,\dot{\xi}^k - \dot{\xi}^i\dot{\xi}^j\,\Gamma_{ij}^k(\xi)\right)\frac{\partial}{\partial \dot{\xi}^k} + \frac{\varrho^2}{2}\,(\dot{\xi}^k\dot{\xi}^\ell - g^{k\ell}(\xi))\,\frac{\partial^2}{\partial \dot{\xi}^k\partial \dot{\xi}^\ell}\,. \qquad (10)$$

The relativistic diffusion $(\xi_s, \dot{\xi}_s)$ is parametrized by proper time $s \ge 0$ (since $g_{\xi_s}(\dot{\xi}_s, \dot{\xi}_s) = 1$), possibly till some positive explosion time.

As in the Riemannian case (recall Remark 3.3), on the one hand this construction uses the Cartan moving frame method and provides the stochastic parallel transport (Ψ_s) along the relativistic Brownian curve $(\xi_s, \dot{\xi}_s)$ together with the curve itself, and on the other hand, the relativistic Brownian motion $(\xi_s, \dot{\xi}_s)$ is covariant with respect to the isometries of $\mathcal{M}$: for any such (Lorentzian) isometry f, the process $\big(f\circ(\xi_s, \dot{\xi}_s)\big)$ is another relativistic Brownian motion, starting at $f(\xi_0, \dot{\xi}_0)$.

In local coordinates (ξ^i, e_j^k), setting $\Psi_s = (\xi_s^i, e_j^k(s))$, Equation $(**)$ becomes locally equivalent to the following system of Itô equations:

$$d\xi_s^k = \dot{\xi}_s^k\,ds = e_0^k(s)\,ds\,;\quad d\dot{\xi}_s^k = -\Gamma_{i\ell}^k(\xi_s)\,\dot{\xi}_s^i\,\dot{\xi}_s^\ell\,ds + \varrho\sum_{i=1}^d e_i^k(s)\,dw_s^i + \frac{d\,\varrho^2}{2}\,\dot{\xi}_s^k\,ds\,,\quad \text{and}$$

$$de_j^k(s) = -\Gamma_{i\ell}^k(\xi_s)\,e_j^\ell(s)\,\dot{\xi}_s^i\,ds + \varrho\,\dot{\xi}_s^k\,dw_s^j + \frac{\varrho^2}{2}\,e_j^k(s)\,ds\,,\quad \text{for } 1\le j\le d,\ 0\le k\le d\,.$$

Furthermore, on $T^1\mathcal{M}$ we have:

$$\sum_{j=1}^{d} V_j^2\mathcal{E} = 2(d+1)\,\mathcal{E} - 2\,\mathrm{Tr}(T) = 2(d+1)\,\mathcal{E} + (d-1)\,R\,.$$

As an application, a direct computation yields the following evolution of the energy.

Remark 6.3 The random energy process $\mathcal{E}_s = \mathcal{E}(\xi_s, \dot{\xi}_s)$ associated to the basic relativistic diffusion $\pi_1(\Psi_s) = (\xi_s, \dot{\xi}_s)$ satisfies the following equation (where $\nabla_V := V^j\nabla_j$):

$$d\mathcal{E}_s = \nabla_{\dot{\xi}_s}\mathcal{E}_s\,ds + \varrho^2\left[(d+1)\mathcal{E}_s + \tfrac{d-1}{2}\,R(\xi_s)\right]ds + dM_s^{\mathcal{E}}\,,$$

with the quadratic variation of its martingale part $dM_s^{\mathcal{E}}$ given by:

$$[d\mathcal{E}_s, d\mathcal{E}_s] = [dM_s^{\mathcal{E}}, dM_s^{\mathcal{E}}] = 4\varrho^2\,[\mathcal{E}_s^2 - \langle\tilde{T}\dot{\xi}_s, \tilde{T}\dot{\xi}_s\rangle]\,ds\,.$$

Note that generally the energy $\mathcal{E}_s$ is not a Markov process.

6.1 Example: The Schwarzschild Solution (After [13])

This space-time is commonly used in physics to model the complement of a spherical body, star or black hole; see for example ([21], Sect. 97). It is a basic example of space-time, i.e., of exact solution to the Einstein equations. This is actually the unique such solution which is both radial and empty (the latter amounts to having a vanishing Ricci tensor); see ([23], Theorems 3.1, 3.4, 3.7) for more specific statements in this direction.

Take $\mathcal{M} = \mathcal{S}_0 := \left\{\xi = (t, r, \theta) \in \mathbb{R}\times[R, +\infty[\times\mathbb{S}^2\right\}$, where $R \in \mathbb{R}_+$ is a parameter of the central body, endowed with the radial pseudo-metric:

$$(1-\tfrac{R}{r})\,dt^2 - (1-\tfrac{R}{r})^{-1}dr^2 - r^2|d\theta|^2\,.$$

The coordinate t represents the absolute time, and r the distance from the origin. The Ricci tensor vanishes, the space $\mathcal{S}_0$ being empty. A theorem by Birkhoff asserts that there is no other radial pseudo-metric in $\mathcal{S}_0$ which satisfies this constraint. Take as local coordinates the global spherical coordinates: $\xi \equiv (\xi^0, \xi^1, \xi^2, \xi^3) := (t, r, \varphi, \psi)$. According to the above, the system of Itô stochastic differential equations governing the relativistic diffusion $(\xi_s, \dot{\xi}_s)$ becomes:

$$dt_s = e_0^0(s)\,ds\,,\ dr_s = e_0^1(s)\,ds\,,\ d\varphi_s = e_0^2(s)\,ds\,,\ d\psi_s = e_0^3(s)\,ds\,,$$

$$de_0^0(s) = \tfrac{3\sigma^2}{2}\,e_0^0(s)\,ds - \tfrac{R}{r_s(r_s-R)}\,e_0^0(s)\,e_0^1(s)\,ds + dM_s^0\,,$$

$$de_0^2(s) = \tfrac{3\sigma^2}{2}\, e_0^2(s)\, ds - \tfrac{2}{r_s}\, e_0^1(s)\, e_0^2(s)\, ds + \sin\varphi_s \cos\varphi_s\, e_0^3(s)^2\, ds + dM_s^2\,,$$

$$de_0^3(s) = \tfrac{3\sigma^2}{2}\, e_0^3(s)\, ds - \tfrac{2}{r_s}\, e_0^1(s)\, e_0^3(s)\, ds - 2\,\mathrm{cotg}\,\varphi_s\, e_0^2(s)\, e_0^3(s)\, ds + dM_s^3\,,$$

where the martingale $M_s := (M_s^0, M_s^1, M_s^2, M_s^3)$ has the following rank 3 quadratic covariation matrix: $K_s = \sigma^2\,(e_0(s)\,{}^t e_0(s) - g^{-1}(\xi_s))$.

Let us introduce the angular momentum $\vec{b} := r^2\,\theta \wedge \dot{\theta}$, the energy $a := (1 - \frac{R}{r})\,\dot{t}$, and the norm of $\vec{b}$: $b := |\vec{b}| = r^2\, U$, with $U := |\dot{\theta}|$. Let us also set $T := \dot{r}$, and accordingly

$$T_s := \dot{r}_s = e_0^1(s)\,, \quad U_s := |\dot{\theta}_s| = \sqrt{e_0^2(s)^2 + \sin^2\varphi_s\, e_0^3(s)^2}\,, \quad \text{and} \quad D := \min\{s > 0 \,|\, r_s = R\}\,.$$

Standard stochastic calculus computations yield the following:

Proposition 6.1.1 *(i) The unit pseudo-norm relation (which expresses that the parameter s is precisely the arc length, i.e., the so-called proper time) is given by*

$$T_s^2 = a_s^2 - (1 - R/r_s)(1 + b_s^2/r_s^2)\,.$$

(ii) The process (r_s, a_s, b_s, T_s) is a degenerate diffusion, with lifetime D, which solves the following system of stochastic differential equations:

$$dr_s = T_s\, ds\,, \quad dT_s = dM_s^T + \tfrac{3\sigma^2}{2}\, T_s\, ds + (r_s - \tfrac{3}{2}R)\,\frac{b_s^2}{r_s^4}\, ds - \frac{R}{2r_s^2}\, ds\,,$$

$$da_s = dM_s^a + \tfrac{3\sigma^2}{2}\, a_s\, ds\,, \quad db_s = dM_s^b + \tfrac{3\sigma^2}{2}\, b_s\, ds + \frac{\sigma^2\, r_s^2}{2\, b_s}\, ds\,,$$

with quadratic covariation matrix of the local martingale (M^a, M^b, M^T) given by

$$K_s' := \sigma^2 \begin{pmatrix} a_s^2 - 1 + \frac{R}{r_s} & a_s\, b_s & a_s\, T_s \\ a_s\, b_s & b_s^2 + r_s^2 & b_s\, T_s \\ a_s\, T_s & b_s\, T_s & T_s^2 + 1 - \frac{R}{r_s} \end{pmatrix}.$$

We get in particular the following statement, in which the dimension is reduced.

Corollary 6.1.2 *The process (r_s, b_s, T_s) is a diffusion, with lifetime D and infinitesimal generator*

$$\mathcal{G}' := T\,\frac{\partial}{\partial r} + \frac{\sigma^2}{2}\,(b^2 + r^2)\,\frac{\partial^2}{\partial b^2} + \frac{\sigma^2}{2b}\,(3b^2 + r^2)\,\frac{\partial}{\partial b} + \sigma^2 bT\,\frac{\partial^2}{\partial b\partial T}$$

$$+ \frac{\sigma^2}{2}\left(T^2 + 1 - \tfrac{R}{r}\right)\frac{\partial^2}{\partial T^2} + \left(\frac{3\sigma^2}{2}\, T + (r - \tfrac{3}{2}R)\,\frac{b^2}{r^4} - \frac{R}{2r^2}\right)\frac{\partial}{\partial T}\,.$$

In the geodesic case $\sigma = 0$ five types of behaviour can occur, owing to the trajectory of (r_s) ; it can be:

- running from R to $+\infty$, or in the opposite direction;
- running from R to R in finite proper time;
- running from $+\infty$ to $+\infty$;
- running from R to some R_1 or from R_1 to $+\infty$, or idem in the opposite direction;
- running endlessly in a bounded region away from R.

Though the stochastic case $\sigma \neq 0$ can be seen as a perturbation of the geodesic case $\sigma = 0$, the asymptotic behaviour classification regarding it is quite different.

Theorem 6.1.3 *([13]) (i) For any initial condition, the radial process (r_s) almost surely hits R within a finite proper time D or goes to $+\infty$ as $s \to +\infty$.*
(ii) Both events in (1) occur with positive probability, from any initial condition.
(iii) Conditionally on the event $\{D = \infty\}$ of non-hitting the central body, the Schwarzschild relativistic diffusion $(\xi_s, \dot{\xi}_s)$ goes almost surely to infinity in some random asymptotic direction of $\mathbb{R}^3$, asymptotically with the velocity of light.

In particular, the relativistic diffusion almost surely cannot explode before a finite proper time D.

The Schwarzschild relativistic diffusion has been analyzed further in [13], using the (Kruskal-Szekeres) maximal extension of the Schwarzschild spacetime (also considered by [23, 24]): not only its behaviour till the hitting of the singularity can be thoroughly specified, but also a continuation of the diffusion thereafter makes sense, at least mathematically, and can be analyzed for proper time running the whole $\mathbb{R}_+$.

6.2 Example: The Gödel Universe (After [11])

The Gödel universe was intended by K. Gödel to object to the Einstein general theory of 1915: while being an exact solution to the Einstein equations, it presents the striking particularity of excluding global causality, since it has closed future-directed timelike continuous paths, which makes theoretically possible to access to one's own past after a long travel. That particular feature soon made this universe famous, and it is still the object of numerous developments.

The Gödel universe G is the manifold $\mathbb{R}^4$, endowed with coordinates $\xi := (t, x, y, z)$, and with the pseudo-metric g defined for some positive constant ω, by:

$$ds^2 := dt^2 - dx^2 + \tfrac{1}{2}\, e^{2\sqrt{2}\omega x}\, dy^2 + 2\, e^{\sqrt{2}\omega x}\, dt\, dy - dz^2 .$$

Along any timelike curve (t_s, x_s, y_s, z_s), the unit pseudo-norm relation, defining proper time s , is:

$$1 + \dot{t}_s^2 + \dot{x}_s^2 + \dot{z}_s^2 = \tfrac{1}{2} \left[e^{\sqrt{2}\omega x_s}\, \dot{y}_s + 2\, \dot{t}_s \right]^2 .$$

Gödel's universe can be viewed as a matrix group, on which Gödel's metric g happens to be the left-invariant metric generated by the Lorentz metric g^0 on the Lie algebra $\mathcal{G}$: $\langle \mathcal{L}_A, \mathcal{L}_A \rangle_g = \langle A, A \rangle_{g^0}$ for any $A \in \mathcal{G}$. This group structure is given by the following: for any $\xi_0 = (t_0, x_0, y_0, z_0)$, $\xi = (t, x, y, z) \in G$, $\xi_0 \times \xi = (t + t_0, x + x_0, y\, e^{-\sqrt{2}\omega x_0} + y_0, z + z_0)$.

Proposition 6.2.1 *The Gödel universe G is piece-wise geodesically transitive: any two points of it can be linked by a piece-wise geodesic future-directed timelike continuous path.*

The relativistic diffusion $(\xi_s, \dot{\xi}_s)$, in coordinates $\xi = (t, x, y, z)$, solves the following system of stochastic differential equations:

$$dt_s = \dot{t}_s\, ds\ ;\quad dx_s = \dot{x}_s\, ds\ ;\quad dy_s = \dot{y}_s\, ds\ ;\quad dz_s = \dot{z}_s\, ds\ ;$$

$$d\dot{t}_s = -2\sqrt{2}\,\omega\, \dot{t}_s\, \dot{x}_s\, ds - \sqrt{2}\,\omega\, e^{\sqrt{2}\omega x_s}\, \dot{x}_s\, \dot{y}_s\, ds + \tfrac{3\sigma^2}{2}\, \dot{t}_s\, ds + \sigma\, dM_s^t\ ;$$

$$d\dot{x}_s = -\sqrt{2}\,\omega\, e^{\sqrt{2}\omega x_s}\, \dot{t}_s\, \dot{y}_s\, ds - (\omega/\sqrt{2})\, e^{2\sqrt{2}\omega x_s}\, \dot{y}_s^2\, ds + \tfrac{3\sigma^2}{2}\, \dot{x}_s\, ds + \sigma\, dM_s^x\ ;$$

$$d\dot{y}_s = 2\sqrt{2}\,\omega\, e^{-\sqrt{2}\omega x_s}\, \dot{t}_s\, \dot{x}_s\, ds + \tfrac{3\sigma^2}{2}\, \dot{y}_s\, ds + \sigma\, dM_s^y\ ;$$

$$d\dot{z}_s = \tfrac{3\sigma^2}{2}\, \dot{z}_s\, ds + \sigma\, dM_s^z\ ;$$

where the $\mathbb{R}^4$-valued martingale $M_s := (M_s^t, M_s^x, M_s^y, M_s^z)$ has (rank 3) quadratic covariation matrix:

$$((K_s^{ij})) := \frac{\langle dM_s^i, dM_s^j \rangle}{ds} = \begin{pmatrix} \dot{t}_s^2 + 1 & \dot{t}_s\, \dot{x}_s & \dot{t}_s\, \dot{y}_s - 2\, e^{-\sqrt{2}\omega x_s} & \dot{t}_s\, \dot{z}_s \\ \dot{t}_s\, \dot{x}_s & \dot{x}_s^2 + 1 & \dot{x}_s\, \dot{y}_s & \dot{x}_s\, \dot{z}_s \\ \dot{t}_s\, \dot{y}_s - 2\, e^{-\sqrt{2}\omega x_s} & \dot{x}_s\, \dot{y}_s & \dot{y}_s^2 + 2\, e^{-2\sqrt{2}\omega x_s} & \dot{y}_s\, \dot{z}_s \\ \dot{t}_s\, \dot{z}_s & \dot{x}_s\, \dot{z}_s & \dot{y}_s\, \dot{z}_s & \dot{z}_s^2 + 1 \end{pmatrix}.$$

The following quantities, as $\dot{z}_s$, are constant along each geodesic:

$$a_s := \dot{t}_s + e^{\sqrt{2}\omega x_s}\, \dot{y}_s \quad \text{and} \quad b_s := e^{\sqrt{2}\omega x_s}\, (2\, \dot{t}_s + e^{\sqrt{2}\omega x_s}\, \dot{y}_s)\,.$$

The quantity a_s^2 represents an energy. Then we have:

$$da_s = \tfrac{3\sigma^2}{2}\, a_s\, ds + \sigma\, dM_s^a = \tfrac{3\sigma^2}{2}\, a_s\, ds + \sigma\, (dM_s^t + e^{\sqrt{2}\omega x_s}\, dM_s^y)\ ;$$

and

$$db_s = \tfrac{3\sigma^2}{2}\, b_s\, ds + \sigma\, dM_s^b = \tfrac{3\sigma^2}{2}\, b_s\, ds + \sigma\, e^{\sqrt{2}\omega x_s}\, (2\, dM_s^t + e^{\sqrt{2}\omega x_s}\, dM_s^y)\,.$$

Moreover we have:

$$d\dot{x}_s = (\omega/\sqrt{2})\, e^{-2\sqrt{2}\,\omega\, x_s}\, b_s^2\, ds - \sqrt{2}\,\omega\, e^{-\sqrt{2}\,\omega\, x_s}\, a_s\, b_s\, ds + \tfrac{3\sigma^2}{2}\, \dot{x}_s\, ds + \sigma\, dM_s^x\,,$$

and the $\mathbb{R}^4$-valued martingale $\tilde{M}_s := (M_s^a, M_s^b, M_s^x, M_s^{\dot{z}})$ has (rank 3) quadratic covariation matrix:

$$((\tilde{K}_s^{ij})) = \begin{pmatrix} a_s^2 - 1 & a_s\, b_s - 2\, e^{\sqrt{2}\,\omega\, x_s} & a_s\, \dot{x}_s & a_s\, \dot{z}_s \\ a_s\, b_s - 2\, e^{\sqrt{2}\,\omega\, x_s} & b_s^2 - 2\, e^{2\sqrt{2}\,\omega\, x_s} & b_s\, \dot{x}_s & b_s\, \dot{z}_s \\ a_s\, \dot{x}_s & b_s\, \dot{x}_s & \dot{x}_s^2 + 1 & \dot{x}_s\, \dot{z}_s \\ a_s\, \dot{z}_s & b_s\, \dot{z}_s & \dot{x}_s\, \dot{z}_s & \dot{z}_s^2 + 1 \end{pmatrix}.$$

From this, we deduce the following, which allows the asymptotic study, as proper time s goes to infinity, of relativistic paths.

Corollary 6.2.2 *The (7-dimensional) relativistic diffusion $(\xi_s, \dot{\xi}_s)$ admits the following sub-diffusions: (a_s) ; $(\dot{z}_s)$; $(a_s, \dot{z}_s)$; $(x_s, \dot{x}_s, a_s, b_s)$.*

The unit pseudo-norm relation can be written as: $1 + \dot{x}_s^2 + \dot{z}_s^2 + \frac{1}{2}\,(2\, a_s - e^{-\sqrt{2}\,\omega\, x_s}\, b_s)^2 = a_s^2$.
Hence the phase space $\mathcal{E}$ of the relativistic diffusion $(\xi_s, \dot{\xi}_s)$ can be written equivalently:

$$\mathcal{E} = \left\{ (t, x, y, z, a, b, \dot{x}, \dot{z}) \in \mathbb{R}^8 \,\middle|\, 1 + \dot{x}^2 + \dot{z}^2 + \tfrac{1}{2}\,(2\, a - e^{-\sqrt{2}\,\omega\, x}\, b)^2 = a^2 \right\},$$

in which the particular phase subspace $\mathcal{E}_0$ has to be distinguished:

$$\mathcal{E}_0 = \mathcal{E} \cap \left\{ a^2 = 1 + \dot{z}^2 \;;\; 2\, a = e^{-\sqrt{2}\,\omega\, x}\, b \;;\; \dot{x} = 0 \right\} = \mathcal{E} \cap \left\{ a^2 = 1 + \dot{z}^2 \right\}.$$

Remark 6.2.3 The phase space $\mathcal{E}$ splits into two connected components: $\mathcal{E} = \mathcal{E}^+ \sqcup \mathcal{E}^-$, with $\mathcal{E}^+ := \mathcal{E} \cap \{a \geq 1,\, b > 0\}$ and $\mathcal{E}^- := \mathcal{E} \cap \{a \leq -1,\, b < 0\}$. Similarly, $\mathcal{E}_0 = \mathcal{E}_0^+ \sqcup \mathcal{E}_0^-$, with $\mathcal{E}_0^+ := \mathcal{E}_0 \cap \mathcal{E}^+$ and $\mathcal{E}_0^- := \mathcal{E}_0 \cap \mathcal{E}^-$. Note that since $2\,\dot{t}_s + e^{\sqrt{2}\,\omega\, x_s}\, \dot{y}_s = e^{-\sqrt{2}\,\omega\, x_s}\, b_s$, the paths in $\mathcal{E}^+$ are always future-directed. Since the symmetry $(a, b) \mapsto (-a, -b)$ exchanges $(\mathcal{E}^+, \mathcal{E}_0^+)$ and $(\mathcal{E}^-, \mathcal{E}_0^-)$, from now on, we can restrict the phase space of the relativistic diffusion $(\xi_s, \dot{\xi}_s)$ to $\mathcal{E}^+$ (its behaviour on $\mathcal{E}^-$ being trivially related).

Recall that in a strongly causal space-time, it seems natural to use the causal boundary, in the sense of Penrose (see the conformal compactification considered in Sect. 2 of [24]), to classify lightlike geodesics by gathering in an equivalence class, called a *beam*, all geodesics which converge to a given causal boundary point (having asymptotically the same past, see ([18], Sect. 6.8)). On the contrary, in the present setting (recall Proposition 6.2.1) such a classification is totally inoperative. It seems that no alternative classification has been proposed before [11], which is relevant in a non-causal setting. Now, owing to the analysis of lightlike geodesics of G, the following alternative classification of lightlike geodesics into beams was adopted in [11], viewing then the 3-dimensional space of beams as an alternative notion of (non-causal, however conformal) boundary, as follows.

Definition 6.2.4 Let us call beam, or boundary point, of Gödel's universe, any equivalence class of lightlike geodesics, identifying those which have the same impact parameter $B = (\ell, \varrho, Y) \in \mathcal{B} = [-1, 1] \times \mathbb{R}_+^* \times \mathbb{R}$. Thus $\mathcal{B}$ will be the boundary of Gödel's universe.

Here the definition of the impact parameter B is exactly as in the following main result, though it is of course by far easier to obtain in the case of a lightlike geodesic.

Theorem 6.2.5 (i) *The relativistic diffusion is irreducible on its phase space* $\mathcal{E}^+ \setminus \mathcal{E}_0^+$.

(ii) *Almost surely, the relativistic diffusion path possesses a 3-dimensional asymptotic random variable* $B = (\ell, \varrho, Y) \in \mathcal{B}$. *Namely, it converges to this beam* B *in the sense that, as proper time* s *goes to infinity, we almost surely have:*

$$\dot{z}_s/a_s \longrightarrow \ell \in]-1, 0[\cup]0, 1[\ ; \quad b_s/a_s \longrightarrow \varrho \in]0, \infty[\ ; \quad Y_s := \frac{\sqrt{2}\,\dot{x}_s}{\omega\, b_s} + y_s \longrightarrow Y \in \mathbb{R}\,;$$

$$\left[\frac{\varrho}{2}\, e^{-\sqrt{2}\,\omega\, x_s} - 1\right]^2 + \left[\frac{\omega\,\varrho}{2}\,(y_s - Y)\right]^2 \longrightarrow \tfrac{1}{2}(1 - \ell^2)\,.$$

(iii) *The asymptotic random variable* (ℓ, ϱ, Y) *can be arbitrarily close to any given* $(\ell_0, \varrho_0, y) \in]-1, 1[\times]0, \infty[\times\mathbb{R}$, *with positive probability. Hence, the whole boundary (space of beams)* $\mathcal{B}$ *is the support of beams the relativistic diffusion can converge to.*

It remains an open question to establish whether the limiting random variable $B = (\ell, \varrho, Y)$ contains all asymptotic information, i.e., generates the invariant or the tail σ-field of the whole Gödel relativistic diffusion, and thereby is enough to represent all bounded harmonic functions (i.e., the *Poisson boundary*, recall Theorem 4.3) of the Gödel universe.

7 Covariant Ξ-relativistic Diffusions

We present here other intrinsic Lorentz-covariant diffusions, taking advantage of the curvature tensor (recall Sect. 5). They could be seen as maybe more physical than the basic relativistic diffusion presented till now, as their quadratic variation is locally determined by their velocity and the curvature of the space, and vanishes in flat or in Ricci-flat (empty) regions.

Let Ξ denote a non-negative smooth function on $G(\mathcal{M})$, invariant under the right action of $\mathrm{SO}(d)$ (so that it is identified with a function on $T^1\mathcal{M}$).

Our basic non-constant examples will be $\Xi = -\sigma^2 R$ and $\Xi = \sigma^2 \mathcal{E}$ (for a constant $\sigma > 0$).

We start with the following Stratonovitch stochastic differential equation on $G(\mathcal{M})$ (for a given $\mathbb{R}^d$-valued Brownian motion (w_s^j)):

$$d\Phi_s = H_0(\Phi_s)\,ds + \tfrac{1}{4}\sum_{j=1}^{d} V_j\,\Xi\,(\Phi_s)V_j(\Phi_s)\,ds + \sum_{j=1}^{d}\sqrt{\Xi(\Phi_s)}\,V_j(\Phi_s)\circ dw_s^j\,. \tag{11}$$

Note that all coefficients in this equation are clearly smooth, except $\sqrt{\Xi}$ on its vanishing set $\Xi^{-1}(0)$. However, $\sqrt{\Xi}$ is a locally Lipschitz function; see ([20], Proposition IV.6.2). Hence, Equation (11) does define a unique $G(\mathcal{M})$-valued diffusion (Φ_s). We have the following theorem, which defines the Ξ-*relativistic diffusion* (or Ξ-*diffusion*) (Φ_s) on $G(\mathcal{M})$ and $(\xi_s, \dot{\xi}_s)$ on $T^1\mathcal{M}$, possibly till some positive explosion time.

Theorem 7.1 (see [14]) *(i) The Stratonovitch stochastic differential equation (11) has a unique solution $(\Phi_s) = (\xi_s;\, \dot{\xi}_s\,, e_1(s), \ldots, e_d(s))$, possibly defined till some positive explosion time. This is a $G(\mathcal{M})$-valued covariant diffusion process, with generator*

$$\mathcal{H}_\Xi := H_0 + \tfrac{1}{2}\sum_{j=1}^{d} V_j\,\Xi\,V_j\,. \tag{12}$$

(ii) Its projection $\pi_1(\Phi_s) = (\xi_s, \dot{\xi}_s)$ defines a covariant diffusion on $T^1\mathcal{M}$, with SO(d)*-invariant generator*

$$\mathcal{H}^1_\Xi := \mathcal{L}_0 + \tfrac{1}{2}\,\nabla^v\,\Xi\,\nabla^v\,, \tag{13}$$

∇^v denoting the gradient on $T^1_\xi\mathcal{M}$ equipped with the hyperbolic metric induced by $g(\xi)$.
(iii) Moreover, the adjoint of $\mathcal{H}_\Xi$ with respect to the Liouville measure of $G(\mathcal{M})$ is
$\mathcal{H}^*_\Xi := -H_0 + \frac{1}{2}\sum_{j=1}^{d} V_j\,\Xi\,V_j\,.$ In particular, if there is no explosion, then the Liouville measure is invariant. Furthermore, if Ξ does not depend on $\dot{\xi}$, i.e., is a function on $\mathcal{M}$, then the Liouville measure is preserved by the stochastic flow defined by Eq. (11).

We specify right away how this looks in a local chart.

Corollary 7.2 *The $T^1\mathcal{M}$-valued Ξ-diffusion $(\xi_s, \dot{\xi}_s)$ satisfies $d\xi_s = \dot{\xi}_s\,ds$, and in any local chart, the following Itô stochastic differential equations: for $0 \le k \le d$, (denoting $\Xi_s \equiv \Xi(\xi_s, \dot{\xi}_s)$)*

$$d\dot{\xi}_s^k = dM_s^k - \Gamma_{ij}^k(\xi_s)\,\dot{\xi}_s^i\,\dot{\xi}_s^j\,ds + \tfrac{d}{2}\,\Xi_s\,\dot{\xi}_s^k\,ds + \tfrac{1}{2}\left[\dot{\xi}_s^k\,\dot{\xi}_s^\ell - g^{k\ell}(\xi_s)\right]\frac{\partial \Xi}{\partial \dot{\xi}^\ell}(\xi_s, \dot{\xi}_s)\,ds\,, \tag{14}$$

with the quadratic covariation matrix of the martingale term (dM_s) *given by:*

$$\left[d\dot{\xi}_s^k, d\dot{\xi}_s^\ell\right] = \left[\dot{\xi}_s^k\,\dot{\xi}_s^\ell - g^{k\ell}(\xi_s)\right]\Xi_s\,ds\,, \quad \textit{for } 0 \le k, \ell \le d\,.$$

Remark 7.3 (i) The vertical terms could be seen as an effect of the matter or the radiation present in the space-time $\mathcal{M}$. The Ξ-diffusion (Φ_s) reduces to the geodesic flow in the regions of the space where Ξ vanishes, which happens in particular for empty space-times $\mathcal{M}$ in the cases $\Xi = -\sigma^2 R(\xi)$, or $\Xi = \sigma^2\,\mathcal{E}(\xi, \dot{\xi})$, or also $\Xi = -\sigma^2 R(\xi)\,e^{\kappa\,\mathcal{E}(\xi,\dot{\xi})/R(\xi)}$ (for any positive constant κ) for example.

(ii) As for the basic relativistic diffusion, the law of the Ξ-relativistic diffusion is covariant with respect to any isometry of $(\mathcal{M}, g)$. The basic relativistic diffusion corresponds to $\Xi \equiv \sigma^2 > 0$, and the geodesic flow to $\Xi \equiv 0$.

(iii) In [5] is considered a general model for relativistic diffusions, which may be covariant or not. Up to enlarging it by allowing the "rest frame" (denoted by z in [5]) to have space vectors of non-unit norm, this model includes the generic Ξ-diffusion (compare the above Eq. (11) to (2.5), (3.3) in [5]).

7.1 Example 1: The R-diffusion

We assume here that the scalar curvature $R = R(\xi)$ is everywhere non-positive on $\mathcal{M}$, which is physically relevant: this is the *strong energy condition*, in the case where the energy-momentum tensor T has a timelike eigenvector (the so-called *type I*, e.g., a perfect fluid); see ([18] page 95); this is also the *dominant energy condition* in the terminology used by ([19], Sect. 2.2); up to the convention used for the sign. Consider the particular case corresponding to $\Xi = -\sigma^2 R(\xi)$, with a constant positive parameter σ.
In this case, as its central term clearly vanishes, Eq. (11) takes on the simple form:

$$d\Phi_s = H_0(\Phi_s)\,ds + \sigma \sum_{j=1}^{d} \sqrt{-R(\Phi_s)}\,V_j(\Phi_s) \circ dw_s^j\,.$$

7.2 Example 2: The $\mathcal{E}$-diffusion

We assume that the Weak Energy Condition (recall Sect. 5) holds (everywhere on $T^1\mathcal{M}$), which is physically relevant (see ([21], (94, 10)), [18]: this means that the energy has to be non-negative everywhere in the space-time), and consider the particular case corresponding to $\Xi = \sigma^2\,\mathcal{E} = \sigma^2\,\mathcal{E}(\xi, \dot{\xi}) = \sigma^2\,T_{00}$.

We call *energy relativistic diffusion* or *$\mathcal{E}$-diffusion* the $G(\mathcal{M})$-valued diffusion process (Φ_s) we get in this way, as well as its $T^1\mathcal{M}$-valued projection $\pi_1(\Phi_s)$.
The following computational lemma implies that the central drift term in Eq. (11) is a function of the Ricci tensor alone when Ξ is.

Lemma 7.2.1 *We have* $V_j R_i^k = \delta_{0i}\, R_j^k - \eta_{ij}\, R_0^k + \delta_0^k\, R_{ij} - \delta_j^k\, R_{0i}$ *, for* $0 \le i, k \le d$ *and* $1 \le j \le d$ *. In particular,* $V_j R = 0$*, and* $V_j \mathcal{E} = V_j T_{00} = V_j R_{00} = 2R_{0j}$ *.*

Lemma 7.2.1 and some more computation lead to the following, to be compared with Corollary 6.3. The notation $(\tilde{T}\dot{\xi})^k \equiv \tilde{T}_m^k \dot{\xi}^m$ below has the meaning of a matrix product.

Remark 7.2.2 The random energy $\mathcal{E}_s := \mathcal{E}(\xi_s, \dot{\xi}_s)$ associated to the $\mathcal{E}$-diffusion (Φ_s) satisfies the following equation (where $\nabla_V := V^j \nabla_j$):

$$d\mathcal{E}_s = \nabla_{\dot{\xi}_s} \mathcal{E}(\xi_s, \dot{\xi}_s)\, ds + (d+2)\, \sigma^2\, \mathcal{E}_s^2\, ds - 2\sigma^2\, g(\tilde{T}\dot{\xi}_s, \tilde{T}\dot{\xi}_s)\, ds + 2\sigma\, dM_s^{\mathcal{E}} \,,$$

with the quadratic variation of its martingale part $dM_s^{\mathcal{E}}$ given by:

$$[d\mathcal{E}_s, d\mathcal{E}_s] = 4\sigma^2 \left[dM_s^{\mathcal{E}}, dM_s^{\mathcal{E}}\right] = 4\sigma^2 \left[\mathcal{E}_s^2 - g(\tilde{T}\dot{\xi}_s, \tilde{T}\dot{\xi}_s)\right] \mathcal{E}_s\, ds.$$

Remark 7.2.3 The case of Einstein Lorentz manifolds.

The Lorentz manifold $\mathcal{M}$ is said to be Einstein if its Ricci tensor is proportional to its metric tensor. Bianchi's contracted identities (see for example [18]), which entail the conservation equations $\nabla_k \tilde{T}^{jk} = 0$, force the proportionality coefficient $\tilde{p}$ to be constant on $\mathcal{M}$. Hence: $\tilde{R}_{\ell m}(\xi) = \tilde{p}\, g_{\ell m}(\xi)$, for any ξ in $\mathcal{M}$ and $0 \le \ell, m \le d$.

Then the scalar curvature is $R(\xi) = (d+1)\tilde{p}$, and by Einstein's Equations (5) we have:

$$\tilde{T}_{\ell m}(\xi) = (\Lambda - \tfrac{d-1}{2}\, \tilde{p})\, g_{\ell m}(\xi) \;=:\; -p\, g_{\ell m}(\xi)\,.$$

Hence Eq. (6) holds, with $q = 0$: we are in a limiting case of a perfect fluid. Moreover, R and $\mathcal{E}$ are constant, so that in an Einstein Lorentz manifold, the R-diffusion and the $\mathcal{E}$-diffusion coincide with the basic relativistic diffusion (of Sect. 6).

8 Example of Robertson-Walker (R-W) Manifolds

These important manifolds are intended to model an expanding (or shrinking) universe resulting from a "Big-Bang", as ours is believed to be. They admit an absolute time coordinate t ; in the classical terminology used in particular by ([19], Sect. 2), they are "globally hyperbolic".

They are particular cases of warped product: they can be written as $\mathcal{M} = I \times M$, where I is an open interval of $\mathbb{R}_+$ and $M \in \{\mathbb{S}^3, \mathbb{R}^3, \mathbb{H}^3\}$, with spherical coordinates $\xi \equiv (t, r, \varphi, \psi)$ (which are global in the case of $\mathbb{R}^3$, $\mathbb{H}^3$, and are defined separately on two hemispheres in the case of $\mathbb{S}^3$), and are endowed with the pseudo-norm:

$$g(\dot\xi,\dot\xi) := \dot t^2 - \alpha(t)^2\Big(\frac{\dot r^2}{1-kr^2} + r^2\dot\varphi^2 + r^2\sin^2\!\varphi\,\dot\psi^2\Big), \tag{15}$$

where the constant scalar spatial curvature k belongs to $\{-1,0,1\}$ (note that $r\in[0,1]$ for $k=1$ and $r\in\mathbb{R}_+$ for $k=0,-1$), and the expansion factor α is a positive C^2 function on I. The so-called *Hubble function* is: $H(t) := \alpha'(t)/\alpha(t)$.
Note that we necessarily have $\dot t \ge 1$ everywhere on $T^1\mathcal{M}$. The curvature operator is given (denoting by X,Y,A,Z vectors over M and by h the metric tensor of M) by:

$$\Big\langle \mathcal{R}\left((u\partial_t + X)\wedge(v\partial_t + Y)\right), (a\partial_t + A)\wedge(w\partial_t + Z)\Big\rangle_\eta$$

$$= \alpha\alpha''\,h(uY - vX, aZ - wA) - \alpha^2(\alpha'^2 + k)\,[h(X,A)h(Y,Z) - h(X,Z)h(Y,A)].$$

The Ricci tensor $((\tilde R_{\ell m}))$ is diagonal, with diagonal entries:

$$\Big(-3\,\frac{\alpha''(t)}{\alpha(t)}, \frac{A(t)}{1-kr^2}, A(t)\,r^2, A(t)\,r^2\sin^2\!\varphi\Big), \quad \text{where } A(t) := \alpha(t)\,\alpha''(t) + 2\,\alpha'(t)^2 + 2k,$$

and the scalar curvature is $R = -6\,[\alpha(t)\,\alpha''(t) + \alpha'(t)^2 + k]\,\alpha(t)^{-2}$.
The Einstein energy-momentum tensor $\tilde R_{\ell m} - \frac{1}{2}\,R\,g_{\ell m} = \tilde T_{\ell m}$ is diagonal as well, with diagonal entries:

$$\Big(3\,\frac{\alpha'(t)^2 + k}{\alpha(t)^2}, \frac{-\tilde A(t)}{1-kr^2}, -\tilde A(t)\,r^2, -\tilde A(t)\,r^2\sin^2\!\varphi\Big), \quad \text{with } \tilde A(t) := 2\alpha(t)\alpha''(t) + \alpha'(t)^2 + k.$$

Hence, we have

$$\tilde T_{\ell m} - \alpha(t)^{-2}\tilde A(t)\,g_{\ell m} = 2\,[k\,\alpha(t)^{-2} - H'(t)]\,1_{\{\ell=m=0\}}.$$

Thus, we have here an example of perfect fluid: Eq. (6) holds, with

$$U_j \equiv \delta_j^0, \quad -p(\xi) = k\,\alpha(t)^{-2} + 2H'(t) + 3H^2(t), \quad q(\xi) = 2\,[k\,\alpha(t)^{-2} - H'(t)], \tag{16}$$

$$\tilde p(\xi) = -2\,[2k\,\alpha(t)^{-2} + H'(t) + 3H^2(t)]/(d-1).$$

Note that

$$\mathcal{A}_s = U_i(\xi_s)\,\dot\xi_s^i = \dot t_s \quad \text{and} \quad \mathcal{E}_s = 2\,[k\,\alpha(t_s)^{-2} - H'(t_s)]\,\dot t_s^2 - p(\xi_s). \tag{17}$$

The weak energy condition is equivalent to: $\alpha'^2 + k \ge (\alpha\,\alpha'')^+$.

We shall consider only eternal Robertson-Walker space-times, which have their future-directed half-geodesics complete. This amounts to $I = \mathbb{R}_+^*$, together with $\int^\infty \frac{\alpha}{\sqrt{1+\alpha^2}} = \infty$. In the case of the basic relativistic diffusion (within such a Robertson-Walker model), we have in particular:

$$d\dot{t}_s = \sigma\sqrt{\dot{t}_s^2 - 1}\, dw_s + \tfrac{3\sigma^2}{2}\, \dot{t}_s\, ds - H(t_s)[\dot{t}_s^2 - 1]\, ds\,. \tag{18}$$

8.1 Ξ-relativistic Diffusions in an Einstein-De Sitter-Like Manifold

We consider henceforth the particular case $I =]0, \infty[$, $k = 0$, and $\alpha(t) = t^c$, with exponent $c > 0$. Note that such expansion functions α can be obtained by solving a proportionality relation between p and q (see [18] or [21]).
Thus $q = 2c\, t^{-2}$, $p = (2-3c)c\, t^{-2}$, $R = -6\, c\, (2c-1)\, t^{-2}$, $\mathcal{E} = c\, t^{-2}\, (2\, \dot{t}^2 + 3c - 2)$.
Note that the weak energy condition holds. The scalar curvature is non-positive if and only if $c \geq 1/2$, and the pressure p is non-negative if and only if $c \leq 2/3$.

Note that the particular case $c = \frac{2}{3}$ corresponds to a vanishing pressure p, and is precisely known as that of *Einstein-de Sitter universe* (see for example [18]). And the analysis of [21] shows precisely both limiting cases $c = \frac{2}{3}$ and $c = \frac{1}{2}$.

8.1.1 Basic Relativistic Diffusion in an Einstein-De Sitter-Like Manifold

In order to compare with the other relativistic diffusions, we mention, first for the basic relativistic diffusion (of Sect. 6), the stochastic differential equations satisfied by the main coordinates $\dot{t}_s$ and $\dot{r}_s$, appearing in the 4-dimensional sub-diffusion $(t_s, \dot{t}_s, r_s, \dot{r}_s)$. By (18), we have, for independent standard real Brownian motions $w, \tilde{w}$:

$$d\dot{t}_s = \sigma\sqrt{\dot{t}_s^2 - 1}\, dw_s + \tfrac{3\sigma^2}{2}\, \dot{t}_s\, ds - \frac{c}{t_s}\, (\dot{t}_s^2 - 1)\, ds\,; \tag{19}$$

$$d\dot{r}_s = \frac{\sigma\, \dot{t}_s\, \dot{r}_s}{\sqrt{\dot{t}_s^2 - 1}}\, dw_s + \sigma\sqrt{\frac{1}{t_s^{2c}} - \frac{\dot{r}_s^2}{\dot{t}_s^2 - 1}}\, d\tilde{w}_s + \tfrac{3\sigma^2}{2}\, \dot{r}_s\, ds + \left[\frac{\dot{t}_s^2 - 1}{t_s^{2c}} - \dot{r}_s^2\right]\frac{ds}{r_s} - \frac{2c}{t_s}\, \dot{t}_s\, \dot{r}_s\, ds\,. \tag{20}$$

Almost surely (see [1]), $\lim_{s\to\infty} \dot{t}_s = \infty$; moreover x_s/r_s and $\dot{x}_s/|\dot{x}_s|$ converge in $\mathbb{S}^2$, to the same random limit.

Further results are established in [1], where the whole diffusion is thoroughly considered. In particular, the Poisson boundary is determined in some sub-cases of interest, yielding in such a curved framework an analogue of Theorem 4.3.

8.1.2 *R*-diffusion in an Einstein-De Sitter-Like Manifold

With the above, Sect. 7.1 gives here, for the R-relativistic diffusion, when $c \geq 1/2$:

$$d\dot{\xi}_s = \sigma\, dM_s + 9c\,(2c-1)\,\sigma^2\, t_s^{-2}\, \dot{\xi}_s\, ds - \Gamma_{ij}^{\cdot}(\xi_s)\, \dot{\xi}_s^i\, \dot{\xi}_s^j\, ds\,, \tag{21}$$

with the quadratic covariation matrix of the martingale part dM_s given by:

$$\sigma^{-2}\left[d\dot{\xi}_s^k, d\dot{\xi}_s^\ell\right] = 6c\,(2c-1)\left[\dot{\xi}_s^k\, \dot{\xi}_s^\ell - g^{k\ell}(\xi_s)\right] t_s^{-2}\, ds\,, \quad \text{for } 0 \leq k, \ell \leq d\,.$$

In particular, we have for independent standard real Brownian motions w, $\tilde{w}$:

$$d\dot{t}_s = \frac{\sigma}{t_s}\sqrt{6\,c\,(2c-1)(\dot{t}_s^2 - 1)}\, dw_s + \frac{9\,\sigma^2 c\,(2c-1)}{t_s^2}\, \dot{t}_s\, ds - \frac{c}{t_s}(\dot{t}_s^2 - 1)\, ds\,; \tag{22}$$

$$d\dot{r}_s = \frac{\sigma\sqrt{6c\,(2c-1)}}{t_s}\left[\frac{\dot{t}_s\, \dot{r}_s}{\sqrt{\dot{t}_s^2 - 1}}\, dw_s + \sqrt{\frac{1}{t_s^{2c}} - \frac{\dot{r}_s^2}{\dot{t}_s^2 - 1}}\, d\tilde{w}_s\right] \tag{23}$$

$$+ \frac{9\,\sigma^2 c\,(2c-1)}{t_s^2}\, \dot{r}_s\, ds + \left[\frac{\dot{t}_s^2 - 1}{t_s^{2c}} - \dot{r}_s^2\right]\frac{ds}{r_s} - \frac{2c}{t_s}\, \dot{t}_s\, \dot{r}_s\, ds\,.$$

As the scalar curvature $R_s = 6c\,(1-2c)/t_s^2$ vanishes asymptotically, we expect that almost surely the R-diffusion behaves eventually as a timelike geodesic, and in particular that $\lim\limits_{s\to\infty} \dot{t}_s = 1$.

8.1.3 $\mathcal{E}$-diffusion in an Einstein-De Sitter-Like Manifold

Similarly, using (16) and (17), we have here $\mathcal{E}\,\dot{\xi} - \tilde{T}\dot{\xi} = 2(0 - H')(\dot{t}^2\dot{\xi} - \dot{t}\, U)$, so that Sect. 7.2 reads here, for the $\mathcal{E}$-diffusion:

$$d\dot{\xi}_s = \sigma\, dM_s + \tfrac{3\sigma^2 c}{2}\, t_s^{-2}\,(2\,\dot{t}_s^2 + 3c - 2)\, \dot{\xi}_s\, ds + 2\sigma^2 c\, t_s^{-2}(\dot{t}_s\, \dot{\xi}_s - U_s)\, \dot{t}_s\, ds - \Gamma_{ij}^{\cdot}(\xi_s)\, \dot{\xi}_s^i\, \dot{\xi}_s^j\, ds\,, \tag{24}$$

with the quadratic covariation matrix of the martingale part dM_s given by:

$$\sigma^{-2}\,[d\dot{\xi}_s^k, d\dot{\xi}_s^\ell] = c\,[\dot{\xi}_s^k\, \dot{\xi}_s^\ell - g^{k\ell}(\xi_s)]\,(2\,\dot{t}_s^2 + 3c - 2)\, t_s^{-2}\, ds\,, \quad \text{for } 0 \leq k, \ell \leq d\,.$$

In particular, we have for some standard real Brownian motion w:

$$d\dot{t}_s = \frac{\sigma\sqrt{c}}{t_s}\sqrt{(2\,\dot{t}_s^2 - 2 + 3c)(\dot{t}_s^2 - 1)}\, dw_s + c\left[5\sigma^2(\dot{t}_s^2 - 1 + \tfrac{9c}{10})\,\frac{\dot{t}_s}{t_s^2} - \frac{\dot{t}_s^2 - 1}{t_s}\right] ds\,; \tag{25}$$

$$d\dot{r}_s = \frac{\sigma\sqrt{c}}{t_s}\sqrt{2\,\dot{t}_s^2 - 2 + 3c}\left[\frac{\dot{t}_s\,\dot{r}_s}{\sqrt{\dot{t}_s^2-1}}\,dw_s + \sqrt{\frac{1}{t_s^{2c}} - \frac{\dot{r}_s^2}{\dot{t}_s^2-1}}\,d\tilde{w}_s\right] \tag{26}$$

$$+\,\sigma^2 c\,(5\,\dot{t}_s^2 - 3 + \tfrac{9c}{2})\,\frac{\dot{r}_s}{t_s^2}\,ds - \frac{2c}{t_s}\,\dot{t}_s\,\dot{r}_s\,ds + \left[\frac{\dot{t}_s^2-1}{t_s^{2c}} - \dot{r}_s^2\right]\frac{ds}{r_s}\,.$$

Remark 8.1.4 Comparison of Ξ-diffusions in an Einstein-de Sitter-like manifold. Along the preceding Sects. 8.1.1, 8.1.2, 8.1.3, we specified the various Ξ-diffusions we considered successively in Sects. 6, 7.1, 7.2 to an Einstein-de Sitter-like manifold. Restricting to the only equation related to the hyperbolic angle $\mathcal{A}_s = \dot{t}_s$, or in other words, to the simplest sub-diffusion $(t_s, \dot{t}_s)$, this yields Equations (19), (22), (25) respectively. We observe that even in this simple case, all these covariant relativistic diffusions differ notably, having pairwise distinct minimal sub-diffusions (with 3 non-proportional diffusion factors).

8.2 Asymptotic Behavior of the *R*-diffusion in an Einstein-De Sitter Manifold

We present here the asymptotic study of the R-diffusion of an Einstein-de Sitter-like manifold (recall Sects. 8.1, 8.1.2). We will focus our attention on the simplest sub-diffusion $(t_s, \dot{t}_s)$, and on the space component $x_s \in \mathbb{R}^3$. Recall from (17) that $\dot{t}_s = \mathcal{A}_s$ equals the hyperbolic angle, measuring the gap between the ambient fluid and the velocity of the diffusing particle. Recall also that, by the unit pseudo-norm relation, $\dot{t}_s$ controls the behavior of the whole velocity $\dot{\xi}_s$. We get as a consequence the asymptotic behavior of the energy $\mathcal{E}_s$. As quoted in Sect. 8.1.2, we must here have $c \geq \frac{1}{2}$.

Note that for $c = \frac{1}{2}$, the scalar curvature vanishes, and the R-diffusion reduces to the geodesic flow, whose equations are easily solved and whose time coordinate satisfies (for constants a and s_0):

$$s - s_0 = \sqrt{t_s\,(t_s + a^2)} - a^2 \log[\sqrt{t_s} + \sqrt{t_s + a^2}\,]\,, \quad \text{whence} \quad t_s \sim s\,.$$

The proofs in this section (and in the following one) repeatedly use the elementary fact that almost surely a continuous local martingale cannot go to infinity.

The following confirms a conjecture stated at the end of Sect. 8.1.2.

Proposition 8.2.1 *The process $\dot{t}_s$ goes almost surely to 1, and $\mathcal{E}_s \to 0$, as $s \to \infty$.*

The following reveals the asymptotic behavior of the space component (x_s) for $c > \frac{1}{2}$.

Proposition 8.2.2 *For $c > \frac{1}{2}$, the space component converges almost surely (as $s \to \infty$):*

$$x_s \to x_\infty \in \mathbb{R}^3.$$

This does not hold in the geodesic flow limiting case $c = \frac{1}{2}$, since then we have

$$r_s = \sqrt{b^2/a^2 + (a + o(1)) \log s} \sim \sqrt{a \log s} \quad \text{as } s \to \infty .$$

To compare the R-diffusion with geodesics, note that (as is easily seen; see for example [1]) along any timelike geodesic, we have $x_s = x_1 + \frac{\dot{x}_1}{|\dot{x}_1|} \int_1^s \frac{a\, d\tau}{t_\tau^{2c}}$ (and $\frac{\dot{x}_s}{|\dot{x}_s|} = \frac{\dot{x}_1}{|\dot{x}_1|}$), which converges precisely for $c > \frac{1}{2}$; and along any lightlike geodesic, we have $x_s = x_1 + \frac{\dot{x}_1}{|\dot{x}_1|} \int_{t_1}^{t_s} \frac{d\tau}{\tau^c} \sim V \times s^{\frac{1-c}{1+c}}$ (and $\frac{\dot{x}_s}{|\dot{x}_s|} = \frac{\dot{x}_1}{|\dot{x}_1|}$), which converges only for $c > 1$.

On the other hand, for $c \leq 1$, the behavior of the basic relativistic diffusion is shown to satisfy (see [1]): $r_s \underset{s \to \infty}{\sim} \int_1^s \frac{a_\tau\, d\tau}{t_\tau^{2c}} \longrightarrow \infty$ (exponentially fast, at least for $c < 1$).

Hence, the R-diffusion behaves asymptotically more like a (timelike) geodesic than like the basic relativistic diffusion. However due to other facts, the asymptotic behavior of the R-diffusion seems to be somehow intermediate between those of the geodesic flow and of the basic relativistic diffusion.

8.3 Asymptotic Energy of the $\mathcal{E}$-diffusion in an Einstein-De Sitter Manifold

We consider here the case of Sect. 8.1.3, dealing with the energy diffusion in an Einstein-de Sitter-like manifold, and more precisely, with its absolute-time minimal sub-diffusion $(t_s, \dot{t}_s)$ satisfying Eq. (25), and with the resulting random energy:

$$\mathcal{E}_s = c\, t_s^{-2} \,(2\, \dot{t}_s^2 + 3c - 2) = 2c\, (\dot{t}_s / t_s)^2 + \mathcal{O}(s^{-2}).$$

Let us denote by ζ the explosion time: $\zeta := \sup\{s > 0 \mid \dot{t}_s < \infty\} \in]0, \infty]$.

Lemma 8.3.1 *We have almost surely: either $\lim_{s \to \zeta} \dot{t}_s = 1$ and $\zeta = \infty$, or $\lim_{s \to \zeta} \dot{t}_s = \infty$.*

The asymptotic behavior can, with positive probability, be partly opposite to that of the preceding R-diffusion:

Proposition 8.3.2 *From any starting point $(t_{s_0}, \dot{t}_{s_0})$, there is a positive probability that both $\mathcal{A}_s \equiv \dot{t}_s$ and the energy $\mathcal{E}_s$ explode. This happens with arbitrarily large probability, starting with $\dot{t}_{s_0}/t_{s_0}$ sufficiently large and t_0 bounded away from zero.*

On the other hand, there is also a positive probability that the hyperbolic angle $\mathcal{A}_s = \dot{t}_s$ does not explode and goes to 1, and then that the random energy $\mathcal{E}_s$ goes to 0. This happens actually with arbitrarily large probability, starting with sufficiently large $t_{s_0}/\dot{t}_{s_0}$.

9 Sectional Relativistic Diffusion

We turn now our attention towards a different class of intrinsic relativistic generators on $G(\mathcal{M})$, whose expressions derive directly from the commutation relations of Sect. 5, on canonical vector fields of $TG(\mathcal{M})$. They all project on the unit tangent bundle $T^1\mathcal{M}$ onto a unique relativistic generator $\mathcal{H}^1_{curv}$, whose expression involves the curvature tensor. Semi-ellipticity of $\mathcal{H}^1_{curv}$ requires the assumption of non-negativity of timelike sectional curvatures. Note that in general $\mathcal{H}^1_{curv}$ does not induce the geodesic flow in an empty space.

We shall actually consider, among these generators, those which are invariant under the action of $SO(d)$ on $G(\mathcal{M})$. To this aim, we introduce the following dual vertical vector fields, by lifting indexes: $V^{ij} := \eta^{im}\,\eta^{jn}\,V_{e_m\wedge e_n}$, so that $V^j \equiv V^{0j} = -V_j$ and $V^{ij} = V_{e_i\wedge e_j}$ for $1 \le i, j \le d$. We again fix a positive parameter σ.

Theorem 9.1 *(i) The following four* SO(d)*-invariant differential operators define the same operator $\mathcal{H}^1_{curv}$ on $T^1\mathcal{M}$:*

$$H_0 - \tfrac{\sigma^2}{2}\sum_{j=1}^{d}\Big([H_0, H_j]V^j + V^j[H_0, H_j]\Big)\,; \quad H_0 + \sigma^2\sum_{j=1}^{d}[H_j, H_0]V^j\,;$$

$$H_0 + \sigma^2\sum_{j=1}^{d} R_0^j\, V_j - \sigma^2\sum_{1\le j,k\le d}\mathcal{R}_0{}^{j0k}\, V_j V_k\,; \quad H_0 - \tfrac{\sigma^2}{4}\sum_{1\le i,j\le d}\Big([H_i, H_j]\,V^{ij} + V^{ij}\,[H_i, H_j]\Big);$$

(ii) $(\mathcal{H}^1_{curv} - \mathcal{L}_0)$ is self-adjoint with respect to the Liouville measure of $T^1\mathcal{M}$.
(iii) In local coordinates, the so-defined second order operator $\mathcal{H}^1_{curv}$ on $T^1\mathcal{M}$ is given by:

$$\mathcal{H}^1_{curv} = \dot\xi^j\frac{\partial}{\partial\xi^j} - \dot\xi^i\dot\xi^j\,\Gamma^k_{ij}\frac{\partial}{\partial\dot\xi^k} + \tfrac{\sigma^2}{2}\,\dot\xi^n\,\tilde{R}^k_n\frac{\partial}{\partial\dot\xi^k} - \tfrac{\sigma^2}{2}\,\dot\xi^p\dot\xi^q\,\widetilde{\mathcal{R}}_p{}^k{}_q{}^\ell\,\frac{\partial^2}{\partial\dot\xi^k\partial\dot\xi^\ell}$$

$$= \dot\xi^j\frac{\partial}{\partial\xi^j} - \dot\xi^i\dot\xi^j\,\Gamma^k_{ij}\frac{\partial}{\partial\dot\xi^k} + \tfrac{\sigma^2}{2}\,\dot\xi^m\,\widetilde{\mathcal{R}}_{mnpq}\left(g^{nq}\,g^{pk}\frac{\partial}{\partial\dot\xi^k} - \dot\xi^p\,g^{nk}g^{q\ell}\,\frac{\partial^2}{\partial\dot\xi^k\partial\dot\xi^\ell}\right).$$

The generator $\mathcal{H}^1_{curv}$ defined on $T^1\mathcal{M}$ by Theorem 9.1 is covariant with respect to any Lorentz isometry of $(\mathcal{M}, g)$. Hence, it is a candidate to generate a covariant "sectional" relativistic diffusion on $T^1\mathcal{M}$. Now, a necessary and sufficient condition, in order that such an operator be the generator of a well-defined diffusion, is that it be subelliptic.
We are thus led to consider the following negativity condition on the curvature:

$$\big\langle \mathcal{R}(u \wedge v), u \wedge v \big\rangle_\eta \le 0\,, \quad \text{for any timelike } u \text{ and any spacelike } v\,. \tag{27}$$

This condition is equivalent to the following lower bound on sectional curvatures of timelike planes $\mathbb{R}u + \mathbb{R}v$:

$$\frac{\langle \mathcal{R}(u \wedge v), u \wedge v \rangle_\eta}{g(u \wedge v, u \wedge v)} \ge 0\,,$$

since $g(u \wedge v, u \wedge v) := g(u,u)g(v,v) - g(u,v)^2 < 0$ for such planes.

When this negativity condition is fulfilled, we call the resulting covariant diffusion on $T^1\mathcal{M}$, which has generator $\mathcal{H}^1_{curv}$ given by Theorem 9.1, the *sectional relativistic diffusion*. Note that the sectional curvature classically plays a significant role in Lorentzian geometry, see for example ([23], Theorems 2.2 and 2.3).

Remark 9.2 Consider a Lorentz manifold $(\mathcal{M}, g)$ having the warped product form, for example a Robertson-Walker one. Then the sign condition (27) is equivalent to: $\alpha'' \le 0$ on I, together with the following lower bound on sectional curvatures of the Riemannian factor (M, h) :

$$\inf_{X,Y \in TM} \frac{\langle \mathcal{K}(X \wedge Y), X \wedge Y \rangle}{h(X,X)\,h(Y,Y) - h(X,Y)^2} \ge \sup_I \{\alpha\,\alpha'' - \alpha'^2\}.$$

In an Einstein-de Sitter-like manifold (recall Sect. 8.1), the sign condition (27) holds if and only if $\alpha'' \le 0$, i.e., if and only if $c \le 1$. The generator $\mathcal{H}^1_{curv}$ is fully computable, but has a complicated expression, even in such a simple example.

References

1. J. Angst, Asymptotic behavior of a relativistic Brownian diffusion in Robertson-Walker spacetimes. Ann. Inst. H. Poincaré Prob. Stat. **52**(1), 376–411 (2016)
2. M. Arnaudon, K.A. Coulibaly, A. Thalmaier, Brownian motion with respect to a metric depending on time; definition, existence and applications to Ricci flow. C. R. Acad. Sci. Paris **346**, 773–778 (2008)
3. M. Arnaudon, A. Thalmaier, Li-Yau type gradient estimates and Harnack inequalities by stochastic analysis. Advanc. Stud. Pure Math. **57**, 29-48 (2010)
4. I. Bailleul, Poisson boundary of a relativistic diffusion. Probab. Theory Relat. Fields **141**, 283–329 (2008)
5. I. Bailleul, A stochastic approach to relativistic diffusions. Ann. I.H.P. **46**(3), 760–795 (2010)

6. I. Bailleul, J. Franchi, Non-explosion criteria for relativistic diffusions. Ann. Prob. **40**(5), 2168–2196 (2012)
7. F. Debbasch, A diffusion process in curved space-time. J. Math. Phys. **45**(7), 2744–2760 (2004)
8. R.M. Dudley, Lorentz-invariant Markov processes in relativistic phase space. Arkiv för Matematik **6**(14), 241–268 (1965)
9. D. Elworthy, Geometric aspects of diffusions on manifolds, in *École d'Été de Probabilités de Saint-Flour (1985–87)*, vol. XV-XVII. Lecture Notes in Math, vol. 1362 (Springer, Berlin, 1988), pp. 277–425
10. M. Émery, On two transfer principles in stochastic differential geometry, in *Séminaire de Probabilités*, vol. XXIV (Springer, Berlin, 1990), pp. 407–441
11. J. Franchi, Relativistic diffusion in Gödel's Universe. Comm. Math. Phys. **290**(2), 523–555 (2009)
12. J. Franchi, Small time asymptotics for an example of strictly hypoelliptic heat kernel, in *Séminaire de Probabilités*, vol. 46 (2014)
13. J. Franchi, Y. Le Jan, Relativistic Diffusions and Schwarzschild Geometry. Comm. Pure Appl. Math. **LX**(2), 187–251 (2007)
14. J. Franchi, Y. Le Jan, Curvature diffusions in general relativity. Comm. Math. Phys. **307**(2), 351–382 (2011)
15. J. Franchi, Y. Le Jan, Hyperbolic dynamics and Brownian motion, in *Oxford Mathematical Monographs* (Oxford Science Publications, 2012)
16. A. Grigor'yan, Heat kernels on weighted manifolds and applications. Contemp. Math. **398**, 93–191 (2006)
17. E.P. Hsu, *Stochastic Analysis on Manifolds*. Graduate studies in Mathematics, vol. 38 (A.M.S., Providence, 2002)
18. S.W. Hawking, G.F.R. Ellis, The Large-scale Structure of Space-Time (Cambrige University Press, 1973)
19. A. Hermann, E. Humbert, On the positive mass theorem for closed Riemanniann manifolds. (This volume, p. 515)
20. N. Ikeda, S. Watanabe, Stochastic Differential Equations and Diffusion Processes (North-Holland Kodansha, 1981)
21. L. Landau, E. Lifchitz, Physique théorique, tome II : Théorie des champs. Éditions MIR de Moscou (1970)
22. P. Malliavin, Géométrie différentielle stochastique (Les Presses de l'Université de Montréal, Montréal, 1978)
23. M. Mars, On local characterization results in geometry and gravitation. (This volume, p. 541)
24. J.-P. Nicolas, The conformal approach to asymptotic analysis. (This volume, p. 571)
25. M.K. von Renesse, K.T. Sturm, Transport inequalities, gradient estimates, entropy and Ricci curvature. Comm. Pure Appl. Math. **68**, 923–940 (2005)

Part V
Relativity

On the Positive Mass Theorem for Closed Riemannian Manifolds

Andreas Hermann and Emmanuel Humbert

Abstract The Positive Mass Conjecture for asymptotically flat Riemannian manifolds is a famous open problem in geometric analysis. In this article we consider a variant of this conjecture, namely the Positive Mass Conjecture for closed Riemannian manifolds. We explain why the two positive mass conjectures are equivalent. After that we explain our proof of the following result: If one can prove the Positive Mass Conjecture for one closed simply-connected non-spin manifold of dimension $n \geq 5$ then the Positive Mass Conjecture is true for all closed manifolds of dimension n.

1 Introduction

General Relativity is a geometrical theory: the background is a 4-Lorentzian manifold $(\mathcal{M}, G)$ whose metric satisfies:

$$\mathrm{Ric}_G - \frac{1}{2} s_G G = 8\pi T. \tag{1}$$

where Ric_G and s_G denote the Ricci curvature and the scalar curvature of the metric G respectively. The tensor T is the *energy momentum tensor* and contains all the information on the matter and Formula (1) explains how the physical objects (the tensor T) are related to the geometrical background (the curvature). After Gauss' work on the curvature of surfaces which resulted in his famous theorema egregium, it was Riemann who generalized the notion of curvature to higher dimensions by introducing a quantity which today is called the Riemann curvature tensor. In this way he founded the language of modern differential geometry which was later used

A. Hermann (✉)
Institut für Mathematik, Universität Potsdam, Karl-Liebknecht-Str. 24-25, 14476 Potsdam, Germany
e-mail: hermanna@uni-potsdam.de

E. Humbert
LMPT, Université de Tours, Parc de Grandmont, 37200 Tours, France
e-mail: emmanuel.humbert@lmpt.univ-tours.fr

L. Ji et al. (eds.), *From Riemann to Differential Geometry and Relativity*,
DOI 10.1007/978-3-319-60039-0_17

by Einstein to formulate the field Eq. (1). Many physical laws can be read through this formula: in particular, it describes gravitation in terms of curvature which at the beginning of the 20th century was the real breakthrough of the theory. This formulation completely denies the existence of a global time which nevertheless has a strong meaning for each human being. One way to overcome this difficulty is to consider an appropriate space-type hypersurface M, which is a Riemannian manifold when equipped with the restriction of the metric G. This manifold M is called a *Cauchy hypersurface* and can be interpreted as a picture of the universe at some fixed time. General Relativity is a deterministic theory: specifying enough data on the chosen Cauchy hypersurface allows to reconstruct the whole spacetime: this is a result by Choquet-Bruhat and Geroch [7]. This is one reason for which many problems can be formulated in terms of Riemannian geometry.

In this survey, we present the Positive Mass Theorem which illustrates what is explained above: the physical question of whether a local positive energy density of an isolated system gives rise to a positive total energy can be formulated as a problem of pure Riemannian geometry. So it is not surprising that in this context, geometrical tools were crucial to (partially) answer this question. On the other hand it is really surprising that the Positive Mass Theorem was the key point in the final argument of R. Schoen for the solution of the Yamabe problem, which was a challenging question in Riemannian geometry and partial differential equations for more than 20 years and which has nothing to do with General Relativity.

In the paper, we will start by introducing the ADM mass of an asymptotically flat manifold and we will explain the reason why it can be interpreted as the total energy of an isolated system. The approach we choose in this text is inspired by the book of Vaugon and the second author [14] where the reader will find many more details. The ADM mass measures how an isolated system affects the trajectories of massless particles evolving far from the system. Note that in the chapter written by Jacques Franchi [8], it is explained how such a system also perturbs Brownian motions (see e.g. the study on Schwarzschild solution). A famous open problem is to know if, under natural physical assumptions, the ADM mass is non-negative together with a rigidity result (for many other interesting rigidity results, the reader may refer to the chapter written by Marc Mars [12]): it should vanish if and only if the spacetime does not contain any matter nor gravity fields and hence is flat.

We will formulate the Positive Mass Conjecture for asymptotically flat manifolds which has been solved in the context of General Relativity (that is, in dimension 3), but which is still considered as open in dimension greater or equal to 8 if the manifold is not spin. Then we will study the mass of closed manifolds, more precisely of manifolds which are compactifications of some asymptotically flat manifolds. As in the chapter written by Jean-Philippe Nicolas [13], the situation appears much clearer in the compact setting giving rise to new applications. We will explain how Schoen defined the mass of a closed Riemannian manifold which can be regarded as the ADM mass of a blown-up manifold. This will give rise to a Positive Mass Conjecture for closed manifolds which was the main point in the last argument for solving the Yamabe problem. It is well known that the Positive Mass Conjectures for closed and for asymptotically flat manifolds are actually equivalent: the proof of this equivalence is essentially contained in the work of Schoen, Lohkamp or Lee-Parker [16, 17, 19]. But, even if all the ingredients of the proof are in the literature, it is quite difficult to find a reference where this equivalence is explicitly stated and proved. We do it in this paper.

In the last paragraph we present our recent results which reduce the Positive Mass Conjecture to proving that it holds for every metric on one fixed closed non spin and simply connected manifold in each dimension.

2 ADM Mass in General Relativity

The goal of this paragraph is to explain how we can define the energy of an isolated system in General Relativity. We will not be very precise in this section: our goal is mainly to explain the physical considerations which lead to the definition of the ADM Mass. For more information on this subject, the reader may refer to [14].

2.1 Modeling Isolated Systems in General Relativity

An *isolated system* is a physical system contained in a "bounded region". Obviously, the notion of "bounded region" should be understood in the sense "bounded in space" and not in time. The general idea is that the gravitational field induced by the system decreases as the distance to the system increases and vanishes at infinity.

It is generally considered as natural that the spacetime $(\mathcal{M}, G)$ (a 4-Lorentzian manifold satisfying Eq. 1) is *globally hyperbolic*, i.e. possesses a *Cauchy hypersurface M*, which means that any inextensible causal curve of $\mathcal{M}$ meets M exactly once. The manifold $\mathcal{M}$ is then homeomorphic to $M \times \mathbb{R}$ and the hypersurfaces $M \times \{t\}$ can be interpreted as "pictures of the universe" at some time t as was proved by Geroch [9]. Bernal-Sánchez [6] were able to strengthen this result by showing that $\mathcal{M}$ is diffeomorphic to $M \times \mathbb{R}$. With this assumption, it is natural to impose that:

- the energy-momentum tensor T is compactly supported in each hypersurface $M \times \{t\}$ which means that the system is contained in a bounded region of space;
- the metric g becomes flat at infinity, in a sense to be made precise, which means that the gravitational field decreases far from the system.

The first assumption is in general too strong. For example it is not satisfied for the Reissner-Nordström metric. Actually, in all physical applications, we just need to consider the second assumption: it does not imply the first one but implies that the energy-momentum tensor T vanishes at infinity. A good way to take into account these considerations is to assume that the spacetime $(\mathcal{M}, G)$ is *asymptotically flat*: there exists a coordinate system (t, x, y, z) such that the hypersurfaces $t =$ constant correspond to Cauchy hypersurfaces and such that in these coordinates we have

$$\lim_{r\to\infty} |G_{ab} - \eta_{ab}| = O(r^{-1})$$
$$\lim_{r\to\infty} |\partial_p G_{ab}| = O(r^{-2})$$
$$\lim_{r\to\infty} |\partial_{pq} G_{ab}| = O(r^{-3})$$

where $r^2 = x^2 + y^2 + z^2$ and η is the Minkowski metric on $\mathbb{R}^4$. We should be more precise on the domain where these coordinates are defined but this will not be crucial in what follows. Note that the convergence speeds we require in this definition are chosen to make the gravitational field energy density decay in a sensible way. As explained above, this assumption can be weakened.

For the problems presented in this text, we will restrict to Cauchy hypersurfaces and we will omit the possible contributions of the second fundamental forms (which cannot be neglected in some other contexts). More precisely, we will consider *asymptotically flat Cauchy hypersurfaces*:

Definition 2.1 A Cauchy hypersurface M of $(\mathcal{M}, G)$ is called *asymptotically flat* if there exists a compact subset $K \subset M$ and a diffeomorphism

$$\eta : \left| \begin{array}{ccc} M\backslash K & \to & \mathbb{R}^3\backslash\overline{B} \\ p & \mapsto & (x, y, z) \end{array} \right.$$

where B is the standard unit ball in $\mathbb{R}^3$, such that for the restriction of the metric G to TM, which is denoted by g (and which is a Riemannian metric on M), we have

$$\lim_{r\to\infty} |g_{ab} - \delta_{ab}| = O(r^{-1})$$
$$\lim_{r\to\infty} |\partial_p g_{ab}| = O(r^{-2})$$
$$\lim_{r\to\infty} |\partial_{pq} g_{ab}| = O(r^{-3})$$

where again $r^2 = x^2 + y^2 + z^2$ and where δ is the Euclidean metric on $\mathbb{R}^3$.

Note that this definition imposes that each Cauchy hypersurface has the topology of $\mathbb{R}^3\backslash B$ outside a ball which is quite restrictive. One can also consider a more general definition of asymptotically flat Cauchy hypersurfaces with several "asymptotically flat ends". Note also that the definition above is intrinsic and does not involve the metric G which allows to talk about *asymptotically flat manifolds* of dimension 3. For the mathematical problems studied in this text, it will be convenient not to restrict to dimension 3 and to weaken the speeds of convergence. This leads to the following definition:

Definition 2.2 A Riemannian manifold (M, g) of dimension n is called *asymptotically flat of order* $\tau > 0$ if there exists a compact subset $K \subset M$ and a diffeomorphism

$$\eta : \left| \begin{array}{ccc} M\backslash K & \to & \mathbb{R}^n\backslash\overline{B} \\ p & \mapsto & (x_1, \cdots, x_n) \end{array} \right.$$

where B is the standard unit ball in $\mathbb{R}^n$, such that in these coordinates we have

$$\lim_{r\to\infty} |g_{ab} - \delta_{ab}| = O(r^{-\tau})$$
$$\lim_{r\to\infty} |\partial_p g_{ab}| = O(r^{-\tau-1})$$
$$\lim_{r\to\infty} |\partial_{pq} g_{ab}| = O(r^{-\tau-2})$$

where again $r^2 = x_1^2 + \cdots + x_n^2$ and where δ is the Euclidean metric on $\mathbb{R}^n$.

The speeds of convergence will be chosen the smallest possible to ensure that all the quantities studied are well defined. It may depend on the situation but the right assumption for the setting of this paper is to ensure that

$$\tau > \frac{n-2}{2}$$

as we will explain in the next section. The simplest example of such a manifold is $\mathbb{R}^n$ equipped with its canonical Euclidean metric ξ^n. Another fundamental example (but which does not fit in this form to the previous definition since the manifold has a non-empty boundary) is the Schwarzschild metric which is $(\mathbb{R}^3 \backslash B(0, \frac{m}{2}), g)$ where $m > 0$, $B(0, \frac{m}{2})$ is the Euclidean ball of radius $\frac{m}{2}$ centered at 0 and where g is a metric conformal to the Euclidean metric ξ^n on $\mathbb{R}^3$ and is defined by

$$g := \left(1 + \frac{m}{2|x|}\right)^4 \xi^n. \tag{2}$$

In General Relativity, the Schwarzschild space is a model for a static black hole of mass m whose horizon coincides with the boundary.

2.2 ADM Mass and Positive Mass Conjecture

We start by giving the

Definition 2.3 Let (M, g) be an asymptotically flat manifold of dimension $n \geq 3$, of order $\tau > \frac{n-2}{2}$ and such that $s_g \in L^1(M)$. The *ADM mass* of (M, g) is defined by

$$m_{\mathrm{ADM}}(M, g) := \lim_{r\to\infty} \frac{1}{4\omega_{n-1}} \int_{S_r} (\partial_i g_{ij} - \partial_j g_{ii})\nu^j \, ds_{\xi^n}.$$

Here the coordinates are given by the definition of asymptotic flatness, ω_{n-1} is the volume of the $(n-1)$-dimensional unit sphere in $\mathbb{R}^n$ with the standard metric, i.e. with the Riemannian metric induced by the Euclidean metric on $\mathbb{R}^n$, S_r is the sphere of radius $r > 0$ in $\mathbb{R}^n$ and ν is the outward unit normal vector field on S_r.

It is known from a result by Bartnik that the conditions $\tau > \frac{n-2}{2}$ and $s_g \in L^1(M)$ ensure that the limit in the definition of $m_{\mathrm{ADM}}(M, g)$ exists and is independent of the choice of asymptotic coordinates ([5, Theorem 4.2], see also [16, Theorem 9.6]). In his article Bartnik also quotes results showing that the bound on τ is optimal.

In General Relativity, the ADM mass represents the energy of an isolated system. In the following paragraphs we will explain the origin of this definition. First we formulate the

Positive Mass Conjecture: *Let (M, g) be an asymptotically flat manifold of dimension $n \geq 3$ and of order $\tau > \frac{n-2}{2}$. We assume that the scalar curvature is integrable and non-negative.* Then

$$m_{\mathrm{ADM}}(M, g) \geq 0$$

with equality if and only if (M, g) is isometric to $(\mathbb{R}^n, \xi^n)$.

The assumption of non-negativity of the scalar curvature is the reformulation in this particular situation of the dominant energy condition, which is a natural hypothesis in General Relativity preventing information to travel faster than light. Under this condition, the energy of an isolated system should be non-negative and it should vanish if and only if the spacetime is empty (i.e. $(M, G) = (\mathbb{R}^{n+1}, \eta)$). This conjecture is actually a theorem in the context of General Relativity, i.e. in dimension 3: indeed, the conjecture has been proven by Schoen and Yau in [20] if $3 \leq n \leq 7$. Witten [25] found another proof for spin manifolds of any dimension greater or equal to 3. Note that every oriented manifold of dimension 3 is spin. The general statement is still a conjecture.

2.3 *The Origin of the ADM Mass*

The notion of ADM mass arises naturally in the Hamiltonian formulation of General Relativity. The basic idea is to assume that the spacetime has the special form $M = M \times I$ where I is a real interval. Solving the Eq. (1), i.e. finding a metric G solving (1) can be done by setting $g_t := G|_{M\times\{t\}}, t \in I$, and then studying $(g_t)_{t\in I}$ as a dynamical system. This approach yields some conserved quantities with respect to time t: one of these, namely the ADM mass, is scalar and hence is interpreted as the energy of the system under consideration. Its definition is not physical but many physical reasons indicate that this interpretation is sensible (see the references at the end of this section).

2.3.1 Lagrangian Formulation of General Relativity

Obtaining a Hamiltonian formulation of a problem is a general principle in physics. A first step, much easier, is to find a *Lagrangian formulation* of the problem.

Lagrangian formulation of a physical problem: Assume for instance that we have to find a tensor ψ over M satisfying some equation (E). The Lagrangian formulation of the problem consists in finding a functional, called the *action functional*, having the form

$$L : \left| \begin{array}{ll} \mathcal{T} \to & \mathbb{R} \\ \psi \mapsto & \int_M \mathcal{L}(\psi, \nabla\psi) dv_G \end{array} \right. \tag{3}$$

where $\mathcal{T}$ is the space of tensors in which we seek ψ, G is a metric on the domain of definition of ψ and where L is such that ψ satisfies equation (E) if and only if ψ is a critical point of L. The function $\mathcal{L}$, called the *Lagrangian*, can be generally found by physical considerations. Let us consider for instance the problem of describing the position $x(t)$ of a particle of mass m in $\mathbb{R}^3$ subject to a potential V. Then its Lagrangian is given by

$$\mathcal{L}(x, x') = \frac{1}{2} m (x')^2 - V(x).$$

Here, x is a function from an interval $[a, b]$ with values in $\mathbb{R}^3$. This choice is natural: the particle will try to spend the smallest possible amount of energy and hence will tend to minimize its kinetic energy and maximize the contribution of V.

In the context of General Relativity, the tensor ψ is the metric itself and the equation (E) is the Einstein equation (1). The Lagrangian formulation of General Relativity is constructed from the following calculation: let $(g_a)_a$ be a family of metrics such that $g_0 = g$ and $\frac{d}{da}|_{a=0} g_a = h$. Let Ω be a relatively compact domain of M. Then, one can compute that

$$\frac{d}{da}\Big|_{a=0} \int_\Omega s_{g_a} dv_{g_a} = \int_\Omega h_{kl} E_g^{kl} dv_g + \int_{\partial\Omega} K(h) dv_g \tag{4}$$

where $E_g := \mathrm{Ric}_g - \frac{1}{2} s_g g$ is the Einstein tensor of the metric g and where $K(h)$ is a quantity constructed from h and its derivatives. If now h is compactly supported in Ω, then the boundary term above vanishes. Hence, g is a solution of the vacuum Einstein equation (i.e. Einstein equation with $T = 0$) if and only if g is a critical point of the functional $g' \mapsto \int_M s_{g'} dv_{g'}$ defined on the space of all metrics on $\mathcal{M}$ with integrable scalar curvature. This Lagrangian formulation does not have the form (3): indeed, the volume element depends on the metric g which is the unknown variable. This could appear as a minor point since Eq. (4) allows to recover General Relativity (in the vacuum context). However, we recall that the goal here is to obtain a Hamiltonian formulation of General Relativity for which the exact form (3) is needed in order to apply the procedure described in Sect. 2.3.2. So the trick consists in fixing any metric G on M (actually, as explained in Sect. 2.3.2, a suitable choice of G will simplify the situation) and setting

$$\mathcal{L}(g, \nabla g, \nabla^2 g) = s_g \frac{\sqrt{-det(g)}}{\sqrt{-det(G)}}$$

and

$$L(g) = \int_M \mathcal{L}(g, \nabla g, \nabla^2 g) dv_G$$

which has the desired form (3). Obviously,

$$L(g) = \int_M s_g dv_g$$

and formula (4) implies that g is a critical point of L if and only if g is a solution of the vacuum Einstein equation.

The same procedure could be applied to get a Lagrangian formulation of General Relativity with a non-vanishing T by a convenient modification of the Lagrangian.

2.3.2 Hamiltonian Formulation of General Relativity

This paragraph is mainly inspired by Wald's famous book [24]. The goal here is to explain how, once a Lagrangian formulation for a problem is found, there is a natural procedure to deduce the Hamiltonian formulation, which consists in constructing some quantities whose conservation with time is equivalent to the Einstein equation. We explain it very briefly.

We assume that M is a product $M \times I$, where I is a real interval and M is a Cauchy hypersurface. Let us denote $M_t := M \times \{t\}$, $t \in I$. This formulation will help in considering General Relativity as an evolution problem: instead of seeking g solution of the Einstein equation on M, we will seek its restriction g_t, $t \in I$ on M_t as the solution of a dynamical system described by a suitable Hamiltonian formulation.

Let us now describe the procedure. We recall that our goal here is to present the general method to build a Hamiltonian formulation from a Lagrangian formulation but not to describe the explicit computations in the special context of General Relativity, which would require to be much more precise than in what follows. So the situation is the following: we want to find a tensor ψ satisfying some equation (E) on M. We have to keep in mind that in the context of General Relativity, ψ would be the metric itself while (E) is the Einstein Equation. As above, instead of seeking ψ, we will describe the evolution of its restriction on M_t. A Hamiltonian formulation of the problem consists in constructing some quantities whose conservation with respect to t is equivalent to solving Equation (E).

We proceed in the following way: we first seek a functional, called the *Hamiltonian* depending on two tensorial variables π, q of the form

$$H(\pi, q) = \int_{M_t} \mathcal{H}(\pi(x), q(x)) dv_{G_0}$$

where π, q do depend on t (this dependence is omitted in the notation), where G_0 is a fixed metric independent of t when considered as a metric on $M \equiv M_t$ and where $\mathcal{H}$ is a real function called the *Hamiltonian density* and chosen such that

$$q' = \frac{\partial H}{\partial \pi}(\pi, q) \text{ and } \pi' = -\frac{\partial H}{\partial q}(\pi, q), \tag{5}$$

where the derivatives are taken with respect to t.

The variables π and q are in fact tensor fields on M_t such that if q is a tensor of type (k, l) then π is a tensor of type (l, k). This allows to give a meaning to Formula (5): q' and $\frac{\partial H}{\partial \pi}(\pi, q)$ are both considered as linear forms on the space of tensor fields of type (l, k) via the formulas

$$q'(\eta) = \int_{M_t} (q')^{j_1 \cdots j_l}_{i_1 \cdots i_k} \eta^{i_1 \cdots i_k}_{j_1 \cdots j_l} dv_{G_0} \tag{6}$$

and

$$\frac{\partial H}{\partial \pi}(\pi, q)(\eta) = \frac{d}{da}\Big|_{a=0} H(\pi + a\eta, q).$$

Such a formulation can be constructed from a Lagrangian formulation of the problem. Indeed, let us assume that the action functional

$$L : \left| \begin{array}{ll} \mathcal{T} \to & \mathbb{R} \\ \psi \mapsto & \int_M \mathcal{L}(\psi, \nabla\psi) dv_G \end{array} \right.$$

is given where the notation is the same as in the previous paragraph. Here, it is convenient to assume that the metric G has the form $G := G_0 - dt^2$. Note that it is not a restriction in the context of General Relativity since the metric G is arbitrary. The expression ∇ is not precise in this general context: it means that $\mathcal{L}$ depends on the derivatives of ψ. When the situation is explicit, these notations should be precisely specified: in particular, in many situations, a convenient choice is to assume that ∇ denotes the Levi-Civita connection associated to the metric G but it can also denote the derivatives of ψ with respect to some system of local coordinates. Then for a fixed $\psi \in \mathcal{T}$, we set:

$$q = \psi|_{M_t}$$
$$\pi = \frac{\partial \mathcal{L}}{\partial q'}.$$

We then obtain a Hamiltonian formulation of the problem by defining

$$\mathcal{H}(\pi, q) = \pi q' - \mathcal{L}(q, q'). \tag{7}$$

This expression needs some explanation: In the definition $\mathcal{L}(\psi, \nabla\psi)$ above, the derivatives of ψ in tangential directions to M_t can be considered as operators on q while the derivatives of ψ with respect to t are contained in the variable q' and hence $\mathcal{L}$ can be considered as a function of q and q' only. Now, we prove:

Proposition 2.4 *The tensor ψ is a critical point of L if and only if Eq. (5) are satisfied.*

Proof First, let us assume that ψ is a critical point of L. Let us set

$$H(\pi, q) = \int_{M_t} \mathcal{H}(\pi, q)\, dv_{G_0}.$$

Then, recalling the definition (7) of $\mathcal{H}$:

$$\frac{\partial H}{\partial \pi} = \int_{M_t} q' \cdot dv_{G_0}$$

since $\mathcal{L}$ is independent of π. From (6), the linear map $\eta \mapsto \int_{M_t} q'\eta dv_{G_0}$ is equal to the quantity q' in the Eq. (5). As a consequence,

$$\frac{\partial H}{\partial \pi} = q'.$$

Since ψ is a critical point of L we have for all compactly supported β on $\mathcal{M}$

$$\begin{aligned} 0 &= \frac{d}{da}\Big|_{a=0} \int_{\mathcal{M}} \mathcal{L}(\psi + a\beta, \psi' + a\beta')\, dv_G \\ &= \int_{\mathcal{M}} \Big(\frac{\partial \mathcal{L}}{\partial q}\beta + \frac{\partial \mathcal{L}}{\partial q'}\beta'\Big)\, dv_G \\ &= \int_{\mathcal{M}} \Big(\frac{\partial \mathcal{L}}{\partial q} - \frac{d}{dt}\frac{\partial \mathcal{L}}{\partial q'}\Big)\beta\, dv_G. \end{aligned}$$

It follows that

$$\frac{\partial \mathcal{L}}{\partial q} = \frac{d}{dt}\frac{\partial \mathcal{L}}{\partial q'} = \pi'$$

and thus

$$\frac{\partial H}{\partial q} = -\int_{M_t} \frac{\partial \mathcal{L}}{\partial q}\, dv_{G_0} = -\int_{M_t} \pi'\, dv_{G_0}.$$

Conversely, let us assume that Eq. (5) hold. More exactly let us consider a family $(\psi_a)_{a\in(-\varepsilon,\varepsilon)}$ such that $\delta\psi_a$ is compactly supported in $]t_1, t_2[\times M$, where δ means $\frac{d}{da}|_{a=0}$. Let us denote by π_a and q_a the associated quantities. We assume that π_0 and q_0 satisfy (5). Then we set

$$J(\psi_a) = \int_{t_1}^{t_2} \int_{M_t} \mathcal{H}(\pi_a, q_a)\, dv_{G_0}\, dt = \int_{t_1}^{t_2} \int_{M_t} (\pi_a q_a' - \mathcal{L}(q_a, q_a'))\, dv_{G_0}\, dt.$$

Since $\delta\mathcal{L}(q_a, q_a')$ is compactly supported in $]t_1, t_2[\times M$ it follows that

$$\delta J(\psi_a) = \int_{t_1}^{t_2} \int_{M_t} (\pi_a \delta q_a' + (\delta\pi_a) q_a')\, dv_{G_0}\, dt - \delta L(q_a, q_a').$$

Integrating by parts and using Eq. (5), it follows that:

$$\begin{aligned}\delta J(\psi_a) &= \int_{t_1}^{t_2} \int_{M_t} (-\pi_a' \delta q_a + (\delta\pi_a) q_a')\, dv_{G_0}\, dt - \delta L(q_a, q_a') \\ &= \int_{t_1}^{t_2} \int_{M_t} \delta(\mathcal{H}(\pi_a, q_a))\, dv_{G_0}\, dt - \delta L(q_a, q_a').\end{aligned}$$

Let us notice that the first term on the right is exactly equal to $\delta J(\psi_a)$. We then get $\delta L(q_a, q_a') = 0$ which proves that ψ_0 is a critical point of L. This ends the proof of Proposition 2.4. □

In the particular context of General Relativity, the Eq. (5) are complicated: we do not write them here but they can be computed explicitly (see e.g. [24, p. 465]).

Remark 2.5 The construction of the Hamiltonian formulation of a problem provides in a natural way some conserved quantities: the ADM mass introduced in the next paragraph is an example of this fact.

2.3.3 ADM Mass of an Isolated System

We are now ready to define the ADM mass of an isolated system. Again we assume that $\mathcal{M} = M \times I$ where I is a real interval and where for all t, $M_t := M \times \{t\}$ is a Cauchy hypersurface. As explained in Sect. 2.1, we also make the assumption that each M_t is an asymptotically flat Riemannian manifold of dimension 3.

Let us consider a domain outside the system i.e. a domain where the energy momentum tensor T vanishes. Following the last paragraph, one can construct the Hamiltonian H as well as the quantities π and q so that Eq. (5) hold. Choose a family of metrics (g_a) satisfying $g_0 = g$ and whose variation at $a = 0$ is compactly supported. Let us denote the corresponding quantities by q_a and π_a. Equations (5) tell us that

$$\delta H(g_a) = \frac{\partial H}{\partial \pi} \delta\pi_a + \frac{\partial H}{\partial q} \delta q_a = q_a' \delta\pi_a - \pi_a' \delta q_a.$$

In the computation of the Hamiltonian should appear some boundary terms which actually turn out to vanish when integrating $\mathcal{H}$ on M_t since their support is compact. Now we assume that the family (g_a) describes some metrics on each slice: we denote it by (g_t) instead of (g_a) to indicate that g_t is a metric on M_t. To fit the model we constructed for an isolated system where each M_t is asymptotically flat, there is no reason to assume that its first variation at $t = 0$ is compactly supported. But we may assume that it has the property to preserve the asymptotic flatness. For this type of variations, the boundary terms no longer vanish. A (quite long) computation (see [24, p. 469] and the reference given there) leads to

$$\delta H(g_t) = q_t' \delta\pi_t - \pi_t' \delta q_t - \delta C_t$$

where, in the coordinates (x^1, x^2, x^3) given by the asymptotic flatness of M_t,

$$C_t = \lim_{R\to\infty} \int_{S_R} (\partial_i (g_t)_{ij} - \partial_j (g_t)_{ii}) \nu^j \, ds_{\xi^n}.$$

Here S_R denotes the sphere $S_R = \{(x^1, x^2, x^3) | r = R\}$ and ν is the outward unit normal on S_R. Furthermore ds_{ξ^n} is the volume element induced by the Euclidean metric $\xi^n := (dx^1)^2 + (dx^2)^2 + (dx^3)^2$ on S_R.

This indicates that we should construct a new Hamiltonian $\tilde{H}$ by setting

$$\tilde{H} = H + C_t$$

which is not of the form $\int_{M_t} \mathcal{H} dv_g$ but this turns out not to be a problem. This choice allows the following formula to be true:

$$\delta \tilde{H}(g_t) = q_t' \delta\pi_t - \pi_t' \delta q_t$$

even for this type of variations and leads again to Eq. (5). Unfortunately, until now, we have neglected the matter, i.e. we are only allowed to consider variations which are supported in the domain where T vanishes. To take into account this problem, we should modify the Hamiltonian density $\mathcal{H}$ where T does not vanish. Finally, we will work with the Hamiltonian

$$\overline{H} = \int_{M_t} \overline{\mathcal{H}} dv_g + C_t$$

where $\overline{\mathcal{H}}$ is the new Hamiltonian density which allows all variations preserving the asymptotic flatness.

We are now able to show:

Proposition 2.6 *Let G be an asymptotically flat metric on $M = M \times I$, solution of Einstein's equation (1) in presence of an isolated system, then $t \mapsto C_t$ is a constant function.*

Proof We first compute $\frac{d}{dt}\overline{H}(g_t)$. Note that this variation preserves asymptotic flatness. Let us fix $t_0 \in I$. From Eq. (5):

$$\frac{d}{dt}\Big|_{t=t_0} \overline{H}(g_t) = \frac{\partial \overline{H}}{\partial \pi} \pi' + \frac{\partial \overline{H}}{\partial q} q' = q'\pi' - \pi' q' = 0. \tag{8}$$

We also have

$$\frac{d}{dt}\Big|_{t=t_0} \overline{H}(g_t) = \frac{d}{ds}\Big|_{s=0} \overline{H}(g_{t_0} + s g_{t_0}').$$

We write $g'_{t_0} = h_r + h'_r$, where $h_r = \eta g'_{t_0}$, $h'_r = (1-\eta)g'_{t_0}$, $\eta \in [0, 1]$ being a cut-off function equal to 1 on $M_{t_0} \setminus B_{2r}$ and vanishing on B_r where B_r is a ball with large radius r. By linearity of the differential,

$$\frac{d}{dt}\Big|_{t=t_0} \overline{H}(g_t) = \frac{d}{ds}\Big|_{s=0} \overline{H}(g_{t_0} + sh_r) + \frac{d}{ds}\Big|_{s=0} \overline{H}(g_{t_0} + sh'_r).$$

Let us recall that

$$\overline{H} = \int_{M_t} \overline{\mathcal{H}} dv_G + C_t.$$

One checks that

$$\lim_{r\to\infty} \frac{d}{ds}\Big|_{s=0} \overline{H}(g_{t_0} + sh_r) = \frac{d}{dt}\Big|_{t=t_0} C_t + o(1)$$

where $o(1)$ tends to 0 as r tends to $+\infty$. This comes from the fact that matter is almost entirely supported outside the support of h_r and hence, only the boundary term remains when r tends to $+\infty$. Since h'_r is compactly supported,

$$\lim_{r\to\infty} \frac{d}{ds}\Big|_{s=0} \overline{H}(g_{t_0} + sh'_r) = 0.$$

This last equality is obtained in the same way as (8). Finally, with (8),

$$\frac{d}{dt}\Big|_{t=t_0} C_t = 0$$

which ends the proof of Proposition 2.6. □

This proposition shows that C_t or at least a multiple of it is a good candidate to be the energy of the system, or at least a multiple of it, but there exist other reasons: for instance, its value coincides with the Komar mass which is a *physical* definition of the energy in the stationary case (see for instance [15] for the definition of the Komar mass and [3] for its comparison with the ADM mass). This explains the definition of the ADM mass. The constant $\frac{1}{4\omega_{n-1}}$ in Definition 2.3 is a normalization constant which makes the ADM mass coincide with the energy of some particular isolated systems for which the exact energy is known: in particular, a static black hole of mass m modeled by the Schwarzschild metric (2) has ADM mass exactly equal to m.

3 The Mass of a Closed Manifold

In this section, we introduce the mass of a closed manifold and make the link with the last section.

3.1 Yamabe Operator and Yamabe Problem

Let (M, g) be a closed Riemannian manifold of dimension $n \geq 3$. The operator

$$L_g = \frac{4(n-1)}{n-2}\Delta_g + s_g$$

is called the *conformal Laplace operator* or *Yamabe operator* for the metric g. Here Δ_g denotes the Laplace operator with non-negative eigenvalues and s_g is the scalar curvature for the metric g. If φ is a smooth positive function on M and $h := \varphi^{4/(n-2)}g$ is a metric on M which is conformal to g, then for all $u \in C^\infty(M)$ we have

$$L_h(\varphi^{-1}u) = \varphi^{-(n+2)/(n-2)}L_g(u) \tag{9}$$

(see e.g. [16], p. 43). This equation is interesting in many aspects. First of all, choosing $u = \varphi$ and noticing that $L_h(1) = s_h$, we obtain

$$L_g(\varphi) = s_h\varphi^{(n+2)/(n-2)}. \tag{10}$$

In particular, any positive solution φ of the *Yamabe equation*

$$L_g(\varphi) = \mu\varphi^{(n+2)/(n-2)} \tag{11}$$

provides a metric $h := \varphi^{4/(n-2)}g$ conformal to g with constant scalar curvature μ. This problem - finding a metric with constant scalar curvature in a given conformal class - is known as the *Yamabe problem* and was an open challenging problem for more than twenty years. It was solved by the works of Yamabe [26], Trudinger [23], Aubin [4] and Schoen [18]. Let us describe briefly the most famous method for solving this problem: it suffices to find a minimizer φ of the *Yamabe functional*

$$Y_g(u) = \frac{\int_M uL_gudv_g}{\|u\|^2_{L^{\frac{2n}{n-2}}}}.$$

Indeed, any such minimizer is a solution of the Euler equation associated to Y_g which is exactly Eq. (11) with

$$\mu = Y_g(\varphi)\|\varphi\|^{-\frac{4}{n-2}}_{L^{\frac{2n}{n-2}}}.$$

Standard elliptic theory finally implies that u is smooth and positive. So the problem reduces to finding a minimizer u of Y_g. Aubin [4] showed in 1976 that this is always possible if

$$\mu(M, g) < \mu(S^n, \sigma^n) \tag{12}$$

where μ is a conformal invariant called the *Yamabe invariant* and defined by

$$\mu(M, g) = \inf\{Y_g(u) | u \in C^\infty(M), u \not\equiv 0\}$$

and where (S^n, σ^n) denotes the n-dimensional unit sphere with the standard metric. Solving the Yamabe problem is now reduced to showing Inequality (12) except when (M, g) is conformally equivalent to the sphere with the standard metric where of course this inequality is an equality. In this special case, the Yamabe problem becomes trivial since the conformal class contains the standard metric of the sphere which has constant scalar curvature. Many cases were solved by Aubin but the case where both conditions

- $n \in \{3, 4, 5\}$ or (M, g) is locally conformally flat
- L_g has only positive eigenvalues

hold was much more difficult. This case was solved by Schoen in 1984 using the Positive Mass Theorem as explained in the next paragraph.

The condition that L_g has only positive eigenvalues is actually equivalent to each of the following conditions:

- The operator $L_g : C^\infty(M) \to C^\infty(M)$ is positive and invertible.
- There exists a metric h in the conformal class of g with positive scalar curvature.

3.2 Positive Mass Conjecture for Closed Manifolds

Let us start by defining the Green function for L_g. We keep the same notation as in the previous paragraphs.

Definition 3.1 Let $p \in M$. We say that $G \in C^\infty(M\backslash\{p\}) \cap L^1_{\text{loc}}(M)$ is a Green function for L_g at p if in the sense of distributions, $L_g G = \delta_p$ where δ_p is the Dirac distribution at p. This means that for any $u \in C_c^\infty(M)$,

$$\int_M G L_g u dv_g = u(p).$$

Note that, in the sense of smooth functions,

$$L_g G = 0 \tag{13}$$

on $M\backslash\{p\}$.

Example 3.2 Consider $\mathbb{R}^n$ with the Euclidean metric ξ^n and let r denote the function which gives the distance to 0. Then $\frac{1}{4(n-1)\omega_{n-1}r^{n-2}}$ is a (actually unique up to a constant) Green function for $L_{\xi^n} = \frac{4(n-1)}{n-2}\Delta_{\xi^n}$ at 0.

The following proposition is a crucial result

Proposition 3.3 *Assume that M is closed and that L_g has only positive eigenvalues. Then the following holds.*

1. *At every point $p \in M$ there exists a unique Green function G for L_g. Moreover G is strictly positive on $M \setminus \{p\}$.*
2. *Let $p \in M$ and assume that there exists an open neighborhood U of p such that g is flat on U. Then the function G has the following expansion as $x \to p$*

$$G(x) = \frac{1}{4(n-1)\omega_{n-1}r^{n-2}} + A + o(1),$$

where $r := \mathrm{d}_g(p, \cdot)$ is the distance function to p, ω_{n-1} is the volume of the $(n-1)$-dimensional unit sphere with the standard metric and A is a real number called the mass of (M, g) at p.

Proof 1. The proof is classical and we omit it here.
2. Let η be a smooth function on M such that $\eta \equiv \frac{1}{4(n-1)\omega_{n-1}}$ on $B(p, \delta)$ and $\mathrm{supp}(\eta) \subset U$. The function F_η: $M \to \mathbb{R}$ defined by

$$F_\eta(x) = \begin{cases} \Delta_g(\eta r^{2-n})(x), & x \neq p \\ 0, & x = p \end{cases}$$

is smooth on M. Since L_g has only positive eigenvalues, L_g is invertible on $C^\infty(M)$. Let $v := L_g^{-1}(F_\eta)$. The function $G := \eta r^{2-n} - v$ is smooth on $M\backslash\{p\}$, is in $L^1(M)$ and satisfies $L_g G = 0$ on $M\backslash\{p\}$. Moreover, near p we have

$$G(x) = \frac{1}{4(n-1)\omega_{n-1}r^{n-2}} - v(x)$$

and $L_g v = \frac{4(n-1)}{n-2}\Delta_g v = 0$. Since there is an open neighborhood of p in M which is flat and thus isometric to a neighborhood of 0 in $\mathbb{R}^n$ and since the Green function for $\frac{4(n-1)}{n-2}\Delta_{\xi^n}$ on $\mathbb{R}^n$ at 0 is $\frac{1}{4(n-1)\omega_{n-1}r^{n-2}}$, we get that $L_g v = \delta_p$ and thus G is a Green function for L_g. This proves the existence.

If now G and G' are Green functions for L_g then $L_g(G - G') = 0$ in the sense of distributions. By standard regularity theorems, $G - G'$ is smooth and hence, by invertibility of L_g we obtain $G = G'$. □

We come back to the Yamabe problem. Schoen used the Green function at some point p to construct a family of test functions $(u_\varepsilon)_{\varepsilon>0}$ such that as $\varepsilon \to 0$

$$Y_g(u_\varepsilon) = \mu(S^n, \sigma^n) - c_n A\varepsilon^2 + o(\varepsilon^2)$$

where c_n is an explicit positive constant depending only on n and A is the real number appearing in the expansion of G at p given by the last proposition. Hence proving Inequality (12) and thus solving the Yamabe problem reduces to proving that $A > 0$. Now it was Schoen's very interesting observation that by setting $M' := M\backslash\{p\}$,

$g' := G^{\frac{4}{n-2}} g$ one obtains an asymptotically flat manifold (M', g') with vanishing scalar curvature (compare 10 and 13) whose ADM-mass is exactly A up to a positive multiplicative constant. Since for $3 \leq n \leq 7$ the Positive Mass Conjecture had been proved by Schoen and Yau [20], Schoen was able to show that for $n \in \{3, 4, 5\}$ we have $A \geq 0$ with equality if and only if (M', g') is isometric to $(\mathbb{R}^n, \xi^n)$. The latter condition is equivalent to the condition that (M, g) is conformally diffeomorphic to (S^n, σ^n). Moreover for locally conformally flat manifolds of any dimension Schoen and Yau were able to show an analogous result without using a Positive Mass Theorem for asymptotically flat manifolds [22]. This concluded the solution of the Yamabe problem.

This explains why the number A denoted from now on by $m(M, g)$, or $m(g)$ when there is no ambiguity on M, is called the *mass of* (M, g) at p. So the expansion of G at p now reads

$$G(x) = \frac{1}{4(n-1)\omega_{n-1} r^{n-2}} + m(g) + o(1).$$

Since $m(g)$ is exactly equal to an ADM-mass, it is natural to state a Positive Mass Conjecture for closed manifolds:

Conjecture 3.4 (Positive Mass Conjecture for closed manifolds) *Let (M, g) be a closed Riemannian manifold of dimension $n \geq 3$.*

(1) Assume that g is flat on an open neighborhood of a point $p \in M$ and that L_g has only positive eigenvalues. Then the mass $m(M, g)$ at p is non-negative.
(2) If there exist a point $p \in M$ and a metric g on M which is flat on an open neighborhood of p such that L_g has only positive eigenvalues and the mass $m(M, g)$ at p is zero, then (M, g) is conformally diffeomorphic to (S^n, σ^n).

As was mentioned above this conjecture is a theorem if (M, g) is locally conformally flat by the work of Schoen and Yau [22]. Moreover by considering the blown-up manifold $(M \setminus \{p\}, G^{\frac{4}{n-2}} g)$ one has a proof of Conjecture 3.4 whenever the Positive Mass Theorem for asymptotically flat manifolds is available. This includes for example the cases where $3 \leq n \leq 7$ by the work of Schoen-Yau [20] or where M is a spin manifold by Witten's result [25]. Note that simple proofs of the Positive Mass Theorem for spin manifolds can also be found in [1] or [11].

At a first glance it would seem that the Positive Mass Conjecture for asymptotically flat manifolds is stronger than Conjecture 3.4 since not every asymptotically flat manifold can be obtained by blowing up a closed Riemannian manifold at a point. In addition, the assumption in Conjecture 3.4 that the metric is flat on an open neighborhood of a point is very strong. However it turns out that both conjectures are in fact equivalent, as we will prove in the next section. For this reason it is not restrictive to consider the mass of a closed Riemannian manifold at a point p only for metrics which are flat on an open neighborhood of p.

4 Equivalence of the Two Positive Mass Conjectures

In this paragraph, we prove

Proposition 4.1 *The Positive Mass Conjecture for closed manifolds is equivalent to the Positive Mass Conjecture for asymptotically flat manifolds.*

This result is well known: it is essentially contained in the work of Schoen, Lohkamp or Lee-Parker. However, it is quite difficult to find it in the literature stated as above and proven with all details. This is what we do here.

Proof It remains to show that under the assumption that Conjecture 3.4 is true the Positive Mass Conjecture for asymptotically flat manifolds follows. Assume that there exists an asymptotically flat Riemannian manifold (M, g) of order $\tau > \frac{n-2}{2}$ with $s_g \geq 0$ and $s_g \in L^1(M)$ and such that $m_{\mathrm{ADM}}(M, g)$ is negative. For example by Proposition 4.1 in [19] or Sect. 5 in [17] there exists a Riemannian metric $\tilde{g}$ on M and there exists a smooth positive function u on M such that the following holds:

(a) There exists a compact subset $K \subset M$ such that $\tilde{g}$ is flat on $M \setminus K$,
(b) $L_{\tilde{g}} u = 0$ and $u \to 1$ as $|x| \to \infty$, in particular $s_{u^{4/(n-2)}\tilde{g}} = 0$ on M,
(c) $(M, u^{4/(n-2)}\tilde{g})$ is asymptotically flat and $m_{\mathrm{ADM}}(M, u^{4/(n-2)}\tilde{g}) < 0$.

Let p be a formal point and denote by $\overline{M} := M \sqcup \{p\}$ the one-point-compactification of M with the usual topology. Then $\overline{M}$ has the structure of a smooth manifold and by (a) there exists a smooth positive function φ on $\overline{M}$ such that the Riemannian metric $\varphi^{4/(n-2)}\tilde{g}$ on M extends to a metric h on $\overline{M}$. Moreover we can choose φ in such a way that $(\overline{M}, h)$ is flat on an open neighborhood of p and such that

$$\varphi(x) \sim (1 + |x|^2)^{(2-n)/2} \quad \text{as } |x| \to \infty.$$

Then there exists a positive constant C such that for the Riemannian distance on $(\overline{M}, h)$ we have

$$\mathrm{d}_h(x, p) \sim C|x|^{-1} \quad \text{as } |x| \to \infty.$$

Since $u \to 1$ as $|x| \to \infty$ we conclude that

$$\frac{u(x)}{\varphi(x)} \sim (1 + |x|^2)^{(n-2)/2} \sim |x|^{n-2} \sim C^{n-2}\mathrm{d}_h(x, p)^{2-n} \quad \text{as } |x| \to \infty. \tag{14}$$

Moreover by the conformal transformation law (9) of $L_{\tilde{g}}$ and by (b) we obtain

$$L_h(\varphi^{-1}u) = \varphi^{-(n+2)/(n-2)} L_{\tilde{g}} u = 0 \tag{15}$$

on $M \subset \overline{M}$. From (14) and (15) we conclude that there exists a positive constant C such that

$$L_h(\varphi^{-1}u) = C\delta_p \tag{16}$$

in the sense of distributions, where δ_p is the Dirac distribution at p. By (14) it is clear that $\varphi^{-1}u$ is integrable on $(\overline{M}, h)$.

We now prove that all eigenvalues of L_h are positive. Let λ be the first eigenvalue of L_h and let w be a corresponding eigenfunction. It is a classical result that w does not change its sign, so we may assume that w is strictly positive on $\overline{M}$. From (16) we obtain

$$\lambda \int_{\overline{M}} \varphi^{-1} u w \, dv^h = \int_{\overline{M}} \varphi^{-1} u L_h w \, dv^h = C w(p) > 0.$$

Since the integral on the left hand side is positive we conclude that $\lambda > 0$.

Thus all eigenvalues of L_h are positive and $\varphi^{-1}u$ is a positive multiple of the Green function of L_h at p. It follows that $(M, u^{4/(n-2)}\tilde{g})$ is up to a constant rescaling the blow-up of $(\overline{M}, h)$ at p by the Green function of L_h and thus by Schoen's result [18] the mass $m(\overline{M}, h)$ at p is a positive multiple of $m_{\mathrm{ADM}}(M, u^{4/(n-2)}\tilde{g})$. By (c) we conclude that $m(\overline{M}, h)$ is negative which contradicts the assertion (1) of Conjecture 3.4. We conclude that the ADM mass of an asymptotically flat manifold satisfying the assumptions of the Positive Mass Conjecture is always non-negative. The equality case of the Positive Mass Conjecture for asymptotically flat manifolds can now be proven exactly as in the proof of Lemma 10.7 in the article [16]. Namely assuming that $m_{\mathrm{ADM}}(M, g) = 0$ one first shows that (M, g) is Ricci flat and then uses the Bochner formula for 1-forms. This proof uses the non-negativity of $m_{\mathrm{ADM}}(M, g)$ but it doesn't require any further restrictions on (M, g). □

5 Some Recent Results on the Positive Mass Conjecture

In this section, we present some results we recently obtained on the Positive Mass Conjecture. We start by defining:

Definition 5.1 We say that a closed manifold M of dimension $n \geq 3$ satisfies PMT (for "Positive Mass Theorem") if for every Riemannian metric g on M such that L_g is a positive operator and for every point $p \in M$ such that g is flat on an open neighborhood of p we have $m(M, g) \geq 0$ at p.

With this definition, proving the Positive Mass Conjecture is thus equivalent to proving that each closed manifold of dimension n satisfies PMT. In the following all simply-connected manifolds are understood to be connected. We are now ready to state

Theorem 5.2 ([11]) *Assume that there exists a closed simply-connected non-spin manifold of dimension $n \geq 5$ satisfying PMT. Then every closed manifold of dimension n satisfies PMT.*

This theorem reduces the Positive Mass Conjecture to finding for each dimension n greater or equal to 8 (since the conjecture is already proven for $3 \leq n \leq 7$ by Schoen and Yau [20]) a closed manifold M satisfying PMT and which is

(1) non spin,
(2) simply connected.

The condition that M is spin only depends on the topology of M. Namely M is spin if and only if its second Stiefel-Whitney class vanishes. This can also be interpreted as a kind of orientability condition of order 2. Indeed a manifold is oriented if the normal bundle of any embedded closed curve (and thus any 1-sphere) is trivial. This should be compared with: a simply-connected manifold is spin if the normal bundle of any embedded 2-sphere is trivial.

A first glance to the statement of Theorem 5.2 leads to two natural questions:

(1) Does the Theorem apply if we find a manifold M which is simply connected, not spin, and which does not possess any metric g such that L_g has only positive eigenvalues?
(2) Is it possible to find even one example of manifold satisfying PMT?

The first question is far from being naive: one easily shows that a closed manifold M possesses a metric such that L_g is positive if and only if it possesses a metric with positive scalar curvature but the classification of such manifolds, despite the fact that it has attracted many interests, is still an open question. Nevertheless, a result of Schoen-Yau [21] and Gromov-Lawson [10] shows that the property of possessing a metric of positive scalar curvature is preserved by surgery of dimension $k \leq n - 3$ (see the next paragraph) with the striking consequence that any closed non-spin simply connected manifold of dimension $n \geq 5$ possesses such a metric.

For the second question, the answer is yes. More precisely, it is easy to find in any dimension a manifold which satisfies PMT and which is non spin or simply connected but of course the question of finding even one manifold satisfying PMT and both conditions is much more difficult. Let us make it precise:

- The sphere S^n is simply connected, spin and satisfies PMT. The latter follows since the Positive Mass Conjecture is true for spin manifolds.
- Let $M = \mathbb{R}P^k \times S^m$ with $k = 1 \bmod 4$. Then M is non-spin due to the restriction on k and satisfies PMT, but it is not simply connected.

The fact that such M satisfies PMT comes from the following proposition using arguments which can be found in [22]. Since the proof is instructive, we give it here.

Proposition 5.3 *Let M be a closed manifold of dimension $n \geq 3$ satisfying PMT. Then every finite quotient of M satisfies PMT.*

Proof Let N be a finite quotient of M and let $\pi \colon M \to N$ be the quotient map. Let h be a Riemannian metric on N which is flat on an open neighborhood of a point $p \in N$

and such that L_h is a positive operator. Let g be the Riemannian metric on M such that π is a Riemannian covering. Since the first eigenvalue λ_0 of L_g is simple and the corresponding eigenfunctions do not change their sign, λ_0 is also an eigenvalue of L_h. It follows that L_g is a positive operator. Now if we write $\pi^{-1}(p) = \{p_1, ..., p_k\}$ and if $G_1, ..., G_k$ denote the Green functions for L_g at $p_1, ..., p_k$ respectively, then for the Green function G of L_h at p we have $G \circ \pi = \sum_{j=1}^k G_j$. In particular if $m^{p_1}(M, g)$ denotes the mass of (M, g) at p_1, then for the mass of L_h at p we have $m(N, h) = m^{p_1}(M, g) + \sum_{j=2}^k G_j(p_1) \geq 0$. □

6 An Idea of the Proof of Theorem 5.2

We now sketch the proof of Theorem 5.2 given in [11]. It is based on surgery arguments.

6.1 *Proofs by Surgery*

Let us describe briefly what a proof by surgery is. First, surgery is a procedure to construct a new manifold from a given manifold M. As an example, "adding a handle" as in the figure below is a surgery (a more precise definition will be given in the following paragraphs).

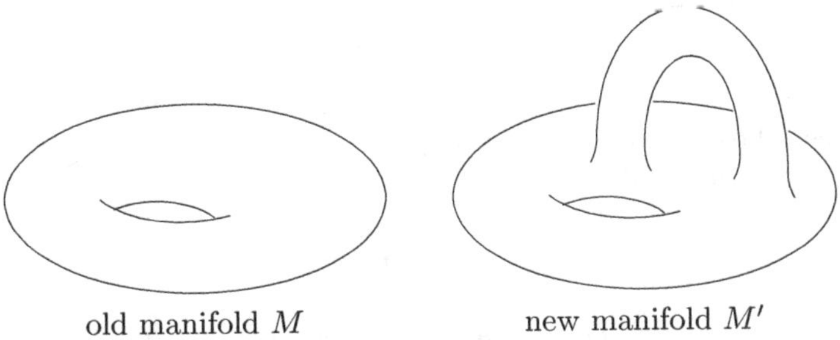

To give an example, assume that we want to prove that any oriented surface M possesses some property P, then, it suffices to

(1) prove it for the sphere S^2;
(2) prove that the property P is preserved by adding a handle.

Indeed, the first assumption says that P holds on the surfaces of genus 0 while the second one ensures that if P holds on surfaces of genus $g \in \mathbb{N}$, then P holds on surfaces of genus $g + 1$. Then we finish the proof by induction on the genus.

This type of proof can be carried over to dimension greater or equal to 3, but two things essentially are different:

- some generalizations of "adding a handle" have to be considered: they are called *surgeries* and will be defined in the next paragraph.
- Any oriented surface of genus g is obtained by g surgeries from the sphere. In dimension greater or equal to 3, this is no longer true but the cobordism theory we will not define here allows to obtain any manifold by a finite number of surgeries from some particular manifolds on which some properties are known.

So again, if a property P is preserved by surgery, some important conclusions can be obtained. An example is the result of Schoen-Yau and Gromov-Lawson described in the last paragraph: they prove that the property

$$P : \text{possesses a metric with positive scalar curvature}$$

is preserved by surgeries of dimensions $k \leq n - 3$. As a conclusion, they obtain that any non-spin simply connected manifold M of dimension $n \geq 5$ possesses such a metric. The proof for this corollary runs exactly as described above: the assumptions allow to say that M can be obtained by a finite number of surgeries of dimension $k \leq n - 3$ from a manifold which possesses the property P.

6.2 Surgery

Let M be a manifold of dimension n, let $k \in \{0, ..., n - 1\}$ and assume that there exists an embedding

$$S^k \times \overline{B}^{n-k} \hookrightarrow M$$

where S^k is the sphere of dimension k and B^{n-k} is the open unit ball of dimension $n - k$. In the following we will not distinguish $S^k \times B^{n-k}$ and its image under the embedding. Then the manifold

$$M \backslash (S^k \times B^{n-k})$$

has boundary $S^k \times S^{n-k-1}$. Observe that the manifold $\overline{B}^{k+1} \times S^{n-k-1}$ has the same boundary $S^k \times S^{n-k-1}$. This allows us to glue these two manifolds along their boundaries to obtain a new manifold M'. It is not obvious but true that a differentiable structure can be constructed on M' around the gluing part in a canonical way.

Definition 6.1 We say that M' is obtained from M by *surgery of dimension* k.

Examples:

• If $k = 0$, since $S^0 = \{\pm 1\}$, then $S^k \times B^{n-k}$ is two copies of an n-dimensional ball. Then, to obtain M', we first remove two balls and then glue along the boundary

$$\overline{B}^1 \times S^{n-1} = [-1, 1] \times S^{n-1}$$

which is a cylinder of section S^{n-1}. This corresponds exactly to what we called "adding a handle" in the last paragraph.

Note that if M_1, M_2 are two manifolds, the connected sum $M_1 \# M_2$ is obtained from the disjoint union $M_1 \cup M_2$ by surgery of dimension 0. We just consider the embedding

$$S^0 \times \overline{B}^n \hookrightarrow M_1 \cup M_2$$

by taking one ball $(\{+1\} \times B^n)$ in M_1 and the other one $(\{-1\}\} \times B^n)$ in M_2.

• If $k = n - 1$, then $S^{n-1} \times B^1$ is removed from M and replaced by two balls. In other words, a surgery of dimension $n - 1$ consists in "removing a handle". More generally, a surgery of dimension k is canceled by a surgery of dimension $n - 1 - k$.

Remark 6.2 It is useful to recall that the subset which we remove while performing surgery (i.e. $S^k \times B^{n-k}$) is in fact a tubular neighborhood of S^k in M. Also, notice that any embedding

$$S^k \times \overline{B}^{n-k} \hookrightarrow M$$

provides an embedding

$$S^k \times \overline{B}^{n-k}(a) \hookrightarrow M$$

for all small $a > 0$ where $B^{n-k}(a)$ is the open ball of radius a in $\mathbb{R}^{n-k}$. Again, one can apply the same procedure: we get a new manifold by removing $S^k \times B^{n-k}(a)$ from M and gluing $\overline{B}^{k+1} \times S^{n-k-1}$ along the boundaries (since the manifolds $M \backslash (S^k \times B^{n-k})$ and $M \backslash (S^k \times B^{n-k}(a))$ have the same boundaries). It is obvious that the new manifold M'_a obtained in this way is diffeomorphic to the original surgery manifold M' but removing a smaller and smaller neighborhood of S^k will be important for what follows.

Remark 6.3 In many results presented in this paper, the property of being spin or not plays a crucial role. We do not give here the exact reason, but just try to give a feeling of why this is important for surgery techniques: as explained above, for a simply connected manifold, being spin is equivalent to the fact that each embedded 2-sphere has a trivial normal bundle. This means that every 2-sphere in M admits a tubular neighborhood diffeomorphic to $S^2 \times B^{n-2}$ and can be used as a surgery sphere. Note that this seems to go in the wrong direction compared to the statement of Theorem 5.2 since it could appear that being spin helps while being non-spin does not help. It is actually the property of being non-spin which helps: the reason is that forming a connected sum of two 2-spheres with non trivial normal bundles gives a 2-sphere with a trivial normal bundle which allows to perform surgeries.

7 Preservation of Mass by Surgery

As explained before, Schoen-Yau and Gromov-Lawson proved that the property of admitting a metric with positive L_g (or equivalently, with positive scalar curvature) is preserved by surgery of dimension less or equal to $n - 3$ with the consequence that any non-spin closed simply connected manifold admits such a metric. Actually, we need a more precise statement. Let (M, g) be a closed manifold such that L_g is a positive operator and perform now a surgery by removing a smaller and smaller neighborhood of some embedded S^k as explained in Remark 6.2. Then, the surgery manifold $M' = M'_a$ is equal to

$$M' = (M \backslash (S^k \times B^{n-k}(a)) \cup (\overline{B}^{k+1} \times S^{n-k-1}))/\#$$

where # indicates that we glue the boundaries. Therefore, we may consider metrics $g(a)$ on M' which are equal to g on $M \backslash (S^k \times B^{n-k}(a))$ and glued to some suitable metric $h(a)$ on $\overline{B}^{k+1} \times S^{n-k-1}$. We are now in a position to give a more precise statement of Schoen-Yau and Gromov-Lawson's result:

Theorem 7.1 ([10, 21]) *With this notation, there exists a sequence of metrics $g(a)$ such that for a small enough, $L_{g(a)}$ is a positive operator.*

Actually, this theorem has been also proven by Ammann-Dahl and the second author in [2] with another sequence of metrics $g(a)$ to improve the consequences we can get from this result. We will not describe these consequences here but from now on, the metrics $g(a)$ will refer to the metrics introduced in [2].

Now, assume that p is a fixed point outside the surgery sphere and hence in $M \backslash (S^k \times B^{n-k}(a))$ when a is small enough and assume also that the metric g is flat around p. Then, the mass $m(M', g(a))$ is well defined. The natural question is then: can we compare the mass $m(M', g(a))$ with the mass $m(M, g)$? The answer is yes:

Theorem 7.2 ([11]) *We have:*

$$m(M', g(a)) \longrightarrow m(M, g)$$

as a goes to 0.

In particular, if $m(M, g) > 0$ then the mass $m(M', g(a)) > 0$ for small a. It is natural to first think of this obvious application but much more interesting is the following: if $m(M, g) < 0$ then the mass $m(M', g(a)) < 0$ for small a. Indeed, it proves that if there exists a metric on M with negative mass preventing M from satisfying PMT, then the same holds on M'. This can be summarized in the following corollary:

Corollary 7.3 *Consider for a closed manifold M the property*

$$P : \textit{ does not satisfy PMT.}$$

Then, the property P is preserved by surgery of dimension less or equal to $n - 3$.

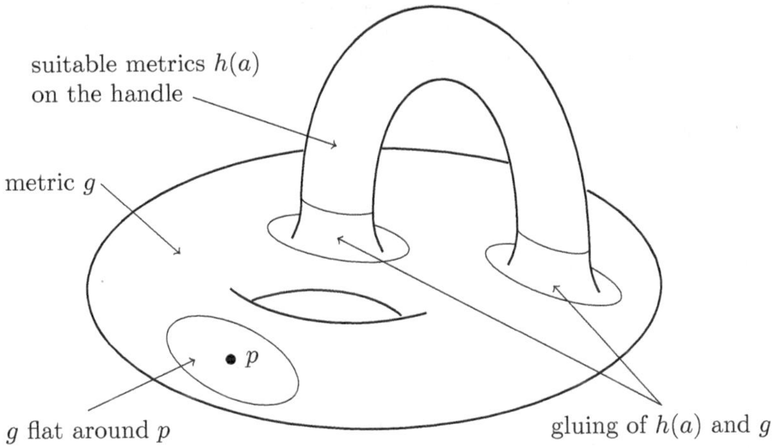

the metrics $g(a)$ on M'

Then, together with some tricks using cobordism theory, we obtained Theorem 5.2. Corollary 7.3 could also be obtained by working on an asymptotically flat manifold. However the above formulation says more: it describes precisely the behavior of the Green function of $L_{g(a)}$ as a tends to 0.

References

1. B. Ammann, E. Humbert, Positive mass theorem for the Yamabe problem on spin manifolds. Geom. Funct. Anal. **15**(3), 567–576 (2005)
2. B. Ammann, M. Dahl, E. Humbert, Smooth Yamabe invariant and surgery. J. Differ. Geom. **94**(1), 1–58 (2013)
3. A. Ashtekar, A. Magnon-Ashtekar, On conserved quantities in general relativity. J. Math. Phys. **20**, 793–800 (1979)
4. T. Aubin, Équations différentielles non linéaires et problème de Yamabe concernant la courbure scalaire. J. Math. Pures Appl. (9) **55**(3), 269–296 (1976)
5. R. Bartnik, The mass of an asymptotically flat manifold. Commun. Pure Appl. Math. **39**(5), 661–693 (1986)
6. A.N. Bernal, M. Sánchez, On smooth Cauchy hypersurfaces and Geroch's splitting theorem. Commun. Math. Phys. **243**(3), 461–470 (2003)
7. Y. Choquet-Bruhat, R. Geroch, Global aspects of the Cauchy problem in general relativity. Commun. Math. Phys. **14**, 329–335 (1969)
8. J. Franchi, From Riemannian to relativistic diffusions. (This volume)
9. R. Geroch, Domain of dependence. J. Math. Phys. **11**, 437–449 (1970)
10. M. Gromov, H. B. Lawson, The classification of simply connected manifolds of positive scalar curvature. Ann. Math. (2) **111**(3), 423–434 (1980)

11. A. Hermann, E. Humbert, About the mass of certain second order elliptic operators. Adv. Math. **294**, 596–633 (2016)
12. M. Mars, On local characterization results in geometry and gravitation. (This volume)
13. J.-P. Nicolas, The conformal approach to asymptotic analysis. (This volume)
14. E. Humbert, M. Vaugon, La relativité générale expliquée aux mathématiciens. Editions Ellipses (2014)
15. A. Komar, Covariant conservation laws in general relativity. Phys. Rev. (2) **113**, 934–936 (1959)
16. J. M. Lee, T. H. Parker, The Yamabe problem. Bull. Am. Math. Soc. (N. S.) **17**, 37–91 (1987)
17. J. Lohkamp, Scalar curvature and hammocks. Math. Ann. **313**(3), 385–407 (1999)
18. R. Schoen, Conformal deformation of a Riemannian metric to constant scalar curvature. J. Differ. Geom. **20**(2), 479–495 (1984)
19. R. Schoen, Variational theory for the total scalar curvature functional for Riemannian metrics and related topics, in *Topics in Calculus of Variations*, Lecture Notes in Mathematics, vol. 1365 (Springer, Berlin, 1989), pp. 120–154. (Montecatini Terme, 1987)
20. R. Schoen, S.-T. Yau, On the proof of the positive mass conjecture in General Relativity. Commun. Math. Phys. **65**(1), 45–76 (1979)
21. R. Schoen, S.-T. Yau, On the structure of manifolds with positive scalar curvature. Manuscripta Math. **28**(1–3), 159–183 (1979)
22. R. Schoen, S.-T. Yau, Conformally flat manifolds, Kleinian groups and scalar curvature. Invent. Math. **92**(1), 47–71 (1988)
23. N. S. Trudinger, Remarks concerning the conformal deformation of Riemannian structures on compact manifolds. Ann. Scuola Norm. Sup. Pisa (3) **22**, 265–274 (1968)
24. R.M. Wald, *General Relativity* (University of Chicago Press, Chicago, 1984)
25. E. Witten, A new proof of the positive energy theorem. Commun. Math. Phys. **80**(3), 381–402 (1981)
26. H. Yamabe, On a deformation of Riemannian structures on compact manifolds. Osaka Math. J. **12**, 21–37 (1960)

On Local Characterization Results in Geometry and Gravitation

Marc Mars

Abstract An important problem in differential geometry and in gravitation is to identify metrics in a fully coordinate independent manner. In fact, the very foundation of Riemannian geometry is based on the existence of a tensor, the Riemann or curvature tensor, which vanishes if and only if the metric is locally flat. Many other such local characterizations of metrics are known. The aim of this article is to present a brief selection of them as an example of the fruitful interplay between differential geometry and gravity.

1 Introduction

Albert Einstein's great insight was that the gravitational interaction was a manifestation of the curvature of spacetime. With this radically new point of view, Einstein gave physical reality to non-euclidean geometries, introduced in Bernhard Riemann's inaugural lecture "Über die Hypothesen, welche der Geometrie zu Grunde liegen" in 1854 [47], about sixty years before Einstein's theory of general relativity was founded. It is most remarkable that Riemann himself, in the last paragraphs of his dissertation, had already realized that the grounds for the metric relations in physical space had to come from binding forces that act upon it, in a view that was already radically different from Newton's conception of space and time as a fixed and invariable entity completely unaffected by physical processes occurring within it. Building on Riemann's mathematical ideas, Einstein could transform his outstanding physical intuition into one of the most elegant physical theories of all times. One of the fundamental consequences of this theory is that differential geometry is not only a fundamental branch of mathematics but also a basic tool for doing physics, and the

Work partially supported by the projects FIS2012-30926, FIS2015-65140-P (Spanish MINECO/FEDER) and P09-FQM-4496 (Junta de Andalucía/FEDER).

M. Mars (✉)
Instituto de Física Fundamental y Matemáticas, IUFFyM, Universidad de Salamanca, Plaza de la Merced s/n, 37008 Salamanca, Spain
e-mail: marc@usal.es

L. Ji et al. (eds.), *From Riemann to Differential Geometry and Relativity*,
DOI 10.1007/978-3-319-60039-0_18

interplay between these two aspects has given a tremendous boost to this scientific discipline.

One of the basic questions that differential geometry has to address is: when are two Riemannian manifolds[1] isometric? Obviously this is also a fundamental question in general relativity, as it addresses the question: when two spacetimes are physically equivalent? This question can be stated, and it is relevant, also at the local level: When are two Riemannian manifolds locally isometric? In fact, answering this question in the case of Euclidean space led Riemann to the introduction of his curvature tensor and to the statement that the vanishing of the curvature at every point is the necessary and sufficient condition for the metric being (locally) flat. The vanishing of the Riemann curvature tensor as a local characterization of Euclidean space is one of the most fundamental results in Riemannian geometry. A natural question is whether similar local characterizations for other geometries also exist. A satisfactory local characterization of a geometry is a very useful tool both in geometry and in general relativity (or any other geometric theory of gravitation). Many results along those lines are known, some coming from pure geometry and some from gravitational theory. The purpose of this chapter is to review a selection of such characterizations as an example of the fruitful interplay between differential geometry and gravity. It must be emphasized that I have no intention of presenting a comprehensive review of known results in this area of research. This would require much more space and a more knowledgeable author. The selection is based solely on my own taste, so many interesting local characterizations will be left out, for which I apologize in advance.

2 Classical Characterizations

Local characterizations involve the notion of local isometry, in the following standard sense. Two Riemannian manifolds $(\mathcal{M}_1, g_1)$ and $(\mathcal{M}_2, g_2)$ are locally isometric if given any point $p \in \mathcal{M}_1$, there exist a connected, open neighbourhood $\mathcal{U}_p$ of p and a smooth map $\Phi_p : (\mathcal{U}_p, g_1|_{\mathcal{U}_p}) \longrightarrow (\mathcal{M}_2, g_2)$ which is an isometry onto its image. Often it is necessary to extend this definition to a situation when $(\mathcal{M}_2, g_2)$ is not a single space but belongs to a class $\{(\mathcal{M}_2^{\mathfrak{a}}, g_2^{\mathfrak{a}})\}$. In this case, the target of the local isometry Φ_p is allowed to depend on p. All manifolds considered here are finite dimensional, smooth, Hausdorff and connected.

As already mentioned, the most important local characterization of a geometry is that of Euclidean space or, more generally, of flat space of signature (p, q), $\mathbb{M}^{p,q} := (\mathbb{R}^{p+q}, \eta^{(p,q)} = -\sum_{i=1}^{p}(dx^i)^2 + \sum_{i=p+1}^{p+q}(dx^i)^2)$. Depending on the point of view, the Riemann tensor can be defined in several ways. The modern approach is to define the curvature operator as

[1]My convention here is that a Riemannian metric is not necessarily positive definite, but can have any non-degenerate signature. The term *strictly Riemannian* is reserved for the positive definite case.

$$\mathrm{Curv}^g(X,Y)Z = \left(\nabla_X\nabla_Y - \nabla_Y\nabla_X - \nabla_{[X,Y]}\right)Z, \qquad X,Y,Z \in \mathfrak{X}(\mathcal{M})$$

where ∇ is the metric-compatible, torsion-free connection associated to g

$$\mathrm{Riem}^g(W,Z,X,Y) = \langle W, \mathrm{Curv}^g(X,Y)Z\rangle, \qquad W,X,Y,Z \in \mathfrak{X}(\mathcal{M})$$

where $\langle\, ,\, \rangle$ denotes scalar product with g. In terms of this tensor Riemann's characterization of flat space is

Theorem 2.1 (Local characterization of flat space, Riemann [47]) *A Riemannian manifold* $(\mathcal{M}, g)$ *is locally isometric to the flat space with the same dimension and signature if and only if* $\mathit{Riem}^g = 0$.

Being intended to all faculty members, Riemann's lecture contains hardly any formula (see Chap. 4 in Spivak's comprehensive treatise [73] for a detailed exposition of Riemann's lecture) so no detailed proof of this claim is given there. Not even an explicit form of the Riemann tensor in a general coordinate system appears there. Following Spivak [73], such an explicit expression, and a proof of the fact that its non-vanishing is an obstruction for a metric being locally flat, appeared first in an unpublished paper, known as the *Commentatio*, submitted in 1861 to the Paris Academy and which appeared in the second edition of the collected works of Riemann [62]. However, as emphasized in [26] (see also [34]), this unpublished paper was primarily concerned with a problem of heat conduction proposed by the Academy in 1858, so its primary concern was not in Geometry, and it is debatable whether our modern understanding of the field might be biasing the interpretation of its results (see the interesting discussion on this *Commentatio* in [34] and references therein). In any case, an explicit expression of what we now know as the Riemann tensor appeared for the first time in this unpublished paper by Riemann. The first explicit proof that the vanishing of this tensor is sufficient for a metric being (locally) flat was given by Elwin Bruno Christoffel [17]. The proof of Theorem 2.1 is so central in differential geometry that many different proofs have been given. Spivak's treatise includes seven explicit proofs, which very nicely illustrate how differential geometry has evolved over the years from a discipline where the use of coordinates was central, to a mature theory where geometric objects fully independent of coordinates are the basic entities.

Another fundamental local characterization of Riemannian manifolds involves manifolds of constant sectional curvature. Recall that the sectional curvature of a two-plane $\Pi \subset T_p\mathcal{M}$, $p \in (\mathcal{M}, g)$ with non-degenerate induced first fundamental form h is defined as

$$K(p;\Pi) = \frac{\mathrm{Riem}^g(X,Y,X,Y)}{h(X,X)h(Y,Y) - h(X,Y)^2}, \qquad \{X,Y\} \text{ any basis of } \Pi.$$

The sectional curvature $K(p;\Pi)$ agrees with the Gauss curvature at p of the surface ruled by geodesics starting at p with tangent vector $X \in \Pi$. A Riemannian manifold has constant curvature k if $K(p;\Pi) = k$ for all $p \in \mathcal{M}$ and all non-degenerate plane $\Pi \in T_p\mathcal{M}$. A well-known lemma by Friedrich Schur [70] asserts that any Riemannian manifold with sectional curvature $K(p, \Pi) = k(p)$ independent of Π is in fact of constant curvature.

For any $k \in \mathbb{R}$, define the manifold $\mathbb{M}_k^{(p,q)}$ as the subset $B(K) \subset \mathbb{R}^{p+q}$

$$B(K) := \{x \in \mathbb{R}^{p+q}; 1 + \frac{k}{4}(\eta_{ij}^{(p,q)} x^i x^j) > 0\},$$

endowed with the metric g_k given by

$$g_k := \frac{1}{\left(1 + \frac{k}{4}(\eta_{ij} x^i x^j)\right)^2} \eta_{ij}^{(p,q)} dx^i dx^j,$$

$$\eta^{(p,q)} := \mathrm{diag}(\underbrace{-1, \ldots, -1}_{p}, \underbrace{1, \ldots, 1}_{q})$$

(it is easily seen that $B(K)$ is connected for all values of k and all signatures (p, q)). Spaces of constant curvature are all isometric to this space. This is already mentioned without proof in Riemann's inaugural lecture [47] and it was also known to Eugenio Beltrami [4].

Theorem 2.2 *Let $(\mathcal{M}, g)$ be a Riemannian manifold of constant sectional curvature k and signature (p, q). Then $(\mathcal{M}, g)$ is locally isometric to $\mathbb{M}_k^{(p,q)}$.*

Riemannian manifolds of constant curvature are also locally characterized by the condition that the Killing algebra of $(\mathcal{M}, g)$ has maximal dimension. Recall that a Killing vector is a vector field $\xi \in \mathfrak{X}(\mathcal{M})$ satisfying $\mathcal{L}_\xi g = 0$ where $\mathcal{L}$ denotes Lie derivative. The set of Killing vectors $\mathcal{K}$ in a manifold defines a finite-dimensional algebra (with respect to the Lie bracket) called Killing algebra. If the manifold has dimension n, the Killing algebra has dimension $\dim(\mathcal{K}) \leq \frac{n(n+1)}{2}$ and whenever equality is attained, the manifold is called maximally symmetric. A classical result by Luigi Bianchi [6] asserts that maximally symmetric spaces are precisely those of constant sectional curvature.

Theorem 2.3 (Bianchi [6]) *A Riemannian manifold $(\mathcal{M}, g)$ is maximally symmetric if and only if it has constant sectional curvature.*

Another important class of Riemannian manifolds singled out by the properties of its isometry group is the class of Riemannian symmetric spaces. These are defined as Riemannian manifolds $(\mathcal{M}, g)$ such that, at any point $p \in \mathcal{M}$ there exists an isometry $\Psi_p : \mathcal{M} \to \mathcal{M}$, $\Psi_p^\star(g) = g$ satisfying $\Psi_p(p) = p$ and $d\Psi_p|_p = -\mathrm{Id}|_p$, i.e. this isometry leaves p invariant and reverses any vector at p. Riemannian symmetric spaces can be locally characterized in terms of the Riemann tensor, as first proved by Élie Cartan.

Theorem 2.4 (Locally symmetric spaces, Cartan [12]) *A Riemannian manifold $(\mathcal{M}, g)$ is locally isometric to a Riemannian symmetric space if and only if the curvature tensor satisfies*

$$\nabla Riem^g = 0.$$

The Riemann tensor can be decomposed in its trace and its trace-free parts. The trace part defines the Ricci tensor $\mathrm{Ric}^g(X,Y) = \mathrm{tr}_g \mathrm{Curv}^g(X,\cdot)Y$ from which the scalar curvature is defined as $\mathrm{Scal}^g = \mathrm{tr}_g \mathrm{Ric}^g$. Useful tensors defined in terms of the Ricci tensor are the Schouten tensor Sch^g (in dimension $n > 2$) and the Einstein tensor Ein^g:

$$\mathrm{Sch}^g := \frac{1}{n-2}\left(\mathrm{Ric}^g - \frac{1}{2(n-1)}\mathrm{Scal}^g g\right), \qquad \mathrm{Ein}^g = \mathrm{Ric}^g - \frac{1}{2}\mathrm{Scal}^g.$$

The Weyl tensor is the trace-free part of the Riemann tensor, defined as

$$\mathrm{Weyl}_g = \mathrm{Riem}^g - \mathrm{Sch}^g \odot g$$

where $\odot$ is the Kulkarni-Nomizu product on symmetric tensors

$$\begin{aligned}(A \odot B)(X_1, X_2, X_3, X_4) =& A(X_1, X_3)B(X_2, X_4) - A(X_1, X_4)B(X_2, X_3)\\ &+ A(X_2, X_4)B(X_1, X_3) - A(X_2, X_3)B(X_1.X_4).\end{aligned}$$

Hermann Weyl introduced [79] the tensor named after him in an attempt to find a theory of gravitation invariant under scale (conformally invariant). The Weyl tensor has the fundamental property that $\mathrm{Weyl}_{\Omega^2 g} = \mathrm{Weyl}_g$ where $\Omega \in C^\infty(\mathcal{M}, \mathbb{R}^+)$. Thus, it vanishes automatically in any locally conformally flat space. Schouten [66] proved that the vanishing of this tensor is also sufficient in dimension $n \geq 4$ (in dimension $n = 2$ the Weyl tensor is not defined and in $n = 3$ it vanishes identically). As is well-known, in dimension $n = 2$ any space is locally conformally flat. In dimension three, the necessary and sufficient condition for local conformal flatness was obtained by Émile Cotton [16] in terms of the vanishing of the so-called Cotton tensor, which in dimension $n > 2$ is defined as

$$\mathrm{Cott}(X, Y, Z) = (n-2)\left((\nabla_Z \mathrm{Sch})(X, Y) - (\nabla_Y \mathrm{Sch})(X, Z)\right),$$

where $X, Y, Z \in \mathfrak{X}(\mathcal{M})$. This tensor is conformally invariant in any space with vanishing Weyl tensor (hence always in dimension three). Summarizing the discussion above, we have

Theorem 2.5 (Locally conformally flat spaces, Cotton, Weyl and Schouten) *A Riemannian manifold $(\mathcal{M}^n, g)$ $n \geq 3$ of signature (p, q) is locally isometric to a conformally flat space $(\mathbb{R}^n, \Omega^2 \eta^{(p,q)})$, $\Omega \in C^\infty(\mathbb{R}^n, \mathbb{R}^+)$ if and only if*

(i) If $n = 3$, the Cotton tensor vanishes.
(ii) If $n \geq 4$, the Weyl tensor vanishes.

3 Local Characterizations of the Schwarzschild and Kruskal Spacetimes

Among indefinite Riemannian manifolds, the Lorentzian ones (i.e. with signature $(-, +, \ldots, +)$) play a fundamental role in physics, as they describe spacetimes. For the purposes of this article by **spacetime** we simply mean a Lorentzian manifold[2] $(\mathcal{M}, g)$ of dimension $n \geq 2$. In Einstein's theory of General Relativity the spacetime geometry is linked to the energy-momentum contents of the non-gravitational fields via the Einstein field equations

$$\mathrm{Ein}^g = \chi T$$

where χ is a coupling constant depending on Newton's gravitational constant G and T is the energy-momentum tensor of the fields. In particular, a spacetime is *vacuum* whenever its Einstein tensor (or equivalently its Ricci tensor) vanishes.

With the exception of the flat Minkowski spacetime, perhaps the most important spacetime in general relativity is the Schwarzschild spacetime of mass $M \in \mathbb{R}$ found by Karl Schwarschild [67] shortly after the Einstein field equations were proposed (see [77] for the corresponding solution in higher dimensions). To be specific, by this solution we mean the manifold $\mathcal{M}_{\mathrm{Sch}} := \mathbb{R} \times (r_0, \infty) \times \mathbb{S}^{m-1}$ $m \geq 3$, where $r_0 = 0$ if $M \leq 0$ and $r_0 = (2M)^{\frac{1}{m-2}}$ if $M > 0$, endowed with the metric (in global coordinates $(t, r) \in \mathbb{R} \times (r_0, \infty)$)

$$g_{\mathrm{Sch}} = -\left(1 - \frac{2M}{r^{m-2}}\right) dt^2 + \left(1 - \frac{2M}{r^{m-2}}\right)^{-1} dr^2 + r^2 \gamma^{\mathbb{S}^{m-1}} \tag{1}$$

in units where $8\pi G = (m-1)\omega_{m-1}$ and $\gamma^{\mathbb{S}^k}$ is the standard metric of the k-sphere $\mathbb{S}^k$ and ω_k the corresponding total volume. The Schwarzschild spacetime is spherically symmetric and vacuum. Spherically symmetric means that the spacetime admits an action of $SO(m)$ as a group of isometries with orbits being either spacelike codimension-two surfaces or points. The fact that the metric (1) is defined also in another manifold, namely $\mathbb{R} \times (0, r_0) \times \mathbb{S}^2$ was found (for $m = 3$) independently by Droste [20] and Hilbert [40].

The Schwarzschild spacetime of mass $M \neq 0$ admits a maximal extension that keeps the properties of being spherically symmetric and vacuum. This is the Kruskal spacetime, which is defined as follows. Let $\mathcal{M}_{\mathrm{Kr}} = \mathcal{U} \times \mathbb{S}^{m-1}$ where $\mathcal{U} \subset \mathbb{R}^2$ is an open set defined by the inequality $\mathcal{U} := \{(u, v) \in \mathbb{R}^2; r(u \cdot v) > 0\}$ where $r := (a, b) \subset \mathbb{R} \to \mathbb{R}$ is the maximal solution of the differential equation

$$\frac{dr}{dx} = \frac{1}{(m-2)\,x} (2|M|)^{\frac{1}{m-2}} \left(1 - \frac{2M}{r^{m-2}}\right). \tag{2}$$

[2]Usually, the notion of spacetime requires also time-orientability, but this is irrelevant for the local characterization theorems we are describing here.

For $M < 0$, the maximal domain of definition does not include $x = 0$. However, for $M > 0$, the solution to this equation extends smoothly through $x = 0$ and has a non-zero gradient at $x = 0$. The Kruskal metric g_{Kr} is defined on $M_{\mathrm{Kr}} = \mathcal{U} \times \mathbb{S}^{m-1}$ and reads

$$g_{\mathrm{Kr}} = \frac{4}{(m-2)^2}|2M|^{\frac{2}{m-2}} \frac{1 - 2Mr(u \cdot v)^{2-m}}{uv} du dv + r(u \cdot v)^2 \gamma^{\mathbb{S}^{m-1}}.$$

The case $m = 3$ is best known, as it corresponds to the original extension by Martin Kruskal [46] (also found independently by Georges Szekeres [75]) of the Schwarzschild spacetime of dimension $n = 4$. In this case the ODE (2) for $M > 0$ can be integrated in simple terms as

$$uv = \left(1 - \frac{r}{2M}\right) e^{\frac{r}{2M}} \quad \Longrightarrow \quad g_{\mathrm{Kr}} = -\frac{32M^3}{r(uv)} e^{-\frac{r}{2M}} du dv + r^2(uv) \gamma^{\mathbb{S}^{m-1}}.$$

The original Schwarzschild spacetime corresponds to the domain $\{v > 0\} \cap \{u < 0\}$ and describes the exterior of the black hole spacetime. The conformal diagram of this exterior domain can be found in [56] along with interesting results on scattering and peeling of test fields on this and other backgrounds. The process of relativistic diffusion on the Schwarzschild spacetime is discussed in [30].

The most important local characterization theorem in gravitational theory is the Birkhoff theorem [7]. This theorem appears in the literature in two different guises, either as a local characterization theorem of the Kruskal spacetime, or as a statement that under suitable circumstances, a spherically symmetric spacetime must admit an additional Killing vector.[3] The two guises appear because the Schwarzschild spacetime was obtained under the assumptions of vacuum, spherically symmetry, time-independence and the implicit assumption that ∇r is spacelike. Birkhoff's proof that any vacuum, spherically symmetric spacetime admits an additional Killing vector and the fact that this Killing is timelike whenever ∇r is spacelike, provides therefore a local characterization theorem for the Schwarzschild metric among vacuum spherically symmetric spacetimes. In fact, the first result concerning the existence of an additional static symmetry is due to Jebsen [42] in a work virtually forgotten for a long time despite the fact that due credit was given e.g. in [33, 69]. This work has been brought back to light in more recent times, first in [64] and later in [19], see [43] and references therein for further details on the discovery of the Schwarzschild spacetime and Birkhoff's theorem.

Historical remarks aside, the fact of the matter is that the Schwarzschild and Kruskal spacetimes are unique among vacuum spherically symmetric n-dimensional Lorentzian spacetimes. This result has enormous physical significance as it means that, like in Newtonian theory, a spherically symmetric gravitational field outside its sources depends on a single constant, namely the total mass of the configuration. It

[3]The second guise is sometimes stated that spherically symmetric, say vacuum, spacetimes are static, but this is clearly not true in this generality as the Kruskal extension admits no global timelike Killing vector, see [22] where this fact is discussed in some more detail.

also states that spherical sources cannot emit gravitational radiation, even if they are themselves evolving. The precise local characterization result is as follows.

Theorem 3.1 (Local characterization of the Kruskal spacetime) *Let $(\mathcal{M}^n, g)$ be a Lorentzian n-dimensional space such that the group $SO(n-1)$ acts as a group of isometries with orbits which are either codimension-two spacelike submanifolds or points. If $(\mathcal{M}, g)$ is vacuum, then there exists $M \in \mathbb{R}$ such that $(\mathcal{M}, g)$ is locally isometric to the Kruskal spacetime $(\mathcal{M}_{Kr}, g_{Kr})$ of mass M. Let $r : \mathcal{M} \to \mathbb{R}$ be defined as*

$$r(p) = \left(\frac{|\mathcal{O}_p|}{\omega_{n-2}}\right)^{\frac{1}{n-2}}$$

where $\mathcal{O}_p$ is the orbit of $SO(n-1)$ containing p, $|\mathcal{O}_p|$ its $(n-2)$-volume and ω_{n-2} the $(n-2)$-volume of the round unit sphere $\mathbb{S}^{n-2}$. If $\langle \nabla r, \nabla r \rangle_p \neq 0$, then $(\mathcal{M}, g)$ is locally isometric near p to the Schwarzschild spacetime $(\mathcal{M}_{Sch}, g_{Sch})$ of mass M.

Analogous local characterization theorems of spherically symmetry spacetimes exist for many other matter models, including Λ-vacuum (i.e. $\mathrm{Ric}^g = (n-1)\Lambda g$, $\Lambda \in \mathbb{R}$), electrovacuum, Einstein-Maxwell-dilaton and many others. A detailed description of all such result is beyond the scope of this note, see [68] and references therein for results in this direction.

3.1 *Characterization of Spherically Symmetric Spacetimes with an Additional Killing Vector*

As already mentioned, the Birkhoff theorem for the vacuum spherically symmetric spacetime can be viewed as a local characterization result of the Kruskal spacetime or as a statement that spherically symmetric vacuum spacetimes admit an additional local isometry with generator orthogonal to the spherically symmetric orbits and which is timelike on regions where ∇r is spacelike. Viewed from this perspective, the Birkhoff theorem is a particular case of a result proved by Eiesland in [23] (and announced at the Eastern Meeting of the American Mathematical Society at Chicago in 1921 [24] shortly before the result by Birkhoff appeared). The theorem by Eiesland, although correct, uses a rather strange convention of the Einstein tensor. In terms of the usual definition $\mathrm{Ein}^g = \mathrm{Ric}^g - \frac{1}{2}\mathrm{Scal}^g g$ this theorem has been recently quoted in [76]. We follow the notation in this reference (with a couple of typo corrections).

Theorem 3.2 (Eiesland [23]) *The necessary and sufficient conditions that a locally spherically symmetric Lorentzian manifold with line-element*

$$ds^2 = -A^2 dt^2 + B^2 dr^2 + C^2 \left(d\theta^2 + \sin^2\theta d\phi^2\right)$$

A, B and C being arbitrary, non-zero functions of t and r, and C not a constant, shall admit a one-parameter group of isometries generated by a vector field $K = k_0(t,r)\partial_t + k_1(t,r)\partial_r$ are

$$A^2 Ein^t_r = \frac{2\Psi'}{C}\partial_r C \partial_t C,$$
$$A^2 B^2 \left(Ein^t_t - Ein^r_r\right) = \frac{2\Psi'}{C}\left[A^2(\partial_r C)^2 + B^2(\partial_t C)^2\right],$$

where $\Psi = \Psi(C)$ is some differentiable function of C, or a constant, and Ein^r_t, Ein^t_t etc. denote the components of the Einstein tensor (with one index raised) in the coordinate basis $\{\partial_t, \partial_r, \partial_\theta, \partial_\phi\}$. Under these conditions the Killing field K is, up to a constant multiple, given by

$$k_0(t,r) = -\frac{1}{e^\Psi AB}\partial_r C, \qquad k_1(t,r) = \frac{1}{e^\Psi AB}\partial_t C.$$

This theorem has been extended in [65] to Einstein spaces which are locally warped products $g = h + r^2\gamma$, where h is a Lorentzian metric and r a positive non-constant function on a two-dimensional space $\mathbb{Q}$, and γ is a metric on an m-dimensional manifold Σ^m, $m \geq 2$. Let us state and prove a theorem that generalizes Eiesland's result to arbitrary warped space with two-dimensional base.

Theorem 3.3 *Let $(\mathcal{M}, g)$ be locally isometric to a warped product space $(\mathbb{Q} \times \Sigma^m, g)$ $(m \geq 2)$, i.e. $g = h + r^2\gamma$, where r is a smooth positive function and h a metric on a two-dimensional manifold $\mathbb{Q}$ and (Σ, γ) is a (pseudo-)Riemannian manifold. Let $\epsilon = 1$ if h is Lorentzian and $\epsilon = -1$ if h is positive or negative definite.*

Assume that there is no open set in $\mathbb{Q}$ where $|\nabla r|^2_h$ vanishes identically. Then $(\mathcal{M}, g)$ admits a Killing vector $K \in \mathfrak{X}(\mathbb{Q})$ if and only if, away from the points where $|\nabla r|^2_h = 0$, there exists a smooth, nowhere-zero, function $H(r)$ such that

$$Ric^g|_{\mathbb{Q}} - \frac{1}{2}tr_h\left(Ric^g|_{\mathbb{Q}}\right)h = \frac{(m-1)H'}{2Hr}\left(dr \otimes dr + \epsilon dr^\star \otimes dr^\star\right)$$

where $Ric^g|_{\mathbb{Q}}$ is the restriction of the Ricci tensor of g to the tangent planes of $\mathbb{Q}$ and $\star$ is the Hodge dual in $(\mathbb{Q}, h)$. Moreover, the one-form $\boldsymbol{K} := h(K, \cdot)$ is $\boldsymbol{K} = H(r)dr^\star$

Proof It is immediate to check that a vector field K tangent to the $\mathbb{Q}$ factor is Killing if and only if K is Killing for $(\mathbb{Q}, h)$ and $K(r) = 0$. On the open set where dr is not null for the metric h (which is dense on $\mathbb{Q}$ by assumption), dr and $dr^\star$ are a basis of the cotangent space of any point in $\mathbb{Q}$. We work on this open (possibly disconnected) set from now on. Note that $\langle dr, dr^\star\rangle = 0$ and $\langle dr, dr\rangle = -\epsilon\langle dr^\star, dr^\star\rangle$ where $\langle\cdot,\cdot\rangle$ denotes scalar product with h. The condition $K(r) = 0$ is equivalent to the existence of a function $H \in C^\infty(\mathbb{Q}, \mathbb{R})$ such that $\boldsymbol{K} = H dr^\star$ or, in index notation $K_i = H\eta_{ij}\nabla^j r$ where $\boldsymbol{\eta}$ is the volume form of h and ∇ denotes the Levi-Civita derivative of h. It is known (and follows e.g. by explicit computation) that

$$\mathrm{Ric}^g|_{\mathbb{Q}} = K^h h - \frac{m-1}{r}\mathrm{Hess}_h r$$

where the Hessian of a functions is, as usual, defined by $(\mathrm{Hess} f)(X,Y) = \langle \nabla_X \mathrm{grad}\, f, Y\rangle$ and where K^h is the Gauss curvature of h. Let us denote by $\mathring{\mathrm{Ric}}^g_{\mathbb{Q}}$ the trace-free part of $\mathrm{Ric}^g|_{\mathbb{Q}}$ so that we have

$$\mathrm{Hess}_h r = -\frac{r}{m-1}\mathring{\mathrm{Ric}}^g_{\mathbb{Q}} + fh \tag{3}$$

where f is a scalar function whose form does not concern us. The Killing equations for K read

$$\begin{aligned} 0 = \nabla_i K_j + \nabla_j K_i &= \eta_{jl}\nabla_i H \nabla^l r + H\eta_j{}^l \nabla_i \nabla_l r + \eta_{il}\nabla_j H\nabla^l r + H\eta_i{}^l \nabla_j\nabla_l r \\ &= \eta_{jl}\left(\nabla_i H\nabla^l r - \frac{Hr}{m-1}(\mathring{\mathrm{Ric}}^g_{\mathbb{Q}})^l{}_i\right) + \eta_{il}\left(\nabla_j H\nabla^l r - \frac{Hr}{m-1}(\mathring{\mathrm{Ric}}^g_{\mathbb{Q}})^l{}_i\right) \end{aligned}$$

after inserting (3) and noticing that the term proportional to f drops. An endomorphism A^i_j satisfying $\eta_{jl}A^l_i + \eta_{il}A^l_j = 0$ must be proportional to the identity. Thus, there exists a function F on $\mathbb{Q}$ such that

$$\nabla_i H \nabla_j r - \frac{Hr}{m-1}(\mathring{\mathrm{Ric}}^g_{\mathbb{Q}})_{ij} = F h_{ij} \qquad \Longrightarrow \qquad A = \frac{1}{2}\langle \nabla H, \nabla r\rangle \tag{4}$$

where the implication follows simply by taking trace. The antisymmetric part of (4) implies that there exists a function $\hat{H} : I \to \mathbb{R}$ such that $H = \hat{H}\circ r$. We make the usual abuse of notation of writing simply $H(r)$. Note that H cannot vanish anywhere (on the open set $|\nabla r|^2_h \neq 0$) because if $H = 0$ at one point p, then (4) implies that H' also vanishes at that point, and the Killing K vanishes at p together with its covariant derivative, which can only happen if K is identically zero.

A basis for trace-free symmetric tensors is given by $dr\otimes dr + \epsilon dr^\star \otimes dr^\star$ and $dr \otimes dr^\star + dr^\star \otimes dr$. Expanding $\mathring{\mathrm{Ric}}^g_{\mathbb{Q}}$ in this basis and using $h = \frac{1}{|\nabla r|^2_h}(dr\otimes dr - \epsilon dr^\star \otimes dr^\star)$ it follows that

$$\mathring{\mathrm{Ric}}^g_{\mathbb{Q}} = \frac{(m-1)H'}{2Hr}\left(dr\otimes dr + \epsilon dr^\star\otimes dr^\star\right),$$

as stated in the theorem. □

Eiesland's Theorem 3.2 follows from this result because on open connected sets where $|\nabla r|^2_h \neq 0$, we can choose $H > 0$ (possible after reversing the Killing K). Writing $H = e^{-\Psi}$, Theorem 3.2 is recovered after a straightforward calculation.

3.2 Characterization of Schwarzschild Within Static Vacuum Spacetimes

Let us define a static, vacuum spacetime as a product manifold $\mathbb{R} \times \Sigma^m$ endowed with a metric $g = -N^2 dt^2 + h$, where h is a strictly Riemannian metric on Σ and $N > 0$ is a smooth positive function on Σ, called static potential, satisfying

$$N\mathrm{Ric}_h = \mathrm{Hess}_h N, \qquad \Delta_h N = 0. \tag{5}$$

Static spacetimes in this sense can be equivalently defined in terms of the data (Σ^m, h, N). The data corresponding to the Schwarzschild spacetime of mass M (and dimension $m+1$) can be written as

$$\Sigma_{\mathrm{Sch}} = \mathbb{R}^m \setminus B\left((|M|/2)^{\frac{1}{m-2}}\right), \qquad g_{\mathrm{Sch}} = \left(1 + \frac{M}{2|x|^{m-2}}\right)^{\frac{4}{m-2}} g_E,$$
$$N_{\mathrm{Sch}} = \frac{1 - \frac{M}{2|x|^{m-2}}}{1 + \frac{M}{2|x|^{m-2}}}. \tag{6}$$

where $B(a)$ is a centered closed ball of radius a.

Note that the Eq. (5) admit a trivial rescaling $N' = cN$, where $c \neq 0$ is a constant. Thus, two vacuum, static data will be said to be isometric if the corresponding Riemannian manifolds are isometric and, under the isometry, the static potentials are transformed to each other except for a non-zero multiplicative constant.

In this section we discuss two local characterizations of the Schwarzschild static data. Both have played a role in establishing the fundamental black hole uniqueness result that static, asymptotically flat static initial data with a totally geodesic boundary are isometric to Schwarzschild data (proved so far for dimensions $3 \leq m \leq 7$, as the proof relies on the positive energy theorem). The first characterization involves the conformal flatness of the metric h with a specific conformal factor constructed from N. As far as I know, it has been stated (and used) in dimension three and in the asymptotically flat context, but it can be generalized to a truly local statement in any dimension (we add a sketch of proof, since it does not seem to have appeared in the literature).

Theorem 3.4 *Let $(\Sigma^m, h, N > 0)$ be static vacuum initial data. The data is isometric to a domain of Schwarzschild data of mass M in an open, connected neighbourhood U_p of $p \in \Sigma$ if and only if the metric*

$$\hat{h} := \Omega^{\frac{4}{m-2}} h$$

is flat in U_p, where $\Omega = \frac{1+cN}{2}$ for some constant $c > 0$ with the property $|1 - cN| > 0$ on U_p.

Sketch of proof. The "only if" part is immediate from (6) with the choice $c = 1$ if $M \neq 0$ and $c \neq 1$ if $M = 0$. For the "if" part we work on U_p and scale N so that $c = 1$. The fact that $\text{Ric}_{\hat{h}} = 0$ and the properties of how the Ricci tensor and Hessian transform under a conformal rescaling turn out to imply the following equation on U_p

$$\text{Hess}_{\hat{h}} \left(\left| \frac{1+N}{1-N} \right| \right)^{\frac{2}{m-2}} = 2F(N)|\nabla N|^2_{\hat{h}} \hat{h}$$

where $F(N)$ is a manifestly positive and explicit function of N which we do not write. It follows that either N is constant (and the data is Minkowski, hence Schwarzschild of mass $M = 0$), or $F(N)|\nabla N|^2_{\hat{h}} = (\frac{2}{|M|})^{\frac{2}{m-2}}$ for a positive constant $|M|$. The solution of the hessian equation

$$\text{Hess}_{\hat{h}} \left(\left| \frac{1+N}{1-N} \right| \right)^{\frac{2}{m-2}} = 2 \left(\frac{2}{|M|} \right)^{\frac{2}{m-2}} \hat{h}$$

is, after a suitable choice of Cartesian coordinates $\{x\}$ for $\hat{h}$:

$$\left| \frac{1+N}{1-N} \right|^{\frac{2}{m-2}} = A + \left(\frac{2}{|M|} \right)^{\frac{2}{m-2}} |x|^2$$

where A is a constant. Inserting this expression into $F(N)|\nabla N|^2_{\hat{h}} = (\frac{2}{|M|})^{\frac{2}{m-2}}$, one concludes that $A = 0$, so the solution is

$$\left| \frac{1+N}{1-N} \right| = \frac{2}{|M|} |x|^{m-2}.$$

Choosing now the sign of M (which is still free) to satisfy $\text{sign}(M) = \text{sign}(1 - N)$, the Schwarzschild data (6) follows readily. □

The second local characterization of Schwarzschild among static vacuum initial data has been obtained by Reiris in [60] and provided a new way of showing the uniqueness of static, asymptotically flat black holes with connected horizon, via comparison geometry. The characterization involves integrable geodesic congruences, meaning congruences $\mathcal{F} = \{\gamma(s)\}$ parametrized by arc-length s and such that $s =$ const defines a smooth surface Σ_s orthogonal to the geodesics γ. Recall that the expansion θ of a congruence $\mathcal{F}$ is equal to the mean curvature of the surfaces Σ_s and coincides with Δs, where s is viewed as a scalar function on the domain covered by $\mathcal{F}$.

Given any static vacuum data set $(\Sigma^3, h, N > 0)$, it is often convenient to consider the naturally conformally rescaled space $(\Sigma^3, \hat{h} = N^2 h)$. Integrable, geodesic congruences $\mathcal{F}$ in $(\Sigma^3, \hat{h})$ have the following interesting property [60]: at each $\gamma \in \mathcal{F}$ and for any real number a, the function $\mathcal{M}_a(s)$ along $\gamma(s)$, defined by

$$\mathcal{M}_a(s) = \big(\frac{\theta}{2}(s+a)^2 - (s+a)\big)N^2, \qquad (7)$$

($\theta = \theta(\gamma(s))$ and $N = N(\gamma(s))$, is monotonically decreasing as s increases. For radial geodesics in Schwarzschild this function is constant (and equal to the mass for a suitable choice of a). This property is also true for the so-called Levi-Civita static solutions [49], labelled as classes A1, A2 and A3 in [21], and which can be collectively written as $(\Sigma^3_{\kappa,M}, h_{\kappa,M}, N_{\kappa,M})$, $\kappa \in \{1,-1,0\}$, $M \in \mathbb{R}$

$$\Sigma^3_{\kappa,M} = I \times S^2_\kappa, \quad N^2_{\kappa,M} = \kappa - \frac{2M}{r}, \quad h_\kappa = \frac{dr^2}{N^2_{\kappa,M}} + r^2\gamma_\kappa$$

and $I \subset \mathbb{R}$ is an open interval and $(S^2_\kappa, \gamma_\kappa)$ is either the standard sphere ($\kappa = 1$), the Euclidean plane ($\kappa = 0$) or the hyperbolic plane ($\kappa = -1$). Remarkably, the constancy of $\mathcal{M}_a$ characterizes locally these metrics [60, 61].

Theorem 3.5 (Reiris) *Let $(\Sigma^3, h, N > 0)$ be a static, vacuum, three-dimensional initial data set and let $\hat{h} = N^2 h$. Then, (Σ, h, N) is locally isometric to a Levi-Civita static solution if and only if around every point there exists an integrable geodesic congruence $\mathcal{F}$ of $(\Sigma, \hat{h})$ and a (not-necessarily continuous) function $a : \mathcal{F} \to \mathbb{R}$ such that $\mathcal{M}_{a(\gamma)}(s)$ is constant along each geodesic $\gamma(s)$.*

The data is locally isometric to the Schwarzschild data (i.e. to the subcase $\kappa = 1$ of the Levi-Civita static class) if and only if, in addition, $\theta > 1/(s+a)$ in at least one point.

3.3 Local Characterization of Kruskal Without Spherical Symmetry

The Kruskal and Schwarzschild spacetime are locally unique among vacuum spherically symmetric Lorentzian n−dimensional spacetimes. An interesting issue that has been raised in the literature is whether the assumption of spherical symmetry can be replaced by suitable local conditions on the curvature which, a posteriori, ensure that the space is locally spherically symmetric, i.e. that a local isometric group action of $SO(n-1)$ with codimension-two spacelike orbits (or points) exists. As far as I know, this has only been achieved in dimension $n = 4$ [28]. Before stating the theorem proved in this reference, we introduce some notation. We note that the definitions used here differ slightly from the definitions in [28].

For a double two-form, i.e. a four-covariant tensor $W_{\alpha\beta\mu\nu}$ satisfying $W_{\alpha\beta\mu\nu} = W_{[\alpha\beta]\mu\nu} = W_{\alpha\beta[\mu\nu]}$ we write $^\star W$ for the Hodge dual with respect to the first pair of indices. The square of W is defined as $W^2{}_{\alpha\beta\mu\nu} = W_{\alpha\beta\rho\sigma}W^{\rho\sigma}{}_{\mu\nu}$. This obviously defines a double two-form, so any power W^k can be defined iteratively. For any pair of two-forms F and H, $F \otimes H$ is clearly a double two-form. The total trace of F and of W are defined respectively as

$$\mathrm{Tr}(F^2) := F_{\alpha\beta}F^{\alpha\beta}, \qquad \mathrm{Tr}(W) := W_{\alpha\beta}{}^{\alpha\beta}.$$

A *symmetric* double two-form satisfies in addition $W_{\alpha\beta\mu\nu} = W_{\mu\nu\alpha\beta}$. Given one such W and a symmetric two-covariant tensor B, the product $W \cdot B$ is the symmetric tensor

$$(W \cdot B)_{\alpha\beta} := W_{\alpha\mu\beta\nu}B^{\mu\nu}.$$

Finally, for any covariant tensor $U_{\alpha_1\cdots\alpha_p}$ we write $|U|_g^2 := U_{\alpha_1\cdots\alpha_p}U^{\alpha_1\cdots\alpha_p}$.

Theorem 3.6 (Ferrando and Sáez [28]) *Let* $(\mathcal{M}, g)$ *be a* 4*-dimensional spacetime with Weyl tensor* $Weyl_g$ *satisfying*

$$\rho := -\left(\frac{1}{96}\mathrm{Tr}((Weyl_g)^3)\right)^{\frac{1}{3}} \neq 0.$$

Define the Riemann tensor concomitants

$$\begin{aligned} S &:= \frac{1}{3\rho}Weyl_g - \frac{1}{6}g \odot g \\ B &:= (Ric^g) - S \cdot Ric^g - \frac{Scal^g}{2}g, \\ F &:= -2\rho + \frac{1}{12}Scal^g + \frac{1}{4}|\Phi|_g^2 + \frac{\epsilon}{4}\sqrt{|B|_g^2} \end{aligned}$$

where $\Phi^\alpha := \frac{1}{2}S^{\alpha\beta}{}_{\mu\nu}\nabla^\rho S_{\rho\beta}{}^{\mu\nu}$, $\epsilon = 0$ *if* $B = 0$ *and* $\epsilon = -\frac{B(u,u)}{|B(u,u)|}$ *if* $B \neq 0$, u *being an arbitrary unit timelike vector field. Then* $(\mathcal{M}, g)$ *is locally spherically symmetric if and only if it satisfies:*

$$\begin{aligned} &S^2 + 2S = 0, \\ &2\nabla_\alpha S^\alpha{}_{\beta\mu\nu} + \frac{3}{2}S_{\mu\nu}{}^{\rho\sigma}\nabla^\alpha S_{\alpha\beta\rho\sigma} - g_{\beta\mu}\Phi_\nu + g_{\beta\nu}\Phi_\mu = 0, \\ &{}^\star S(\cdot, \Phi, \cdot, \Phi) = 0, \qquad 2S(\Phi, u, \Phi, u) - |\Phi|_g^2 > 0, \\ &\Phi_\alpha \textit{is an exact one-form}. \end{aligned}$$

In combination with the Birkhoff theorem, a local characterization of the Kruskal spacetime directly in terms of curvature concomitants can be obtained [28, 29].

Theorem 3.7 (Ferrando and Sáez [28, 29]) *Let* $(\mathcal{M}, g)$ *be a* 4*-dimensional spacetime with Weyl tensor* $Weyl_g$ *satisfying*

$$\rho := -\left(\frac{1}{96}\mathrm{Tr}((Weyl_g)^3)\right)^{\frac{1}{3}} \neq 0.$$

Let $S := \frac{1}{3\rho}Weyl_g - \frac{1}{6}g \odot g$. *Then* $(\mathcal{M}, g)$ *is locally isometric to the Kruskal spacetime if and only if*

$$
\begin{aligned}
&Ric^g = 0, \qquad S^2 + 2S = 0,\\
&\frac{1}{9\rho^2}|\nabla\rho|_g^2 - 2\rho > 0,\\
&S^\star(\cdot, \nabla\rho, \cdot, \nabla\rho) = 0, \qquad 2S(\nabla\rho, u, \nabla\rho, u) - |\nabla\rho|_g^2 > 0
\end{aligned}
$$

where u *is an arbitrary timelike unit vector field.*

The general idea of this theorem can be understood as follows. The equalities involving S in the theorem restrict the Weyl tensor pointwise. This, in combination with the Ricci flat condition, provices sufficient information on the Riemann tensor so as to show that the local isometry group of the spacetime is three dimensional with two-dimensional orbits. The inequality conditions of the theorem are then used to make the Lie algebra of Killing vectors isomorphic to the Lie algebra of the rotation group (as opposed to the the Lie algebra of isometries of the Euclidean or hyperbolic planes). Thus, the spacetime is locally spherically symmetric and the Birkhoff theorem implies that it is locally isometric to the Kruskal spacetime.

This local characterization theorem has been used in [32] to derive a local characterization theorem for initial data sets of Kruskal, i.e. triples (Σ, γ, K), where (Σ^3, γ) is a strictly Riemannian manifold and K a symmetric $(0, 2)-$tensor such that there exists an isometric embedding $\Phi : (\Sigma, \gamma) \mapsto (\mathcal{M}_{\mathrm{Kr}}, g_{\mathrm{Kr}})$ with second fundamental form equal to K.

4 Local Characterization of pp-Waves and Related Spacetimes

Brinkmann spaces are, by definition, Riemannian spaces admitting a parallel (also called covariantly constant) null vector field, i.e. a vector field ξ satisfying $\langle\xi, \xi\rangle = 0$ and $\nabla\xi = 0$. They were considered by Brinkmann [11] in his study of Einstein spaces $(\mathcal{M}, g)$ admitting a non-constant positive function $\Omega \in C^\infty(\mathcal{M}, \mathbb{R}^+)$, $d\Omega \not\equiv 0$ for which $(\mathcal{M}, \Omega^2 g)$ is also Einstein.[4] In this work a local coordinate system for any such spaces was also found. This provides a local characterization of spaces, which can be phrased a follows:

In order to discuss the local characterization results of Brinkmann spaces, it is useful to consider the manifold $\mathcal{M}_B := \mathbb{R} \times \mathbb{R} \times \Sigma$, where Σ is a $k \geq 1$ manifold, endowed with a Riemannian metric g_B of the form

$$g_B = -2du\,(dv + H_u du + W_u) + g_u.$$

[4] The term *Einstein space* to denote Riemannian manifolds of dimension $m \geq 3$ with Ricci tensor of pure trace appears to originate in Brinkmann's earlier work [10], see [5].

Here (u, v) are global coordinates on $(\mathbb{R} \times \mathbb{R})$, and H_u, W_u and g_u are, respectively, a u-dependent function, one-form and Riemannian metric on Σ. The signature of the space is $(p+1, q+1)$ if the signature of g_u is (p, q). The vector field $\xi := \partial_v$ is null and parallel, hence this Riemannian manifold is a Brinkmann space. The value of (M_B, g_B) is that, as found by Brinkmann [11], the converse is also locally true

Theorem 4.1 (Brinkmann [11]) *A Brinkmann space is locally isometric to a Riemannian manifold* (M_B, g_B).

In fact, a stronger result was found by Schimming [63].

Theorem 4.2 (Schimming [63]) *A Brinkmann space is locally isometric to a Riemannian manifold* (M_B, g_B) *with* $W_u = 0$ *and* $H_u = 0$.

An important class of spacetimes is the class of pp-waves, or plane fronted waves with parallel rays, which physically describe waves far away from bounded sources. There is no universally accepted definition of pp-wave, although all of them require the space to admit a parallel null vector field. Some authors do not require any extra condition, and hence identify pp-waves with Brinkmann spaces. Other authors add an algebraic condition of the curvature tensor. Here we follow [63] and define pp-waves as Brinkmann spaces of Lorentzian signature (and arbitrary dimension) with the Riemann tensor satisfying the trace condition

$$\mathrm{Tr}((\mathrm{Riem}^g)^2) = 0.$$

Schimming [63] found the following local characterization of pp-waves.

Theorem 4.3 (pp-waves, Schimming [63]) *A Lorentzian spacetime* $(\mathcal{M}^n, g)$ *of dimension* $n \geq 3$ *is a pp-wave if and only if it is locally isometric to* $(\mathcal{M}_B = \mathbb{R} \times \mathbb{R} \times \mathbb{R}^{n-2}, g_B)$ *with metric*

$$g_B = -2du dv + H(u, x^i) du^2 + \sum_{i=1}^{n-2} dx_i^2.$$

The algebraic conditions on the Riemann tensor that define pp-waves can be stated in several equivalent forms, as proved by Schimming [63] and Leistner [48].

Theorem 4.4 (Equivalent algebraic conditions for pp-waves) *A Lorentzian Brinkmann space* $(\mathcal{M}, g)$ *is a pp-wave if and only if one of the following conditions holds:*

(i) *[63]* $\xi_\alpha Riem^g_{\beta\gamma\mu\nu} + \xi_\alpha Riem^g_{\beta\gamma\mu\nu} + \xi_\alpha Riem^g_{\beta\gamma\mu\nu} = 0.$
(ii) *[63] There exists a symmetric 2-covariant tensor* V *satisfying* $V(\xi, \cdot) = 0$ *such that*

$$Riem^g_{\alpha\beta\mu\nu} = \xi_\alpha V_{\beta\mu} \xi_\nu - \xi_\beta V_{\alpha\mu} \xi_\nu + \xi_\beta V_{\alpha\nu} \xi_\mu - \xi_\alpha V_{\beta\nu} \xi_\mu.$$

(iii) [63] There exists a function $f \in C^\infty(\mathcal{M}, \mathbb{R})$ *such that*

$$(Riem^g)^\alpha{}_{\beta\mu\nu}(Riem^g)^\nu{}_{\rho\sigma\alpha} = f\xi_\beta\xi_\mu\xi_\rho\xi_\sigma.$$

(iv) [48] For all $U, V \in \mathfrak{X}(\mathcal{M})$ *and any pair of vectors* $X, Y \in \mathfrak{X}(\mathcal{M})$ *orthogonal to* ξ*:*

$$Curv^g(U, V)X \in span(\xi) \quad \text{or, equivalently} \quad Riem^g(\cdot, \cdot, X, Y) = 0,$$

where ξ *is the parallel null vector of* $(\mathcal{M}, g)$*.*

The Ricci tensor of any pp-wave with parallel null vector ξ is of the form $\mathrm{Ric}^g = \Phi\boldsymbol{\xi} \otimes \boldsymbol{\xi}$, where $\Phi \in C^\infty(\mathcal{M}, g)$ and $\boldsymbol{\xi} := g(\xi, \cdot)$. Thus, a pp-wave is an Einstein space only if it is vacuum. The case of dimension $n = 4$ is particularly interesting for physics. In this case the condition $\mathrm{Ric}^g = \Phi\boldsymbol{\xi} \otimes \boldsymbol{\xi}$ turns out to be equivalent to a Brinkmann space being a pp-wave. This follows from the fact that $\mathrm{Riem}^g(\xi, \cdot, \cdot, \cdot) = 0$ (which is a direct consequence of $\nabla\boldsymbol{\xi} = 0$), implies $\mathrm{Weyl}_g(\xi, \cdot, \cdot, \cdot) = 0$ when $\mathrm{Ric}^g = \Phi\boldsymbol{\xi} \otimes \boldsymbol{\xi}$. In dimension $n = 4$ this means that the Weyl tensor is of Petrov type N (from Bel's characterization [3] of the Petrov type [59]) and the Riemann tensor necessarily satisfies item (iii) of Theorem 4.4.

According to Ehlers and Kundt [21], the notion of pp-wave in general relativity was discovered independently by Robinson (unpublished work) in 1956 and by Hély [37] and Peres [58], Robinson being the first to discuss its physical significance. In four dimensions, additional characterization results exist, see [21, 74, 81] for further details. Here we mention only one for pp-waves that is genuinely four-dimensional [21].

Theorem 4.5 (Ehlers and Kundt [21]) *A four-dimensional vacuum spacetime* $(\mathcal{M}, g)$ *is a pp-wave if and only if it admits a covariantly constant null two-form, i.e. a two-form* $\boldsymbol{F}$ *satisfying*

$$\nabla F = 0, \qquad \mathrm{Tr}(F^2) = 0 \qquad \mathrm{Tr}(F \otimes F^\star) = 0$$

where $F^\star$ *is the Hodge dual of* $\boldsymbol{F}$*.*

As already said, in four dimensions vacuum pp-waves are simply vacuum Brinkmann spaces. Sufficient conditions similar to this theorem ensuring the existence of a parallel null vector exist in arbitrary dimension, see Lemma 3.1 and Remark iv in [71].

Brinkmann spaces have arisen also in connection to generalizations of locally symmetric spaces. As discussed before, locally symmetric spaces are defined by the condition that the Riemann tensor is covariantly constant. It is natural to consider k-th order symmetric spaces $k \geq 1$, as those with a curvature tensor satisfying

$$\underbrace{\nabla \cdots \nabla}_{k} \mathrm{Riem}^g = 0. \tag{8}$$

However, in the strictly Riemannian case (g positive definite) a classical theorem asserts that these spaces are automatically locally symmetric (see [50, 57] under additional restrictions and [78] for the general case, attributed there to unpublished work by Nomizu). Moreover, Tanno [78] also proves under a non-degeneracy condition for the Riemann tensor that condition (8) $k \geq 2$ implies locally symmetric in arbitrary signature. However, k-th order symmetric ($k \geq 2$) indefinite Riemannian spaces which are not locally symmetric do exist and their systematic study in the Lorentzian case and for $k = 2$ was initiated by Senovilla [71] as late as 2008, where the nomenclature of k-th order symmetric spaces was introduced and the following result was proved.

Theorem 4.6 (Senovilla [71]) *Let $(\mathcal{M}, g)$ be a proper second order symmetric space, i.e. an n-dimensional Lorentzian space satisfying*

$$\nabla\nabla Riem^g = 0 \qquad \nabla Riem^g \not\equiv 0.$$

Then $(\mathcal{M}, g)$ admits a parallel null vector (i.e. it is a Brinkmann space).

Obviously, not all Brinkmann spaces are proper second order symmetric, so it is natural to determine them. This has been achieved recently in [8] (announced in [9]), where the following local characterization result is proven.

Theorem 4.7 (Second order symmetric Lorentzian spaces [8]) *Let $(\mathcal{M}^n, g)$ be an n-dimensional proper second order symmetric Lorentzian space. Then $(\mathcal{M}, g)$ is locally isometric to a direct product $(\mathcal{M}_1 \times \mathbb{R}^{d+2}, g_1 \oplus g_2)$, $(d \geq 0)$ where $(\mathcal{M}_1, g_1)$ is a non-flat symmetric space with positive definite metric and g_2 is, in Cartesian coordinates for $\mathbb{R}^{d+2}$,*

$$g_2 = -2dudv + \left(\sum_{i,j=2}^{d+1} p_{ij}(u) x^i x^j \right) du^2 + \sum_{i=2}^{d+1} (dx^i)^2$$

where $p_{ij} = (H_1)_{ij} u + (H_0)_{ij}$ and $(H_0)_{ij}$, $(H_1)_{ij}$ are symmetric real matrices, with H_1 not identically zero.

This classification result has also been proved independently and about the same time in [2] using an approach fully based on holonomy groups. This holonomy method has been extended recently in [31], where the following classification result for third order Lorentzian spaces is proved.

Theorem 4.8 (Third order symmetric Lorentzian spaces [31]) *Let $(\mathcal{M}^n, g)$ be an n-dimensional third order symmetric Lorentzian space which is not second order symmetric. Then the same conclusion as in Theorem 4.7 holds, except that the metric g_2 is now*

$$g_2 = -2dudv + \left(\sum_{i,j=2}^{d+1} p_{ij}(u)x^i x^j \right) du^2 + \sum_{i=2}^{d+1} (dx^i)^2$$

where $p_{ij} = (H_2)_{ij}u^2 + (H_1)_{ij}u + (H_0)_{ij}$ *and* $(H_a)_{ij}$, $(a = 0, 1, 2)$ *are symmetric, real matrices, with* H_2 *not identically zero.*

5 Local Characterizations of the Kerr, Kerr-Newman and Kerr-De Sitter Metrics

The class of Kerr spacetimes is one of the most important classes in general relativity. It was discovered by Roy Kerr [44] as the first family of stationary and axially symmetric vacuum spacetimes and depends on two real parameters a and m. For a suitable subset of the parameters, it represents a stationary asymptotically flat black hole. A fundamental conjecture in general relativity states that the exterior region of the Kerr black hole class is the unique class of stationary vacuum, asymptotically flat exterior black hole regions. This conjecture is known to be true under suitable conditions, mainly of technical nature (see [14, 39]). Thus, the Kerr class of spacetimes enjoys a privileged position among the very many possible stationary and asymptotically flat vacuum spacetimes. Obviously, the condition of being a black hole is essentially of global nature, and it is therefore of interest to find local characterizations of the Kerr class of spacetimes.

To be explicit, by Kerr spacetime we mean the maximal analytic extension of the Kerr exterior region, as obtained by Carter [13]. For parameter values $a \cdot m \neq 0$, the manifold outside a countable collection of smooth closed, codimension-two surfaces, which correspond to the set of zeros of a Killing vector, can be covered by a countable union of patches of Kerr-Schild type, each one of them isometric to $(M_a, g_{m,a})$ given by

$$\begin{aligned}
\mathcal{M}_a &:= \mathbb{R} \times \left(\mathbb{R}^3 \setminus \{x^2 + y^2 \leq a^2, z = 0\}\right) \\
g_{m,a} &:= -dt^2 + dx^2 + dy^2 + dz^2 + \frac{2mr^3}{r^4 + a^2 z^2} \boldsymbol{\ell} \otimes \boldsymbol{\ell}, \\
\boldsymbol{\ell} &:= dt + \frac{r}{r^2 + a^2}(xdx + ydy) + \frac{a}{r^2 + a^2}(ydx - xdy) + \frac{zdz}{r}, \\
r &: \mathcal{M}_a \mapsto \mathbb{R}^+ \quad \text{defined by} \quad \frac{x^2 + y^2}{r^2 + a^2} + \frac{z^2}{r^2} = 1
\end{aligned}$$

where t is the coordinate in the $\mathbb{R}$ factor and (x, y, z) are Cartesian coordinates in the other factor. The metric is of Kerr-Schild type [45] because it is locally of the form $g = \eta + \boldsymbol{\ell} \otimes \boldsymbol{\ell}$ where η is the Minkowski metric and $\boldsymbol{\ell}$ is a null one-form with respect to η (and also g). The parameters m and a are called respectively *mass* and *specific angular momentum*. The mass parameter m is the ADM mass of the spacetime, see [38] for the definition of ADM mass and for a thorough discussion on its positivity properties and the interesting relationship it has with a concept of mass in closed manifolds by means of a conformal blow-up of a point.

Natural generalizations of the Kerr class exist to the case of electrovacuum spacetimes with cosmological constant, i.e. spacetimes $(\mathcal{M}, g)$ with a two-form H (the electromagnetic field) and a constant Λ (the cosmological constant) satisfying the field equations

$$\mathrm{Ein}^g + \Lambda g = 2\,\mathrm{tr}_{2,4}(H \otimes H) - \frac{1}{2}\mathrm{Tr}(H^2)g, \qquad dH = 0, \quad dH^\star = 0. \tag{9}$$

The pure electrovacuum (i.e. with $\Lambda = 0$) generalization is called *Kerr-Newman* and also plays a fundamental role in the classification of stationary black holes in four dimensions [14, 39]. The spacetime depends on three parameters $\{m, a, q\}$ where q is called *electric charge*. Similarly as for the Kerr spacetime above, whenever $a \cdot m \neq 0$, the Kerr-Newman spacetime can be covered outside a countable, disjoint, union of closed, spacelike, totally geodesic codimension-two surfaces by Kerr-Schild patches $(\mathcal{M}_a, g_{m,a,q})$ where $\mathcal{M}_a$ is as before and the metric is given by

$$g_{m,a,q} := -dt^2 + dx^2 + dy^2 + dz^2 + \frac{r^2\left(2mr - q^2\right)}{r^4 + a^2 z^2}\boldsymbol{\ell} \otimes \boldsymbol{\ell},$$

with $\boldsymbol{\ell}$ and r defined exactly as in the Kerr class.

The Einstein space generalization (i.e. with vanishing electromagnetic field H and $\Lambda \neq 0$) is called *Kerr-NUT-(anti-)de Sitter* spacetime and depends on four real parameters $\{\Lambda, m, a, l\}$. The spacetime can no longer be covered by Kerr-Schild patches. However, it does admit a *double Kerr-Schild* form [15], i.e. the metric can be locally written as $g = g_0 + f_1 \ell \otimes \ell + f_2 k \otimes k$, where k and ℓ are null in the base metric g_0 which in this case is the spacetime of constant curvature $\frac{\Lambda}{3}$. For the purposes of this article, we describe the spacetime in advanced Eddington-Finkelstein coordinates as follows. The *Kerr-NUT-(A) de Sitter spacetime* of parameters $\{\Lambda, m, a, l\}$ is the maximal spacetime $(\mathcal{M}_{KNdS}, g_{\mathrm{KNdS}})$ for which any point $p \in \mathcal{M}_{\mathrm{KNdS}}$ where no Killing vector vanishes admits an open, connected neighbourhood U_p and local coordinates $\{u, r, \theta, \phi\}$ such that g_{KNdS} takes the form

$$\begin{aligned} g_{\mathrm{KNdS}} = &- \frac{\Delta - a^2 \sin^2\theta \Delta_\theta}{\rho^2}\left(du - (a\sin^2\theta + 4l\sin^2(\theta/2))d\phi\right)^2 \\ &+ 2\left(dr - a\sin^2\theta\Delta_\theta d\phi\right)\left(du - (a\sin^2\theta + 4l\sin^2(\theta/2))d\phi\right) \\ &+ \rho^2\left(\frac{d\theta^2}{\Delta_\theta} + \Delta_\theta \sin^2\theta d\phi^2\right) \end{aligned}$$

where

$$\rho^2 := r^2 + (l + a\cos\theta)^2,$$
$$\Delta := a^2 - l^2 - 2mr + r^2 - \frac{\Lambda}{3}(3l^2(a^2 - l^2) + (a^2 + 6l^2)r^2 + r^4).$$

For other expressions of the local form of this metric see [35, 36]. The case $\Lambda = 0$ of this metric defines the *Kerr-NUT spacetime*. The case $\Lambda = l = 0$ corresponds precisely to the Kerr spacetime.

Note that all the spacetimes discussed in this section admit a Killing vector $\xi = \partial_t$ which turns out to have very distinctive features. The characterizations we shall describe in this section exploit essentially the existence of this Killing vector. This fits well with the problem of stationary black hole spacetimes and in fact these characterizations have had useful applications in the black hole uniqueness context ([1] and references therein). The characterization is also specific to four dimensions, which again fits well with the fact that the classification of black holes in higher dimensions is necessarily of a very different nature than in four dimensions, given the plethora of already known examples, starting with the celebrated black ring of Emparan and Reall [25], see [14, 41] for results on stationary black hole in higher dimensions.

Consider any four-dimensional spacetime $(\mathcal{M}, g)$ admitting a Killing vector ξ. Assume $(\mathcal{M}, g)$ to be orientable with volume form η^g and let $\star$ denote the corresponding Hodge operator. For a real two-form U, we define $\mathcal{U} := U + iU^\star$, which satisfies $\mathcal{U}^\star = -i\mathcal{U}$ and it is hence called *self-dual* (the term *anti self-dual* is also common in the literature). Conversely, any (necessarily complex) self-dual two form $\mathcal{U}$ can be written as $\mathcal{U} = \mathrm{Re}(\mathcal{U}) + i(\mathrm{Re}(\mathcal{U}))^\star$ where Re denotes the real part. A two-form is called non-degenerate at $p \in \mathcal{M}$ if $\mathcal{U}^2 := \mathrm{Tr}(\mathcal{U} \otimes \mathcal{U})$ satisfies $\mathcal{U}^2|_p \neq 0$. A non-degenerate two-form at p admits precisely two linearly independent real eigenvectors $\ell_\pm \in T_p\mathcal{M}$, i.e. solutions of $(\mathcal{U} - \lambda_\pm g)|_p(\ell_\pm, \cdot) = 0$ with $\lambda_\pm \in \mathbb{R}$. The eigenvectors $\ell_\pm$ are necessarily null and the two-dimensional timelike plane they span $\mathcal{T}_\mathcal{U} := \mathrm{span}\{\ell_+, \ell_-\}$ is called the timelike eigenplane of $\mathcal{U}$ at p.

Given a Killing vector ξ, the tensor $F := \nabla\boldsymbol{\xi}$ defines a two-form. The corresponding two-form $\mathcal{F} := F + iF^\star$ is called self-dual Killing form of ξ and the so-called Ernst one-form of ξ is defined by $\boldsymbol{\chi} := 2\mathcal{F}(\xi, \cdot)$.

For a tensor W with the same symmetries as the Weyl-tensor, define the left-Hodge dual $W^\star$ as the Hodge dual in the second pair of indices and the *self-dual of* $\mathcal{W}$ as $\mathcal{W} = W + iW^\star$. The self-dual of the Weyl tensor is denoted by $\mathcal{W}eyl_g$. The space of self-dual two-forms admits a natural metric, denoted by $\mathcal{I}_g$ and defined by

$$\mathcal{I}_g = \frac{1}{8}\left(g \odot g + i\,(g \odot g)^\star\right).$$

The Kerr spacetime has the algebraic property that the self-dual Weyl tensor and the self-dual Killing form are related to each other by the following simple relation

$$\mathcal{W}eyl_g = \mathcal{Q}\left(\mathcal{F} \otimes \mathcal{F} - \frac{1}{3}\mathcal{F}^2\mathcal{I}_g\right) \tag{10}$$

where $\mathcal{Q}$ is a smooth complex function on $\mathcal{M}$. It turns out that this algebraic property is useful to characterize locally the Kerr class among all vacuum spacetimes admitting a Killing vector ξ. Although not explicitly stated in this form, the following local characterization theorem is proved in [51]. The idea of the theorem stems from a previous characterization result for the Kerr metric in the so-called quotient formalism by Simon [72].

Theorem 5.1 (Mars [51]) *Let $(\mathcal{M}, g)$ be a smooth, vacuum, four-dimensional spacetime admitting a Killing vector ξ. Assume the following two conditions hold:*

(i) There exists a smooth complex function $\mathcal{Q} : \mathcal{M} \mapsto \mathbb{C}$ such that the self-dual Weyl tensor $\mathcal{W}eyl_g$ of g and self-dual Killing form $\mathcal{F}$ of ξ satisfy

$$\mathcal{W}eyl_g = \mathcal{Q}\left(\mathcal{F} \otimes \mathcal{F} - \frac{1}{3}\mathcal{F}^2\mathcal{I}_g\right). \tag{11}$$

(ii) There exist $p \in \mathcal{M}$ where $\mathcal{Q}\mathcal{F}^2|_p \neq 0$.

Then, the Ernst one-form $\boldsymbol{\chi}$ is exact $\boldsymbol{\chi} = d\chi$ and $\mathcal{Q}$, $\mathcal{F}^2$ take the form $\mathcal{Q} = -6/(c - \chi)$, $\mathcal{F}^2 = A(c - \chi)^4$ where $A \neq 0$ and c are complex constants.

If, in addition, $Re(c) > 0$, A is real and negative and ξ is somewhere not orthogonal to the timelike eigenplane $\mathcal{T}_{\mathcal{F}}$ of $\mathcal{F}$, then the spacetime $(\mathcal{M}, g)$ is locally isometric to a Kerr spacetime.

This result was quoted in [52] with the unfortunate omission of one of the hypotheses, namely the condition that the Killing vector ξ is somewhere not orthogonal to the timelike eigenplane of $\mathcal{F}$. In the proof of [51] this condition was automatically true as the Killing vector was assumed to be timelike somewhere. See [55] for a detailed discussion of this issue.

The main assumption (11) imposes algebraic restrictions on the Weyl tensor at each point. Recall that the self-dual Weyl tensor $\mathcal{W}eyl_g$ defines an endomorphism on the space of self-dual two-forms. The algebraic classification of this endomorphism leads to the so-called Petrov classification [3, 59] of spacetimes, which plays an important role in studying the geometric properties of spacetimes. Details can be found e.g. in [74]. The Petrov type of a Weyl tensor can be I, II, III, N, D or 0 and Condition (11) restricts the Weyl tensor, at each point, to be of Petrov types D, N or O, with $\mathcal{F}$ being an eigenvector of $\mathcal{W}eyl_g$. Thus, Theorem 5.1 identifies the Kerr spacetime essentially as the only vacuum spacetime satisfying this Petrov-type restriction everywhere and, moreover, being of Petrov type exactly D at least at one point (by assumption (ii)). The proof proceeds by solving the Einstein equations and Bianchi identities in a frame constructed geometrically from $\mathcal{F}$ and the Killing vector ξ. The word "essentially" above refers to the fact that, besides ξ being somewhere not orthogonal to $\mathcal{T}_{\mathcal{F}}$, conditions on the constants A and c need to be imposed in order to select the Kerr spacetime among the larger class of Kerr-NUT spacetimes with either spherical, plane and hyperbolic topologies, which is the class that arises when the conditions on A and c are dropped. In fact a complete classification can be obtained [55] when Condition (ii) is dropped and, in addition, no a priori assumption on ξ at any point is made.

In the case of the Kerr-Newman spacetime, the spacetimes does not only satisfy the alignment condition (10) but, in addition, the self-dual electromagnetic field $\mathcal{H} := H + iH^\star$ is proportional to the self-dual Killing form $\mathcal{F}$. As proved by W. Wong in [80], this double alignment, together with suitable additional restrictions, characterizes the Kerr-Newman class of spacetimes (for a complex number A, $\overline{A}$ is the complex conjugate, $|A|^2 = A\overline{A}$ and $A = \mathrm{Re}(A) + i\mathrm{Im}(A)$, with $\mathrm{Re}(A), \mathrm{Im}(A) \in \mathbb{R}$)

Theorem 5.2 (Wong [80]) *Let $(\mathcal{M}, g, H)$ be a simply connected, smooth electrovacuum spacetime admitting a Killing vector ξ satisfying $\mathcal{L}_\xi H = 0$. Assume that ξ is timelike somewhere and that the self-dual two form $\mathcal{H} := H + iH^\star$ satisfies $\mathcal{H}^2 := \mathrm{Tr}(\mathcal{H} \otimes \mathcal{H}) \neq 0$ everywhere. Assume that*

(i) The self-dual Killing form $\mathcal{F}$ is proportional to $\mathcal{H}$: $\mathcal{F} = \overline{u}\mathcal{H}$, where $u \in C^\infty(\mathcal{M}, \mathbb{C})$
(ii) The function u satisfies $du = \mathcal{H}(\xi, \cdot)$.
(iii) There exists a non-zero complex constant C_1 such that the self-dual Weyl tensor satisfies

$$\mathcal{W}eyl_g = 3P\left(\frac{1}{2}(\mathcal{F}\otimes\mathcal{H} + \mathcal{H}\otimes\mathcal{F}) - \frac{1}{3}\mathrm{Tr}(\mathcal{F}\otimes\mathcal{H})\mathcal{I}_g\right)$$

where $P := \left(\frac{-4}{C_1^2\mathcal{H}^2}\right)^{\frac{1}{4}}$.

Then, there exists a complex constant C_2 such that $P^{-1} - u = C_2$ and a real constant C_4 such that $\langle \xi, \xi\rangle + |u|^2 = C_4$.

If C_2 further satisfies that $C_1\overline{C_2}$ is real and $C_4 = |C_2|^2 - 1$, then there exists a non-negative real constant $\mathfrak{U}$ such that $|C_1|^2|P|^2\langle Im(C_1\nabla P), Im(C_1\nabla P)\rangle + (Im(C_1P))^2 - \mathfrak{U} = 0$ everywhere and $(\mathcal{M}, g)$ is locally isometric to the Kerr-Newman spacetime of charge $q = |C_1|$, specific angular momentum $\sqrt{\mathfrak{U}}$ and mass $C_1\overline{C_2}$.

Remark 5.3 Note that this theorem does not immediately admit a vacuum limit, since $\mathcal{H}^2$ is assumed to be non-zero by hypotheses. Also, the proportionality functions Ξ and P in items (i) and (iii) are linked by hypothesis to the electromagnetic field $\mathcal{H}$. Comparing with the characterization theorem in vacuum Theorem 5.1, where the proportionality function between the self-dual Weyl tensor and the self-dual Killing form is not fixed a priori, this raises the question whether a stronger characterization result exists with weaker a priori conditions on Ξ and P. In view of the result in vacuum and in the case of cosmological constant below, we conjecture that such a stronger characterization should indeed exist. Interesting steps towards finding this generalized characterization have been obtained in [18].

A local characterization result for the Kerr-(A)de Sitter class among Λ-vacuum spacetimes (i.e. satisfying (9) with $H = 0$) and admitting a Killing vector ξ along the same lines as before has been recently obtained in [54]. The method of proof is much more geometric than for the two results mentioned above (strongly based on a tetrad construction) and is based on the presence of a second Killing vector constructed in terms of geometric information arising solely from ξ and an underlying geometric structure in terms of a Riemannian submersion of a conformal metric. The following theorem has been proved in [54]. Unfortunately, the statement of the result as given in [54] contains a few typos and, more importantly, has the same missing hypothesis as in the vacuum case discussed above. We provide the correct version here.

Theorem 5.4 (Mars and Senovilla [54]) *Let $(\mathcal{M}, g)$ be a Λ-vacuum spacetime admitting a Killing vector ξ with self-dual Killing form $\mathcal{F}$. Assume there exists $Q \in C^\infty(\mathcal{M}, \mathbb{C})$ such that the self-dual Weyl tensor of g satisfies*

$$\begin{aligned}
&(i) \quad \mathcal{W}eyl_g = Q\left(\mathcal{F} \otimes \mathcal{F} - \frac{1}{3}\mathcal{F}^2 \mathcal{I}_g\right) \\
&(ii) \quad \text{There exists } p \in \mathcal{M} \text{ such that } Q\mathcal{F}^2|_p \neq 0, \\
&(iii) \quad \text{There exists } p' \in \mathcal{M} \text{ such that } (Q\mathcal{F}^2 - 4\Lambda)|_{p'} \neq 0.
\end{aligned}$$

Then $\mathcal{F}^2 \neq 0$ and $Q\mathcal{F}^2 - 4\Lambda \neq 0$ everywhere. Assuming further that (iv) ξ is somewhere not orthogonal to the timelike eigenplane $\mathcal{T}_{\mathcal{F}}$ of $\mathcal{F}$, then there exist constants $b_1, b_2, c, k \in \mathbb{R}$ such that

$$36Q(\mathcal{F}^2)^{\frac{5}{2}} + (b_2 - ib_1)(Q\mathcal{F}^2 - 4\Lambda)^3 = 0 \tag{12}$$

$$\langle \xi, \xi \rangle_g + Re\left(\frac{6\mathcal{F}^2 (Q\mathcal{F}^2 + 2\Lambda)}{(Q\mathcal{F}^2 - 4\Lambda)^2}\right) + c = 0$$

$$-k + \left|\frac{36\mathcal{F}^2}{(Q\mathcal{F}^2 - 4\Lambda)^2}\right|^2 |\nabla Z|^2_g - b_2 Z + cZ^2 + \frac{\Lambda}{3} Z^4 = 0$$

where $Z := Im\left(\frac{6i\sqrt{\mathcal{F}^2}}{Q\mathcal{F}^2 - 4\Lambda}\right)$. Moreover, $|\nabla Z|^2_g \geq 0$ everywhere.

If these constants are such that the polynomial $V(\zeta) := k + b_2\zeta - c\zeta^2 - \frac{\Lambda}{3}\zeta^4$ can be factored as

$$V(\zeta) = \hat{V}(\zeta)(\zeta - \zeta_0)(\zeta_1 - \zeta) \tag{13}$$

with $\zeta_0 \leq \zeta_1$ and $\hat{V}(\zeta) > 0$ on $[\zeta_0, \zeta_1]$ and $Z : \mathcal{M} \to [\zeta_0, \zeta_1]$ then $(\mathcal{M}, g)$ is locally isometric to the Kerr-NUT-(A)dS spacetime with parameters $\{\Lambda, m, a, l\}$ where

$$m = \frac{b_1}{2v_0\sqrt{v_0}}, \qquad a = \frac{\zeta_1 - \zeta_0}{2\sqrt{v_0}}, \qquad l = \frac{\zeta_1 + \zeta_0}{2\sqrt{v_0}}$$

and $v_0 := \hat{V}(\frac{\zeta_0 + \zeta_1}{2})$.

The statement of this theorem is somewhat long. However, the core of the hypotheses lies in items (i), (ii) and (iii) which are fully analogous to the vacuum case before. All the rest of conditions deal with identifying a number of constants, and then selecting the subset of values for which the spacetime is locally isometric to Kerr-NUT-de Sitter. A full classification of spacetimes satisfying (i), (ii) can be found in [54].

It should be emphasized that this theorem includes vacuum as a particular case. In fact, the proof in [54] applies equally well to the case $\Lambda = 0$. It is instructive to see how the vacuum characterization Theorem 5.1 follows from this theorem. First note that when $\Lambda = 0$, assumption (iii) follows from (ii). Thus, under (i), (ii) and (iv) in Theorem 5.4, the function $V \circ Z : \mathcal{M} \mapsto \mathbb{R}$ is necessarily non-negative everywhere (because $|\nabla Z|^2_g \geq 0$). If we assume $c > 0$ and given that V must be non-negative somewhere (in order to fulfill the condition $V \circ Z \geq 0$), it follows that $V(\zeta)$ must have two real zeroes $\zeta_0 \leq \zeta_1$ and $Z \in [\zeta_0, \zeta_1]$. The polynomial $\hat{V}$ is simply $\hat{V} = c > 0$, so all required conditions hold and the following corollary follows (the only if part holds because the Kerr-NUT class does have $c > 0$).

Corollary 5.5 *Let $(\mathcal{M}, g)$ be a vacuum spacetime admitting a Killing vector ξ with self-dual two-form $\mathcal{F}$. Assume that hypotheses (i), (ii) and (iv) in Theorem 5.4 hold and define the constants b_1, b_2, k and c as in Theorem 5.4). Then the spacetime is locally isometric to the Kerr-NUT spacetime if and only if $c > 0$.*

This result was proved in [53] by exploiting the local characterization Theorem 5.1 for Kerr and the action of a so-called Ehlers transformation group on vacuum spacetimes admitting a Killing vector ξ which, as shown in [53], happens to preserve conditions (i) and (ii).

Concerning vacuum with vanishing NUT parameter, the condition $l = 0$ is equivalent to $\zeta_0 + \zeta_1 = 0$ which, in turn, is equivalent to $b_2 = 0$. Since Eq. (12) with $\Lambda = 0$ is

$$Q^4 \mathcal{F}^2 (b_2 - i b_1)^2 = 36^2$$

it follows that $b_2 = 0$ is equivalent to $Q^4 \mathcal{F}^2$ being real and negative. Thus, the conditions $c > 0$, $\mathrm{Re}(Q^4 \mathcal{F}^2) < 0$, $\mathrm{Im}(Q^4 \mathcal{F}^2) = 0$ and (iv) characterize locally the Kerr metric, and we recover Theorem 5.1.

As already mentioned above, there exists a spacetime that naturally generalizes both the Kerr-NUT-(anti) de Sitter and the Kerr-Newman spacetimes into a single five parametric family of Λ-electrovacuum spacetimes, called *Kerr-Newman-NUT-(anti) de Sitter spacetime*. In view of the various characterization theorems of this section, we conjecture that a local characterization result along these lines also exists for the Kerr-Newman-(A)de Sitter spacetime, which includes both Theorems 5.1, 5.2 and 5.4 as special cases.

5.1 Local Characterization of Kerr Without Killing Vector

The local characterization of the Kerr spacetime and its charged and cosmological constant in the previous section all require the existence of a Killing vector ξ. In Sect. 3.3 a characterization of the Kruskal spacetime without assuming a priori spherical symmetry has been discussed. Similarly, it is of interest to obtain local characterizations of the Kerr spacetime without the need of assuming a priori the existence of a Killing vector. This has been achieved in [27]. In order to state the theorem, we need to introduce some additional notation concerning double two-forms.

We have introduced above the self-dual Weyl tensor $\mathcal{W}eyl_g = \mathrm{Weyl}_g + i\mathrm{Weyl}^\star_g$. A spacetime $(\mathcal{M}, g)$ is of Petrov type D [3, 59] at a point p if there exists a non-zero constant ω such that

$$\mathcal{W}eyl_g^2 + 4\omega\mathcal{W}eyl_g - 32\omega^2\mathcal{I}_g|_p = 0.$$

This equation is equivalent to the existence of a self-dual two-form $\mathcal{U}$ at p satisfying $\mathcal{U}^2 = -2$ and such that

$$\mathcal{W}eyl_g\big|_p = 6\,\omega\left(\mathcal{U}\otimes\mathcal{U} + \frac{2}{3}\mathcal{I}_g\right)\bigg|_p.$$

The two-form $\mathcal{U}$ can be computed algorithmically from $\mathcal{W}eyl_g$ simply by contracting $\mathcal{W}eyl_g - 4\omega\mathcal{I}_g$ with a self-dual two-form $\mathcal{W}$ not lying in its kernel. For a two-form V, its divergence is defined as $(\mathrm{div}V)_\beta := \nabla^\alpha V_{\alpha\beta}$.

Theorem 5.6 (Ferrando and Sáez [27]) *Let $(\mathcal{M}, g)$ be a smooth, vacuum, four-dimensional spacetime. Assume that there exists a smooth, nowhere vanishing function $\omega : \mathcal{M} \to \mathbb{C}$ such that*

$$\mathcal{W}eyl_g^2 + 4\omega\mathcal{W}eyl_g - 32\omega^2\mathcal{I}_g = 0.$$

Let $\mathcal{U}$ be the unique self-dual two-form satisfying

$$\mathcal{U}^2 = -2 \quad \text{and} \quad \mathcal{W}eyl_g = 6\omega\left(\mathcal{U}\otimes\mathcal{U} + \frac{2}{3}\mathcal{I}_g\right). \tag{14}$$

Let $\zeta := \mathrm{div}(\mathcal{U})$ and decompose it in real and imaginary parts $\zeta = \zeta_R + i\zeta_I$. Let $\widehat{\mathcal{M}} := \{p \in \mathcal{M};\ \zeta_R \text{ is non-null}\}$ and assume that $\mathcal{M}\setminus\widehat{\mathcal{M}}$ has empty interior. On $\widehat{\mathcal{M}}$ define $\lambda := \langle\zeta_R, \zeta_I\rangle\langle\zeta_R, \zeta_R\rangle^{-1}$. If the following conditions are satisfied

$$
\begin{aligned}
&(i) && \zeta_R \wedge \zeta_I = 0, \\
&(ii) && Im\left(\overline{\omega}\mathcal{W}eyl_g(\cdot, \zeta_R, \cdot, \zeta_R, \cdot)\right) \neq 0 \quad \textit{on a dense set}, \\
&(iii) && Im(\omega(1 - i\lambda)^3) = 0 \quad \textit{somewhere on } \widehat{\mathcal{M}}, \\
&(iv) && \frac{2Re(\omega)}{3\lambda^2 - 1} - \frac{1}{4}\langle \zeta_R, \zeta_R \rangle > 0 \quad \textit{somewhere on} \widehat{\mathcal{M}},
\end{aligned}
$$

then $(\mathcal{M}, g)$ *is locally isometric to a Kerr spacetime* $(\mathcal{M}_{m,a}, g_{m,a})$ *with non-vanishing specific angular momentum a.*

Note the similarity between (14) and (10). The fundamental difference is that $\mathcal{U}$ is not a priori related to any Killing vector. This condition is replaced by Condition (i), which is of algebraic nature, so much more natural from a local characterization point of view. On the other hand, the assumption $\omega \neq 0$ and (14) impose that the Petrov type is D everywhere. In view of the characterization result in Theorem 5.1, which a priori allows for Petrov degenerations (namely N and 0), it seems plausible that the restriction $\omega \neq 0$ in the hypotheses can be relaxed. Determining whether this is possible, and to which extent, is an interesting open problem that, in my opinion, deserves consideration.

Acknowledgements I would like to thank José M.M. Senovilla for useful comments on the manuscript and to Martín Reiris for various clarifications concerning Theorem 3.5 and for providing the detailed statement of this result.

References

1. S. Alexakis, A.D. Ionescu, S. Klainerman, Uniqueness of smooth stationary black holes in vacuum: small perturbations of the Kerr spaces. Commun. Math. Phys. **299**, 89127 (2010)
2. D.V. Alekseevskii, A.S. Galaev, Two-symmetric Lorentzian manifolds. J. Geom. Phys. **61**, 2331–2340 (2011)
3. Ll. Bel, "Les états de radiation et le problème de l'énergie en relativité générale", *Cahiers de Physique*, **16**, 59–80 (1962). (Reprinted and translated to English in "Radiation States and the Problem of Energy in General Relativity" *General Relativity and Gravitation*, **32**, 2047–2078 (2000))
4. E. Beltrami, Teoria fondamentale degli spazii di curvatura costante. Annali di Matematica Pura ed Applicata **2**, 232–255 (1868)
5. G. Berg, On the Origin of the Concept of an Einstein Space in *Studies in the History of General Relativity*, eds. by J. Eisendtaedt, A.J. Kox. Einstein Studies, vol. 3 (Birkhäuser, 1992), pp. 336–343
6. L. Bianchi, Lezioni di Geometria Differenzialle, vol. I. (Enrico Spoerri, Libraio-Editore, Seconda edizione, Pisa, 1902), p. 348
7. G.D. Birkhoff, *Relativity and Modern Physics* (Harvard University Press, Cambridge, 1923)
8. O.F. Blanco, M. Sánchez, J.M.M. Senovilla, Structure of second-order symmetric Lorentzian manifolds. J. Eur. Math. Soc. (JEMS) **15**, 595–634 (2013)
9. O.F. Blanco, M. Sánchez, J.M.M. Senovilla, Second-order symmetric Lorentzian manifolds II: structure and global properties. J. Phys. Conf. Ser. **314**, 012021 (2011)
10. H.W. Brinkmann, Riemann Spaces Conformal to Einstein Spaces. Mathematische Annalen **91**, 269–278 (1924)

11. H.W. Brinkmann, Einstein spaces which are mapped conformally on each other. Mathematische Annalen **94**, 119–145 (1925)
12. É. Cartan, Sur une classe remarquable d'espaces de Riemann. II. Bulletin de la Société Mathématique de France **114–134**, 55 (1927)
13. B. Carter, Global structure of the Kerr family of gravitational fields. Phys. Rev. **174**, 1559–1571 (1968)
14. P.T. Chruściel, J.L. Costa, M. Heusler, Stationary black holes: uniqueness and beyond. Living Rev. Relat. **15**, 7 (2012), http://www.livingreviews.org/lrr-2012-7, (cited on 03/2015)
15. Z.W. Chong, G.W. Gibbons, H. Lü, C.N. Pope, Separability and Killing tensors in Kerr-Taub-NUT-de Sitter metrics in higher dimensions. Phys. Lett. B **609**, 124–132 (2005)
16. É. Cotton, Sur les variétés à trois dimensions. Annales de la Faculté des Sciences de Toulouse. **II**(1), 385–438 (1899)
17. E.B. Christoffel, Über die Transformation der homogenen Differentialausdrücke zweiten Grades. Journal fur die reine und angewandte Mathematik (Crelle) **70**, 46–70 (1869)
18. M.J. Cole, J.A. Valiente Kroon, Killing spinors as a characterisation of rotating black hole spacetimes. Class. Quantum Gravity **33**, 125019 (21pp) (2016)
19. S. Deser, J. Franklin, Schwarzschild and Birkhoff a la Weyl. Am. J. Phys. **73**, 261–264 (2005)
20. J. Droste, The field of a single centre in Einstein's Theory of gravitation and the motion of a particle in that field, in *Koninklijke Nederlandsche Akademie van Wetenshappen Proceedings*, **19**, 197–221 (1917). (Reprinted in *General Relativity and Gravitation*, **34**, 1545–1563 (2002))
21. J. Ehlers, W. Kundt, Exact solutions of the Gravitational field equations, in *Gravitation: An Introduction to Current Research*, ed. by L. Witten. (Wiley, New York, 1962)
22. J. Ehlers, A. Krasiński, Comment on the paper by J.T. Jebsen reprinted in Gen. Rel. Grav. **37**, 2253–2259. Gen. Relativ. Gravit. **38**, 1329–1330 (2006)
23. J. Eiesland, The group of motions of an Einstein space. Trans. Am. Math. Soc. **27**, 213–245 (1925)
24. J. Eiesland, The group of motions of an Einstein space, paper presented at the Easter Meeting at Chicago, March 25th and 26th, 1921, summarized by A. Dresden in *Bulletin of the American Mathematical Society, 1921* vol. 27, Classic Reprint Series by Americal Mathematical Society, (Published by Forgotten Books 2013) p. 410
25. R. Emparan, H.S. Reall, A rotating black ring in five dimensions. Phys. Rev. Lett. **88**, 101101 (2002)
26. R. Farwell, C. Knee, The missing link: Riemann's "Commentatio", differential geometry and tensor analysis. Historia Mathematica **17**, 223–255 (1990)
27. J.J. Ferrando, D. Sáez, An intrinsic characterization of the Kerr metric. Class. Quantum Gravity **26**, 075013 (2009)
28. J.J. Ferrando, D. Sáez, An intrinsic characterization of spherically symmetric spacetimes. Class. Quantum Gravity **27**, 205024 (2010)
29. J.J. Ferrando, D. Sáez, An intrinsic characterization of the Schwarzschild metric. Class. Quantum Gravity **15**, 1323–1330 (1998)
30. J. Franchi, From Riemannian to Relativistic diffusions. (This volume)
31. A.S. Galaev, Classification of third-order symmetric Lorentzian manifolds (2014). arXiv:1407.3594
32. A. García-Parrado Gómez-Lobo, J. Valiente Kroon, Initial data sets for the Schwarzschild spacetime. Phys. Rev. D, **75**, 024027 (2007)
33. H. Goenner, Einstein tensor and generalizations of Birhoff's theorem. Commun. Math. Phys. **16**, 34–47 (1970)
34. T. Gray, Riemann on geometry, physics, and philosophy-some remarks. (This volume)
35. J.B. Griffiths, J. Podolský, A note on the parameters of the Kerr-NUT-(anti) de Sitter spacetime. Class. Quantum Gravity **24**, 1687–1689 (2007)
36. J.B. Griffiths, J. Podolský, *Exact Space-Times in Einstein's General Relativity*, Cambridge Monographs on Mathematical Physics (Cambridge University Press, Cambridge, 2009)
37. J. Hély, Modèles d'univers en état de radiation pure. Comptes Rendus de l'Acad?ie des Sciences (Paris) **249**, 1867–1868 (1959)

38. A. Hermann, E. Humbert, On the positive mass theorem for closed Riemannian manifolds. (This volume)
39. M. Heusler, *Black Hole Uniqueness Theorems*. Cambridge Lecture Notes in Physics, vol. 6, (Cambrige University Press, 1996)
40. D. Hilbert, Die Grundlagen der Physik. (Zweite Mitteilung), *Königliche Gesellschaft der Wissenschaften zu Göttingen. Mathematisch-Physikalische Klasse. Nachrichten* (1917), pp. 53–76
41. S. Hollands, A. Ishibashi, Black hole uniqueness theorems in higher dimensional spacetimes. Class. Quantum Gravity **29**, 163001 (2012)
42. J.T. Jebsen, Über die allgemeinen kugelsymmetrischen Lösungen der Einsteinschen Gravitationsgleichungen im Vakuum, *Arkiv för Matematik, Astronomi och Fysik (Stockholm)*, **15**, nr. 18 (1921). (English translation in: J. T. Jebsen, "On the general spherically symmetric solutions of Einstein's gravitational equations in vacuo", *General Relativity and Gravitation*, **37**, 2253–2259 (2005))
43. N.V. Johansen, F. Ravndal, On the discovery of Birkhoff's theorem. Gen. Relativ. Gravit. **38**, 537–540 (2006)
44. R.P. Kerr, Gravitational field of a spinning mass as an example of algebraically special metrics. Phys. Rev. Lett. **11**, 237–238 (1963)
45. R.P. Kerr, A. Schild, A new class of vacuum solutions of the Einstein field equations, in *Atti del Congregno Sulla Relativita Generale*, (Galileo Centenario, 1965)
46. M.D. Kruskal, Maximal extension of Schwarzschild metric. Phys. Rev. **119**, 1743–1745 (1960)
47. Lecture published posthumously in 1866 by R. Dedekind in *Abhandlungen der Königlichen Gesellschaft der Wissenschaften zu Göttingen*, **13** (1867)
48. T. Leistner, Screen bundles of Lorentzian manifolds and generalisations of pp-waves. J. Geom. Phys. **56**, 2117–2134 (2006)
49. T. Levi-Civita, ds^2 Einsteiniani in Campi Newtoniani III. Rendiconti della Reale Accademia dei Lincei **27**, 183 (1917)
50. A. Lichnerowicz, Courbure, nombres de Betti, et espaces symétriques, in *Proceedings of the International Congress of Mathematicians*, Cambridge, Mass, 1950, vol. 2 (American Mathematical Society, Providence, R.I. 1952), pp. 216–223
51. M. Mars, A spacetime characterization of the Kerr metric. Class. Quantum Gravity **16**, 2507–2523 (1999)
52. M. Mars, Uniqueness properties of the Kerr metric. Class. Quantum Gravity **17**, 3353–3374 (2000)
53. M. Mars, Spacetime Ehlers group: transformation law for the Weyl tensor. Class. Quantum Gravity **18**, 719–738 (2001)
54. M. Mars, J.M.M. Senovilla, A spacetime characterization of the Kerr-NUT-(A)de Sitter and related metrics. Annales Henri Poincaré **16**, 1509–1550 (2015)
55. M. Mars, J.M.M. Senovilla, Spacetime characterizations of Λ-vacuum metrics with a null Killing 2-form. Class. Quantum Gravity **33**, 195004 (2016)
56. J.-P. Nicolas, The conformal approach to asymptotic analysis. (This volume)
57. K. Nomizu, H. Ozeki, A theorem on curvature tensor fields. Proc. Natl. Acad. Sci. (USA) **48**, 206–207 (1962)
58. A. Peres, Some gravitational waves. Phys. Rev. Lett. **3**, 571–572 (1959)
59. A.Z. Petrov, Klassifikacya prostranstv opredelyayushchikh polya tyagoteniya, Uchenye Zapiski Kazanskogo Gosudarstvennogo Universiteta im. V.I. Ulyanova-Lenina [Scientific Proceedings of Kazan State University, named after V.I. Ulyanov-Lenin], **114**, 55–69 (1954). (Reprinted and translated to English in "The Classification of Spaces Defining Gravitational Fields" *General Relativity and Gravitation*, **32**, 1665–1685 (2000))
60. M. Reiris, Static solutions from the point of view of comparison geometry. J. Math. Phys. **53**, 012501 (2012)
61. M. Reiris, unpublished
62. B. Riemann, *Collected Works of Bernhard Riemann (Gesammelte Mathematische Werke)*, 2nd edn. and the Supplement, (ed.) by H. Weber, (Dover Publications, New York, 1953)

63. R. Schimming, Riemannsche Räume mit ebenfrontiger und mit ebener Symmetrie. Mathematische Nachrichten **59**, 129–162 (1974)
64. H.-J. Schmidt, A new proof of Birkhoff's theorem. Gravit. Cosmol. **3**, 185–190 (1997)
65. H.-J. Schmidt, The tetralogy of Birkhoff theorems. Gen. Relativ. Gravit. **45**, 395410 (2013)
66. J.A. Schouten, Über die konforme Abbildung n-dimensionaler Mannig- faltigkeiten mit quadratischer Massbestimmung auf eine Mannigfaltigkeit mit euklidischer Massbestimmung. Mathematische Zeitschrift **11**, 58–88 (1921)
67. K. Schwarzschild, Über das Gravitationsfeld eines Massenpunktes nach der Einsteinschen Theorie, *Sitzungsberichte der Königlich Preussischen Akademie der Wissenschaften*, **3**, 189–196 (1916). (English translation in K. Schwarzschild, "On the Gravitational Field of a Mass Point According to Einstein's Theory", *General Relativity and Gravitation*, **35**, 951–959 (2003))
68. K. Schleich, D.M. Witt, A simple proof of Birkhoff's theorem for cosmological constant. J. Math. Phys. **51**, 112502 (2010)
69. E. Schmutzer, *Relativistische Physik* (Teubner-Verlag, Leipzig, 1968)
70. F. Schur, Über den Zusammenhang der Raume konstanten Krummungs- masses mit den projektiven Raumen. Mathematische Annalen **27**, 537–567 (1886)
71. J.M.M. Senovilla, Second-order symmetric Lorentzian manifolds. I. Characterization and general results. Class. Quantum Gravity **25**, 245010 (2008)
72. W. Simon, Characterizations of the Kerr metric. Gen. Relativ. Gravit. **16**, 465–476 (1984)
73. M. Spivak, *A Comprehensive Introduction to Differential Geometry*, vol. 2, 3rd edn. (Publish or Perish Inc., Houston, Texas, 1999)
74. H. Stephani, D. Kramer, M. Maccallum, C. Hoenselaers, E. Herlt, *Exact Solutions of Einsteins Field Equations*, 2nd Edn. (Cambridge University Press, 2003)
75. G. Szekeres, On the singularities of a Riemannian manifold, *Publicationes Mathematicae Debrecen*, **7**, 285–301 (1960). (Reprinted in *General Relativity and Gravitation* **34**, 2001–2016 (2002))
76. A.S. Tahvildar-Zadeh, On the static spacetime of a single point charge. Rev. Math. Phys. **23**, 309–346 (2011)
77. F.R. Tangherlini, Schwarzschild field in n dimensions and the dimensionality of space problem. Il Nuovo Cimento **27**, 636–651 (1963)
78. S. Tanno, Curvature tensors and covariant derivatives. Annali di Matematica Pura et Applicata **96**, 233–241 (1972)
79. H. Weyl, Reine Infinitesimalgeometrie. Mathematische Zeitschrift **2**, 384–411 (1918)
80. W.W-Y. Wong, A space-time characterization of the Kerr-Newman metric. Annales Henri Poincaré, **10**, 453–484 (2009)
81. V.D. Zakharov, *Gravitational waves in Einstein's theory*, (Hallsted Press, John Wiley & Sons Inc., New York, 1973). (Chapter 10 and references therein)

The Conformal Approach to Asymptotic Analysis

Jean-Philippe Nicolas

Abstract Albert Einstein's general theory of relativity is a geometric theory of gravity, using the framework of Lorentzian geometry: an extension of Riemannian geometry in which space and time are united in a real 4-dimensional manifold endowed with an indefinite metric of signature $(1,3)$ or $(3,1)$. The metric allows to distinguish between timelike and spacelike directions in an intrinsic manner and, provides a description of gravity via its curvature. The introduction by Minkowski in 1908 of the notion of spacetime was a decisive change of viewpoint which opened the road for Einstein to develop the geometrical framework for the fully covariant theory he was after. Instead of discussing the history of this development and the crucial influence of Riemannian geometry through the help of Marcel Grossmann, this essay explores Roger Penrose's approach to general relativity which bears a remarkable kindred of spirit with Einstein's and perpetuates the geometrical view of the universe initiated by Riemann and Einstein. More specifically, Penrose's approach to asymptotic analysis in general relativity, which is based on conformal geometric techniques, is presented through historical and recent aspects of two specialized topics: conformal scattering and peeling. Other essays in this volume are related to general relativity: Jacques Franchi [15] discusses relativistic analogues of the Brownian motion on various Lorentzian manifolds; Andreas Hermann and Emmanuel Humbert [23] discuss the positive mass theorem, which is closely related to the Yamabe problem in Riemannian geometry; Marc Mars [28] presents some local intrinsic ways of characterizing a spacetime.

1 Introduction

Hermann Minkowski, in his famous speech at the 80th Assembly of German Natural Scientists and Physicians in Köln in 1908, cast in a rather emphatic way the mould for what would, from then on, be the framework of relativistic mathematical physics:

J.-P. Nicolas (✉)
Department of Mathematics, University of Brest, 6 avenue Victor Le Gorgeu, 29200 Brest, France
e-mail: Jean-Philippe.Nicolas@univ-brest.fr

L. Ji et al. (eds.), *From Riemann to Differential Geometry and Relativity*,
DOI 10.1007/978-3-319-60039-0_19

"The views of space and time which I wish to lay before you have sprung from the soil of experimental physics, and therein lies their strength. They are radical. Henceforth space by itself, and time by itself, are doomed to fade away into mere shadows, and only a kind of union of the two will preserve an independent reality." The last sentence is the founding principle of the geometrical description of special relativity and by extension one of the founding principles of general relativity. The new framework was to retain the essential features of Riemannian geometry, but to incorporate time into the picture and provide an indefinite metric as an intrinsic geometrical way of comparing the speed of a particle to that of light. Minkowski was not merely attempting to introduce a convenient geometrical framework for special relativity, one which was to become known as the Minkowski spacetime. He was advocating, *in view of physical evidence*, to give up the notion of simultaneity; to give up the picture of the universe as a space that changes with time; to replace it by a spacetime that is not to be understood as the succession of instants glued together, but as a global object.

When in 1917 Karl Schwarzschild discovered his famous static spherically symmetric solution of the Einstein vacuum equations, it was misunderstood as having a spherical singularity. It is precisely the change of viewpoint urged by Minkowski that allowed to replace the Schwarzschild coordinate system, tied in with a foliation by spacelike slices in the exterior region, by the Eddington-Finkelstein coordinates, based on principal null geodesics,[1] thus leading to our present understanding of the nature of the event horizon. But of course, the nature of a truly 4-dimensional reality goes against our intuition and our experience of space and time. This is probably the reason why even though we use spacetimes every day in relativity, we are also tempted to do away with a founding principle by breaking them into a succession of spaces, what we call a $3+1$, or an $n+1$, decomposition. We are attached to such notions of simultaneity. The Cauchy problem, the related notion of global hyperbolicity, the constraints for the Einstein equations or for other overdetermined systems of equations such as Maxwell, Rarita-Schwinger, are all based on this type of decomposition. In fact, and I mention this at the risk of appearing dogmatic, even the choice of signature $-+++$, as opposed to $+---$, is prompted by similar motivations: the reason why people choose the latter is usually because the metric allows to measure the proper time along causal curves (i.e. curves that are timelike or null at each point), which is in the spirit of Minkowski's geometrical approach; the standard reason for choosing the former is that the induced metric on a spacelike slice is positive, rather than negative, definite.

The objects that our senses, our eyes in particular, allow us to observe in nature,[2] are causal and in most cases in fact null. Of course we have access to spacelike structures but as cuts of causal objects, a topological spacelike sphere as a cut of

[1] These coordinate systems are now called Eddington-Finkelstein coordinates, as they were discovered first by Eddington in 1924 then re-discovered by Finkelstein in 1958.

[2] A question that is related to that of observability is whether or not it is possible to identify a spacetime by local observations; some spacetimes can be locally or pseudo-locally recognized, see the discussion and examples in the article by Mars [28] in this volume.

a light-cone for example, and these cuts have a degree of arbitrariness. In a truly 4-dimensional approach to general relativity, it seems that causal objects should be given a certain preference. As an illustration, the Cauchy problem can be replaced by the Goursat problem. The Cauchy problem in the framework of general relativity is the usual one for hyperbolic partial differential equations, where the data is set on a Cauchy hypersurface, i.e. a spacelike hypersurface such that any inextendible timelike curve intersects it exactly once, in other words, a slice of simultaneity. The Goursat problem is similar to the Cauchy problem with the important difference that the data are now set on a null hypersurface, typically a null cone. One must be wary of a "perversion" of this notion whereby the initial null hypersurface is the union of two intersecting null slices meeting on a spacelike submanifold. This type of hypersurface, insofar as it is based on a spacelike structure, is not more natural than a Cauchy hypersurface. The reason why this type of problem is considered at all is because of its relative simplicity compared to the light-cone case; the singularity of the hypersurface is very mild by comparison and the situation has advantageous similarities with the $1 + 1$ dimensional case. The drawback of the Goursat problem on a lightcone is that it is usually a local problem in the neighbourhood of the vertex, lightcones in generic spacetimes tending to develop caustics. Scattering theory is a global problem that can be understood as an analogue of the Goursat problem for a light-cone at infinity. The whole evolution of the field is then summarized in an operator acting between null asymptotic structures, by-passing the Cauchy problem as an unnecessary intermediate stage.

Some of us are less than others prone to thinking in $3 + 1$ dimensions. Roger Penrose certainly seems to have remained very faithful to Minkowski's viewpoint. His view of relativity appears to be truly 4-dimensional. He has not systematically avoided $3 + 1$ decompositions and indeed has had major inputs within this approach, particularly in relation to the notion of mass (see the survey by Hermann and Humbert in this volume [23] which discusses Penrose's positive mass theorem and its remarkable role in the resolution of the a priori unrelated Yamabe problem in Riemannian geometry), but a very significant proportion of his research is concerned with the light-cone structure (i.e. the conformal structure) of spacetime. His geometrical ideas have yielded new methods for analysis in general relativity. In this paper, I will focus on his notion of conformal compactification and how it can be used to study two types of asymptotic analytic questions: peeling and scattering.

This paper is organized as follows. Section 2 is devoted to the description of the principles of conformal compactification, the explicit treatment of the case of Minkowski spacetime and how it provides asymptotic information on a large class of solutions of conformally invariant field equations. Section 3 starts with a historical presentation of the notion of peeling, proposes a different way of looking at the question and an alternative approach to studying it, which are strongly inspired by the ideas of Penrose, and finally presents the recent results in the field. Scattering, or rather a version of scattering based on conformal compactifications, is the object of Sect. 4: the history of the topic is described from the founding idea by Penrose to the first actual construction by F. G. Friedlander; a new approach is proposed which, by giving up some of the analytic niceties of Friedlander's results, allows to extend the construction to much more general situations; finally the recent results following this approach are reviewed. Section 5 contains concluding remarks.

2 Conformal Compactification

2.1 The Principles of Conformal Compactification

The notion of conformal compactification in general relativity was introduced by Penrose in a short note [35] in Physical Review Letters in 1963. Its usage as a tool for studying asymptotic properties is clearly mentioned but not developed. The year after, in Les Houches, he gave a series of three lectures [36] explaining the technique in details and the differences depending on the sign of the cosmological constant Λ. In 1965, specializing to the case where $\Lambda = 0$, he published a long and thorough study of the asymptotic behaviour of zero rest-mass fields by means of the conformal technique [37]. Another reference where a clear and detailed description of the method can be found is Spinors and Spacetime vol. 2 [38].

There are two essential ingredients. The first is a geometrical construction: the conformal compactification itself. It can be presented in a very general manner as follows.

- The "physical" spacetime is the spacetime on which we wish to study asymptotic properties, of test fields for example. It is a smooth, 4-dimensional, real, Lorentzian manifold $(\mathcal{M}, g)$.
- The "unphysical," or "compactified," spacetime is a smooth manifold $\bar{\mathcal{M}}$ with boundary $\mathcal{B}$ and interior $\mathcal{M}$.
- The link between the boundary and the physical spacetime is provided by a boundary defining function Ω; it is a positive function on $\mathcal{M}$, smooth on $\bar{\mathcal{M}}$, such that $\Omega|_{\mathcal{B}} = 0$ and $\mathrm{d}\Omega|_{\mathcal{B}} \neq 0$.
- The metric $\hat{g} := \Omega^2 g$ extends as a smooth non degenerate Lorentzian metric on $\bar{\mathcal{M}}$ (hence the name "conformal factor" for Ω).

This conformal "compactification"[3] is not always possible. The property for a spacetime to admit a smooth conformal compactification can be characterized in terms of the decay of the Weyl curvature at infinity. When such a compactification exists, the boundary $\mathcal{B}$ will have a structure: different parts corresponding to different ways of going to infinity in the physical spacetime. Different parts of the boundary will play a role in relation to different types of asymptotic properties: for example timelike decay and scattering, when studied by means of the conformal method, will not involve the same parts of the conformal boundary.

[3]The word compactification is a little misleading since in general the unphysical spacetime will not be compact, there will be holes in the boundary. Only in exceptional cases such as Minkowski spacetime will the rescaled spacetime be compact.

The second ingredient is the equation we wish to study on the physical spacetime. It is important that it admits some rather explicit transformation law under conformal rescalings, so that we can study it on the rescaled spacetime and gain information on its behaviour in the physical spacetime. Conformally invariant equations are the natural class to consider, but not the only possible class.

2.2 Conformal Compactification of Minkowski Spacetime

Let us now present the conformal method in more detail on an explicit example: the simple case of the wave equation on flat spacetime. The contents of this section, and much more, can be found in [37].

2.2.1 The Geometrical Construction

The Minkowski metric in spherical coordinates is expressed as

$$\eta = \mathrm{d}t^2 - \mathrm{d}r^2 - r^2 \mathrm{d}\omega^2 \,, \; \mathrm{d}\omega^2 = \mathrm{d}\theta^2 + \sin^2\theta \, \mathrm{d}\varphi^2 \,.$$

We choose the advanced and retarded coordinates

$$u = t - r \,, \; v = t + r \,. \tag{1}$$

The metric η in terms of these coordinates takes the form

$$\eta = \mathrm{d}u \mathrm{d}v - \frac{(v-u)^2}{4} \mathrm{d}\omega^2 \,, \; v - u \geq 0 \,.$$

Then we introduce new null coordinates that allow to describe the whole of Minkowski spacetime as a bounded domain:

$$p = \arctan u \,, \; q = \arctan v \,. \tag{2}$$

The expression of the Minkowski metric in the coordinates (p, q, ω) is given by

$$\eta = (1+u^2)(1+v^2) \mathrm{d}p \, \mathrm{d}q - \frac{(v-u)^2}{4} \mathrm{d}\omega^2 \,.$$

From p and q we can define new time and space coordinates as follows,

$$\begin{aligned} \tau &= p + q = \arctan(t-r) + \arctan(t+r) \,, \\ \zeta &= q - p = \arctan(t+r) - \arctan(t-r) \,, \end{aligned} \tag{3}$$

and we get

$$\eta = \frac{(1+u^2)(1+v^2)}{4}\left(\mathrm{d}\tau^2 - \mathrm{d}\zeta^2\right) - \frac{(v-u)^2}{4}\mathrm{d}\omega^2 \,.$$

We choose the conformal factor

$$\Omega = \frac{2}{\sqrt{(1+u^2)(1+v^2)}} = \frac{2}{\sqrt{(1+\tan^2 p)(1+\tan^2 q)}} = 2\cos p \cos q \,, \qquad (4)$$

i.e. we rescale the metric by $1/r^2$ in null directions, $1/r^4$ in spacelike directions and $1/t^4$ in timelike directions. We obtain

$$\begin{aligned}
\mathfrak{e} := \Omega^2\eta &= \mathrm{d}\tau^2 - \mathrm{d}\zeta^2 - \frac{(v-u)^2}{(1+u^2)(1+v^2)}\mathrm{d}\omega^2 \\
&= \mathrm{d}\tau^2 - \mathrm{d}\zeta^2 - ((\tan q - \tan p)\cos p \cos q)^2 \,\mathrm{d}\omega^2 \\
&= \mathrm{d}\tau^2 - \mathrm{d}\zeta^2 - (\sin q \cos p - \sin p \cos q)^2 \,\mathrm{d}\omega^2 \\
&= \mathrm{d}\tau^2 - \mathrm{d}\zeta^2 - (\sin(q-p))^2 \,\mathrm{d}\omega^2 \\
&= \mathrm{d}\tau^2 - \mathrm{d}\zeta^2 - (\sin\zeta)^2 \,\mathrm{d}\omega^2 \\
&= \mathrm{d}\tau^2 - \sigma^2_{S^3} \,,
\end{aligned}$$

where $\sigma^2_{S^3}$ is the euclidean metric on the 3-sphere. Minkowski spacetime is now described as the diamond

$$\mathbb{M} = \{|\tau| + \zeta < \pi \,,\ \zeta \geq 0 \,,\ \omega \in S^2\} \,.$$

The metric $\mathfrak{e}$ is the Einstein metric; it extends analytically to the whole Einstein cylinder $\mathfrak{E} = \mathbb{R}_\tau \times S^3_{\zeta,\theta,\varphi}$. The full conformal boundary of Minkowski spacetime can be defined in this framework. It is described as

$$\partial\mathbb{M} = \{|\tau| + \zeta = \pi \,,\ \zeta \geq 0 \,, \omega \in S^2\} \,.$$

Several parts can be distinguished.

- Future and past null infinities:

$$\mathscr{I}^+ = \left\{(\tau\,,\ \zeta\,,\ \omega)\,;\ \tau + \zeta = \pi\,,\ \zeta \in]0, \pi[\,,\ \omega \in S^2\right\} \,,$$
$$\mathscr{I}^- = \left\{(\tau\,,\ \zeta\,,\ \omega)\,;\ \zeta - \tau = \pi\,,\ \zeta \in]0, \pi[\,,\ \omega \in S^2\right\} \,.$$

Proposition 2.1 *The hypersurfaces $\mathscr{I}^\pm$ are smooth null hypersurfaces for $\mathfrak{e}$. Their null generators are respectively the vector fields*

$$\partial_\tau - \partial_\zeta \text{ for } \mathscr{I}^+ \text{ and } \partial_\tau + \partial_\zeta \text{ for } \mathscr{I}^- \,.$$

Proof They are clearly smooth hypersurfaces since $\mathfrak{e}$ is analytic up to $\mathscr{I}^{\pm}$ and does not degenerate there: its determinant

$$\det(\mathfrak{e}) = -\sin^4\zeta\,.\sin^2\theta$$

does not vanish on $\mathscr{I}^{\pm}$ (except for the usual singularity due to spherical coordinates). Now the vector fields $\partial_\tau - \partial_\zeta$ and $\partial_\tau + \partial_\zeta$ are null and tangent respectively to $\mathscr{I}^+$ and $\mathscr{I}^-$. They are orthogonal to the two other generators of $\mathscr{I}^{\pm}$: ∂_θ and ∂_φ. They are therefore normal to $\mathscr{I}^+$ and $\mathscr{I}^-$ respectively. This proves the proposition. □

- Future and past timelike infinities:

$$i^{\pm} = \left\{(\tau = \pm\pi\,,\ \zeta = 0\,,\ \omega)\,;\ \omega \in S^2\right\}\,.$$

  They are smooth points for $\mathfrak{e}$ (2-spheres whose area is zero because they correspond to $\zeta = 0$).
- Spacelike infinity:

$$i^0 = \left\{(\tau = 0\,,\ \zeta = \pi\,,\ \omega)\,;\ \omega \in S^2\right\}\,.$$

  It is also a smooth point for $\mathfrak{e}$.

Note that the Einstein spacetime $(\mathfrak{E}, \mathfrak{e})$ is static: in the coordinates $\tau, \zeta, \theta, \varphi$, it is obvious that ∂_τ is a global timelike Killing vector field,[4] orthogonal to the level hypersurfaces of τ, which are 3-spheres.

In Fig. 1, the conformal boundary of Minkowski spacetime is represented with its different parts. In Fig. 2, we display the Penrose diagram of compactified Minkowski spacetime, i.e. a representation of $\overline{\mathbb{M}} = \mathbb{M} \cup \partial\mathbb{M}$ quotiented by the group of isometries inherited from the group of rotations in $\mathbb{M}$: the spherical degrees of freedom do not appear, the advantage is that the causal structure is clearly readable on the resulting 2-dimensional diagram.

2.2.2 An Application for a Conformally Invariant Equation

Let us consider a simple example of conformally invariant equation, the conformal wave equation

$$(\Box_g + \frac{1}{6}\mathrm{Scal}_g)\phi = 0\,. \tag{5}$$

Its conformal invariance can be expressed precisely as follows: we consider a spacetime $(\mathcal{M}, g)$ and a metric $\hat{g}$ in the conformal class of g with conformal factor Ω, i.e. $\hat{g} = \Omega^2 g$. Then we have the equality of operators acting on scalar fields on $\mathcal{M}$

[4] A Killing vector field is a vector field K^a whose flow is an isometry, i.e. $\mathcal{L}_K g_{ab} = 0$, which is equivalent to the Killing equation $\nabla^{(a}K^{b)} = 0$.

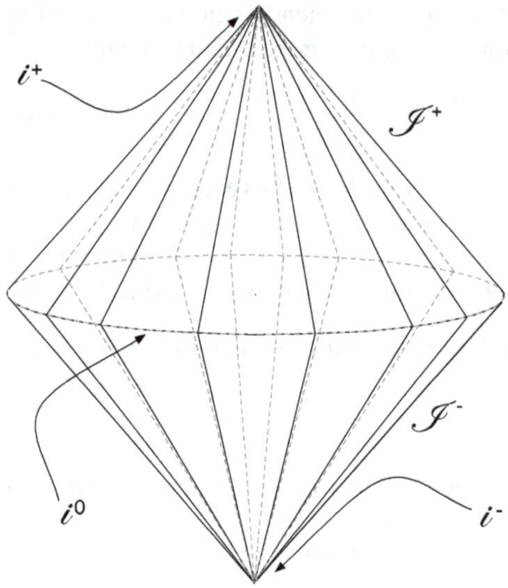

Fig. 1 Compactified Minkowski spacetime, i^0 is merely a point, just like $i^\pm$

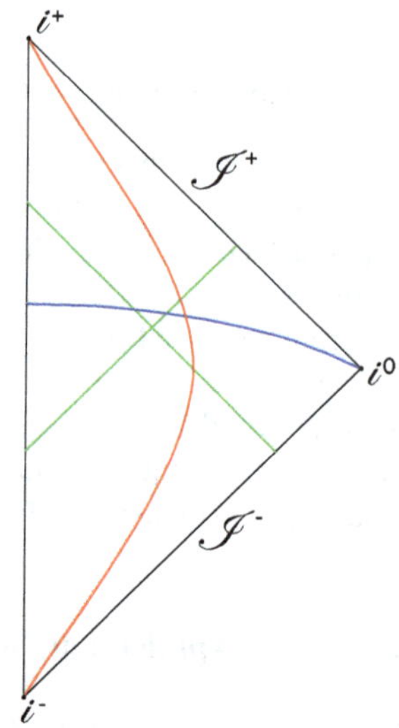

Fig. 2 Penrose diagram of compactified Minkowski spacetime with spacelike, timelike and null geodesics

$$\Box_g + \frac{1}{6}\mathrm{Scal}_g = \Omega^3 \left(\Box_{\hat{g}} + \frac{1}{6}\mathrm{Scal}_{\hat{g}} \right) \Omega^{-1} , \qquad (6)$$

which also entails the expression of the scalar curvature of $\hat{g}$ in terms of that of g

$$\mathrm{Scal}_{\hat{g}} = \Omega^{-2}\mathrm{Scal}_g + 6\Omega^{-3}\Box_g \Omega . \qquad (7)$$

Minkowski spacetime is flat, its scalar curvature vanishes, whereas the scalar curvature of the metric $\mathfrak{e}$ is equal to 6. Hence, a distribution $\phi \in \mathcal{D}'(\mathbb{R}^4)$ satisfies the wave equation

$$\partial_t^2 \phi - \Delta \phi = 0 \,, \tag{8}$$

if and only if $\tilde{\phi} := \Omega^{-1}\phi$ (Ω defined by (4)) satisfies

$$\Box_{\mathfrak{e}} \tilde{\phi} + \tilde{\phi} = 0 \,, \tag{9}$$

where

$$\Box_{\mathfrak{e}} = \partial_\tau^2 - \Delta_{S^3} \,.$$

Since the Einstein cylinder is globally hyperbolic,[5] for smooth data on S^3, the Cauchy problem for (9) has a unique smooth solution on the whole of $\mathbb{R}_\tau \times S^3$ (see Leray [27]). Let us consider data ϕ_0 , ϕ_1 at $t = 0$ for the Cauchy problem for (8), i.e.

$$\phi_0 = \phi|_{t=0} \,, \; \phi_1 = \partial_t \phi|_{t=0} \,. \tag{10}$$

From these, we can easily calculate the corresponding data at $\tau = 0$ for $\tilde{\phi}$, i.e.

$$\tilde{\phi}_0 = \tilde{\phi}|_{\tau=0} \,, \; \tilde{\phi}_1 = \partial_\tau \tilde{\phi}|_{\tau=0} \,, \tag{11}$$

using the fact that on $\mathbb{M}$, $t = 0$ is equivalent to $\tau = 0$. First it is immediate that

$$\tilde{\phi}_0 = (\Omega|_{t=0})^{-1}\, \phi_0 = \frac{1+r^2}{2}\phi_0 \,.$$

Concerning the other part of the data,

$$\partial_t \phi = (\partial_t \Omega)\, \tilde{\phi} + \Omega \frac{\partial \tau}{\partial t} \partial_\tau \tilde{\phi} + \Omega \frac{\partial \zeta}{\partial t} \partial_\zeta \tilde{\phi}$$

and

$$\frac{\partial \Omega}{\partial t}|_{t=0} = 0 \,, \; \frac{\partial \tau}{\partial t}|_{t=0} = \frac{2}{1+r^2} \,, \; \frac{\partial \zeta}{\partial t}|_{t=0} = 0 \,,$$

which gives

$$\partial_t \phi_{|_{t=0}} = \frac{4}{(1+r^2)^2} \partial_\tau \tilde{\phi}_{|_{\tau=0}} = \frac{4}{(1+r^2)^2} \tilde{\phi}_1 \,.$$

Hence the relation between the data (11) for the rescaled field and (10) for the physical field:

$$\tilde{\phi}_0 = \frac{1+r^2}{2}\phi_0 \,, \; \tilde{\phi}_1 = \frac{(1+r^2)^2}{4}\phi_1 \,. \tag{12}$$

[5] A spacetime is said to be globally hyperbolic if it admits a Cauchy hypersurface.

Let us make the assumption that $\tilde{\phi}_0$ and $\tilde{\phi}_1$, which are naturally defined only on the 3-sphere at $\tau = 0$ with the point i^0 removed, extend as smooth functions on S^3. Then the rescaled solution $\tilde{\phi} = \Omega^{-1}\phi$ extends as a smooth function on $\overline{\mathbb{M}}$ and by simple explicit calculations, we can infer precise pointwise decay rates of the unrescaled field in all causal directions. All we need is to use the continuity of $\tilde{\phi}$ at the boundary of $\overline{\mathbb{M}}$ and the behaviour of the conformal factor Ω along null and timelike geodesics (easily obtained from (4)).

1. **Decay along null directions**. There exist smooth functions $\tilde{\phi}^{\pm} \in \mathcal{C}^\infty(\mathbb{R} \times S^2)$ such that for all u, v, ω,

$$\lim_{r\to+\infty} r\phi(t = r + u, r, \omega) = \frac{1}{\sqrt{1+u^2}}\tilde{\phi}^+(u,\omega)\,, \tag{13}$$

$$\lim_{r\to+\infty} r\phi(t = -r + v, r, \omega) = \frac{1}{\sqrt{1+v^2}}\tilde{\phi}^-(v,\omega)\,. \tag{14}$$

   The functions $\tilde{\phi}^{\pm}$ are simply the restrictions of $\tilde{\phi}$ on $\mathscr{I}^{\pm}$; the two limits above are referred to as the future and past asymptotic profiles, or radiation fields, of ϕ.
2. **Decay along timelike directions**. There exist two constants $C^{\pm}$ such that for all r, ω,

$$\lim_{t\to\pm\infty} t^2\phi(t, r, \omega) = 2C^{\pm}\,.$$

   These constants are $C^{\pm} = \tilde{\phi}(i^{\pm})$ (recall that $i^{\pm}$ are points on the Einstein cylinder, not 2-spheres).

In other words, the physical solution ϕ decays like $1/r$ along radial null geodesics and like $1/t^2$ along the integral lines of ∂_t. These decay rates are valid for solutions ϕ of the wave equation on Minkowski spacetime such that $\tilde{\phi} = \Omega^{-1}\phi$ extends as a smooth function on $\mathfrak{E}$. Implicit in this hypothesis are some requirements on the fall-off of initial data for ϕ. Using (12), the smoothness of $\tilde{\phi}_0$ and $\tilde{\phi}_1$ on S^3 entails that there exist two constants C_0, C_1 such that for all ω,

$$\lim_{r\to+\infty} r^2\phi(0, r, \omega) = 2C_0\,,$$

$$\lim_{r\to+\infty} r^4\partial_t\phi(0, r, \omega) = 4C_1\,,$$

the constants C_0 and C_1 being the respective values of $\tilde{\phi}_0$ and $\tilde{\phi}_1$ at i^0 (which, like $i^{\pm}$ is a point on $\mathfrak{E}$).

The crucial observation which allowed us to derive the above decay rates is that the information on the pointwise decay of ϕ at infinity is equivalent to the continuity of the rescaled field at the conformal boundary. It is possible to go further. In the next two sections, we present two refinements of this first use of conformal compactification for asymptotic analysis.

3 Peeling

The peeling, or peeling-off of principal null directions, is a generic asymptotic behaviour discovered by R. Sachs for spin 1 and 2 fields in the flat case [39] and in the asymptotically flat case [40]. A zero-rest-mass field of spin s is described as a symmetric spinor of rank $2s$. Such an object possesses $2s$ principal null directions at each point, which are analogous to the roots of a polynomial of degree $2s$. An outgoing zero rest-mass field of spin s, along a null geodesic going out to infinity, can be expressed as an expansion in powers of $1/r$. This expansion is such that the part of the field falling-off like r^{-k}, $1 \leq k \leq 2s$, has $2s - k$ of its principal null directions aligned along the null geodesic. The notion was explored further by Newman and Penrose [33] and by Penrose [35, 37], using the spin-coefficient formalism (now referred to as the Newman-Penrose formalism) and the conformal method. In [37], the conformal method is used in order to show that the peeling-off of principal null directions is equivalent to a very simple property: the boundedness of the rescaled field at null infinity. The question I will focus my attention on is that of the genericity of the peeling behaviour. It is a delicate question which remained controversial for some years. The reason for this controversy was the logarithmic divergence observed when comparing the asymptotic structures of Minkowski and Schwarzschild spacetimes. This is expected to be generic since a physically relevant asymptotically flat spacetime ought to be a short-range perturbation of a Schwarzschild spacetime. The question was however answered in the affirmative in [30, 31] by treating the Schwarzschild case. The key idea was to reformulate the peeling property in terms of energy estimates instead of pointwise behaviour along outgoing null geodesics. This section is a description of these results and of the path that led to them.

3.1 A New Approach to the Peeling

As we saw above, the peeling in its original form is equivalent to the boundedness of the rescaled field at null infinity. It is naturally tempting to define a notion of peeling of higher order, corresponding to higher degrees of regularity at the conformal boundary. This is exactly what Penrose did when he defined k-asymptotically simple spacetimes as a generic model for asymptotic flatness (see [38] Vol. 2). A 4-dimensional, globally hyperbolic, Lorentzian space-time $(\mathscr{M}, g)$, $\mathscr{M} \simeq \mathbb{R}^4$, is called k asymptotically simple if there exists a globally hyperbolic C^{k+1} Lorentzian manifold $(\hat{\mathscr{M}}, \hat{g})$ with boundary $\mathscr{I}$ and a scalar field Ω on $\hat{\mathscr{M}}$ such that:

(i) $\mathscr{M}$ is the interior of $\hat{\mathscr{M}}$;
(ii) $\hat{g}_{ab} = \Omega^2 g_{ab}$ on $\mathscr{M}$;
(iii) Ω and $\hat{g}$ are C^k on $\hat{\mathscr{M}}$;
(iv) $\Omega > 0$ on $\mathscr{M}$; $\Omega = 0$ and $\mathrm{d}\Omega \neq 0$ on $\mathscr{I}$;
(v) every null geodesic in $\mathscr{M}$ acquires a future endpoint on $\mathscr{I}^+$ and a past endpoint on $\mathscr{I}^-$.

However, when defining peeling of higher order for zero rest-mass fields, it is not convenient to use $\mathcal{C}^k$ spaces to characterize their regularity at $\mathscr{I}$. The reason is that the Cauchy problem for hyperbolic equations is not well-posed in $\mathcal{C}^k$ spaces.

In a large portion of the litterature concerned with the peeling and particularly its genericity, what seems to have been lacking is a precise definition of what one is really trying to prove or disprove. For example, saying that on a given asymptotically flat spacetime there is no peeling would not make much sense, because, unless the asymptotic flatness is too week to even define $\mathscr{I}$ as a conformal boundary, the chances are that there will always be conformally rescaled fields that extend continuously at $\mathscr{I}$; smooth compactly supported initial data would usually guarantee such a behaviour. Also finding examples of data for which the rescaled field does not extend continuously at $\mathscr{I}$ is no proof that the asymptotic structure of the spacetime is radically different from that of Minkowski spacetime. Indeed, in the flat case, if we take for the wave equation smooth initial data that are, say, exponentially increasing at spacelike infinity, the rescaled field will not even be bounded at $\mathscr{I}$. An important information however, would be, on a given asymptotically flat spacetime, to have data such that the associated solution does not peel, but whose regularity and decay at spacelike infinity is enough to entail peeling in Minkowski spacetime. As we shall see below, such a situation is however very unlikely, certainly it is not possible on the Schwarzschild metric for the wave equation, Dirac or Maxwell fields. Here is a precise way of addressing the question of the genericity of the peeling, or rather its higher order version, in the form of a scheme in two steps:

Step 1. Characterize, on a given asymptotically flat spacetime, the class of data that ensures a given regularity of the rescaled field at $\mathscr{I}$.
Step 2. Compare such classes between different spacetimes, in particular, compare them with the corresponding classes in the case of Minkowski spacetime.

It remains to decide how to measure the regularity of the rescaled field at $\mathscr{I}$ and how to proceed to obtain the optimal classes of data ensuring such regularity. I shall adopt four main guiding principles to do so.

1. **Work on the compactified spacetime**: formulate the peeling in terms of regularity at $\mathscr{I}$, not in terms of an asymptotic expansion along null directions. The reason for this is that pointwise regularity at $\mathscr{I}$ can be precisely controlled, without loss of information, in terms of regularity and decay of initial data. This is less true of the asymptotic behaviour of fields described in terms of finite asymptotic expansions, particularly when we are after a one to one correspondance between an order of expansion and a class of data.
2. **Work in a neighbourhood of spacelike infinity**. Once the regularity is established at $\mathscr{I}$ near spacelike infinity, it can easily be inferred further up $\mathscr{I}$ provided the solution is smooth enough in the bulk. What we wish to avoid is singularities creeping up $\mathscr{I}$ due to insufficient decay assumptions on the data.
3. **Avoid the use of $\mathcal{C}^k$ spaces**. The first natural idea is to consider that a field peels at order k if the rescaled field is $\mathcal{C}^k$ at $\mathscr{I}$. This is a perfectly valid definition but, because of the ill-posedness of the Cauchy problem in $\mathcal{C}^k$ spaces, it makes

it difficult to characterize this behaviour by a class of initial data. Instead, I will characterize the order of peeling using Sobolev spaces whose norms are energy fluxes and for which the Cauchy problem is naturally well-posed. The optimal class of data for which fields peel at a given order can then be studied using geometric energy estimates.

4. **Avoid Sobolev embeddings**. A common approach in the study of decay of fields is to use integrated energy estimates and to turn them into pointwise estimates via Sobolev embeddings: by means of geometric energy estimates, one controls weighted Sobolev norms in the bulk, then Sobolev inequalities, by providing an embedding of the relevant weighted Sobolev space into a weighted $\mathcal{C}^k$ space, give locally uniform pointwise decay rates. The problem is that Sobolev embeddings lose derivatives. Besides, they are valid for an open set of regularities so estimates in fact always lose a little more derivatives than necessary. The method is therefore not adapted to finding the optimal class of data ensuring a certain regularity at the conformal boundary. The solution is to define the peeling of a given order as a Sobolev regularity of the rescaled field at $\mathscr{I}$. It is precise, does not involve conversion with loss between Sobolev and $\mathcal{C}^k$ regularities and allows optimal control in terms of the regularity of the data by means of energy estimates. The kind of estimates we will use here are not integrated estimates, but estimates between the energy fluxes on $\mathscr{I}$ and on a Cauchy hypersurface. Their optimality will be ensured by imposing that they be valid both ways. This essentially means that we prove the equivalence of the energies on $\mathscr{I}$ and the Cauchy hypersurface. The fact that we work in a neighbourhood of spacelike infinity and not on the whole spacetime merely requires to have an added term in the estimates, corresponding to the flux of energy leaving the neighbourhood of i^0.

3.2 A First Natural Framework in the Flat Case

For a first approach, we use the complete regular conformal compactification on Minkowski spacetime to perform global energy estimates. These estimates provide us with a definition and characterization of the peeling at any order for the wave equation.

Let us consider the stress energy tensor for Eq. (9)

$$\tilde{T}_{ab} = \partial_a\tilde{\phi}\partial_b\tilde{\phi} - \frac{1}{2}\mathfrak{e}_{ab}\mathfrak{e}^{cd}\partial_c\tilde{\phi}\partial_d\tilde{\phi} + \frac{1}{2}\tilde{\phi}^2\mathfrak{e}_{ab}\,. \tag{15}$$

It is symmetric and divergence-free when $\tilde{\phi}$ is a solution of (9) since

$$\tilde{\nabla}^a\tilde{T}_{ab} = (\Box_{\mathfrak{e}}\tilde{\phi} + \tilde{\phi})\partial_b\tilde{\phi}\,,$$

where $\tilde{\nabla}$ denotes the Levi-Civita connection associated with the Einstein metric $\mathfrak{e}$. Hence, contracting $\tilde{T}_{ab}$ with the Killing vector field $K = \partial_\tau$, we have the conservation law

$$\tilde{\nabla}^a \left(K^b \tilde{T}_{ab} \right) = 0 . \tag{16}$$

The vector field $J^a := K^b \tilde{T}_b^a$ is the energy current that we shall use for the estimates. On a given oriented piecewise $\mathcal{C}^1$ hypersurface S, the flux of J is given by

$$\mathcal{E}_{K,S}(\tilde{\phi}) = \int_S J_a n^a (l \lrcorner \mathrm{dVol}) ,$$

where l^a is a vector field transverse to S compatible with the orientation of S and n^a a normal vector field to S such that $l_a n^a = 1$.

For instance, denoting $X_\tau = \{\tau\} \times S^3$ the level hypersurfaces of the function τ,

$$\mathcal{E}_{K,X_\tau}(\tilde{\phi}) = \frac{1}{2} \int_{X_\tau} \left((\partial_\tau \tilde{\phi})^2 + \left| \nabla_{S^3} \tilde{\phi} \right|^2 + \tilde{\phi}^2 \right) \mathrm{d}\mu_{S^3} = \frac{1}{2} \left(\| \tilde{\phi} \|^2_{H^1(X_\tau)} + \| \partial_\tau \tilde{\phi} \|^2_{L^2(X_\tau)} \right) . \tag{17}$$

Also, parametrizing $\mathscr{I}^+$ as $\tau = \pi - \zeta$,

$$\begin{aligned} \mathcal{E}_{K,\mathscr{I}^+}(\tilde{\phi}) &= \frac{1}{\sqrt{2}} \int_{\mathscr{I}^+} \left(-2 \partial_\tau \tilde{\phi} \, \partial_\zeta \tilde{\phi} + (\partial_\tau \tilde{\phi})^2 + \left| \nabla_{S^3} \tilde{\phi} \right|^2 + \tilde{\phi}^2 \right) \mathrm{d}\mu_{S^3} \\ &= \frac{1}{\sqrt{2}} \int_{\mathscr{I}^+} \left(\left| \partial_\tau \tilde{\phi} - \partial_\zeta \tilde{\phi} \right|^2 + \frac{1}{\sin^2 \zeta} \left| \nabla_{S^2} \tilde{\phi} \right|^2 + \tilde{\phi}^2 \right) \mathrm{d}\mu_{S^3} . \end{aligned} \tag{18}$$

This is a natural H^1 norm of $\tilde{\phi}$ on $\mathscr{I}^+$, involving only the tangential derivatives of $\tilde{\phi}$ along $\mathscr{I}^+$.

Now consider a smooth solution $\tilde{\phi}$ of (9) on $\mathfrak{E}$. The conservation law (16) tells us that the flux of J across the closed hypersurface made of the union of X_0 and $\mathscr{I}^+$ is zero. Hence,

$$\mathcal{E}_{K,\mathscr{I}^+}(\tilde{\phi}) = \mathcal{E}_{K,X_0}(\tilde{\phi}) . \tag{19}$$

Moreover, since ∂_τ is a Killing vector, for any $k \in \mathbb{N}$, $\partial_\tau^k \tilde{\phi}$ satisfies Eq. (9), whence

$$\mathcal{E}_{K,\mathscr{I}^+}(\partial_\tau^k \tilde{\phi}) = \mathcal{E}_{K,X_0}(\partial_\tau^k \tilde{\phi}) .$$

Using Eq. (9), for $k = 2p$, $p \in \mathbb{N}$, we have

$$\begin{aligned} 2\mathcal{E}_{K,X_0}(\partial_\tau^k \tilde{\phi}) &= \| \partial_\tau^{2p} \tilde{\phi} \|^2_{H^1(X_0)} + \| \partial_\tau^{2p+1} \tilde{\phi} \|^2_{L^2(X_0)} \\ &= \| (1 - \Delta_{S^3})^p \tilde{\phi} \|^2_{H^1(X_0)} + \| (1 - \Delta_{S^3})^p \partial_\tau \tilde{\phi} \|^2_{L^2(X_0)} \\ &\simeq \| \tilde{\phi} \|^2_{H^{2p+1}(X_0)} + \| \partial_\tau \tilde{\phi} \|^2_{H^{2p}(X_0)} , \end{aligned} \tag{20}$$

and for $k = 2p + 1$, $p \in \mathbb{N}$,

$$\begin{aligned}
2\mathcal{E}_{K,X_0}(\partial_\tau^k\tilde{\phi}) &= \|\partial_\tau^{2p+1}\tilde{\phi}\|^2_{H^1(X_0)} + \|\partial_\tau^{2p+2}\tilde{\phi}\|^2_{L^2(X_0)} \\
&= \|(1-\Delta_{S^3})^p\partial_\tau\tilde{\phi}\|^2_{H^1(X_0)} + \|(1-\Delta_{S^3})^{p+1}\tilde{\phi}\|^2_{L^2(X_0)} \\
&\simeq \|\tilde{\phi}\|^2_{H^{2p+2}(X_0)} + \|\partial_\tau\tilde{\phi}\|^2_{H^{2p+1}(X_0)}\,. \qquad (21)
\end{aligned}$$

Hence, we have for each $k \in \mathbb{N}$:

$$\|\tilde{\phi}\|^2_{H^{k+1}(X_0)} + \|\partial_\tau\tilde{\phi}\|^2_{H^k(X_0)} \simeq \mathcal{E}_{K,X_0}(\partial_\tau^k\tilde{\phi}) = \mathcal{E}_{K,\mathscr{I}^+}(\partial_\tau^k\tilde{\phi}) \simeq \|\partial_\tau^k\tilde{\phi}\|^2_{H^1(\mathscr{I}^+)}$$

and using the fact that the H^k norm controls all the lower Sobolev norms, this gives us the apparently stronger equivalence

$$\|\tilde{\phi}\|^2_{H^{k+1}(X_0)} + \|\partial_\tau\tilde{\phi}\|^2_{H^k(X_0)} \simeq \sum_{p=0}^{k} \|\partial_\tau^p\tilde{\phi}\|^2_{H^1(\mathscr{I}^+)}\,. \qquad (22)$$

Remark 3.1 In principle, this equivalence should not be understood as providing a solution to a Goursat problem on $\mathscr{I}^+$. Indeed, in Lars Hörmander's paper on the Goursat problem for the wave equation [22], it is made very clear that such an equivalence only provides us with a trace operator on $\mathscr{I}^+$ that is a partial isometry. It is then necessary to prove the surjectivity of this operator in order to solve the Goursat problem. However, we know from the same paper that the Goursat problem for Eq. (9) with data $\tilde{\phi}_{|\mathscr{I}^+} \in H^1(\mathscr{I}^+)$ is well posed and gives rise to solutions $\tilde{\phi} \in \mathcal{C}^0(\mathbb{R}_\tau\,;\, H^1(S^3)) \cap \mathcal{C}^1(\mathbb{R}_\tau\,;\, L^2(S^3))$. Hence (22) indeed provides us with a regularity result for the Goursat problem: data on $\mathscr{I}^+$ for which the norm on the right-hand side is finite give rise to solutions that are in $\mathcal{C}^l(\mathbb{R}_\tau\,;\, H^{k+1-l}(S^3))$ for all $0 \le l \le k+1$. This is however stronger than the information we are interested in. We simply extract from (22) the fact that for smooth solutions, the control of the transverse regularity on $\mathscr{I}^+$ described by $\mathcal{E}_{\mathscr{I}^+}(\partial_\tau^p\tilde{\phi})$, $0 \le p \le k$, is equivalent to that of the H^{k+1} norm of the restriction of $\tilde{\phi}$ to X_0 and the H^k norm of the restriction of $\partial_\tau\tilde{\phi}$ to X_0. By a standard density argument, this shows that if we wish to guarantee, by means of some control on the initial data, that the restriction to $\mathscr{I}^+$ of $\partial_\tau^p\tilde{\phi}$, $0 \le p \le k$, is in $H^1(\mathscr{I}^+)$, the optimal condition to impose is that $\tilde{\phi}_0 \in H^{k+1}(X_0)$ and $\tilde{\phi}_1 \in H^k(X_0)$. This is our first definition of a peeling of order k and its characterization by a function space of initial data.

Definition 3.1 A solution ϕ of (8) is said to peel at order $k \in \mathbb{N}$ if the traces on $\mathscr{I}^+$ of $\partial_\tau^p\tilde{\phi}$, $0 \le p \le k$, are in $H^1(\mathscr{I}^+)$. The optimal function space of initial data $(\tilde{\phi}_0\,,\,\tilde{\phi}_1)$ giving rise to solutions that peel at order k is $H^{k+1}(S^3) \times H^k(S^3)$.

Expressing the space of data in terms of the physical field ϕ using (12) gives us the exact function space of physical data giving rise to solutions of (8) that peel at order k.

3.3 The "Correct" Version in the Flat Case

The previous construction is very natural but its drawback is that very few spacetimes admit such a complete and regular compactification. As a consequence, we may have a valid definition of the peeling at any order in the flat case, but we will not be able to compare with other asymptotically flat spacetimes, whose natural compactifications are associated with much weaker conformal factors. On the Schwarzschild spacetime for instance, we cannot hope to compactify the exterior of the black hole in as complete a manner as Minkowski spacetime. Timelike and spacelike infinities will be singularities of the conformal structure. The natural compactification is associated to $\Omega = 1/r$, which also has the pleasant property that we have the same symmetries before and after rescaling. Since the conformal factors $1/r$ and (4) are not uniformly equivalent at infinity in the flat case, it would not make sense to compare the peeling on the Schwarzschild metric constructed with $1/r$, to the version above on Minkowski spacetime constructed with (4). We must therefore redefine the peeling in the flat case using $1/r$ as a conformal factor. Then we can perform similar constructions on the Schwarzschild metric and compare the results. Besides, on most asymptotically flat spacetime, it will be possible to define an analogous conformal factor that will allow to construct $\mathscr{I}$. This choice of conformal factor will therefore make the comparison with Minkowski spacetime relatively easy for a large class of asymptotically flat geometries.

We perform the construction on $\mathbb{M}$ purely for future null infinity. The analogous construction for $\mathscr{I}^-$ is obtained by a straightforward modification. We write the Minkowski metric in terms of the variables $u = t - r$, r, and ω:

$$\eta = \mathrm{d}u^2 + 2\mathrm{d}u\mathrm{d}r - r^2\mathrm{d}\omega^2 .$$

Then putting $R = 1/r$, we get

$$\hat{\eta} := R^2\eta = R^2\mathrm{d}u^2 - 2\mathrm{d}u\mathrm{d}R - \mathrm{d}\omega^2 . \tag{23}$$

This rescaled metric extends analytically on $\mathbb{R}_u \times [0, +\infty[_R \times S^2_\omega$, which is Minkowski spacetime minus the $r = 0$ coordinate line, with the added boundary $\mathbb{R}_u \times \{0\}_R \times S^2_\omega$. This boundary can easily be seen to be $\mathscr{I}^+$ since for u and ω constant, we move on an outgoing radial null geodesic and the boundary is reached as $r \to +\infty$ along such lines. In the new coordinates u, R, ω, the vector field ∂_u is the timelike Killing vector field for η that used to be ∂_t in the t, r, ω coordinates. It turns out that ∂_u is still Killing for the rescaled metric $\hat{\eta}$. The scalar curvature of $\hat{\eta}$ vanishes

$$\frac{1}{6}\mathrm{Scal}_{\hat{\eta}} = R^{-3}\Box_\eta R = 0 , \tag{24}$$

so $\phi \in \mathcal{D}'(\mathbb{M})$ satisfies (8) if and only if $\hat{\phi} := R^{-1}\phi = r\phi$ is a solution of

$$\Box_{\hat{\eta}}\hat{\phi} = 0\,,\ \Box_{\hat{\eta}} = -2\partial_u\partial_R - \partial_R R^2\partial_R - \Delta_{S^2}\,. \tag{25}$$

We work in a neighbourhood of i^0 of the following form

$$\Omega_{u_0} = \{(u, R, \omega)\,;\ u \leq u_0\,,\ R \geq 0\,,\ t \geq 0\}\,,$$

for $u_0 << -1$. The boundary of Ω_{u_0} is made of three parts:

$$\begin{aligned} \mathscr{I}^+_{u_0} &= \mathscr{I}^+ \cap \Omega_{u_0}\,, \\ \Sigma_0^{u_0} &= \Sigma_0 \cap \Omega_{u_0}\,, \\ \mathcal{S}_{u_0} &= \{u = u_0\} \cap \Omega_{u_0}\,. \end{aligned}$$

The usual stress-energy tensor for the wave equation

$$\hat{T}_{ab} = \partial_a\hat{\phi}\,\partial_b\hat{\phi} - \frac{1}{2}\hat{\eta}_{ab}\hat{\eta}^{cd}\partial_c\hat{\phi}\,\partial_d\hat{\phi} \tag{26}$$

is symmetric and divergence-free for solutions of (25) since

$$\hat{\nabla}_a\hat{T}_{ab} = (\Box_{\hat{\eta}}\hat{\phi})\partial_b\hat{\phi}\,,$$

where $\hat{\nabla}$ denotes the Levi-Civita connection associated with the metric $\hat{\eta}$.

We need to find an analogue of ∂_τ on the Einstein cylinder with which to define an energy current: a vector field that extends smoothly at the conformal boundary as a transverse vector field to $\mathscr{I}$ and that is as close as possible to being Killing. We could simply re-express ∂_τ in terms of the coordinates u, R, ω but its flow would not be an isometry. It turns out that a modification of ∂_τ by a multiple of ∂_t is a Killing vector field for $\hat{\eta}$ and is transversal to $\mathscr{I}$: it is the Morawetz vector field introduced by Kaithleen Morawetz in 1961 [32] in order to prove decay estimates for the wave equation on Minkowski spacetime. It is defined in terms of the variables u and v as

$$T = u^2\partial_u + v^2\partial_v,\ \text{i.e.}\ T = (t^2 + r^2)\partial_t + 2tr\partial_r\,.$$

and is therefore timelike everywhere on $\mathbb{M}$ except on the lightcone of the origin, where it is null. It is future-oriented on $\mathbb{M}$, except at the origin where it vanishes. In terms of the variables u and R, it reads

$$T = u^2\partial_u - 2(1 + uR)\partial_R\,. \tag{27}$$

It extends smoothly at $\mathscr{I}^+$ where it takes the expression $u^2\partial_u - 2\partial_R$ which is the sum of two future-oriented null vector fields, one tangent to $\mathscr{I}$ and the other transverse. Moreover, T satisfies the Killing equation for $\hat{\eta}$:

$$\hat{\nabla}^{(a}T^{b)} = 0\,,$$

so we immediately get the following energy identity, for a smooth solution ϕ of (8) with compactly supported initial data,

$$\mathcal{E}_{T,\mathscr{I}^+_{u_0}}(\hat{\phi}) + \mathcal{E}_{T,\mathcal{S}_{u_0}}(\hat{\phi}) = \mathcal{E}_{T,\Sigma_0^{u_0}}(\hat{\phi})\,, \tag{28}$$

where

$$\begin{aligned}\mathcal{E}_{T,\Sigma_0^{u_0}}(\hat{\phi}) &= \int_{\Sigma_0^{u_0}} \Big(u^2(\partial_u\hat{\phi})^2 + R^2u^2\partial_u\hat{\phi}\partial_R\hat{\phi} \\ &\quad + R^2\left(\frac{(2+uR)^2}{2} - (1+uR)\right)(\partial_R\hat{\phi})^2 + \left(\frac{u^2R^2}{2}+1+uR\right)\left|\nabla_{S^2}\hat{\phi}\right|^2\Big)\mathrm{d}u\mathrm{d}^2\omega \\ &\simeq \int_{\Sigma_0^{u_0}} \left(u^2(\partial_u\hat{\phi})^2 + R^2(\partial_R\hat{\phi})^2 + \left|\nabla_{S^2}\hat{\phi}\right|^2\right)\mathrm{d}u\mathrm{d}^2\omega\,, \end{aligned}\tag{29}$$

$$\mathcal{E}_{T,\mathscr{I}^+_{u_0}}(\hat{\phi}) = \int_{\mathscr{I}^+_{u_0}} \left(u^2(\partial_u\hat{\phi})^2 + |\nabla_{S^2}\hat{\phi}|^2\right)\mathrm{d}u\mathrm{d}^2\omega\,, \tag{30}$$

$$\mathcal{E}_{T,\mathcal{S}_u}(\phi) = \int_{\mathcal{S}_u} \frac{1}{2}\left((2+uR)^2(\partial_R\hat{\phi})^2 + u^2|\nabla_{S^2}\hat{\phi}|^2\right)\mathrm{d}R\mathrm{d}^2\omega\,. \tag{31}$$

The next thing to work out is how to obtain higher order estimates from (28). We could simply commute T into the equation; since it is Killing, we would immediately obtain identitites similar to (28) for $T^k\hat{\phi}$. Although this would again be a perfectly valid definition, it would not extend naturally to other spacetimes such as Schwarzschild, because in these spacetimes, the Killing form of T induces terms that are delicate to handle in the higher order estimates (see [30] for details). Another vector field that we can use is ∂_R. It is a null vector field, as can be seen from (23), it is transverse to $\mathscr{I}^+$ and has the following expression in terms of the variables t, r (which we shall use later on):

$$\partial_R = -r^2(\partial_t + \partial_r)\,. \tag{32}$$

The vector field ∂_R is not Killing for $\hat{\eta}$ but it is nonetheless easy to control the error terms in the higher order estimates, since they merely involve polynomials in R. Moreover, this vector field will turn out to be just as easy to use in the Schwarzschild framework, for the same reason. Using Gronwall's inequality, we obtain the following estimates both ways: for each $k \in \mathbb{N}$, there exists a positive constant C_k such that, for any smooth solution $\hat{\phi}$ of (25) with compactly supported initial data,

$$\mathcal{E}_{T,\mathscr{I}^+_{u_0}}(\partial_R^k\hat{\phi}) \leq C_k\sum_{p=0}^{k}\mathcal{E}_{T,\Sigma_0^{u_0}}(\partial_R^p\hat{\phi})\,, \tag{33}$$

$$\mathcal{E}_{T,\Sigma_0^{u_0}}(\partial_R^k\hat{\phi}) \leq C_k\sum_{p=0}^{k}\left(\mathcal{E}_{T,\mathscr{I}^+_{u_0}}(\partial_R^p\hat{\phi}) + \mathcal{E}_{T,\mathcal{S}_{u_0}}(\partial_R^p\hat{\phi})\right)\,. \tag{34}$$

Using the spherical symmetry, we can also add angular derivatives in the estimates above. These higher order estimates give us a definition of the peeling at any order on Minkowski spacetime, that is different from the one above, but that extends to other asymptotically flat spacetimes.

Definition 3.2 We say that a solution $\hat{\phi}$ of (25) peels at order $k \in \mathbb{N}$ if for all polynomial P in ∂_R and ∇_{S^2} of order lower than or equal to k, we have $\mathcal{E}_{T,\mathscr{I}^+_{u_0}}(P\hat{\phi}) < +\infty$. This means than for all $p \in \{0, 1, ..., k\}$ we have for all $q \in \{0, 1, ..., p\}$, $\mathcal{E}_{T,\mathscr{I}^+_{u_0}}(\partial_R^q \nabla_{S^2}^{p-q} \hat{\phi}) < +\infty$.

In view of Estimates (33), (34) the condition on initial data that guarantees peeling at order k is therefore that

$$\forall p \in \{0, 1, ..., k\}\,,\ \forall q \in \{0, 1, ..., p\}\,,\ \mathcal{E}_{T,\Sigma_0^{u_0}}(\partial_R^q \nabla_{S^2}^{p-q} \hat{\phi}) < +\infty\,.$$

This can easily be re-expressed, using the equation, purely in terms of initial data. First, note that Equation (25) can be written in terms of the variables (t, r, ω) as

$$(\partial_t + \partial_r)(\partial_t - \partial_r)\hat{\phi} - \frac{1}{r^2}\Delta_{S^2}\hat{\phi} = 0\,.$$

Whence,

$$\begin{aligned} \partial_R \begin{pmatrix} \hat{\phi} \\ \partial_t \hat{\phi} \end{pmatrix} &= -r^2\,(\partial_t + \partial_r) \begin{pmatrix} \hat{\phi} \\ \partial_t \hat{\phi} \end{pmatrix} \\ &= -r^2 \begin{pmatrix} \partial_r & 1 \\ \partial_r^2 + \frac{1}{r^2}\Delta_{S^2} & \partial_r \end{pmatrix} \begin{pmatrix} \hat{\phi} \\ \partial_t \hat{\phi} \end{pmatrix} =: L \begin{pmatrix} \hat{\phi} \\ \partial_t \hat{\phi} \end{pmatrix}. \end{aligned}$$

The operator L purely involves spacelike derivatives. Now, we can express the spaces of initial data that entail peeling at a given order.

Definition 3.3 Given $\hat{\phi}_0, \hat{\phi}_1 \in \mathcal{C}_0^\infty([-u_0, +\infty\,[_{r_*} \times S^2_\omega)$, we define the following squared norm of order k:

$$\left\| \begin{pmatrix} \hat{\phi}_0 \\ \hat{\phi}_1 \end{pmatrix} \right\|_k^2 := \sum_{p=0}^{k} \sum_{q=0}^{p} \mathcal{E}_{T,\Sigma_0^{u_0}} \left(L^q \nabla_{S^2}^{p-q} \begin{pmatrix} \hat{\phi}_0 \\ \hat{\phi}_1 \end{pmatrix} \right), \tag{35}$$

where we have denoted by $\mathcal{E}_{T,\Sigma_0^{u_0}} \begin{pmatrix} \hat{\phi}_0 \\ \hat{\phi}_1 \end{pmatrix}$ the energy $\mathcal{E}_{T,\Sigma_0^{u_0}}(\hat{\phi})$, given in (29), where $\hat{\phi}$ is replaced by $\hat{\phi}_0$ and $\partial_t \hat{\phi} = \partial_u \hat{\phi}$ is replaced by $\hat{\phi}_1$.

Theorem 1 *The space of initial data (on $[-u_0, +\infty\,[_r \times S^2_\omega$) for which the associated solution peels at order k is the completion of $\mathcal{C}_0^\infty([-u_0, +\infty\,[_r \times S^2_\omega) \times \mathcal{C}_0^\infty([-u_0, +\infty\,[_r \times S^2_\omega)$ in the norm (35). The fact that we have estimates both ways at all orders guarantees that this setting is optimal for our definition.*

3.4 Results on the Schwarzschild Metric

The Schwarzschild metric expressed in terms of Schwarzschild coordinates is

$$g = \left(1 - \frac{2m}{r}\right) \mathrm{d}t^2 - \left(1 - \frac{2m}{r}\right)^{-1} \mathrm{d}r^2 - r^2 \mathrm{d}\omega^2$$

where $m > 0$ is the mass of the black hole and $\mathrm{d}\omega^2$ is the Euclidean metric on the unit sphere S^2. We work on the exterior of the black hole

$$\mathcal{M} = \mathbb{R}_t \times]2m, +\infty[_r \times S^2_\omega .$$

The associated d'Alembertian is

$$\Box_g = \left(1 - \frac{2m}{r}\right)^{-1} \frac{\partial^2}{\partial t^2} - \frac{1}{r^2}\frac{\partial}{\partial r} r^2 \left(1 - \frac{2m}{r}\right) \frac{\partial}{\partial r} - \frac{1}{r^2}\Delta_{S^2} ,$$

where Δ_{S^2} is the Laplacian on S^2 endowed with the Euclidean metric. We perform a conformal rescaling that is similar to the partial compactification of Minkowski spacetime: we consider the variables

$$R = 1/r , \; u = t - r_* , \; \text{with } r_* = r + 2m \log(r - 2m) ,$$

and rescale the metric g as follows

$$\hat{g} = R^2 g = R^2(1 - 2mR)\mathrm{d}u^2 - 2\mathrm{d}u\mathrm{d}R - \mathrm{d}\omega^2 .$$

In these coordinates,

$$\mathscr{I}^+ = \{0\}_R \times \mathbb{R}_u \times S^2_\omega .$$

The scalar curvature of $\hat{g}$ is given by

$$\mathrm{Scal}_{\hat{g}} = 12mR$$

and the conformally invariant wave equation on the metric $\hat{g}$ has the following form

$$\left(\Box_{\hat{g}} + 2mR\right) \hat{\phi} = \left(-2\partial_u \partial_R - \partial_R R^2 \left(1 - 2mR\right) \partial_R - \Delta_{S^2} + 2mR\right) \hat{\phi} = 0 . \quad (36)$$

As in the case of flat spacetime, the two following properties are equivalent:

1. $\phi \in \mathcal{D}'(\mathcal{M})$ satisfies $\Box_g \phi = 0$;
2. $\hat{\phi} := R^{-1}\phi = r\phi$ satisfies (36) on $\mathcal{M}$.

The situation is now slightly more complicated because Equation (36) does not admit a conserved stress-energy tensor. We choose to use the stress-energy tensor for the free wave equation

$$\hat{T}_{ab} = \hat{T}_{(ab)} = \partial_a \hat{\phi}\, \partial_b \hat{\phi} - \frac{1}{2} \hat{g}_{ab} \hat{g}^{cd} \partial_c \hat{\phi}\, \partial_d \hat{\phi}\,, \tag{37}$$

which satisfies

$$\nabla^a \hat{T}_{ab} = \Box_{\hat{g}} \hat{\phi}\, \partial_b \hat{\phi} = -2mR \hat{\phi}\, \partial_b \hat{\phi}\,. \tag{38}$$

Then we adapt the flat spacetime Morawetz vector field to construct a timelike vector field transverse to $\mathscr{I}^+$: this is a classic construction (it was first introduced by Inglese and Nicolò [24] and then used by other authors, for instance by Dafermos and Rodnianski in [10]), one chooses a coordinate system in Minkowski spacetime, obtains the expression of the Morawetz vector field in these coordinates, then keeps the expression in an analogous coordinate system on the Schwarzschild spacetime. So in a sense, there are several Morawetz vector fields on $\mathcal{M}$, depending on the coordinate system one chooses. The usual choice is to work with $u = t - r$, $v = t + r$, ω and then to transpose the expression in the coordinates $u = t - r_*$, $v = t + r_*$, ω on $\mathcal{M}$. Instead, we use the expression (27) and define on $\mathcal{M}$ the vector field

$$T := u^2 \partial_u - 2(1 + uR)\partial_R \tag{39}$$

in the coordinate system $u = t - r_*$, $R = 1/r$, ω. This is now no longer a Killing vector field, but its Killing form has a rather fast decay at infinity:

$$\hat{\nabla}_{(a} T_{b)} \mathrm{d}x^a \mathrm{d}x^b = 4mR^2(3 + uR)\mathrm{d}u^2\,.$$

The associated energy current will satisfy an approximate conservation law, with error terms coming both from the equation and the Killing form of T. More precisely,

$$J^a := T_b \hat{T}^{ab}\,;\ \hat{\nabla}_a J^a = -2mR \nabla_T \hat{\phi} + \nabla_{(a} T_{b)} \hat{T}^{ab}\,.$$

Then, it simply remains to apply the method we have developed in the flat case and to check that all error terms can be controlled via a priori estimates of Gronwall type. The details can be found in [30] for the wave equation and [31] for Dirac and Maxwell fields. We now express the definition of the peeling at any order on the Schwarzschild metric and its characterization in terms of classes of initial data. As before, we work on a domain $\{u \leq u_0\}$ for $u_0 << -1$ and we consider the energy fluxes through the three parts of its boundary (with the same notation as in the flat case)

$$\begin{aligned}
\mathcal{E}_{T,\Sigma_0^{u_0}}(\hat{\phi}) &= \int_{\Sigma_0^{u_0}} \Big(u^2(\partial_u\hat{\phi})^2 + R^2(1-2mR)u^2\partial_u\hat{\phi}\partial_R\hat{\phi} \\
&\quad + R^2(1-2mR)\left(\frac{(2+uR)^2}{2} - mu^2R^3 - (1+uR)\right)(\partial_R\hat{\phi})^2 \\
&\quad + \left(\frac{u^2R^2(1-2mR)}{2} + 1 + uR\right)\left|\nabla_{S^2}\hat{\phi}\right|^2\Big)\,\mathrm{d}u\mathrm{d}^2\omega \\
&\simeq \int_{\Sigma_0^{u_0}} \left(u^2(\partial_u\hat{\phi})^2 + \frac{R}{|u|}(\partial_R\hat{\phi})^2 + \left|\nabla_{S^2}\hat{\phi}\right|^2\right)\mathrm{d}u\mathrm{d}^2\omega\,, \\
\mathcal{E}_{T,\mathscr{I}^+_{u_0}}(\hat{\phi}) &= \int_{\mathscr{I}^+_{u_0}} \left(u^2(\partial_u\hat{\phi})^2 + |\nabla_{S^2}\hat{\phi}|^2\right)\mathrm{d}u\mathrm{d}^2\omega\,, \\
\mathcal{E}_{T,\mathcal{S}_u}(\phi) &= \int_{\mathcal{S}_u} \frac{1}{2}\left(\left((2+uR)^2 - 2mu^2R^3\right)(\partial_R\hat{\phi})^2 + u^2|\nabla_{S^2}\hat{\phi}|^2\right)\mathrm{d}R\mathrm{d}^2\omega\,.
\end{aligned}$$

We obtain Estimates (33) and (34) in the Schwarzschild framework and the spherical symmetry once again allows us to get analogous estimates for angular derivatives. Hence the following definition of the peeling at any order on the Schwarzschild metric:

Definition 3.4 We say that a solution $\hat{\phi}$ of (36) peels at order $k \in \mathbb{N}$ if for all polynomials P in ∂_R and ∇_{S^2} of order lower than or equal to k, we have $\mathcal{E}_{T,\mathscr{I}^+_{u_0}}(P\hat{\phi}) < +\infty$. This means than for all $p \in \{0, 1, ..., k\}$ we have for all $q \in \{0, 1, ..., p\}$, $\mathcal{E}_{T,\mathscr{I}^+_{u_0}}(\partial_R^q \nabla_{S^2}^{p-q}\hat{\phi}) < +\infty$. This condition can be re-expressed as the finiteness of the following norm of the data for $\hat{\phi}$ at $t = 0$:

$$\left\| \begin{pmatrix} \hat{\phi}_0 \\ \hat{\phi}_1 \end{pmatrix} \right\|_k^2 := \sum_{p=0}^{k}\sum_{q=0}^{p} \mathcal{E}_{T,\Sigma_0^{u_0}}\left(L^q \nabla_{S^2}^{p-q} \begin{pmatrix} \hat{\phi}_0 \\ \hat{\phi}_1 \end{pmatrix}\right), \tag{40}$$

the operator L now reading

$$L = -\frac{r^3}{r-2m}\begin{pmatrix} \partial_{r_*} & 1 \\ \partial_{r_*}^2 - \frac{2m(r-2m)}{r^4} + \frac{r-2m}{r^3}\Delta_{S^2} & \partial_{r_*} \end{pmatrix}.$$

Theorem 2 *The space of initial data (on $[-u_0, +\infty[_{r_*} \times S^2_\omega$) for which the associated solution peels at order k is the completion of $\mathcal{C}_0^\infty([-u_0, +\infty[_{r_*} \times S^2_\omega) \times \mathcal{C}_0^\infty([-u_0, +\infty[_{r_*} \times S^2_\omega)$ in the norm (40). This is optimal for our definition.*

The remaining task is to compare the characterizations of peeling at the same given order between the flat case and the Schwarzschild case. After the care we took to make sure that our constructions would be as close to one another as possible, it turns out that not only is comparison between our classes of data natural, but also these classes are almost trivially equivalent in the following sense: the classes are defined by weighted Sobolev norms; the weights intervening in the flat and Schwarzschild cases have equivalent behaviours at infinity when using r for the radial variable in

$\mathbb{M}$ and r_* in $\mathcal{M}$.[6] The essential reason for this is that the norms in the Schwarzschild situation are uniform in the mass m of the black hole on any given bounded interval $]0, M]$. The details can be found in [30, 31].

4 Conformal Scattering

Scattering theory is a way of summarizing the whole evolution of solutions to a certain equation by a scattering operator that, to their asymptotic behaviour in the distant past, associates their asymptotic behaviour in the distant future. These asymptotic behaviours are usually solutions to a simpler equation, a comparison dynamics. A complete scattering theory will not only show the existence of a scattering operator but will also establish that the solutions are completely and uniquely characterized by their past (resp. future) asymptotic behaviours, which entails in particular the invertibility of the scattering operator. Different choices of comparison dynamics are always possible, giving different scattering operators that are not necessarily defined on the same function spaces. Some simplified dynamics can be described geometrically as transport equations along congruences of null geodesics defining null infinity; the fields that they propagate then correspond to functions defined on $\mathscr{I}$. Using such comparison dynamics for scattering theories on asymptotically flat spacetimes means that the scattering data, i.e. the large time asymptotic behaviours, are merely radiation fields.

The idea of using a conformal compactification in order to obtain a time-dependent scattering theory formulated in terms of radiation fields is due to Penrose. His discussion of the topic in [37] clearly indicates that it was one of the main motivations of the conformal technique. The first actual conformal scattering theory appeared fifteen years later in Friedlander's founding paper [19] as a combination of Friedlander's own work on radiation fields [16–18] and the Lax-Phillips approach to time-dependent scattering [26]. This first paper treated the case of the conformal wave equation. The geometrical background was a static asymptotically flat spacetime with a fast decay at infinity, too fast for allowing for the presence of energy when considering solutions of the Einstein vacuum equations, but fast enough to ensure a smooth conformal compactification including at spacelike and timelike infinities. The principle of the construction was first to re-interpret the scattering theory as the well-posedness of the Goursat problem for the rescaled equation at null infinity, then to solve this Goursat problem. Friedlander's main goal was then, it seems to me, to extend the results of Lax and Phillips, in all their analytic precision, to a curved situation. In particular, he wanted to recover the fundamental structure in the Lax-Phillips theory: the so-called "translation representer" of the solution. The existence of such an object requires a timelike Killing vector field that extends as the null generator of $\mathscr{I}$. This is probably

[6] In fact, the choice of r_* in the Schwarzschild situation is not crucial for the comparison of the asymptotic behaviour of the weights, simply because $r_* \simeq r$ at infinity. This choice is however the natural one because the radial derivatives appearing in the norms are ∂_r on $\mathbb{M}$ and ∂_{r_*} on $\mathcal{M}$.

what motivated the choice of a static background, even though the conformal scattering construction itself can be performed on non stationary geometries. Friedlander's method was then applied to nonlinear equations, but still for static geometries, by J. C. Baez, I. E. Segal and Zhou Z. F. in 1989–1990 [1–5]. Friedlander himself at the end of his life came back to conformal scattering in a posthumously published note [20]. Lars Hörmander published in 1990 a short paper, in the form of a remark prompted by [4], entitled "A remark on the characteristic Cauchy problem" [22], in which he presented a general method for solving the initial value problem on a weakly spacelike hypersurface, for a general wave equation on spatially compact spacetimes. The method is entirely based on energy estimates and compactness methods. Using this approach, a conformal scattering theory on generically non stationnary backgrounds was developed by L.J. Mason and the author [29]. Then, more recently, J. Joudioux obtained the first result for a non linear wave equation [25] in non stationary situations. I came back to the topic a couple of years ago to propose an extension of these methods to black hole spacetimes [34], describing the construction in the Schwarzschild spacetime, which is static, and discussing the case of the Kerr metric and the associated difficulties.

This section starts by a brief description of the Lax-Phillips theory in a simplified setting. Then we describe the conformal scattering construction for the wave equation in the flat case and move on to extensions to non stationary and black hole situations.

4.1 A Simple Overview of Lax-Phillips Theory

The Lax-Phillips theory describes the scattering of a massless scalar field by an obstacle. We present here a version of the theory without obstacle, i.e. for the free wave equation. Spectral analysis is used to construct a translation representer of the free wave equation, which is then re-interpreted geometrically as a radiation field. We describe the construction of the translation representer and its geometrical reinterpretation. It is usual to consider that in the case of a free equation, there is no scattering. Indeed the existence of the translation representer can be understood in this manner. But this statement is not invariant, it merely corresponds to chosing the equation itself as comparison dynamics. The geometrical re-interpretation of the translation representer describes asymptotic behaviours as radiation fields. For this choice of comparison dynamics, the scattering process is non trivial even on flat spacetime. Moreover the resulting scattering operator can be defined geometrically without resorting to the spectral analytic part of the Lax-Phillips theory. The construction thus modified can be easily generalized to a large class of curved situations. This will be the object of Sects. 4.2, 4.3 and 4.4.

4.1.1 Finite-Dimensional Case

Let us first describe the essential ideas on a finite-dimensional toy model. Consider the following equation for a time-dependent vector in $\mathbb{C}^n$:

$$\partial_t V(t) = iAV(t) \tag{41}$$

where A is an $n \times n$ Hermitian matrix A with n distinct eigenvalues σ_1, ..., σ_n. Let $\{e_1, \dots, e_n\}$ be an orthonormal basis of eigenvectors of A. The Cauchy problem for (41) is solved by the propagator e^{itA}, i.e. if V is a solution of (41),

$$V(t) = e^{i(t-s)A}V(s)\,,\ \forall t, s \in \mathbb{R}$$

and the matrices e^{itA} are unitary.

Instead of considering V as an element of $\mathbb{C}^n$, we represent it as a function on the spectrum of A, which is square integrable for the natural spectral measure $\mu = \sum_{i=1}^n \delta_{\sigma_i}$:

$$\begin{array}{rcl} \mathbb{C}^n & \to & L^2(\mathbb{R};\, \mathrm{d}\mu) \\ V & \mapsto & \tilde{V}(\sigma_i) = \langle V, e_i\rangle\,. \end{array}$$

This is a spectral representation in the sense that the action of A is now described simply by multiplication by the spectral parameter:

$$\widetilde{AV}(\sigma) = \sigma\tilde{V}(\sigma)\,.$$

Then taking the Fourier transform,

$$\hat{\tilde{V}} := \mathcal{F}_\sigma \tilde{V}\,,$$

we obtain a new representation for which the evolution is described by a simple translation of t:

$$\mathcal{F}_\sigma\left(\widetilde{e^{itA}V}\right)(s) = \mathcal{F}_\sigma(e^{it\sigma}\tilde{V})(s) = \hat{\tilde{V}}(s-t)\,.$$

This is called a translation representer of the solution of (41). A similar construction can be performed for the wave equation on Minkowski spacetime and is at the heart of the Lax-Phillips theory.

4.1.2 The Wave Equation

We now consider the wave equation on Minkowski spacetime:

$$\partial_t^2\phi - \Delta\phi = 0\,. \tag{42}$$

It can be written formally as Eq. (41) in the following manner:

$$\partial_t U = iAU \,, \; U := \begin{pmatrix} \phi \\ \partial_t \phi \end{pmatrix} , \; A = -i \begin{pmatrix} 0 & 1 \\ \Delta & 0 \end{pmatrix} .$$

The operator A is self-adjoint on $\mathcal{H} = \dot{H}^1(\mathbb{R}^3) \times L^2(\mathbb{R}^3)$, where $\dot{H}^1(\mathbb{R}^3)$ is the first-order homogeneous Sobolev space on $\mathbb{R}^3$, completion of $\mathcal{C}_0^\infty(\mathbb{R}^3)$ in the norm

$$\|\psi\|^2_{\dot{H}^1(\mathbb{R}^3)} = \int_{\mathbb{R}^3} |\nabla\psi|^2 \mathrm{d}^3 x \,.$$

The spectrum of A is the whole real axis and is purely absolutely continuous. In particular the point spectrum of A is empty. The equation $AU = \sigma U$ for $\sigma \in \mathbb{R}$, which reads

$$\begin{cases} u_2 = i\sigma u_1 \,, \\ \Delta u_1 = i\sigma u_2 \,, \\ \quad\;\; = -\sigma^2 u_1 \,, \end{cases} \tag{43}$$

does not have any finite energy solution, by which we mean a solution in $\mathcal{H}$. However, for each $\sigma \in \mathbb{R}^*$, A has a whole 2-sphere worth of generalized eigenfunctions which are plane waves:

$$e_{\sigma,\omega}(x) = \begin{pmatrix} e^{-i\sigma x.\omega} \\ i\sigma e^{-i\sigma x.\omega} \end{pmatrix} , \; \omega \in S^2 \,.$$

For $\sigma = 0$, the 2-sphere collapses to a point and the only solution is

$$e_{0,\omega}(x) = e_0(x) = \begin{pmatrix} 1 \\ 0 \end{pmatrix} .$$

We now proceed exactly as in the finite dimensional case. Consider $U \in \mathcal{C}_0^\infty(\mathbb{R}^3) \times \mathcal{C}_0^\infty(\mathbb{R}^3)$, we represent it as a function on $\mathbb{R}_\sigma \times S^2_\omega$ by taking its inner product with plane waves, suitably normalized:

$$\begin{aligned} \tilde{U}(\sigma,\omega) &:= \frac{1}{(2\pi)^{3/2}} \langle U, e_{\sigma,\omega} \rangle_{\mathcal{H}} \\ &= \frac{1}{(2\pi)^{3/2}} \int_{\mathbb{R}^3} (\nabla u_1 \overline{\nabla e^{-i\sigma x.\omega}} + u_2 \overline{i\sigma e^{-i\sigma x.\omega}}) \mathrm{d}^3 x \\ &= \frac{1}{(2\pi)^{3/2}} \int_{\mathbb{R}^3} (u_1 (\overline{-\Delta e^{-i\sigma x.\omega}}) + u_2 \overline{i\sigma e^{-i\sigma x.\omega}}) \mathrm{d}^3 x \\ &= \frac{1}{(2\pi)^{3/2}} \int_{\mathbb{R}^3} (\sigma^2 u_1 - i\sigma u_2) e^{i\sigma x.\omega} \mathrm{d}^3 x \\ &= \sigma^2 \hat{u}_1(-\sigma\omega) - i\sigma \hat{u}_2(-\sigma\omega) \,, \end{aligned}$$

where "^" denotes the Fourier transform on $\mathbb{R}^3$. Although the intermediate calculations do not, the final formula extends to $\mathcal{H}$ and the map that to U associates $\tilde{U}$ extends as an isometry from $\mathcal{H}$ onto $L^2(\mathbb{R}_\sigma \times S^2)$. This provides a spectral representation of A and its propagator:

$$\widetilde{AU} = \sigma\tilde{U}\,,\ \widetilde{e^{itA}U} = e^{it\sigma}\tilde{U}\,.$$

Then we take the Fourier transform in σ to obtain the new representation $\mathcal{R}$:

$$\mathcal{R}U(r,\omega) := \mathcal{F}_\sigma(\tilde{U}(.,\omega))(r)\,.$$

Then just as in the finite-dimensional case, $\mathcal{R}$ is a translation representation; we have

$$\mathcal{R}(e^{itA}U)(r,\omega) = (\mathcal{R}U)(r-t,\omega)\,.$$

This representation is of course also an isometry from $\mathcal{H}$ onto $L^2(\mathbb{R} \times S^2)$.

In a sense, the existence of a translation representer can be interpreted as the fact that there is no scattering; the evolution merely corresponds to the translation, without deformation, of a function representing the initial data. The Lax-Phillips theory does not stop there however. The representation $\mathcal{R}$ can be expressed as follows

$$\mathcal{R}U = \frac{1}{4\pi}(-\partial_s^2 Ru_1 + \partial_s Ru_2)(s,\omega)\,,$$

where R is the Radon transform R defined for $f \in \mathcal{C}_0^\infty(\mathbb{R}^3)$ by

$$Rf(s,\omega) = \int_{x.\omega=s} f(x)\mathrm{d}^2\sigma(x)\,.$$

This observation and the knowledge of the inverse Radon transform

$$R^*\psi(x) = \int_{S^2} \psi(x.\omega,\omega)\mathrm{d}^2\omega$$

give an explicit converse map to $\mathcal{R}$:

$$\mathcal{I}k = \frac{1}{2\pi}(R^*k\,,\quad - R^*\partial_s k)\,.$$

Lax and Phillips then use this to establish the *asymptotic profile* property which essentially re-interprets the translation representer as a future radiation field:

$$\mathcal{R}U(s,\omega) = -\lim_{r\to+\infty} r\partial_t\phi(r,(r+s)\omega)\,. \tag{44}$$

In addition, the fact that $\mathcal{I}$ is the inverse of $\mathcal{R}$ gives immediately an integral formula for the solution in terms of its translation representer

$$\phi(t,x) = \frac{1}{2\pi}\int_{S^2} \mathcal{R}U(x.\omega + t, \omega)\mathrm{d}^2\omega\,.$$

This is precisely Whittaker's formula from 1903 [42] for the solutions to the wave equation, which the Lax-Phillips theory allowed to reinterpret as providing the solution to the Goursat problem for the wave equation at null infinity.

Performing a similar construction for $-A$ instead of A, we obtain new representations $\check{\mathcal{R}}$ and $\check{\mathcal{I}}$ which relate the solution to its past radiation field. The scattering operator, which to the past radiation field associates the future radiation field, is then given by

$$S = \mathcal{R}\check{\mathcal{I}}\,.$$

4.2 Conformal Scattering on Minkowski Spacetime

The Lax Phillips theory provides a scattering operator turning past radiation fields into future radiation fields and which is therefore naturally understood as acting on compactified Minkowski spacetime. However the construction leading to this operator is entirely performed on $\mathbb{M}$, not on the compactified spacetime. Besides the techniques used require a static, or at least a stationary, background. We provide here an alternative construction of a similar scattering operator described in terms of the radiation field for ϕ instead of that for $\partial_t\phi$. Our construction is performed on the compactified spacetime using essentially the regularity of the conformal boundary. It follows the approach of [22] for the resolution of the Goursat problem and is done in three main steps. We adopt the full compactification of Minkowski spacetime and the notation of Sect. 3.2, i.e. the rescaled field is denoted $\tilde{\phi}$.

Step 1. We define the trace operator T^+ that to the data for $\tilde{\phi}$ at $\tau = 0$ associates the trace of $\tilde{\phi}$ on $\mathscr{I}^+$. This operator is well defined for data that extend as smooth functions on S^3, in particular for $\tilde{\phi}_0, \tilde{\phi}_1 \in \mathcal{C}_0^\infty(\mathbb{R}^3)$. The image of such data is a smooth function on $\mathscr{I}^+$.[7]

Step 2. We prove energy estimates both ways between the data and their image through T^+. In the case of Minkowski spacetime, we have a stronger result which is the energy equality (19), saying that for $\tilde{\phi}_0, \tilde{\phi}_1 \in \mathcal{C}^\infty(S^3)$,

$$\mathcal{E}_{K,\mathscr{I}^+}(\tilde{\phi}) = \mathcal{E}_{K,X_0}(\tilde{\phi})\,.$$

This implies that T^+ extends as a partial isometry from $H^1(S^3)\times L^2(S^3)$ into $H^1(\mathscr{I}^+)$ (see (17) and (18) for the energy norms on X_τ and $\mathscr{I}^+$) and is one-to-one with closed range.

[7]Note that in the case of Minkowski spacetime, we could straight away consider data $\tilde{\phi}_0, \tilde{\phi}_1 \in \mathcal{C}^\infty(S^3)$ to define T^+. However, this cannot be extended to more general situations. Hence we prefer to use a more flexible approach that is very common in scattering theory: to start with smooth compactly supported data and extend the operators by density using uniform estimates.

Step 3. In order to prove that the trace operator is onto, we only need to establish that its range is dense in $H^1(\mathscr{I}^+)$ since we already know that it is a closed subspace of $H^1(\mathscr{I}^+)$. This is done by solving the Goursat problem for a dense subset of $H^1(\mathscr{I}^+)$. We must be able to find a solution of the rescaled equation for which we have access to the trace on $\mathscr{I}^+$ in the strong sense and to check that this trace is the data we started from. For the Goursat problem on $\mathscr{I}^+$ for the rescaled wave equation on the Einstein cylinder, this follows from [22] for data in $H^1(\mathscr{I}^+)$, so we prove directly the surjectivity without resorting to the closed range property.

After these three steps, we have established that the operator T^+ is an isometry from $H^1(S^3) \times L^2(S^3)$ onto $H^1(\mathscr{I}^+)$. A similar construction can be performed in the past for the trace operator T^-. Then the scattering operator that maps past radiation fields to future radiation fields[8] is simply given by

$$S := T^+(T^-)^{-1} .$$

The above construction does not use any of the symmetries of Minkowski spacetime. All it requires is a regular conformal compactification. This is in fact quite strong. When dealing with asymptotically flat solutions to the Einstein vacuum equations, one does not expect that the conformal metric will be smooth at i^0 unless the ADM mass is zero and the spacetime is flat. In the case of black hole spacetimes, the situation is even worse since timelike infinities are rather strong singularities of the conformal metric. The method needs to be modified in order to deal with these singularities. The treatment of the difficulty at i^0 can be found in [29] and a conformal scattering on the Schwarzschild metric was developed in [34] with a way of dealing with timelike infinities. We give in the next two subsections the essential ingredients of the extension on the conformal scattering construction to spacetimes with singular i^0 and to black hole spacetimes.

4.3 The Case of Asymptotically Simple Spacetimes

4.3.1 Geometrical Background

We work on smooth globally hyperbolic asymptotically simple spacetimes that contain energy (the ADM energy is not zero) and such that $i^\pm$ are regular points of the conformal structure. The fact that the ADM energy does not vanish means in particular that the conformal structure is singular at i^0. More precisely, the type of spacetimes

[8]Here we call radiation fields the traces of the rescaled solution at $\mathscr{I}^\pm$. This is not quite the way the radiation fields were defined in (13) and (14). The two notions of radiation fields only differ by the presence of factors independent of the solution, which are the limits at $\mathscr{I}^\pm$ of the ratio of the two conformal factors $1/r$ and Ω. We shall use in this section the more flexible definition as traces of the rescaled field for the conformal factor we choose to work with.

we work on are as follows: a 4-dimensional, globally hyperbolic, Lorentzian space-time $(\mathscr{M}, g)$, $\mathscr{M} \simeq \mathbb{R}^4$, such that there exists another globally hyperbolic, Lorentzian space-time $(\hat{\mathscr{M}}, \hat{g})$ and a smooth scalar function Ω on $\hat{\mathscr{M}}$ satisfying:

(i) $\mathscr{M}$ is the interior of $\hat{\mathscr{M}}$, its boundary is the union of two points i^- and i^+ and a smooth null hypersurface $\mathscr{I}$, which is the disjoint union of the past light-cone $\mathscr{I}^+$ of i^+ and of the future light-cone $\mathscr{I}^-$ of i^-;

(ii) $\Omega > 0$, $\hat{g} = \Omega^2 g$ on $\mathscr{M}$, $\Omega = 0$ and $\mathrm{d}\Omega \neq 0$ on $\partial\mathscr{M}$;

(iii) every inextendible null geodesic in $\mathscr{M}$ acquires a future endpoint on $\mathscr{I}^+$ and a past endpoint on $\mathscr{I}^-$.
Since null unparametrized geodesics are conformally invariant objects and since from any point of $\mathscr{I}$ we can find a null geodesic for $\hat{g}$ that enters the spacetime, the definition above implies that $\mathscr{I}^+$ (resp. $\mathscr{I}^-$) is the set of future (resp. past) end-points of null geodesics. Spacelike infinity, i^0, is the boundary of Cauchy hypersurfaces in $\mathscr{M}$; it does not belong to $\hat{\mathscr{M}}$ therefore the definition remains somewhat abstract but we shall not need to make it more precise here.
In the results that have appeared sofar (i.e. [25, 29]), the following additional symmetry assumption was imposed:

(iv) $\mathscr{M}$ is diffeomorphic to Schwarzschild's spacetime outside the domain of influence of a given compact subset K of a Cauchy hypersurface Σ_0. In particular, we assume that outside the domain of influence of K, $\Omega = 1/r$ where r is the radial variable in the Schwarzschild coordinates.

The reason for this additional hypothesis is that (i)–(iii) do not give enough control on the geometry of $(\hat{\mathscr{M}}, \hat{g})$ near i^0 to perform energy estimates in that region. In order to gain this control, we can either impose some symmetry, or specify the decay in spacelike and null directions of $\hat{g}$ and its derivatives of order up to 2. The first solution has the advantage of simplicity. Besides, there are large classes of solutions to the Einstein vacuum equations whose behaviour is generically non stationary within the domain of influence of K and which satisfy assumptions (i)–(iv) (see [6–9]).

4.3.2 Dealing with Spacelike Infinity

On the class of asymptotically simple spacetimes defined above, we adopt the same strategy as on Minkowski spacetime in order to construct a conformal scattering theory. We describe the construction towards the future. A similar one can be performed towards the past and, put together, they provide the scattering operator. We choose a Cauchy hypersurface Σ_0 and denote by $\hat{\mathscr{M}}^+$ the part of $\hat{\mathscr{M}}$ in the future of Σ_0.

Step 1. We take smooth compactly supported initial data on Σ_0: $\hat{\phi}_0, \hat{\phi}_1 \in \mathcal{C}_0^\infty(\Sigma_0)$ and consider $\hat{\phi} \in \mathcal{C}^\infty(\mathscr{M})$, the associated solution of

$$\left(\Box_{\hat{g}} + \frac{1}{6}\mathrm{Scal}_{\hat{g}}\right)\hat{\phi} = 0 . \tag{45}$$

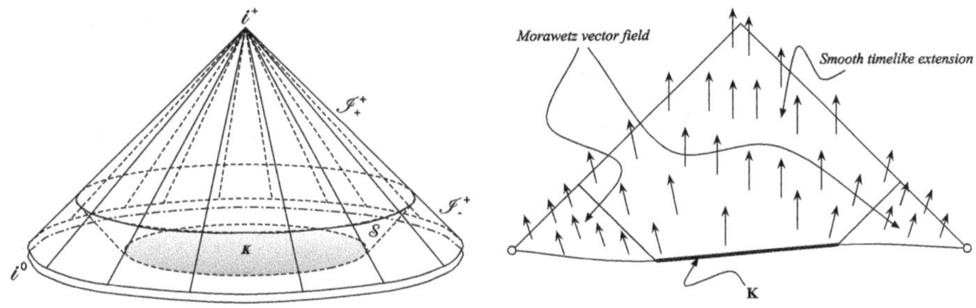

Fig. 3 Future of Σ_0 in $\hat{\mathscr{M}}$, then a 2D-cut with a choice of global timelike vector field

As a simple consequence of the finite propagation speed (which implies that the support of $\hat{\phi}$ remains away from i^0) and of the regularity of $\hat{g}$ up to the boundary of $\hat{\mathscr{M}}$, $\hat{\phi}$ extends as a smooth function on $\hat{\mathscr{M}}$. We can therefore define the trace operator T^+ that to data $\hat{\phi}_0$, $\hat{\phi}_1 \in \mathcal{C}_0^\infty(\Sigma_0)$ associates the trace of the associated solution $\hat{\phi}$ on $\mathscr{I}^+$.

Step 2. We establish estimates both ways between the energy of the data and that of their image through T^+. First, we need to choose a timelike vector field with which to define the energies. We use the symmetry assumption (iv). Outside the domain of influence of K, we have a Morawetz vector field associated to the Schwarzschild metric by the construction done in Sect. 3.3 and we extend it as a smooth timelike vector field T^a over $\hat{\mathscr{M}}$ (the decomposition of the future of Σ_0 into the domain of influence of K and its complement as well as the construction of T^a are shown in Fig. 3). Let us denote by $\mathscr{S}$ the boundary of the future domain of influence of K in $\mathscr{M}$, by $\mathscr{I}^+_+$ the part of $\mathscr{I}^+$ in the future of $\mathscr{S}$ and by $\mathscr{I}^+_-$ the part of $\mathscr{I}^+$ in the past of $\mathscr{S}$. Thanks to the regularity of the conformal metric at $\mathscr{I}^+$ and i^+, the estimates both ways between $\mathscr{I}^+_+$ and $K \cup \mathscr{S}$ are straightforward, simply requiring standard Gronwall estimates. We have

$$\mathcal{E}_{T,\mathscr{I}^+_+}(\hat{\phi}) \simeq \mathcal{E}_{T,\mathscr{S}}(\hat{\phi}) + \mathcal{E}_{T,K}(\hat{\phi}), \tag{46}$$

the constants in the equivalence depending only on the geometry and not on the solution considered. Outside the domain of influence of K, we have estimates both ways between the energies on $\mathscr{I}^+_- \cup \mathscr{S}$ and on $\Sigma_0 \setminus K$ using the peeling results:

$$\mathcal{E}_{T,\mathscr{I}^+_-}(\hat{\phi}) + \mathcal{E}_{T,\mathscr{S}}(\hat{\phi}) \simeq \mathcal{E}_{T,\Sigma_0 \setminus K}(\hat{\phi}). \tag{47}$$

Putting (46) and (47) together, we obtain the estimates both ways between the energies on $\mathscr{I}^+$ and Σ_0:

$$\mathcal{E}_{T,\mathscr{I}^+}(\hat{\phi}) \simeq \mathcal{E}_{T,\Sigma_0}(\hat{\phi}). \tag{48}$$

This implies that T^+ extends in a unique manner as a linear bounded operator defined on the completion $\mathcal{H}_{\Sigma_0}$ of $\mathcal{C}_0^\infty(\Sigma_0) \times \mathcal{C}_0^\infty(\Sigma_0)$ in the norm $\sqrt{\mathcal{E}_{T,\Sigma_0}(\hat{\phi})}$ with values in the completion $\mathcal{H}_{\mathscr{I}^+}$ of $\mathcal{C}_0^\infty(\mathscr{I}^+)$ in the norm $\sqrt{\mathcal{E}_{T,\mathscr{I}^+}(\hat{\phi})}$; the resulting operator is one-to-one and has closed range. Here $\mathcal{C}_0^\infty(\mathscr{I}^+)$ denotes smooth functions on $\mathscr{I}^+$ supported away from both i^+ and i_0. Since $\mathscr{I}^+$ is of dimension 3 and i^+ is merely a point, assuming the functions supported away from i^+ does not impose that the elements of the completion vanish at i^+.

Step 3. In order to show that T^+ is onto, we merely need to establish that its range is dense in $\mathcal{H}_{\mathscr{I}^+}$. We do this by solving the Goursat problem from $\mathscr{I}^+$ for data $\hat{\phi}_\infty \in \mathcal{C}_0^\infty(\mathscr{I}^+)$. We know from [22] that (45) has a unique solution $\hat{\phi} \in \mathcal{C}^\infty(\hat{\mathscr{M}}^+)$ whose restriction to $\mathscr{I}^+$ is $\hat{\phi}_\infty$. The difficulty is to see that

$$(\hat{\phi}_0\,,\ \hat{\phi}_1) := (\hat{\phi}|_{\Sigma_0}\,,\ \nabla_T\hat{\phi}|_{\Sigma_0}) \in \mathcal{H}_{\Sigma_0}\,, \tag{49}$$

which will then automatically entail that $\hat{\phi}_\infty = T^+(\hat{\phi}_0\,,\ \hat{\phi}_1)$ as well as the density of the range of T^+. The idea is to choose a spacelike hypersurface $\mathcal{S}$ in $\hat{\mathscr{M}}$, crossing $\mathscr{I}^+$ in the past of the support of $\hat{\phi}_\infty$ (see Fig. 4). The restrictions to $\mathcal{S}$ of $\hat{\phi}$ and $\nabla_T\hat{\phi}$ are smooth and the crucial observation is that, due to the location of $\mathcal{S}$ below the support of the data on $\mathscr{I}^+$, $\hat{\phi}|_{\mathcal{S}}$ vanishes at the boundary of $\mathcal{S}$, i.e. at $\mathcal{S} \cap \mathscr{I}^+$. The energy norm on $\mathcal{S}$ associated with the vector field T is equivalent to the natural $H^1 \times L^2$ norm on $\mathcal{S}$ for the rescaled metric $\hat{g}$. Therefore, $\hat{\phi}|_{\mathcal{S}} \in H_0^1(\mathcal{S})$. It follows that $(\hat{\phi}|_{\mathcal{S}}\,,\ \nabla_T\hat{\phi}|_{\mathcal{S}})$ can be approached, in the energy norm, by a pair of sequences $(\hat{\phi}_0^n\,,\ \hat{\phi}_1^n)$ of smooth functions on $\mathcal{S}$, supported away from $\mathscr{I}^+$. We denote by $\hat{\phi}^n$ the solution of (45) in $\mathcal{C}^\infty(\hat{\mathscr{M}})$ such that

$$\hat{\phi}^n|_{\mathcal{S}} = \hat{\phi}_0^n\,,\ \nabla_T\hat{\phi}^n|_{\mathcal{S}} = \hat{\phi}_1^n\,.$$

Since by finite propagation speed $\hat{\phi}^n$ is supported away from i^0, we have estimates both ways between the energies of $\hat{\phi}^n$ on $\mathcal{S}$ and on Σ_0, uniformly in n. The convergence of $(\hat{\phi}^n|_{\mathcal{S}}\,,\ \nabla_T\hat{\phi}^n|_{\mathcal{S}})$ towards $(\hat{\phi}|_{\mathcal{S}}\,,\ \nabla_T\hat{\phi}|_{\mathcal{S}})$ in the energy norm on $\mathcal{S}$ together with the uniform energy estimates imply that $(\hat{\phi}^n|_{\Sigma_0}\,,\ \nabla_T\hat{\phi}^n|_{\Sigma_0})$ is a Cauchy sequence in $\mathcal{H}_{\Sigma_0}$. This entails that $(\hat{\phi}_0\,,\ \hat{\phi}_1) \in \mathcal{H}_{\Sigma_0}$.

Therefore the trace operator T^+ extends as an isomorphism from $\mathcal{H}_{\Sigma_0}$ onto $\mathcal{H}_{\mathscr{I}^+}$. We can construct T^- in a similar manner and the scattering operator $S = T^+(T^-)^{-1}$ is then an isomorphism from $\mathcal{H}_{\mathscr{I}^-}$ onto $\mathcal{H}_{\mathscr{I}^+}$.

Compared to the case of Minkowski spacetime, the essential change is just the loss of the regularity of the conformal metric at i^0, the loss of symmetry is of no importance for our construction. This loss of regularity at i^0 is dealt with in a simple manner using essentially the finite propagation speed.

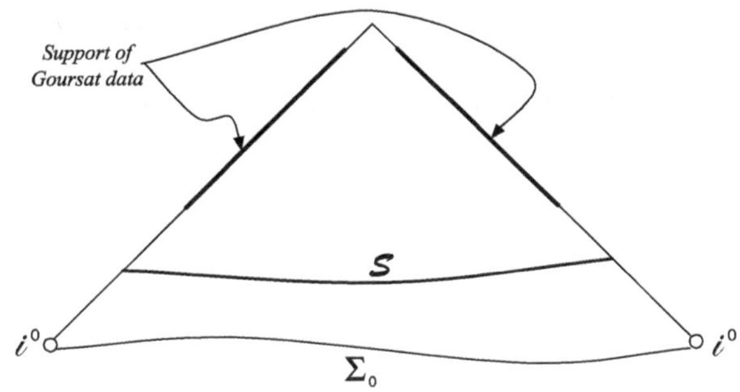

Fig. 4 A choice of intermediate hypersurface $\mathcal{S}$ for estimating the energy on Σ_0 of the solution to the Goursat problem

4.4 Conformal Scattering on the Schwarzschild Metric

For spacetimes describing isolated black holes in an asymptotically flat universe, the conformal compactification of the exterior is more complicated than for asymptotically simple spacetimes. The singularity at i^0 is the same as in the asymptotically simple case, but now the conformal metric is also singular at timelike infinities. This is a much more serious difficulty than the singularity at spacelike infinity. The reason is that finite propagation speed cannot help us here, whatever hypothesis we may make on the supports of the data, i^+ will be in their domain of influence. The crucial step is to establish estimates both ways between the energy of the data and that of the trace of the rescaled solution at the conformal boundary. Since the solutions propagate into i^+ and the conformal structure is singular there, the regularity of the rescaled field at i^+, which allowed to apply Stokes's theorem for the energy current, needs to be replaced by asymptotic information on the behaviour of the field at timelike infinity, typically a sufficiently fast and uniform decay rate in timelike directions. Once this is done, the rest of the construction is mostly unchanged because the behaviour of the conformal metric at i^+ has no influence on the propagation of solutions from their future scattering data into the spacetime. We give in this subsection the essential features of the conformal scattering theory developed on the Schwarzschild metric in [34]. For more technical details as well as a discussion of the additional difficulties in the case of the Kerr metric, see [34] and references therein, in particular the recent work by Dafermos et al. [11].

When we considered the conformal compactification of the exterior of the Schwarzschild metric in Sect. 3.4, we were only interested in constructing $\mathscr{I}^+$ and in working in the neighbourhood of i^0. We are now attempting to develop a conformal scattering theory; this requires to understand the global geometry of the exterior of the black hole. We still perform the compactification using the simplest conformal factor $\Omega = 1/r$, but we look at the rescaled metric using different coordinate systems in order to construct all the components of the boundary. Recall that the Schwarzschild metric is given on $\mathbb{R}_t \times]0, +\infty[_r \times S^2_\omega$ by

$$g = F\mathrm{d}t^2 - F^{-1}\mathrm{d}r^2 - r^2\mathrm{d}\omega^2\,,\; F = 1 - 2m/r$$

and the change of radial variable $r_* = r + 2m\log(r - 2m)$ maps the exterior of the black hole $\mathbb{R}_t \times]2m, +\infty[_r \times S^2_\omega$ to the domain $\mathbb{R}_t \times \mathbb{R}_{r_*} \times S^2_\omega$ with the new expression for g

$$g = F(\mathrm{d}t^2 - \mathrm{d}r_*^2) - r^2\mathrm{d}\omega^2\,.$$

Using coordinates $u = t - r_*$, $R = 1/r$, ω, the rescaled metric $\hat{g} = R^2 g$ reads

$$\hat{g} = R^2(1 - 2mR)\mathrm{d}u^2 - 2\mathrm{d}u\mathrm{d}R - \mathrm{d}\omega^2\,.$$

In this coordinate system, $\mathscr{I}^+$ and the past horizon $\mathscr{H}^-$ appear as the smooth null hypersurfaces

$$\mathscr{I}^+ = \mathbb{R}_u \times \{0\}_R \times S^2_\omega\,,\; \mathscr{H}^- = \mathbb{R}_u \times \{1/2m\}_R \times S^2_\omega\,.$$

Similarly, using the coordinates $v = t + r_*$, R, ω, the metric $\hat{g}$ takes the form

$$\hat{g} = R^2(1 - 2mR)\mathrm{d}v^2 + 2\mathrm{d}v\mathrm{d}R - \mathrm{d}\omega^2\,.$$

We now have access to past null infinity $\mathscr{I}^-$ and to the future horizon $\mathscr{H}^+$, appearing respectively as the smooth null hypersurfaces

$$\mathscr{I}^- = \mathbb{R}_v \times \{0\}_R \times S^2_\omega\,,\; \mathscr{H}^+ = \mathbb{R}_v \times \{1/2m\}_R \times S^2_\omega\,.$$

At the past and future horizons, not only the rescaled metric, but also the physical metric g, extends analytically as a non degenerate metric; $\mathscr{H}^+$ and $\mathscr{H}^-$ meet at a 2-sphere S^2_c, called the crossing sphere, at which both g and $\hat{g}$ extend analytically and are non degenerate. The construction of S^2_c can be done using Kruskal-Szekeres coordinates (see for example [21] or [41]). The Penrose diagram of the compactified exterior is given in Fig. 5. Instead of two null hypersurfaces diffeomorphic to $\mathbb{R} \times S^2$, the boundary of our compactified spacetime now contains four such hypersurfaces.

Remark 4.1 Note that the different components of the boundary of the rescaled exterior of the black hole are of two quite different natures. Null infinities $\mathscr{I}^\pm$ on the one hand are genuinely part of the conformal boundary of the spacetime, describing "points at infinity." The horizons $\mathscr{H}^\pm$ on the other hand do not describe points at infinity for g but the finite boundary of the exterior of the black hole. They are part of the boundary of our compactified spacetime only because we restrict our study to the exterior of the black hole. This is justified by the fact that our scattering theory is assumed to reflect the point of view of an observer static at infinity, whose perception does not go beyond the horizon.

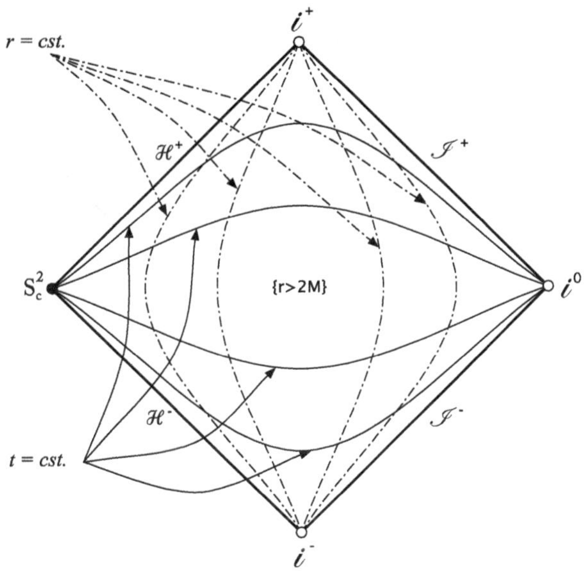

Fig. 5 The conformal compactification of the exterior of the *black hole*

We consider the Cauchy hypersurface

$$\Sigma_0 = \{0\}_t \times \mathbb{R}_{r_*} \times S^2_\omega .$$

In order to establish estimates both ways between Σ_0 and $\mathscr{I}^+ \cup \mathscr{H}^+$, we proceed in two steps.

1. For $T > 0$, we consider the three hypersurfaces[9] (see Fig. 6)

$$S_T = \left\{ (t, r_*, \omega) \in \mathbb{R} \times \mathbb{R} \times S^2 \, ; \; t = T + \sqrt{1 + r_*^2} \right\} , \tag{50}$$

$$\mathscr{I}^+_T = \mathscr{I}^+ \cap \{u \leq T\} =]-\infty, T]_u \times \{0\}_R \times S^2_\omega , \tag{51}$$

$$\mathscr{H}^+_T = S^2_c \cup (\mathscr{H}^+ \cap \{v \leq T\}) = S^2_c \cup (]-\infty, T]_v \times \{1/2m\}_R \times S^2_\omega) . \tag{52}$$

We establish energy estimates both ways between Σ_0 and $\mathscr{H}^+_T \cup S_T \cup \mathscr{I}^+_T$, uniformly in $T > 0$. We use the stress-energy tensor (37) for the wave equation (36) associated with $\hat{g}$. We recall its conservation law (38) for $\hat{\phi}$ solution of (36):

$$\hat{\nabla}^a \hat{T}_{ab} = -2mR\hat{\phi}\hat{\nabla}_b\hat{\phi} .$$

[9]We give here an explicit choice of hypersurface S_T, but the only important properties that S_T needs to satisfy are that it is achronal for the rescaled metric and that $\Sigma_0 \cup \mathscr{H}^+_T \cup S_T \cup \mathscr{I}^+_T$ forms a closed hypersurface (except where $\mathscr{I}^+$ and Σ_0 meet i^0).

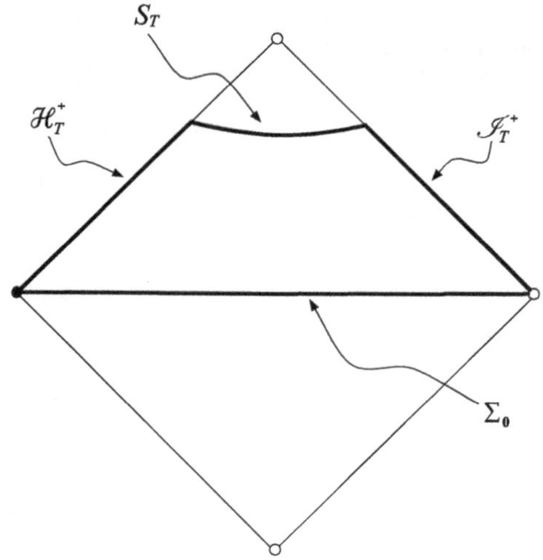

Fig. 6 The main hypersurfaces represented on the compactified exterior

This entails that the energy current

$$\hat{J}_a = K^b \hat{T}_{ab}\,,\ K = \partial_t\,,$$

is not conserved but satisfies instead

$$\hat{\nabla}^a \hat{J}_a = -2mR\hat{\phi}\partial_t\hat{\phi}\,.$$

However, thanks to the symmetries of Schwarzschild's spacetime, this equation can easily be seen as the exact conservation law

$$\hat{\nabla}_a \left(\hat{J}^a + V^a \right) = 0\,,\ \text{with } V = mR\hat{\phi}^2\partial_t\,. \tag{53}$$

Since V is causal and future oriented, we still have that the flux of the modified current $J + V$ defines a positive definite (resp. non negative) quadratic form on spacelike (resp. achronal) hypersurfaces. We introduce the modified energy on an oriented hypersurface S

$$\hat{\mathcal{E}}_{\partial_t, S} = \int_S (\hat{J}_a + V_a) n^a (l \lrcorner \mathrm{dVol})\,,$$

where l^a is a vector field transverse to S compatible with the orientation of S and n^a a normal vector field to S such that $\hat{g}(l, n) = 1$. For any $T > 0$, we have

$$\hat{\mathcal{E}}_{\partial_t, \Sigma_0} = \hat{\mathcal{E}}_{\partial_t, \mathscr{I}_T^+} + \hat{\mathcal{E}}_{\partial_t, \mathscr{H}_T^+} + \hat{\mathcal{E}}_{\partial_t, S_T}\,. \tag{54}$$

2. We take T to $+\infty$. The modified energies on $\mathscr{H}_T^+$ and $\mathscr{I}_T^+$ tend to $\hat{\mathcal{E}}_{\partial_t,\mathscr{H}^+}$ and $\hat{\mathcal{E}}_{\partial_t,\mathscr{I}^+}$ respectively. The last thing we need in order to conclude is to show that $\hat{\mathcal{E}}_{\partial_t,S_T}$ tends to zero. On the Schwarzschild metric, we know enough on the decay of solutions to the wave equation to infer this (see Dafermos and Rodnianski [10]). Generally for this type of approach, the expected generic decay known as Price's law will be sufficient to establish that $\hat{\mathcal{E}}_{\partial_t,S_T}$ tends to zero.

As mentioned above, the rest of the construction is essentially unchanged.

5 Concluding Remarks

We have described two approaches to asymptotic analysis making a fundamental use of conformal compactifications. Both constructions in fact rely on a choice of spacelike hypersurface as an intermediate tool. In the case of conformal scattering, the Cauchy hypersurface is used to construct the trace operators $T^{\pm}$, which play the role of inverse wave operators. The object that the theory aims to construct is the scattering operator, mapping the past radiation field to the future radiation field. This operator is independent of the choice of spacelike hypersurface from which the trace operators are defined and the theory is in fact truly covariant. For our approach of the peeling, the choice of Cauchy hypersurface is more fundamental since we study the asymptotic properties of Cauchy data that entail a certain transverse regularity of the rescaled solution at null infinity. However, provided we only work with asymptotically flat Cauchy hypersurfaces, these asymptotic properties ought to be independent of the choice of Cauchy hypersurface and in this sense the theory could also be understood as covariant. An interesting alternative (and much more delicate) approach to the peeling would be to characterize the transverse regularity at $\mathscr{I}^+$ in terms of the function space of past scattering data.

Acknowledgements The author would like to thank the organizer of the conference, Athanase Papadopoulos, for the invitation. The research presented in this essay was partly supported by the ANR funding ANR-12-BS01-012-01.

References

1. J.C. Baez, Scattering and the geometry of the solution manifold of $\Box f + \lambda f^3 = 0$. J. Funct. Anal. **83**, 317–332 (1989)
2. J.C. Baez, Scattering for the Yang-Mills equations. Trans. Am. Math. Soc. **315**(2), 823–832 (1989)
3. J.C. Baez, Conserved quantities for the Yang-Mills equations. Adv. Math. **82**(1), 126–131 (1990)
4. J.C. Baez, I.E. Segal, Zhou Z. F., The global Goursat problem and scattering for nonlinear wave equations. J. Funct. Anal. **93**(2), 239–269 (1990)

5. J.C. Baez, Z.F. Zhou, The global Goursat problem on $\mathbb{R} \times S^1$. J. Funct. Anal. **83**, 364–382 (1989)
6. P. Chrusciel, E. Delay, Existence of non trivial, asymptotically vacuum, asymptotically simple space-times. Class. Quantum Grav. **19**, L71–L79, (2002). (erratum Class. Quantum Grav. **19**(2002), 3389 (2002))
7. P. Chrusciel, E. Delay, On mapping properties of the general relativistic constraints operator in weighted function spaces, with applications. Mém. Soc. Math. Fr. (N.S.) **94**, vi+103 (2003)
8. J. Corvino, Scalar curvature deformation and a gluing construction for the Einstein constraint equations. Commun. Math. Phys. **214**, 137–189 (2000)
9. J. Corvino, R.M. Schoen, On the asymptotics for the vacuum Einstein constraint equations, gr-qc 0301071 (2003). J. Differ. Geom. **73**(2), 185–217 (2006)
10. M. Dafermos, I. Rodnianski, The redshift effect and radiation decay on black hole space-times. Commun. Pure Appl. Math. **62**(7), 859–919 (2009)
11. M. Dafermos, I. Rodnianski, Y. Shlapentokh-Rothman, A scattering theory for the wave equation on Kerr black hole exteriors, arXiv:1412.8379
12. J. Dimock, Scattering for the wave equation on the Schwarzschild metric. Gen. Relat. Grav. **17**(4), 353–369 (1985)
13. J. Dimock, B.S. Kay, Classical and quantum scattering theory for linear scalar fields on the Schwarzschild metric. II. J. Math. Phys. **27**(10), 2520–2525 (1986)
14. J. Dimock, B.S. Kay, Classical and Quantum Scattering theory for linear scalar fields on the Schwarzschild metric I. Ann. Phys. **175**, 366–426 (1987)
15. J. Franchi, From Riemannian to relativistic diffusions, to appear as. (This volume)
16. F.G. Friedlander, On the radiation field of pulse solutions of the wave equation. Proc. R. Soc. Ser. A **269**, 53–65 (1962)
17. F.G. Friedlander, On the radiation field of pulse solutions of the wave equation II. Proc. R. Soc. Ser. A **279**, 386–394 (1964)
18. F.G. Friedlander, On the radiation field of pulse solutions of the wave equation III. Proc. R. Soc. Ser. A **299**, 264–278 (1967)
19. F.G. Friedlander, Radiation fields and hyperbolic scattering theory. Math. Proc. Camb. Philos. Soc. **88**, 483–515 (1980)
20. F.G. Friedlander, Notes on the Wave Equation on Asymptotically Euclidean Manifolds. J. Funct. Anal. **184**, 1–18 (2001)
21. S.W. Hawking, G.F.R. Ellis, *The large scale structure of space-time*, Cambridge monographs in mathematical physics (Cambridge University Press, 1973)
22. L. Hörmander, A remark on the characteristic Cauchy problem. J. Funct. Anal. **93**, 270–277 (1990)
23. E. Humbert, On the positive mass theorem for closed Riemannian manifolds, to appear as. (This volume)
24. W. Inglese, F. Nicolò, Asymptotic properties of the electromagnetic field in the external Schwarzschild spacetime. Ann. Henri Poincaré **1**(5), 895–944 (2000)
25. J. Joudioux, Conformal scattering for a nonlinear wave equation. J. Hyperbolic Differ. Equ. **9**(1), 1–65 (2012)
26. P.D. Lax, R.S. Phillips, *Scattering Theory* (Academic Press, 1967)
27. J. Leray, *Hyperbolic Differential Equations*, Lecture notes (Princeton Institute for Advanced Studies, 1953)
28. M. Mars, On local characterization results in geometry and gravitation, to appear as. (This volume)
29. L.J. Mason, J.-P. Nicolas, Conformal scattering and the Goursat problem. J. Hyperbolic Differ. Equ. **1**(2), 197–233 (2004)
30. L.J. Mason, J.-P. Nicolas, Regularity at space-like and null infinity. J. Inst. Math. Jussieu **8**(1), 179–208 (2009)
31. L.J. Mason, J.-P. Nicolas, Peeling of Dirac and Maxwell fields on a Schwarzschild background. J. Geom. Phys. **62**, 867–889 (2012)

32. C.S. Morawetz, The decay of solutions of the exterior initial-boundary value problem for the wave equation. Commun. Pure Appl. Math. **14**, 561–568 (1961)
33. E.T. Newman, R. Penrose, An approach to gravitational radiation by a method of spin coefficients. J. Math. Phys. **3**, 566–768 (1962)
34. J.-P. Nicolas, Conformal scattering on the Schwarzschild metric, arXiv:1312.1386
35. R. Penrose, Asymptotic properties of fields and space-times. Phys. Rev. Lett. **10**, 66–68 (1963)
36. R. Penrose, *Conformal treatment of infinity*, 1964 Relativité, Groupes et Topologie (Lectures, Les Houches, 1963 Summer School of Theoret. Phys., Univ. Grenoble) 563–584 Gordon and Breach, New York
37. R. Penrose, Zero rest-mass fields including gravitation: asymptotic behaviour. Proc. R. Soc. Lond. A **284**, 159–203 (1965)
38. R. Penrose, W. Rindler, *Spinors and Space-Time*, vol. I, (Cambridge University Press, 1984). (vol. 2, (1986))
39. R. Sachs, Gravitational waves in general relativity VI, the outgoing radiation condition. Proc. R. Soc. Lond. A **264**, 309–338 (1961)
40. R. Sachs, Gravitational waves in general relativity VIII, waves in asymptotically flat space-time. Proc. R. Soc. Lond. A **270**, 103–126 (1962)
41. R. Wald, *General Relativity* (University of Chicago Press, 1984)
42. E.T. Whittaker, On the partial differential equations of mathematical physics. Mathematische Annalen **57**, 333–355 (1903)

Part V
Concluding Chapter

Bernhard Riemann and His Work

Lizhen Ji

Abstract This chapter attempts to give a brief summary of the life and work of Riemann. It tries to address the following issues: the influence of his education and early life on his work, and a summary of his major works and an overview of his work, and the general impact of his work through the concepts and terminology named after him.

1 Introduction

Bernhard Riemann (1826–1866) was one of the most original mathematicians in the history of mathematics. Though he had written only a small number of papers, he changed the way we view and do mathematics. For example, to get a taste of his impact, it suffices to think of several concepts named after him: Riemann surfaces, Riemann-Roch theorem, Riemann zeta function, and Riemannian geometry.

Specifically, before Riemann, complex functions were studied on domains of the complex plane $\mathbb{C}$, and the issue of analytic continuation and the ensuing problem of multi-valuedness were complicated. Now after the work of Riemann, the proper way to understand a complex analytic function and its analytic continuation, in particular to understand properly an algebraic function defined by an algebraic equation, is to consider its associated Riemann surface. One can even turn things around so that Riemann surfaces become the more basic spaces, and holomorphic functions on them can be viewed as maps between Riemann surfaces. (In the language of categories, Riemann surfaces become objects and holomorphic maps become morphisms between the spaces.) This thinking led to the Riemann mapping theorem, which in turn led to the uniformization theorem for Riemann surfaces, which is one of the most important theorems in mathematics. Consequently, geometry and analysis are now intimately connected. Riemann proved the crucial part of Riemann's inequality in the Riemann-Roch Theorem. In doing this, he initiated the theory of algebraic topology and made topology an essential part of geometric function theory and geometric

L. Ji (✉)
Department of Mathematics, University of Michigan, Ann Arbor, MI 48109, USA
e-mail: lji@umich.edu

L. Ji et al. (eds.), *From Riemann to Differential Geometry and Relativity*,
DOI 10.1007/978-3-319-60039-0_20

analysis. The Riemann-Roch theorem also motivated the important index theories for higher dimensional manifolds and algebraic varieties.

Similarly, the Riemann zeta function brought the methods of complex analysis into analytic number theory, and it is now hard to imagine what remains if we remove ideas of complex analysis, for example, the Riemann zeta function and generalized Riemann zeta functions, from analytic number theory. Generalized Riemann zeta functions include L-functions which are also crucial in algebraic number theory and arithmetic geometry. For example, without them, it is almost impossible to formulate the celebrated Langlands program.

In differential geometry, the idea of Riemannian metric, Riemannian connections and their non-positive definite generalizations in the theory of relativity are indispensable to most people working in the subject, and higher dimensional generalizations of the Riemannian mapping theorem can also be expressed in terms of Riemannian geometry.

Besides his geometric and global view, Riemann's work has several other features. Riemann's intuitive reasoning played a crucial role in his work. It might fall somewhat short of a rigorous proof, but the brilliant ideas in his works are so much clearer, because his work is not overly filled with lengthy computations (though heavy computations were important for him to gain his understanding and intuition). One example of Riemann's intuitive reasoning is his application of Dirichlet's principle to various problems such as the Riemann mapping theorem and the Riemann-Roch theorem.

Integration between mathematics and physics runs through a large part of Riemann's work. According to Klein [8, p. 167], this integration is the source of inspiration for Riemann:

> he endeavored again and again to find a general mathematical formulation for the laws underlying all natural phenomena.... *these physical views are the mainspring of Riemann's purely mathematical investigations.*

Riemann had a bigger perspective and was a better philosopher than most mathematicians. For example, his paper on the hypothesis which lies at the basis of geometry contains a substantial philosophic part. Not only the intrinsic geometry is important to him, its relation to reality, or to the real space we live in, is also important to him. According to Freudenthal [4], Riemann was

> one of the most profound and imaginative mathematicians of all time, he had a strong inclination to philosophy, indeed, was a great philosopher. Had he lived and worked longer, philosophers would acknowledge him as one of them.

Riemann was also capable of doing very extensive and heavy calculations. This can be confirmed by some existing calculating sheets or scratching papers used by him at the Nachlass of Riemann in the Historical Library of the University of Göttingen. This is consistent with Riemann's ability and joy in doing computation when he was a young boy. Riemann resembled Gauss in the aspect of powerful computation. Computation was crucial to Riemann, and his intuition, deep insights and understanding might result from, but is certainly supported by, his extensive computations.

Besides his originality, Riemann was also broad. His work covered almost all major subjects in mathematics. It is important and interesting to get an overview of the work of Riemann. This task is probably beyond the ability of any single person, though several people including Klein [8], the editor of the Russian edition of Riemann's Collected works, and the author of the book [9], had tried. Narasimhan edited the 1990 edition of Riemann's collected works [16]. In the preface, he also summarized and commented on some of Riemann's major works after writing "No one person is capable of a full analysis of Riemann's work, its history, its development and its influence on mathematics."

In this book, there are several chapters devoted to summarizing and analyzing some aspects of Riemann's works, for example, [5, 12–14, 18]. On the other hand, in this chapter, we want to give a brief overview of all major aspects of Riemann's work and life. Given the long lasting impact of Riemann on the development of mathematics, it is worthwhile to view Riemann and his work from different perspectives. This chapter should complement other chapters of this book. We also hope to understand better, or give some explanation for, his work from the perspective of his education and interaction with others.

2 Riemann's Work I: His Best Known Works

As we mentioned earlier, Riemann made deep contributions to many different subjects. In the opinion of most mathematicians, he is probably best known for his work in the following four subjects:

1. Complex analysis.
2. Real analysis.
3. Riemannian geometry.
4. Number theory.

We will start with the most famous result of his work during his life time, and then summarize some of his major results in the above four subjects.

1.a. *Abelian functions and the Jacobian inversion problem.*
Given what most people know about Riemann, it might be surprising to point out that during Riemann's life time, he first became famous and probably was best known for his solution of the Jacobi inversion problem for Riemann surfaces (or algebraic functions) of higher genus.[1] Though this particular result of Riemann is

[1] As it is known, the theory of elliptic integrals motivated the theory of Abelian integrals. One basic insight in the theory of elliptic integrals is to consider the inverse of an elliptic integral. The general framework for Abelian integrals on a fixed compact Riemann surface is the map via integration of holomorphic 1-forms from the group of divisors of degree 0 to the Jacobian variety of the Riemann surface. To invert this map, we need to identify its image. One version of the Jacobi inversion problem, as presented in most modern textbooks on Riemann surfaces, asks to prove that this map is surjective. The full version asks, in addition, for a precise description of the inverse images of this map.

still important, it is probably fair to say that it is not the most noticeable achievement of Riemann to many mathematicians today. According to Klein [8, p. 172]:

> It must always be regarded as one of the greatest achievements of Jacobi that, by a sort of inspiration, he established for these integrals a problem of inversion which furnishes single-valued functions just as the simple inversion does in the case of the elliptic integrals. The actual solution of this problem of inversion is the central task performed at the same time, but by different methods, by Weierstrass and Riemann. The great memoir on the Abelian functions in which Riemann published his theory in 1857 has always been recognized as the most brilliant of all the achievements of his genius. Indeed, the result is here reached, not by laborious calculations, but in the most direct way, by a proper combination of the geometrical considerations just referred to....
>
> The second half, which is concerned with the theta-series, is perhaps still more remarkable. The important result is here deduced that the theta-series required for the solution of Jacobi's problem of inversion are not the general theta-series; and this leads to the new problem of determining the position of the general theta-series in our theory.

This solution to the famous inversion problem of Jacobi was contained in Riemann's paper *Theory of Abelian functions* in 1857, which continued where his doctoral dissertation had left off and developed further the idea of Riemann surfaces and their topological properties. He examined multi-valued functions as single valued over a special Riemann surface and used these results to solve general inversion problems which had been solved for the special case of elliptic integrals by Abel and Jacobi.

1.b. *Complex analysis and Riemann surfaces.*
The title of Riemann's thesis, "Foundations for a general theory of functions of a complex variable", is very appropriate, and this paper is most important to people in complex analysis. He made systematic use of the Cauchy-Riemann equation, explained the special feature of functions of complex variables versus real variables, introduced Riemann surfaces as the natural domains of holomorphic functions, pointed out the right geometric meaning of such functions as conformal mappings, the relevance of topology in analysis, and proposed the Riemann mapping theorem and outlined a proof. If we take a moment and think of how many new concepts, directions, perspectives Riemann introduced and approaches to understand them, it is totally astonishing. It is hard to imagine modern mathematics without them.

Six years after his thesis, Riemann published his paper on Abelian functions, which was instantly considered or recognized as a masterpiece, as mentioned in the quote from Klein above. It was the result of work carried out over several years and contained in a lecture course he gave to only three people in 1855–56, including Dedekind. This paper also contained many groundbreaking ideas and results: for example, Riemann's inequality on the dimension of spaces of meromorphic functions with prescribed singularities which is a crucial part of the Riemann-Roch theorem, basic results on Jacobian varieties and Riemann theta functions, the Riemann-Hurwitz formula, a solution of the Jacobi inversion problem on Abelian integrals (as mentioned at the beginning of this section), the birational geometry of algebraic curves and the notion of moduli of Riemann surfaces (or algebraic curves).

Most of the results which were proved and of the problems raised in this paper are still very central in mathematics. For example, Riemann's moduli space is arguably the most important space in algebraic geometry. For more details about the history and development of moduli space of Riemann surfaces, see [1, 6]. The extensive and influential subject of index theories was motivated by the Riemann-Roch theorem, and various index theorems can be viewed as higher dimensional generalizations of the Riemann-Roch theorem. See the informal and historical discussions in [19] and the references there.

It is perhaps helpful to point out that one of the main motivations for Weierstrass to develop complex analysis was to solve the Jacobi inversion problem. According to Klein [7, p. 263],

> Weierstrass now had a life goal: Through rigorous, methodical work on power series (also of several variables) to master the inversion problem for hyperelliptic integrals of arbitrarily high rank, as it had been set by the divinatory Jacobi–perhaps even for the most general Abelian integrals.
>
> It is on this path that what is called Weierstrass's theory of analytic functions appeared, to to speak, as a mere by-product.

Klein continued [7, p. 264]:

> when Weierstrass submitted a first treatment of general Abelian functions to the Berlin Academy in 1857, Riemann's paper on the same theme appeared in Crelle's Journal, Volume 54. It contained so many unexpected, new concepts that Weierstrass withdrew his paper and in fact published no more. It must have caused Weierstrass extraordinary agitation.
>
> In any case, in the winter of 1859/60 he showed traces of overwork, and these were followed in 1861 by a complete nervous breakdown, ...

The Dirichlet principle was a crucial tool in both Riemann's thesis and his paper on Abelian functions. But that is probably also the most famous shortcoming in Riemann's work: a non-rigorous application of Dirichlet principle. The problem can be described as follows. It is well-known that if a functional (or function) has a lower bound, its infimum may not be realized. Weierstrass showed that a minimizing function was not guaranteed by the Dirichlet Principle, and this made people doubtful about Riemann's methods, though not about his results. For example, Weierstrass firmly believed in Riemann's results, and asked his student Hermann Schwarz to try to find other proofs of Riemann's existence theorems which do not use the Dirichlet Principle. Schwarz succeeded in 1869–70. Finally, in 1901 Hilbert mended Riemann's approach by giving the correct form of the Dirichlet Principle through the so-called direct methods of calculus of variations. Thus, he succeeded in making Riemann's proofs rigorous, and at the same time, he contributed substantially to development of the theory of the calculus of variation.

It is also important to note that the search for a rigorous proof of Riemann's theorems had been fruitful in other subjects too. For example, many important ideas in algebraic geometry were discovered by Clebsch, Dedekind, Gordan, Brill and Max Noether, Weber while they tried to reprove Riemann's results using algebraic methods. One striking publication is the long paper by Dedekind and Weber [2].

2. *Fourier series and Riemann integral.*
In his Habilitation dissertation, the degree which would allow him to become a lecturer, Riemann studied the problem of the representability of functions by trigonometric series. To do this, he introduced the notion of Riemann integral and gave the necessary and sufficient conditions in terms of the subset of discontinuity for a function to have an integral. This is now called the condition of Riemann integrability.

After this preparation, Riemann studied conditions on functions so that they can be represented by a Fourier series. He wrote [15, pp. 234–235]:

> The previous work on this topic served the purpose of proving the Fourier series for the cases occurring in nature. Hence the proofs could start for an arbitrary function, and later for the purposes of the proof one could impose arbitrary restrictions on the function, when they did not impair the goal. For our purposes we only impose conditions necessary for the representation of the function. Hence we must first look for necessary conditions for the representation, and from these select sufficient conditions for the representation. While the previous work showed: 'If a function has this or that property then it is represented by a Fourier series', we must start from the converse question: If a function is represented by a Fourier series, what are the consequences for the function, regarding the changes of its values with a continuous change of the argument?

This paper of Riemann had a huge impact on the subject of Fourier series. (See the book [20] for a systematic exposition of the subject, where Riemann's work is clearly visible.). One particular example is the Riemann-Lebesgue Lemma which appeared in Sect. 9 of [15, p. 243]. This paper also led to the work of Cantor on the uniqueness of the representation by trigonometric series, which in turn led to Cantor's famous set theory.

3. *Riemannian geometry.*
Though Riemann wrote only few papers on geometry, his geometric way of thinking permeated all his work. To complete his Habilitation, Riemann had to give a trial lecture. As it will be discussed below, Gauss picked the unexpected third topic: *On the hypotheses that lie at the foundations of geometry*. There were two parts to Riemann's lecture. In the first part he posed the problem of how to define an n-dimensional space and he ended up giving a definition of what today we call a Riemannian space. One of the main points of this part of Riemann's lecture was the definition of the curvature tensor. The second part of Riemann's lecture posed deep questions about the relationship of geometry to the world we live in. For example, he asked what the dimension of real space was and whether geometry described real space.

As we all know, there is not a big jump from calculus of functions in two variables to calculus of functions of more variables. In some sense, Riemann's paper on the foundation of geometry is less original in view of Gauss' work on the intrinsic geometry of surfaces in space. On the other hand, besides the technical difficulties in higher dimension, for example, arriving at the right formulation of Riemannian curvature, Riemann changed people's ways of thinking of spaces and geometry. Manifolds are not necessarily sitting in some standard ambient spaces. This perspective and his emphasis on the relation between mathematical geometry and reality is also profound. (One should note that when Riemann entered the subject, the newly discovered non-Euclidean geometry weighed heavily on people. Riemann changed people's mode of thinking of such matters.) All these aspects have had huge impact on both mathematics and physics, in particular on the theory of general relativity of Einstein. It is helpful to note that the philosophical nature of his lecture was consistent with the whole perspective of Riemann. According to Klein [7, p. 233],

> I would like to direct particular attention to the beginning of paragraph 3."My main work" Riemann says there. Therefore he himself values his natural-philosophical speculations significantly higher than his, to us classical, works on the theory of complex functions $f(x+iy)$.

4. *Riemann zeta function.*
The best known open problem in mathematics now is the Riemann hypothesis concerning the nontrivial zeros of Riemann zeta function. This came from his single paper in number theory. This paper on the zeta function might seem incidental, but it fits into his mathematics world. As a newly elected member of the Berlin Academy of Sciences, Riemann had to report on his most recent research, and he sent a report titled *On the number of primes less than a given magnitude.* This is another masterpiece of Riemann which has been a pillar of number theory since its inception and will probably continue to be so.

Though this zeta function

$$\zeta(s) = \sum_{n\geq 1} \frac{1}{n^s} = \prod_p (1-p^{-s})^{-1}$$

was considered before by Euler as a function of the real variable $s \in \mathbb{R}$, Riemann considered a very different question and its implication. He studied the zeta function as a complex function of the complex variable $s \in \mathbb{C}$ rather than a real one and brought his results in and understanding of complex analysis to bear on the problem. In particular, he extended $\zeta(s)$ to a function of one complex variable, proved the existence of a meromorphic continuation and found the functional equation, and established a relation between its zeros and the distribution of prime numbers.

The Riemann zeta function is the most basic example of L-functions in number theory (in analytic and algebraic number theory, and in arithmetic geometry), which are crucial for many applications. For example, the basic reciprocity laws in class field theory and the vast generalization via the Langlands program make essential use of L-functions.

3 Riemann's Work II: Some Little Known or Even Unknown Works

The previous section listed four major areas of Riemann's work, which clearly cannot cover the broad contribution of Riemann to many subjects of mathematics and sciences. Besides his less known work in mathematics, Riemann's results in physics and philosophy are probably less known to mathematicians in general.

To complement the discussion in the previous section, we highlight in this section some deep works of Riemann which are probably not so well-known to many mathematicians.

1. Riemann initiated the study of birational algebraic geometry of algebraic curves in the paper he wrote in 1857 on Abelian functions. Many basic problems and their solutions in algebraic geometry, for example, relations between affine varieties and commutative algebras were considered and solved by Riemann in this paper. See [3] for some explanation.
2. The notion of manifolds was formally defined by Weyl in his classical book on Riemann surfaces. But Weyl was influenced by Klein's understanding of Riemann surfaces as abstract spaces rather than coverings of the complex plane or the complex sphere, and Klein believed that Riemann had the abstract notion of manifolds already. As pointed out before, this was one important difference between the works of Gauss and of Riemann on geometry.
3. His paper on the propagation of waves of finite amplitude in a compressible two-dimensional medium started the now comprehensive theory of shock waves and the theory of hyperbolic partial differential equations.
4. After Riemann introduced and counted the number of moduli for Riemann surfaces, a proper definition and an understanding of moduli spaces of Riemann surfaces (or algebraic curves) has been pursued intensively by many people. The impact of this subject has gone much beyond the subject of complex analysis and algebraic geometry, for example, through Teichmüller space and mapping class groups and their applications in low dimensional geometry, topology and geometric group theory, developed in the twentieth century.
5. One approach to understand Riemann's moduli space is to use the period (or Jacobian) map from this moduli space to the moduli space of principally polarized Abelian varieties. One famous problem on characterizing the locus of Jacobian varieties was often attributed to Schottky, the so-called Schottky problem. Actually, Riemann raised this question and discussed it in special cases when he studied the Jacobi inversion problem via theta functions.
6. Riemann's work on minimal surfaces is still yielding most recent new results: every properly embedded minimal planar domain in $\mathbb{R}^3$ is either a minimal surface constructed by Riemann, a catenoid, a helicoid or a plane. (See [10, 18] for detail.)
7. Riemann made deep contributions to electrodynamics. We quote a good summary from [9, pp. 269–270]:

   > ... at the dawn of the 20th century, in 1905, before the appearance of the fundamental papers of Einstein and Plank, prominent physicists took Riemann seriously as one of their own. This comes through with particular clarity in the articles in the second half of the fifth volume (physics) of *Encyklopädie der Mathematischen Wissenschaften*. Issue 1 appeared on 16 June 1904 and was devoted to electricity. The article V. 12, by R. Reiff and A. Sommerfeld, presented an historical account titled "The standpoint of action at a distance. The elementary laws" (pp. 3–62)....
   >
   > The subsection devoted to Gauss and Riemann begins with the words (p. 45): "While Weber consistently championed the standpoint of action at a distance, diametrically

> opposite tendencies were put forward in his immediate milieu by his teacher Gauss and by his student Riemann"....
> the paper "Ein Beitrag zur Elektrodynamik", presented by Riemann to the Göttingen Society of Sciences on 10 February 58, made him a forerunner of Maxwell, and that the "recent electron theory" had, in a certain sense, led back to Riemann's form of a (retarded) elementary potential.
> In fact, Riemann was the first to formulate the differential equation
>
> $$\frac{1}{c^2}\frac{\partial^2 U}{\partial t^2} = \Delta U + 4\pi\rho$$
>
> for the potential U and the charge density ρ that was later deduced from Maxwell's theory, and to observe that his results agreed with experience if c was taken to be the velocity of light....

The book then quoted Riemann's own words [9, p. 270]:

> I have found that the electrodynamic actions of galvanic currents can be explained if one assumes that the action of one electrical mass on others is not instantaneous but propagates itself towards them with constant speed (equal, within the limits of observational errors, to that of the speed of light). Under this assumption, the differential equation of the electrical force is the same as that for the propagation of light and radiant heat.

It may be interesting to note that Riemann gave his talk in 1858, and his paper was posthumously published in 1867. In the meantime, Maxwell published his comprehensive paper titled "*A dynamical theory of the electromagnetic field*" in 1865. One could ask whether Maxwell might have known Riemann's result.

4 Riemann's Publications and his Impact

In the previous two sections, we summarized some major works of Riemann and explained how they affected many subjects in mathematics. Surprisingly, Riemann had only a small number of publications. In his lifetime, he published formally **nine** papers. If we can also add his thesis and a report for a conference, then the list of **eleven papers** is as follows:

1. *Grundlagen für eine allgemeine Theorie der Functionen einer veränderlichen complexen Grösse*, (Inauguraldissertation, Göttingen, 1851).
   Foundations for a general theory of functions of a complex variable, (Inaugural dissertation, Göttingen, 1851).
   In this paper, Riemann sets up the foundation of complex analysis. For example, a holomorphic function can be characterized by the Cauchy-Riemann equation and can be viewed as a conformal map. These approaches and results of Riemann fit well the modern points of view of differential equations and geometric function theory. The proper domain of definition of a holomorphic function is a Riemann surface, instead of being a domain of $\mathbb{C}$. Riemann stated and outlined a proof of

the Riemann mapping theorem, and introduced the Dirichlet principle to produce harmonic functions on domains in Riemann surfaces.

2. *Ueber die Gesetze der Vertheilung von Spannungselectricität in ponderabeln Körpern, wenn diese nicht als vollkommene Leiter oder Nichtleiter, sondern als dem Enthalten von Spannungselectricität mit endlicher Kraft widerstrebend betrachtet werden*, (Amtlicher Bericht über die 31. Versammlung deutscher Naturforscher und Aerzte zu Göttingen im September 1854).
About the laws of distribution of electric electricity in ponderable bodies, if these are not considered perfect conductors or insulators, rather than may be viewed as resisting the holding of electric charge with finite power, (Official Report on the 31st meeting of German natural scientists and physicians to Göttingen in September 1854)
This is a summary of Riemann's first work in mathematical physics. Its long title explains its contents to a certain extent. A detailed version was later published in his collected works with the title *New theory of residual charge in apparatus for static charge.*
3. *Zur Theorie der Nobili'schen Farbenringe*, (Annalen der Physik und Chemie, 95 (1855), 130–139).
On the theory of Nobili's color rings, (Annals of Physics and Chemistry, 95 (1855), 130–139).
The side of Riemann as a physicist is not so well-known to mathematicians. This paper shows clearly that Riemann was well versed in experimental physics and used mathematics effectively to understand experiments. This paper [15, pp. 49–50] started with the following words: "Nobili's color rings represent a valuable tool for the experimental study of the laws of current flow in a body made conducting by decomposition...."
4. *Beiträge zur Theorie der durch die Gauss'sche Reihe $F(\alpha, \beta, \gamma, x)$ darstellbaren Functionen*, (Abhandlungen der Königlichen Gesellschaft der Wissenschaften zu Göttingen, 7 (1857), 3–32).
Contributions to the theory of represented by the Gaussian series $F(\alpha, \beta, \gamma, x)$ functions, (Memoirs of the Royal Society of Sciences in Göttingen, 7 (1857), 3–32).
Hypergeometric functions are important special functions. They were studied by Euler, Gauss and Kummer before Riemann. In this paper, Riemann brought in a completely new perspective: the monodromy group of differential equations with algebraic coefficients which are singular. In [15, p. 57], Riemann writes:
"In the present work I have treated these transcendental functions by a new method, which essentially applies to any function that satisfies a linear differential equation with algebraic coeffficients. The method yields results almost directly from the definition, that were formerly obtained only after somewhat troublesome calculations."
This paper is also one of the first papers where the topology has played a crucial role.
5. *Selbstanzeige: Beiträge zur Theorie der durch die Gauss'sche Reihe $F(\alpha, \beta, \gamma, x)$ darstellbaren Functionen*, (Göttinger Nachrichten, 1857, 6–8)

Voluntary disclosure: contributions to the theory of represented by the Gaussian series $F(\alpha, \beta, \gamma, x)$ functions, (Göttingen News, 1857, 6–8)
This paper is a short announcement of results in the previous paper. It also gave some motivations of that paper. For example, Riemann wrote [15, p. 77]:
"This memoir deals with a class of functions which are used for solving many of the problems of mathematical physics. The series formed from them perform the same roles in the more difficult problems as are served in the easier ones by the trigonometrical series, now so frequently employed, which proceed in terms of sines and cosines of multiples of a variable.
These applications, particularly in astronomy, appear to have led Gauss—following Euler who had already frequently concerned himself with these functions because of their theoretical interest—to undertake his researches into the series which he denoted by $F(\alpha, \beta, \gamma, x)$."

6. *Theorie der Abel'schen Functionen*, (Journal für die reine und angewandte Mathematik, 54 (1857), 101–155).
Theory of Abelian functions, (Crelle's Journal, 54 (1857), 101–155).
This is one of the most important papers written by Riemann. The reason why this paper is titled *Theory of Abelian functions* is that the essential point of this paper is to understand Abelian integrals, i.e., integrals of rational functions on algebraic curves, which were first proposed by Abel. This paper contains many new ideas which have had long lasting impact. For example, it establishes a direct correspondence between plane algebraic curves and compact Riemann surfaces, introduces the idea of homology groups of surfaces to produce meromorphic functions, and proves the Riemann inequality in the Riemann-Roch theorem. It also introduces the notion of moduli spaces of compact Riemann surfaces and algebraic curves, and the Riemann theta function to solve the Jacobi inversion problem. The Riemann-Hurwitz formula also appears here. It is probably less known to people in complex analysis that this paper is also the starting point of algebraic geometry and birational geometry of algebraic varieties.
7. *Ueber die Anzahl der Primzahlen unter einer gegebenen Grösse*, (Monatsberichte der Berliner Akademie, November 1859, 671–680).
The number of primes below a given size, (Monthly reports of the Berlin Academy, November 1859, 671–680).
This is probably the most famous paper of Riemann. It deals with the Riemann zeta function and the distribution of prime numbers. Though this paper is concerned with problems in number theory, Riemann's work on complex analysis is a crucial ingredient in this work.
8. *Ueber die Fortpflanzung ebener Luftwellen von endlicher Schwingungsweite*, (Abhandlungen der Königlichen Gesellschaft der Wissenschaften zu Göttingen, 8 (1860), 43–65).
Concerning propagation of plane air waves of finite amplitude (Memoirs of the Royal Society of Sciences in Göttingen, 8 (1860), 43–65).
Waves play an important role in our life. A particular type of waves, called shock waves, which occur when the waves move faster than the local speed of sound in the fluid. It may not be well-known to people who do not work in nonlinear

differential equations and wave theory that Riemann was one of the originators of theory of shock waves and he introduced some fundamental notions and problems in this paper. See [17] for more detail and references.

9. *Selbstanzeige: Ueber die Fortpflanzung ebener Luftwellen von endlicher Schwingungsweite*, (Göttinger Nachrichten, 1859, 192–197)
Voluntary disclosure: concerning propagation of plane air waves of finite amplitude, (Göttingen News, 1859, 192–197)
This announcement of results of the previous paper also explains Riemann's perspective on differential equations. For example, he wrote [15, p. 167]:
"For the solution of linear partial differential equations, the most fruitful methods have not been found by developing the general idea of the problem, but rather from the treatment of special physical problems. In the same way, the theory of nonlinear partial differential equations seems generally to demand a thorough treatment of particular physical problems, taking into account all the secondary factors. Indeed, the solution of the quite special problem that is the subject of this work requires new methods and concepts, and leads to results that will probably play a role in more general problems."
10. *Ein Beitrag zu den Untersuchungen über die Bewegung eines flüssigen gleichartigen Ellipsoides*, (Abhandlungen der Königlichen Gesellschaft der Wissenschaften zu Göttingen, 9 (1860), 3–36)
A contribution to the studies of the motion of a homogeneous liquid ellipsoid, (Memoirs of the Royal Society of Sciences in Göttingen, 9 (1860), 3–36)
This paper was motivated by the problem on the shape of the earth. Due to its rotation, the earth is not a sphere, but rather an ellipsoid. This problem was studied before by Newton, Jacobi, Dirichlet et al. The beginning of this paper [15, p. 171] explains the paper and the problem addressed:
"Dirichlet investigated the motion of a homogeneous fluid ellipsoid, whose elements are attracted to one another by the law of gravity. His approach is surprising and opens a new path. The continuation of this fine research has a special appeal for mathematicians, quite apart from the question of the form of heavenly bodies which was the occasion for the investigation. Dirichlet himself carried through the solution of the problem completely only in the simplest cases. For the continuation of the investigation, it is convenient to give a form of the differential equation for the motion of a fluid body that it independent of the time-origin chosen."
11. *Ueber das Verschwinden der Theta-Functionen*, (Journal für die reine und angewandte Mathematik, 65 (1866), 161–172)
About the vanishing of theta functions, (Crelle's Journal, 65 (1866), 161–172)
This paper is a continuation of the paper 6 on Abelian functions above. Riemann introduced his theta functions earlier. By understanding better the vanishing of his theta functions, he used them in this paper to solve the Jacobi inversion problem.

Seven more papers of Riemann based on his manuscripts and extracts of Riemann's correspondence were posthumously published:

12. *Ueber die Darstellbarkeit einer Function durch eine trigonometrische Reihe.* (Habilitationsschrift, 1854, Abhandlungen der Königlichen Gesellschaft der Wissenschaften zu Göttingen, 13 (1868))
Concerning the representability of a function by a trigonometric series. (Habilitationsschrift, 1854 Memoirs of the Royal Society of Sciences in Göttingen, 13 (1868))
This paper studies the question of when periodic functions can be written as sums of sine and cosine functions. To do this, Riemann needed to introduce the so-called Riemann integrals and give sufficient and necessary conditions for a function to be integrable. This is one of the long papers by Riemann. It consists of two parts: a history of Fourier series in three periods (from Euler to Fourier, from Fourier to Dirichlet, after Dirichlet), and then his own new work.
13. *Ueber die Hypothesen, welche der Geometrie zu Grunde liegen.* (Habilitationsschrift, 1854, Abhandlungen der Königlichen Gesellschaft der Wissenschaften zu Göttingen, 13 (1868))
On the Hypotheses which lie at the bases of Geometry. (Habilitationsschrift, 1854 Memoirs of the Royal Society of Sciences in Göttingen, 13 (1868))
This is the famous article which set the foundation for Riemannian geometry. Since it was the written version of an address towards the general faculty of Göttingen University, it is not a technical mathematical paper, and does not contain formulas and precise definitions. It also contains a philosophical discussion on geometry and its applications in reality.
14. *Ein Beitrag zur Elektrodynamik.* (1858, Annalen der Physik und Chemie, 131 (1867), 237–243)
A contribution to electrodynamics. (1858, Annals of Physics and Chemistry, 131 (1867), 237–243).
One important conclusion in this paper is the importance of the propagation at the speed of light. It also includes one equation in Maxewell's systems of equations for electromagnetism.
15. *Beweis des Satzes, dass eine einwerthige mehr als 2nfach periodische Function von n Veränderlichen unmöglich ist.* (26. October 1859, Journal für die reine und angewandte Mathematik, 71 (1870), 197–200)
A Proof of the proposition that a single-valued periodic function of n variables cannot be more than 2n-fold periodic. (Crelle's Journal, 71 (1870), 197–200)
The title describes the content. This is one part of a letter from Riemann to Weierstrass on October 26, 1859.
16. *Estratto di una lettera scritta in lingua Italiana il di 21 Gennaio 1864 al Sig. Professore Enrico Betti.* (Annali di Matematica, 7 (Ser. 1, 1865), 281–283)
Extract from a letter written in Italian on the day January 21, 1864 to Mr Professor Enrico Betti. (Annals of Mathematics, 7 (Ser. 1, 1865), 281–283)
The letter starts with [15, p. 283]: "... To find the attraction due to any homogeneous right ellipsoidal cylinder, I consider the infinite cylinder whose points, in rectangular Cartesian co-ordinates x, y, z are those satisfying the inequality ....". Riemann went on to compute the partial derivatives of the potential.

17. *Ueber die Fläche vom kleinsten Inhalt bei gegebener Begrenzung.* (Abhandlungen der Königlichen Gesellschaft der Wissenschaften zu Göttingen, 13 (1868))
    On the surface of least area with a given boundary. (Memoirs of the Royal Society of Sciences in Göttingen, 13 (1868))
    One may note that this paper on minimal surfaces was used recently in the paper [10] to finish the classification of properly embedded noncompact, connected and genus zero minimal surfaces in $\mathbb{R}^3$, which are also called properly embedded minimal planar domains.
18. *Mechanik des Ohres*, (Aus Henle und Pfeuffer's Zeitschrift für rationelle Medicin, dritte Reihe. BD. 29)
    Mechanics of the ear, (From Henle and Pfeuffer's magazine for rational medicine, third vol. Vol. 29)

Various editions of Riemann's collected works contain additional papers or notes based on his writings. The English translation of Riemann's collected works in [15] contains almost all the papers of Riemann in the 1892 edition of Riemann's collected works edited by Weber with the help of Dedekind.

Among the 11 papers published in his lifetime, the papers 1, 4, 5, 6, 7, 11 in our list above are in mathematics, and the papers 2, 3, 8, 9, 10 are in physics.

Among the posthumously published papers, the papers 12, 13, 15, 16, 17 are in mathematics, the paper 14 is in physics, and the paper 18 is special and belongs to medicine.

Topics affected by Riemann's work.

We have emphasized that Riemann contributed to many different subjects. Whatever subject he touched, he changed the subject and people's perspective of it. Let us summarize a list of of topics to which Riemann made substantial contribution:

1. Analysis: integration theories and trigonometric series.
2. Functions of one complex variable.
3. The Riemann mapping theorem, uniformization theorem for Riemann surfaces and generalizations.
4. Riemann surfaces and complex manifolds.
5. Moduli spaces of Riemann surfaces and related varieties.
6. Birational geometry of algebraic curves and varieties.
7. The Riemann-Roch theorem and index theories.
8. Topology of surfaces and the Riemann-Hurwitz formula.
9. Hypergeometric functions and generalizations.
10. The Riemann zeta functions and analytic number theory.
11. Riemannian geometry and general relativity.
12. Calculus of variations, in particular the Dirichlet principle.
13. Partial differential equations: shock waves.
14. Differential equations: the Riemann-Hilbert problem.
15. Monodromy groups and the Riemann–Hilbert correspondence.
16. Physics: electrodynamics.
17. Physics: motion of a homogeneous liquid ellipsoid.

18. Philosophy.

The book [9] contains some overview and development of Riemann's mathematics. It concentrates on more classical topics and does not discuss much of more recent development.

Concepts, methods and results named after Riemann.

Given the depth of Riemann's work, it is not surprising that many things in mathematics are named after Riemann:

1. The Riemann sphere.
2. Riemann surfaces.
3. Riemann's moduli space.
4. The Cauchy-Riemann equations.
5. The tangential Cauchy-Riemann equation.
6. The tangential Cauchy-Riemann equation complex.
7. The Riemann mapping.
8. The Measurable Riemann mapping theorem.
9. Riemann's theorem on removable singularities, or Riemann extension theorem.
10. The Riemann theta function.
11. The Riemann vanishing theorem.
12. The Riemann-Siegel theta function.
13. The Riemann bilinear relations.
14. The Riemann form.
15. The Riemann matrix.
16. The Riemann singularity theorem on the theta divisor.
17. The Riemann-Hilbert correspondence.
18. The Riemann zeta function.
19. The Riemann Xi function, a variant of the Riemann zeta function, and is defined so as to have a particularly simple functional equation.
20. The Riemann hypothesis.
21. The generalized Riemann hypothesis.
22. The grand Riemann hypothesis.
23. The Riemann's explicit formula, an explicit formula for the normalized prime-counting function of prime numbers.
24. The Riemann-Siegel formula, an asymptotic formula for the error of the approximate functional equation of the Riemann zeta function.
25. The Riemann-von Mangoldt formula on the distribution of the zeros of the Riemann zeta function.
26. The Riemann operator in a spectral theory approach to Riemann hypothesis.
27. The Riemann hypothesis for curves over finite fields.
28. The Riemann integral.
29. Riemann integrability.
30. The Riemann sum.
31. The generalized Riemann integral.
32. The Riemann-Stieltjes integral.

33. The Riemann multiple integral.
34. The Riemann-Lebesgue lemma.
35. The Riemann-Liouville integral.
36. The Riemann series theorem.
37. The Riemann-Hurwitz formula.
38. The Riemann-Roch theorem.
39. The arithmetic Riemann-Roch theorem.
40. The Riemann-Roch theorem for smooth manifolds.
41. The Grothendieck-Hirzebruch-Riemann-Roch theorem.
42. The Hirzebruch-Riemann-Roch theorem.
43. The Zariski-Riemann space.
44. The Riemann geometry.
45. Riemannian manifolds.
46. The Riemann curvature tensor also called Riemann tensor.
47. The Riemann-Cartan geometry.
48. The Riemannian metric.
49. The Riemannian distance.
50. The Riemannian bundle metric on vector bundles over manifolds.
51. The Riemannian connection.
52. The Riemannian volume form.
53. The Fundamental theorem of Riemannian geometry.
54. The Riemannian holonomy.
55. Riemannian submanifolds.
56. The Riemannian submersion.
57. Sub-Riemannian manifolds.
58. Pseudo-Riemannian manifolds.
59. Riemannian symmetric spaces.
60. Pseudo-Riemannian symmetric spaces.
61. The Riemannian circle in metric geometry.
62. The Riemannian-Penrose inequality.
63. The Riemann-Hilbert problem.
64. The Riemann initial value problem.
65. The Riemann's differential equation, a generalization of the hypergeometric differential equation.
66. The Riemann's existence theorem on ramified coverings of a compact Riemann surface.
67. Riemann's minimal surface.
68. The Riemann invariant for systems of conservation equations.
69. The free Riemann gas, also called primon gas.
70. The Riemann solver, a numerical method to solve a Riemann problem.
71. The Riemann problem for initial value problems of conservation equations.
72. The Riemann-Silberstein vector, a complex vector that combines the electric field and the magnetic field in electromagnetism.

5 How Riemann Developed

There is no book length biography about Riemann,[2] and the best and relatively short biography about him was written by his friend Dedekind for the first edition of Riemann's collected works. All subsequent descriptions of Riemann's life such as those in the books [7, 9, 11] and the article [4] follow this biography by Dedekind.

Let us give a short summary of some of the main points of his life and education to see how they might have affected his mathematics or made Riemann the mathematician as we know.

Riemann was born on September 17, 1826, in a small village in the kingdom of Hanover. At that time, his father was a local Lutheran pastor. His mother, Charlotte Ebell, died when Riemann was 20. Riemann was the second of six children. The family was loving and happy. Due to his sheltered upbringing, Riemann was shy and had a fear of speaking in public. He suffered from numerous nervous breakdowns throughout his life, and he took refuge in solitude and the world of his imagination, where he displayed the greatest boldness and open-mindedness.

Riemann's father acted as a teacher to his children and taught Riemann until he was ten years old. At the age of five, Riemann was very interested in history. But soon after that Riemann exhibited exceptional calculation abilities. There was nothing he liked better than to discover by himself hard problems and then solve them. On the other hand, he was not too good with writing and expressing himself. This might explain that Riemann had written and published relatively little later in life.

Between the age of ten and thirteen and a half, Riemann was taught by a tutor, who gave him a solid training in arithmetic and geometry. But soon the tutor found that Riemann surpassed him. Riemann surprised him with solutions to problems he assigned.

In school, Riemann did well in all subjects, but he excelled in mathematics. In high school, his exceptional talent for mathematics was recognized by the headmaster, who allowed him to study mathematics texts from his own library. It is possible he lent Riemann Legendre's book on the theory of numbers. Riemann read the 900 page book and understood it. This might have buried the seed of Riemann's future work on prime numbers.

Riemann also read Euler, and gained solid knowledge of advanced analysis and great skill in computation and manipulative ingenuity.

In high school, Riemann also studied the Bible intensively, but he was often distracted by mathematics. He was religious and later saw his life as a mathematician as another way to serve God.

At the age of nineteen and a half, in the spring of 1846, Riemann enrolled at the University of Göttingen. His father had encouraged him to study theology and so he entered the theology faculty. At the same time, he attended some mathematics

---

[2]One explanation was given by Klein [8, p. 167]: "The outward life of Riemann may perhaps appeal to your sympathy; but it was too uneventful to arouse particular interest. Riemann was one of those retiring men of learning who allow their profound thoughts to mature slowly in the seclusion of their study."

lectures of Stern on numerical solution of equations, lectures on terrestrial magnetism by Goldschmidt, lectures of Gauss on the method of least square. Soon he found that he could not resist the attraction of mathematics and asked his father if he could transfer to the faculty of philosophy so that he could study mathematics. Riemann was always very close to his family and he would never have changed courses without his father's permission. Since Riemann's parents always regarded the proper education of their children as their main duty, his father granted his permission.

It may be thought that Riemann was in just the right place to study mathematics at Göttingen, since Gauss was the acknowledged greatest mathematician in the world for the past half century and was teaching there. Gauss did lecture to Riemann but he was only giving elementary courses. It is not clear whether at this time he recognized Riemann's genius. Another teacher of Riemann, Moritz A. Stern did realize Riemann's talent and said that at this time Riemann "already sang like a canary."

Since Riemann had taught himself quite advanced mathematics already and felt that he could not learn much new mathematics in Göttingen, he moved from Göttingen to Berlin University in the spring of 1847. During his time of studying there, Jacobi, Lejeune Dirichlet, and Steiner were teaching their newest results and their struggles in obtaining them, and hence they attracted a large number of students from all over Germany. He also met and interacted with Eisenstein, who was regarded by Gauss as one of the few geniuses in mathematics. They discussed how complex numbers should be introduced into the theory of functions and held very different points: Eisenstein emphasized the formal algorithmic approach, while Riemann emphasized understanding holomorphic functions by the partial differential equation, which was the basic and starting point of his thesis in 1851 on the foundation of complex analysis of one variable.

Riemann was attracted to Dirichlet's approach to mathematics and his style of thinking. According to Klein [7, pp. 234–235],

> Riemann was bound to Dirichlet by the strong inner sympathy of a like mode of thought. Dirichlet loved to make things clear to himself on an intuitive level; along with this he would give acute, logical analyses of foundational questions and would avoid long computations as much as possible. His manner suited Riemann, who adopted it and worked according to Dirichlet's methods.

In 1849 Riemann returned to Göttingen, and things had changed in Göttingen. Wilhelm Weber had returned to a chair of physics at Göttingen from Leipzig during the time that Riemann was in Berlin, and Listing had been appointed as a professor of physics in 1849.

Before he submitted his Ph.D. thesis, supervised by Gauss in 1851, Riemann attended courses on experimental physics given by Weber, and joined in the Fall of 1850 the recently formed seminar on mathematical physics led by Weber, Ulrich, Stern and Listing. He took part in the experimental work. Riemann was the assistant to Weber for eighteen months. Listing was one of the founders of topology and Riemann interacted with him. All these scientists had a huge impact on Riemann's future work.

Due to his various interests, it was not until November 1851, at the age of 26, that Riemann submitted his thesis, which was grounding breaking for complex analysis. This has been proved both by the ultimate test of time and Gauss' evaluation[3]:

> The dissertation submitted by Herr Riemann offers convincing evidence of the author's thorough and penetrating investigations in those parts of the subject treated in the dissertation, of a creative, active, truly mathematical mind, and of a gloriously fertile originality. The presentation is perspicuous and concise and, in places, beautiful. The majority of readers would have preferred a greater clarity of arrangement. The whole is a substantial, valuable work, which not only satisfies the standards demanded for doctoral dissertations, but far exceeds them.

At that time, Riemann thought that he could finish his habilitation thesis fairly soon, but it took him two and half years to finish it. After he finished and submitted his habilitation thesis in early December 1853, he wrote to his brother on December 28 about the trial lecture [15, p. 523]:

> My work is now in a reasonable state; I handed in my habilitation thesis at the beginning of December, and I must now propose three subjects for the trial lecture, one of which is then chosen by the faculty. I had already prepared the first two, and I had hoped that one of these would be chosen, but Gauss chose the third, and so I am again in something of a tight spot, as I now have to do some more work on it.

The topic Gauss picked is on the hypothesis which lies at the foundation of geometry. One reason is that Gauss wanted to see how the talented Riemann could handle this difficult issue. The first two topics were associated with his investigations on electricity which Riemann had been working on.

Riemann tried to prepare his lecture so that members of the faculty without mathematical training could understand it and hence suppressed all the detailed mathematical computation. The lecture was well beyond Gauss' expectations. Gauss was completely astonished and excitedly told Weber so after Riemann's lecture.

From the letter of Riemann to his brother, we may imagine that Riemann worked hard on this lecture and the written notes. This is supported by the reworking and correction on his manuscripts from his Nachlass at the University of Göttingen.

The success of the trial lecture was followed by a successful presentation on the distribution of electricity in non-conductors at a meeting of the Association for Scientific Research in September 1854. He wrote to his father [15, p. 526]:

> The fact that I had spoken in public to this particular gathering before gave me a bit more courage to give my lecture, but I now see that there is a vast difference between, on the one hand, having thought about a subject for a long time and having sorted everything out, and on the other having only just prepared the subject-matter immediately before the lecture.

This was followed by his successful first lectures which attracted a large number of students. Then several tragedies followed. In 1855, his father and one of his

---

[3]It might be interesting to compare this with Gauss' evaluation on Dedekind's thesis: "The paper submitted by Mr. Dedekind deals with problems in calculus which are by no means commonplace. The author not only shows very good knowledge in this field but also an independence which indicates favorable promise for his future achievements. As paper for admission to the examination this text is fully sufficient".

sisters died. Near the end of 1857, his brother died and he assumed responsibility of providing for his three sisters. In March 1858, another of his sister died.

Though Riemann's talents were recognized right away after his thesis in 1951, it was only in 1857 that he became an extraordinary (or associate) professor. In 1859, at the age of 33, Riemann became a full professor and got the chair used to be occupied by Gauss and then Dirichlet, after the death of Dirichlet in 1859. A few days later he was elected to the Berlin Academy of Sciences. He had been proposed by three of the Berlin mathematicians, Kummer, Borchardt and Weierstrass. Their proposal read [11, p. 51]:

> Prior to the appearance of his most recent work [Theory of Abelian functions], Riemann was almost unknown to mathematicians. This circumstance excuses somewhat the necessity of a more detailed examination of his works as a basis of our presentation. We considered it our duty to turn the attention of the Academy to our colleague whom we recommend not as a young talent which gives great hope, but rather as a fully mature and independent investigator in our area of science, whose progress he in significant measure has promoted.

In June 1862, Riemann married Elise Koch who was a friend of his sister. They had one daughter. In the autumn 1862, Riemann caught a heavy cold which turned to tuberculosis. Riemann tried to fight the illness by going to the warmer climate of Italy.

The winter of 1862–63 was spent in Sicily, and he then traveled through Italy, spending time with Betti and other Italian mathematicians who had visited Göttingen. He returned to Göttingen in June 1863 but his health soon deteriorated and once again he returned to Italy. Having spent from August 1864 to October 1865 in northern Italy, Riemann returned to Göttingen for the winter of 1865–66, then returned to Selasca on the shores of Lake Maggiore on June 16, 1866. Riemann died on July 20, 1866 in Selasca, and was buried in Italy. According to Dedekind [15, p. 533]:

> His strength declined rapidly, and he himself felt that his end was near. But even on the day before his death, he was working quietly under a fig-tree, filled with a great joy looking out over the beautiful landscape, on his last, and alas unfinished work. His life ended very peacefully, without any struggle or fear of death; ... His wife had to give him bread and wine; he gave his blessing to those he loved at home and said to her: kiss our child. She recited the Lord's Prayer, but he could not speak; at the words "forgive us our trespasses"his eyes looked upwards, she felt his hand in hers grow colder, and after a few breaths his pure and noble heart ceased to beat.

6 People Who Influenced Riemann

As we have emphasized, Riemann's work has influenced many generations of mathematicians. It is interesting to see who had influenced him and in what ways. Though Riemann is usually described as shy and keeping to himself, his interaction with others had been crucial to his development and achievement, in spite of his originality.

1. Dirichlet. Among all teachers of Riemann, Dirichlet was the most inspiring to Riemann. Riemann took courses in the theory of numbers and analysis with him

in Berlin. Dirichlet's interests in mathematical physics also influenced Riemann. Riemann named the powerful guiding principle, the Dirichlet principle, after him. Dirichlet was also personally close to Riemann. Indeed, in 1852, in a letter to his father, Riemann wrote [15, p. 523]:

> The other morning, Dirichlet was with me for about two hours; he gave me some notes which I needed for my habilitation thesis, which were so comprehensive that it will significantly lighten my work; otherwise I might have had to spend a lot of time in the library searching for all kinds of things. He also went through my dissertation with me and, in general, was extremely friendly towards me, which I scarcely expected because of the great gap in standing. I have hopes that he will not forget me later.

2. Gauss. Gauss' influence on Riemann probably did not come from his courses. According to Klein [7, p. 233],

> Quite wonderful and almost engimatic to us is Riemann's close relation to Gauss in his scientific ideas. He cannot have attended many lectures by the then 70 year old Gauss, who lectured little anyway. And surely the young, shy student could not form social relations with Gauss. Gauss taught unwillingly, had little interest in most of his auditors, and was otherwise quite inaccessible. Nevertheless, we call Riemann a pupil of Gauss; indeed, he is Gauss's only true pupil, entering into his inner ideas....

   The fact that Gauss picked the third topic of the trial lecture made Riemann to work and think more about geometry, which was an extension of Gauss' work. Riemann's work on prime numbers was also influenced by Gauss' work on the prime number theorem. They also shared many similarities: their points of view connecting holomorphic functions with conformal maps and harmonic functions, and their work on hypergeometric functions. Furthermore, to both of them, mathematics was always connected with physics.
3. Eisenstein. He was only 3 years older than Riemann but was an established mathematician when Riemann met him in Berlin as a student. The fact that Riemann discussed with him and held his own different perspective about functions of one complex variable before he developed his own theory gave Riemann confidence. This encounter might have been very important for a shy person like Riemann.
4. Weber. Riemann contributed to both mathematics and physics, and the interaction between these two subjects gave Riemann a lot of motivation and inspiration. Weber contributed substantially to Riemann's training both as an experimental and theoretical physicist. Furthermore, according to Klein [7, p. 235], "In Weber, Riemann found a patron and fatherly friend. Weber recognized Riemann's genius and drew the shy student to him."
5. Dedekind. He was both a friend and colleague of Riemann and probably one of the very few people with whom Riemann could chat and discuss mathematics, and who really appreciated Riemann's mathematics. When Riemann lectured on his masterpiece on Abelian functions, there were only three listeners: two students plus Dedekind. This is not surprising, given the difficulty of the subject. Dedekind's biography of Riemann is essentially the only reliable source about Riemann and his life, and he was one of the first editors of Riemann's collected works.

6. Jacobi. He was one of the inspiring teachers of Riemann during his stay in Berlin. Riemann's solution of the Jacobi inversion problem immediately established Riemann as a first rate mathematician.
7. Stern. He was the first mathematics university teacher whose many lectures motivated Riemann to major in mathematics. He noticed very early Riemann's talent.
8. Listing. He came up with the name of "topology" and published in 1847 the first book on topology titled *Vorstudien zur Topologie*. He probably influenced Riemann to introduce highly original and powerful topological methods into the theory of functions of one complex variable. According to Klein [7, p. 234],

   > We cannot regard this otherwise than that the Göttingen atmosphere was then saturated with these geometric interests and exerted a compelling force on the very gifted and receptive Riemann. How important is a man's spiritual environment, influencing him more strongly than the facts and concrete knowledge offered him.

9. Friedrich Herbart. He was a philosopher and education-scientist. Riemann's philosophy was greatly influenced by his teaching, even though they never met. Herbart was born in 1776, and spent a major portion of his life in Göttingen. He died in 1841, five years before Riemann arrived there. He led the renewed 19th-century interest in realism and is considered as one of the founders of modern scientific pedagogy. Riemann wrote essays on philosophy. His interest in this subject is also reflected, for example, in his famous paper on the foundation of geometry. As Riemann wrote [7, p. 233], "My main work concerns a new conception of the known laws of nature.... I was led to this mainly through studying the works of Newton, Euler, and–from another aspect–Herbart."

References

1. N. A'Campo, L. Ji, A. Papadopoulos, On the early history of moduli and Teichmüller spaces, in *Lipman Bers, a Life in Mathematics*, ed. by L. Keen, I. Kra, R. Rodriguez. American Mathematical Society (Providence, RI, 2015), pp. xiii+329, 175–262
2. R. Dedekind, H. Weber, *Theory of algebraic functions of one variable.* Translated from the 1882 German original and with an introduction, bibliography and index by John Stillwell. History of Mathematics, 39. American Mathematical Society (Providence, RI; London Mathematical Society, London, 2012), pp. viii+152
3. J. Dieudonné, Algebraic geometry. Adv. Math. **3**, 233–321 (1969)
4. H. Freudenthal, Riemann, Georg Friedrich Bernhard, in *Dictionary of Scientific Biography*, vol. 11, 1975, New York, pp. 447-456
5. J. Gray, Riemann on geometry, physics, and philosophy–some remarks. (In this book)
6. L. Ji, The story of Riemann's moduli space. ICCM Not. **3**(2), 46–73 (2015)
7. F. Klein, *Development of mathematics in the 19th century.* With a preface and appendices by Robert Hermann. Translated from the German by M. Ackerman. Lie Groups: History, Frontiers and Applications, IX. mathematics Sci Press, Brookline, Mass., 1979. ix+630, pp. 175–180
8. F. Klein, Riemann and his significance for the development of modern mathematics. Bull. Amer. Math. Soc. **1**(7), 165–180 (1895)
9. D. Laugwitz, *Bernhard Riemann 1826–1866. Turning points in the conception of mathematics* (Birkhäuser Boston, Inc., Boston, MA, 2008), pp. xviii+357

10. W. Meeks, J.Pérez, A. Ros, Properly embedded minimal planar domains. Ann. Math. **2**(2, 181) 473–546 (2015)
11. M. Monastyrsky, in *Riemann, Topology, and Physics*, ed. by Freeman J. Dyson (Birkhäuser Boston, Inc., Boston, MA, 2008), pp. xvi+215
12. K. Ohshika, The origin of the notion of manifold: from Riemann's Habilitationsvortrag onward. (In this book)
13. A. Papadopoulos, Physics in Riemann's mathematical papers. (In this book)
14. A. Plotnitsky, Comprehending the Connection of Things: Bernhard Riemann and the Architecture of Mathematical Concepts. (In this book)
15. B. Riemann, *Collected Papers*. Translated from the 1892 German edition by Roger Baker, Charles Christenson and Henry Orde. (Kendrick Press, Heber City, UT, 2004), pp. x+555
16. B. Riemann, Gesammelte mathematische Werke, wissenschaftlicher Nachlass und Nachträge. Based on the edition by Heinrich Weber and Richard Dedekind. Edited and with a preface by Raghavan Narasimhan. (Springer-Verlag, Berlin; BSB B. G. Teubner Verlagsgesellschaft, Leipzig, 1990), pp. vi+911
17. J. Smoller, *Shock waves and reaction-diffusion equations*. Second edition. Grundlehren der Mathematischen Wissenschaften, 258 (Springer-Verlag, New York, 1994), pp. xxiv+632
18. S. Yamada, Riemann on minimal surfaces. (In this book)
19. S.T. Yau, *The founders of index theory: reminiscences of and about Sir Michael Atiyah, Raoul Bott, Friedrich Hirzebruch, and I. M. Singer.* Second edition (International Press, Somerville, MA, 2009), pp. lii+393
20. A. Zygmund, in *Trigonometric Series*. Cambridge Mathematical Library, Vol. I, II. 3rd edn. ed. by R.A. Fefferman (Cambridge University Press, Cambridge, 2002), pp. xii. (Vol. I: pp. xiv+383, Vol. II: pp. viii+364)

Index

L. Ji et al. (eds.), *From Riemann to Differential Geometry and Relativity*,
DOI 10.1007/978-3-319-60039-0

T

Printed in the USA
CPSIA information can be obtained
at www.ICGtesting.com
LVHW010952070424
776681LV00003B/97

9 783319 600383